Log on to  **fl.msmath3.net**

Florida Online Book

- Complete Student Edition
- Links to Online Study Tools

Florida Online Study Tools

- Extra Examples
- Self-Check Quizzes
- Vocabulary Review
- Chapter Test Practice
- FCAT Practice

Online Activities

- WebQuest Projects
- Game Zone Activities
- Career Links
- Data Updates

Parent & Student Study Guide Workbook

- Printable Worksheets

Graphing Calculator Keystrokes

- Calculator Keystrokes for other calculators

GLENCOE MATHEMATICS

Florida Edition

Mathematics

Applications and Concepts

Course 3

Contents

McGraw Hill Glencoe

New York, New York   Columbus, Ohio   Chicago, Illinois   Peoria, Illinois   Woodland Hills, California

ISBN: 0-07-860139-8 *(Florida Student Edition)*

Sunshine State Standards and Grade 8 Expectations

Strand A NUMBER SENSE, CONCEPTS, AND OPERATIONS

Standard 1: *The student understands the different ways numbers are represented and used in the real world.*

Benchmark MA.A.1.3.1: The student associates verbal names, written word names, and standard numerals with integers, fractions, decimals; numbers expressed as percents; numbers with exponents; numbers in scientific notation; radicals; absolute value; and ratios.

1	knows word names and standard numerals for integers, fractions, decimals, numbers expressed as percents, numbers with exponents, numbers expressed in scientific notation, absolute value, radicals, and ratios.

Benchmark MA.A.1.3.2: The student understands the relative size of integers, fractions, and decimals; numbers expressed as percents; numbers with exponents; numbers in scientific notation; radicals; absolute value; and ratios.

1	compares and orders fractions, decimals, integers, and radicals using graphic models, number lines, and symbols.
2	compares and orders numbers expressed in absolute value, scientific notation, integers, percents, numbers with exponents, fractions, decimals, radicals, and ratios.

Benchmark MA.A.1.3.3: The student understands concrete and symbolic representations of rational numbers and irrational numbers in real-world situations.

1	knows examples of rational and irrational numbers in real-world situations.
2	describes the meanings of rational and irrational numbers using physical or graphical displays.
3	constructs models to represent rational and irrational numbers.

Benchmark MA.A.1.3.4: The student understands that numbers can be represented in a variety of equivalent forms, including integers, fractions, decimals, percents, scientific notation, exponents, radicals, and absolute value.

1	knows the relationships among fractions, decimals, and percents given a real-world context.
2	simplifies expressions using integers, exponents, and radicals.
3	knows equivalent forms of large and small numbers in scientific and standard notation.
4	identifies and explains the absolute value of a number.

Standard 2: *The student understands number systems.*

Benchmark MA.A.2.3.1: The student understands and uses exponential and scientific notation.

1	expresses rational numbers in exponential notation including negative exponents (for example, $2^{-3} = \frac{1}{2^3} = \frac{1}{8}$).
2	expresses numbers in scientific or standard notation including decimals between 0 and 1.
3	evaluates numerical or algebraic expressions that contain exponential notation.

Benchmark MA.A.2.3.2: The student understands the structure of number systems other than the decimal number system.

1	expresses base ten numbers as equivalent numbers in different bases, such as base two, base five, and base eight.
2	discusses the application of the binary (base two) number system in computer technology.
3	expresses non-base ten numbers as equivalent numbers in base ten.

Standard 3: *The student understands the effects of operations on numbers and the relationships among these operations, selects appropriate operations, and computes for problem solving.*

Benchmark MA.A.3.3.1: The student understands and explains the effects of addition, subtraction, multiplication, and division on whole numbers, fractions, including mixed numbers, and decimals, including the inverse relationships of positive and negative numbers.

1	knows the effects of the four basic operations on whole numbers, fractions, mixed numbers, decimals, and integers.
2	knows the inverse relationship of positive and negative numbers.
3	applies the properties of real numbers to solve problems (commutative, associative, distributive, identity, equality, inverse, and closure).

Benchmark MA.A.3.3.2: The student selects the appropriate operation to solve problems involving addition, subtraction, multiplication, and division of rational numbers, ratios, proportions, and percents, including the appropriate application of the algebraic order of operations.

1	knows the appropriate operations to solve real-world problems involving integers, ratios, rates, proportions, numbers expressed as percents, decimals, and fractions.
2	solves real-world problems involving integers, ratios, proportions, numbers expressed as percents, decimals, and fractions in two- or three-step problems.
3	solves real-world problems involving percents including percents greater than 100% (for example percent of change, commission).
4	writes and simplifies expressions from real-world situations using the order of operations.

Benchmark MA.A.3.3.3: The student adds, subtracts, multiplies, and divides whole numbers, decimals, and fractions, including mixed numbers, to solve real-world problems, using appropriate methods of computing, such as mental mathematics, paper and pencil, and calculator.

1	solves multi-step real-world problems involving fractions, decimals, and integers using appropriate methods of computation, such as mental computation, paper and pencil, and calculator.

Standard 4: *The student uses estimation in problem solving and computation.*

Benchmark MA.A.4.3.1: The student uses estimation strategies to predict results and to check the reasonableness of results.

1	knows appropriate estimation techniques for a given situation using real numbers.
2	estimates to predict results and to check reasonableness of results.

Standard 5: *The student understands and applies theories related to numbers.*

Benchmark MA.A.5.3.1: The student uses concepts about numbers, including primes, factors, and multiples, to build number sequences.

1	knows if numbers are relatively prime.
2	applies number theory concepts to determine the terms in a real number sequence.
3	applies number theory concepts, including divisibility rules, to solve real-world or mathematical problems.

Strand B MEASUREMENT

Standard 1: *The student measures quantities in the real world and uses the measures to solve problems.*

Benchmark MA.B.1.3.1: The student uses concrete and graphic models to derive formulas for finding perimeter, area, surface area, circumference, and volume of two- and three-dimensional shapes, including rectangular solids and cylinders.

1	uses concrete and graphic models to explore and derive formulas for surface area and volume of three-dimensional regular shapes, including pyramids, prisms, and cones.
2	solves and explains real-world problems involving surface area and volume of three-dimensional shapes.

Benchmark MA.B.1.3.2: The student uses concrete and graphic models to derive formulas for finding rates, distance, time, and angle measures.

1	applies formulas for finding rates, distance, time and angle measures.
2	describes and uses rates of change (for example, temperature as it changes throughout the day, or speed as the rate of change in distance over time) and other derived measures.

Benchmark MA.B.1.3.3: The student understands and describes how the change of a figure in such dimensions as length, width, height, or radius affects its other measurements such as perimeter, area, surface area, and volume.

1	knows how a change in a figure's dimensions affects its perimeter, area, circumference, surface area, or volume.
2	knows how changes in the volume, surface area, area, or perimeter of a figure affect the dimensions of the figure.
3	solves real-world or mathematical problems involving the effects of changes either to the dimensions of a figure or to the volume, surface area, area, perimeter, or circumference of figures.

Benchmark MA.B.1.3.4: The student constructs, interprets, and uses scale drawings such as those based on number lines and maps to solve real-world problems.

1	interprets and applies various scales including those based on number lines, graphs, models, and maps. (Scale may include rational numbers.)
2	constructs and uses scale drawings to recreate a given situation.

Standard 2: *The student compares, contrasts, and converts within systems of measurement (both standard/nonstandard and metric/customary).*

Benchmark MA.B.2.3.1: The student uses direct (measured) and indirect (not measured) measures to compare a given characteristic in either metric or customary units.

1	finds measures of length, weight or mass, and capacity or volume using proportional relationships and properties of similar geometric figures.

Benchmark MA.B.2.3.2: The student solves problems involving units of measure and converts answers to a larger or smaller unit within either the metric or customary system.

1	solves problems using mixed units within each system, such as feet and inches, hours and minutes.
2	solves problems using the conversion of measurements within the customary system.

| 3 | solves problems using the conversions of measurement within the metric system. |

Standard 3: *The student estimates measurements in real-world problem situations.*

Benchmark MA.B.3.3.1: The student solves real-world and mathematical problems involving estimates of measurements including length, time, weight/mass, temperature, money, perimeter, area, and volume, in either customary or metric units.

| 1 | knows a variety of strategies to estimate, describe, make comparisons, and solve real-world and mathematical problems involving measurements. |

Standard 4: *The student selects and uses appropriate units and instruments for measurement to achieve the degree of precision and accuracy required in real-world situations.*

Benchmark MA.B.4.3.1: The student selects appropriate units of measurement and determines and applies significant digits in a real-world context. (Significant digits should relate to both instrument precision and to the least precise unit of measurement).

1	selects the appropriate unit of measure for a given situation.
2	knows the precision of different measuring instruments.
3	determines the appropriate precision unit for a given situation.
4	identifies the number of significant digits as it relates to the least precise unit of measure.
5	determines the greatest possible error of a given measurement and the possible actual measurements of an object.

Benchmark MA.B.4.3.2: The student selects and uses appropriate instruments, technology, and techniques to measure quantities in order to achieve specified degrees of accuracy in a problem situation.

| 1 | applies significant digits in the real-world context. |
| 2 | selects and uses appropriate instruments, technology, and techniques to measure quantities and dimensions to a specified degree of accuracy. |

Strand C   GEOMETRY AND SPATIAL SENSE

Standard 1: *The student describes, draws, identifies, and analyzes two- and three-dimensional shapes.*

Benchmark MA.C.1.3.1: The student understands the basic properties of, and relationships pertaining to, regular and irregular geometric shapes in two- and three-dimensions.

1	determines and justifies the measures of various types of angles based upon geometric relationships in two- and three-dimensional shapes.
2	compares regular and irregular polygons and two- and three-dimensional shapes.
3	draws and builds three-dimensional figures from various perspectives (for example, flat patterns, isometric drawings, nets).
4	knows the properties of two- and three-dimensional figures.

Standard 2: *The student visualizes and illustrates ways in which shapes can be combined, subdivided, and changed.*

Benchmark MA.C.2.3.1: The student understands the geometric concepts of symmetry, reflections, congruency, similarity, perpendicularity, parallelism, and transformations, including flips, slides, turns, and enlargements.

1	use the properties of parallelism, perpendicularity, and symmetry in solving real-world problems.
2	identifies congruent and similar figures in real-world situations and justifies the identification.
3	identifies and performs the various transformations (reflection, translation, rotation, dilation) of a given figure on a coordinate plane.

Benchmark MA.C.2.3.2: The student predicts and verifies patterns involving tessellations (a covering of a plane with congruent copies of the same pattern with no holes and no overlaps, like floor tiles).

| 1 | continues a tessellation pattern using the needed transformations. |
| 2 | creates an original tessellating tile and tessellation pattern using a combination of transformations. |

Standard 3: *The student uses coordinate geometry to locate objects in both two- and three-dimensions and to describe objects algebraically.*

Benchmark MA.C.3.3.1: The student represents and applies geometric properties and relationships to solve real-world and mathematical problems.

| 1 | observes, explains, makes and tests conjectures regarding geometric properties and relationships (among regular and irregular shapes of two and three dimensions). |
| 2 | applies the Pythagorean Theorem in real-world problems (for example, finds the relationship among sides in 45°-45° and 30°-60° right triangles). |

Benchmark MA.C.3.3.2: The student identifies and plots ordered pairs in all four quadrants of a rectangular coordinate system (graph) and applies simple properties of lines.

1	given an equation or its graph, finds ordered-pair solutions (for example, $y = 2x$).
2	given the graph of a line, identifies the slope of the line (including the slope of vertical and horizontal lines).
3	given the graph of a linear relationship, applies and explains the simple properties of lines on a graph, including parallelism, perpendicularity, and identifying the x- and y-intercepts, the midpoint of a horizontal or vertical line segment, and the intersection point of two lines.

Strand D ALGEBRAIC THINKING

Standard 1: *The student describes, analyzes, and generalizes a wide variety of patterns, relations, and functions.*

Benchmark MA.D.1.3.1: The student describes a wide variety of patterns, relationships, and functions through models, such as manipulatives, tables, graphs, expressions, equations, and inequalities.

1	reads, analyzes, and describes graphs of linear relationships.
2	uses variables to represent unknown quantities in real-world problems.
3	uses the information provided in a table, graph, or rule to determine if a function is linear and justifies reasoning.
4	finds a function rule to describe tables of related input-output variables.
5	predicts outcomes based upon function rules.

Benchmark MA.D.1.3.2: The student creates and interprets tables, graphs, equations, and verbal descriptions to explain cause-and-effect relationships.

| 1 | interprets and creates tables and graphs (function tables). |

2	writes equations and inequalities to express relationships.
3	graphs equations and inequalities to explain cause-and-effect relationships.
4	interprets the meaning of the slope of a line from a graph depicting a real-world situation.

Standard 2: *The student uses expressions, equations, inequalities, graphs, and formulas to represent and interpret situations.*

Benchmark MA.D.2.3.1: The student represents and solves real-world problems graphically, with algebraic expressions, equations, and inequalities.

1	translates verbal expressions and sentences into algebraic expressions, equations, and inequalities.
2	translates algebraic expressions, equations, or inequalities representing real-world relationships into verbal expressions or sentences.
3	solves single- and multiple-step linear equations and inequalities in concrete or abstract form.
4	graphs linear equations on the coordinate plane using tables of values.
5	graphically displays real-world situations represented by algebraic equations or inequalities.
6	evaluates algebraic expressions, equations, and inequalities by substituting integral values for variables and simplifying the results.
7	simplifies algebraic expressions that represent real-world situations by combining like terms and applying the properties of real numbers.

Benchmark MA.D.2.3.2: The student uses algebraic problem-solving strategies to solve real-world problems involving linear equations and inequalities.

1	simplifies algebraic expressions with a maximum of two variables.
2	solves single- and multi-step linear equations and inequalities that represent real-world situations.

Strand E DATA ANALYSIS AND PROBABILITY

Standard 1: *The student understands and uses the tools of data analysis for managing information.*

Benchmark MA.E.1.3.1: The student collects, organizes, and displays data in a variety of forms, including tables, line graphs, charts, bar graphs, to determine how different ways of presenting data can lead to different interpretations.

1	reads and interprets data displayed in a variety of forms including histograms.
2	constructs and interprets displays of data, (including circle, line, bar, and box-and-whisker graphs) and explains how different displays of data can lead to different interpretations.

Benchmark MA.E.1.3.2: The student understands and applies the concepts of range and central tendency (mean, median, and mode).

1	finds the mean, median, and mode of a set of data using raw data, tables, charts, or graphs.
2	interprets measures of dispersion (range) and of central tendency.
3	determines appropriate measures of central tendency for a given situation or set of data.

Benchmark MA.E.1.3.3: The student analyzes real-world data by applying appropriate formulas for measures of central tendency and organizing data in a quality display, using appropriate technology, including calculators and computers.

1	determines the mean, median, mode, and range of a set of real-world data using appropriate technology.
2	organizes, graphs and analyzes a set of real-world data using appropriate technology.

Standard 2: *The student identifies patterns and makes predictions from an orderly display of data using concepts of probability and statistics.*

Benchmark MA.E.2.3.1: The student compares experimental results with mathematical expectations of probabilities.

1	compares and explains the results of an experiment with the mathematically expected outcomes.
2	calculates simple mathematical probabilities for independent and dependent events.

Benchmark MA.E.2.3.2: The student determines odds for and odds against a given situation.

1	predicts the mathematical odds for and against a specified outcome in a given real-world situation.

Standard 3: *The student uses statistical methods to make inferences and valid arguments about real-world situations.*

Benchmark MA.E.3.3.1: The student formulates hypotheses, designs experiments, collects and interprets data, and evaluates hypotheses by making inferences and drawing conclusions based on statistics (range, mean, median, and mode) and tables, graphs, and charts.

1	formulates a hypothesis and designs an experiment.
2	performs the experiment and collects, organizes, and displays the data.
3	evaluates the hypothesis by making inferences and drawing conclusions based on statistical results.

Benchmark MA.E.3.3.2: The student identifies the common uses and misuses of probability or statistical analysis in the everyday world.

1	knows appropriate uses of statistics and probability in real-world situations.
2	knows when statistics and probability are used in misleading ways.
3	identifies and uses different types of sampling techniques (for example, random, systematic, stratified).
4	knows whether a sample is biased.

How To...

Prepare for FCAT

Countdown To FCAT

Pages FL4–FL24 of this text include a section called **Countdown to FCAT**. Each page contains 7 problems that are just like those on FCAT. You should plan to complete one page each week to help you prepare for the test.

Plan to spend a few minutes each day working on the FCAT problem(s) for that day unless your teacher asks you to do otherwise. Each day of the week has the same type of problem(s). If you have difficulty with any problem, you can refer to the lesson that is referenced in parentheses after the problem.

Monday	Extended Response
Tuesday	Multiple Choice, Gridded Response
Wednesday	Multiple Choice, Gridded Response
Thursday	Short Response
Friday	Multiple Choice or Gridded Response

Your teacher can provide you with an answer sheet to record your work and your answers for each week. A printable worksheet is also available at fl.msmath3.net. At the end of the week, your teacher may want you to turn in the answer sheet.

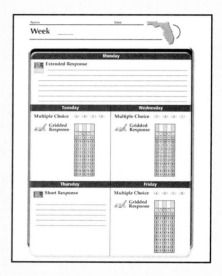

FCAT Practice and Sample Test Workbook

The **FCAT Practice and Sample Test Workbook, Grade 8,** contains a diagnostic test, practice for each of the Sunshine State Standards, and a sample test.

As you practice and master each objective, you can record your progress in the Student Recording Chart in your workbook.

Your teacher may also ask you to take a sample test at various points throughout the year to see if you're ready to take the real FCAT.

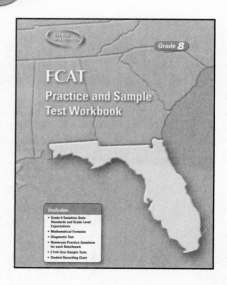

Your Textbook

Your textbook contains many opportunities for you to get ready for FCAT every day. Take advantage of these so you don't need to cram before the test.

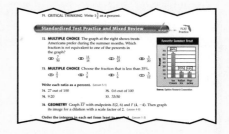

- **Each lesson** contains at least two FCAT practice problems. You can use these problems every day to keep your FCAT skills sharp. The **Mid-Chapter Practice Test** and the **Chapter Practice Test** also include FCAT practice problems.

- **Worked-out examples** in each chapter show you step-by-step solutions of FCAT problems. Just like the practice problems, these include multiple choice, gridded response, short response, and extended response. **Test-Taking Tips** are also included.

- Two pages of **FCAT Practice** are included at the end of each chapter. These problems may cover any of the content up to and including the chapter they follow.

Test-Taking Tips

- ✓ Go to bed early the night before the test. You will think more clearly after a good night's rest.

- ✓ Read each problem carefully and think about ways to solve the problem before you try to answer the question.

- ✓ Relax. Most people get nervous when taking a test. It's natural. Just do your best.

- ✓ Answer questions you are sure about first. If you do not know the answer to a question, skip it and go back to that question later.

- ✓ Think positively. Some problems may seem hard to you, but you may be able to figure out what to do if you read each question carefully.

- ✓ If no figure is provided, draw one. If one is furnished, mark it up to help you solve the problem.

- ✓ When you have finished each problem, reread it to make sure your answer is reasonable.

- ✓ Become familiar with common formulas and when they should be used. Use the FCAT Reference sheet found at the back of this book.

- ✓ Make sure that the number of the question on the answer sheet matches the number of the question on which you are working in your test booklet.

Countdown to FCAT

Week 1

Monday

 Extended Response Study the pattern at the right. (Lesson 1-1)

Part A Make a table showing the number of tiles in the first eight figures.

Part B Describe the pattern.

Part C Predict how many tiles would be in the 10th figure.

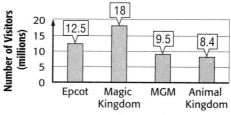

1st   2nd   3rd   4th

Tuesday

Multiple Choice Evaluate $6 + 5 \times 4 - (9 \div 3 + 2)$. (Lesson 1-2)

- Ⓐ $42\frac{1}{5}$
- Ⓑ 39
- Ⓒ 21
- Ⓓ $7\frac{4}{5}$

 Gridded Response The next figure in the pattern will have what fraction of its area shaded? Write your answer as a fraction in lowest terms. (Lesson 1-1)

Wednesday

Multiple Choice Which statement below is true? (Lesson 1-2)

- Ⓕ The division operation is commutative.
- Ⓖ The subtraction operation is associative.
- Ⓗ The subtraction operation is commutative.
- Ⓘ The addition operation is associative.

 Gridded Response Evaluate $\frac{rt}{s} + r^2$ if $r = 3$, $s = 2$, and $t = 4$. (Lesson 1-2)

Thursday

Short Response Draw and label a number line and graph the integers below. Use the number line to write the integers in order from **least** to **greatest**. (Lesson 1-3)

$$0, -1, 4, -3, -5, 3$$

Friday

Gridded Response The graph shows the average number of visitors per year to theme parks in Orlando. In millions, how many more people visit Magic Kingdom in a year than Animal Kingdom and MGM combined? (Lesson 1-2)

Theme Park Visitors

Week 2

Monday

Extended Response If Derek saves half his allowance for 12 weeks, he will have enough money saved to purchase a $60 skateboard. (Lesson 1-9)

Part A Write an algebraic equation for this situation.

Part B Then solve your equation, showing all steps.

Part C How much allowance does Derek get each week?

Tuesday

Multiple Choice Write *six less than two times a number is fourteen* as an algebraic equation. (Lesson 1-7)

Ⓐ $6 - 2x = 14$ Ⓑ $6 + 2x = 14$

Ⓒ $2x - 6 = 14$ Ⓓ $2x + 6 = 14$

Gridded Response Evaluate $3|a| - |b|$ if $a = -3$ and $b = 5$. (Lesson 1-3)

Wednesday

Multiple Choice The temperature at 6:00 P.M. was 10°. Between 6:00 P.M. and midnight, the temperature dropped three different times—dropping 4° each time. What was the temperature at midnight? (Lesson 1-6)

Ⓕ $-12°$ Ⓖ $-2°$ Ⓗ $0°$ Ⓘ $2°$

Gridded Response Simplify $\dfrac{2(6|-3| - 2|-2| + 1)}{3}$. (Lesson 1-3)

Thursday

Short Response Write a number sentence to represent the transactions in Sarah's checking account. An overdraft charge happened when her balance fell below the minimum allowed. (Lesson 1-4)

Bank Statement		
Date	Type of Transaction	Amount
3/31	Previous Balance	$68
4/6	Check #201	$35
4/12	Check #202	$40
4/13	Overdraft Charge	$25
4/15	Deposit	$82
4/30	Final Balance	$50

Friday

Gridded Response Lake Okeechobee, the largest lake in Florida, is 17 feet deep at its deepest point. The average depth of the lake is 8 feet less than this. What is the average depth of the lake in feet? (Lesson 1-5)

Monday

Extended Response Amanda is planning a rectangular vegetable garden using a roll of border fencing that she was given. The roll of fencing is $45\frac{3}{4}$ feet long. (Lesson 2-6)

Part A If she makes the width of the garden $10\frac{1}{4}$ feet, what must the length be?

Part B Draw and label the sides of the garden. Show all your work.

Tuesday

Multiple Choice Each year, 15.6 million people visit Walt Disney World's Magic Kingdom, whereas only 6.0 million visit Animal Kingdom. How many times as many people visit Magic Kingdom than Animal Kingdom? (Lesson 2-4)

A $\frac{5}{13}$ **B** $2\frac{3}{5}$ **C** 3 **D** $3\frac{1}{5}$

Gridded Response What is $1\frac{1}{4} \cdot 720$? (Lesson 2-3)

Wednesday

Multiple Choice Which of the following is NOT equal to $-\frac{3}{5}$? (Lesson 2-1)

F $-\frac{6}{10}$ **G** -0.6

H -0.60 **I** $-\frac{10}{15}$

Gridded Response What fraction in lowest terms is represented by the shaded portion of the model decimal shown? (Lesson 2-2)

Thursday

Short Response Write the rational numbers in order from **least to greatest**. Explain your strategy. (Lesson 2-2)

$$0.3, \frac{1}{3}, -\frac{2}{3}, -1, -0.3, 0.03$$

Friday

Multiple Choice Find the area of the parallelogram. (Lesson 2-3)

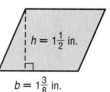

$h = 1\frac{1}{2}$ in.

$b = 1\frac{3}{8}$ in.

A $2\frac{1}{16}$ square inches

B $3\frac{3}{16}$ square inches

C $\frac{11}{12}$ square inches

D $\frac{16}{33}$ square inches

Week 4

Monday

Extended Response Mattie uses colored sand to make decorative sand candles in clear different-sized jars. She makes candles containing $2\frac{1}{4}$ cups sand, $3\frac{3}{8}$ cups sand, and $5\frac{3}{4}$ cups sand. (Lesson 2-6)

Part A Copy and complete the table showing the decimal representation of each fractional amount and the totals as both fractions and decimals.

Part B Which candle accounts for more than half the sand used? Explain.

Candle	Amount of Sand (c)	
	Fraction	Decimal
1	$2\frac{1}{4}$	
2	$3\frac{3}{8}$	
3	$5\frac{3}{4}$	
TOTALS		

Tuesday

Multiple Choice Write $m \cdot m \cdot m \cdot m \cdot n \cdot n \cdot n \cdot n \cdot n$ using exponents. (Lesson 2-8)

Ⓐ $(mn)^9$　　　Ⓑ m^5n^4

Ⓒ m^4n^5　　　Ⓓ $(mn \cdot mn)^5$

 Gridded Response Find the value of b in $b - (-0.38) = 4.1$. (Lesson 2-7)

Wednesday

Multiple Choice Which expression is equivalent to $\frac{1}{4^{-2}}$? (Lesson 2-8)

Ⓕ $\frac{1}{16}$　　Ⓖ -4　　Ⓗ $-\frac{1}{16}$　　Ⓘ 16

 Gridded Response Write 3.7×10^3 in standard form. (Lesson 2-9)

Thursday

Short Response Manatee Springs in Levy County, Florida, produces a water flow of up to 150,000,000 gallons of crystal clear water per day. Write this in scientific notation. How many gallons flow from Manatee Springs in one hour? Write that amount in scientific notation also. (Lesson 2-9)

Friday

Multiple Choice A cube is a three-dimensional "box" in which the length, width, and height are all the same measure. Using the formula for the volume of a cube $V = \ell \cdot w \cdot h$, find the volume of this cube. (Lesson 2-8)

Ⓐ $3x$

Ⓑ x^3

Ⓒ x^2

Ⓓ $3x^3$

x

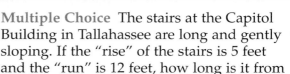
Monday

 Extended Response What is the perimeter of a square whose area is 121 square centimeters? Draw a picture and show all your work. (Lesson 3-1)

Tuesday

Multiple Choice $\sqrt{90}$ is between what two whole numbers? (Lesson 3-2)

- (A) 8 and 9
- (B) 9 and 10
- (C) 10 and 11
- (D) 11 and 12

 Gridded Response $\sqrt{52}$ is closest to what whole number? (Lesson 3-3)

Wednesday

Multiple Choice The stairs at the Capitol Building in Tallahassee are long and gently sloping. If the "rise" of the stairs is 5 feet and the "run" is 12 feet, how long is it from point A to point B? (Lesson 3-4)

- (F) 13 ft
- (G) 12 ft
- (H) 11 ft
- (I) 10 ft

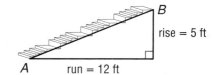

 Gridded Response The base of a ten-foot ladder stands six feet from a house. How many feet up the side of the house does the ladder reach? (Lesson 3-4)

Thursday

 Short Response Place the numbers in the appropriate region of the Venn diagram. Explain your choices. (Lesson 3-3)

$$0, -1, \frac{1}{2}, \sqrt{2}, 0.\overline{81}$$

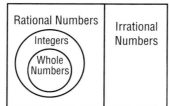

Friday

Multiple Choice Which numbers are ordered correctly from **least** to **greatest**? (Lesson 3-3)

- (A) $1, \sqrt{2}, 2, 3.1, \sqrt{5}$
- (B) $1, \sqrt{2}, 2, \sqrt{5}, 3.1$
- (C) $\sqrt{2}, 1, 2, 3.1, \sqrt{5}$
- (D) $\sqrt{2}, 1, 2, \sqrt{5}, 3.1$

Week 6

Monday

 Extended Response Jack and Mayuko left school at the same time to run errands. Mayuko walked east 1,600 feet to the bookstore. Jack walked south 1,200 feet to the gym. Jack and Mayuko wanted to meet for ice cream halfway between the bookstore and the gym. (Lesson 3-4)

Part A Draw a diagram to represent the problem.

Part B How many feet (starting from school) would each have to walk? Show all your work.

Tuesday

Multiple Choice Find the distance between the points shown. (Lesson 3-6)

- Ⓐ 6 units
- Ⓑ 7.5 units
- Ⓒ 8.1 units
- Ⓓ 9 units

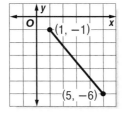

 Gridded Response Find the length in centimeters of the missing side. (Lesson 3-4)

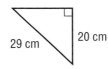

29 cm 20 cm

Wednesday

Multiple Choice Which of the following sets is NOT a Pythagorean triple? (Lesson 3-5)

- Ⓕ 3, 4, 5
- Ⓖ 5, 12, 14
- Ⓗ 8, 6, 10
- Ⓘ 15, 8, 17

 Gridded Response What is the distance in units between the points shown? (Lesson 3-6)

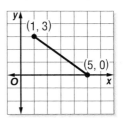

Thursday

 Short Response On a coordinate grid, graph the points at $(-4, 2)$ and $(-1, -1)$. What is the distance between the points? (Lesson 3-6)

Friday

Multiple Choice A right triangle is formed by connecting Tallahassee to Jacksonville to Orlando. The distance from Tallahassee to Jacksonville is 163 miles and the distance from Jacksonville to Orlando is 135 miles. How far, to the nearest mile, is it from Tallahassee to Orlando? (Lesson 3-5)

- Ⓐ 298 miles
- Ⓑ 250 miles
- Ⓒ 212 miles
- Ⓓ 200 miles

Monday

Extended Response The table shows the weight of a puppy each month during the first year. (Lesson 4-2)

Part A Find the rate of change in the puppy's weight from birth to 6 months and from 6 months to 12 months.

Part B Are the rates the same? Why or why not?

Month	Weight (lb)	Month	Weight (lb)
Birth	2.5	7	8
1	2.75	8	8
2	3	9	9
3	4	10	10.2
4	4.5	11	10.8
5	5.75	12	11.4
6	7.5		

Tuesday

Multiple Choice The slope of the line passing through the points at $(0, -2)$ and $(-1, 0)$ can be found with which expression? (Lesson 4-3)

- **A** $\frac{0-2}{-1-0}$
- **B** $\frac{0+2}{-1-0}$
- **C** $\frac{-1+0}{-2+0}$
- **D** $\frac{-1-0}{2+0}$

Gridded Response Find the slope of the line shown. (Lesson 4-3)

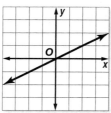

Wednesday

Multiple Choice Between 1990 and 1995, the average salary for a professional baseball player skyrocketed. What is the rate of change representing this increase? (Lesson 4-3)

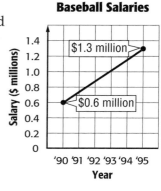

Baseball Salaries

- **F** $1,400 per year
- **G** $14,000 per year
- **H** $140,000 per year
- **I** $14,000,000 per year

Gridded Response Write the ratio 6:27 as a fraction in **lowest** terms. (Lesson 4-1)

Thursday

Short Response Between 1983 and 1993, the unemployment rate dropped from 10.2% to 7%. What was the rate of change for those years? Show your work. (Lesson 4-2)

Friday

Multiple Choice The state of Florida has 120 legislators in its State Congress. If the population of Florida is approximately 16,917,308, which ratio shows the number of people to each legislator? (Lesson 4-1)

- **A** 140,978:1
- **B** 16,917,308:120
- **C** 1:140,978
- **D** 120:16,917,308

Monday

 Extended Response Determine whether the two rectangles are similar. Explain. (Lesson 4-5)

3 in. ⌐ 8 in.

1.5 in. ⌐ 5 in.

Tuesday

Multiple Choice
Triangle *ABC* is similar to △*RST*. Find the length of $\overline{ST}$. (Lesson 4-5)

Ⓐ 2 units   Ⓑ 1.5 units
Ⓒ 1 units   Ⓓ 0.75 unit

 Gridded Response Solve $\frac{3}{x} = \frac{7}{12}$. (Lesson 4-4)

Wednesday

Multiple Choice An architect is building a scale model of the office complex he is designing. If the tallest building of the complex is 160 feet and the scale is 1 inch = 8 feet, how tall is the highest point of the scale model? (Lesson 4-6)

Ⓕ 20 inches   Ⓖ 25 inches
Ⓗ 30 inches   Ⓘ 40 inches

 Gridded Response The triangle formed by the flagpole and its shadow is similar to the given triangle. How many feet tall is the flagpole? (Lesson 4-7)

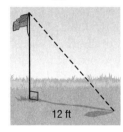

12 ft

6 ft

8 ft

Thursday

 Short Response Raccoons are common to the woods of north central Florida. A female raccoon usually has between 1 and 8 babies. If a den of 3 mother raccoons lives with 14 babies, about how many babies does each mother raccoon have? Write and solve a proportion to reflect the information given. (Lesson 4-4)

Friday

Multiple Choice Using which scale would a scale model of a building appear $\frac{1}{20}$ the size of the actual building? (Lesson 4-6)

Ⓐ 3 inches = 10 feet
Ⓑ 5 inches = 10 feet
Ⓒ 6 inches = 10 feet
Ⓓ 10 inches = 12 feet

Monday

Extended Response (Lesson 5-5)

Part A Explain in detail how you could ESTIMATE 126% of 400.

Part B What is 126% of 400?

Part C How does your estimate compare with the actual value?

Tuesday

Multiple Choice What number is 42% of 300? (Lesson 5-3)

- (A) 12,600
- (B) 1,260
- (C) 126
- (D) 12.6

 Gridded Response Write 105% as a fraction in lowest terms. (Lesson 5-2)

Wednesday

Multiple Choice Write 0.5% as a decimal. (Lesson 5-2)

- (F) 5
- (G) 0.5
- (H) 0.05
- (I) 0.005

Gridded Response What percent of the octagon is shaded? (Lesson 5-1)

Thursday

Short Response In 2001, 68 of the freshmen who enrolled at Florida State University were National Merit Finalists. If there were 27,128 new freshmen in 2001, what percent were National Merit finalists? Show your work. (Lesson 5-3)

Friday

Multiple Choice The example below represents three types of transformations. Which transformations are shown? (Lesson 4-8)

- (A) reflection, translation, rotation
- (B) reflection, dilation, translation
- (C) translation, rotation, dilation
- (D) rotation, dilation, reflection

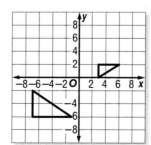

Week 10

Monday

Extended Response Suppose you and a friend go out to eat and the bill is $28.79. You want to leave a 15% tip. (Lesson 5-5)

Part A Explain how you could quickly, and without a calculator, ESTIMATE a 15% tip for the bill.

Part B What would be the exact amount of a 15% tip?

Tuesday

Multiple Choice 18% of 40 is closest to what whole number? (Lesson 5-5)

(A) 10   (B) 9
(C) 8   (D) 4

Gridded Response ESTIMATE the percent shaded to the nearest whole percent. (Lesson 5-5)

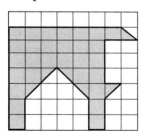

Wednesday

Multiple Choice Twenty-seven is 30% of what number? (Lesson 5-6)

(F) $0.0\overline{1}$ (G) 8.1
(H) 81   (I) 90

Gridded Response A pair of name brand athletic shoes were on sale for $69.99. If the tax is 6%, what is the **total** cost? (Lesson 5-6)

Thursday

Short Response Jose went shopping during the After Holiday Sale at the local mall. He found shirts that were originally $39.98 on sale for $23.49. What was the percent markdown on the shirts? Show your work. (Lesson 5-7)

Friday

Multiple Choice Write 4,830,000,000 in scientific notation. (Lesson 2-9)

(A) 483×10^7 (B) 48.3×10^8
(C) 4.83×10^9 (D) 0.483×10^{10}

Monday

 Extended Response A Florida map shows the coordinates of four locations: Lake City at (0, 7), Jacksonville Beach at (3, 7), Orlando at (3, 2), and Tampa at (1, 0). (Lesson 6-4)

Part A Plot the locations on a coordinate plane. Then connect the points.

Part B Name the figure. Explain your choice.

Part C If the angle at Tampa measures 74°, the angle at Orlando measures 114°, and the angle at Jacksonville Beach measures 93°, what is the measure of the angle at Lake City?

Tuesday

Multiple Choice Find $m\angle B$. (Lesson 6-2)

(A) 20°
(B) 50°
(C) 70°
(D) 80°

A 42° 68° *C*   *B* ?

Gridded Response Find $m\angle 2$ in degrees. (Lesson 6-1)

2   70°

Wednesday

Multiple Choice Name all angles that are congruent to $\angle 6$ if $a \parallel b$ and $c \parallel d$. (Lesson 6-1)

(F) 1, 3, 8, 9, 11, 14, 16
(G) 2, 4, 5, 7, 10, 12, 13, 15
(H) 7, 10, 11
(I) 1, 2, 5

```
        c      d
      1/2    3/4    a
      5/6    7/8
    9/10   11/12    b
  13/14   15/16
```

Gridded Response What is the measure of $\angle 3$ in degrees? (Lesson 6-1)

3   35°

Thursday

 Short Response Find the lengths of the missing sides. Show all your work. (Lesson 6-3)

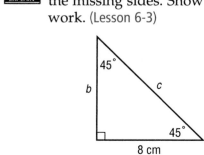

45°

b   *c*

45°
8 cm

Friday

Multiple Choice Which **best** describes the triangle shown? (Lesson 6-2)

(A) acute and isosceles
(B) obtuse and isosceles
(C) acute and equilateral
(D) obtuse and equilateral

Week 12

Monday

 Extended Response Use the triangle shown to draw and label a graph for each transformation. (Lesson 6-8)

Part A Translate the triangle 3 units up and 2 units left.

Part B Then reflect the figure from Part A over the *y*-axis.

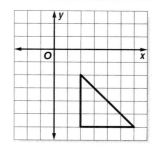

Tuesday

Multiple Choice The orange blossom is Florida's state flower. If the orange blossom shown is translated 4 units right and 2 units up, where will the new point *B* be located? (Lesson 6-8)

- **A** (1, −3)
- **B** (0, −3)
- **C** (−1, −1)
- **D** (−1, −2)

 Gridded Response What is the measure in degrees of ∠1? (Lesson 6-5a)

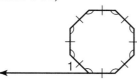

Wednesday

Multiple Choice If △*ABC* ≅ △*RST*, which statement is NOT true? (Lesson 6-5)

- **F** ∠*B* ≅ ∠*S*
- **G** $\overline{AB}$ ≅ $\overline{RS}$
- **H** $\overline{AC}$ ≅ $\overline{ST}$
- **I** ∠*C* ≅ ∠*T*

Gridded Response What is the *y*-coordinate of the point at (2, −3) after it is reflected across the *x*-axis? (Lesson 7-7)

Thursday

 Short Response Reflect the figure over the *y*-axis. Draw and label the vertices of the new figure. (Lesson 6-7)

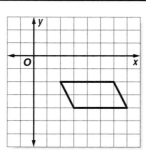

Friday

Multiple Choice How many lines of symmetry are there in the letter E? (Lesson 6-6)

- **A** 1
- **B** 2
- **C** 3
- **D** 4

Monday

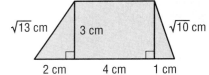

THINK SOLVE EXPLAIN

Extended Response Refer to the trapezoid shown. (Lesson 7-1)

Part A Find the area of the trapezoid using the formula for area of a trapezoid.

Part B Find the area of the trapezoid by finding the areas of the two triangles and the rectangle and add.

Part C State whether the area is the same using both methods.

√13 cm   3 cm   √10 cm
2 cm   4 cm   1 cm

Tuesday

Multiple Choice What is the total area of the figure shown? (Lesson 7-3)

- Ⓐ 92.5 square centimeters
- Ⓑ 64.3 square centimeters
- Ⓒ 56.5 square centimeters
- Ⓓ 36.0 square centimeters

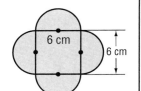

6 cm
6 cm

Gridded Response What is the area of the shaded region, to the nearest hundredth of a square inch? (Lesson 7-2)

4 in.

Wednesday

Multiple Choice The clock shown has a diameter of 10 inches. Which is the **best** estimate for the area of the shaded region? (Lesson 7-2)

- Ⓕ 78.5 square inches
- Ⓖ 32.7 square inches
- Ⓗ 15.7 square inches
- Ⓘ 6.5 square inches

Gridded Response What is the area to the nearest hundredth of a square foot of the seal on a Florida flag measuring 4 feet wide by 6 feet long if the circular seal's diameter is half the width of the flag? (Lesson 7-2)

Thursday

THINK SOLVE EXPLAIN

Short Response Find the area of the shaded triangle. Explain how it compares to the area of the parallelogram. (Lesson 7-1)

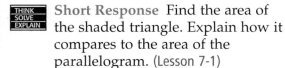

3 cm
10 cm

Friday

Multiple Choice Figure A is an example of which transformation of Figure B? (Lesson 6-9)

- Ⓐ translation
- Ⓑ rotation
- Ⓒ reflection
- Ⓓ dilation

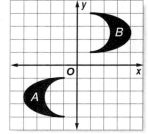

Week 14

Monday

Extended Response Find the volume of the figure. Show your work. (Lesson 7-5)

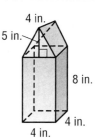

Tuesday

Multiple Choice What is the ratio of the volume of a cone to the volume of a cylinder if the radius and height of the cone are the same as the radius and height of the cylinder? (Lesson 7-6)

Ⓐ 1:3 　　　　Ⓑ 3:1
Ⓒ 1:9 　　　　Ⓓ 9:1

Gridded Response Find the area of the regular hexagon in square centimeters. (Lesson 7-3)

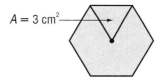

$A = 3 \text{ cm}^2$

Wednesday

Multiple Choice To find the surface area of a square pyramid, you would calculate the area of which shapes making up the square pyramid? (Lesson 7-8)

Ⓕ 1 square and 3 triangles
Ⓖ 1 square and 4 triangles
Ⓗ 5 triangles
Ⓘ 3 squares and 2 triangles

Gridded Response The Dolphin Research Center in Grassy Key, Florida, provides a "Dolphin Encounter" where you can swim with a dolphin in a cylindrical tank 30 feet across and 6 feet deep. What is the volume of water in cubic feet held by the dolphin tank? Answer to the nearest whole number. (Lesson 7-6)

Thursday

Short Response Which shipping container has a **greater** volume? Support your answer mathematically. (Lesson 7-5)

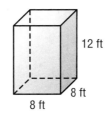

5 ft
8 ft
12 ft
8 ft
8 ft

Friday

Multiple Choice Name the solids that make up the figure shown below. (Lesson 7-4)

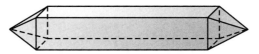

Ⓐ 1 pyramid and 1 prism
Ⓑ 2 pyramids and 1 prism
Ⓒ 1 pyramid and 2 prisms
Ⓓ 2 pyramids and 2 prisms

Monday

Extended Response The 8th grade Student Council at Northrup High School is selling T-shirts as a fund-raiser. The shirts come in yellow (Y), blue (B), or white (W). You can choose from a black and white logo (BW) or a logo in color (C). The sizes are small (S), medium (M), and large (L). (Lesson 8-2)

Part A Draw a tree diagram to show the different T-shirt selections that can be made.

Part B How many different T-shirts are possible?

Tuesday

Multiple Choice How many different 4-digit locker combinations are possible using the digits 1–9 with no digit repeated? (Lesson 8-3)

Ⓐ 3,024   Ⓑ 5,040

Ⓒ 5,184   Ⓓ 6,561

Gridded Response What is the probability that in 3 spins you land on yellow all 3 times? (Lesson 8-1)

Wednesday

Multiple Choice In a bag containing 3 blue, 3 red, 2 green, and 2 yellow marbles, what is the probability of randomly selecting a marble that is NOT blue on your first selection? (Lesson 8-1)

Ⓕ $\frac{3}{10}$ Ⓖ $\frac{5}{10}$

Ⓗ $\frac{6}{10}$ Ⓘ $\frac{7}{10}$

Gridded Response What are the odds of landing on a number divisible by 3? Write as a fraction. (Lesson 8-1)

Thursday

Short Response The Silverwood family is visiting Walt Disney World's Magic Kingdom. The children want to see Fantasyland, Adventureland, Tomorrowland, and Frontierland. In how many orders can they visit the four areas of Magic Kingdom? Show your work. (Lesson 8-3)

Friday

Gridded Response Five points are located randomly within a rectangle so that no three points lie on the same line. How many line segments can be drawn connecting these points? (Lesson 8-4)

Week 16

Monday

 Extended Response From a bag containing 6 red, 8 blue, 4 yellow, and 2 white marbles, you will draw 2 marbles without replacement. (Lesson 8-5)

Part A Explain why the probability of selecting 2 yellow marbles is NOT $\frac{4}{20} \cdot \frac{4}{20}$ or $\frac{1}{25}$.

Part B What **is** the probability of selecting 2 yellow marbles?

Tuesday

Multiple Choice A coin was tossed 50 times, and the results recorded. What is the experimental probability of tossing the coin and landing on tails? (Lesson 8-6)

H	T
29	21

Ⓐ $\frac{21}{29}$ Ⓑ $\frac{29}{50}$

Ⓒ $\frac{1}{2}$ Ⓓ $\frac{21}{50}$

 Gridded Response You toss 3 coins. What is the theoretical probability that all 3 land on heads? (Lesson 8-5)

Wednesday

Multiple Choice Tallahassee is the capital of Florida. Each of the letters spelling TALLAHASSEE is put on a letter card and one card is drawn at random. What is the probability of selecting a vowel on the first draw? (Lesson 8-1)

Ⓕ $\frac{1}{5}$ Ⓖ $\frac{5}{11}$

Ⓗ $\frac{6}{11}$ Ⓘ $\frac{5}{6}$

 Gridded Response A sample of 50 students was asked to name their favorite ice cream. The results are shown here. What percent of the students surveyed prefer chocolate? (Lesson 8-7)

Flavor	Number
vanilla	24
chocolate	16
strawberry	6
Neapolitan	4

Thursday

 Short Response Determine how many ways a president, vice president, secretary, and treasurer can be chosen from a 10-member Student Council. Show your work or explain in words. (Lesson 8-3)

Friday

Multiple Choice Which situation is represented by $6 \cdot 5 \cdot 4$? (Lesson 8-4)

Ⓐ the number of arrangements of 6 people in line

Ⓑ the number of ways to choose a committee of any 3 from 6 people

Ⓒ the number of ways to choose a president, secretary, and treasurer from 6 people

Ⓓ the number of 3-digit locker combinations from the numbers 0–6

Countdown to FCAT

Week 17

Monday

 Extended Response The frequency table shows the grams of carbohydrates in a wide variety of breakfast foods. (Lesson 9-1)

Part A Draw a histogram to represent the data.

Part B Which adjacent intervals contain most of the data?

Breakfast Foods		
Carbohydrates (g)	Tally	Frequency
6–10	III	3
11–15	IIII	4
16–20	I	1
21–25	JHT I	6
26+	JHT	5

Tuesday

Multiple Choice Of the 1,200 students at Rosewood High School, how many can be predicted to receive a bachelor's degree as their highest degree earned? (Lesson 9-2)

Ⓐ 439
Ⓑ 268
Ⓒ 142
Ⓓ 63

Education

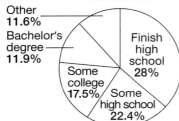

Other 11.6%
Bachelor's degree 11.9%
Some college 17.5%
Some high school 22.4%
Finish high school 28%

 Gridded Response What is the mean of this data set? (Lesson 9-4)

> 84, 87, 78, 91, 89, 90

Wednesday

Multiple Choice Of the 35,462 students enrolled at Florida State University in the fall of 2001, how many more undergraduates were there than graduate students? (Lesson 9-2)

Ⓕ 27,128
Ⓖ 20,816
Ⓗ 6,312
Ⓘ 2,021

Students at Florida State

Graduate students 17.8%
Other 5.7%
Undergraduates 76.5%

Gridded Response What is the median of the data set? (Lesson 9-4)

> 84, 87, 78, 91, 89, 90

Thursday

Short Response Draw a circle graph of the data in the table showing what percent of students prefer each type of music. (Lesson 9-2)

Favorite Type of Music	
Type	Number of Students
rock	73
country	20
classical	3
pop	104

Friday

Multiple Choice In what year did the number of in-line skaters pass the 5-million-skaters mark? (Lesson 9-1)

Ⓐ 1990
Ⓑ 1991
Ⓒ 1992
Ⓓ 1993

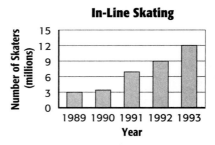

In-Line Skating

Week 18

Monday

Extended Response (Lesson 9-5)

Part A Find the median, range, lower quartile, and upper quartile for the rainfall data. Show all your work.

Part B Draw a box-and-whisker plot for the data.

Part C Write an observation about the data.

Rainfall for Daytona Beach, 2001			
Month	**Rainfall (in.)**	**Month**	**Rainfall (in.)**
Jan.	0.88	July	9.55
Feb.	0.38	Aug.	3.57
Mar.	9.98	Sept.	16.11
Apr.	0.28	Oct.	3.22
May	1.77	Nov.	6.93
June	5.26	Dec.	0.35

Tuesday

Multiple Choice The weight gain for a puppy over the first year of its life could **best** be graphed using which type of graph? (Lesson 9-3)

- Ⓐ circle graph
- Ⓑ histogram
- Ⓒ bar graph
- Ⓓ line graph

Gridded Response Which data point if removed, would have the **greatest** effect on the mean? (Lesson 9-5)

72, 78, 34, 71, 70, 94

Wednesday

Multiple Choice Which statement is true of the data set shown? (Lesson 9-4)

Number of TVs in Your Home		
Number	**Tally**	**Frequency**
1	卌	5
2	卌 卌 II	12
3	卌 卌 卌 IIII	19
4	IIII	4

- Ⓕ The mean and median are the same.
- Ⓖ The median and mode are the same.
- Ⓗ The mean is greater than the median.
- Ⓘ The mode is 19.

Gridded Response What is the range of the data set? (Lesson 9-5)

2.9, 2.4, 2.8, 3.0, 1.1, 1.9, 1.7

Thursday

Short Response Calculate the mean, median, and mode of the bowling scores shown below. Then order the mean, median, and mode from **least to greatest**. (Lesson 9-4)

135	188	188	136
121	143	158	140

Friday

Gridded Response What is the median of the data shown in the box-and-whisker plot? (Lesson 9-6)

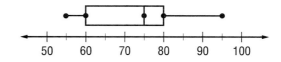

Monday

 Extended Response Manatee Springs, one of Florida's largest and most beautiful natural springs, has a water flow of between 50 million and 150 million gallons of water per day. (Lesson 10-5)

Part A Write an inequality to represent *between 50 million and 150 million*.

Part B Write an inequality that describes how many gallons of water flow from Manatee Springs in one hour.

Tuesday

Multiple Choice Michael is 2 years older than his sister Brooke. Write an expression that represents the sum of Michael and Brooke's ages. (Lesson 10-1)

  A $2(x + 1)$ **B** $x^2 + 2$

  C $x + 2$ **D** $x + 2x$

 Gridded Response What is the least value of x that will solve $3x \geq 12$? (Lesson 10-7)

Wednesday

Multiple Choice A plumber charges $40 for a house call plus 10% over the cost of any materials needed for repair. Which equation represents the total amount t charged by the plumber using m dollars in repair materials? (Lesson 10-3)

  F $t = 0.10m + 40$ **G** $t = 0.10 + 40m$

  H $t = 1.1m + 40$ **I** $t = 1.1 + 40m$

 Gridded Response What is the greatest value of y that will solve $y - 4 \leq -2$? (Lesson 10-6)

Thursday

 Short Response Find the value of x so that the two rectangles have the same perimeter. Show your work. (Lesson 10-4)

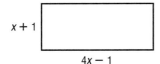

$x + 1$ $4x - 1$ $2x$ $4x - 2$

Friday

Multiple Choice Which inequality represents **no more than** -2? (Lesson 10-5)

  A $x > -2$

  B $x \geq -2$

  C $x < -2$

  D $x \leq -2$

Week 20

Monday

Extended Response (Lesson 11-3)

Part A Graph $y = 3x - 1$.

Part B Make a table that shows the coordinates of four points on the graph of the equation.

Part C Is this a function? Explain.

Tuesday

Multiple Choice Which statement is true of the scatter plot shown? (Lesson 11-6)

- Ⓐ There is a negative correlation.
- Ⓑ There is a positive correlation.
- Ⓒ As x increases, y increases.
- Ⓓ There is no obvious pattern.

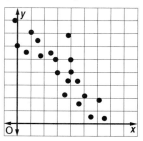

 Gridded Response What is the next term in the sequence shown? (Lesson 11-1)

$$4, 1, \frac{1}{4}, \frac{1}{16}, \dots$$

Wednesday

Multiple Choice Which statement is true? (Lesson 11-3)

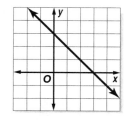

x	y
−3	6
−1	4
0	3
1	1
2	3

$$y = -x + 3$$

- Ⓕ The table does not represent the same function as the graph and the equation.
- Ⓖ The graph does not represent the same function as the table and the equation.
- Ⓗ The equation does not represent the same function as the graph and the table.
- Ⓘ All three models represent different functions.

 Gridded Response If $f(x) = -\frac{x}{2}$, find $f(-8)$. (Lesson 11-2)

Thursday

Short Response Begin with $\frac{1}{2}$. Write the next four terms of a geometric sequence whose common ratio is $\frac{3}{2}$. (Lesson 11-1)

Friday

Multiple Choice The graph shown represents the rise in citrus production in Florida between 1998 and 1999. What is the slope of this line? (Lesson 11-4)

- Ⓐ −2
- Ⓑ −1
- Ⓒ 1
- Ⓓ 2

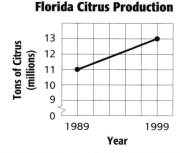

Florida Citrus Production

Monday

 Extended Response The height and weight of the Miami Heat Basketball players are given in the table. (Lesson 11-6)

Part A Make a scatter plot of the data.

Part B Then draw a line that seems to best represent the data.

Part C Describe the relationship.

Player	Height (in.)	Weight (lb)
Allen	82	255
Best	71	185
Butler	79	205
Carter	74	195
Ellis	80	240
Grant	81	254
House	73	180
James	74	188
Mourning	82	261
Stepania	85	255

Tuesday

Multiple Choice Simplify $(3x^2 - 2) - (4x^2 + 7)$. (Lesson 12-5)

- Ⓐ $-x^2 - 9$
- Ⓑ $-x^2 + 5$
- Ⓒ $x^2 - 9$
- Ⓓ $x^2 + 5$

Gridded Response Find $m\angle 1$ in degrees if $m\angle 1 = (10x - 15)°$. (Lesson 12-4)

1 / 3x°

Wednesday

Multiple Choice Which linear inequality is represented by the graph? (Lesson 11-8)

- Ⓕ $y > -1$
- Ⓖ $y < -2$
- Ⓗ $y \geq -2$
- Ⓘ $y \leq -2$

Gridded Response What is $-3x^2 - 2x$ if $x = -2$? (Lesson 12-6)

Thursday

Short Response Graph $y = x^2 - 1$ on a coordinate plane. Show a table of at least four pairs of x and y values to go with your graph. (Lesson 12-2)

Friday

Gridded Response What is the missing term of the sequence? (Lesson 11-1)

$$-\frac{3}{4}, -\frac{1}{4}, \frac{1}{4}, \underline{\quad ? \quad}, 1\frac{1}{4}, \dots$$

Florida Edition

Mathematics

Applications and Concepts

Course 3

Bailey
Day
Frey
Howard
Hutchens
McClain

Moore-Harris
Ott
Pelfrey
Price
Vielhaber
Willard

McGraw Hill Glencoe

New York, New York   Columbus, Ohio   Chicago, Illinois   Peoria, Illinois   Woodland Hills, California

The McGraw·Hill Companies

The Standardized Test Practice features in this book were aligned and verified
by The Princeton Review, the nation's leader in test preparation. Through its
association with McGraw-Hill, The Princeton Review offers the best way to
help students excel on standardized assessments.

The Princeton Review is not affiliated with Princeton University or Educational Testing Service.

The USA TODAY® service mark, USA TODAY Snapshots® trademark and other
content from USA TODAY® has been licensed by USA TODAY® for use for certain
purposes by Glencoe/McGraw-Hill, a Division of The McGraw-Hill Companies, Inc.
The USA TODAY Snapshots® and the USA TODAY® articles, charts, and photographs
incorporated herein are solely for private, personal, and noncommerical use.

Microsoft Excel is a registered trademark of Microsoft Corporation in the United States
and other countries.

Send all inquiries to:
Glencoe/McGraw-Hill
8787 Orion Place
Columbus, OH 43240-4027

ISBN: 0-07-860139-8

1 2 3 4 5 6 7 8 9 10 043/027 10 09 08 07 06 05 04 03

Contents in Brief

Rhonda Bailey
Mathematics Consultant
Mathematics by Design
DeSoto, Texas

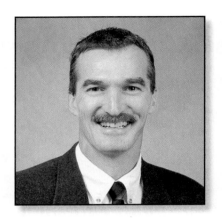

Roger Day, Ph.D.
Associate Professor
Illinois State University
Normal, Illinois

Patricia Frey
Director of Staffing and
Retention
Buffalo City Schools
Buffalo, New York

Arthur C. Howard
Mathematics Teacher
Houston Christian High
School
Houston, Texas

**Deborah T. Hutchens,
Ed.D.**
Assistant Principal
Great Bridge Middle School
Chesapeake, Virginia

Kay McClain, Ed.D.
Assistant Professor
Vanderbilt University
Nashville, Tennessee

Beatrice Moore-Harris
Mathematics Consultant
League City, Texas

Jack M. Ott, Ph.D.
Distinguished Professor of
 Secondary Education
 Emeritus
University of South Carolina
Columbia, South Carolina

Ronald Pelfrey, Ed.D.
Mathematics Specialist
Appalachian Rural Systemic
 Initiative
Lexington, Kentucky

Jack Price, Ed.D.
Professor Emeritus
California State Polytechnic
 University
Pomona, California

Kathleen Vielhaber
Mathematics Specialist
Parkway School District
St. Louis, Missouri

Teri Willard, Ed.D.
Adjunct Instructor and
 Mathematics Consultant
Montana State University
Bozeman, Montana

Contributing Authors

USA TODAY
The USA TODAY Snapshots®, created
by USA TODAY®, help students make the
connection between real life and mathematics.

Dinah Zike
Educational Consultant
Dinah-Might Activities, Inc.
San Antonio, Texas

Content Consultants

Each of the Content Consultants reviewed every chapter and gave suggestions for improving the effectiveness of the mathematics instruction.

Mathematics Consultants

L. Harvey Almarode
Curriculum Supervisor, Mathematics K–12
Augusta County Public Schools
Fishersville, VA

Claudia Carter, MA, NBCT
Mathematics Teacher
Mississippi School for Mathematics and Science
Columbus, MS

Carol E. Malloy, Ph.D.
Associate Professor, Curriculum Instruction,
    Secondary Mathematics
The University of North Carolina at Chapel Hill
Chapel Hill, NC

Melissa McClure, Ph.D.
Mathematics Instructor
University of Phoenix On-Line
Fort Worth, TX

Robyn R. Silbey
School-Based Mathematics Specialist
Montgomery County Public Schools
Rockville, MD

Leon L. "Butch" Sloan, Ed.D.
Secondary Mathematics Coordinator
Garland ISD
Garland, TX

Barbara Smith
Mathematics Instructor
Delaware County Community College
Media, PA

Reading Consultant

Lynn T. Havens
Director
Project CRISS
Kalispell, MT

ELL Consultants

Idania Dorta
Mathematics Educational Specialist
Miami–Dade County Public Schools
Miami, FL

Frank de Varona, Ed.S.
Visiting Associate Professor
Florida International University
    College of Education
Miami, FL

Teacher Reviewers

Each Teacher Reviewer reviewed at least two chapters of the Student Edition, giving feedback and suggestions for improving the effectiveness of the mathematics instruction.

Royallee Allen
Teacher, Math Department Head
Eisenhower Middle School
San Antonio, TX

Dennis Baker
Mathematics Department Chair
Desert Shadows Middle School
Scottsdale, AZ

Rosie L. Barnes
Teacher
Fairway Middle School–KISD
Killeen, TX

Charlie Bialowas
Math Curriculum Specialist
Anaheim Union High School District
Anaheim, CA

Stephanie R. Boudreaux
Teacher
Fontainebleau Jr. High School
Mandeville, LA

Dianne G. Bounds
Teacher
Nettleton Junior High School
Jonesboro, AR

Susan Peavy Brooks
Math Teacher
Louis Pizitz Middle School
Vestavia Hills, AL

Karen Sykes Brown
Mathematics Educator
Riverview Middle School
Grundy, VA

Kay E. Brown
Teacher, 7th Grade
North Johnston Middle School
Micro, NC

Renee Burgdorf
Middle Grades Math Teacher
Morgan Co. Middle
Madison, GA

Kelley Summers Calloway
Teacher
Baldwin Middle School
Montgomery, AL

Carolyn M. Catto
Teacher
Harney Middle School
Las Vegas, NV

Claudia M. Cazanas
Math Department Chair
Fairmont Junior High
Pasadena, TX

David J. Chamberlain
Secondary Math Resource Teacher
Capistrano Unified School District
San Juan Capistrano, CA

David M. Chioda
Supervisor Math/Science
Marlboro Township Public Schools
Marlboro, NJ

Carrie Coate
7th Grade Math Teacher
Spanish Fort School
Spanish Fort, AL

Toinette Thomas Coleman
Secondary Mathematics Teacher
Caddo Middle Career & Technology
   School
Shreveport, LA

Linda M. Cordes
Math Department Chairperson
Paul Robeson Middle School
Kansas City, MO

Polly Crabtree
Teacher
Hendersonville Middle School
Hendersonville, NC

Dr. Michael T. Crane
Chairman Mathematics
B.M.C. Durfee High School
Fall River, MA

Tricia Creech, Ph.D.
Curriculum Facilitator
Southeast Guilford Middle School
Greensboro, NC

Lyn Crowell
Math Department Chair
Chisholm Trail Middle School
Round Rock, TX

B. Cummins
Teacher
Crestdale Middle School
Matthews, NC

Debbie Davis
8th Grade Math Teacher
Max Bruner, Jr. Middle School
Ft. Walton Beach, FL

Diane Yendell Day
Math Teacher
Moore Square Museums Magnet
   Middle School
Raleigh, NC

Wendysue Dodrill
Teacher
Barboursville Middle School
Barboursville, WV

Judith F. Duke
Math Teacher
Cranford Burns Middle School
Mobile, AL

Carol Fatta
Math/Computer Instructor
Chester Jr. Sr. M.S.
Chester, NY

Cynthia Fielder
Mathematics Consultant
Atlanta, GA

Georganne Fitzgerald
Mathematics Chair
Crittenden Middle School
Mt. View, CA

Jason M. Fountain
7th Grade Mathematics Teacher
Bay Minette Middle School
Bay Minette, AL

Sandra Gavin
Teacher
Highland Junior High School
Cowiche, WA

Ronald Gohn
8th Grade Mathematics
Dover Intermediate School
Dover, PA

Larry J. Gonzales
Math Department Chairperson
Desert Ridge Middle School
Albuquerque, NM

Shirley Gonzales
Math Teacher
Desert Ridge Middle School
Albuquerque, NM

Paul N. Hartley, Jr.
Mathematics Instructor
Loudoun County Public Schools
Leesburg, VA

Deborah L. Hewitt
Math Teacher
Chester High School
Chester, NY

Steven J. Huesch
Mathematics Teacher/Department
   Chair
Cortney Jr. High
Las Vegas, NV

Sherry Jarvis
8th Grade Math/Algebra 1 Teacher
Flat Rock Middle School
East Flat Rock, NC

Mary H. Jones
Math Curriculum Coordinator
Grand Rapids Public Schools
Grand Rapids, MI

Vincent D.R. Kole
Math Teacher
Eisenhower Middle School
Albuquerque, NM

Ladine Kunnanz
Middle School Math Teacher
Sequoyah Middle School
Edmond, OK

Barbara B. Larson
Math Teacher/Department Head
Andersen Middle School
Omaha, NE

Judith Lecocq
7th Grade Teacher
Murphysboro Middle School
Murphysboro, IL

Paula C. Lichiello
7th Grade Math and Pre-Algebra
   Teacher
Forest Middle School
Forest, VA

Michelle Mercier Maher
Teacher
Glasgow Middle School
Baton Rouge, LA

Jeri Manthei
Math Teacher
Millard North Middle School
Omaha, NE

Albert H. Mauthe, Ed.D.
Supervisor of Mathematics (Retired)
Norristown Area School District
Norristown, PA

Karen M. McClellan
Teacher & Math Department Chair
Harper Park Middle
Leesburg, VA

Ken Montgomery
Mathematics Teacher
Tri-Cities High School
East Point, GA

Helen M. O'Connor
Secondary Math Specialist
Harrison School District Two
Colorado Springs, CO

Cindy Ostrander
8th Grade Math Teacher
Edwardsville Middle School
Edwardsville, IL

Michael H. Perlin
8th Grade Mathematics Teacher
John Jay Middle School
Cross River, NY

Denise Pico
Mathematics Teacher
Jack Lund Schofield Middle School
Las Vegas, NV

Ann C. Raymond
Teacher
Oak Ave. Intermediate School
Temple City, CA

M.J. Richards
Middle School Math Teacher
Davis Middle School
Dublin, OH

Linda Lou Rohleder
Math Teacher, Grades 7 & 8
Jasper Middle School
Jasper, IN

Dana Schaefer
Pre-Algebra & Algebra I Teacher
Coachman Fundamental Middle
   School
Clearwater, FL

Donald W. Scheuer, Jr.
Coordinator of Mathematics
Abington School District
Abington, PA

Angela Hardee Slate
Teacher, 7th Grade Math/Algebra
Martin Middle School
Raleigh, NC

Mary Ferrington Soto
7th Grade Math
Calhoun Middle School-Ouachita
   Parish Schools
Calhoun, LA

Diane Stilwell
Mathematics Teacher/Technology
   Coordinator
South Middle School
Morgantown, WV

Pamela Ann Summers
K–12 Mathematics Coordinator
Lubbock ISD–Central Office
Lubbock, TX

Marnita L. Taylor
Mathematics Teacher/Department
   Chairperson
Tolleston Middle School
Gary, IN

Susan Troutman
Teacher
Dulles Middle School
Sugar Land, TX

Barbara C. VanDenBerg
Math Coordinator, K–8
Clifton Board of Education
Clifton, NJ

Mollie VanVeckhoven-Boeving
7th Grade Math and Algebra Teacher
White Hall Jr. High School
White Hall, AR

Mary A. Voss
7th Grade Math Teacher
Andersen Middle School
Omaha, NE

Christine Waddell
Teacher Specialist
Jordan School District
Sandy, UT

E. Jean Ware
Supervisor
Caddo Parish School Board
Shreveport, LA

Karen Y. Watts
9th Grade Math Teacher
Douglas High School
Douglas, AL

Lu Wiggs
Supervisor
I.S. 195
New York, NY

Teacher Advisory Board

Glencoe/McGraw-Hill wishes to thank the following teachers for their feedback on *Mathematics: Applications and Concepts*. They were instrumental in providing valuable input toward the development of this program.

Katie Davidson
Legg Middle School
Coldwater, MI

Lynanne Gabriel
Bradley Middle School
Huntersville, NC

Kathleen M. Johnson
New Albany-Plain Local Middle School
New Albany, OH

Ronald C. Myer
Indian Springs Middle School
Columbia City, IN

Mike Perlin
John Jay Middle School
Cross River, NY

Reema Rahaman
Brentwood Middle School
Brentwood, MO

Diane T. Scheuber
Elizabeth Middle School
Elizabeth, CO

Deborah Sykora
Hubert H. Humphrey Middle School
Bolingbrook, IL

DeLynn Woodside
Roosevelt Middle School,
 Oklahoma City Public Schools
Oklahoma City, OK

Field Test Schools

Glencoe/McGraw-Hill wishes to thank the following schools that field-tested pre-publication manuscript during the 2002–2003 school year. They were instrumental in providing feedback and verifying the effectiveness of this program.

Knox Community Middle School
Knox, IN

Roosevelt Middle School
Oklahoma City, OK

Brentwood Middle School
Brentwood, MO

Elizabeth Middle School
Elizabeth, CO

Legg Middle School
Coldwater, MI

Great Hollow Middle School
Nesconset, NY

Student Advisory Board

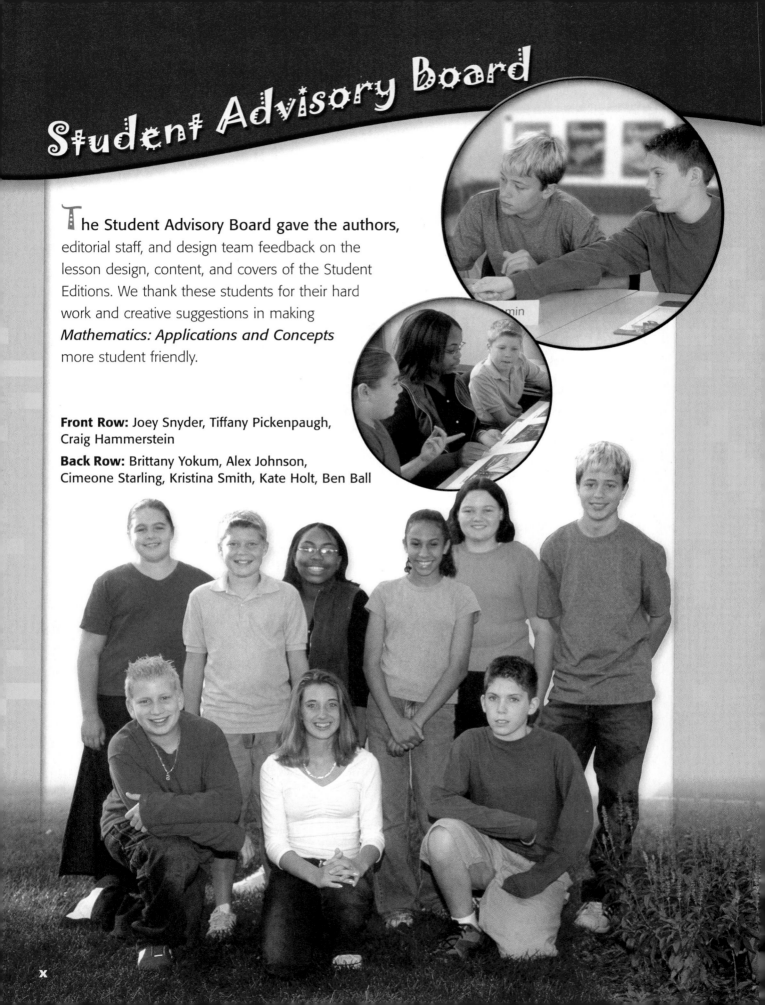

The Student Advisory Board gave the authors, editorial staff, and design team feedback on the lesson design, content, and covers of the Student Editions. We thank these students for their hard work and creative suggestions in making *Mathematics: Applications and Concepts* more student friendly.

Front Row: Joey Snyder, Tiffany Pickenpaugh, Craig Hammerstein

Back Row: Brittany Yokum, Alex Johnson, Cimeone Starling, Kristina Smith, Kate Holt, Ben Ball

Real Numbers and Algebra **2**

Lesson 1-7, p. 40

Table of Contents

UNIT 1

Lesson 2-9, p. 107

UNIT 1

Lesson 3-3, p. 127

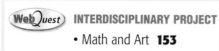
Lesson 4-6, p. 187

Prerequisite Skills
• Getting Started **155**
• Getting Ready for the Next Lesson **159, 164, 169, 173, 182, 187, 191**

 Study Organizer **155**

Study Skills
• Study Tips **160, 167, 170, 179, 184, 188, 194, 195**
• Homework Help **158, 163, 168, 172, 181, 186, 190, 196**

 Snapshots **159, 164**

Reading and Writing Mathematics
• Link to Reading **194**
• Reading Math **156, 161, 171**
• Writing Math **158, 163, 168, 172, 181, 183, 186, 189, 193, 196**

Standardized Test Practice
• Multiple Choice **159, 164, 169, 173, 174, 177, 182, 187, 191, 197, 201, 202**
• Short Response/Grid In **159, 164, 169, 173, 174, 182, 187, 191, 201, 203**
• Extended Response **203**
• Worked-Out Example **180**

UNIT 2

Lesson 5-2, p. 210

UNIT 3

Geometry and Measurement 252

Chapter 6 Geometry 254

Lesson 6-8, p. 299

UNIT 3

Chapter 7 Geometry: Measuring Area and Volume 312

Lesson 7-8, p. 354

Probability and Statistics 370

Lesson 8-5, p. 397

Prerequisite Skills
• Getting Started **373**
• Getting Ready for the Next Lesson **377, 383, 387,
391, 399, 403**

FOLDABLES Study Organizer **373**

Study Skills
• Study Tips **375, 397, 401, 404, 407**
• Homework Help **376, 382, 386, 390, 398, 402, 408**

 Snapshots **399**

Reading and Writing Mathematics
• Reading Math **375, 385, 389**
• Writing Math **376, 382, 386, 390, 392, 393, 398,
402, 408**

Standardized Test Practice
• Multiple Choice **377, 379, 383, 387, 391, 394, 399,
403, 409, 413, 414**
• Short Response/Grid In **383, 391, 394, 399, 403, 415**
• Extended Response **415**
• Worked-Out Example **385**

UNIT 4

Prerequisite Skills
- Getting Started **417**
- Getting Ready for the Next Lesson **424, 429, 433, 438, 445, 449, 453**

FOLDABLES Study Organizer **417**

Study Skills
- Study Tips **443, 450**
- Homework Help **422, 428, 432, 437, 444, 448, 452, 456**

Reading and Writing Mathematics
- Link to Reading **442**
- Reading Math **421, 455**
- Writing Math **422, 428, 432, 434, 437, 444, 448, 451, 456**

Standardized Test Practice
- Multiple Choice **419, 424, 429, 433, 438, 440, 445, 449, 453, 457, 461, 462**
- Short Response/Grid In **429, 433, 440, 445, 449, 463**
- Extended Response **463**
- Worked-Out Example **447**

 Snapshots **426, 433**

 INTERDISCIPLINARY PROJECT **457**

Lesson 9-2, p. 428

UNIT 5

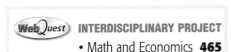
Lesson 10-1, p. 471

Prerequisite Skills
• Getting Started **467**
• Getting Ready for the Next Lesson **473, 477, 481, 487, 495, 499**

FOLDABLES Study Organizer **467**

Study Skills
• Study Tips **470, 471, 475, 479, 485, 493, 497, 501**
• Homework Help **472, 477, 480, 486, 494, 498, 503**

 Snapshots **495, 504**

Reading and Writing Mathematics
• Link to Reading **469**
• Writing Math **472, 476, 480, 482, 483, 486, 494, 498, 503**

Standardized Test Practice
• Multiple Choice **473, 477, 481, 487, 489, 490, 495, 499, 504, 507, 508**
• Short Response/Grid In **473, 477, 481, 487, 490, 509**
• Extended Response **509**
• Worked-Out Example **485**

UNIT 5

Prerequisite Skills

 Study Organizer **511**

Study Skills

Reading and Writing Mathematics

Standardized Test Practice

Snapshots **528**

Lesson 11-7, p. 547

UNIT 5

Lesson 12-2, p. 566

HOW TO...
Use Your Math Book

BEFORE YOU READ

Have a Goal

- What information are you trying to find?
- Why is this information important to you?
- How will you use the information?

Have a Plan

- Read *What You'll Learn* at the beginning of the lesson.
- Look over photos, tables, graphs, and opening activities.
- Locate boldfaced words and read their definitions.
- Find Key Concept and Concept Summary boxes for a preview of what's important.
- Skim the example problems.

Have an Opinion

- Is this information what you were looking for?
  - Do you understand what you have read?
  - How does this information fit with what you already know?

IN CLASS

During class is the opportunity to learn as much as possible about that day's lesson. Take advantage of this time! Ask questions about things that you don't understand, and take notes to help you remember important information.

To help keep your notes in order, try making a Foldables Study Organizer. It's as easy as 1-2-3! Here's a Foldable you can use to keep track of the rules for addition, subtraction, multiplication, and division.

FOLDABLES™ Study Organizer

Operations Make this Foldable to help you organize your notes. Begin with a sheet of 11" × 17" paper.

STEP 1

Fold
Fold the short sides toward the middle.

STEP 2

Fold Again
Fold the top to the bottom.

STEP 3

Cut
Open. Cut along the second fold to make four tabs.

STEP 4

Label
Label each of the tabs as shown.

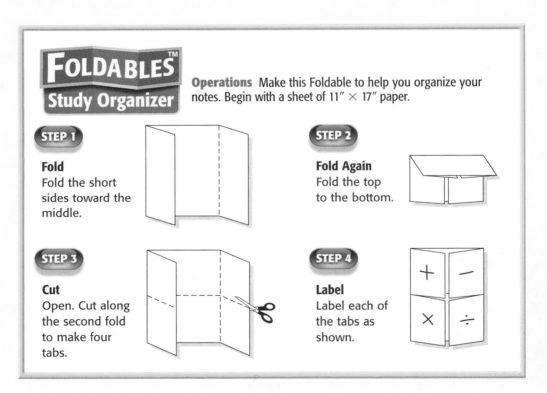

LOOK FOR...

FOLDABLES™

on these pages:

5, 61, 115, 155, 205, 255, 313, 373, 417, 467, 511, and 559.

DOING YOUR HOMEWORK

Regardless of how well you paid attention in class, by the time you arrive at home, your notes may no longer make any sense and your homework seems impossible. It's during these times that your book can be most useful.

- Each lesson has example problems, solved step-by-step, so you can review the day's lesson material.

- A Web site has extra examples to coach you through solving those difficult problems.

- Each exercise set has Homework Help boxes that show you which examples may help with your homework problems.

- Answers to the odd-numbered problems are in the back of the book. Use them to see if you are solving the problems correctly. If you have difficulty on an even problem, do the odd problem next to it. That should give you a hint about how to proceed with the even problem.

LOOK FOR...

The Web site with extra examples on these pages in Chapter 1: 7, 13, 19 25, 29, 35, 39, 47, and 51.

Homework Help boxes on these pages in Chapter 1: 9, 14, 20, 26, 30, 37, 41, 48, and 52.

Selected Answers starting on page 677.

1. $-2y - 7 = 3$

2. $4 + 5R = 3$

3. $4x + 5 = 13$

4. $6 + 3.50x = 20$

5. $P - 14 = 27$

Slope
a $(-2, -4) (1, 5)$
b $(0, 3) (4, -1)$
c $(2, 2), (5, 3)$

HOW TO...
Use Your Math Book

BEFORE A TEST

Admit it! You think there is no way to study for a math test! However, there *are* ways to review before a test. Your book offers help with this also.

- Review all of the new vocabulary words and be sure you understand their definitions. These can be found on the first page of each lesson or highlighted in yellow in the text.

- Review the notes you've taken on your Foldable and write down any questions that you still need answered.

- Practice all of the concepts presented in the chapter by using the chapter Study Guide and Review. It has additional problems for you to try as well as more examples to help you understand. You can also take the Chapter Practice Test.

- Take the self-check quizzes from the Web site.

LOOK FOR...
The Web site with self-check quizzes on these pages in Chapter 1: 9, 15, 21, 27, 31, 37, 41, 49, and 53.

The Study Guide and Review for Chapter 1 on page 54.

LET'S GET STARTED

To help you find the information you need quickly, use the Scavenger Hunt below to learn where things are located in each chapter.

CHAPTER 1

1. What is the title of Chapter 1?

2. How can you tell what you'll learn in Lesson 1-1?

3. What is the key concept presented in Lesson 1-2?

4. Sometimes you may ask "When am I ever going to use this"? Name a situation that uses the concepts from Lesson 1-3?

5. In Lesson 1-3, there is a paragraph that tells you that the absolute value of a number is not the same as the opposite of a number. What is the main heading above that paragraph?

6. What is the web address where you could find extra examples?

7. List the new vocabulary words that are presented in Lesson 1-4.

8. How many Examples are presented in Lesson 1-5?

9. In Lesson 1-8, there is a problem presented that deals with the minimum wage. Where could you find information about the current minimum wage?

10. Suppose you're doing your homework on page 48 and you get stuck on Exercise 18. Where could you find help?

11. There is a Real-Life Career mentioned in Lesson 1-9. What is it?

12. What is the web address that would allow you to take a self-check quiz to be sure you understand the lesson?

13. On what pages will you find the Study Guide and Review?

14. Suppose you can't figure out how to do Exercise 29 in the Study Guide on page 55. Where could you find help?

15. You complete the Practice Test on page 57 to study for your chapter test. Where could you find another test for more practice?

UNIT 1
Real Numbers and Algebra

Your study of math includes many different types of real numbers. In this unit, you will solve equations using integers, rational numbers, and irrational numbers.

WebQuest INTERDISCIPLINARY PROJECT

MATH and GEOGRAPHY

BON VOYAGE!

All aboard! We're setting sail on an adventure that will take us to exotic vacation destinations. Along the way, you'll act as a travel agent for one of three different families, gathering data about the cost of cruise packages, working to meet their vacation needs while still staying within their budget. You will also plan their itinerary and offer choices of activities for them to participate in at their respective destinations. We'll be departing shortly, so pack your problem-solving tool kit and hop on board.

 Log on to msmath3.net/webquest to begin your WebQuest.

Algebra: Integers

""How do you use math in scuba diving?""

Recreational scuba divers dive no more than 130 feet below the water's surface. You can use the integer −130 to describe this depth. In algebra, you will use integers to describe many real-life situations.

You will describe situations using integers in Lesson 1-3.

GETTING STARTED

Take this quiz to see whether you are ready to begin Chapter 1.

▶ Vocabulary Review

Choose the correct term to complete each sentence.

1. To find the product of two numbers, you must (add, multiply).

2. (Division, Addition) and subtraction are opposites because they undo each other.

▶ Prerequisite Skills

Add.

3. $64 + 13$

4. $10.3 + 4.7$

5. $2.5 + 77$

6. $38 + 156$

Subtract.

7. $200 - 48$

8. $59 - 26$

9. $3.3 - 0.7$

10. $73.5 - 0.87$

Multiply.

11. $3 \times 5 \times 2$

12. 2.8×5

13. 12×6

14. $4 \times 9 \times 3$

Divide.

15. $244 \div 0.2$

16. $72 \div 9$

17. $96 \div 3$

18. $100 \div 0.5$

19. $2 \div 5$

20. $0.36 \div 0.3$

Replace each ● with $<$, $>$, or $=$ to make a true sentence.

21. $13 ● 16$

22. $5 ● 0.5$

23. $25 ● 22$

24. $3 ● 3.0$

Integers and Equations
Make this Foldable to help you organize your notes. Begin with a plain piece of 11" x 17" paper.

STEP 1 Fold
Fold the paper in sixths lengthwise.

STEP 2 Open and Fold
Fold a 4" tab along the short side. Then fold the rest in half.

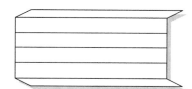

STEP 3 Label
Draw lines along the folds and label as shown.

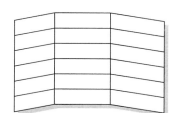

	Words	Example(s)
A Plan for Problem Solving		
+ & − of Integers		
× & ÷ of Integers		
Solving + & − Equations		
Solving × & ÷ Equations		

Reading and Writing As you read and study the chapter, explain each concept in words and give one or more examples.

1-1

A Plan for Problem Solving

Sunshine State Standards
MA.A.3.3.2-1, MA.A.4.3.1-1, MA.A.4.3.1-2

What You'll Learn

Solve problems by using the four-step plan.

➤ *NEW* Vocabulary

conjecture

HANDS-ON Mini Lab

Materials
- blue and white square tiles

Work with a partner.

Suppose you are designing rectangular gardens that are bordered by white tiles. The three smallest gardens you can design are shown below.

Garden 1

Garden 2

Garden 3

1. How many white tiles does it take to border each of these three gardens?

2. Predict how many white tiles it will take to border the next-largest garden. Check your answer by modeling the garden.

3. How many white tiles will it take to border a garden that is 10 tiles long? Explain your reasoning.

In this textbook, you will be solving many kinds of problems. Some, like the problem presented above, can be solved by using one or more problem-solving strategies.

No matter which strategy you use, you can always use the four-step plan to solve a problem.

Explore
- Determine what information is given in the problem and what you need to find.
- Do you have all the information you need?
- Is there too much information?

Plan
- Visualize the problem and select a strategy for solving it. There may be several strategies that you can use.
- Estimate what you think the answer should be.

Solve
- Solve the problem by carrying out your plan.
- If your plan doesn't work, try another.

Examine
- Examine your answer carefully.
- See if your answer fits the facts given in the problem.
- Compare your answer to your estimate.
- You may also want to check your answer by solving the problem again in a different way.
- If the answer is not reasonable, make a new plan and start again.

Some problem-solving strategies require you to make an educated guess or **conjecture**.

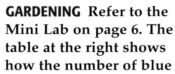

EXAMPLE Use the Four-Step Plan

1 GARDENING Refer to the Mini Lab on page 6. The table at the right shows how the number of blue tiles it takes to represent each garden is related to the number of white tiles needed to border the garden. How many white tiles will it take to border a garden that is 12 blue tiles long?

Blue Tiles	1	2	3	4	5	6
White Tiles	8	10	12	14	16	18

Explore You know the number of white tiles it takes to border gardens up to 6 tiles long. You need to determine how many white tiles it will take to border a garden 12 tiles long.

Plan You might make the conjecture that there is a pattern to the number of white tiles used. One method of solving this problem is to look for a pattern.

Solve First, look for the pattern.

Blue Tiles	1	2	3	4	5	6
White Tiles	8	10	12	14	16	18

+2 +2 +2 +2 +2

Next, extend the pattern.

Blue Tiles	6	7	8	9	10	11	12
White Tiles	18	20	22	24	26	28	30

+2 +2 +2 +2 +2 +2

It would take 30 white tiles to border a garden that was 12 blue tiles long.

Examine It takes 8 white tiles to border a garden that is 1 blue tile wide. As shown below, each additional blue tile added to the shape of the garden needs 2 white tiles to border it, one above and one below.

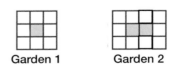

Garden 1   Garden 2

So, to border a garden 12 blue tiles long, it would take 8 white tiles for the first blue tile and 11×2 or 22 for the 11 additional tiles. Since $8 + 22 = 30$, the answer is correct.

Some problems can be solved by adding, subtracting, multiplying, or dividing. Others can require a combination of these operations.

READING Math

Word Problems It is important to read a problem more than once before attempting to solve it. You may discover important details that you overlooked when you read the problem the first time.

EXAMPLE Use the Four-Step Plan

2 **WORK** Refer to the graphic. On average, how many more hours per week did a person in the United States work in 2000 than a person in the United Kingdom?

Explore *What do you know?*
You know the average number of hours worked in the year 2000 by a person in the United States, Japan, Canada, the United Kingdom, and Italy.

What are you trying to find?
You need to find the difference in the number of hours worked *per week* by a person in the United States and in the United Kingdom.

Plan Extra information is given. Use only the number of hours worked for the United States, 1,877, and the United Kingdom, 1,708.

Begin by subtracting to find the annual difference in the number of hours worked in each country. Then divide by the number of weeks in a year to find the weekly difference.

Estimate $1,900 - 1,700 = 200$ and $200 \div 50 = 4$
The number of hours is about 4.

Solve $1,877 - 1,708 = 169$ USA hours in 2000 − UK hours in 2000
$169 \div 52 = 3.25$ There are 52 weeks in a year.

On average, a person in the United States worked 3.25 hours more per week in 2000 than a person in the United Kingdom.

Examine Is your answer reasonable? The answer is close to the estimate, so the answer is reasonable.

1. **Explain** each step in the four-step problem-solving plan.

2. **OPEN ENDED** Describe another method you could use to find the number of white tiles it takes to border a garden 12 green tiles long.

3. **NUMBER SENSE** Find a pattern in this list of numbers 4, 5, 7, 10, 14, 19. Then find the next number in the list.

GUIDED PRACTICE

Use the four-step plan to solve each problem.

4. **SCHOOL SUPPLIES** At the school bookstore, a pen costs $0.45, and a small writing tablet costs $0.85. What combination of pens and tablets could you buy for exactly $2.15?

5. **HOBBIES** Lucero put 4 pounds of sunflower seeds in her bird feeder on Sunday. On Friday, the bird feeder was empty, so she put 4 more pounds of seed in it. The following Sunday, the seeds were half gone. How many pounds of sunflower seeds were eaten that week?

6. **FIELD TRIP** Two 8th-grade teams, the Tigers and the Waves, are planning a field trip to Washington, D.C. There are 123 students and 4 teachers on the Tiger team and 115 students and 4 teachers on the Waves team. If one bus holds 64 people, how many buses are needed for this field trip?

Practice and Applications

Use the four-step plan to solve each problem.

7. **FOOD** Almost 90 million jars of a popular brand of peanut butter are sold annually. Estimate the number of jars sold every second.

HOMEWORK HELP	
For Exercises	See Examples
7–17	1, 2
Extra Practice See pages 616, 648.	

Draw the next two figures in each of the patterns below.

8.

9.

ART For Exercises 10–12, use the following information.
The number of paintings an artist produced during her first four years as a professional is shown in the table at the right.

10. Estimate the total number of paintings the artist has produced.

11. About how many more paintings did she produce in the last two years than in the first two years?

12. About how many more paintings did she produce in the odd years than the even years?

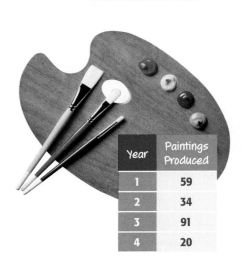

Year	Paintings Produced
1	59
2	34
3	91
4	20

HISTORY For Exercises 13 and 14, use the information below.
In 1803, the United States bought the Louisiana Territory from France for $15 million. The area of this purchase was 828,000 square miles.

13. If one square mile is equal to 640 acres, how many acres of land did the United States acquire through the Louisiana Purchase?

14. About how much did the United States pay for the Louisiana Territory per acre?

15. **BABY-SITTING** Kayla earned $30 baby-sitting last weekend. She wants to buy 3 CDs that cost $7.89, $12.25, and $11.95. Does she have enough money to purchase the CDs, including tax? Explain your reasoning.

16. **MEDICINE** The table shows how the amount of a medicine in the bloodstream is related to the time since it was taken. A doctor wants the concentration to be at least 0.32 milligram per liter at all times. If the medicine takes exactly 30 minutes to be absorbed into the bloodstream after it is taken, what is the maximum amount of time the patient should wait before taking more of the medicine?

Time (hours)	Concentration (mg/L)
0	0
1	0.60
2	0.75
3	0.69
4	0.60
5	0.52
6	0.45

17. **SHOPPING** Miguel went to the store to buy jeans. Each pair costs $24. If he buys two pairs, he can get the second pair for half price. How much will he save per pair if he buys two pairs?

18. **CRITICAL THINKING** Draw the next figure in the pattern shown below. Then predict the number of tiles it will take to create the 10th figure in this pattern. Explain your reasoning.

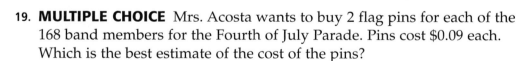

| Figure 1 | Figure 2 | Figure 3 | Figure 4 |

Standardized Test Practice and Mixed Review

FCAT Practice

19. **MULTIPLE CHOICE** Mrs. Acosta wants to buy 2 flag pins for each of the 168 band members for the Fourth of July Parade. Pins cost $0.09 each. Which is the best estimate of the cost of the pins?

    A $8   **B** $20   **C** $30   **D** $50

20. **GRID IN** John stocks the vending machines at Rose Hill Elementary School every 9 school days and Nassaux Intermediate School every 6 school days. In September, he stocked both schools on the 27th. How many school days earlier had he stocked the vending machines at both schools on the same day?

GETTING READY FOR THE NEXT LESSON

BASIC SKILL Add, subtract, multiply, or divide.

21. $15 + 45$ 22. $1{,}287 - 978$ 23. 4×3.6 24. $280 \div 0.4$

Variables, Expressions, and Properties

Sunshine State Standards
MA.A.1.3.4-2, MA.A.3.3.1-3, MA.A.3.3.3-1, MA.D.2.3.1-6

What You'll Learn

Evaluate expressions and identify properties.

➤ NEW Vocabulary

variable
algebraic
    expression
numerical
    expression
evaluate
order of operations
powers
equation
open sentence
property
counterexample

HANDS-ON Mini Lab

Materials
• toothpicks

The figures at the right are formed using toothpicks. If each toothpick is a unit, then the perimeter of the first figure is 4 units.

Figure 1   Figure 2   Figure 3

1. Copy and complete the table below.

Figure Number	1	2	3	4	5	6
Perimeter	4	8				

2. What would be the perimeter of Figure 10?

3. What is the relationship between the figure number and the perimeter of the figure?

You can use the variable n to represent the figure number in the Mini Lab above. A **variable** is a symbol, usually a letter, used to represent a number.

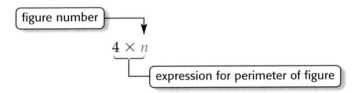

The expression $4 \times n$ is called an **algebraic expression** because it contains a variable, a number, and at least one operation symbol. When you substitute 10 for n, or replace n with 10, the algebraic expression $4 \times n$ becomes the **numerical expression** 4×10.

When you **evaluate** an expression, you find its numerical value. To avoid confusion, mathematicians have agreed on a set of rules called the **order of operations**.

Key Concept   Order of Operations

1. Do all operations within grouping symbols first; start with the innermost grouping symbols.

2. Evaluate all powers before other operations.

3. Multiply and divide in order from left to right.

4. Add and subtract in order from left to right.

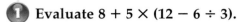**Evaluate a Numerical Expression**

1 **Evaluate $8 + 5 \times (12 - 6 \div 3)$.**

$$8 + 5 \times (12 - 6 \div 3) = 8 + 5 \times (12 - 2) \quad \text{Divide inside parentheses first.}$$

$$= 8 + 5 \times (10) \quad \text{Subtract inside parentheses next.}$$

$$= 8 + 50 \text{ or } 58 \quad \text{Multiply 5 and 10. Then add 8 and 50.}$$

Your Turn Evaluate each expression.

a. $(9 + 6) \div 5 \times 4 + (3 - 2)$ **b.** $2 + 3 \times (18 - 7)$

STUDY TIP

Technology Enter $4 + 3 \times 2$ into your calculator. If it displays 10, then your calculator follows the order of operations. If it displays 14, then it does not.

Algebra has special symbols that are used for multiplication.

$8 \cdot 5$ $3(2)$ $4n$ xy

| 8 times 5 | 3 times 2 | 4 times n | x times y |

Expressions such as 7^2 and 2^3 are called **powers** and represent repeated multiplication.

| 7 squared or 7 times 7 | → 7^2 | 2^3 ← | 2 cubed or 2 times 2 times 2 |

To evaluate an algebraic expression, replace the variable or variables with the known values and then use the order of operations.

EXAMPLES **Evaluate Algebraic Expressions**

Evaluate each expression if $a = 5$, $b = 4$, and $c = 8$.

2 $4a - 3b + 1$

$$4a - 3b + 1 = 4(5) - 3(4) + 1 \quad \text{Replace } a \text{ with 5 and } b \text{ with 4.}$$

$$= 20 - 12 + 1 \quad \text{Do all multiplications first.}$$

$$= 8 + 1 \text{ or } 9 \quad \text{Add and subtract in order from left to right.}$$

3 $\dfrac{c^2}{a - 3}$

The fraction bar is a grouping symbol. Evaluate the expressions in the numerator and denominator separately before dividing.

$$\frac{c^2}{a - 3} = \frac{8^2}{5 - 3} \quad \text{Replace } c \text{ with 8 and } a \text{ with 5.}$$

$$= \frac{64}{5 - 3} \quad \text{Evaluate the power in the numerator. } 8^2 = 8 \cdot 8 \text{ or } 64$$

$$= \frac{64}{2} \text{ or } 32 \quad \text{Subtract in the denominator. Then divide.}$$

STUDY TIP

Parentheses Parentheses around a single number do not necessarily mean that multiplication should be performed first. Remember to multiply or divide in order from left to right.

$20 \div 4(2) = 5(2)$ or 10

Your Turn Evaluate each expression if $m = 9$, $n = 2$, and $p = 5$.

c. $25 - 5p$ **d.** $\dfrac{4 + 6m}{2p - 8}$ **e.** $n^2 + 5n - 6$

A mathematical sentence that contains an equals sign (=) is called an **equation**. Some examples of equations are shown below.

$$7 + 8 = 15 \qquad 3(6) = 18 \qquad x + 2 = 5$$

An equation that contains a variable is an **open sentence**. When a number is substituted for the variable in an open sentence, the sentence is true or false. Consider the equation $x + 2 = 5$.

Replace x with 2.　　　$2 + 2 \stackrel{?}{=} 5$ ✗　　　This equation is false.

Replace x with 3.　　　$3 + 2 \stackrel{?}{=} 5$ ✔　　　This equation is true.

Properties are open sentences that are true for any numbers. You may remember some of the properties in the table below from your previous mathematics courses.

Property	Algebra	Arithmetic
Commutative	$a + b = b + a$ $a \cdot b = b \cdot a$	$6 + 1 = 1 + 6$ $7 \cdot 3 = 3 \cdot 7$
Associative	$a + (b + c) = (a + b) + c$ $a \cdot (b \cdot c) = (a \cdot b) \cdot c$	$2 + (3 + 8) = (2 + 3) + 8$ $3 \cdot (4 \cdot 5) = (3 \cdot 4) \cdot 5$
Distributive	$a(b + c) = ab + ac$ $a(b - c) = ab - ac$	$4(6 + 2) = 4 \cdot 6 + 4 \cdot 2$ $3(7 - 5) = 3 \cdot 7 - 3 \cdot 5$
Identity	$a + 0 = a$ $a \cdot 1 = a$	$9 + 0 = 9$ $5 \cdot 1 = 5$

EXAMPLE　**Identify Properties**

④ **Name the property shown by the statement $2 \cdot (5 \cdot n) = (2 \cdot 5) \cdot n$.**

The grouping of the numbers and variables changed. This is the Associative Property of Multiplication.

You may wonder whether each of the properties applies to subtraction. If you can find a counterexample, the property does not apply. A **counterexample** is an example that shows that a conjecture is false.

EXAMPLE　**Find a Counterexample**

⑤ **State whether the following conjecture is *true* or *false*. If *false*, provide a counterexample.**

Division of whole numbers is commutative.

Write two division expressions using the Commutative Property, and then check to see whether they are equal.

$15 \div 3 \stackrel{?}{=} 3 \div 15$　　　State the conjecture.

$5 \neq \dfrac{1}{5}$　　　Divide.

We found a counterexample. That is, $15 \div 3 \neq 3 \div 15$. So, division is *not* commutative. The conjecture is false.

Skill and Concept Check

Writing Math
Exercises 1 & 2

1. **Compare** the everyday meaning of the term *variable* with its mathematical definition.

2. **Describe** the difference between $3k + 9$ and $3k + 9 = 15$.

3. **OPEN ENDED** Write an equation that illustrates the Commutative Property of Multiplication.

GUIDED PRACTICE

Evaluate each expression.

4. $14 - 3 \cdot 2$

5. $4^2 - 2 \cdot 5 + (8 - 2)$

6. $\dfrac{28}{4^2 - 2}$

Evaluate each expression if $a = 2$, $b = 7$, and $c = 4$.

7. $6b - 5a$

8. $\dfrac{bc}{2}$

9. $b^2 - (8 + 3c)$

Name the property shown by each statement.

10. $3(2 + 5) = 3(2) + 3(5)$

11. $3(12 \cdot 4) = (12 \cdot 4)3$

State whether each conjecture is *true* or *false*. If *false*, provide a counterexample.

12. Subtraction of whole numbers is associative.

13. The sum of two different whole numbers is always greater than either addend.

Practice and Applications

Evaluate each expression.

14. $12 \div 4 + 2$

15. $25 - 15 \div 5$

16. $3(7) - 4 \div 2$

17. $18 + 1(12) \div 6$

18. $16 - 6 + 5 \cdot 2^3$

19. $5^2 - 4 \cdot 6 \div 3$

20. $4^3 \div (16 - 12) \cdot 3$

21. $4^2 - 2 \cdot 5 + (8 - 2)$

22. $\dfrac{36}{3^2 - 3}$

23. $(14 - 8) \cdot 3 + \dfrac{31 - 9}{11}$

24. $\dfrac{(18 - 12)(21 + 4)}{3}$

25. $\dfrac{8 + 2(4 - 1)}{3^2 - 2}$

26. $2[18 - (5 + 3^2) \div 7]$

27. $4 - 3 + 7(12 - 2^2)$

HOMEWORK HELP	
For Exercises	See Examples
14–27	1
28–42	2, 3
43–48	4
51–54	5

Extra Practice
See pages 616, 648.

Evaluate each expression if $w = 2$, $x = 6$, $y = 4$, and $z = 5$.

28. $2x + y$

29. $3z - 2w$

30. $9 + 7x - y$

31. $3y + z - x$

32. wx^2

33. $(wx)^2$

34. $x(3 + y) - z$

35. $2(xy - 9) \div z$

36. $\dfrac{x^2 - 3}{2z + 1}$

37. $\dfrac{wz^2}{y + 6}$

38. $3y^2 + 2y - 7$

39. $2z^2 - 4z + 5$

40. **INSECTS** The number of times a cricket chirps can be used to estimate the temperature in degrees Fahrenheit. Find the approximate temperature if a cricket chirps 140 times in a minute. Use the expression $c \div 4 + 37$ where c is the number of chirps per minute.

PETS **For Exercises 41 and 42, use the information below.**
You can estimate the number of households with pets in your community using the expression $\frac{c}{n} \cdot p$, where c is the population of your community, n is the national number of people per household, and p is the national percent of households owning pets.

41. In 2000, the U.S. Census Bureau estimated that there were 2.62 people per household. Estimate the number of dog-owning households for a community of 50,000 people.

42. Estimate the number of bird-owning households.

National Percent of Households Owning Pets	
Dogs	0.316
Cats	0.273
Birds	0.046
Horses	0.015

Source: U.S. Pet Ownership & Demographics Sourcebook

Name the property shown by each statement.

43. $(6 + 3)2 = 6(2) + 3(2)$

44. $1 \cdot 5abc = 5abc$

45. $5 + (8 + 12) = (8 + 12) + 5$

46. $(3 + 9) + 20 = 3 + (9 + 20)$

47. $(5 + x) + 0 = 5 + x$

48. $5(10 - 4) = 5(10) - 5(4)$

Rewrite each expression using the indicated property.

49. $6(4) + 6(3)$, Distributive Property

50. x, Identity Property

State whether each conjecture is *true* or *false*. If *false*, provide a counterexample.

51. The sum of two even number is always even.

52. The sum of two odd numbers is always odd.

53. Division of whole numbers is associative.

54. Subtraction of whole numbers is commutative.

55. **RESEARCH** Use the Internet or another resource to find out who first introduced a mathematical symbol such as the equals sign ($=$).

56. **CRITICAL THINKING** Decide whether $6 + 7 \cdot 2 + 5 = 55$ is *true* or *false*. If *false*, copy the equation and insert parentheses to make it true.

Standardized Test Practice and Mixed Review

FCAT Practice

57. **MULTIPLE CHOICE** What is the value of $3^2 + 4 \cdot 2 - 6 \div 2$?

   Ⓐ 5.5   Ⓑ 10   Ⓒ 14   Ⓓ 26

58. **MULTIPLE CHOICE** Which is an example of the Associative Property?

   Ⓕ $4 \cdot 6 = 6 \cdot 4$ Ⓖ $5 + (4 + 1) = (4 + 1) + 5$

   Ⓗ $7 + (3 + 2) = 7 + (2 + 3)$ Ⓘ $8(9 \cdot 2) = (8 \cdot 9)2$

59. **DINING** Kyung had $17. His dinner cost $5.62, and he gave the cashier a $10 bill. How much change should he receive? (Lesson 1-1)

GETTING READY FOR THE NEXT LESSON

BASIC SKILL Replace each ● with $<$, $>$, or $=$ to make a true sentence.

60. $4 \, ● \, 9$ 61. $7 \, ● \, 7$ 62. $8 \, ● \, 5$ 63. $3 \, ● \, 2$

Study Skill
HOW TO....
Study Math Vocabulary

WORD MAP

Learning new math vocabulary is more than just memorizing the definition. Try using a word map to really understand the meaning of the word.

New vocabulary terms are clues about important concepts. Your textbook helps you find those clues by highlighting them in yellow, as **integers** is highlighted on the next page.

Whenever you see a highlighted word, stop and ask yourself these questions.
- How does this fit with what I already know?
- How is this alike or different from something I learned earlier?

Organize your answers in a word map like the one below.

Definition from Text

Negative numbers like −86, positive numbers like +125, and zero are members of the set of integers.

In Your Own Words

Integers are whole numbers and negative "whole" numbers, not fractions or decimals.

Word

Integer

Examples

−3, 0, 2, 56, −89

Nonexamples

$\frac{1}{2}$, $3\frac{2}{5}$, 0.5, −1.8

SKILL PRACTICE
Make a word map for each term. The term is defined on the given page.

1. greatest common factor (p. 610)

2. least common multiple (p. 612)

3. perimeter (p. 613)

4. area (p. 613)

1-3

Integers and Absolute Value

Sunshine State Standards
MA.A.1.3.1-1, MA.A.1.3.2-1, MA.A.1.3.2-2, MA.A.1.3.4-4, MA.A.3.3.2-2

What You'll Learn

Graph integers on a number line and find absolute value.

NEW Vocabulary

negative number
integer
coordinate
inequality
absolute value

WHEN am I ever going to use this?

GEOGRAPHY Badwater, in Death Valley, California, is the lowest point in North America, while Mt. McKinley in Alaska, is the highest point. The graph shows their elevations and extreme temperatures.

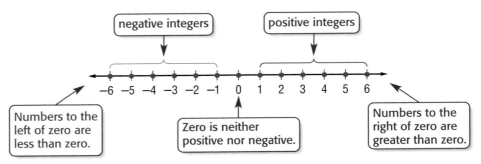

Mt. McKinley
Elevation : 6,184 m
Temperature: −35°F

Badwater
Elevation: −86 m
Temperature: 125°F

1. What does an elevation of −86 meters represent?

2. What does a temperature of −35° represent?

With sea level as the starting point of 0, you can express 86 meters below sea level as $0 - 86$, or -86. A **negative number** is a number less than zero.

Negative numbers like -86, positive numbers like $+125$, and zero are members of the set of **integers**. Integers can be represented as points on a number line.

negative integers   positive integers

$$-6 \ -5 \ -4 \ -3 \ -2 \ -1 \quad 0 \quad 1 \ 2 \ 3 \ 4 \ 5 \ 6$$

Numbers to the left of zero are less than zero.

Zero is neither positive nor negative.

Numbers to the right of zero are greater than zero.

This set of integers can be written as $\{\ldots, -3, -2, -1, 0, 1, 2, 3, \ldots\}$, where . . . means *continues indefinitely*.

EXAMPLES Write Integers for Real-Life Situations

Write an integer for each situation.

1 a 15-yard loss   The integer is -15.

2 3 inches above normal   The integer is $+3$.

Your Turn Write an integer for each situation.

a. a gain of $2 a share   b. 10 degrees below zero

To graph integers, locate the point named by the integers on a number line. The number that corresponds to a point is called the **coordinate** of that point.

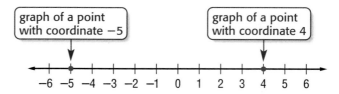

| graph of a point with coordinate −5 |
| graph of a point with coordinate 4 |

$$-6 \ -5 \ -4 \ -3 \ -2 \ -1 \ 0 \ 1 \ 2 \ 3 \ 4 \ 5 \ 6$$

Notice that −5 is to the left of 4 on the number line. This means that −5 is less than 4. A sentence that compares two numbers or quantities is called an **inequality**. They contain symbols like < and >.

| −5 is less than 4. | ➤ $-5 < 4$ | | $4 > -5$ | ◄ | 4 is greater than −5. |

EXAMPLES   Compare Two Integers

Replace each ● with <, >, or = to make a true sentence. Use the integers graphed on the number line below.

$$-6 \ -5 \ -4 \ -3 \ -2 \ -1 \ 0 \ 1 \ 2 \ 3 \ 4 \ 5 \ 6$$

3 **1 ● −6**   1 is greater than −6, since it lies to the right of −6. Write $1 > -6$.

4 **−4 ● −2**   −4 is less than −2, since it lies to the left of −2. Write $-4 < -2$.

Your Turn Replace each ● with <, >, or = to make a true sentence.

c. -3 ● 2 d. -5 ● -6 e. -1 ● 1

Integers are used to compare numbers in many real-life situations.

EXAMPLE   Order Integers

5 **WEATHER** The table below shows the record low temperatures for selected states. Order these temperatures from least to greatest.

State	AL	CA	GA	IN	KY	NC	TN	TX	VA
Temperature (°F)	−27	−45	−17	−36	−37	−34	−32	−23	−30

Graph each integer on a number line.

$$-46 \ -44 \ -42 \ -40 \ -38 \ -36 \ -34 \ -32 \ -30 \ -28 \ -26 \ -24 \ -22 \ -20 \ -18 \ -16$$

Write the numbers as they appear from left to right.

The temperatures $-45°, -37°, -36°, -34°, -32°, -30°, -27°, -23°,$ and $-17°$ are in order from least to greatest.

On the number line, notice that −4 and 4 are on opposite sides of zero and that they are the same distance from zero. In mathematics, we say they have the same absolute value, 4.

The **absolute value** of a number is the distance the number is from 0 on the number line. The symbol for absolute value is two vertical bars on either side of the number.

| The absolute value of 4 is 4. | → $|4| = 4$ | $|-4| = 4$ ← | The absolute value of −4 is 4. |

Since distance cannot be negative, the absolute value of a number is always positive or zero.

EXAMPLES Expressions with Absolute Value

Evaluate each expression.

6 $|-7|$

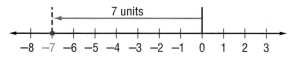

$|-7| = 7$ The graph of −7 is 7 units from 0 on the number line.

7 $|5| + |-6|$

$|5| + |-6| = 5 + |-6|$ The absolute value of 5 is 5.

$\qquad\qquad = 5 + 6$ The absolute value of −6 is 6.

$\qquad\qquad = 11$ Simplify.

Your Turn Evaluate each expression.

f. $|14|$ g. $|-9| + |3|$ h. $|-8| - |-2|$

Since variables represent numbers, you can use absolute value notation with algebraic expression involving variables.

EXAMPLE Expressions with Absolute Value

8 Evaluate $8 + |n|$ if $n = -12$.

$8 + |n| = 8 + |-12|$ Replace *n* with −12.

$\qquad\quad = 8 + 12$ $|-12| = 12$

$\qquad\quad = 20$ Simplify.

Your Turn Evaluate each expression if $a = -5$ and $b = 3$.

i. $|b| + 7$ j. $|a| - 2$ k. $4|a| + b$

Writing Math
Exercise 2

1. **OPEN ENDED** Write two inequalities using the same two integers.

2. **Which One Doesn't Belong?** Identify the phrase that cannot be described by the same integer as the other three. Explain your reasoning.

5° below normal	5 miles above sea level	a loss of 5 pounds	giving away $5

GUIDED PRACTICE

Write an integer for each situation.

3. a 10-yard gain

4. 34 miles below sea level

5. Graph the set of integers {5, −3, 0} on a number line.

Replace each ● with <, >, or = to make a true sentence.

6. −4 ● 3

7. −10 ● −12

8. 7 ● −7

Evaluate each expression.

9. $|-5|$

10. $|20| - |-3|$

11. $|-16| + |-12|$

Evaluate each expression if $x = -10$ and $y = 6$.

12. $3 + |x|$

13. $|x| - y$

14. $3|y|$

Practice and Applications

Write an integer for each situation.

15. a loss of 2 hours

16. earning $45

17. a gain of 4 ounces

18. 13° below zero

19. a $60 deposit

20. spending $25

Graph each set of integers on a number line.

21. {−1, 4, −7}

22. {0, −5, 3, 8}

23. {−2, −8, −4, −9}

24. {−4, 0, −6, −1, 2}

Replace each ● with <, >, or = to make a true sentence.

25. −7 ● −2

26. −9 ● −10

27. −3 ● 0

28. 0 ● 12

29. 4 ● −11

30. −15 ● 14

31. $|-8|$ ● $|8|$

32. $|-13|$ ● $|-6|$

Order the integers in each set from least to greatest.

33. {45, −23, 55, 0, −12, −37}

34. {97, −46, −50, 38, −100}

35. {−17, −2, −5, −11, 6}

36. {21, 8, −47, 3, −1, 0}

Evaluate each expression.

37. $|-14|$

38. $|18|$

39. $|25|$

40. $|0|$

41. $|2| + |-13|$

42. $|-15| + |-6|$

43. $|-20| - |17|$

44. $|31| - |-1|$

45. $-|3|$

46. $-|-10|$

47. $|5 + 9|$

48. $|17 - 8|$

HOMEWORK HELP

For Exercises	See Examples
15–20, 50	1, 2
21–32, 52	3, 4
33–36, 49	5
37–48	6, 7
52–57	8

Extra Practice
See pages 616, 648.

GOLF For Exercises 49 and 50, use the information below.

In golf, a score of 0 is called *even par*. Two under par is written as -2. Two over par is written as $+2$.

49. The final round scores of the top ten Boys' Division finishers in the 2001 Westfield Junior PGA Championship in Westfield Center, Ohio, are shown at the right. Order the scores from least to greatest.

50. Taylor Hall of LaGrange, Georgia, won the Championship, finishing 4 under par. Write an integer to describe Taylor Hall's score.

51. **SCIENCE** Hydrogen freezes at about $-435°F$, and helium freezes at about $-458°F$. Which element has the lower freezing point?

Name	Final Round
Taylor Hall	+1
Casey Wittenberg	−2
Brady Schnell	−5
Kevin Silva	−1
Gabriel Borrud	−2
Chris Brown	0
Troy Hawkins	−1
Erik Olson	+2
Ethan Swift	−2
John Browndorf	+1

Source: www.pga.com

Evaluate each expression if $a = 5$, $b = -8$, and $c = -3$.

52. $|b| + 7$

53. $a - |c|$

54. $|a| + |b|$

55. $|4a|$

56. $6|b|$

57. $|16 - a|$

58. **WRITE A PROBLEM** Write about a real-life situation that requires comparing absolute values. Then write the integer that describes it.

CRITICAL THINKING Determine whether the following statements are *always*, *sometimes*, or *never* true. Explain your reasoning.

59. The absolute value of a positive integer is a negative integer.

60. The absolute value of a negative integer is a positive integer.

61. If a and b are integers and $a > b$, then $|a| > |b|$.

Standardized Test Practice and Mixed Review

62. **MULTIPLE CHOICE** Which is in order from least to greatest?
    Ⓐ $-4, 2, 8$ Ⓑ $4, -1, -6$ Ⓒ $-1, 2, -4$ Ⓓ $0, -1, -4$

63. **MULTIPLE CHOICE** If $a = -3$ and $b = 3$, then which statement is false?
    Ⓕ $|a| > 2$ Ⓖ $|a| = |b|$ Ⓗ $|b| < 2$ Ⓘ $|a| = b$

Evaluate each expression. (Lesson 1-2)

64. $3[14 - (8 - 5)^2] + 20$

65. $6 \cdot (18 - 14) + \dfrac{22 - 4}{9}$

66. $\dfrac{45 - 9}{3^2 + 3}$

67. **CHARITY WALK** Krystal knows that she can walk about 1.5 meters per second. If she can maintain that pace, about how long should it take her to complete a 10-kilometer charity walk? (Lesson 1-1)

GETTING READY FOR THE NEXT LESSON

BASIC SKILL Add or subtract.

68. $9 + 14$

69. $100 - 57$

70. $47 - 19$

71. $18 + 34 + 13$

What You'll Learn
Graph and interpret data.

Materials
- metric measuring tape
- masking tape
- cardboard
- 2 metersticks
- tennis ball

Sunshine State Standards
MA.E.1.3.1-2, MA.E.3.3.1-2, MA.E.3.3.1-3

Graphing Data

INVESTIGATE *Work in groups of 4.*

In this Lab, you will investigate the relationship between the height of a chute and the distance an object travels as it leaves the chute.

STEP 1 Make a meter-long chute for the ball out of cardboard. Reinforce the chute by taping it to one of the metersticks.

STEP 2 Use the tape measure to mark off a distance of 3 meters on the floor. Make a 0-meter mark and a 3-meter mark using tape.

STEP 3 Place the end of your chute at the edge of the 0-meter mark. Raise the back of the chute to a height of 5 centimeters.

STEP 4 Let a tennis ball roll down the chute. When the ball stops, measure how far it is from the 3-meter mark.

STEP 5 Copy the table shown and record your results. If the ball stops short of the 3-meter mark, record the distance as a negative number. If the ball passes the 3-meter mark, record the distance as a positive number.

STEP 6 Raise the chute by 5 centimeters and repeat the experiment. Continue until the chute is 40 centimeters high.

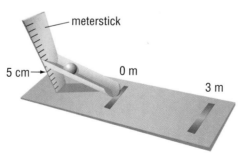
meterstick
5 cm
0 m
3 m

Height h of Chute (cm)	Distance d from 3-meter Mark (cm)
5	
10	
15	

Writing Math

Work with a partner.

1. **Graph** the ordered pairs (h, d) on a coordinate grid.

2. **Describe** how the points appear on your graph.

3. **Describe** how raising the chute affects the distance the ball travels.

4. Use your graph to **predict** how far the ball will roll when the chute is raised to the 50-centimeter mark. Then check your prediction.

Adding Integers

Sunshine State Standards
MA.A.1.3.4-2, MA.A.3.3.1-2, MA.A.3.3.1-3, MA.A.3.3.2-1, MA.A.3.3.2-2, MA.D.2.3.1-6

What You'll Learn
Add integers.

⇨NEW Vocabulary
opposites
additive inverse

⟳REVIEW Vocabulary
addend: numbers that are added together
sum: the result when two or more numbers are added together

1. Write an integer that describes the game show host's statement.
2. Write an addition sentence that describes this situation.

The equation $-3,200 + (-7,400) + (-2,600) = -13,200$ is an example of adding integers with the same sign. Notice that the sign of the sum is the same as the sign of each addend.

EXAMPLE   Add Integers with the Same Sign

1 Find $-4 + (-2)$.

Method 1 Use a number line.

- Start at zero.
- Move 4 units left.
- From there, move 2 units left.

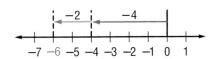

So, $-4 + (-2) = -6$.

Method 2 Use counters.

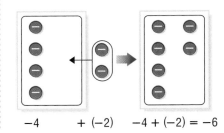

-4 $+ (-2)$ $-4 + (-2) = -6$

Your Turn Add using a number line or counters.

a. $-3 + (-2)$ b. $1 + 5$ c. $-5 + (-4)$

These examples suggest a rule for adding integers with the same sign.

Key Concept	Adding Integers with the Same Sign
Words	To add integers with the same sign, add their absolute values. Give the result the same sign as the integers.
Examples	$-7 + (-3) = -10$ $5 + 4 = 9$

EXAMPLE   Add Integers with the Same Sign

2 **Find** $-13 + (-18)$.

$$-13 + (-18) = -31$$ Add $|-13|$ and $|-18|$.
Both numbers are negative, so the sum is negative.

Your Turn Add.

d. $-43 + -11$ e. $15 + 22$ f. $-28 + 0$

Models can also help you add integers with different signs.

EXAMPLES   Add Integers with Different Signs

3 **Find** $5 + (-2)$.

Method 1 Use a number line.

- Start at zero.
- Move 5 units right.
- From there, move 2 units left.

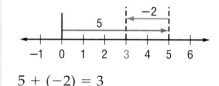

$5 + (-2) = 3$

Method 2 Use counters.

Remove zero pairs.

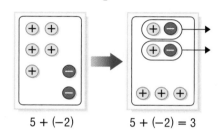

$5 + (-2)$ $5 + (-2) = 3$

STUDY TIP

Adding Integers on an Integer Mat
When one positive counter is paired with one negative counter, the result is called a *zero pair*.

4 **Find** $-4 + 3$.

Method 1 Use a number line.

- Start at zero.
- Move 4 units left.
- From there, move 3 units right.

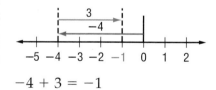

$-4 + 3 = -1$

Method 2 Use counters.

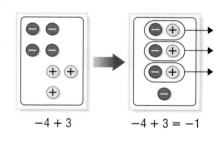

$-4 + 3$ $-4 + 3 = -1$

Your Turn Add using a number line or counters.

g. $7 + (-5)$ h. $-6 + 4$ i. $-1 + 8$

These examples suggest a rule for adding integers with different signs.

Key Concept	Adding Integers with Different Signs
Words	To add integers with different signs, subtract their absolute values. Give the result the same sign as the integer with the greater absolute value.
Examples	$8 + (-3) = 5$ $-8 + 3 = -5$

Add Integers with Different Signs

5 **Find −14 + 9.**

$-14 + 9 = -5$ To find $-14 + 9$, subtract $|9|$ from $|-14|$.
The sum is negative because $|-14| > |9|$.

Your Turn Add.

j. $-20 + 4$ k. $17 + (-6)$ l. $-8 + 27$

Two numbers with the same absolute value but different signs are called **opposites**. For example, -2 and 2 are opposites. An integer and its opposite are also called **additive inverses**.

Key Concept   **Additive Inverse Property**

Words   The sum of any number and its additive inverse is zero.

Symbols   Arithmetic   Algebra
                     $7 + (-7) = 0$ $x + (-x) = 0$

The Commutative, Associative, and Identity Properties, along with the Additive Inverse Property, can help you add three or more integers.

EXAMPLES **Add Three or More Integers**

STUDY TIP

Mental Math
• One way to add a group of integers mentally is to look for addends that are opposites.
• Another way is to group the positive addends together and the negative addends together. Then add.

6 **Find −4 + (−12) + 4.**

$$-4 + (-12) + 4 = -4 + 4 + (-12) \quad \text{Commutative Property}$$
$$= 0 + (-12) \quad \text{Additive Inverse Property}$$
$$= -12 \quad \text{Identity Property of Addition}$$

7 **Find −9 + 8 + (−2) + 16.**

$$-9 + 8 + (-2) + 16 = -9 + (-2) + 8 + 16 \quad \text{Commutative Property}$$
$$= [-9 + (-2)] + (8 + 16) \quad \text{Associative Property}$$
$$= -11 + 24 \text{ or } 13 \quad \text{Simplify.}$$

EXAMPLE **Use Integers to Solve a Problem**

8 **MONEY** The starting balance in a checking account is $75. What is the balance after checks for $12 and $20 are written?

Writing a check *decreases* your account balance, so integers for this situation are -12 and -20. Add these integers to the starting balance to find the new balance.

$$75 + (-12) + (-20) = 75 + [-12 + (-20)] \quad \text{Associative Property}$$
$$= 75 + (-32) \quad -12 + (-20) = -32$$
$$= 43 \quad \text{Simplify.}$$

The balance is now $43.

1. **Explain** how to add integers that have different signs.

2. **OPEN ENDED** Give an example of a positive and a negative integer whose sum is negative. Then find their sum.

3. **Which One Doesn't Belong?** Identify the pair of numbers that does not have the same characteristic as the other three. Explain your reasoning.

| −16 and 16 | 22 and −22 | 45 and 54 | −3 and 3 |

GUIDED PRACTICE

Add.

4. $-4 + (-5)$ 5. $10 + (-6)$ 6. $7 + (-18)$

7. $-21 + 8$ 8. $11 + (-3) + 9$ 9. $-14 + 2 + (-15) + 7$

10. **GOLF** Suppose a player shot −5, +2, −3, and −2 in four rounds of a tournament. What was the player's final score?

Practice and Applications

Add.

11. $-18 + (-8)$ 12. $7 + 16$ 13. $14 + 8$

14. $-14 + (-6)$ 15. $-3 + (-12)$ 16. $-5 + (-31)$

17. $20 + (-5)$ 18. $-15 + 8$ 19. $45 + (-4)$

20. $-19 + 2$ 21. $-10 + 34$ 22. $17 + (-18)$

23. $13 + (-43)$ 24. $-21 + 30$ 25. $-7 + (-25)$

26. $-54 + (-14)$ 27. $36 + (-47)$ 28. $-41 + 33$

HOMEWORK HELP	
For Exercises	See Examples
11–16, 25–26	1, 2
17–24, 27–28	3–5
33–38	6, 7
43–44	8

Extra Practice
See pages 617, 648.

Write an addition expression to describe each situation. Then find each sum.

29. **FOOTBALL** A team gains 8 yards on one play, then loses 5 yards on the next.

30. **SCUBA DIVING** A scuba diver dives 125 feet. Later, she rises 46 feet.

31. **ELEVATOR** You get on an elevator in the basement of a building, which is one floor below ground level. The elevator goes up 7 floors.

32. **WEATHER** The temperature outside is −2°F. The temperature drops by 9°.

Add.

33. $8 + (-6) + 5$ 34. $-7 + 2 + (-9)$ 35. $-5 + 11 + (-4)$

36. $3 + (-15) + 1$ 37. $-13 + 6 + (-8) + 13$ 38. $9 + (-4) + (-12) + (-9)$

Evaluate each expression if $a = -5$, $b = 2$, and $c = -3$.

39. $8 + a$ 40. $|b| + c$ 41. $a + b + c$ 42. $|a + b|$

MUSIC TRENDS For Exercises 43 and 44, use the table below that shows the change in music sales to the nearest percent from 1997 to 2000.

43. What is the percent of music sold in 2000 for each of these musical categories?

44. What was the total percent change in the sale of these types of music?

 Data Update What percent of music sold last year was rock, rap/hip hop, pop, or country? Visit msmath3.net/data_update to learn more.

Style of Music	Percent of Music Sold in 1997	Percent Change as of 2000
Rock	33	−8
Rap/Hip Hop	10	+3
Pop	9	+2
Country	14	−3

Source: Recording Industry Assoc. of America

45. **CRITICAL THINKING** Determine whether the following statement is *always*, *sometimes*, or *never* true.

If x and y are integers, then $|x + y| = |x| + |y|$.

Standardized Test Practice and Mixed Review

FCAT Practice

46. **MULTIPLE CHOICE** A stock on the New York Stock Exchange opened at $52 on Monday morning. The table shows the change in the value of the stock for each day that week. What was the stock worth at the close of business on Friday?

Day	Change
Monday	−$2
Tuesday	+$1
Wednesday	+$3
Thursday	−$1
Friday	−$4

   Ⓐ $41   Ⓑ $49   Ⓒ $57   Ⓓ $63

47. **MULTIPLE CHOICE** Simplify $-6 + |5| + (-3) + 4$.

   Ⓕ 0   Ⓖ −4   Ⓗ −10   Ⓘ −18

Replace each ● with <, >, or = to make a true sentence. (Lesson 1-3)

48. $-6 ● -11$ 49. $5 ● -5$ 50. $5 ● |8|$ 51. $|-7| ● -7$

52. **WEATHER** The time t in seconds between seeing lightning and hearing thunder can be used to estimate a storm's distance in miles. How far away is a storm if this time is 15 seconds? Use the expression $\frac{t}{5}$. (Lesson 1-2)

TELEVISION For Exercises 53 and 54, use the information below and the graph at the right.
The graph shows the number of prime-time television viewers in millions for different age groups. (Lesson 1-1)

53. Estimate the total number of viewers for all the age groups given.

54. About how many more people 65 years and over watch prime-time television than 18 to 24-year-olds?

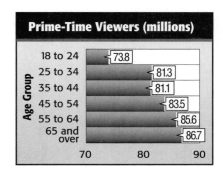

Prime-Time Viewers (millions)

Age Group	
18 to 24	73.8
25 to 34	81.3
35 to 44	81.1
45 to 54	83.5
55 to 64	85.6
65 and over	86.7

70   80   90

GETTING READY FOR THE NEXT LESSON

PREREQUISITE SKILL Evaluate each expression if $x = 3$, $y = 9$, and $z = 5$. (Lesson 1-2)

55. $x + 14$ 56. $z - 2$ 57. $y - z$ 58. $x + y - z$

Subtracting Integers

Sunshine State Standards
MA.A.1.3.4-2, MA.A.3.3.1-2, MA.A.3.3.1-3, MA.A.3.3.2-1, MA.A.3.3.2-2, MA.D.2.3.1-6

What You'll Learn
Subtract integers.

HANDS-ON Mini Lab

Materials
- counters
- integer mat

Work with a partner.

You can also use counters to model the subtraction of two integers. Follow these steps to model 3 − 5. Remember that subtract means *take away* or *remove*.

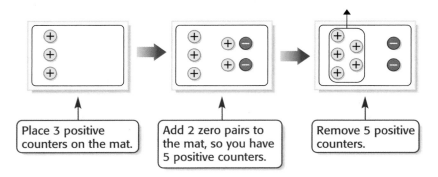

| Place 3 positive counters on the mat. | Add 2 zero pairs to the mat, so you have 5 positive counters. | Remove 5 positive counters. |

Since 2 negative counters remain, 3 − 5 = −2.

1. How does this result compare with the result of 3 + (−5)?
2. Use counters to find −4 − 2.
3. How does this result compare to −4 + (−2)?
4. Use counters to find each difference and sum. Compare the results in each group.

   a. 1 − 5; 1 + (−5)   b. −6 − 4; −6 + (−4)

When you subtract 3 − 5, as you did using counters in the Mini Lab, the result is the same as adding 3 + (−5). When you subtract −4 − 2, the result is the same as adding −4 + (−2).

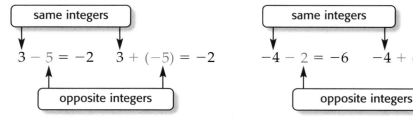

same integers

$$3 - 5 = -2 \qquad 3 + (-5) = -2$$

opposite integers

same integers

$$-4 - 2 = -6 \qquad -4 + (-2) = -6$$

opposite integers

These and other examples suggest a method for subtracting integers.

Key Concept   Subtracting Integers

Words	To subtract an integer, add its opposite or additive inverse.
Symbols	**Arithmetic** **Algebra**
	$4 - 7 = 4 + (-7)$ or -3 $a - b = a + (-b)$

Subtract a Positive Integer

1 **Find 9 − 12.**

$9 - 12 = 9 + (-12)$ To subtract 12, add −12.

$= -3$ Add.

2 **Find −6 − 8.**

$-6 - 8 = -6 + (-8)$ To subtract 8, add −8.

$= -14$ Add.

Your Turn Subtract.

a. $3 - 8$ **b.** $-5 - 4$ **c.** $10 - 7$

In Examples 1 and 2, you subtracted a positive integer by adding its opposite, a negative integer. To subtract a negative integer, you also add its opposite, a positive integer.

EXAMPLES **Subtract a Negative Integer**

3 **Find 7 − (−15).**

$7 - (-15) = 7 + 15$ To subtract −15, add 15.

$= 22$ Add.

4 **Find −30 − (−20).**

$-30 - (-20) = -30 + 20$ To subtract −20, add 20.

$= -10$ Add.

Your Turn Subtract.

d. $6 - (-7)$ **e.** $-5 - (-19)$ **f.** $-14 - (-2)$

Use the rule for subtracting integers to evaluate expressions.

EXAMPLES **Evaluate Algebraic Expressions**

Evaluate each expression if $a = 9$, $b = -8$, and $c = -2$.

5 $14 - b$

$14 - b = 14 - (-8)$ Replace b with −8.

$= 14 + 8$ To subtract −8, add 8.

$= 22$ Add.

6 $c - a$

$c - a = -2 - 9$ Replace c with −2 and a with 9.

$= -2 + (-9)$ To subtract 9, add −9.

$= -11$ Add.

Your Turn Evaluate each expression if $x = -5$ and $y = 7$.

g. $x - (-8)$ **h.** $-3 - y$ **i.** $y - x - 3$

STUDY TIP

Common Error
In Example 5, a common error is to replace b with 8 instead of its correct value of −8. Prevent this error by inserting parentheses before replacing b with its value.

$14 - b = 14 - (\ \ )$
$= 14 - (-8)$

Writing Math
Exercise 2

1. **OPEN ENDED** Write an expression involving the subtraction of a negative integer. Then write an equivalent addition expression.

2. **FIND THE ERROR** Anna and David are finding $-5 - (-8)$. Who is correct? Explain.

Anna
$-5 - (-8) = 5 + 8$
$= 13$

David
$-5 - (-8) = -5 + 8$
$= 3$

GUIDED PRACTICE

Subtract.

3. $8 - 13$ 4. $-4 - 10$ 5. $5 - 24$

6. $7 - (-3)$ 7. $-2 - (-6)$ 8. $-18 - (-7)$

Evaluate each expression if $n = 10$, $m = -4$, and $p = -12$.

9. $n - 17$ 10. $m - p$ 11. $p + n - m$

Practice and Applications

Subtract.

12. $5 - 9$ 13. $1 - 8$ 14. $12 - 15$

15. $4 - 16$ 16. $-6 - 3$ 17. $-8 - 8$

18. $-3 - 14$ 19. $-7 - 13$ 20. $2 - (-8)$

21. $9 - (-5)$ 22. $10 - (-2)$ 23. $5 - (-11)$

24. $-5 - (-4)$ 25. $-18 - (-7)$ 26. $-3 - (-6)$

27. $-7 - (-14)$ 28. $2 - |12|$ 29. $|-6| - |8|$

HOMEWORK HELP

For Exercises	See Examples
12–33, 42–44	1–4
34–41	5, 6

Extra Practice
See pages 617, 648.

GEOGRAPHY For Exercises 30–33, use the table at the right.

30. How far below the surface is the deepest part of Lake Huron?

31. How far below the surface is the deepest part of Lake Superior?

32. Find the difference between the deepest part of Lake Erie and the deepest part of Lake Superior.

33. How does the deepest part of Lake Michigan compare with the deepest part of Lake Ontario?

Great Lake	Deepest Point (m)	Surface Elevation (m)
Erie	−64	174
Huron	−229	176
Michigan	−281	176
Ontario	−244	75
Superior	−406	183

Source: National Ocean Service

Evaluate each expression if $a = -3$, $b = 14$, and $c = -8$.

34. $b - 20$ 35. $a - c$ 36. $a - b$ 37. $c - 15$

38. $a + b$ 39. $c + b$ 40. $b - a - c$ 41. $a - c - b$

42. SPACE On Mercury, the temperatures range from 805°F during the day to −275°F at night. Find the drop in temperature from day to night.

WEATHER For Exercises 43 and 44, use the following information and the table at the right.
The wind makes the outside temperature feel colder than the actual temperature.

43. How much colder does a temperature of 0°F with a 30-mile-per-hour wind feel than the same temperature with a 10-mile-per-hour wind?

44. How much warmer does 20°F feel than −10°F when there is a 30-mile-per-hour wind blowing?

Wind Chill Temperature			
Wind (miles per hour)			
Calm	10	20	30
20°	9°	4°	1°
10°	−4°	−9°	−12°
0°	−16°	−22°	−26°
−10°	−28°	−35°	−39°

Temperature (F)

Source: National Weather Service

45. WRITE A PROBLEM Write a problem about a real-life situation involving subtraction of integers for which the answer is −4.

CRITICAL THINKING For Exercises 46 and 47, determine whether the statement is *true* or *false*. If *false*, give a counterexample.

46. If x is a positive integer and y is a positive integer, then $x - y$ is a positive integer.

47. Subtraction of integers is commutative.

Standardized Test Practice and Mixed Review

FCAT Practice

48. MULTIPLE CHOICE Use the thermometers at the right to determine how much the temperature increased between 8:00 A.M. and 12:00 P.M.

   Ⓐ 14°F   Ⓑ 15°F   Ⓒ 30°F   Ⓓ 31°F

8:00 A.M. °F   12:00 P.M. °F   23°

−8°

49. MULTIPLE CHOICE Find the distance between A and B.

   Ⓕ −7 units   Ⓖ −3 units   Ⓗ 3 units   Ⓘ 7 units

50. BASEBALL The table at the right shows the money taken in (income) of several baseball teams in 2001. What was the total income of all of these teams? (*Hint:* A gain is positive income and a loss is negative income.) (Lesson 1-4)

Team	Income (thousands)
Atlanta Braves	−$14,360
Chicago Cubs	$4,797
Florida Marlins	−$27,741
New York Yankees	$40,859

Source: www.mlb.com

Evaluate each expression. (Lesson 1-3)

51. $|-14| + |3|$

52. $|20| - |-5|$

53. Name the property shown by $12n = 12n + 0$. (Lesson 1-2)

GETTING READY FOR THE NEXT LESSON

BASIC SKILL Multiply.

54. $4 \cdot 13$ **55.** $9 \cdot 15$ **56.** $2 \cdot 7 \cdot 6$ **57.** $3 \cdot 9 \cdot 4 \cdot 5$

CHAPTER 1 Mid-Chapter Practice Test

Vocabulary and Concepts

1. **OPEN ENDED** Write an equation that illustrates the Associative Property of Addition. (Lesson 1-2)

2. **Explain** how to determine the absolute value of a number. (Lesson 1-3)

Skills and Applications

3. **TRAVEL** A cruise ship has 148 rooms, with fifty on the two upper decks and the rest on the two lower decks. An upper deck room costs $1,100, and a lower deck room costs $900. Use the four-step plan to find the greatest possible room sales on one trip. (Lesson 1-1)

4. Evaluate $6 + 2(5 - 6 \div 2)$. (Lesson 1-2)

5. Find the value of $x^2 + y^2 - z^2$ if $x = 3$, $y = 6$, and $z = 2$. (Lesson 1-2)

Replace each ● with <, >, or = to make a true sentence. (Lesson 1-3)

6. $-2 ● 3$

7. $5 ● -6$

8. $|-4| ● |4|$

Evaluate each expression if $x = -7$ and $y = 3$. (Lesson 1-3)

9. $|-2|$

10. $|3| + |-6|$

11. $5 + |x|$

12. $|x| + |y|$

Add or subtract. (Lessons 1-4 and 1-5)

13. $6 + (-1)$

14. $-5 + (-8)$

15. $2 - 6$

16. $-2 - 3$

17. $-7 + 2$

18. $-1 - 7$

Standardized Test Practice

19. **GRID IN** You plant bushes in a row across the back and down two sides of a yard. A bush is planted at each of the four corners and every 4 meters. How many bushes are planted? (Lesson 1-1)

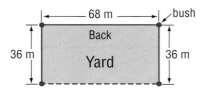

20. **MULTIPLE CHOICE** Naya recorded the low temperature for each of four days. Which list shows these temperatures in order from coldest to warmest? (Lesson 1-3)

Ⓐ $-2.3°C, 1.4°C, -1.2°C, -0.7°C$

Ⓑ $-0.7°C, -1.2°C, -2.3°C, 1.4°C$

Ⓒ $-0.7°C, -1.2°C, 1.4°C, -2.3°C$

Ⓓ $-2.3°C, -1.2°C, -0.7°C, 1.4°C$

The Game Zone

Math Skill

Comparing Integers

A Place To Practice Your Math Skills

Absolutely!

- ### GET READY!

  Players: two
  Materials: scissors, 14 index cards

- ### GET SET!

  - Cut each index card in half, making 28 cards.
  - Copy the integers below, one integer onto each of 24 cards.

    $$-17 \quad -3 \quad 0 \quad 19 \quad 16 \quad 5 \quad 25 \quad -10$$
    $$3 \quad -2 \quad -8 \quad -7 \quad 7 \quad 6 \quad 9 \quad 22$$
    $$11 \quad 12 \quad 1 \quad 14 \quad -20 \quad -13 \quad -16 \quad -18$$

  - Write "absolute value" on the 4 remaining cards and place these cards aside.
  - Shuffle the integer cards and deal them facedown to each player. Each player gets 2 "absolute value" cards.

- ### GO!

  - Each player turns the top card from his or her pile faceup. The player with the greater card takes both cards and puts them facedown in a separate pile. When there are no more cards in the original pile, shuffle the cards in the second pile and use them.
  - Twice during the game, each player can use an "absolute value" card after the two other cards have been played. When an absolute value card is played, players compare the absolute values of the integers on the cards. The player with the greater absolute value takes both cards. If there is a tie, continue play.
  - **Who Wins?** The player who takes all of the cards is the winner.

Multiplying and Dividing Integers

Sunshine State Standards
MA.A.1.3.4-2, MA.A.3.3.1-3, MA.A.3.3.2-2, MA.D.2.3.1-6

What You'll Learn
Multiply and divide integers.

⟲ REVIEW Vocabulary

factor: numbers that are multiplied together
product: the result when two or more numbers are multiplied together

WHEN am I ever going to use this?

OCEANOGRAPHY A deep-sea submersible descends 120 feet each minute to reach the bottom of Challenger Deep in the Pacific Ocean, the deepest point on Earth's surface. A descent of 120 feet is represented by -120. The table shows the submersible's depth after various numbers of minutes.

Time (min)	Depth (ft)
1	-120
2	-240
⋮	⋮
9	$-1,800$
10	$-1,200$

1. Write two different addition sentences that could be used to find the submersible's depth after 3 minutes. Then find their sum.

2. Write a multiplication sentence that could be used to find this same depth. Explain your reasoning.

3. Write a multiplication sentence that could be used to find the submersible's depth after 10 minutes. Then find the product.

Multiplication is repeated addition. So, $3(-120)$ means that -120 is used as an addend 3 times.

$$3(-120) = -120 + (-120) + (-120)$$
$$= -360$$

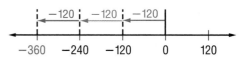

By the Commutative Property of Multiplication, $3(-120) = -120(3)$. This example suggests the following rule.

Key Concept — Multiplying Two Integers with Different Signs

Words   The product of two integers with different signs is negative.

Examples $2(-5) = -10$ $-5(2) = -10$

EXAMPLES — Multiply Integers with Different Signs

1 Find $6(-8)$.

$6(-8) = -48$ The factors have different signs. The product is negative.

2 Find $-9(2)$.

$-9(2) = -18$ The factors have different signs. The product is negative.

Your Turn Multiply.

a. $5(-3)$ b. $-8(6)$ c. $-2(4)$

The product of two positive integers is positive. For example, $3 \cdot 2 = 6$. What is the sign of the product of two negative integers? Look for a pattern to find $-3 \cdot (-2)$.

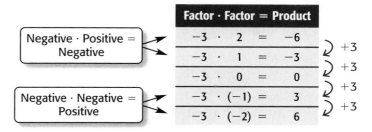

Factor · Factor = Product
$-3 \cdot 2 = -6$
$-3 \cdot 1 = -3$
$-3 \cdot 0 = 0$
$-3 \cdot (-1) = 3$
$-3 \cdot (-2) = 6$

Negative · Positive = Negative

Negative · Negative = Positive

This example suggests the following rule.

> **Key Concept**   **Multiplying Two Integers with the Same Sign**
>
> **Words**   The product of two integers with the same sign is positive.
>
> **Examples** $2(5) = 10$ $-2(-5) = 10$

EXAMPLE   **Multiply Integers with the Same Sign**

3 Find $-4(-3)$.

$-4(-3) = 12$ The factors have the same sign. The product is positive.

Your Turn Multiply.

d. $-3(-7)$ e. $6(4)$ f. $(-5)^2$

To multiply more than two integers, group factors using the Associative Property of Multiplication.

EXAMPLE   **Multiply More than Two Integers**

4 Find $-2(3)(-9)$.

$$-2(3)(-9) = [-2(3)](-9) \quad \text{Associative Property}$$
$$= -6(-9) \quad -2(3) = -6$$
$$= 54 \quad -6(-9) = 54$$

You know that multiplication and division are opposite operations. Examine the following multiplication sentences and their related division sentences.

READING Math

Division In a division sentence like $12 \div 3 = 4$, the number you are dividing, 12, is called the *dividend*. The number you are dividing by, 3, is called the *divisor*. The result is called the *quotient*.

Multiplication Sentence	Related Division Sentences	
$4(3) = 12$	$12 \div 3 = 4$	$12 \div 4 = 3$
$-4(3) = -12$	$-12 \div 3 = -4$	$-12 \div (-4) = 3$
$4(-3) = -12$	$-12 \div (-3) = 4$	$-12 \div 4 = -3$
$-4(-3) = 12$	$12 \div (-3) = -4$	$12 \div (-4) = -3$

These examples suggest that the rules for dividing integers are similar to the rules for multiplying integers.

Words	The quotient of two integers with different signs is negative. The quotient of two integers with the same sign is positive.
Examples	$16 \div (-8) = -2$ $-16 \div (-8) = 2$

EXAMPLES **Divide Integers**

5 **Find** $-24 \div 3$**.**   The dividend and the divisor have different signs.

$-24 \div 3 = -8$ The quotient is negative.

6 **Find** $-30 \div (-15)$**.**   The signs are the same.

$-30 \div (-15) = 2$ The quotient is positive.

Your Turn **Divide.**

   g. $-28 \div (-7)$ **h.** $\dfrac{36}{-2}$ **i.** $\dfrac{-40}{8}$

You can use all of the rules you have learned for adding, subtracting, multiplying, and dividing integers to evaluate algebraic expressions. Remember to follow the order of operations.

EXAMPLE **Evaluate Algebraic Expressions**

7 **Evaluate** $-2a - b$ **if** $a = -3$ **and** $b = -5$**.**

$$-2a - b = -2(-3) - (-5)$$ Replace a with -3 and b with -5.

$$= 6 - (-5)$$ The product of -2 and -3 is positive.

$$= 6 + 5$$ To subtract -5, add 5.

$$= 11$$ Add.

REAL-LIFE MATH

CARD GAMES In the game of *Hearts*, the object is to avoid scoring points. Each heart is worth one penalty point, the queen of spades is worth 13, and the other cards have no value.

Source: www.pagat.com

EXAMPLE **Find the Mean of a Set of Integers**

8 **CARD GAMES** **In a certain card game, you can gain or lose points with each round played. Atepa's change in score for each of five rounds is shown. Find Atepa's mean (average) point gain or loss per round.**

Atepa
−10
−30
−20
10
20

To find the mean of a set of numbers, find the sum of the numbers. Then divide the result by how many numbers there are in the set.

$$\frac{-10 + (-30) + (-20) + 10 + 20}{5} = \frac{-30}{5}$$ Find the sum of the set of numbers. Divide by the number in the set.

$$= -6$$ Simplify.

Atepa lost an average 6 points per round of cards.

1. **State** whether each product or quotient is positive or negative.

   a. $-8(-6)$ b. $16 \div (-4)$ c. $5(-7)(9)$

2. **OPEN ENDED** Name two integers whose quotient is -7.

3. **NUMBER SENSE** Find the sign of each of the following if n is a negative number. Explain your reasoning.

   a. n^2 b. n^3 c. n^4 d. n^5

GUIDED PRACTICE

Multiply.

4. $-4 \cdot 5$ 5. $-2(-7)$ 6. $(-3)^2$ 7. $-4(5)(-7)$

Divide.

8. $-25 \div (-5)$ 9. $-16 \div 4$ 10. $\dfrac{-49}{7}$ 11. $\dfrac{30}{-10}$

Evaluate each expression if $a = -5$, $b = 8$, and $c = -12$.

12. $4a + 9$ 13. $\dfrac{b - c}{a}$ 14. $3b - a^2$

Practice and Applications

Multiply.

15. $7(-8)$	**16.** $-5 \cdot 8$	**17.** $-4(-6)$	
18. $-14(-2)$	**19.** $12 \cdot 5$	**20.** $-3(-9)$	
21. $8(-9)$	**22.** $4(7)$	**23.** $(-8)^2$	
24. $(-7)^2$	**25.** $6(-2)(-7)$	**26.** $-3(-3)(-4)$	
27. $(-5)^3$	**28.** $(-3)^3$	**29.** $-2(4)(-3)(-10)$	
30. $-4(-8)(-2)(-5)$	**31.** $-2(-4)^2$	**32.** $(2)^2 \cdot (-6)^2$	

HOMEWORK HELP

For Exercises	See Examples
15–34	1–4
35–48	5, 6
49–56	7
57–60	8

Extra Practice
See pages 617, 648.

33. **HIKING** For every 1-kilometer increase in altitude, the temperature drops $7°C$. Find the temperature change for a 5-kilometer altitude increase.

34. **LIFE SCIENCE** Most people lose 100 to 200 hairs per day. If you were to lose 150 hairs per day for 10 days, what would be the change in the number of hairs you have?

Divide.

35. $50 \div (-5)$	**36.** $28 \div 7$	**37.** $-60 \div 3$	**38.** $-84 \div (-4)$
39. $45 \div 9$	**40.** $64 \div (-8)$	**41.** $-34 \div (-2)$	**42.** $-72 \div 6$
43. $\dfrac{-108}{12}$	**44.** $\dfrac{-39}{-13}$	**45.** $\dfrac{-42}{-6}$	**46.** $\dfrac{121}{-11}$

47. **WEATHER** A weather forecaster says that the temperature is changing at a rate of $-8°$ per hour. At that rate, how long will it take for the temperature change to be $-24°$?

48. OCEANOGRAPHY In 1960, a submersible named *Trieste* descended into Challenger Deep at a rate of 125 feet per minute to measure its depth. The bottom of the Challenger Deep has been estimated at −36,000 feet. How long did it take for the *Trieste* to reach the bottom?

Evaluate each expression if $w = -2$, $x = 3$, $y = -4$, and $z = -5$.

49. $x + 6y$ **50.** $9 - wz$ **51.** $\dfrac{w - x}{z}$ **52.** $\dfrac{8y}{x - 5}$

53. $\dfrac{6z}{x} - y$ **54.** $\dfrac{-42}{y - x} + w$ **55.** $-z^2$ **56.** $4(3w + 2)^2$

Find the mean of each set of integers.

57. $-2, -7, -6, 5, -10$ **58.** $-14, -17, -20, -16, -13$

59. $-23, -21, -28, -27, -25, -26$ **60.** $-15, 19, -13, 17, -12, 16$

61. CRITICAL THINKING Explain how you can use the number of negative factors to determine the sign of the product when multiplying more than two integers.

EXTENDING THE LESSON

The sum or product of any two whole numbers (0, 1, 2, 3, . . .) is always a whole number. So, the set of whole numbers is said to be *closed* under addition and multiplication. This is an example of the *Closure Property*. State whether each statement is *true* or *false*. If *false*, give a counterexample.

62. The set of whole numbers is closed under subtraction.

63. The set of integers is closed under multiplication.

64. The set of integers is closed under division.

Standardized Test Practice and Mixed Review

FCAT Practice

65. MULTIPLE CHOICE A glacier receded at a rate of 350 feet per day for two consecutive weeks. How much did the glacier's position change in all?

    A −336 ft   **B** −348 ft   **C** −700 ft   **D** −4,900 ft

66. SHORT RESPONSE On six consecutive days, the low temperature in a city was −5°C, −4°C, 6°C, 3°C, −1°C, and −8°C. What was the average low temperature for the six days?

Subtract. (Lesson 1-5)

67. $12 - 18$ **68.** $-5 - (-14)$ **69.** $-3 - 20$

Add. (Lesson 1-4)

70. $-9 + 2 + (-8)$ **71.** $-24 + (-11) + 24$ **72.** $-7 + 12 + (-3) + 6$

GETTING READY FOR THE NEXT LESSON

BASIC SKILL Give an example of a word or phrase that could indicate each operation.

Example: addition → the sum of

73. addition   **74.** subtraction   **75.** multiplication   **76.** division

Writing Expressions and Equations

 Sunshine State Standards
MA.A.3.3.2-4, MA.D.1.3.1-2, MA.D.1.3.2-2, MA.D.2.3.1-1, MA.D.2.3.1-2

What You'll Learn

Write algebraic expressions and equations from verbal phrases and sentences.

→ NEW Vocabulary

defining a variable

WHEN am I ever going to use this?

PARTY PLANNING It costs $8 per guest to hold a birthday party at the Community Center, as shown in the table.

1. What is the relationship between the number of guests and the cost of the party?

2. Write an expression representing the cost of a party with g guests.

Number of Guests	Party Cost
5	5 · 8 or 40
7	7 · 8 or 56
10	10 · 8 or 80
12	12 · 8 or 96
g	?

An important skill in algebra is writing verbal expressions as algebraic expressions. The steps in this process are given below.

```
❶ Write a model of the     ❷ Define a variable.     ❸ Write an algebraic
   situation using words.                               expression.
```

When you choose a variable and an unknown quantity for the variable to represent, this is called **defining a variable**. In the example above, g is defined as the *unknown number of guests*.

Algebraic expressions are made up of variables and operation symbols. The following table lists some common words and phrases that usually indicate the four operations.

Addition or Subtraction				Multiplication or Division			
plus	increased by	minus	subtract	times	each	divided	rate
sum	in all	less	decreased by	product	of	quotient	ratio
total	more than	less than	difference	multiplied	factors	an, in, or per	separate

STUDY TIP

Defining a Variable Any letter can be used as a variable, but it is often helpful to select letters that can be easily connected to the quantity they represent.

Example: age → a

EXAMPLE Write an Algebraic Expression

❶ Write *five years older than her brother* as an algebraic expression.

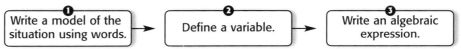

Words	five years older than her brother
Variable	Let a represent her brother's age.
Expression	five years older than her brother's age
	5 + a

The expression is $5 + a$.

 **EXAMPLE** Write an Algebraic Expression

2 Write *six dollars an hour times the number of hours* as an algebraic expression.

Words	six dollars an hour times the number of hours
Variable	Let h represent the number of hours.
Expression	six dollars an hour times the number of hours 6 · h

The expression is $6 \cdot h$ or $6h$.

You can also translate a verbal sentence into an equation. Some key words that indicate an equation are *equals* and *is*.

EXAMPLE Write an Algebraic Equation

3 Write *a number less 8 is 22* as an algebraic equation.

Words	A number less 8 is 22.
Variable	Let n represent the number.
Equation	A number less 8 is 22. $n - 8$ = 22

The equation is $n - 8 = 22$.

Your Turn Write each verbal sentence as an algebraic equation.

a. 4 inches shorter than Ryan's height is 58 inches

b. 30 is 6 times a number.

REAL-LIFE MATH

GEOGRAPHY More than 6 million cubic feet of water go over the crestline of Niagara Falls every minute during peak daytime tourist hours.

Source: www.infoniagara.com

EXAMPLE Write an Equation to Solve a Problem

4 **GEOGRAPHY** Niagara Falls is one of the most visited waterfalls in North America, but it is not the tallest. Yosemite Falls is 2,249 feet taller. If Yosemite Falls is 2,425 feet high, write an equation to find the height of Niagara Falls.

Words	Yosemite's height is 2,249 feet taller than Niagara's height.
Variable	Let n represent the height of Niagara Falls.
Equation	Yosemite's height is 2,249 ft taller than Niagara's height. 2,425 = 2,249 + n

The equation is $2,425 = 2,249 + n$.

1. **OPEN ENDED** Write two different verbal phrases that could be represented by the algebraic expression $x + 4$.

2. **FIND THE ERROR** Regina and Kamilah are translating the verbal phrase *6 less than a number* into an algebraic expression. Who is correct? Explain.

> Regina
> $n - 6$

> Kamilah
> $6 - n$

GUIDED PRACTICE

Write each verbal phrase as an algebraic expression.

3. 18 seconds faster than Tina's time

4. the difference between 7 and a number

5. the quotient of a number and 9

Write each verbal sentence as an algebraic equation.

6. The sum of 6 and a number is 2.

7. When the people are separated into 5 committees, there are 3 people on each committee.

Practice and Applications

Write each verbal phrase as an algebraic expression.

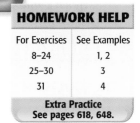

HOMEWORK HELP

For Exercises	See Examples
8–24	1, 2
25–30	3
31	4

Extra Practice
See pages 618, 648.

8. a $4 tip added to the bill

9. a number decreased by 6

10. half of Jessica's allowance

11. the sum of a number and -9

12. 5 points less than the average

13. a number divided by -3

14. 16 pounds more than his sister's weight

15. 20 fewer people than the number expected

16. 65 miles per hour for a number of hours

17. 4 more touchdowns than the other team scored

18. **LIFE SCIENCE** An adult cat has 2 fewer teeth than an adult human. Define a variable and write an expression for the number of teeth in an adult cat.

19. **HISTORY** Tennessee became a state 4 years after Kentucky. Define a variable and write an expression for the year Tennessee became a state.

20. **HEALTH** You count the number of times your heart beats in 15 seconds. Define a variable and write an expression for the number of times your heart beats in a minute.

21. **TRAVEL** Define a variable and write an expression for the number of miles Travis's car gets per gallon of gasoline if he drives 260 miles.

Write an algebraic expression that represents the relationship in each table.

22.

Age Now	Age in 12 years
5	17
8	20
12	24
16	28
a	■

23.

Number of Servings	Total Calories
2	300
5	750
7	1,050
12	1,800
n	■

24.

Regular Price	Sale Price
$8	$6
$12	$9
$16	$12
$20	$15
p	■

Write each verbal sentence as an algebraic equation.

25. 8 less than some number is equal to 15.

26. -30 is the product of -5 and a number.

27. -14 is twice a number.

28. 10 batches of cookies is 4 fewer than she made yesterday

29. $10 less the amount she spent is $3.50.

30. 3 pairs of jeans at d each is $106.50.

31. **MUSIC** A musician cannot be inducted into the Rock and Roll Hall of Fame until 25 years after their first album debuted. If an artist was inducted this year, write an equation to find the latest year y the artist's first album could have debuted.

32. **CRITICAL THINKING** Write an expression to represent *the difference of twice x and 3*. Then find the value of your expression if $x = -2$.

Standardized Test Practice and Mixed Review

FCAT Practice

33. **MULTIPLE CHOICE** Javier is 4 years older than his sister Rita. If Javier is y years old, which expression represents Rita's age?

   Ⓐ $y + 4$ Ⓑ $y - 4$ Ⓒ $4y$ Ⓓ $y \div 4$

34. **SHORT RESPONSE** Write an expression for the perimeter of a figure in the pattern at the right that contains x triangles. The sides of each triangle are 1 unit in length.

Figure 1   Figure 2   Figure 3

Multiply or divide. (Lesson 1-6)

35. $-9(10)$ 36. $-5(-14)$ 37. $34 \div (-17)$ 38. $\dfrac{-105}{-5}$

39. **BUSINESS** The formula $P = I - E$ is used to find the profit P when income I and expenses E are known. One month, a business had income of $18,600 and expenses of $20,400. What was the business's profit that month? (Lesson 1-5)

GETTING READY FOR THE NEXT LESSON

PREREQUISITE SKILL Add. (Lesson 1-4)

40. $-11 + 11$ 41. $-14 + 5$ 42. $6 + (-23)$ 43. $-7 + (-20)$

Problem-Solving Strategy
A Preview of Lesson 1-8

🟠 **Sunshine State Standards**
MA.A.3.3.2-1, MA.A.3.3.2-2, MA.D.1.3.2-1, MA.E.1.3.1-1

What You'll Learn
Solve problems by using the work backward strategy.

Work Backward

The closing day activities at the Junior Camp must be over by 2:45 P.M. We need $1\frac{1}{2}$ hours for field competitions, another 45 minutes for the awards ceremony, and an hour and 15 minutes for the cookout.

We also need an hour for room checkout. So how early do we need to get started? Let's **work backward** to figure it out.

Explore	We know the time that the campers must leave. We know the time it takes for each activity. We need to determine the time the day's activities should begin.
Plan	Let's start with the ending time and work backward.
Solve	The day is over at 2:45 P.M. **2:45 P.M.** **Go back** 1 hour for checkout. ➡ **1:45 P.M.** **Go back** 1 hour and 15 minutes for the cookout. ➡ **12:30 A.M.** **Go back** 45 minutes for the awards ceremony. ➡ **11:45 A.M.** **Go back** $1\frac{1}{2}$ hours for the field competitions. ➡ **10:15 A.M.** So, the day's activities should start no later than 10:15 A.M.
Examine	Assume that the day starts at 10:15 A.M. After $1\frac{1}{2}$ hours of field competitions, it is 11:45 A.M. After a 45-minute awards ceremony, it is 12:30 P.M. After the 1 hour and 15 minute cookout, it is 1:45 P.M., and after one hour for checkout, it is 2:45 P.M. So starting at 10:15 A.M. gives us enough time for all activities.

Analyze the Strategy

1. **Tell** why the work backward strategy is the best way to solve this problem.

2. **Explain** how you can examine the solution when you solve a problem by working backward.

3. **Write** a problem that could be solved by working backward. Then write the steps you would take to find the solution to your problem.

Solve. Use the work backward strategy.

4. **FAMILY** Mikal's great-grandmother was 6 years old when her family came to the United States, 73 years ago. If the year is 2003, when was her great-grandmother born?

5. **MONEY** The cash-in receipts in Brandon's cash drawer total $823.27, and his cash-out receipts total $734.87. If he currently has $338.40 in his drawer, what was his opening balance?

Mixed Problem Solving

Solve. Use any strategy.

TRAVEL For Exercises 6 and 7, use the graph at the right.

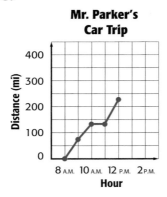

Mr. Parker's Car Trip

6. What may have happened between 10:00 A.M. and 11:00 A.M.?

7. Mr. Parker's total trip will cover 355 miles. If he maintains the speed set between 11:00 A.M. and noon, about what time should he reach his final destination?

8. **GRADES** Amelia wants to maintain an average of at least 90 in science class. So far her grades are 94, 88, 93, 85, and 91. What is the minimum grade she can make on her next assignment to maintain her average?

9. **CARS** Ms. Calzada will pay $375 a month for five years in order to buy her new car. The bank loaned her $16,800 to pay for the car. How much extra will Ms. Calzada end up paying for the loan?

10. **USE A MODEL** Suppose you had 100 sugar cubes. What is the largest cube you could build with the sugar cubes?

JEANS For Exercises 11 and 12, use the following information.
A store tripled the price it paid for a pair of jeans. After a month, the jeans were marked down $5. Two weeks later, the price was divided in half. Finally, the price was reduced by $3, and the jeans sold for $14.99.

11. How much did the store pay for the jeans?

12. Did the store make or lose money on the sale of the jeans?

13. **SPORTS** The graph shows the number of injuries for the top seven summer recreational activities. About how many injuries were there in all for these activities?

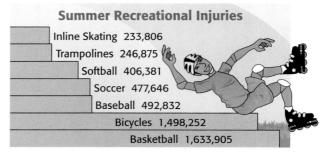

Summer Recreational Injuries

Inline Skating 233,806
Trampolines 246,875
Softball 406,381
Soccer 477,646
Baseball 492,832
Bicycles 1,498,252
Basketball 1,633,905

Source: American Academy of Orthopedic Surgeons

14. **STANDARDIZED TEST PRACTICE** FCAT Practice
Find the next three numbers in the pattern 5, 2, −1, −4 . . . .
Ⓐ −1, 2, 5
Ⓑ −7, −10, −13
Ⓒ −5, −6, −7
Ⓓ −6, −8, −10

You will use the work backward strategy in the next lesson.

1-8 Solving Addition and Subtraction Equations

Sunshine State Standards MA.A.3.3.2-1, MA.A.3.3.2-2, MA.A.3.3.3-1, MA.D.1.3.1-2, MA.D.1.3.2-2, MA.D.2.3.1-1, MA.D.2.3.1-3, MA.D.2.3.2-2

What You'll Learn
Solve equations using the Subtraction and Addition Properties of Equality.

⇒ NEW Vocabulary
solve
solution
inverse operations

 HANDS-ON Mini Lab

Materials
- cups
- counters
- equation mat

Work with a partner.

When you **solve** an equation, you are trying to find the values of the variable that makes the equation true. These values are called the **solutions** of the equation. You can use cups, counters, and an equation mat to solve $x + 4 = 6$.

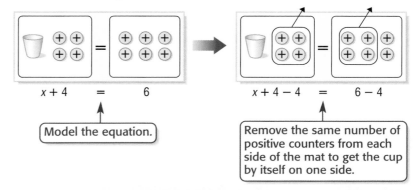

$x + 4 = 6$

Model the equation.

$x + 4 - 4 = 6 - 4$

Remove the same number of positive counters from each side of the mat to get the cup by itself on one side.

The number of positive counters remaining on the right side of the mat represents the value of x. So when $x = 2$, $x + 4 = 6$ is true.

Solve each equation using cups and counters.

1. $x + 1 = 4$　　　2. $x + 3 = 7$　　　3. $x + (-4) = -5$

4. Explain how you would find a value of x that makes $x + (-3) = -8$ true without using models.

In the Mini Lab, you solved the equation $x + 4 = 6$ by *removing*, or subtracting, the same number of positive counters from each side of the mat. This suggests the **Subtraction Property of Equality**.

Key Concept　　　　　Subtraction Property of Equality

Words	If you subtract the same number from each side of an equation, the two sides remain equal.	
Symbols	**Arithmetic**	**Algebra**
	$7 = 7$	$x + 4 = 6$
	$7 - 3 = 7 - 3$	$x + 4 - 4 = 6 - 4$
	$4 = 4$	$x = 2$

You can use this property to solve any addition equation. Remember to check your solution by substituting it back into the original equation.

Lesson 1-8 Solving Addition and Subtraction Equations　**45**

EXAMPLE Solve an Addition Equation

1 Solve $x + 5 = 3$. Check your solution.

Method 1 Vertical Method

$$x + 5 = 3 \quad \text{Write the equation.}$$

$$\begin{array}{r} x + 5 = 3 \\ -5 = -5 \\ \hline x = -2 \end{array} \quad \begin{array}{l} \text{Subtract 5 from} \\ \text{each side.} \end{array}$$

$5 - 5 = 0$ and $3 - 5 = -2$. x is by itself.

Method 2 Horizontal Method

$$x + 5 = 3$$

$$x + 5 - 5 = 3 - 5$$

$$x = -2$$

The solution is -2.

Check $x + 5 = 3$ Write the original equation.

$-2 + 5 \stackrel{?}{=} 3$ Replace x with -2. Is this sentence true?

$3 = 3$ ✔   The sentence is true.

Your Turn Solve each equation. Check your solution.

a. $a + 6 = 2$ b. $y + 3 = -8$ c. $5 = n + 4$

Addition and subtraction are called **inverse operations** because they "undo" each other. For this reason, you can use the **Addition Property of Equality** to solve subtraction equations like $x - 7 = -5$.

Key Concept   **Addition Property of Equality**

Words   If you add the same number to each side of an equation, the two sides remain equal.

Symbols   Arithmetic   Algebra

$$7 = 7 \qquad\qquad x - 5 = 6$$

$$7 + 3 = 7 + 3 \qquad x - 5 + 5 = 6 + 5$$

$$10 = 10 \qquad\qquad x = 11$$

EXAMPLE Solve a Subtraction Equation

2 Solve $y - 7 = -6$.

Method 1 Vertical Method

$$y - 7 = -6 \quad \text{Write the equation.}$$

$$\begin{array}{r} y - 7 = -6 \\ +7 = +7 \\ \hline y = 1 \end{array} \quad \text{Add 7 to each side.}$$

$-7 + 7 = 0$ and $-6 + 7 = 1$. y is by itself.

Method 2 Horizontal Method

$$y - 7 = -6$$

$$y - 7 + 7 = -6 + 7$$

$$y = 1$$

The solution is 1.   Check the solution.

Your Turn Solve each equation. Check your solution.

d. $x - 8 = -3$ e. $b - 4 = -10$ f. $7 = p - 12$

EXAMPLE **Write and Solve an Equation**

3 **MULTIPLE-CHOICE TEST ITEM** What value of n makes the sum of n and 25 equal -18?

Ⓐ -43 Ⓑ -7 Ⓒ 7 Ⓓ 43

Read the Test Item To find the value of n, write and solve an equation.

Solve the Test Item

The sum of n and 25,	equals	$-18.$	
$n + 25$	$=$	-18	Write the equation.
$n + 25 - 25 = -18 - 25$			Subtract 25 from each side.
$n = -43$			$-18 - 25 = -18 + (-25)$

The answer is A.

Test-Taking Tip

The Princeton Review

Backsolving In some instances, it may be easier to try each choice than to write and solve an equation.

Skill and Concept Check

Writing Math

Exercises 1 & 3

1. **Tell** what you might say to the boy in the cartoon to explain why the solution is correct.

"Hey, wait. That can't be right; yesterday we said x equals 3."

2. **OPEN ENDED** Write one addition equation and one subtraction equation that each have -3 as a solution.

3. **Which One Doesn't Belong?** Identify the equation that cannot be solved using the same property of equality as the other three. Explain.

| $g + 4 = 2$ | $a + 5 = -3$ | $m - 6 = 4$ | $x + 1 = -7$ |

GUIDED PRACTICE

Solve each equation. Check your solution.

4. $a + 4 = 10$ 5. $z + 7 = 2$ 6. $x + 9 = -3$

7. $y - 2 = 5$ 8. $n - 5 = -6$ 9. $d - 11 = -8$

Write and solve an equation to find each number.

10. The sum of a number and 8 is 1.

11. If you decrease a number by 20, the result is -14.

Practice and Applications

Solve each equation. Check your solution.

HOMEWORK HELP

For Exercises	See Examples
12–31	1, 2
32–44	3

Extra Practice
See pages 618, 648.

12. $x + 5 = 18$

13. $p + 11 = 9$

14. $a + 7 = 1$

15. $y + 12 = -3$

16. $w + 8 = -6$

17. $n + 3 = 20$

18. $g - 2 = -13$

19. $m - 15 = 3$

20. $b - 9 = -8$

21. $r - 20 = -4$

22. $k - 4 = 17$

23. $t - 6 = -16$

24. $28 + n = 34$

25. $52 + x = -7$

26. $-49 = c - 18$

27. $62 = f - 14$

28. $35 = -19 + d$

29. $-22 = -14 + q$

30. Find the value of x if $x + (-5) = -7$.

31. If $a - (-2) = 10$, what is the value of a?

Write and solve an equation to find each number.

32. If you increase a number by 12, the result is 7.

33. If you decrease a number by 8, the result is -14.

34. The difference of a number and 24 is -10.

35. The sum of a number and 30 is 9.

36. **GEOMETRY** Two angles are complementary if the sum of their measures is 90°. Angles A and B, shown at the right, are complementary. Write and solve an addition equation to find the measure of angle A.

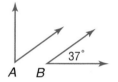

37. **BANKING** After you deposit $50 into your savings account, the balance is $124. Write and solve an addition equation to find your balance before this deposit.

38. **WEATHER** After falling 10°F, the temperature was -8°F. Write and solve a subtraction equation to find the starting temperature.

39. **GOLF** After four rounds of golf, Lazaro's score was 5 under par or -5. Lazaro had improved his overall score during the fourth round by decreasing it by 6 strokes. Write and solve a subtraction equation to find Lazaro's score after the third round.

BASKETBALL For Exercises 40 and 41, use the information below and in the table.
Katie Smith averaged 3.5 points per game more than Lisa Leslie during the 2001 WNBA regular season.

40. Write and solve an addition equation to find Lisa Leslie's average points scored per game.

41. Yolanda Griffith of the New York Liberty averaged 0.6 fewer points than Chamique Holdsclaw that season. Write and solve an equation to find how many points Yolanda Griffith averaged per game.

2001 WNBA Regular Season Points Leaders	
Player (Team)	AVG
Katie Smith (Minnesota Lynx)	23.1
Lisa Leslie (Los Angeles Sparks)	?
Tina Thompson (Houston Comets)	19.3
Janeth Arcain (Houston Comets)	18.5
Chamique Holdsclaw (Washington Mystics)	16.8

MINIMUM WAGE For Exercises 42 and 43, use the information in the table.

Year	Action
1996	A subminimum wage of $4.25 an hour is established for employees under 20 years of age during their first 90 days of employment.
1997	Congress raises the minimum wage to $5.15 an hour.

42. Write and solve an addition equation to find the increase in pay a teenager who started out at the subminimum wage would receive after their first 90 days of work.

43. In 1997, the minimum wage was increased by $0.40 per hour. Write and solve an addition equation to find the minimum wage before this increase.

 Data Update What is the current minimum wage? Visit msmath3.net/data_update to learn more.

44. **MULTI STEP** Suppose you go to the store and buy some supplies for school. You buy a pencil for $1.25, a notebook for $6.49, and some paper. The total cost of the supplies before tax was $8.79. Write an equation that can be used to find the cost c in dollars of the paper. Then solve your equation to find the cost of the paper.

45. **WRITE A PROBLEM** Write a problem about a real-life situation that can be answered by solving the equation $x + 60 = 20$. Then solve the equation to find the answer to your problem.

46. **CRITICAL THINKING** Solve $|x| + 5 = 7$.

Standardized Test Practice and Mixed Review

FCAT Practice

47. **MULTIPLE CHOICE** Dante paid $42 for a jacket, which included $2.52 in sales tax. Which equation could be used to find the price of the jacket before tax?

    Ⓐ $x - 2.52 = 42$ Ⓑ $x + 2.52 = 42$

    Ⓒ $x - 42 = 2.52$ Ⓓ $x + 42 = 2.52$

48. **MULTIPLE CHOICE** The record low temperature for the state of Virginia is 7°F warmer than the record low for West Virginia. If the record low for Virginia is −30°F, what is West Virginia's record low?

    Ⓕ −37°F   Ⓖ −23°F   Ⓗ 23°F   Ⓘ 37°F

Write each verbal phrase as an algebraic expression. (Lesson 1-7)

49. 7 inches per minute for a number of minutes

50. 5 degrees warmer than yesterday's high temperature

51. **MULTI STEP** Experts estimate that there may have been 100,000 tigers living 100 years ago. Now there are only about 6,000. Find the average change in the tiger population per year for the last 100 years. (Lesson 1-6)

GETTING READY FOR THE NEXT LESSON

PREREQUISITE SKILL Multiply. (Lesson 1-6)

52. 3(9)   53. −2(18)   54. −5(−11)   55. 4(−15)

Solving Multiplication and Division Equations

Sunshine State Standards MA.A.3.3.2-1, MA.A.3.3.2-2, MA.D.1.3.1-2, MA.D.1.3.2-2, MA.D.2.3.1-1, MA.D.2.3.1-3, MA.D.2.3.2-2

What You'll Learn

Solve equations by using the Division and Multiplication Properties of Equality.

⟳ REVIEW Vocabulary

Identity Property (×): the product of a number and 1 is that same number

WHEN am I ever going to use this?

PLANTS Some species of a bamboo can grow 35 inches per day. That is as many inches as the average child grows in the first 10 years of his or her life!

1. If d represents the number of days the bamboo has been growing, write a multiplication equation you could use to find how long it would take for the bamboo to reach a height of 210 inches.

Bamboo Growth	
Day	Height (in.)
1	35(1) = 35
2	35(2) = 70
3	35(3) = 105
⋮	⋮
d	?

The equation $35d = 210$ models the relationship described above. To undo the multiplication of 35, divide each side of the equation by 35.

EXAMPLE Solve a Multiplication Equation

1 Solve $35d = 210$.

$$35d = 210 \qquad \text{Write the equation.}$$
$$\frac{35d}{35} = \frac{210}{35} \qquad \text{Divide each side of the equation by 35.}$$
$$1d = 6 \qquad 35 \div 35 = 1 \text{ and } 210 \div 35 = 6$$
$$d = 6 \qquad \text{Identity Property; } 1d = d$$

The solution is 6.   Check the solution.

Your Turn Solve each equation. Check your solution.

a. $8x = 72$ b. $-4n = 28$ c. $-12 = -6k$

In Example 1, you used the **Division Property of Equality** to solve a multiplication equation.

Key Concept

Division Property of Equality

Words If you divide each side of an equation by the same nonzero number, the two sides remain equal.

Symbols

Arithmetic	Algebra
$12 = 12$	$5x = -60$
$\dfrac{12}{4} = \dfrac{12}{4}$	$\dfrac{5x}{5} = \dfrac{-60}{5}$
$3 = 3$	$x = -12$

Division Expressions
Remember, $\frac{a}{-3}$ means *a divided by* -3.

You can use the **Multiplication Property of Equality** to solve division equations like $\frac{a}{-3} = -7$.

Key Concept **Multiplication Property of Equality**

Words   If you multiply each side of an equation by the same number, the two sides remain equal.

Symbols

Arithmetic	Algebra
$5 = 5$	$\frac{x}{2} = 8$
$5(-4) = 5(-4)$	$\frac{x}{2}(2) = 8(2)$
$-20 = -20$	$x = 16$

EXAMPLE **Solve a Division Equation**

2 Solve $\frac{a}{-3} = -7$.

$\frac{a}{-3} = -7$ Write the equation.

$\frac{a}{-3}(-3) = -7(-3)$ Multiply each side by -3 to undo the division in $\frac{a}{-3}$.

$a = 21$ $-7 \cdot (-3) = 21$

The solution is 21   Check the solution.

Your Turn Solve each equation. Check your solution.

d. $\frac{y}{-4} = -8$ e. $\frac{m}{5} = -9$ f. $30 = \frac{b}{-2}$

EXAMPLE **Use an Equation to Solve a Problem**

3 **REPTILES** A Nile crocodile grows to be 4,000 times as heavy as the egg from which it hatched. If an adult crocodile weighs 2,000 pounds, how much does a crocodile egg weigh?

Words	The weight of an adult crocodile is 4,000 times as heavy as the weight of a crocodile egg.
Variable	Let g = the weight of the crocodile egg.
Equation	Weight of adult is 4,000 times the egg's weight.

Weight of adult is   4,000 times the egg's weight.
2,000   =   4,000g

$2,000 = 4,000g$ Write the equation.

$\frac{2,000}{4,000} = \frac{4,000g}{4,000}$ Divide each side by 4,000.

$0.5 = g$ $2,000 \div 4,000 = 0.5$

The crocodile egg weighs 0.5 pound.   Check this solution.

Skill and Concept Check

1. **State** what property you would use to solve $-4a = 84$. Explain your reasoning.

2. **OPEN ENDED** Write a division equation whose solution is -10.

3. **NUMBER SENSE** Without solving the equation, tell what you know about the value of x in the equation $\frac{x}{25} = 300$.

GUIDED PRACTICE

Solve each equation. Check your solution.

4. $5b = 40$
5. $-7k = 14$
6. $-3n = -18$
7. $-20 = 4x$
8. $\frac{p}{9} = 9$
9. $\frac{a}{12} = -3$
10. $\frac{m}{-2} = -22$
11. $7 = \frac{z}{-8}$

Write and solve an equation to find each number.

12. The product of -9 and a number is 45.

13. When you divide a number by 4, the result is -16.

Practice and Applications

Solve each equation. Check your solution.

14. $4c = 44$
15. $9b = 72$
16. $34 = -2x$
17. $-36 = 18y$
18. $-8d = 32$
19. $-5n = -35$
20. $-52 = -4g$
21. $-90 = 6w$
22. $\frac{k}{12} = 2$
23. $10 = \frac{m}{7}$
24. $6 = \frac{u}{-9}$
25. $\frac{h}{-3} = 33$
26. $\frac{c}{12} = -8$
27. $\frac{r}{24} = -3$
28. $\frac{q}{-5} = -20$
29. $\frac{t}{-4} = -15$
30. $\frac{10}{x} = -5$
31. $\frac{-126}{a} = -21$

HOMEWORK HELP

For Exercises	See Examples
14–31	1, 2
32–44	3

Extra Practice
See pages 618, 648.

Write and solve an equation to find each number.

32. The product of a number and 8 is 56.

33. When you multiply a number by -3, the result is 39.

34. When you divide a number by -5, the result is -10.

35. The quotient of a number and 7 is -14.

MEASUREMENT For Exercises 36–39, refer to the table. Write and solve an equation to find each quantity.

36. the number of yards in 18 feet

37. the number of feet in 288 inches

38. the number of yards in 540 inches

39. the number of miles in 26,400 feet

Customary System Conversions (length)
1 foot = 12 inches
1 yard = 3 feet
1 yard = 36 inches
1 mile = 5,280 feet
1 mile = 1,760 yards

40. LAWN SERVICE Josh charges $15 to mow an average size lawn in his neighborhood. Write and solve a multiplication equation to find how many of these lawns he needs to mow to earn $600.

41. ANIMALS An African elephant can eat 500 pounds of vegetation per day. Write and solve a multiplication equation to find how many days a 3,000-pound supply of vegetation will last for one elephant.

POPULATION For Exercises 42–44 use the information in the graphic at the right.

42. Write a multiplication equation that could be used to find how many hours it would take the world's population to increase by 1 million.

43. Solve the equation. Round to the nearest hour.

44. There are 24 hours in one day. Write and solve a multiplication equation to determine how many days it would take the world's population to increase by 1 million. Round to the nearest day.

45. CRITICAL THINKING If an object is traveling at a rate of speed r, then the distance d the object travels after a time t is given by $d = rt$. Write an expression for the value of t.

USA TODAY Snapshots®

Population grows by the hour
The world's 6.1 billion population increases by nearly 9,000 people each hour:

Born: **15,020**
— Die: **6,279**

Increase: **8,741**

Source: Census Bureau

By Marcy E. Mullins, USA TODAY

Standardized Test Practice and Mixed Review

FCAT Practice

46. SHORT RESPONSE The base B of a triangular prism has an area of 24 square inches. If the volume V of the prism is 216 cubic inches, use the formula $V = Bh$ to find the height of the prism in inches.

h

B

47. MULTIPLE CHOICE Luis ran 2.5 times the distance that Mark ran. If Mark ran 3 miles, which equation can be used to find the distance d in miles that Luis ran?

Ⓐ $d = 2.5 + 3$ Ⓑ $d + 2.5 = 3$ Ⓒ $d = 2.5(3)$ Ⓓ $2.5d = 3$

48. ARCHITECTURE William G. Durant wanted the Empire State Building to be taller than the building being built by his competitor, Walter Chrysler. He secretly had a 185-foot spire built inside the building and then hoisted to the top of the building upon its completion. Write and solve an addition equation that could be used to find the height of the Empire State Building without its spire. (Lesson 1-8)

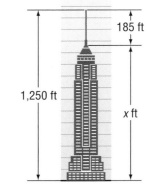

185 ft

1,250 ft

x ft

Write each verbal sentence as an algebraic equation. (Lesson 1-7)

49. A number increased by 10 is 4.

50. 8 feet longer than she jumped is 15 feet.

Study Guide and Review

Vocabulary and Concept Check

absolute value (p. 19)	evaluate (p. 11)	order of operations (p. 11)
additive inverse (p. 25)	inequality (p. 18)	powers (p. 12)
algebraic expression (p. 11)	integer (p. 17)	property (p. 13)
conjecture (p. 7)	inverse operations (p. 46)	solution (p. 45)
coordinate (p. 18)	negative number (p. 17)	solve (p. 45)
counterexample (p. 13)	numerical expression (p. 11)	variable (p. 11)
defining a variable (p. 39)	open sentence (p. 13)	
equation (p. 13)	opposites (p. 25)	

Choose the letter of the term that best matches each statement or phrase.

1. an integer and its opposite
2. a number less than zero
3. value of the variable that makes the equation true
4. a sentence that compares two numbers
5. contains a variable, a number, and at least one operation symbol
6. to find the value of an expression
7. a mathematical sentence that contains an equals sign
8. an open sentence that is true for any number
9. the distance a number is from zero
10. operations that "undo" each other

a. algebraic expression
b. evaluate
c. absolute value
d. equation
e. additive inverses
f. property
g. inverse operations
h. solution
i. negative number
j. inequality

Lesson-by-Lesson Exercises and Examples

1-1 **A Plan for Problem Solving** (pp. 6–10)

Use the four-step plan to solve each problem.

11. **SCIENCE** A chemist pours table salt into a beaker. If the beaker plus the salt has a mass of 84.7 grams and the beaker itself has a mass of 63.3 grams, what was the mass of the salt?

12. **SPORTS** In a basketball game, the Sliders scored five 3-point shots, seven 2-point shots, and 15 1-point shots. Find the total points scored.

Example 1 At Smart's Car Rental, it costs $57 per day plus $0.10 per mile to rent a certain car. How much will it cost to rent the car for 1 day and drive 180 miles?

Multiply the number of miles by the cost per mile. Then add the daily cost.

$0.10 \times 180 = \$18$

$\$18 + \$57 = \$75$

The cost is $75.

54 **Chapter 1** Algebra: Integers

 msmath3.net/vocabulary_review

1-2 Variables, Expressions, and Properties (pp. 11–15)

Evaluate each expression.

13. $3^2 - 2 \cdot 3 + 5$ 14. $4 + 2(5 - 2)$

15. $\dfrac{25}{6^2 - 11}$ 16. $\dfrac{22 - 6}{4} + 10(2)$

Evaluate each expression if $a = 6$, $b = 2$, and $c = 1$.

17. $3a - 2b + c$ 18. $\dfrac{(a + 2)^2}{bc}$

Example 2 Evaluate $x^2 + yx - z^2$ if $x = 4$, $y = 2$, and $z = 1$.

$x^2 + yx - z^2$ Write the expression.

$= 4^2 + (2)(4) - (1)^2$ $x = 4, y = 2,$ and $z = 1$

$= 16 + (2)(4) - 1$ Evaluate powers first.

$= 16 + 8 - 1$ Multiply.

$= 23$ Add and subtract.

1-3 Integers and Absolute Value (pp. 17–21)

19. **MONEY** Kara made an $80 withdrawal from her checking account. Write an integer for this situation.

Replace each ● with $<$, $>$, or $=$ to make a true sentence.

20. -8 ● 7 21. -2 ● -6

22. Order the set of integers $\{-7, 8, 0, -3, -2, 5, 6\}$ from least to greatest.

Evaluate each expression.

23. $\left|-5\right|$ 24. $\left|-12\right| - \left|4\right|$

Example 3 Replace the ● in -3 ● -7 with $<$, $>$, or $=$ to make a true sentence. Graph the integers on a number line.

-3 lies to the right of -7, so $-3 > -7$.

Example 4 Evaluate $\left|-3\right|$.
Since the graph of -3 is 3 units from 0 on the number line, the absolute value of -3 is 3.

1-4 Adding Integers (pp. 23–27)

Add.

25. $-54 + 21$ 26. $100 + (-75)$

27. $-14 + (-20)$ 28. $38 + (-46)$

29. $-14 + 37 + (-20) + 2$

30. **WEATHER** At 8:00 A.M., it was $-5°$F. By noon, it had risen $34°$. Write an addition statement to describe this situation. Then find the sum.

Example 5 Find $-16 + (-11)$.

$-16 + (-11)$

$= -27$ Add $\left|-16\right|$ and $\left|-11\right|$. Both numbers are negative, so the sum is negative.

Example 6 Find $-7 + 20$.

$-7 + 20$

$= 13$ To find $-7 + 20$, subtract $\left|-7\right|$ from $\left|20\right|$. The sum is positive because $\left|20\right| > \left|-7\right|$.

1-5 Subtracting Integers (pp. 28–31)

Subtract.

31. $-2 - (-5)$ 32. $-30 - 13$

33. $11 - 15$ 34. $25 - (-11)$

Example 7 Find $-27 - (-6)$.

$-27 - (-6) = -27 + 6$ To subtract -6, add 6.

$= -21$ Add.

1-6 **Multiplying and Dividing Integers** (pp. 34–38)

Multiply or divide.

35. $-4(-25)$ 36. $-7(3)$

37. $-15(-4)(-1)$ 38. $180 \div (-15)$

39. $-170 \div (-5)$ 40. $-88 \div 8$

41. **GOLF** José scored -2 on each of six golf holes. What was his overall score for these six holes?

Example 8 Find $3(-20)$.

$3(-20) = -60$ The factors have different signs. The product is negative.

Example 9 Find $-48 \div (-12)$.

$-48 \div (-12) = 4$ The dividend and the divisor have the same sign. The quotient is positive.

1-7 **Writing Expressions and Equations** (pp. 39–42)

Write each verbal phrase or sentence as an algebraic expression or equation.

42. Six divided by a number is $\frac{1}{2}$.

43. the sum of a number and 7

44. A number less 10 is 25.

45. Four times a number is 48.

Example 10 Write *nine less than a number is 5* **as an algebraic equation.**

Let n represent the number.

Nine less than a number   is   5.

$n - 9 \qquad = \qquad 5$

The equation is $n - 9 = 5$.

1-8 **Solving Addition and Subtraction Equations** (pp. 45–49)

Solve each equation. Check your solution.

46. $n + 40 = 90$ 47. $x - 3 = 10$

48. $c - 30 = -18$ 49. $9 = a + 31$

50. $d + 14 = -1$ 51. $27 = y - 12$

52. **CANDY** There are 75 candies in a bowl after you remove 37. Write and solve a subtraction equation to find how many candies were originally in the bowl.

Example 11 Solve $5 + k = 18$.

$\quad 5 + k = 18$ Write the equation.

$5 - 5 + k = 18 - 5$ Subtract 5 from each side.

$\qquad n = 13$ $18 - 5 = 13$

Example 12 Solve $n - 13 = -62$.

$\quad n - 13 = -62$ Write the equation.

$n - 13 + 13 = -62 + 13$ Add 13 to each side.

$\qquad n = -49$ $-62 + 13 = -49$

1-9 **Solving Multiplication and Division Equations** (pp. 50–53)

Solve each equation. Check your solution.

53. $15x = -75$ 54. $-72 = -6f$

55. $-4x = 52$ 56. $\frac{s}{7} = 42$

57. $-3 = \frac{d}{24}$ 58. $\frac{y}{-10} = -15$

Example 13

Solve $60 = 5t$.

$60 = 5t$

$\dfrac{60}{5} = \dfrac{5t}{5}$

$12 = t$

Example 14

Solve $\dfrac{m}{-2} = 8$.

$\dfrac{m}{-2} = 8$

$\left(\dfrac{m}{-2}\right)(-2) = 8(-2)$

$m = -16$

Vocabulary and Concepts

1. **Determine** whether the following statement is *true* or *false*. Explain.
   The absolute value of a positive number is negative.

2. **Write** two different verbal phrases for the algebraic expression $8 - n$.

Skills and Applications

Evaluate each expression if $a = 3$, $b = 2$, and $c = 5$.

3. $a + 15 \div a$

4. $(2c + b) \div a - 3$

5. $4a^2 - 5a - 12$

Replace each ● with $<$, $>$, or $=$ to make a true sentence.

6. $-6 ● 3$

7. $-8 ● -11$

8. $|8| ● -8$

9. Order the set of integers $\{-3, 0, -5, 1, -2\}$ from least to greatest.

10. Find the value of $|y| - |x|$ if $x = -4$ and $y = -9$.

Add or subtract.

11. $-27 + 8$

12. $-12 + 60$

13. $-9 + (-11)$

14. $10 - 24$

15. $-41 - 13$

16. $-4 - (-35)$

Multiply or divide.

17. $-5(-13)$

18. $8(-9)$

19. $7(-10)(-4)$

20. $-105 \div 15$

21. $\dfrac{-70}{-5}$

22. $\dfrac{36}{-4}$

Solve each equation. Check your solution.

23. $k - 10 = 65$

24. $x + 15 = -3$

25. $-7 = a - 11$

26. $3d = 24$

27. $\dfrac{n}{-2} = 16$

28. $-96 = 8y$

29. **CARDS** After losing two rounds in a card game, Eneas' score was -40. After winning the third round, her score was 5. Write and solve an addition equation to find the number of points scored in the third round.

Standardized Test Practice

30. **MULTIPLE CHOICE** What is the distance between the airplane and the submarine?

    Ⓐ 524 ft

    Ⓑ 536 ft

    Ⓒ 1,156 ft

    Ⓓ 1,176 ft

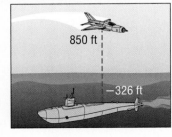

🔶 FCAT Practice

PART 1 Multiple Choice

Record your answers on the answer sheet provided by your teacher or on a sheet of paper.

1. The table shows how much Miranda spent on her school lunch during one week.

Day	Amount
Monday	$3.15
Tuesday	$3.25
Wednesday	$3.85
Thursday	$2.95
Friday	$4.05

Which of the following is a good estimate for the total amount Miranda spent on lunches that week? (Prerequisite Skill, p. 600)

   Ⓐ $15   Ⓑ $16

   Ⓒ $17   Ⓓ $18

2. What is the order of operations for the expression $27 \div (4 + 5) \times 2$? (Lesson 1-2)

   Ⓕ $\div, +, \times$ Ⓖ $\div, \times, +$

   Ⓗ $+, \div, \times$ Ⓘ $+, \times, \div$

3. Which of the following properties is illustrated by the equation $6 + (2 + 5) = (2 + 5) + 6$? (Lesson 1-2)

   Ⓐ Associative Property

   Ⓑ Commutative Property

   Ⓒ Distributive Property

   Ⓓ Inverse Property

4. Find $-15 - (-9)$. (Lesson 1-5)

   Ⓕ -24 Ⓖ -6 Ⓗ 4 Ⓘ 6

5. Solve the equation $y = -8(-4)(-2)$.
(Lesson 1-6)

   Ⓐ -64 Ⓑ -32 Ⓒ 32 Ⓓ 64

TEST-TAKING TIP

Question 6 Be careful when a question involves adding, subtracting, multiplying, or dividing negative integers. Check to be sure that you have chosen an answer with the correct sign.

6. What is -54 divided by -6? (Lesson 1-6)

   Ⓕ -60 Ⓖ -48 Ⓗ -9 Ⓘ 9

7. Which of these expresses the equation below in words? (Lesson 1-7)

   $$3(x - 4) = 7x + 5$$

   Ⓐ Three times a number minus four is seven times that number plus five.

   Ⓑ Three times a number minus four is seven times the sum of that number and five.

   Ⓒ Three times the difference of a number and four is seven times that number plus five.

   Ⓓ Three times the difference of a number and four is seven times the sum of that number and five.

8. Amy had 20 jellybeans. Tariq gave her 18 more. Amy ate the jellybeans as she walked home. When she got home, she had 13 left. Which equation shows how many jellybeans Amy ate? (Lesson 1-7)

   Ⓕ $20 + 18 + x = 13$

   Ⓖ $(20 - 18) - x = 13$

   Ⓗ $(20 - 18) + x = 13$

   Ⓘ $(20 + 18) - x = 13$

9. If $x - 12 = -15$, find the value of $4 - x$. (Lesson 1-8)

   Ⓐ -13 Ⓑ -1 Ⓒ 1 Ⓓ 7

PART 2 Short Response/Grid In

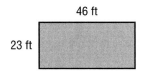

Record your answers on the answer sheet provided by your teacher or on a sheet of paper.

10. What is the area of the empty lot shown below in square feet? (Prerequisite Skill, p. 613)

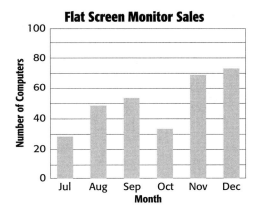

46 ft

23 ft

11. The graph shows the number of flat screen computer monitors sold during the last 6 months of the year at Marvel Computers.

Flat Screen Monitor Sales

Estimate the average number of computers sold per month during the last 6 months to the nearest ten. (Prerequisite Skill, p. 602)

12. One pound of coffee makes 100 cups. If 300 cups of coffee are served at each football game, how many pounds are needed for 7 games? (Lesson 1-1)

13. What is the value of the following expression? (Lesson 1-2)

$$\frac{5 - 3}{8 + 6 \div 2 \times 4}$$

14. Which points on the following number line have the same absolute value? (Lesson 1-3)

15. The table shows Mr. Carson's weight change during the first 3 months of his diet. If he started his diet at 245 pounds, how much did he weigh at the end of month 3? (Lesson 1-4)

Month	1	2	3
Weight Change	−7	−9	−5

16. The eighth grade ordered 216 hot dogs for their end-of-the-year party. If 8 hot dogs come in a single package, how many packages did they buy? (Lesson 1-9)

PART 3 Extended Response

Record your answers on a sheet of paper. Show your work.

17. To raise money for a charity, an 8th-grade science class asked a student group to perform a benefit concert in the school's 400-seat auditorium. Tickets for the 180 seats near the stage sold for $30 each. Tickets for other seats were sold at a lower price. The concert sold out, raising a total of $9,360. (Lesson 1-1)

   a. How many seats are not in the section near the stage?

   b. Write an equation for the price p of each ticket in the section not near the stage.

   c. Find the price of the tickets in the section not near the stage.

18. In the table below, n, p, r, and t each represent a different integer. If $n = -4$ and $t \neq 1$, find each of the following values. Explain your reasoning using the properties of integers. (Lesson 1-2)

$$n \times p = n$$
$$t \times r = r$$
$$n + t = r$$

   a. p b. r c. t

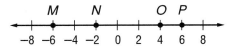

2 Algebra: Rational Numbers

❝What do roller coasters have to do with math?❞

A ride on the roller coaster called *The Beast* takes 3 minutes and 40 seconds. You can write this time as $3\frac{40}{60}$ or $3\frac{2}{3}$ minutes. You can also write this mixed number as the decimal $3.\overline{6}$.

You will order fractions and mixed numbers by writing them as decimals in Lesson 2-2.

GETTING STARTED

Take this quiz to see whether you are ready to begin Chapter 2. Refer to the lesson or page number in parentheses if you need more review.

▶ Vocabulary Review

Complete each sentence.

1. Two numbers with the same absolute value but different signs are called __?__ or __?__ __?__. (Lesson 1-4)

2. The value of a variable that makes an equation true is called the __?__ of the equation. (Lesson 1-8)

▶ Prerequisite Skills

Add. (Lesson 1-4)

3. $-13 + 4$
4. $28 + (-9)$
5. $-18 + 21$
6. $4 + (-16)$

Subtract. (Lesson 1-5)

7. $-8 - 6$
8. $23 - (-15)$
9. $-17 - 11$
10. $-5 - (-10)$

Multiply or divide. (Lesson 1-6)

11. $6(-14)$
12. $36 \div (-4)$
13. $-86 \div (-2)$
14. $-3(-9)$

Solve each equation. (Lessons 1-8 and 1-9)

15. $-12x = 144$
16. $a + 9 = 37$
17. $-18 = y - 42$
18. $25 = \dfrac{n}{5}$

Find the least common multiple (LCM) of each set of numbers. (Page 612)

19. $10, 5, 6$
20. $3, 7, 9$
21. $12, 16$
22. $24, 9$

Rational Numbers Make this Foldable to help you organize your notes. Begin with five sheets of $8\frac{1}{2}$" by 11" paper.

STEP 1 **Stack Pages**
Place 5 sheets of paper $\frac{3}{4}$ inch apart.

STEP 2 **Roll Up Bottom Edges**
All tabs should be the same size.

STEP 3 **Crease and Staple**
Staple along the fold.

STEP 4 **Label**
Label the tabs with the lesson numbers.

Reading and Writing As you read and study the chapter, write examples from each lesson under each tab.

Fractions and Decimals

Sunshine State Standards
MA.A.1.3.3-1, MA.A.1.3.4-1, MA.A.3.3.2-1

WHEN am I ever going to use this?

WHALE WATCHING The top ten places in the Northern Hemisphere to watch whales are listed below.

Viewing Site	Location	Type Seen
Sea of Cortez	Baja California, Mexico	Blue, Finback, Sei, Sperm, Minke, Pilot, Orca, Humpback, Gray
Dana Point	California	Gray
Monterey Bay	California	Gray
San Ignacio Lagoon	Baja California, Mexico	Gray
Churchill River Estuary	Manitoba, Canada	Beluga
Stellwagen Bank National Marine Sanctuary	Massachusetts	Humpback, Finback, Minke
Lahaina	Hawaii	Humpback
Silver Bank	Dominican Republic	Humpback
Mingan Island	Quebec, Canada	Blue
Friday Harbor	Washington	Orca, Minke

1. What fraction of the sites are in the United States?
2. What fraction of the sites are in Canada?
3. At what fraction of the sites might you see gray whales?
4. What fraction of the humpback viewing sites are in Mexico?

Numbers such as $\frac{1}{2}$, $\frac{1}{5}$, $\frac{2}{5}$, and $\frac{1}{10}$ are called **rational numbers**.

Key Concept   Rational Numbers

Words   A rational number is any number that can be expressed in the form $\frac{a}{b}$, where a and b are integers and $b \neq 0$.

Since -7 can be written as $\frac{-7}{1}$ and $2\frac{2}{3}$ can be written as $\frac{8}{3}$, -7 and $2\frac{2}{3}$ are rational numbers. All integers, fractions, and mixed numbers are rational numbers.

Mental Math
It is helpful to memorize these commonly used fraction-decimal equivalencies.

$\frac{1}{2} = 0.5$ $\quad$ $\frac{1}{3} = 0.\overline{3}$

$\frac{1}{4} = 0.25$ $\quad$ $\frac{1}{5} = 0.2$

$\frac{1}{8} = 0.125$

$\frac{1}{10} = 0.1$

Any fraction can be expressed as a decimal by dividing the numerator by the denominator.

EXAMPLE **Write a Fraction as a Decimal**

1 Write $\frac{5}{8}$ as a decimal.

$\frac{5}{8}$ means $5 \div 8$.

```
   0.625
8)5.000      Add a decimal point and zeros to the dividend: 5 = 5.000
  4 8
   20
   16
   40
   40
    0        Division ends when the remainder is 0.
```

You can also use a calculator. $\quad$ 5 ÷ 8 ENTER **0.625**

The fraction $\frac{5}{8}$ can be written as 0.625.

A decimal like 0.625 is called a **terminating decimal** because the division ends, or terminates, when the remainder is 0.

EXAMPLE **Write a Mixed Number as a Decimal**

2 Write $1\frac{2}{3}$ as a decimal.

$1\frac{2}{3}$ means $1 + \frac{2}{3}$. To change $\frac{2}{3}$ to a decimal, divide 2 by 3.

```
   0.666...     The three dots means the six keeps repeating.
3)2.000
  1 8
   20
   18
   20
   18
    2          The remainder after each step is 2.
```

You can also use a calculator. $\quad$ 2 ÷ 3 ENTER **0.666666667**

The mixed number $1\frac{2}{3}$ can be written as $1 + 0.666...$ or $1.666...$.

Your Turn Write each fraction or mixed number as a decimal.

a. $\frac{3}{4}$ $\qquad$ b. $-\frac{3}{5}$ $\qquad$ c. $2\frac{1}{9}$ $\qquad$ d. $5\frac{1}{6}$

STUDY TIP

Bar Notation
The bar is placed above the repeating part. To write 8.636363... in bar notation, write $8.\overline{63}$, not $8.\overline{6}$ or $8.\overline{636}$. To write 0.3444... in bar notation, write $0.3\overline{4}$, not $0.\overline{34}$.

A decimal like 1.666... is called a **repeating decimal**. Since it is not possible to write all of the digits, you can use **bar notation** to show that the 6 repeats.

$$1.666... = 1.\overline{6}$$

Repeating decimals often occur in real-life situations. However, they are usually rounded to a certain place-value position.

EXAMPLE **Round a Repeating Decimal**

3 BASEBALL In a recent season, Kansas City pitcher Kris Wilson won 6 of the 11 games he started. To the nearest thousandth, find his winning average.

To find his winning average, divide the number of wins, 6, by the number of games, 11.

6 ÷ 11 ENTER 0.5454545

Look at the digit to the right of the thousandths place. Round down since 4 < 5.

Kris Wilson's winning average was 0.545.

Terminating and repeating decimals are also rational numbers because you can write them as fractions.

EXAMPLE **Write a Terminating Decimal as a Fraction**

4 Write 0.45 as a fraction.

$$0.45 = \frac{45}{100} \quad \text{0.45 is 45 hundredths.}$$

$$= \frac{9}{20} \quad \text{Simplify. Divide by the greatest common factor of 45 and 100, 5.}$$

The decimal 0.45 can be written as $\frac{9}{20}$.

You can use algebra to change repeating decimals to fractions.

READING Math

Repeating Decimals
Read 0.5 as *point five repeating.*

EXAMPLE **Write a Repeating Decimal as a Fraction**

5 ALGEBRA Write $0.\overline{5}$ as a fraction in simplest form.

Let $N = 0.\overline{5}$ or 0.555... . Then $10N = 5.555...$.

Multiply N by 10 because 1 digit repeats.

Subtract $N = 0.555...$ to eliminate the repeating part, 0.555... .

$$\begin{array}{ll} 10N = 5.555... & \\ \underline{-1N = 0.555...} & N = 1N \\ 9N = 5 & 10N - 1N = 9N \\ \dfrac{9N}{9} = \dfrac{5}{9} & \text{Divide each side by 9.} \\ N = \dfrac{5}{9} & \text{Simplify.} \end{array}$$

The decimal $0.\overline{5}$ can be written as $\frac{5}{9}$.

Your Turn Write each decimal as a fraction or mixed number in simplest form.

e. -0.14 f. 8.75 g. $0.\overline{3}$ h. $-1.\overline{4}$

Writing Math
Exercises 1 & 3

1. **OPEN ENDED** Give an example of a repeating decimal where two digits repeat. Explain why your number is a rational number.

2. **Write** 5.321321321... using bar notation.

3. **Which One Doesn't Belong?** Identify the fraction that cannot be expressed as the same type of decimal as the other three. Explain.

$$\frac{4}{11} \qquad \frac{1}{2} \qquad \frac{1}{9} \qquad \frac{1}{3}$$

GUIDED PRACTICE

Write each fraction or mixed number as a decimal.

4. $\frac{4}{5}$ 5. $4\frac{3}{8}$ 6. $-\frac{1}{3}$ 7. $7\frac{5}{33}$

Write each decimal as a fraction or mixed number in simplest form.

8. 0.6 9. -1.55 10. $-0.\overline{5}$ 11. $2.\overline{1}$

BIOLOGY For Exercises 12 and 13, use the figure at the right.

12. Write the length of the ant as a fraction.

13. Write the length of the ant as a decimal.

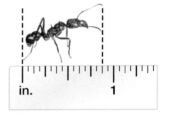

Practice and Applications

Write each fraction or mixed number as a decimal.

14. $\frac{1}{4}$ 15. $\frac{1}{5}$ 16. $-\frac{13}{25}$ 17. $-\frac{11}{50}$

18. $2\frac{1}{8}$ 19. $5\frac{5}{16}$ 20. $-\frac{5}{6}$ 21. $-\frac{2}{9}$

22. $-\frac{4}{33}$ 23. $-\frac{6}{11}$ 24. $6\frac{4}{11}$ 25. $7\frac{8}{33}$

HOMEWORK HELP	
For Exercises	See Examples
14–27	1, 2
28–33, 41–44	4
34–39	5
40	3

Extra Practice
See pages 619, 649.

26. Write $\frac{10}{33}$ as a decimal using bar notation.

27. Write $\frac{2}{45}$ as a decimal using bar notation.

Write each decimal as a fraction or mixed number in simplest form.

28. 0.4 29. 0.5 30. -0.16 31. -0.35

32. 5.55 33. 7.32 34. $-0.\overline{2}$ 35. $-0.\overline{4}$

36. $3.\overline{6}$ 37. $2.\overline{7}$ 38. $-4.\overline{21}$ 39. $-3.\overline{72}$

40. **BASEBALL** In a recent season, Sammy Sosa had 189 hits during his 577 at-bats. What was Sammy Sosa's batting average? Round to the nearest thousandth.

41. Write 0.38 and 0.383838 as fractions.

BIOLOGY For Exercises 42–44, use the information at the right.

42. Write the weight of a queen bee as a fraction.

43. Write the weight of a hummingbird as a fraction.

44. Write the weight of a hamster as a mixed number.

Animal	Weight (ounces)
Queen Bee	0.004
Hummingbird	0.11
Hamster	3.5

Source: *Animals as Our Companions*

THEATER For Exercises 45 and 46, use the following information.
The Tony Award is given to exceptional plays and people involved in making them. The award weighs 1 pound 10 ounces.

45. Write the weight of the Tony Award in pounds using a mixed number in simplest form.

46. Write the weight of the Tony Award in pounds using decimals.

47. CRITICAL THINKING A *unit fraction* is a fraction that has 1 as its numerator.

   a. Write the four greatest unit fractions that are terminating decimals. Write each fraction as a decimal.

   b. Write the four greatest unit fractions that are repeating decimals. Write each fraction as a decimal.

Standardized Test Practice and Mixed Review

FCAT Practice

48. MULTIPLE CHOICE Janeth Arcain of the Houston Comets in the WNBA made 0.9 of her free throws in the 2001 season. Write this decimal as a fraction in simplest form.

  A $\dfrac{8}{9}$ **B** $\dfrac{9}{10}$ **C** $\dfrac{4}{5}$ **D** $\dfrac{3}{5}$

49. MULTIPLE CHOICE A survey asked Americans to name the biggest problem with home improvement. The results are shown in the table. What decimal represents the fraction of people surveyed who chose procrastination?

  F 0.15   **G** 0.32

  H 0.11   **I** 0.42

Reason	Fraction of Respondents
Lack of Time	$\dfrac{21}{50}$
Procrastination	$\dfrac{8}{25}$
Lack of Know-How	$\dfrac{3}{20}$
Lack of Tools	$\dfrac{11}{100}$

Source: Impulse Research for Ace Hardware

50. The product of two integers is 72. If one integer is -18, what is the other integer? (Lesson 1-9)

Solve each equation. Check your solution. (Lesson 1-8)

51. $t + 17 = -5$ **52.** $a - 5 = 14$ **53.** $5 = 9 + x$ **54.** $m - 5 = -14$

GETTING READY FOR THE NEXT LESSON

PREREQUISITE SKILL Find the least common multiple for each pair of numbers. (Page 612)

55. 15, 5   **56.** 6, 9   **57.** 8, 6   **58.** 3, 5

Comparing and Ordering Rational Numbers

 **Sunshine State Standards**
MA.A.1.3.2-1, MA.A.1.3.2-2, MA.A.1.3.4-1, MA.A.3.3.1-3

What You'll Learn
Compare and order rational numbers.

MATH Symbols

< less than

> greater than

WHEN am I ever going to use this?

RECYCLING The table shows the portion of some common materials and products that are recycled.

1. Do we recycle more or less than half of the paper we produce? Explain.

2. Do we recycle more or less than half of the aluminum cans? Explain.

3. Which items have a recycle rate less than one half?

Material	Fraction Recycled
Paper	$\frac{5}{11}$
Aluminum Cans	$\frac{5}{8}$
Glass	$\frac{2}{5}$
Scrap Tires	$\frac{3}{4}$

Source: http://envirosystemsinc.com

4. Which items have a recycle rate greater than one half?

5. Using this estimation method, can you order the rates from least to greatest?

Sometimes you can use estimation to compare rational numbers. Another method is to compare two fractions with common denominators. Or you can also compare decimals.

EXAMPLE Compare Rational Numbers

1 Replace ● with <, >, or = to make $\frac{5}{8} \bullet \frac{3}{4}$ a true sentence.

Method 1 Write as fractions with the same denominator.

For $\frac{5}{8}$ and $\frac{3}{4}$, the least common denominator is 8.

$\frac{5}{8} = \frac{5 \cdot 1}{8 \cdot 1}$ or $\frac{5}{8}$

$\frac{3}{4} = \frac{3 \cdot 2}{4 \cdot 2}$ or $\frac{6}{8}$

Since $\frac{5}{8} < \frac{6}{8}$, $\frac{5}{8} < \frac{3}{4}$.

Method 2 Write as decimals.

Write $\frac{5}{8}$ and $\frac{3}{4}$ as decimals. Use a calculator.

5 ÷ 8 ENTER = **0.625**   3 ÷ 4 ENTER = **0.75**

$\frac{5}{8} = 0.625$ $\frac{3}{4} = 0.75$

Since $0.625 < 0.75$, $\frac{5}{8} < \frac{3}{4}$.

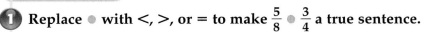

EXAMPLE **Compare Negative Rational Numbers**

2 Replace ● with <, >, or = to make -5.2 ● $-5\frac{1}{4}$ a true sentence.

Write $-5\frac{1}{4}$ as a decimal.

$\frac{1}{4} = 0.25$, so $-5\frac{1}{4} = -5.25$.

Since $-5.2 > -5.25$, $-5.2 > -5\frac{1}{4}$.

Check Use a number line to check the answer.

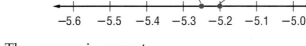

The answer is correct.

STUDY TIP

Number Lines
A number to the left is always less than a number to the right.

Your Turn Replace each ● with <, >, or = to make a true sentence.

a. $\frac{5}{6}$ ● $\frac{7}{9}$ b. $-\frac{5}{7}$ ● -0.7 c. $2\frac{3}{5}$ ● $2.\overline{6}$

You can order rational numbers by writing any fractions as decimals. Then order the decimals.

EXAMPLE **Order Rational Numbers**

REAL-LIFE MATH

ROLLER COASTERS The *Dragon Fire* is a double looping coaster with a corkscrew. The track is 2,160 feet long.

Source: Paramount

3 **ROLLER COASTERS** The ride times for nine roller coasters are shown in the table. Order the times from least to greatest.

Coaster	Ride Time (min)
Dragon Fire	$2\frac{1}{6}$
Mighty Canadian Minebuster	$2.\overline{6}$
Wilde Beast	2.5
Ghoster Coaster	$1\frac{5}{6}$
SkyRider	$2\frac{5}{12}$
Thunder Run	1.75
The Bat	$1\frac{5}{6}$
Vortex	1.75
TOP GUN	$2\frac{5}{12}$

Source: Paramount

$2\frac{1}{6} = 2.1\overline{6}$ $1\frac{5}{6} = 1.8\overline{3}$ $2\frac{5}{12} = 2.41\overline{6}$

From least to greatest, the times are 1.75, 1.75, $1\frac{5}{6}$, $1\frac{5}{6}$, $2\frac{1}{6}$, $2\frac{5}{12}$, $2\frac{5}{12}$, 2.5, and $2.\overline{6}$. So, Vortex and Thunder Run have the shortest ride times, and Mighty Canadian Minebuster has the longest ride time.

68 Chapter 2 Algebra: Rational Numbers

1. **Explain** why 0.28 is less than $0.\overline{28}$.

2. **OPEN ENDED** Name two fractions that are less than $\frac{1}{2}$ and two fractions that are greater than $\frac{1}{2}$.

3. **NUMBER SENSE** Are the fractions $\frac{5}{11}$, $\frac{5}{12}$, $\frac{5}{13}$, and $\frac{5}{14}$ arranged in order from least to greatest or from greatest to least? Explain.

GUIDED PRACTICE

Replace each ● with <, >, or = to make a true sentence.

4. $\frac{3}{4}$ ● $\frac{7}{12}$ 5. $-\frac{4}{5}$ ● $-\frac{7}{9}$ 6. $3\frac{5}{8}$ ● 3.625   7. $-2\frac{4}{9}$ ● -2.42

Order each set of rational numbers from least to greatest.

8. $\frac{4}{5}$, 0.5, $\frac{1}{3}$, 0.65   9. $-\frac{2}{3}$, 0.7, -0.68, $\frac{3}{4}$ 10. $-1\frac{2}{3}$, -1.23, -1.45, $-1\frac{1}{2}$

11. **CARPENTRY** Rondell has some drill bits marked $\frac{7}{16}$, $\frac{3}{8}$, $\frac{5}{32}$, $\frac{9}{16}$, and $\frac{1}{4}$. If these are all measurements in inches, how should he arrange them if he wants them from least to greatest?

Practice and Applications

Replace each ● with <, >, or = to make a true sentence.

			HOMEWORK HELP

For Exercises	See Examples
12–23, 33	1, 2
24–32, 34–35	3

Extra Practice
See pages 619, 649.

12. $\frac{2}{3}$ ● $\frac{7}{9}$ 13. $\frac{3}{5}$ ● $\frac{5}{8}$ 14. $-\frac{3}{11}$ ● $-\frac{1}{3}$

15. $-\frac{8}{11}$ ● $-\frac{7}{9}$ 16. -2.3125 ● $-2\frac{5}{16}$ 17. -5.2 ● $-5\frac{3}{11}$

18. $0.\overline{38}$ ● $\frac{4}{11}$ 19. $0.\overline{26}$ ● $\frac{4}{15}$ 20. $-4.\overline{37}$ ● -4.37

21. $-3.1\overline{6}$ ● $-3.\overline{16}$ 22. $\frac{3}{7}$ ● $0.\overline{42}$ 23. $12\frac{5}{6}$ ● $12.8\overline{3}$

Order each set of rational numbers from least to greatest.

24. 1.8, 1.07, $1\frac{8}{9}$, $1\frac{1}{2}$ 25. $7\frac{1}{5}$, 6.8, 7.6, $6\frac{3}{4}$ 26. $\frac{1}{9}$, 0.1, $-\frac{1}{3}$, -0.25

27. $-\frac{3}{5}$, 0.45, -0.5, $\frac{4}{7}$ 28. $-3\frac{2}{5}$, -3.68, -3.97, $-4\frac{3}{4}$ 29. -2.9, -2.95, $-2\frac{9}{11}$, $-2\frac{13}{14}$

30. Which is least: $\frac{7}{11}$, 0.6, $\frac{2}{3}$, $0.6\overline{3}$, or $\frac{8}{13}$?

31. Which is greatest: $\frac{3}{8}$, 0.376, 0.367, $\frac{2}{5}$, or $0.\overline{37}$?

32. **STATISTICS** If you order a set of numbers from least to greatest, the middle number is the *median*. Find the median of 23.2, 22.45, 21.63, $22\frac{5}{8}$, and $21\frac{3}{5}$.

33. **PHOTOGRAPHY** The shutter time on Diego's camera is set at $\frac{1}{250}$ second. If Diego wants to increase the shutter time, should he set the time at $\frac{1}{500}$ second or $\frac{1}{125}$ second?

34. Match each number with a point on the number line.

   a. 0.425 b. $\frac{3}{8}$ c. $\frac{13}{16}$ d. $0.\overline{15}$

35. **MULTI STEP** The table shows the regular season records of five college baseball teams during the 2002 season. Which team had the best record?

Team	Games Won	Games Played
University of Alabama	48	61
University of Notre Dame	44	59
University of Southern California	34	56
Florida State University	56	68
Rice University	47	58

36. **CRITICAL THINKING** Are there any rational numbers between $0.\overline{2}$ and $\frac{2}{9}$? Explain.

Standardized Test Practice and Mixed Review

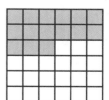

FCAT Practice

37. **MULTIPLE CHOICE** Determine which statement is *not* true.

   Ⓐ $\frac{3}{4} < 0.\overline{7}$ Ⓑ $-\frac{2}{3} = -0.\overline{6}$ Ⓒ $0.81 > \frac{4}{5}$ Ⓓ $-0.58 > -\frac{5}{12}$

38. **SHORT RESPONSE** Is the fraction represented by the shaded part of the square at the right greater than, equal to, or less than 0.41?

39. **HISTORY** During the fourteenth and fifteenth centuries, printing presses used type cut from wood blocks. Each block was $\frac{7}{8}$ inch thick. Write this fraction as a decimal.

(Lesson 2-1)

Solve each equation. Check your solution. (Lesson 1-9)

40. $\frac{y}{7} = 22$ 41. $4p = -60$ 42. $20 = \frac{t}{15}$ 43. $81 = -3d$

GETTING READY FOR THE NEXT LESSON

PREREQUISITE SKILL Multiply. (Lesson 1-6)

44. $-4(-7)$ 45. $8(-12)$ 46. $17(-3)$ 47. $-23(-5)$

Multiplying Rational Numbers

Sunshine State Standards
MA.A.3.3.1-1, MA.A.3.3.2-1, MA.A.3.3.1-3, MA.D.2.3.1-6, MA.D.2.3.1-7

What You'll Learn
Multiply fractions.

NEW Vocabulary

dimensional analysis

REVIEW Vocabulary

greatest common factor (GCF): the greatest of the common factors or two or more numbers (Page 610)

HANDS-ON Mini Lab

Materials
• paper
• colored pencils

Work with a partner.

To multiply $\frac{1}{3}$ and $\frac{2}{5}$, you can use an area

model to find $\frac{1}{3}$ of $\frac{2}{5}$.

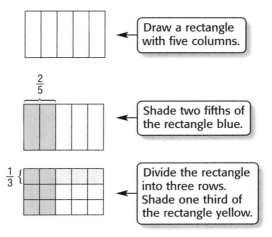

Draw a rectangle with five columns.

Shade two fifths of the rectangle blue.

Divide the rectangle into three rows. Shade one third of the rectangle yellow.

The overlapping green area represents the product of $\frac{1}{3}$ and $\frac{2}{5}$.

1. What is the product of $\frac{1}{3}$ and $\frac{2}{5}$?

2. Use an area model to find each product.

   a. $\frac{3}{4} \cdot \frac{1}{2}$ **b.** $\frac{2}{5} \cdot \frac{2}{3}$ **c.** $\frac{1}{4} \cdot \frac{3}{5}$ **d.** $\frac{2}{3} \cdot \frac{4}{5}$

3. What is the relationship between the numerators of the factors and the numerator of the product?

4. What is the relationship between the denominators of the factors and the denominator of the product?

The Mini Lab suggests the rule for multiplying fractions.

Key Concept	Multiplying Fractions
Words	To multiply fractions, multiply the numerators and multiply the denominators.

Symbols	**Arithmetic**	**Algebra**
	$\frac{2}{3} \cdot \frac{4}{5} = \frac{8}{15}$	$\frac{a}{b} \cdot \frac{c}{d} = \frac{ac}{bd}$, where $b \neq 0,\ d \neq 0$

Multiply Fractions

1 Find $\frac{4}{9} \cdot \frac{3}{5}$. **Write in simplest form.**

$$\frac{4}{9} \cdot \frac{3}{5} = \frac{4}{\overset{}{\underset{3}{9}}} \cdot \frac{\overset{1}{3}}{5} \qquad \text{Divide 9 and 3 by their GCF, 3.}$$

$$= \frac{4 \cdot 1}{3 \cdot 5} \qquad \begin{array}{l} \leftarrow \text{Multiply the numerators.} \\ \leftarrow \text{Multiply the denominators.} \end{array}$$

$$= \frac{4}{15} \qquad \text{Simplify.}$$

Use the rules for multiplying integers to determine the sign of the product.

Multiply Negative Fractions

2 Find $-\frac{5}{6} \cdot \frac{3}{8}$. **Write in simplest form.**

$$-\frac{5}{6} \cdot \frac{3}{8} = \frac{-5}{\overset{}{\underset{2}{6}}} \cdot \frac{\overset{1}{3}}{8} \qquad \text{Divide 6 and 3 by their GCF, 3.}$$

$$= \frac{-5 \cdot 1}{2 \cdot 8} \qquad \begin{array}{l} \leftarrow \text{Multiply the numerators.} \\ \leftarrow \text{Multiply the denominators.} \end{array}$$

$$= -\frac{5}{16} \qquad \text{The fractions have different signs, so the product is negative.}$$

STUDY TIP

Negative Fractions
$-\frac{5}{6}$ can be written as $\frac{-5}{6}$ or as $\frac{5}{-6}$.

Your Turn **Multiply. Write in simplest form.**

a. $\frac{8}{9} \cdot \frac{3}{4}$

b. $-\frac{3}{5} \cdot \frac{7}{9}$

c. $\left(-\frac{1}{2}\right)\left(-\frac{6}{7}\right)$

To multiply mixed numbers, first rename them as improper fractions.

Multiply Mixed Numbers

3 Find $4\frac{1}{2} \cdot 2\frac{2}{3}$. **Write in simplest form.**

$$4\frac{1}{2} \cdot 2\frac{2}{3} = \frac{9}{2} \cdot \frac{8}{3} \qquad 4\frac{1}{2} = \frac{9}{2}, 2\frac{2}{3} = \frac{8}{3}$$

$$= \frac{\overset{3}{9}}{\underset{1}{2}} \cdot \frac{\overset{4}{8}}{\underset{1}{3}} \qquad \text{Divide out common factors.}$$

$$= \frac{3 \cdot 4}{1 \cdot 1} \qquad \begin{array}{l} \leftarrow \text{Multiply the numerators.} \\ \leftarrow \text{Multiply the denominators.} \end{array}$$

$$= \frac{12}{1} \text{ or } 12 \qquad \text{Simplify.}$$

Check $4\frac{1}{2}$ is less than 5, and $2\frac{2}{3}$ is less than 3. Therefore, $4\frac{1}{2} \cdot 2\frac{2}{3}$ is less than $5 \cdot 3$ or 15. The answer is reasonable.

Your Turn **Multiply. Write in simplest form.**

d. $1\frac{1}{2} \cdot 1\frac{2}{3}$

e. $\frac{5}{7} \cdot 1\frac{3}{5}$

f. $\left(-2\frac{1}{6}\right)\left(-1\frac{1}{5}\right)$

Evaluate an Algebraic Expression

4 **ALGEBRA** Evaluate abc if $a = -\frac{1}{2}$, $b = \frac{3}{5}$, and $c = \frac{5}{9}$.

$$abc = -\frac{1}{2} \cdot \frac{3}{5} \cdot \frac{5}{9}$$ Replace a with $-\frac{1}{2}$, b with $\frac{3}{5}$, and c with $\frac{5}{9}$.

$$= -\frac{1}{2} \cdot \frac{\overset{1}{\cancel{3}}}{\cancel{5}} \cdot \frac{\overset{1}{\cancel{5}}}{\underset{3}{\cancel{9}}}$$ Divide out common factors.

$$= -\frac{1 \cdot 1 \cdot 1}{2 \cdot 1 \cdot 3} \text{ or } -\frac{1}{6}$$ Simplify.

Your Turn Evaluate each expression if $a = \frac{3}{4}$, $b = -\frac{1}{2}$, and $c = \frac{2}{3}$.

g. ac **h.** ab **i.** abc

REAL-LIFE MATH

AIRCRAFT A 757 aircraft has a capacity of 242 passengers and a wingspan of 165 feet 4 inches.

Source: *Continental Traveler*

Dimensional analysis is the process of including units of measurement when you compute. You can use dimensional analysis to check whether your answers are reasonable.

Use Dimensional Analysis

5 **AIRCRAFT** Suppose a 757 aircraft is traveling at its cruise speed. How far will it travel in $1\frac{1}{3}$ hours?

Aircraft	Cruise Speed (mph)
MD-80	505
DC-10	550
757	540
ATR-42	328

Source: *Continental Traveler*

Words	Distance equals the rate multiplied by the time.
Variables	$d = r \cdot t$
Equation	d = 540 miles per hour · $1\frac{1}{3}$ hours

STUDY TIP

Mental Math

$\frac{1}{3}$ of 540 is 180. Using the Distributive Property, $1\frac{1}{3}$ of 540 should equal 540 + 180, or 720.

$$d = \frac{540 \text{ miles}}{1 \text{ hour}} \cdot 1\frac{1}{3} \text{ hours}$$ Write the equation.

$$d = \frac{540 \text{ miles}}{1 \text{ hour}} \cdot \frac{4}{3} \text{ hours}$$ $1\frac{1}{3} = \frac{4}{3}$

$$d = \frac{\overset{180}{\cancel{540} \text{ miles}}}{1 \text{ \cancel{hour}}} \cdot \frac{4}{\underset{1}{\cancel{3}}} \text{ \cancel{hours}}$$ Divide by common factors and units.

$$d = 720 \text{ miles}$$

At its cruising speed, a 757 will travel 720 miles in $1\frac{1}{3}$ hours.

Check The problem asks for the distance. When you divide the common units, the answer is expressed in miles. So, the answer is reasonable.

Skill and Concept Check

1. **NUMBER SENSE** Explain why the product of $\frac{1}{2}$ and $\frac{7}{8}$ is less than $\frac{1}{2}$.

2. **OPEN ENDED** Name two fractions whose product is greater than $\frac{1}{2}$ and less than 1.

3. **FIND THE ERROR** Matt and Enrique are multiplying $2\frac{1}{2}$ and $3\frac{1}{4}$. Who is correct? Explain.

Matt

$$2\frac{1}{2} \cdot 3\frac{1}{4} = 2 \cdot 3 + \frac{1}{2} \cdot \frac{1}{4}$$
$$= 6 + \frac{1}{8}$$
$$= 6\frac{1}{8}$$

Enrique

$$2\frac{1}{2} \cdot 3\frac{1}{4} = \frac{5}{2} \cdot \frac{13}{4}$$
$$= \frac{65}{8}$$
$$= 8\frac{1}{8}$$

GUIDED PRACTICE

Multiply. Write in simplest form.

4. $\frac{3}{5} \cdot \frac{5}{7}$

5. $-\frac{1}{8} \cdot \frac{4}{9}$

6. $1\frac{1}{3} \cdot 5\frac{1}{2}$

7. $\left(-\frac{4}{5}\right)\left(-\frac{4}{5}\right)$

8. **FOOD** The nutrition label is from a can of green beans. How many cups of green beans does the can contain?

9. **ALGEBRA** Evaluate xy if $x = \frac{4}{5}$ and $y = \frac{1}{2}$.

Nutrition Facts
Serving Size 1/2 cup (121g)
Servings Per Container approx. 3½

Amount Per Serving

Calories 20 Calories from Fat 0

	% Daily Value*
Total Fat 0g	0%
Saturated Fat 0g	0%
Cholesterol 0mg	0%
Sodium 390mg	15%
Total Carbohydrate 4g	1%
Dietary Fiber 2g	6%
Sugars 2g	
Protein 1g	

Practice and Applications

Multiply. Write in simplest form.

10. $\frac{3}{8} \cdot \frac{4}{5}$

11. $\frac{1}{12} \cdot \frac{4}{7}$

12. $-\frac{3}{8} \cdot \frac{4}{9}$

13. $-\frac{9}{10} \cdot \frac{2}{3}$

14. $-3\frac{3}{8} \cdot \left(-\frac{2}{3}\right)$

15. $-\frac{5}{6} \cdot \left(-1\frac{4}{5}\right)$

16. $3\frac{1}{3} \cdot 1\frac{1}{2}$

17. $2\frac{1}{2} \cdot 1\frac{2}{5}$

18. $-3\frac{1}{3} \cdot 2\frac{1}{4}$

19. $-4\frac{1}{4} \cdot 3\frac{1}{3}$

20. $\left(-\frac{3}{7}\right)\left(-\frac{3}{7}\right)$

21. $\left(-\frac{2}{3}\right)\left(-\frac{2}{3}\right)$

22. Find the product of $\frac{1}{3}$, $-\frac{3}{8}$, and $\frac{4}{5}$.

23. What is one half of the product of $\frac{2}{5}$ and $\frac{3}{4}$?

ALGEBRA Evaluate each expression if $r = -\frac{1}{4}$, $s = \frac{2}{5}$, $t = \frac{8}{9}$, **and** $v = -\frac{2}{3}$.

24. rs

25. rt

26. stv

27. rtv

HOMEWORK HELP

For Exercises	See Examples
10–23	1–3
24–27	4
28–29	5

Extra Practice
See pages 619, 649.

28. PHOTOGRAPHY Minh-Thu has a square photograph that measures $3\frac{1}{2}$ inches on each side. She reduces it to $\frac{2}{3}$ of its size. What is the length of a side of the new photograph?

29. BIOLOGY The bee hummingbird of Cuba is the smallest hummingbird in the world. It is $\frac{1}{4}$ the length of the giant hummingbird. Use the information at the right to find the length of a bee hummingbird.

Giant Hummingbird

$8\frac{1}{4}$ in.

30. RESEARCH Use the Internet or other resource to find a recipe for spaghetti sauce. Change the recipe to make $\frac{2}{3}$ of the amount. Then, change the recipe to make $1\frac{1}{2}$ of the amount.

31. CRITICAL THINKING Find the missing fraction. $\frac{3}{4} \cdot ? = \frac{9}{14}$

EXTENDING THE LESSON
MENTAL MATH You can use number properties to simplify computations.

Example: $\frac{3}{4} \cdot \frac{3}{7} \cdot \frac{4}{3} = \left(\frac{3}{4} \cdot \frac{4}{3}\right) \cdot \frac{3}{7}$ Commutative and Associative Properties

$= 1 \cdot \frac{3}{7}$ or $\frac{3}{7}$ Identity Property of Multiplication

Use mental math to find each product.

32. $\frac{2}{5} \cdot \frac{1}{6} \cdot \frac{5}{2}$ **33.** $5 \cdot 3.78 \cdot \frac{1}{5}$ **34.** $\frac{2}{7} \cdot \frac{4}{9} \cdot \frac{3}{5} \cdot 0$

Standardized Test Practice and Mixed Review

FCAT Practice

35. MULTIPLE CHOICE Find the area of the triangle. Use the formula $A = \frac{1}{2}bh$.

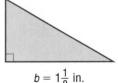

$h = \frac{2}{3}$ in.

$b = 1\frac{1}{8}$ in.

  Ⓐ $\frac{3}{4}$ in²   Ⓑ $\frac{5}{8}$ in²   Ⓒ $\frac{3}{8}$ in²   Ⓓ $\frac{1}{6}$ in²

36. MULTIPLE CHOICE What number will make $\frac{3}{4} \cdot \frac{7}{8} = \frac{7}{8} \cdot n$ true?

  Ⓕ $\frac{4}{8}$ Ⓖ $\frac{3}{4}$ Ⓗ $\frac{10}{12}$ Ⓘ $\frac{7}{8}$

Replace each ● with <, >, or = to make a true sentence. (Lesson 2-2)

37. $\frac{1}{2}$ ● $\frac{4}{7}$ **38.** $\frac{2}{7}$ ● $0.\overline{28}$ **39.** -0.753 ● $-\frac{3}{4}$ **40.** $-\frac{4}{9}$ ● $-0.\overline{4}$

41. HISTORY In 1864, Abraham Lincoln won the presidential election with about 0.55 of the popular vote. Write this as a fraction in simplest form.
(Lesson 2-1)

GETTING READY FOR THE NEXT LESSON

PREREQUISITE SKILL Divide. (Lesson 1-6)

42. $51 \div (-17)$ **43.** $-81 \div (-3)$ **44.** $-92 \div 4$ **45.** $-105 \div (-7)$

2-4 Dividing Rational Numbers

Sunshine State Standards
MA.A.3.3.1-1, MA.A.3.3.1-3, MA.A.3.3.2-1, MA.D.2.3.1-6

What You'll Learn
Divide fractions.

→ NEW Vocabulary

multiplicative
  inverses
reciprocals

↻ REVIEW Vocabulary

additive inverse:
the sum of any
number and its
additive inverse is
zero, $a + (-a) = 0$
(Lesson 1-5)

WHEN am I ever going to use this?

ANIMALS The world's longest
snake is the reticulated python.
It is approximately one-fourth
the length of the blue whale.

1. Find the value of $110 \div 4$.

2. Find the value of $110 \cdot \frac{1}{4}$.

3. Compare the values of $110 \div 4$
   and $110 \cdot \frac{1}{4}$.

4. What can you conclude about
   the relationship between
   dividing by 4 and multiplying by $\frac{1}{4}$?

World's Largest Animals		
Largest Animal	Blue Whale	110 feet long
Largest Reptile	Saltwater Crocodile	16 feet long
Largest Bird	Ostrich	9 feet tall
Largest Insect	Stick Insect	15 inches long

Source: *The World Almanac for Kids*

In Chapter 1, you learned about additive inverses. A similar property
applies to multiplication. Two numbers whose product is 1 are
multiplicative inverses, or **reciprocals**, of each other. For example,
4 and $\frac{1}{4}$ are multiplicative inverses because $4 \cdot \frac{1}{4} = 1$.

Key Concept   **Inverse Property of Multiplication**

Words   The product of a rational number and its multiplicative inverse
is 1.

Symbols   Arithmetic   Algebra

$\frac{3}{4} \cdot \frac{4}{3} = 1$ $\frac{a}{b} \cdot \frac{b}{a} = 1$, where $a, b \neq 0$

EXAMPLE   Find a Multiplicative Inverse

1 Write the multiplicative inverse of $-5\frac{2}{3}$.

$-5\frac{2}{3} = -\frac{17}{3}$ Write $-5\frac{2}{3}$ as an improper fraction.

Since $-\frac{17}{3}\left(-\frac{3}{17}\right) = 1$, the multiplicative inverse of $-5\frac{2}{3}$ is $-\frac{3}{17}$.

Your Turn Write the multiplicative inverse of each number.

a. $-2\frac{1}{3}$ b. $-\frac{5}{8}$ c. 7

Dividing by 4 is the same as multiplying by $\frac{1}{4}$, its multiplicative inverse. This is true for any rational number.

multiplicative inverses

$$110 \div 4 = 27\frac{1}{2} \qquad 110 \cdot \frac{1}{4} = 27\frac{1}{2}$$

same answer

Key Concept   Dividing Fractions

Words   To divide by a fraction, multiply by its multiplicative inverse.

Symbols   **Arithmetic**   **Algebra**

$$\frac{2}{5} \div \frac{3}{4} = \frac{2}{5} \cdot \frac{4}{3} \text{ or } \frac{8}{15} \qquad \frac{a}{b} \div \frac{c}{d} = \frac{a}{b} \cdot \frac{d}{c}, \text{ where } b, c, d \neq 0$$

EXAMPLE   Divide Fractions

2 Find $\frac{7}{8} \div \frac{3}{4}$. Write in simplest form.

$$\frac{7}{8} \div \frac{3}{4} = \frac{7}{8} \cdot \frac{4}{3} \qquad \text{Multiply by the multiplicative inverse of } \frac{3}{4}, \text{ which is } \frac{4}{3}.$$

$$= \frac{7}{\overset{}{\underset{2}{8}}} \cdot \frac{\overset{1}{4}}{3} \qquad \text{Divide 8 and 4 by their GCF, 4.}$$

$$= \frac{7}{6} \text{ or } 1\frac{1}{6} \qquad \text{Simplify.}$$

EXAMPLE   Divide by a Whole Number

STUDY TIP

Dividing By a Whole Number
When dividing by a whole number, always rename it as an improper fraction. Then multiply by its reciprocal.

3 Find $\frac{2}{5} \div 5$. Write in simplest form.

$$\frac{2}{5} \div 5 = \frac{2}{5} \div \frac{5}{1} \qquad \text{Write 5 as } \frac{5}{1}.$$

$$= \frac{2}{5} \cdot \frac{1}{5} \qquad \text{Multiply by the multiplicative inverse of 5, which is } \frac{1}{5}.$$

$$= \frac{2}{25} \qquad \text{Simplify.}$$

EXAMPLE   Divide Negative Fractions

4 Find $-\frac{4}{5} \div \frac{6}{7}$. Write in simplest form.

$$-\frac{4}{5} \div \frac{6}{7} = -\frac{4}{5} \cdot \frac{7}{6} \qquad \text{Multiply by the multiplicative inverse of } \frac{6}{7}, \text{ which is } \frac{7}{6}.$$

$$= \frac{-\overset{2}{4}}{5} \cdot \frac{7}{\underset{3}{6}} \qquad \text{Divide } -4 \text{ and 6 by their GCF, 2.}$$

$$= -\frac{14}{15} \qquad \text{The fractions have different signs, so the quotient is negative.}$$

Your Turn Divide. Write in simplest form.

d. $\frac{3}{4} \div \frac{1}{2}$ **e.** $\frac{3}{5} \div (6)$ **f.** $-\frac{2}{3} \div \left(-\frac{3}{5}\right)$

EXAMPLE **Divide Mixed Numbers**

5 Find $4\frac{2}{3} \div \left(-3\frac{1}{2}\right)$. Write in simplest form.

$$4\frac{2}{3} \div \left(-3\frac{1}{2}\right) = \frac{14}{3} \div \left(-\frac{7}{2}\right) \qquad 4\frac{2}{3} = \frac{14}{3}, \; -3\frac{1}{2} = -\frac{7}{2}$$

$$= \frac{14}{3} \cdot \left(-\frac{2}{7}\right) \qquad \text{The multiplicative inverse of } -\frac{7}{2} \text{ is } -\frac{2}{7}.$$

$$= \frac{\overset{2}{\cancel{14}}}{3} \cdot \left(-\frac{2}{\underset{1}{\cancel{7}}}\right) \qquad \text{Divide 14 and 7 by their GCF, 7.}$$

$$= -\frac{4}{3} \text{ or } -1\frac{1}{3} \qquad \text{Simplify.}$$

Check Since $4\frac{2}{3}$ is about 5 and $-3\frac{1}{2}$ is about -4, you can estimate the answer to be about $5 \div (-4)$, which is $\frac{5}{-4}$ or $-1\frac{1}{4}$. The answer seems reasonable because $-1\frac{1}{3}$ is about $-1\frac{1}{4}$.

Your Turn **Divide. Write in simplest form.**

g. $2\frac{3}{4} \div \left(-2\frac{1}{5}\right)$ **h.** $1\frac{1}{2} \div 2\frac{1}{3}$ **i.** $-3\frac{1}{2} \div \left(-1\frac{1}{4}\right)$

You can use dimensional analysis to check for reasonable answers in division problems as well as multiplication problems.

EXAMPLE **Use Dimensional Analysis**

6 **HOLIDAYS** Isabel and her friends are making ribbons to give to other campers at their day camp on Flag Day. They have a roll with 20 feet of ribbon. How many Flag Day ribbons as shown at the right can they make?

Flag Day   4 in.

Since 4 inches equals $\frac{4}{12}$ or $\frac{1}{3}$ foot, divide 20 by $\frac{1}{3}$.

$$20 \div \frac{1}{3} = \frac{20}{1} \div \frac{1}{3} \qquad \text{Write 20 as } \frac{20}{1}.$$

$$= \frac{20}{1} \cdot \frac{3}{1} \qquad \text{Multiply by the multiplicative inverse of } \frac{1}{3} \text{ which is 3.}$$

$$= \frac{60}{1} \text{ or } 60 \qquad \text{Simplify.}$$

Isabel and her friends can make 60 Flag Day ribbons.

Check Use dimensional analysis to examine the units.

$$\text{feet} \div \frac{\text{feet}}{\text{ribbon}} = \cancel{\text{feet}} \times \frac{\text{ribbon}}{\cancel{\text{feet}}} \qquad \text{Divide out the units.}$$

$$= \text{ribbon} \qquad \text{Simplify.}$$

The result is expressed as ribbons. This agrees with your answer of 60 ribbons.

1. **Explain** how you know if two numbers are multiplicative inverses.

2. **Give a counterexample** to the statement *the quotient of two fractions between 0 and 1 is never a whole number.*

3. **OPEN ENDED** Write a division problem that can be solved by multiplying a rational number by $\frac{6}{5}$.

4. **NUMBER SENSE** Which is greater: $30 \cdot \frac{3}{4}$ or $30 \div \frac{3}{4}$? Explain.

GUIDED PRACTICE

Write the multiplicative inverse of each number.

5. $\frac{5}{7}$

6. -12

7. $-2\frac{3}{4}$

Divide. Write in simplest form.

8. $\frac{2}{3} \div \frac{3}{4}$

9. $-5\frac{5}{6} \div 4\frac{2}{3}$

10. $-\frac{4}{5} \div (-8)$

11. **BIOLOGY** The 300-million-year-old fossil of a cockroach was recently found in eastern Ohio. The ancient cockroach is shown next to the common German cockroach found today. How many times longer is the ancient cockroach than the German cockroach?

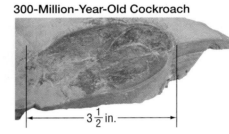

300-Million-Year-Old Cockroach

Common German Cockroach

$\frac{1}{2}$ in.

$3\frac{1}{2}$ in.

Practice and Applications

Write the multiplicative inverse of each number.

12. $-\frac{7}{9}$

13. $-\frac{5}{8}$

14. 15

15. 18

16. $\frac{6}{11}$

17. $\frac{7}{15}$

18. $3\frac{2}{5}$

19. $4\frac{1}{8}$

HOMEWORK HELP

For Exercises	See Examples
12–19	1
20–35	2–5
36–39	6

Extra Practice
See pages 620, 649.

Divide. Write in simplest form.

20. $\frac{2}{5} \div \frac{3}{4}$

21. $\frac{3}{8} \div \frac{2}{3}$

22. $-\frac{3}{8} \div \frac{9}{10}$

23. $-\frac{2}{3} \div \frac{5}{6}$

24. $-5\frac{2}{5} \div \left(-2\frac{1}{10}\right)$

25. $-3\frac{1}{4} \div \left(-8\frac{2}{3}\right)$

26. $3\frac{3}{4} \div 2\frac{1}{2}$

27. $7\frac{1}{2} \div 2\frac{1}{10}$

28. $\frac{4}{5} \div (-6)$

29. $\frac{6}{7} \div (-4)$

30. $-12\frac{1}{4} \div 4\frac{2}{3}$

31. $-10\frac{1}{5} \div 3\frac{3}{15}$

32. What is $\frac{7}{12}$ divided by $\frac{5}{6}$?

33. Divide $\frac{5}{6}$ by $\frac{15}{16}$.

34. **ALGEBRA** Evaluate $x \div y$ if $x = -\frac{5}{12}$ and $y = \frac{5}{8}$.

35. **ALGEBRA** Evaluate $a \div b$ if $a = \frac{3}{4}$ and $b = \frac{5}{6}$.

36. **BIOLOGY** Use the information at the right. How many of the smallest grasshoppers need to be laid end-to-end to have the same length as the largest grasshoppers?

Smallest grasshopper
$\frac{1}{2}$ in.

Largest grasshopper

— 4 in. —

37. **ENERGY** Electricity costs $6\frac{1}{2}$¢ per kilowatt-hour. Of that cost, $3\frac{1}{4}$¢ goes toward the cost of the fuel. What fraction of the cost goes toward the fuel?

GEOGRAPHY For Exercises 38 and 39, use the information at the right.

38. About how many times larger is North America than South America?

39. About how many times larger is Asia than North America?

40. **WRITE A PROBLEM** Write a real-life situation that can be solved by dividing fractions or mixed numbers. Solve the problem.

41. **CRITICAL THINKING** Use mental math to find each value.

a. $\dfrac{43}{594} \cdot \dfrac{641}{76} \div \dfrac{641}{594}$

b. $\dfrac{783}{241} \cdot \dfrac{241}{783} \div \dfrac{72}{53}$

Continent	Fraction of Earth's Landmass
North America	$\frac{1}{6}$
South America	$\frac{1}{8}$
Asia	$\frac{3}{10}$

Source: *The World Almanac*

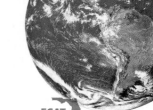

Standardized Test Practice and Mixed Review

FCAT Practice

42. **MULTIPLE CHOICE** A submarine sandwich that is $26\frac{1}{2}$ inches long is cut into $4\frac{5}{12}$-inch mini-subs. How many mini-subs are there?

  A 4   **B** 5   **C** 6   **D** 7

43. **SHORT RESPONSE** What is the multiplicative inverse of $-\frac{1}{a}$?

Multiply. Write in simplest form. (Lesson 2-3)

44. $\dfrac{1}{2} \cdot \dfrac{3}{4}$ 45. $\dfrac{7}{12} \cdot \dfrac{4}{7}$ 46. $1\dfrac{2}{3} \cdot 4\dfrac{1}{5}$ 47. $\dfrac{2}{3} \cdot 3\dfrac{1}{4}$

48. **SCHOOL** In a survey of students at Centerburg Middle School, $\frac{13}{20}$ of the boys and $\frac{17}{25}$ of the girls said they rode the bus to school. Of those surveyed, do a greater fraction of boys or girls ride the bus? (Lesson 2-2)

49. **ALGEBRA** Write an algebraic expression to represent *eight million less than four times the population of Africa*. (Lesson 1-7)

50. Write an integer to describe *10 candy bars short of his goal*. (Lesson 1-3)

GETTING READY FOR THE NEXT LESSON

PREREQUISITE SKILL Add or subtract. (Lessons 1-4 and 1-5)

51. $-7 + 15$ 52. $-9 + (-4)$ 53. $-3 - 15$ 54. $12 - (-17)$

TWO-COLUMN NOTES

Have you ever written a step-by-step solution to a problem, but couldn't follow the steps later? Try using two-column notes. You may like this method of taking notes so well, you'll want to use it for your other classes.

To take two-column notes, first fold your paper lengthwise into two columns. Make the right-hand column about 3 inches wide.

When your teacher solves a problem in class, write all of the steps in the left-hand column. In the right-hand column, add notes in your own words that will help you remember how to solve the problem. Add a ★ by any step that you especially want to remember.

Here's a sample.

← 3 in. →

How to Divide Fractions	My Notes
$\frac{3}{4} \div 1\frac{1}{2} = \frac{3}{4} \div \frac{3}{2}$	Write $1\frac{1}{2}$ as a fraction.
$= \frac{3}{4} \cdot \frac{2}{3}$	★Use the inverse of the second fraction.
$= \frac{\overset{1}{\cancel{3}}}{\underset{2}{\cancel{4}}} \cdot \frac{\overset{1}{\cancel{2}}}{\underset{1}{\cancel{3}}}$	Then multiply. This is important.
$= \frac{1}{2}$	Cancel and multiply.

SKILL PRACTICE

Use the method above to write notes for each step-by-step solution.

1. $1\frac{1}{3} \div 3 = \frac{4}{3} \div \frac{3}{1}$

$= \frac{3}{4} \cdot \frac{1}{3}$

$= \frac{\overset{1}{\cancel{3}}}{4} \cdot \frac{1}{\underset{1}{\cancel{3}}}$

$= \frac{1}{4}$

2. $1\frac{1}{2} \cdot 1\frac{2}{3} = \frac{3}{2} \cdot \frac{5}{3}$

$= \frac{3}{2} \cdot \frac{5}{3}$

$= \frac{5}{2}$

$= 2\frac{1}{2}$

3. $x + 8 = -6$

$\underline{ - 8 = -8}$

$x = -14$

4. $5 - 12 = 5 + (-12)$

$ = -7$

Adding and Subtracting Like Fractions

Sunshine State Standards
MA.A.3.3.1-1, MA.A.3.3.2-1, MA.A.3.3.2-2, MA.A.3.3.3-1, MA.D.2.3.1-6

What You'll Learn
Add and subtract fractions with like denominators.

NEW Vocabulary
like fractions

WHEN am I ever going to use this?

BAKING A bread recipe calls for the ingredients at the right together with small amounts of sugar, oil, yeast, and salt.

Bread	
$1\frac{1}{3}$	cups of whole wheat flour (sifted)
$2\frac{1}{3}$	cups of white flour (sifted)
$\frac{1}{3}$	cup oatmeal
$\frac{1}{3}$	cup apricots (diced)
$\frac{1}{3}$	cup hazelnuts (chopped)
$1\frac{1}{3}$	cups of warm water

1. What is the sum of the whole-number parts of the amounts?

2. How many $\frac{1}{3}$ cups are there?

3. Since $\frac{1}{3} + \frac{1}{3} + \frac{1}{3} = 1$, how many cups do all the $\frac{1}{3}$ cups make?

4. What is the total number of cups of the ingredients listed?

The fractions above have like denominators. Fractions with like denominators are called **like fractions**.

Key Concept
Adding Like Fractions

Words To add fractions with like denominators, add the numerators and write the sum over the denominator.

Symbols

Arithmetic	Algebra
$\frac{1}{3} + \frac{1}{3} = \frac{2}{3}$	$\frac{a}{c} + \frac{b}{c} = \frac{a+b}{c}$, where $c \neq 0$

You can use the rules for adding integers to determine the sign of the sum of fractions.

EXAMPLE Add Like Fractions

1 Find $\frac{5}{8} + \left(-\frac{7}{8}\right)$. **Write in simplest form.**

$\frac{5}{8} + \left(-\frac{7}{8}\right) = \frac{5 + (-7)}{8}$ ← Add the numerators.
                                          ← The denominators are the same.

$= \frac{-2}{8}$ or $-\frac{1}{4}$ Simplify.

STUDY TIP

Look Back You can review **adding integers** in Lesson 1-4.

Your Turn Add. **Write in simplest form.**

a. $\frac{5}{9} + \frac{7}{9}$
b. $-\frac{5}{6} + \frac{1}{6}$
c. $-\frac{1}{6} + \left(-\frac{5}{6}\right)$

Subtracting like fractions is similar to adding them.

Key Concept	Subtracting Like Fractions
Words	To subtract fractions with like denominators, subtract the numerators and write the difference over the denominator.

Symbols   Arithmetic   Algebra

$$\frac{5}{7} - \frac{3}{7} = \frac{5-3}{7} \text{ or } \frac{2}{7} \qquad \frac{a}{c} - \frac{b}{c} = \frac{a-b}{c}, \text{ where } c \neq 0$$

EXAMPLE   Subtract Like Fractions

2 Find $-\frac{8}{9} - \frac{7}{9}$. Write in simplest form.

$$-\frac{8}{9} - \frac{7}{9} = \frac{-8-7}{9} \qquad \begin{array}{l} \leftarrow \text{Subtract the numerators.} \\ \leftarrow \text{The denominators are the same.} \end{array}$$

$$= \frac{-15}{9} \text{ or } -1\frac{2}{3} \qquad \text{Rename } \frac{-15}{9} \text{ as } -1\frac{6}{9} \text{ or } -1\frac{2}{3}.$$

To add mixed numbers, add the whole numbers and the fractions separately. Then simplify.

EXAMPLE   Add Mixed Numbers

3 Find $5\frac{7}{9} + 8\frac{4}{9}$. Write in simplest form.

$$5\frac{7}{9} + 8\frac{4}{9} = (5 + 8) + \left(\frac{7}{9} + \frac{4}{9}\right) \qquad \begin{array}{l} \text{Add the whole numbers} \\ \text{and fractions separately.} \end{array}$$

$$= 13 + \frac{7+4}{9} \qquad \text{Add the numerators.}$$

$$= 13\frac{11}{9} \text{ or } 14\frac{2}{9} \qquad \frac{11}{9} = 1\frac{2}{9}$$

STUDY TIP

Alternative Method You can also add the mixed numbers vertically.

$5\frac{7}{9}$

$+ 8\frac{4}{9}$

$13\frac{11}{9}$ or $14\frac{2}{9}$

One way to subtract mixed numbers is to write the mixed numbers as improper fractions.

EXAMPLE   Subtract Mixed Numbers

4 **HEIGHTS** Jasmine is $60\frac{1}{4}$ inches tall. Amber is $58\frac{3}{4}$ inches tall. How much taller is Jasmine than Amber?   **Estimate** $60 - 59 = 1$

$$60\frac{1}{4} - 58\frac{3}{4} = \frac{241}{4} - \frac{235}{4} \qquad \begin{array}{l} \text{Write the mixed numbers} \\ \text{as improper fractions.} \end{array}$$

$$= \frac{241 - 235}{4} \qquad \begin{array}{l} \leftarrow \text{Subtract the numerators.} \\ \leftarrow \text{The denominators are the same.} \end{array}$$

$$= \frac{6}{4} \text{ or } 1\frac{1}{2} \qquad \text{Rename } \frac{6}{4} \text{ as } 1\frac{2}{4} \text{ or } 1\frac{1}{2}.$$

Jasmine is $1\frac{1}{2}$ inches taller than Amber.

Writing Math

Exercise 3

1. **Draw** a model to show the sum of $\frac{1}{5}$ and $\frac{3}{5}$.

2. **OPEN ENDED** Write a subtraction problem with a difference of $\frac{2}{9}$.

3. **FIND THE ERROR** Allison and Wesley are adding $\frac{1}{7}$ and $\frac{3}{7}$. Who is correct? Explain.

Allison

$$\frac{1}{7} + \frac{3}{7} = \frac{1+3}{7}$$
$$= \frac{4}{7}$$

Wesley

$$\frac{1}{7} + \frac{3}{7} = \frac{1+3}{7+7}$$
$$= \frac{4}{14} \text{ or } \frac{2}{7}$$

GUIDED PRACTICE

Add or subtract. Write in simplest form.

4. $\frac{2}{5} + \frac{2}{5}$

5. $-\frac{3}{4} + \frac{1}{4}$

6. $-5\frac{4}{9} + \left(-2\frac{2}{9}\right)$

7. $\frac{3}{8} - \frac{7}{8}$

8. $8 - 6\frac{1}{6}$

9. $-1\frac{3}{7} - \left(-2\frac{2}{7}\right)$

10. **SPORTS** One of the track and field events is the triple jump. In this event, the athlete takes a running start and makes three jumps without stopping. Find the total length of the 3 jumps for the athlete below.

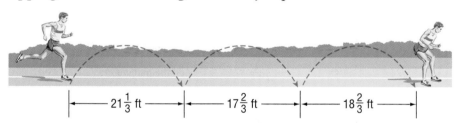

$\longleftarrow 21\frac{1}{3}$ ft $\longrightarrow$ $\longleftarrow 17\frac{2}{3}$ ft $\longrightarrow$ $\longleftarrow 18\frac{2}{3}$ ft $\longrightarrow$

Practice and Applications

Add or subtract. Write in simplest form.

11. $\frac{3}{7} + \frac{3}{7}$

12. $\frac{1}{9} + \frac{1}{9}$

13. $-\frac{5}{12} + \frac{7}{12}$

14. $-\frac{8}{9} + \frac{5}{9}$

15. $-\frac{7}{8} + \left(-\frac{7}{8}\right)$

16. $-\frac{5}{9} + \left(-\frac{7}{9}\right)$

17. $\frac{1}{12} - \frac{7}{12}$

18. $\frac{2}{9} - \frac{8}{9}$

19. $-\frac{4}{5} - \frac{3}{5}$

20. $-\frac{2}{3} - \frac{2}{3}$

21. $3\frac{5}{8} + 7\frac{5}{8}$

22. $9\frac{5}{9} + 4\frac{7}{9}$

23. $8\frac{1}{10} - 2\frac{9}{10}$

24. $8\frac{5}{12} - 5\frac{11}{12}$

25. $-1\frac{5}{6} - 3\frac{5}{6}$

26. $-3\frac{3}{4} - 7\frac{3}{4}$

27. $7 - 5\frac{2}{5}$

28. $9 - 6\frac{3}{7}$

29. $-8 - \left(-3\frac{5}{8}\right)$

30. $-7 - \left(-2\frac{3}{5}\right)$

31. **ALGEBRA** Find $a - b$ if $a = 5\frac{1}{3}$ and $b = -2\frac{1}{3}$.

HOMEWORK HELP

For Exercises	See Examples
11–20, 32	1, 2
21–31, 34–36	3, 4

Extra Practice
See pages 620, 649.

32. ALGEBRA Find $x + y$ if $x = -\dfrac{5}{12}$ and $y = -\dfrac{1}{12}$.

33. MENTAL MATH Explain how to use the Distributive Property to find $\dfrac{1}{2} \cdot \dfrac{3}{4} + \dfrac{1}{2} \cdot \dfrac{1}{4}$.

34. GEOMETRY Find the perimeter of the rectangle at the right.

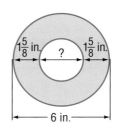

$12\dfrac{1}{4}$ in.

$25\dfrac{3}{4}$ in.

35. CLOTHING Hat sizes are determined by the distance across a person's head. How much wider is a person's head who wears a hat size of $7\dfrac{3}{4}$ inches than someone who wears a hat size of $6\dfrac{1}{4}$ inches?

36. MULTI STEP Quoits was one of five original games in the ancient Greek Pentathlon. Find the distance across the hole of the quoit shown at the right.

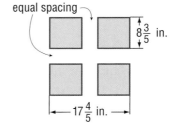

$1\dfrac{5}{8}$ in.   ? $1\dfrac{5}{8}$ in.

6 in.

37. CRITICAL THINKING Explain how to use mental math to find the following sum. Then find the sum.

$$3\dfrac{2}{3} + 4\dfrac{2}{5} + 2\dfrac{1}{6} + 2\dfrac{5}{6} + 1\dfrac{1}{3} + \dfrac{3}{5}$$

Standardized Test Practice and Mixed Review

A B C D

FCAT
Practice

38. MULTIPLE CHOICE Find $\dfrac{7}{8} - \left(-\dfrac{3}{8}\right)$.

   A $-1\dfrac{1}{4}$ **B** $-\dfrac{1}{2}$ **C** $\dfrac{1}{2}$ **D** $1\dfrac{1}{4}$

39. MULTIPLE CHOICE The equal-sized square tiles on a bathroom floor are set as shown. What is the width of the space between the tiles?

equal spacing

$8\dfrac{3}{5}$ in.

$17\dfrac{4}{5}$ in.

   F $\dfrac{3}{5}$ in.   **G** $\dfrac{1}{5}$ in.

   H $\dfrac{3}{10}$ in.   **I** $\dfrac{2}{5}$ in.

Divide. Write in simplest form. (Lesson 2-4)

40. $\dfrac{3}{5} \div \dfrac{6}{7}$ **41.** $\dfrac{7}{8} \div 2\dfrac{4}{5}$ **42.** $-3\dfrac{1}{4} \div 2\dfrac{1}{2}$

43. Find the product of $-\dfrac{7}{8}$ and $-\dfrac{6}{7}$. (Lesson 2-3)

44. FOOD On a typical day, 2 million gallons of ice cream are produced in the United States. About how many gallons are produced each year? (Lesson 1-1)

GETTING READY FOR THE NEXT LESSON

PREREQUISITE SKILL Find the least common multiple (LCM) of each set of numbers. (Page 612)

45. 14, 21   **46.** 18, 9, 6   **47.** 6, 4, 9   **48.** 5, 10, 20

Vocabulary and Concepts

1. **Name** three numbers that are between $\frac{1}{2}$ and $\frac{3}{4}$. (Lesson 2-2)

2. **Define** reciprocals and give the reciprocal of $\frac{2}{3}$. (Lesson 2-4)

3. **OPEN ENDED** Write an addition problem with a sum of $2\frac{2}{3}$. (Lesson 2-5)

Skills and Applications

4. Write $-\frac{2}{9}$ as a decimal. (Lesson 2-1)

5. Write -2.65 as a mixed number in simplest form. (Lesson 2-1)

6. Write $0.\overline{5}$ as a fraction in simplest form. (Lesson 2-1)

Replace each ● with <, >, or = to make a true sentence. (Lesson 2-2)

7. $\frac{1}{3}$ ● $\frac{1}{4}$ 8. $-\frac{2}{5}$ ● $-\frac{3}{10}$ 9. $0.\overline{12}$ ● $\frac{4}{33}$ 10. $-\frac{5}{6}$ ● $-\frac{4}{5}$

Multiply, divide, add, or subtract. Write in simplest form. (Lessons 2-3, 2-4, and 2-5)

11. $-\frac{1}{3} \cdot \frac{2}{3}$ 12. $\frac{1}{2} \div \frac{3}{4}$ 13. $-1\frac{1}{3} \div \left(-\frac{1}{4}\right)$

14. $2\frac{3}{4} \cdot \frac{1}{5}$ 15. $\frac{3}{10} + \left(-\frac{7}{10}\right)$ 16. $-\frac{7}{9} - \frac{8}{9}$

17. **GEOMETRY** Find the area of the rectangle at the right. Use the formula $A = \ell w$. (Lesson 2-3)

$\frac{1}{5}$ unit

$\frac{5}{6}$ unit

18. **CARPENTRY** A board that is $25\frac{1}{2}$ feet long is cut into equal pieces that are each $1\frac{1}{2}$ feet long. Into how many pieces is the board cut? (Lesson 2-4)

Standardized Test Practice

FCAT Practice

Ⓐ Ⓑ Ⓒ Ⓓ

19 **MULTIPLE CHOICE** One centimeter is about 0.392 inch. What fraction of an inch is this?
(Lesson 2-1)

Ⓐ $\frac{49}{500}$ in.   Ⓑ $\frac{49}{125}$ in.

Ⓒ $\frac{98}{125}$ in.   Ⓓ $\frac{392}{100}$ in.

20. **SHORT RESPONSE** A bag of candy weighs 12 ounces. Each individual piece of candy weighs $\frac{1}{6}$ ounce. Write a division problem that you could use to determine the number of candies in the bag. How many candies are in the bag? (Lesson 2-4)

The GameZone

A Place To Practice Your Math Skills

Plug It In

● GET READY!

Players: two
Materials: 1 piece of paper, 9 index cards, scissors, marker

● GET SET!

- Write the following fractions on a piece of paper.

$$-\frac{8}{9}, \ -\frac{7}{9}, \ -\frac{5}{9}, \ -\frac{4}{9}, \ -\frac{2}{9}, \ -\frac{1}{9}, \ \frac{1}{9}, \ \frac{2}{9}, \ \frac{4}{9}, \ \frac{5}{9}, \ \frac{7}{9}, \ \frac{8}{9}$$

- Cut the index cards in half, making 18 cards.

- Write one of the following expressions on each of the cards.

$a + b$	$a - b$	$b - a$	ab	$\frac{1}{2}a$
$a \div b$	$b \div a$	$a + 1$	$b + 1$	$\frac{1}{2}b$
$1 - a$	$1 - b$	$a - 1$	$b - 1$	
a	b	$1 \div a$	$1 \div b$	

● GO!

- The cards are shuffled and dealt facedown to each player.

- One player chooses the value for a from the list of fractions on the paper. The other player chooses the value for b from the same list.

- Each player turns over the top card from his or her pile and evaluates the expression. The person whose expression has the greatest value wins a point. If the values are equal, no points are awarded.

- The players choose new values for a and b. Each player turns over a new card. The play continues until all the cards are used.

- **Who Wins?** The person with the most points wins the game.

Adding and Subtracting Unlike Fractions

Sunshine State Standards
MA.A.3.3.1-1, MA.A.3.3.2-1, MA.A.4.3.1-1, MA.A.4.3.1-2, MA.D.2.3.1-6

What You'll Learn
Add and subtract fractions with unlike denominators.

→ NEW Vocabulary
unlike fractions

↻ REVIEW Vocabulary
least common denominator (LCD): the least common multiple (LCM) of the denominators
(Page 612)

WHEN am I ever going to use this?

FOOD Marta and Brooke are sharing a pizza. Marta eats $\frac{1}{4}$ of the pizza and Brooke eats $\frac{3}{8}$ of the pizza.

1. What are the denominators of the fractions?
2. What is the least common multiple of the denominators?
3. Find the missing value in $\frac{1}{4} = \frac{?}{8}$.
4. What fraction of the pizza did the two girls eat?

The fractions $\frac{1}{4}$ and $\frac{3}{8}$ have different or unlike denominators. Fractions with unlike denominators are called **unlike fractions**. To add or subtract unlike fractions, you must use a common denominator.

Key Concept — Adding and Subtracting Unlike Fractions

Words To find the sum or difference of two fractions with unlike denominators, rename the fractions with a common denominator. Then add or subtract and simplify, if necessary.

Examples
$$\frac{1}{4} + \frac{1}{6} = \frac{1}{4} \cdot \frac{3}{3} + \frac{1}{6} \cdot \frac{2}{2} \qquad \frac{2}{3} - \frac{4}{9} = \frac{2}{3} \cdot \frac{3}{3} - \frac{4}{9}$$
$$= \frac{3}{12} + \frac{2}{12} \text{ or } \frac{5}{12} \qquad\qquad = \frac{6}{9} - \frac{4}{9} \text{ or } \frac{2}{9}$$

EXAMPLE Subtract Unlike Fractions

1 Find $-\frac{2}{3} - \left(-\frac{3}{8}\right)$. Write in simplest form.

$$-\frac{2}{3} - \left(-\frac{3}{8}\right) = -\frac{2}{3} \cdot \frac{8}{8} - \left(-\frac{3}{8}\right) \cdot \frac{3}{3} \qquad \text{The LCD is } 3 \cdot 2 \cdot 2 \cdot 2 \text{ or } 24.$$

$$= -\frac{16}{24} - \left(-\frac{9}{24}\right) \qquad\qquad \text{Rename each fraction using the LCD.}$$

$$= -\frac{16}{24} + \frac{9}{24} \qquad\qquad \text{Subtract } -\frac{9}{24} \text{ by adding its inverse, } \frac{9}{24}.$$

$$= \frac{-16 + 9}{24} \qquad\qquad \text{Add the numerators.}$$

$$= -\frac{7}{24} \qquad\qquad \text{Simplify.}$$

EXAMPLE Add Mixed Numbers

2 Find $-6\frac{2}{9} + 4\frac{5}{6}$. Write in simplest form.

$$-6\frac{2}{9} + 4\frac{5}{6} = -\frac{56}{9} + \frac{29}{6} \qquad \text{Write the mixed numbers as fractions.}$$

$$= -\frac{56}{9} \cdot \frac{2}{2} + \frac{29}{6} \cdot \frac{3}{3} \qquad \text{The LCD is } 3 \cdot 3 \cdot 2 \text{ or } 18.$$

$$= -\frac{112}{18} + \frac{87}{18} \qquad \text{Rename each fraction using the LCD.}$$

$$= \frac{-112 + 87}{18} \qquad \text{Add the numerators.}$$

$$= \frac{-25}{18} \text{ or } -1\frac{7}{18} \qquad \text{Simplify.}$$

Your Turn Add or subtract. Write in simplest form.

a. $-\frac{1}{3} - \left(-\frac{3}{4}\right)$ **b.** $-\frac{5}{6} + \left(-\frac{1}{2}\right)$ **c.** $-\frac{1}{2} + \frac{7}{8}$

d. $-3\frac{1}{2} + 8\frac{1}{3}$ **e.** $-1\frac{2}{5} + \left(-3\frac{1}{3}\right)$ **f.** $2\frac{3}{4} - 6\frac{1}{3}$

EXAMPLE Estimate the Sum of Mixed Numbers

3 **MULTIPLE-CHOICE TEST ITEM** Four telephone books are $2\frac{1}{8}$, $1\frac{15}{16}$, $1\frac{3}{4}$, and $2\frac{3}{8}$ inches thick. If these books were stacked one on top of another, what is the total height of the books?

   Ⓐ $5\frac{3}{16}$ in.   Ⓑ $8\frac{3}{16}$ in.   Ⓒ $11\frac{3}{16}$ in.   Ⓓ $15\frac{3}{16}$ in.

Read the Test Item You need to find the sum of four mixed numbers.

Solve the Test Item It would take some time to change each of the fractions to ones with a common denominator. However, notice that all four of the numbers are about 2. Since 2×4 equals 8, the answer will be about 8. Notice that only one of the choices is close to 8. The answer is B.

EXAMPLE Evaluate Expressions

4 **ALGEBRA** Find the value of $a - b$ if $a = \frac{5}{7}$ and $b = -\frac{3}{5}$.

$$a - b = \frac{5}{7} - \left(-\frac{3}{5}\right) \qquad \text{Replace } a \text{ with } \frac{5}{7} \text{ and } b \text{ with } -\frac{3}{5}.$$

$$= \frac{25}{35} - \left(-\frac{21}{35}\right) \qquad \text{Rename each fraction using the LCD, 35.}$$

$$= \frac{25 - (-21)}{35} \qquad \text{Subtract the numerators.}$$

$$= \frac{46}{35} \text{ or } 1\frac{11}{35} \qquad \text{Simplify.}$$

1. **Describe** the first step in adding unlike fractions.

2. **OPEN ENDED** Write a subtraction problem with unlike fractions with a least common denominator of 12. Find the answer.

3. **NUMBER SENSE** Without doing the computation, determine whether $\frac{4}{7} + \frac{5}{9}$ is greater than, less than, or equal to 1. Explain.

GUIDED PRACTICE

Add or subtract. Write in simplest form.

4. $\frac{3}{4} + \frac{1}{6}$

5. $\frac{7}{8} - \frac{3}{4}$

6. $-\frac{1}{7} - \left(-\frac{4}{5}\right)$

7. $-\frac{2}{5} + \left(-\frac{5}{6}\right)$

8. $3\frac{5}{8} - 1\frac{1}{3}$

9. $-4\frac{2}{3} - \left(-3\frac{4}{5}\right)$

10. **MUSIC** A waltz is written in $\frac{3}{4}$ time. This means the quarter note gets one beat and the total value of each measure is $\frac{3}{4}$. What type of note must be used to finish the last measure of the waltz below?

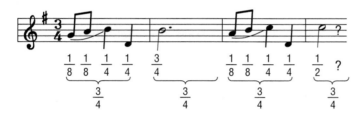

Practice and Applications

Add or subtract. Write in simplest form.

11. $\frac{3}{8} + \frac{5}{6}$

12. $\frac{7}{8} + \frac{3}{12}$

13. $\frac{3}{4} - \frac{1}{6}$

14. $\frac{4}{5} - \frac{2}{15}$

15. $-\frac{6}{7} - \left(-\frac{1}{3}\right)$

16. $-\frac{4}{5} - \left(-\frac{2}{3}\right)$

17. $8\frac{3}{7} - \left(-6\frac{1}{2}\right)$

18. $7\frac{3}{4} - \left(-1\frac{1}{8}\right)$

19. $-4\frac{3}{4} - 5\frac{5}{8}$

20. $-8\frac{1}{3} - 4\frac{5}{6}$

21. $9\frac{1}{6} - 4\frac{1}{2}$

22. $9\frac{1}{3} - 2\frac{1}{2}$

23. $3\frac{1}{5} + \left(-8\frac{1}{2}\right)$

24. $1\frac{1}{6} + \left(-6\frac{2}{3}\right)$

25. $-15\frac{5}{8} + 11\frac{2}{3}$

26. $-22\frac{2}{5} + 15\frac{5}{6}$

27. $\frac{65}{187} - \frac{9}{136}$

28. $\frac{45}{152} - \frac{13}{209}$

29. Subtract $-6\frac{1}{4}$ from 9.

30. What is $2\frac{3}{8}$ less than $-8\frac{1}{5}$?

31. What is the sum of $-\frac{5}{8}$ and $-\frac{1}{2}$?

32. Find the sum of $-\frac{4}{9}$ and $-\frac{2}{3}$.

33. **ALGEBRA** Evaluate $c - d$ if $c = -\frac{3}{4}$ and $d = -12\frac{7}{8}$.

34. **ALGEBRA** Evaluate $r - s$ if $r = -\frac{5}{8}$ and $s = 2\frac{5}{6}$.

HOMEWORK HELP

For Exercises	See Examples
11–32, 35	1–3
33–34	4

Extra Practice
See pages 620, 649.

35. HISTORY In the 1824 presidential election, Andrew Jackson, John Quincy Adams, Henry Clay, and William H. Crawford received electoral votes. Use the information at the right to determine what fraction of the votes William H. Crawford received.

Candidate	Fraction of Vote
Andrew Jackson	$\frac{3}{8}$
John Quincy Adams	$\frac{1}{3}$
Henry Clay	$\frac{1}{7}$

Source: *The World Almanac*

WATER MANAGEMENT For Exercises 36–40, use the following information.
Suppose a bucket is placed under two faucets.

36. If one faucet is turned on alone, the bucket will be filled in 5 minutes. Write the fraction of the bucket that will be filled in 1 minute.

37. If the other faucet is turned on alone, the bucket will be filled in 3 minutes. Write the fraction of the bucket that will be filled in 1 minute.

38. Write the fraction of the bucket that will be filled in 1 minute if both faucets are turned on.

39. Divide 1 by the sum in Exercise 38 to determine the number of minutes it will take to fill the bucket if both faucets are turned on.

40. How many seconds will it take to fill the bucket if both faucets are turned on?

41. CRITICAL THINKING Write an expression for each statement. Then find the answer.

a. $\frac{3}{4}$ of $\frac{2}{3}$

b. $\frac{3}{4}$ more than $\frac{2}{3}$

c. $\frac{3}{4}$ less than $\frac{2}{3}$

d. $\frac{3}{4}$ divided into $\frac{2}{3}$

Standardized Test Practice and Mixed Review

FCAT Practice

42. MULTIPLE CHOICE Teresa worked on homework $\frac{2}{3}$ of an hour on Monday and $1\frac{1}{2}$ hours on Tuesday. How much more time did she spend working on homework on Tuesday than on Monday?

Ⓐ $\frac{1}{6}$ h   Ⓑ $\frac{1}{4}$ h   Ⓒ $\frac{5}{6}$ h   Ⓓ $\frac{13}{6}$ h

43. SHORT RESPONSE Show each step in finding $5\frac{1}{6} + 4\frac{2}{9}$.

Add or subtract. Write in simplest form. (Lesson 2-5)

44. $-\frac{7}{11} + \frac{5}{11}$

45. $-\frac{7}{15} - \frac{4}{15}$

46. $5\frac{4}{5} - 7\frac{1}{5}$

47. ALGEBRA Find $a \div b$ if $a = 3\frac{1}{2}$ and $b = -\frac{7}{8}$. (Lesson 2-4)

GETTING READY FOR THE NEXT LESSON

PREREQUISITE SKILL Solve each equation. Check your solution. (Lessons 1-8 and 1-9)

48. $d - 13 = -44$ **49.** $-18t = 270$ **50.** $-34 = y + 22$ **51.** $-5 = \frac{a}{16}$

Solving Equations with Rational Numbers

Sunshine State Standards MA.A.3.3.3-1, MA.A.3.3.2-1, MA.D.1.3.1-2, MA.D.1.3.2-2, MA.D.2.3.1-1, MA.D.2.3.1-3, MA.D.2.3.2-2

What You'll Learn

Solve equations involving rational numbers.

⟳ REVIEW Vocabulary

equation: a mathematical sentence that contains an equals sign (Lesson 1-8)

WHEN am I ever going to use this?

BIOLOGY An elephant can run $\frac{5}{6}$ as fast as a grizzly bear. If s represents the speed of a grizzly bear, you can write the equation $25 = \frac{5}{6}s$.

1. Multiply each side of the equation by 6. Write the result.

2. Divide each side of the equation in Exercise 1 by 5. Write the result.

3. Multiply each side of the original equation $25 = \frac{5}{6}s$ by the multiplicative inverse of $\frac{5}{6}$. Write the result.

4. What is the speed of a grizzly bear?

5. Which method of solving the equation seems most efficient? Explain.

You used the Multiplication and Division Properties of Equality to solve $25 = \frac{5}{6}s$. You can also use the Addition and Subtraction Properties of Equality to solve equations with rational numbers.

EXAMPLES Solve by Using Addition or Subtraction

1 Solve $p - 7.36 = -2.84$. Check your solution.

$$p - 7.36 = -2.84 \qquad \text{Write the equation.}$$
$$p - 7.36 + 7.36 = -2.84 + 7.36 \qquad \text{Add 7.36 to each side.}$$
$$p = 4.52 \qquad \text{Simplify.}$$

2 Solve $\frac{1}{2} = t + \frac{3}{4}$.

$$\frac{1}{2} = t + \frac{3}{4} \qquad \text{Write the equation.}$$
$$\frac{1}{2} - \frac{3}{4} = t + \frac{3}{4} - \frac{3}{4} \qquad \text{Subtract } \frac{3}{4} \text{ from each side.}$$
$$\frac{1}{2} - \frac{3}{4} = t \qquad \text{Simplify.}$$
$$\frac{2}{4} - \frac{3}{4} = t \qquad \text{Rename } \frac{1}{2}.$$
$$-\frac{1}{4} = t \qquad \text{Simplify.}$$

Solve by Using Multiplication or Division

3 Solve $\frac{4}{7}b = 16$. Check your solution.

$$\frac{4}{7}b = 16 \qquad \text{Write the equation.}$$

$$\frac{7}{4}\left(\frac{4}{7}b\right) = \frac{7}{4}(16) \qquad \text{Multiply each side by } \frac{7}{4}.$$

$$b = 28 \qquad \text{Simplify.}$$

Check $\qquad \frac{4}{7}b = 16 \qquad$ Write the original equation.

$$\frac{4}{7}(28) \stackrel{?}{=} 16 \qquad \text{Replace } b \text{ with 28.}$$

$$16 = 16 \checkmark \qquad \text{Simplify.}$$

4 Solve $58.4 = -7.3m$.

$$58.4 = -7.3m \qquad \text{Write the equation.}$$

$$\frac{58.4}{-7.3} = \frac{-7.3m}{-7.3} \qquad \text{Divide each side by } -7.3.$$

$$-8 = m \qquad \text{Simplify. Check the solution.}$$

Your Turn Solve each equation. Check your solution.

a. $r - 7.81 = 4.32$ **b.** $7.2v = -36$ **c.** $-\frac{2}{3}n = -\frac{3}{5}$

STUDY TIP

Multiply a Rational Number and a Whole Number
Recall that
$\frac{7}{4}(16) = \frac{7}{4} \cdot \frac{16}{1}$.

REAL-LIFE MATH

BASKETBALL During her rookie season for the WNBA, Sue Bird's field goal average was 0.379, and she made 232 field goal attempts.

Source: WNBA.com

You can write equations with rational numbers to solve real-life problems.

Write an Equation to Solve a Problem

5 **BASKETBALL** In basketball, a player's field goal average is determined by dividing the number of field goals made by the number of field goals attempted. Use the information at the left to determine the number of field goals Sue Bird made in her rookie season.

Words	Field goal average equals goals divided by attempts.
Variables	$f \qquad = \qquad \dfrac{g}{a}$
Equation	$0.379 \qquad = \qquad \dfrac{g}{232}$

$$0.379 = \frac{g}{232} \qquad \text{Write the equation.}$$

$$232(0.379) = 232\left(\frac{g}{232}\right) \qquad \text{Multiply each side by 232.}$$

$$87.928 = h \qquad \text{Simplify.}$$

Sue Bird made 88 field goals during her rookie season.

Skill and Concept Check

1. **OPEN ENDED** Write an equation with rational numbers that has a solution of $\frac{1}{4}$.

2. **Which One Doesn't Belong?** Identify the expression that does not have the same value as the other three. Explain your reasoning.

$$\frac{4}{3}\left(\frac{3}{4}x\right)$$ $$-\frac{3}{2}\left(-\frac{2}{3}x\right)$$ $$2\left(\frac{1}{2}x\right)$$ $$-\frac{1}{3}\left(\frac{1}{3}x\right)$$

GUIDED PRACTICE

Solve each equation. Check your solution.

3. $t + 0.25 = -4.12$

4. $a - \frac{3}{4} = -\frac{3}{8}$

5. $-45 = \frac{5}{6}d$

6. $-26.5 = -5.3w$

7. $\frac{5}{8}z = \frac{2}{9}$

8. $p - (-0.03) = 3.2$

SPACE **For Exercises 9 and 10, use the following information.**
The planet Jupiter takes 11.9 years to make one revolution around the Sun.

9. Write a multiplication equation you can use to determine the number of revolutions Jupiter makes in 59.5 years. Let r represent the number of revolutions.

10. How many revolutions does Jupiter make in 59.5 years?

Practice and Applications

Solve each equation. Check your solution.

11. $q + 0.45 = 1.29$

12. $a - 1.72 = 5.81$

13. $-\frac{1}{2} = m - \frac{2}{3}$

14. $-\frac{5}{9} = f + \frac{1}{3}$

15. $-\frac{4}{7}b = 16$

16. $-\frac{2}{9}p = -8$

17. $-1.92 = -0.32s$

18. $-8.4 = 1.2t$

19. $\frac{3}{4}z = -\frac{5}{6}$

20. $-\frac{2}{5}d = \frac{4}{9}$

21. $g - (-1.5) = 2.35$

22. $-1.3 = n - (-6.12)$

23. $\frac{t}{3.2} = -4.5$

24. $-\frac{a}{1.6} = 7.5$

25. $-5\frac{3}{4} = -2\frac{1}{2}x$

26. $4\frac{1}{6} = -3\frac{1}{3}c$

27. $3.5g = -\frac{7}{8}$

28. $-7.5r = -3\frac{1}{3}$

HOMEWORK HELP	
For Exercises	See Examples
11–30	1–4
31–33	5

Extra Practice
See pages 621, 649.

29. Find the solution of $v - \frac{2}{5} = -2$.

30. What is the solution of $-4.2 = \frac{c}{7}$?

31. **MONEY** The currency of Egypt is called a pound. The equation $3\frac{3}{4}d = 21$ can be used to determine how many U.S. dollars d equal 21 Egyptian pounds. Solve the equation.

RECREATION For Exercises 32 and 33, use the graph.

32. Let v equal the number of additional visitors that the Golden Gate National Recreation Area needed in the year 2000 to equal the number of visitors to the Blue Ridge Parkway. Write an addition equation to represent the situation.

33. How many more visitors did the Golden Gate National Recreation Area need to equal the number of visitors to the Blue Ridge Parkway?

34. **CRITICAL THINKING** What is the solution of $\frac{1}{2}y + 3 = 15$? Check your solution.

USA TODAY Snapshots®

Most popular national parks

The most-visited U.S. national park in 2000 was the Blue Ridge Parkway, a scenic roadway and series of parks that stretches 469 miles along the Appalachian Mountains in Virginia and North Carolina. Number of visitors, in millions, at the most popular national parks last year:

Blue Ridge Parkway

19.0

Golden Gate National Recreation Area

14.5

Great Smokey Mountains National Park

10.1

=1 million

Source: National Park Service

By William Risser and Robert W. Ahrens, USA TODAY

Standardized Test Practice and Mixed Review

A B C D **FCAT Practice**

35. **MULTIPLE CHOICE** Find the value of t in $t - (-4.36) = 7.2$.
   A 2.84   B 11.56   C −2.84   D −11.56

36. **MULTIPLE CHOICE** If the area of the rectangle at the right is $22\frac{3}{4}$ square inches, what is the width of the rectangle?

   F $\frac{4}{13}$ in.   G $2\frac{1}{2}$ in.

   H $3\frac{1}{4}$ in.   I $3\frac{3}{4}$ in.

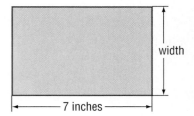

width

7 inches

Add or subtract. Write in simplest form. (Lesson 2-6)

37. $\frac{1}{6} + \frac{1}{7}$ 38. $\frac{7}{8} - \frac{1}{6}$ 39. $-5\frac{1}{2} - 6\frac{4}{5}$ 40. $2\frac{1}{2} + 5\frac{2}{3}$

41. **SHIPPING** Plastic straps are often wound around large cardboard boxes to reinforce them during shipping. Suppose the end of the strap must overlap $\frac{7}{16}$ inch to fasten. How long is the plastic strap around the box at the right? (Lesson 2-5)

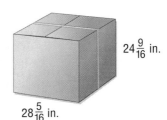

$24\frac{9}{16}$ in.

$28\frac{5}{16}$ in.

42. **ALGEBRA** The sum of two integers is 13. One of the integers is −5. Write an equation and solve to find the other integer. (Lesson 1-8)

43. **ALGEBRA** Write an expression for *17 more than p*. (Lesson 1-7)

GETTING READY FOR THE NEXT LESSON

BASIC SKILL Multiply.
44. $4 \cdot 4 \cdot 4$ 45. $2 \cdot 2 \cdot 2 \cdot 2 \cdot 2$ 46. $3 \cdot 3 \cdot 3 \cdot 3$ 47. $5 \cdot 5 \cdot 5$

Problem-Solving Strategy
A Preview of Lesson 2-8

Sunshine State Standards
MA.A.3.3.1-1, MA.A.3.3.2-1, MA.A.3.3.2-2,
MA.A.3.3.3-1

What You'll Learn
Solve problems using the look for a pattern strategy.

Look for a Pattern

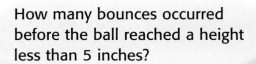

In science class, we dropped a ball from 48 inches above the ground. Each time it hit the ground, it bounced back up $\frac{1}{2}$ of the previous height.

How many bounces occurred before the ball reached a height less than 5 inches?

Explore	We know the original height of the ball. Each time the ball bounced, its height was $\frac{1}{2}$ of the previous height. We want to know the number of bounces before the ball reaches a height less than 5 inches.
Plan	Use a pattern to determine when the ball will reach a height of less than 5 inches.
Solve	<table><tr><th>Bounce</th><th>Height (inches)</th></tr><tr><td>1</td><td>$\frac{1}{2} \cdot 48 = 24$</td></tr><tr><td>2</td><td>$\frac{1}{2} \cdot 24 = 12$</td></tr><tr><td>3</td><td>$\frac{1}{2} \cdot 12 = 6$</td></tr><tr><td>4</td><td>$\frac{1}{2} \cdot 6 = 3$</td></tr></table> After the fourth bounce, the ball will reach a height less than 5 inches.
Examine	Check your pattern to make sure the answer is correct.

Analyze the Strategy

1. **Explain** how Jerome and Haley determined the numbers in the first column.

2. **Describe** how to continue the pattern in the second column. Find the fraction of the height after 7 bounces.

3. **Write** a problem that can be solved by finding a pattern. Describe the pattern.

Solve. Use the look for a pattern strategy.

4. **WATER MANAGEMENT** A tank is draining at a rate of 8 gallons every 3 minutes. If there are 70 gallons in the tank, when will the tank have just 22 gallons left?

5. **MUSIC** The names of musical notes form a pattern. Name the next three notes in the following pattern. whole note, half note, quarter note

Mixed Problem Solving

Solve. Use any strategy.

6. **TRAVEL** Rafael is taking a vacation. His plane is scheduled to leave at 2:20 P.M. He must arrive at the airport at least 2 hours before his flight. It will take him 45 minutes to drive from his house to the airport. When is the latest he should plan to leave for the airport?

7. **GEOMETRY** What is the total number of rectangles, of any size, in the figure below?

8. **TECHNOLOGY** The price of calculators has been decreasing. A calculator sold for $12.50 in 1990. A similar calculator sold for $8.90 in 2000. If the price decrease continues at the same rate, what would be the price in 2020?

9. **FUND-RAISING** Marissa is collecting donations for her 15-mile bike-a-thon. She is asking for pledges between $1.50 and $2.50 per mile. If she has 12 pledges, about how much could she expect to collect?

10. **SCHOOL** Lawanda was assigned some math exercises for homework. She answered half of them in study period. After school, she answered 7 more exercises. If she still has 11 exercises to do, how many exercises were assigned?

11. **SCIENCE** The Italian scientist Galileo discovered a relationship between the time of the back and forth swing of a pendulum and its length. How long is a pendulum with a swing of 5 seconds?

Time of Swing	Length of Pendulum
1 second	1 unit
2 seconds	4 units
3 seconds	9 units
4 seconds	16 units

12. **MULTI STEP** Hiroshi is planning a party. He plans to order 4 pizzas, which cost $12.75 each. If he has a coupon for $1.50 off each pizza, find the total cost of the pizzas.

13. **GEOMETRY** Draw the next two geometric figures in the pattern.

14. **STANDARDIZED TEST PRACTICE** FCAT Practice

Madeline rode her bicycle $\frac{1}{3}$ mile in 2 minutes. If she continues riding at the same rate, how far will she ride in 10 minutes?

Ⓐ $1\frac{2}{3}$ mi   Ⓑ $2\frac{1}{3}$ mi

Ⓒ $2\frac{2}{3}$ mi   Ⓓ $3\frac{1}{3}$ mi

You will use the look for a pattern strategy in the next lesson.

Powers and Exponents

Sunshine State Standards
MA.A.1.3.1-1, MA.A.1.3.4-2, MA.A.2.3.1-1, MA.A.2.3.1-3, MA.A.3.3.1-3, MA.D.2.3.1-6

What You'll Learn
Use powers and exponents in expressions.

NEW Vocabulary
base
exponent
power

REVIEW Vocabulary
evaluate: to find the value of an expression (Lesson 1-2)

WHEN am I ever going to use this?

FAMILY Every person has 2 biological parents. Study the family tree below.

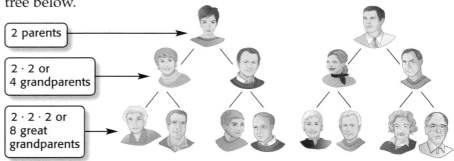

2 parents

2 · 2 or 4 grandparents

2 · 2 · 2 or 8 great grandparents

1. How many 2s are multiplied to determine the number of great grandparents?

2. How many 2s would you multiply to determine the number of great-great grandparents?

An expression like $2 \cdot 2 \cdot 2 \cdot 2$ can be written as the power 2^4.

The **base** is the number that is multiplied. → 2^4 ← The **exponent** tells how many times the base is used as a factor.

The number that is expressed using an exponent is called a **power**.

The table below shows how to write and read powers.

Powers	Words	Repeated Factors
2^1	2 to the first power	2
2^2	2 to the second power or 2 squared	$2 \cdot 2$
2^3	2 to the third power or 2 cubed	$2 \cdot 2 \cdot 2$
2^4	2 to the fourth power	$2 \cdot 2 \cdot 2 \cdot 2$
⋮	⋮	⋮
2^n	2 to the nth power	$\underbrace{2 \cdot 2 \cdot 2 \cdot \ldots \cdot 2}_{n \text{ factors}}$

EXAMPLE Write an Expression Using Powers

1 Write $a \cdot b \cdot b \cdot a \cdot b$ using exponents.

$$a \cdot b \cdot b \cdot a \cdot b = a \cdot a \cdot b \cdot b \cdot b \qquad \text{Commutative Property}$$
$$= (a \cdot a) \cdot (b \cdot b \cdot b) \qquad \text{Associative Property}$$
$$= a^2 \cdot b^3 \qquad \text{Definition of exponents}$$

You can also use powers to name numbers that are less than one. Consider the pattern in the powers of 10.

$$10^3 = 10 \cdot 10 \cdot 10 \text{ or } 1{,}000$$

$$10^2 = 10 \cdot 10 \text{ or } 100$$

$$10^1 = 10$$

$$10^0 = 1$$

$$10^{-1} = \frac{1}{10}$$

$$10^{-2} = \frac{1}{100}$$

$1{,}000 \div 10 = 100$

$100 \div 10 = 10$

$10 \div 10 = 1$

$1 \div 10 = \frac{1}{10}$

$\frac{1}{10} \div 10 = \frac{1}{10^2} \text{ or } \frac{1}{100}$

Negative Exponents
Remember that 10^{-2} equals $\frac{1}{10^2}$, *not* -20 or -100.

The pattern above suggests the following definitions for zero exponents and negative exponents.

Key Concept   Zero and Negative Exponents

Words   Any nonzero number to the zero power is 1. Any nonzero number to the negative n power is 1 divided by the number to the nth power.

Symbols

Arithmetic	Algebra
$5^0 = 1$	$x^0 = 1, x \neq 0$
$7^{-3} = \frac{1}{7^3}$	$x^{-n} = \frac{1}{x^n}, x \neq 0$

EXAMPLES Evaluate Powers

2 Evaluate 5^4.

$5^4 = 5 \cdot 5 \cdot 5 \cdot 5$ Definition of exponents

$\quad = 625$ Simplify.

Check using a calculator. $5 \; \boxed{\wedge} \; 4 \; \boxed{\overset{\text{ENTER}}{=}} \; 625$

3 Evaluate 4^{-3}.

$4^{-3} = \frac{1}{4^3}$ Definition of negative exponents

$\quad = \frac{1}{64}$ Simplify.

4 **ALGEBRA** Evaluate $a^2 \cdot b^4$ if $a = 3$ and $b = 5$.

$a^2 \cdot b^4 = 3^2 \cdot 5^4$ Replace a with 3 and b with 5.

$\quad = (3 \cdot 3) \cdot (5 \cdot 5 \cdot 5 \cdot 5)$ Definition of exponents

$\quad = 9 \cdot 625$ Simplify.

$\quad = 5{,}625$ Simplify.

Your Turn Evaluate each expression.

a. 15^3 **b.** $2^5 \cdot 5^2$ **c.** 5^{-4}

msmath3.net/extra_examples/fcat   **Lesson 2-8** Powers and Exponents   **99**

Skill and Concept Check

Writing Math
Exercise 1

1. **OPEN ENDED** Write an expression with a negative exponent and explain what it means.

2. **NUMBER SENSE** Without evaluating the powers, order 6^{-3}, 6^2, and 6^0 from least to greatest.

GUIDED PRACTICE

Write each expression using exponents.

3. $3 \cdot 3 \cdot 3 \cdot 3 \cdot 3 \cdot 3$

4. $2 \cdot 2 \cdot 2 \cdot 3 \cdot 3 \cdot 3$

5. $r \cdot s \cdot r \cdot r \cdot s \cdot s \cdot r \cdot r$

Evaluate each expression.

6. 7^3

7. $2^3 \cdot 6^2$

8. $4^2 \cdot 5^3$

9. 6^{-3}

10. **ALGEBRA** Evaluate $x^2 \cdot y^4$ if $x = 2$ and $y = 10$.

For Exercises 11–14, use the information below.

11. How many stars can be seen with unaided eyes in an urban area?

12. How many stars can be seen with unaided eyes in a rural area?

13. How many stars can be seen with binoculars?

14. How many stars can be seen with a small telescope?

How Many Stars Can You See?	
Unaided Eye in Urban Area	$3 \cdot 10^2$ stars
Unaided Eye in Rural Area	$2 \cdot 10^3$ stars
With Binoculars	$3 \cdot 10^4$ stars
With Small Telescope	$2 \cdot 10^6$ stars

Source: *Kids Discover*

Practice and Applications

Write each expression using exponents.

15. $8 \cdot 8 \cdot 8$

16. $5 \cdot 5 \cdot 5 \cdot 5$

17. $p \cdot p \cdot p \cdot p \cdot p \cdot p$

18. $d \cdot d \cdot d \cdot d \cdot d$

19. $3 \cdot 3 \cdot 4 \cdot 4 \cdot 4$

20. $2 \cdot 2 \cdot 2 \cdot 5 \cdot 5$

21. $4 \cdot 7 \cdot 4 \cdot 4 \cdot 7 \cdot 7 \cdot 7 \cdot 7$

22. $5 \cdot 5 \cdot 8 \cdot 8 \cdot 5 \cdot 8 \cdot 8$

23. $a \cdot a \cdot b \cdot b \cdot a \cdot b \cdot b \cdot a$

24. $x \cdot y \cdot y \cdot y \cdot x \cdot y \cdot y \cdot y$

25. Write the product $7 \cdot 7 \cdot 7 \cdot 15 \cdot 15 \cdot 7$ using exponents.

26. Write the product $5 \cdot 12 \cdot 12 \cdot 12 \cdot 5 \cdot 5 \cdot 5 \cdot 5$ using exponents.

HOMEWORK HELP

For Exercises	See Examples
15–26	1
27–38	2, 3
39–40	4

Extra Practice
See pages 621, 649.

Evaluate each expression.

27. 2^3

28. 3^4

29. 3^5

30. 9^3

31. $3^2 \cdot 5^2$

32. $3^3 \cdot 4^2$

33. $2^5 \cdot 5^3$

34. $3^2 \cdot 7^3$

35. 5^{-4}

36. 9^{-3}

37. $2^3 \cdot 7^{-2}$

38. $5^2 \cdot 2^{-7}$

39. **ALGEBRA** Evaluate $g^5 \cdot h$ if $g = 2$ and $h = 7$.

40. **ALGEBRA** Evaluate $x^3 \cdot y^4$ if $x = 1$ and $y = 3$.

41. **BIOLOGY** Suppose a bacterium splits into two bacteria every 20 minutes. How many bacteria will there be in 2 hours?

42. **LITERATURE** *The Rajah's Rice* is the story of a young girl named Chandra. She loved elephants and helped take care of the Rajah's elephants. The Rajah was pleased and wanted to reward her. She asked for the following reward.

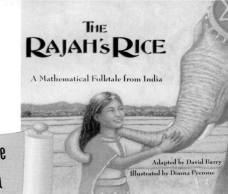

"If Your Majesty pleases, place two grains of rice on the first square of this chessboard. Place four grains on the second square, eight on the next, and so on, doubling each pile of rice till the last square."

Write the number of grains of rice the Rajah should put on the last square using an exponent.

43. **GEOMETRY** To find the volume of a cube, multiply its length, its width, and its depth. Find the volume of each cube.

2 in.   6 in.

44. Continue the following pattern.
$3^4 = 81, 3^3 = 27, 3^2 = 9, 3^1 = 3,$
$3^0 = ?, 3^{-1} = ?, 3^{-2} = ?, 3^{-3} = ?$

45. **CRITICAL THINKING** Write each of the following as a power of 10 or the product of a whole number between 1 and 10 and a power of 10.
   a. 100,000
   b. fifty million
   c. 3,000,000,000
   d. sixty thousand

Standardized Test Practice and Mixed Review

FCAT Practice

46. **MULTIPLE CHOICE** Write $5 \cdot 5 \cdot 7 \cdot 7 \cdot 7 \cdot q \cdot q$ using exponents.
   Ⓐ $5 \cdot 12^2 \cdot q^2$ Ⓑ $5^2 \cdot 7^3 \cdot q^2$ Ⓒ $35^2 \cdot q^2$ Ⓓ $70\, q^2$

47. **SHORT RESPONSE** Write $2^3 \cdot 6^2$ in expanded form. Then find its value.

48. **FOOD** Suppose hamburgers are cut in the shape of a square that is $2\frac{1}{2}$ inches on a side. Write a multiplication equation to determine how many hamburgers can fit across a grill that is 30 inches wide. Solve the equation. (Lesson 2-7)

Add or subtract. Write in simplest form. (Lesson 2-6)

49. $\frac{1}{6} + \frac{4}{9}$ 50. $\frac{2}{5} - \frac{1}{4}$ 51. $1\frac{1}{2} - \left(-\frac{7}{9}\right)$ 52. $-\frac{1}{8} + \frac{5}{6}$

53. **ALGEBRA** Write an algebraic expression for *12 more than a number*. (Lesson 1-7)

GETTING READY FOR THE NEXT LESSON

BASIC SKILL Write each number.

54. two million   55. three hundred twenty   56. twenty six hundred

Sunshine State Standards
MA.A.2.3.2-1, MA.A.2.3.2-2, MA.A.2.3.2-3

What You'll Learn
Use binary numbers.

Materials
- paper and pencil
- grid paper

Binary Numbers

Computers have a language of their own. The digits 0 and 1, also called bits, translate into OFF and ON within the computer's electronic switches system. Numbers that use only the digits 0 and 1 are called **base two numbers** or **binary numbers**. For example, 101001_2 is a binary number. The small 2 after 101001_2 means the number is in base two.

INVESTIGATE

1. Copy and complete the table for the powers of 2.

Power of Two	2^5	2^4	2^3	2^2	2^1
Value	32				

2. Use the pattern in the table to determine the value of 2^0.

Find the value of each expression.

3. $2^3 + 2^2 + 2^0$
4. $2^4 + 2^2$
5. $2^5 + 2^3 + 2^2$
6. $2^5 + 2^2 + 2^0$
7. $2^4 + 2^3 + 2^2 + 2^1$
8. $2^5 + 2^4 + 2^1 + 2^0$

When using binary numbers, use the following rules.

- The digits 0 and 1 are the only digits used in base two.
- The digit 1 represents that the power of two is ON. The digit 0 represents the power is OFF.

Binary numbers can be written in our standard base ten system.

ACTIVITY *Work with a partner.*

1 Write 10011_2 in base ten.

10011_2 is in base two. Each place value represents a power of 2.

1	0	0	1	1
ON	OFF	OFF	ON	ON
2^4 or 16	2^3 or 8	2^2 or 4	2^1 or 2	2^0 or 1

$$10011_2 = (1 \times 2^4) + (0 \times 2^3) + (0 \times 2^2) + (1 \times 2^1) + (1 \times 2^0)$$
$$= (1 \times 16) + (0 \times 8) + (0 \times 4) + (1 \times 2) + (1 \times 1)$$
$$= 16 + 0 + 0 + 2 + 1 \text{ or } 19$$

Therefore, 10011_2 is 19 in base ten.

Your Turn Write each number in base ten.

a. 10101_2
b. 1001_2
c. 110110_2

You can also reverse the process and write base ten numbers in base two.

ACTIVITY *Work with a partner.*

2 **Write 38 in base two.**

STEP 1 Make a base two place-value chart.

2^6 or 64	2^5 or 32	2^4 or 16	2^3 or 8	2^2 or 4	2^1 or 2	2^0 or 1

STEP 2 Find the greatest power of 2 that is less than or equal to 38. Place a 1 in that place value.

	1					
2^6 or 64	2^5 or 32	2^4 or 16	2^3 or 8	2^2 or 4	2^1 or 2	2^0 or 1

STEP 3 Since $38 - 32 = 6$, find the greatest power of 2 that is less than or equal to 6. Place a 1 in that place value.

	1			1		
2^6 or 64	2^5 or 32	2^4 or 16	2^3 or 8	2^2 or 4	2^1 or 2	2^0 or 1

STEP 4 Since $6 - 4 = 2$, find the greatest power of 2 that is less than or equal to 2. Place a 1 in that place value.

	1			1	1	
2^6 or 64	2^5 or 32	2^4 or 16	2^3 or 8	2^2 or 4	2^1 or 2	2^0 or 1

STEP 5 Since $2 - 2 = 0$, place a 0 in any unfilled spaces.

0	1	0	0	1	1	0
2^6 or 64	2^5 or 32	2^4 or 16	2^3 or 8	2^2 or 4	2^1 or 2	2^0 or 1

The zero at the far left is not needed as a placeholder. Therefore, 38 in base ten is equal to 100110 in base two. Or, $38 = 100110_2$.

Your Turn Write each number in base two.

d. 46   e. 70   f. 15

Writing Math

1. **Explain** how to determine the place value of each digit in base two.

2. **Make** a place-value chart of the first four digits in base five.

3. **Identify** the digits you would use in base five.

4. **MAKE A CONJECTURE** Explain how to determine the place values for base n. What digits would you use for base n?

2-9 Scientific Notation

Sunshine State Standards
MA.A.1.3.1-1, MA.A.1.3.2-2, MA.A.1.3.4-3, MA.A.2.3.1-2

What You'll Learn
Express numbers in scientific notation.

NEW Vocabulary

scientific notation

WHEN am I ever going to use this?

LANGUAGES The most frequently spoken languages are listed in the table.

1. All of the values contain 10^8. What is the value of 10^8?

2. How many people speak Mandarin as their native language?

3. How many people speak English as their native language?

Top Five Languages of the World		
Language	**Where Spoken**	**Number of Native Speakers**
Mandarin	China, Taiwan	8.74×10^8
Hindi	India	3.66×10^8
English	U.S.A., Canada, Britain	3.41×10^8
Spanish	Spain, Latin America	3.22×10^8
Arabic	Arabian Peninsula	2.07×10^8

Source: *The World Almanac for Kids*

The number 8.74×10^8 is written in **scientific notation**. Scientific notation is often used to express very large or very small numbers.

Key Concept — Scientific Notation

A number is expressed in scientific notation when it is written as the product of a factor and a power of 10. The factor must be greater than or equal to 1 and less than 10.

Multiplying by a positive power of 10 moves the decimal point right. Multiplying by a negative power of 10 moves the decimal point left.

EXAMPLES — Express Numbers in Standard Form

1 Write 5.34×10^4 in standard form.

$5.34 \times 10^4 = 5.34 \times 10{,}000$ $10^4 = 10 \cdot 10 \cdot 10 \cdot 10$ or 10,000

$ = 53{,}400$ Notice that the decimal point moves 4 places to the right.

2 Write 3.27×10^{-3} in standard form.

$3.27 \times 10^{-3} = 3.27 \times \dfrac{1}{10^3}$ $10^{-3} = \dfrac{1}{10^3}$

$\phantom{3.27 \times 10^{-3}} = 3.27 \times 0.001$ $\dfrac{1}{10^3} = \dfrac{1}{1{,}000}$ or 0.001

$\phantom{3.27 \times 10^{-3}} = 0.00327$ Notice that the decimal point moves 3 places to the left.

Your Turn Write each number in standard form.

a. 7.42×10^5 **b.** 6.1×10^{-2} **c.** 3.714×10^2

To write a number in scientific notation, place the decimal point after the first nonzero digit. Then find the power of 10. If a number is between 0 and 1, the power of ten is negative. Otherwise, the power of ten is positive.

EXAMPLES Write Numbers in Scientific Notation

3 Write 3,725,000 in scientific notation.

$3{,}725{,}000 = 3.725 \times 1{,}000{,}000$ The decimal point moves 6 places.

$\phantom{3{,}725{,}000} = 3.725 \times 10^6$ The exponent is positive.

4 Write 0.000316 in scientific notation.

$0.000316 = 3.16 \times 0.0001$ The decimal point moves 4 places.

$ = 3.16 \times 10^{-4}$ The exponent is negative.

Your Turn Write each number in scientific notation.

d. 14,140,000   e. 0.00876   f. 0.114

You can order numbers written in scientific notation.

EXAMPLE Compare Numbers in Scientific Notation

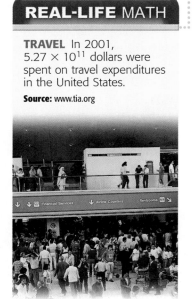

5 **TRAVEL** The number of visitors from various countries to the United States in 2001 are listed in the table. Order the countries according to the number of visitors from greatest to least.

First, order the number according to their exponents. Then order the number with the same exponents by comparing the factors.

International Visitors to the U.S.A.	
Country	**Number of Visitors**
Canada	1.46×10^7
France	1.1×10^6
Germany	1.8×10^6
Japan	5.1×10^6
Mexico	1.03×10^7
United Kingdom	4.7×10^6

Source: International Trade Association

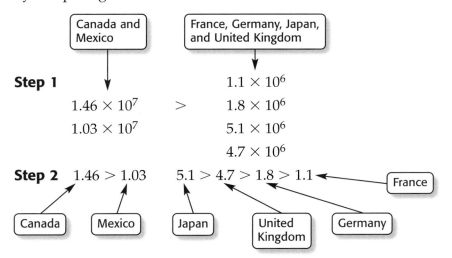

The countries in order are Canada, Mexico, Japan, United Kingdom, Germany, and France.

Skill and Concept Check

1. **Determine** whether a decimal times a power of 10 is *sometimes*, *always*, or *never* scientific notation. Explain.

2. **OPEN ENDED** Write a number in scientific notation that is less than 1 and greater than 0. Then write the number in standard form.

3. **NUMBER SENSE** Determine whether 1.2×10^5 or 1.2×10^6 is closer to one million. Explain.

GUIDED PRACTICE

Write each number in standard form.

4. 7.32×10^4 5. 9.931×10^5 6. 4.55×10^{-1} 7. 6.02×10^{-4}

Write each number in scientific notation.

8. 277,000   9. 8,785,000,000   10. 0.00004955   11. 0.524

12. **CARTOONS** Use scientific notation to write the number of seconds in summer vacation according to the cartoon below.

FoxTrot   by Bill Amend

Practice and Applications

Write each number in standard form.

13. 2.08×10^2 14. 3.16×10^3 15. 7.113×10^7

16. 4.265×10^6 17. 7.8×10^{-3} 18. 1.1×10^{-4}

19. 8.73×10^{-4} 20. 2.52×10^{-5} 21. 1.046×10^6

22. 2.051×10^5 23. 6.299×10^{-6} 24. 5.022×10^{-7}

HOMEWORK HELP

For Exercises	See Examples
13–26	1, 2
27–28, 41	5
29–39	3, 4

Extra Practice
See pages 621, 649.

25. **DINOSAURS** The Giganotosaurus weighed 1.4×10^4 pounds. Write this number in standard form.

26. **HEALTH** The diameter of a red blood cell is about 7.4×10^{-4} centimeter. Write this number using standard form.

27. Which is greater: 6.3×10^5 or 7.1×10^4?   28. Which is less: 4.1×10^3 or 3.2×10^7?

Write each number in scientific notation.

29. 6,700   30. 43,000   31. 52,300,000   32. 147,000,000

33. 0.037   34. 0.0072   35. 0.00000707   36. 0.0000901

37. **TIME** The smallest unit of time is the *yoctosecond*, which equals 0.000000000000000000000001 second. Write this number in scientific notation.

38. **SPACE** The temperature of the Sun varies from 10,900°F on the surface to 27,000,000,000°F at its core. Write these temperatures in scientific notation.

39. **NUMBERS** A googol is a number written as a 1 followed by 100 zeros. Write a googol in scientific notation.

40. **SCIENCE** An oxygen atom has a mass of 2.66×10^{-23} gram. Explain how to enter this number into a calculator.

41. **BASEBALL** The following table lists five Major League Ballparks. List the ballparks from least capacity to greatest capacity.

Ballpark	Team	Capacity
H.H.H. Metrodome	Minnesota Twins	4.8×10^4
Network Associates Coliseum	Oakland Athletics	4.7×10^4
The Ballpark in Arlington	Texas Rangers	4.9×10^4
Wrigley Field	Chicago Cubs	3.9×10^4
Yankee Stadium	New York Yankees	5.5×10^4

Source: www.users.bestweb.net

Data Update What is the capacity of your favorite ballpark? Visit msmath3.net/data_update to learn more.

CRITICAL THINKING Compute and express each value in scientific notation.

42. $\dfrac{(130,000)(0.0057)}{0.0004}$

43. $\dfrac{(90,000)(0.0016)}{(200,000)(30,000)(0.00012)}$

Standardized Test Practice and Mixed Review

FCAT Practice

44. **MULTIPLE CHOICE** The distance from Milford to Loveland is 326 kilometers. If there are 1,000 meters in a kilometer, use scientific notation to write the distance from Milford to Loveland in meters.

Ⓐ 3.26×10^6 m

Ⓑ 32.6×10^5 m

Ⓒ 326×10^5 m

Ⓓ 3.26×10^5 m

45. **SHORT RESPONSE** Name the Great Lake with the second greatest area.

46. **ALGEBRA** Evaluate $a^5 \cdot b^2$ if $a = 2$ and $b = 3$. (Lesson 2-8)

ALGEBRA Solve each equation. Check your solution. (Lesson 2-7)

47. $t + 3\frac{1}{3} = 2\frac{1}{2}$

48. $-\frac{2}{3}y = 14$

49. $\frac{p}{1.3} = 2.4$

50. $-1\frac{3}{4} = n - 4\frac{1}{6}$

Great Lakes	
Lake	Area (square miles)
Erie	9.91×10^3
Huron	2.3×10^4
Michigan	2.23×10^4
Ontario	7.32×10^3
Superior	3.17×10^4

Source: *World Book*

Vocabulary and Concept Check

bar notation (p. 63)	multiplicative inverses (p. 76)	repeating decimal (p. 63)
base (p. 98)	power (p. 98)	scientific notation (p. 104)
dimensional analysis (p. 73)	rational number (p. 62)	terminating decimal (p. 63)
exponent (p. 98)	reciprocals (p. 76)	unlike fractions (p. 88)
like fractions (p. 82)		

Choose the correct term to complete each sentence.

1. The (base, exponent) tells how many times a number is used as a factor.

2. Two numbers whose product is one are called (multiplicative inverses, rational numbers).

3. (Unlike fractions, Like fractions) have the same denominator.

4. A number that is expressed using an exponent is called a (power, base).

5. The (base, exponent) is the number that is multiplied.

6. The number 3.51×10^{-3} is written in (dimensional analysis, scientific notation).

7. The number $\frac{3}{4}$ is a (power, rational number).

8. Bar notation is used to represent a (terminating decimal, repeating decimal).

Lesson-by-Lesson Exercises and Examples

2-1   **Fractions and Decimals** (pp. 62–66)

Write each fraction or mixed number as a decimal.

9. $1\frac{1}{3}$ 10. $-\frac{5}{8}$

11. $5\frac{13}{50}$ 12. $-\frac{5}{6}$

13. $-2\frac{3}{10}$ 14. $\frac{5}{9}$

Write each decimal as a fraction or mixed number in simplest form.

15. 0.3 16. 3.56

17. -2.75 18. -7.14

19. $4.\overline{3}$ 20. $-5.\overline{7}$

Example 1   Write $\frac{3}{5}$ as a decimal.

$\frac{3}{5}$ means $3 \div 5$.

$$\begin{array}{r} 0.6 \\ 5\overline{)3.0} \\ 30 \\ \hline 0 \end{array}$$

The fraction $\frac{3}{5}$ can be written as 0.6.

Example 2   Write 0.25 as a fraction in simplest form.

$0.25 = \frac{25}{100}$ 0.25 is 25 hundredths.

$ = \frac{1}{4}$ Simplify.

The decimal 0.25 can be written as $\frac{1}{4}$.

 msmath3.net/vocabulary_review

2-2 **Comparing and Ordering Rational Numbers** (pp. 67–70)

Replace each ● with $<$, $>$, or $=$ to make a true sentence.

21. $\frac{2}{3}$ ● $\frac{8}{9}$

22. $-0.\overline{24}$ ● $-\frac{8}{33}$

23. $-\frac{1}{2}$ ● $-\frac{55}{110}$

24. $\frac{5}{6}$ ● $\frac{3}{4}$

25. Order $-\frac{1}{2}$, 0.75, $-\frac{3}{4}$, 0 from least to greatest.

Example 3 Replace ● with $<$, $>$, or $=$ to make $\frac{2}{5}$ ● 0.34 a true sentence.

$\frac{2}{5} = 0.4$

Since $0.4 > 0.34$, $\frac{2}{5} > 0.34$.

2-3 **Multiplying Rational Numbers** (pp. 71–75)

Multiply. Write in simplest form.

26. $\frac{3}{5} \cdot 1\frac{2}{3}$

27. $-\frac{2}{3} \cdot \left(-\frac{2}{3}\right)$

28. $\frac{5}{6} \cdot \frac{3}{5}$

29. $\frac{1}{2} \cdot \frac{10}{11}$

30. **COOKING** Crystal is making $1\frac{1}{2}$ times a recipe. The original recipe calls for $3\frac{1}{2}$ cups of milk. How many cups of milk does she need?

Example 4 Find $\frac{2}{3} \cdot \frac{5}{7}$. Write in simplest form.

$\frac{2}{3} \cdot \frac{5}{7} = \frac{2 \cdot 5}{3 \cdot 7}$ ← Multiply the numerators.
← Multiply the denominators.

$= \frac{10}{21}$ Simplify.

2-4 **Dividing Rational Numbers** (pp. 76–80)

Divide. Write in simplest form.

31. $\frac{7}{9} \div \frac{1}{3}$

32. $\frac{7}{12} \div \left(-\frac{2}{3}\right)$

33. $-4\frac{2}{5} \div (-2)$

34. $6\frac{1}{6} \div \left(-1\frac{2}{3}\right)$

Example 5 Find $-\frac{5}{6} \div \frac{3}{5}$. Write in simplest form.

$-\frac{5}{6} \div \frac{3}{5} = -\frac{5}{6} \cdot \frac{5}{3}$ Multiply by the multiplicative inverse.

$= -\frac{25}{18}$ or $-1\frac{7}{18}$ Simplify.

2-5 **Adding and Subtracting Like Fractions** (pp. 82–85)

Add or subtract. Write in simplest form.

35. $\frac{5}{11} + \frac{6}{11}$

36. $\frac{1}{8} + \left(-\frac{3}{8}\right)$

37. $\frac{1}{8} - \frac{7}{8}$

38. $12\frac{4}{5} - 5\frac{3}{5}$

Example 6 Find $\frac{1}{5} - \frac{3}{5}$. Write in simplest form.

$\frac{1}{5} - \frac{3}{5} = \frac{1-3}{5}$ ← Subtract the numerators.
← The denominators are the same.

$= \frac{-2}{5}$ or $-\frac{2}{5}$ Simplify.

2-6 Adding and Subtracting Unlike Fractions (pp. 88–91)

Add or subtract. Write in simplest form.

39. $-\frac{2}{3} + \frac{3}{5}$

40. $\frac{2}{3} + \frac{3}{4}$

41. $-4\frac{1}{2} - 6\frac{2}{3}$

42. $5 - 1\frac{2}{5}$

43. $7\frac{3}{4} + 3\frac{4}{5}$

44. $5\frac{3}{5} - 12\frac{1}{2}$

Example 7 Find $\frac{3}{4} + \frac{1}{3}$. Write in simplest form.

$$\frac{3}{4} + \frac{1}{3} = \frac{9}{12} + \frac{4}{12} \quad \text{Rename the fractions.}$$

$$= \frac{9 + 4}{12} \quad \text{Add the numerators.}$$

$$= \frac{13}{12} \text{ or } 1\frac{1}{12} \quad \text{Simplify.}$$

2-7 Solving Equations with Rational Numbers (pp. 92–95)

Solve each equation. Check your solution.

45. $d - (-0.8) = 4$

46. $\frac{x}{4} = -2.2$

47. $\frac{3}{4}n = \frac{7}{8}$

48. $-7.2 = \frac{r}{1.6}$

49. **AGE** Trevor is $\frac{3}{8}$ of Maria's age. If Trevor is 15, how old is Maria?

Example 8 Solve $t + \frac{1}{3} = \frac{5}{6}$.

$$t + \frac{1}{3} = \frac{5}{6} \quad \text{Write the equation.}$$

$$t + \frac{1}{3} - \frac{1}{3} = \frac{5}{6} - \frac{1}{3} \quad \text{Subtract } \frac{1}{3} \text{ from each side.}$$

$$t = \frac{1}{2} \quad \text{Simplify.}$$

2-8 Powers and Exponents (pp. 98–101)

Write each expression using exponents.

50. $3 \cdot 3 \cdot 3 \cdot 3 \cdot 3$

51. $2 \cdot 2 \cdot 5 \cdot 5 \cdot 5$

52. $x \cdot x \cdot x \cdot x \cdot y$

53. $4 \cdot 4 \cdot 9 \cdot 9$

Evaluate each expression.

54. 5^4

55. $4^2 \cdot 3^3$

56. 5^{-3}

57. $4^2 \cdot 2^3$

Example 9
Write $3 \cdot 3 \cdot 3 \cdot 7 \cdot 7$ using exponents.
$3 \cdot 3 \cdot 3 \cdot 7 \cdot 7 = 3^3 \cdot 7^2$

Example 10
Evaluate 7^3.
$7^3 = 7 \cdot 7 \cdot 7 \text{ or } 343$

2-9 Scientific Notation (pp. 104–107)

Write each number in standard form.

58. 3.2×10^{-3}

59. 6.71×10^4

60. 1.72×10^5

61. 1.5×10^{-2}

Write each number in scientific notation.

62. 0.000064

63. 0.000351

64. $87,500,000$

65. $7,410,000$

Example 11
Write 3.21×10^{-6} in standard form.
$3.21 \times 10^{-6} = 0.00000321$ Move the decimal point 6 places to the left.

Vocabulary and Concepts

1. **Explain** how to write a number in scientific notation.
2. **Write** $3 \cdot 3 \cdot 3 \cdot 3 \cdot 3$ using exponents.

Skills and Applications

Write each fraction or mixed number as a decimal.

3. $1\frac{2}{3}$

4. $\frac{1}{8}$

5. $-\frac{7}{20}$

Write each decimal as a fraction or mixed number in simplest form.

6. 0.78

7. $0.\overline{1}$

8. 2.04

Multiply, divide, add, or subtract. Write in simplest form.

9. $-\frac{2}{3} \cdot \frac{7}{8}$

10. $-6 \div \frac{2}{3}$

11. $-5\frac{1}{4} \cdot \left(-2\frac{1}{3}\right)$

12. $\frac{1}{8} \div \frac{5}{6}$

13. $-\frac{5}{7} + \frac{3}{7}$

14. $1\frac{1}{2} + \frac{2}{3}$

15. $\frac{5}{6} - \frac{1}{2}$

16. $-\frac{7}{8} - \left(-\frac{1}{4}\right)$

17. **BAKING** Madison needs $2\frac{3}{4}$ cups of flour. She has only $1\frac{1}{3}$ cups. How much does she need to borrow from her neighbor Raul?

18. **GEOMETRY** Find the perimeter of the rectangle.

$\frac{3}{4}$ unit

$\frac{2}{3}$ unit

Solve each equation. Check your solution.

19. $x - \frac{5}{6} = \frac{1}{3}$

20. $16 = \frac{2}{3}y$

Write each expression using exponents.

21. $4 \cdot 4 \cdot 4 \cdot 4 \cdot 4 \cdot 5 \cdot 5 \cdot 5$

22. $a \cdot a \cdot a \cdot a \cdot b \cdot b$

23. Write 8.83×10^{-7} in standard form.

24. Write $25{,}000$ in scientific notation.

FCAT Practice

Standardized Test Practice

A   B   C   D

25. **MULTIPLE CHOICE** The table lists four movies and their running times. Which movie is the longest?

Ⓐ Movie A

Ⓑ Movie B

Ⓒ Movie C

Ⓓ Movie D

Movie	Length (h)
Movie A	$2\frac{1}{4}$
Movie B	$2.11\overline{6}$
Movie C	$2\frac{1}{6}$
Movie D	$2.18\overline{3}$

FCAT Practice

PART 1 Multiple Choice

Record your answers on the answer sheet provided by your teacher or on a sheet of paper.

1. Sonia pours 8 ounces of water into a 12-ounce glass. Which of the following fractions represents how full the glass is? (Prerequisite Skill, p. 611)

   Ⓐ $\frac{3}{12}$ Ⓑ $\frac{2}{3}$ Ⓒ $\frac{8}{1}$ Ⓓ $\frac{12}{1}$

2. Which point is graphed at $|-3|$? (Lesson 1-3)

   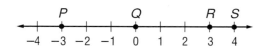

   Ⓕ P Ⓖ Q Ⓗ R Ⓘ S

3. Which of the following is *not* equivalent to $(12)(-9)(-7)(5)$? (Lesson 1-6)

   Ⓐ $12[(-9)(-7)]5$

   Ⓑ $[(12)(-9)](-7)(5)$

   Ⓒ $[(12-9)](-7)(5)$

   Ⓓ $[(12)(-9)][(-7)(5)]$

4. Which decimal can be written as the fraction $\frac{5}{9}$? (Lesson 2-1)

   Ⓕ $0.\overline{5}$ Ⓖ $0.\overline{59}$

   Ⓗ 1.8 Ⓘ 9.500

5. If a whole number greater than one is multiplied by a fraction less than zero, which of the following describes the product? (Lesson 2-3)

   Ⓐ a number greater than the whole number

   Ⓑ a negative number less than the fraction

   Ⓒ a negative number greater than the fraction

   Ⓓ zero

6. What is the length of the rectangle? (Lesson 2-7)

   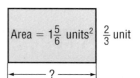

   Ⓕ $\frac{4}{33}$ unit Ⓖ $\frac{4}{11}$ unit

   Ⓗ $\frac{13}{9}$ units Ⓘ $\frac{11}{4}$ units

7. Which of the following represents the expression $12y^4$? (Lesson 2-8)

   Ⓐ $12 \cdot y \cdot 4$

   Ⓑ $12 \cdot 12 \cdot y \cdot y$

   Ⓒ $12 \cdot 12 \cdot 12 \cdot 12 \cdot y$

   Ⓓ $12 \cdot y \cdot y \cdot y \cdot y$

8. What is the same as $(2 \cdot 2 \cdot 2)^3$? (Lesson 2-8)

   Ⓕ 3^2 Ⓖ 2^6 Ⓗ 8^3 Ⓘ 222^3

9. The populations of the three largest countries in the world in 2000 are given below.

Country	Population
China	1,262,000,000
India	1,014,000,000
United States	281,000,000

   Source: *The World Almanac*

   Which of the following does *not* express the population of the United States in another way? (Lesson 2-9)

   Ⓐ 2.81×10^8 Ⓑ 28.1×10^7

   Ⓒ 28.1 million Ⓓ 281 million

10. What is the standard form of 4.673×10^{-5}? (Lesson 2-9)

    Ⓕ 0.00004673 Ⓖ 0.004673

    Ⓗ $46,730$ Ⓘ $467,300$

PART 2   Short Response/Grid In

THINK
SOLVE
EXPLAIN

Record your answers on the answer sheet provided by your teacher or on a sheet of paper.

11. Salvador has finished 28 of the 40 assigned math problems. Write this ratio in a different way. (Prerequisite Skill, p. 611)

12. At a golf tournament, a player scored 3, −4, −7, and −5. What was his total score? (Lesson 1-4)

13. Olivia made a coat rack with seven hooks. She found a board that was $31\frac{1}{2}$ inches long. She divided the board evenly, making the space at the ends of the rack the same as the space between the hooks. Each hook was $\frac{1}{2}$-inch in width. What was the space between each hook? (Lesson 2-5)

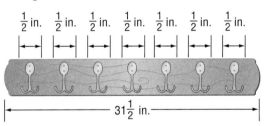

$\frac{1}{2}$ in. $\frac{1}{2}$ in. $\frac{1}{2}$ in. $\frac{1}{2}$ in. $\frac{1}{2}$ in. $\frac{1}{2}$ in. $\frac{1}{2}$ in.

$31\frac{1}{2}$ in.

14. Logan was using 4 tiles of different lengths to make a mosaic. What is the length of the mosaic shown below? (Lesson 2-6)

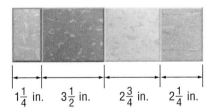

$1\frac{1}{4}$ in. $3\frac{1}{2}$ in. $2\frac{3}{4}$ in. $2\frac{1}{4}$ in.

TEST-TAKING TIP

Questions 13 and 14 You cannot write mixed numbers, such as $2\frac{1}{2}$, on an answer grid. Answers such as these need to be written as improper fractions, such as 5/2, or as decimals, such as 2.5. Choose the method that you like best, so that you will avoid making unnecessary mistakes.

15. During one week, Ms. Ito biked $1\frac{3}{8}$ miles, $1\frac{3}{4}$ miles, and $1\frac{1}{2}$ miles. What was the total distance she biked that week? (Lesson 2-6)

16. Lindsey made $3\frac{3}{4}$ cups of chocolate milk. She poured $1\frac{1}{2}$ cups for her brother. How much did she have left? (Lesson 2-6)

17. Find the value of the expression $4^3 − 3^3$. (Lesson 2-8)

18. Write an expression for the volume of the cube. (Lesson 2-8)

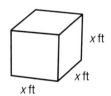

x ft
x ft
x ft

PART 3   Extended Response

Record your answers on a sheet of paper. Show your work.

THINK
SOLVE
EXPLAIN

19. Leo found the value of x in the equation $\frac{5x}{6} − 7 = 3$ to be 30. Is Leo correct or incorrect? Explain. (Lesson 2-7)

20. Masons are making large bricks. The container they are using is 9 inches by 9 inches by 9 inches. They have several boxes measuring 3 inches by 3 inches by 3 inches of cement that they will use to fill the large container. (Lesson 2-8)

    a. Describe how to determine the number of boxes of cement required to fill the container.

    b. Write and simplify an expression to solve the problem.

    c. How many boxes it will take?

CHAPTER 3

Algebra: Real Numbers and the Pythagorean Theorem

"How far can you see from a tall building?"

The Sears Tower in Chicago is 1,450 feet high. You can determine approximately how far you can see from the top of the Sears Tower by multiplying 1.23 by $\sqrt{1{,}450}$. The symbol $\sqrt{1{,}450}$ represents the square root of 1,450.

You will solve problems about how far a person can see from a given height in Lesson 3-3.

GETTING STARTED

Take this quiz to see whether you are ready to begin Chapter 3. Refer to the lesson number in parentheses if you need more review.

▶ Vocabulary Review

State whether each sentence is *true* or *false*. If *false*, replace the underlined word to make a true sentence.

1. The number 0.6 is a <u>rational</u> number.
   (Lesson 2-1)

2. In the number 3^2, the <u>base</u> is 2.
   (Lesson 2-8)

▶ Prerequisite Skills

Graph each point on a coordinate plane.
(Page 614)

3. $A(-1, 3)$ 4. $B(2, -4)$

5. $C(-2, -3)$ 6. $D(-4, 0)$

Evaluate each expression. (Lesson 1-2)

7. $2^2 + 4^2$ 8. $3^2 + 3^2$

9. $10^2 + 8^2$ 10. $7^2 + 5^2$

Solve each equation. Check your solution. (Lesson 1-8)

11. $x + 13 = 45$ 12. $56 + d = 71$

13. $101 = 39 + a$ 14. $62 = 45 + m$

Express each decimal as a fraction in simplest form. (Lesson 2-1)

15. $0.\overline{6}$ 16. 0.35

17. 0.375 18. 0.6

Between which two of the following numbers does each number lie?
1, 4, 9, 16, 25, 36, 49, 64, 81 (Lesson 2-2)

19. 38 20. 74

Real Numbers and the Pythagorean Theorem Make this Foldable to help you organize your notes. Begin with two sheets of $8\frac{1}{2}$" by 11" paper.

STEP 1 Fold and Cut One Sheet
Fold in half from top to bottom. Cut along fold from edges to margin.

STEP 2 Fold and Cut the Other Sheet
Fold in half from top to bottom. Cut along fold between margins.

STEP 3 Assemble
Insert first sheet through second sheet and align folds.

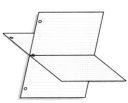

STEP 4 Label
Label each page with a lesson number and title.

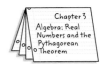

Chapter 3
Algebra: Real Numbers and the Pythagorean Theorem

Reading and Writing As you read and study the chapter, fill the journal with notes, diagrams, and examples for real numbers and the Pythagorean Theorem.

3-1 Square Roots

Sunshine State Standards
MA.A.1.3.1, MA.A.1.3.4-2

What You'll Learn

Find square roots of perfect squares.

NEW Vocabulary

perfect square
square root
radical sign
principal square
 root

REVIEW Vocabulary

exponent: tells the number of times the base is used as a factor **(Lesson 1-7)**

HANDS-ON Mini Lab

Materials
• color tiles

Work with a partner.

Look at the two square arrangements of tiles at the right. Continue this pattern of square arrays until you reach 5 tiles on each side.

1. Copy and complete the following table.

Tiles on a Side	1	2	3	4	5
Total Number of Tiles in the Square Arrangement	1	4			

2. Suppose a square arrangement has 36 tiles. How many tiles are on a side?

3. What is the relationship between the number of tiles on a side and the number of tiles in the arrangement?

Numbers such as 1, 4, 9, 16, and 25 are called **perfect squares** because they are squares of whole numbers. The opposite of squaring a number is finding a **square root**.

Key Concept — Square Root

Words	A square root of a number is one of its two equal factors.
Symbols	**Arithmetic** Since $3 \cdot 3 = 9$, a square root of 9 is 3. Since $(-3)(-3) = 9$, a square root of 9 is -3. **Algebra** If $x^2 = y$, then x is a square root of y.

The symbol $\sqrt{}$, called a **radical sign**, is used to indicate the positive square root. The symbol $-\sqrt{}$ is used to indicate the negative square root.

EXAMPLE — Find a Square Root

 Find $\sqrt{64}$.

$\sqrt{64}$ indicates the *positive* square root of 64.
Since $8^2 = 64$, $\sqrt{64} = 8$.

EXAMPLE — Find the Negative Square Root

2 Find $-\sqrt{121}$.

$-\sqrt{121}$ indicates the *negative* square root of 121.

Since $(-11)(-11) = 121$, $-\sqrt{121} = -11$.

Your Turn Find each square root.

a. $\sqrt{49}$ b. $-\sqrt{225}$ c. $-\sqrt{0.16}$

READING Math

Square Roots A positive square root is called the **principal square root**.

Some equations that involve squares can be solved by taking the square root of each side of the equation. Remember that every positive number has both a positive and a negative square root.

EXAMPLE — Use Square Roots to Solve an Equation

3 **ALGEBRA** Solve $t^2 = \dfrac{25}{36}$.

$t^2 = \dfrac{25}{36}$ Write the equation.

$\sqrt{t^2} = \sqrt{\dfrac{25}{36}}$ or $-\sqrt{\dfrac{25}{36}}$ Take the square root of each side.

$t = \dfrac{5}{6}$ or $-\dfrac{5}{6}$ Notice that $\dfrac{5}{6} \cdot \dfrac{5}{6} = \dfrac{25}{36}$ and $\left(-\dfrac{5}{6}\right)\left(-\dfrac{5}{6}\right) = \dfrac{25}{36}$.

The equation has two solutions, $\dfrac{5}{6}$ and $-\dfrac{5}{6}$.

Your Turn Solve each equation.

d. $y^2 = \dfrac{4}{25}$ e. $196 = a^2$ f. $m^2 = 0.09$

In real-life situations, a negative answer may not make sense.

REAL-LIFE MATH

HISTORY The Great Pyramid at Giza was built in 2600 B.C. It has a square base with an area of about 567,009 square feet.

Source: *World Book*

EXAMPLE — Use an Equation to Solve a Problem

4 **HISTORY** Use the information at the left to determine the length of each side of the base of the Great Pyramid of Giza.

Words	Area	is equal to	the square of the length of a side.
Variables	A	$=$	s^2
Equation	567,009	$=$	s^2

$567,009 = s^2$ Write the equation.

$\sqrt{567,009} = \sqrt{s^2}$ Take the square root of each side.

[2nd] $\sqrt{}$ 567009 [ENTER =]   Use a calculator.

753 or $-753 = s$

The length of a side of the base of the Great Pyramid of Giza is about 753 feet since distance cannot be negative.

Skill and Concept Check

Writing Math

Exercises 1 & 4

1. **Explain** the meaning of $\sqrt{16}$ in the cartoon below.

FoxTrot
by Bill Amend

2. **Write** the symbol for the negative square root of 25.

3. **OPEN ENDED** Write an equation that can be solved by taking the square root of a perfect square.

4. **FIND THE ERROR** Diana and Terrell are solving the equation $x^2 = 81$. Who is correct? Explain.

Diana
$x^2 = 81$
$x = 9$

Terrell
$x^2 = 81$
$x = 9$ or $x = -9$

GUIDED PRACTICE

Find each square root.

5. $\sqrt{25}$ 6. $-\sqrt{100}$ 7. $-\sqrt{\dfrac{16}{81}}$ 8. 0.64

ALGEBRA Solve each equation.

9. $p^2 = 36$ 10. $n^2 = 169$ 11. $900 = r^2$ 12. $t^2 = \dfrac{1}{9}$

13. **ALGEBRA** If $n^2 = 256$, find n.

Practice and Applications

Find each square root.

14. $\sqrt{16}$ 15. $\sqrt{81}$ 16. $-\sqrt{64}$

17. $-\sqrt{36}$ 18. $-\sqrt{196}$ 19. $-\sqrt{144}$

20. $\sqrt{256}$ 21. $\sqrt{324}$ 22. $-\sqrt{\dfrac{16}{25}}$

23. $-\sqrt{\dfrac{9}{49}}$ 24. $\sqrt{0.25}$ 25. $\sqrt{1.44}$

26. Find the positive square root of 169.

27. What is the negative square root of 400?

HOMEWORK HELP

For Exercises	See Examples
14–27	1, 2
28–41	3
42–45	4

Extra Practice
See pages 622, 650.

ALGEBRA Solve each equation.

28. $v^2 = 81$ **29.** $b^2 = 100$ **30.** $y^2 = 225$ **31.** $s^2 = 144$

32. $1{,}600 = a^2$ **33.** $2{,}500 = d^2$ **34.** $w^2 = 625$ **35.** $m^2 = 961$

36. $\dfrac{25}{81} = p^2$ **37.** $\dfrac{9}{64} = c^2$ **38.** $r^2 = 2.25$ **39.** $d^2 = 1.21$

40. ALGEBRA Find a number that when squared equals 1.0404.

41. ALGEBRA Find a number that when squared equals 4.0401.

42. MARCHING BAND A marching band wants to make a square formation. If there are 81 members in the band, how many should be in each row?

GEOMETRY The formula for the perimeter of a square is $P = 4s$, where s is the length of a side. Find the perimeter of each square.

43.

Area = 121 square inches

44.

Area = 25 square feet

45.
Area = 36 square meters

46. MULTI STEP Describe three different-sized squares that you could make at the same time out of 130 square tiles. How many tiles are left?

47. CRITICAL THINKING Find each value.

   a. $\left(\sqrt{36}\right)^2$ **b.** $\left(\sqrt{81}\right)^2$ **c.** $\left(\sqrt{21}\right)^2$ **d.** $\left(\sqrt{x}\right)^2$

48. CRITICAL THINKING *True* or *False*? $\sqrt{-25} = -5$. Explain.

Standardized Test Practice and Mixed Review

FCAT Practice

49. MULTIPLE CHOICE What is the solution of $a^2 = 49$?

   Ⓐ -7 Ⓑ 7 Ⓒ 7 or -7 Ⓓ 7 or 0 or -7

50. SHORT RESPONSE The area of each square is 4 square units. Find the perimeter of the figure.

51. SPACE The Alpha Centauri stars are about 2.5×10^{13} miles from Earth. Write this distance in standard form. (Lesson 2-9)

Write each expression using exponents. (Lesson 2-8)

52. $6 \cdot 6 \cdot 6$ **53.** $2 \cdot 3 \cdot 3 \cdot 2 \cdot 2 \cdot 2$ **54.** $a \cdot a \cdot a \cdot b$ **55.** $s \cdot t \cdot t \cdot s \cdot s \cdot t \cdot s$

56. What is the absolute value of -18? (Lesson 1-3)

GETTING READY FOR THE NEXT LESSON

PREREQUISITE SKILL Between which two perfect squares does each number lie? (Lesson 2-2)

57. 57   **58.** 68   **59.** 33   **60.** 40

Estimating Square Roots

Sunshine State Standards
MA.A.1.3.2-1, MA.A.1.3.2-2, MA.A.1.3.4-2, MA.A.4.3.1-1, MA.A.4.3.1-2

What You'll Learn
Estimate square roots.

***MATH* Symbols**
≈ about equal to

HANDS-ON Mini Lab

Materials
• grid paper

Work with a partner.

On grid paper, draw the largest possible square using no more than 40 small squares.

On grid paper, draw the smallest possible square using at least 40 small squares.

1. How many squares are on each side of the largest possible square using no more than 40 small squares?

2. How many squares are on each side of the smallest possible square using at least 40 small squares?

3. The value of $\sqrt{40}$ is between two consecutive whole numbers. What are the numbers?

Use grid paper to determine between which two consecutive whole numbers each value is located.

4. $\sqrt{23}$ 5. $\sqrt{52}$ 6. $\sqrt{27}$ 7. $\sqrt{18}$

Since 40 is not a perfect square, $\sqrt{40}$ is not a whole number.

The number line shows that $\sqrt{40}$ is between 6 and 7. Since 40 is closer to 36 than 49, the best whole number estimate for $\sqrt{40}$ is 6.

EXAMPLE Estimate Square Roots

Estimate to the nearest whole number.

1 $\sqrt{83}$

• The first perfect square less than 83 is 81.

• The first perfect square greater than 83 is 100.

$$81 < 83 < 100 \qquad \text{Write an inequality.}$$
$$9^2 < 83 < 10^2 \qquad 81 = 9^2 \text{ and } 100 = 10^2$$
$$\sqrt{9^2} < \sqrt{83} < \sqrt{10^2} \qquad \text{Take the square root of each number.}$$
$$9 < \sqrt{83} < 10 \qquad \text{Simplify.}$$

So, $\sqrt{83}$ is between 9 and 10. Since 83 is closer to 81 than 100, the best whole number estimate for $\sqrt{83}$ is 9.

Your Turn Estimate to the nearest whole number.

a. $\sqrt{35}$ b. $\sqrt{170}$ c. $\sqrt{14.8}$

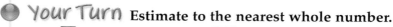

EXAMPLE **Estimate Square Roots**

2 **ART** Many artists believe that the *golden rectangle* is the most pleasing shape to the eye. The Parthenon is one example of a golden rectangle. In a golden rectangle, the length of the longer side divided by the length of the shorter side is equal to $\frac{1 + \sqrt{5}}{2}$. Estimate this value.

2 units

$(1 + \sqrt{5})$ units

First estimate the value of $\sqrt{5}$.

$4 < 5 < 9$ 4 and 9 are perfect squares.

$2^2 < 5 < 3^2$ $4 = 2^2$ and $9 = 3^2$

$2 < \sqrt{5} < 3$ Take the square root of each number.

Since 5 is closer to 4 than 9, the best whole number estimate for $\sqrt{5}$ is 2. Use this to evaluate the expression.

$$\frac{1 + \sqrt{5}}{2} \approx \frac{1 + 2}{2} \text{ or } 1.5$$

In a "golden rectangle," the length of the longer side divided by the length of the shorter side is about 1.5.

STUDY TIP

Technology You can use a calculator to find a more accurate value of $\frac{1 + \sqrt{5}}{2}$.

(1 + 2nd

√ 5)) ÷

2 ENTER 1.618033989

Skill and Concept Check

Writing Math

Exercises 3 & 4

1. **Graph** $\sqrt{78}$ on a number line.

2. **OPEN ENDED** Give two numbers that have square roots between 7 and 8. One number should have a square root closer to 7, and the other number should have a square root closer to 8.

3. **FIND THE ERROR** Julia and Chun are estimating $\sqrt{50}$. Who is correct? Explain.

Julia

$\sqrt{50} \approx 7$

Chun

$\sqrt{50} \approx 25$

4. **NUMBER SENSE** Without a calculator, determine which is greater, $\sqrt{94}$ or 10. Explain your reasoning.

GUIDED PRACTICE

Estimate to the nearest whole number.

5. $\sqrt{28}$ 6. $\sqrt{60}$ 7. $\sqrt{135}$ 8. $\sqrt{13.5}$

9. **ALGEBRA** Estimate the solution of $t^2 = 78$ to the nearest whole number.

Practice and Applications

HOMEWORK HELP

For Exercises	See Examples
10–31	1
34–35	2

Extra Practice
See pages 622, 650.

Estimate to the nearest whole number.

10. $\sqrt{11}$ 11. $\sqrt{15}$ 12. $\sqrt{44}$ 13. $\sqrt{23}$

14. $\sqrt{113}$ 15. $\sqrt{105}$ 16. $\sqrt{82}$ 17. $\sqrt{50}$

18. $\sqrt{15.6}$ 19. $\sqrt{23.5}$ 20. $\sqrt{85.1}$ 21. $\sqrt{38.4}$

22. $\sqrt{200}$ 23. $\sqrt{170}$ 24. $\sqrt{150}$ 25. $\sqrt{130}$

26. $\sqrt{630}$ 27. $\sqrt{925}$ 28. $\sqrt{1300}$ 29. $\sqrt{780}$

30. **ALGEBRA** Estimate the solution of $y^2 = 55$ to the nearest integer.

31. **ALGEBRA** Estimate the solution of $d^2 = 95$ to the nearest integer.

32. Order 7, 9, $\sqrt{50}$, and $\sqrt{85}$ from least to greatest.

33. Order $\sqrt{91}$, 7, 5, $\sqrt{38}$ from least to greatest.

34. **HISTORY** During the first century, the Egyptian mathematician Heron created the formula $A = \sqrt{s(s-a)(s-b)(s-c)}$ to find the area A of a triangle. In this formula, a, b, and c are the measures of the sides, and s is one-half of the perimeter. Use this formula to estimate the area of the triangle at the right.

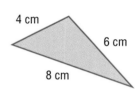

4 cm
6 cm
8 cm

35. **SCIENCE** The formula $t = \dfrac{\sqrt{h}}{4}$ represents the time t in seconds that it takes an object to fall from a height of h feet. If a ball is dropped from a height of 200 feet, estimate how long will it take to reach the ground.

36. **CRITICAL THINKING** If $x^3 = y$, then x is the cube root of y. Explain how to estimate the cube root of 30. What is the cube root of 30 to the nearest whole number?

Standardized Test Practice and Mixed Review

FCAT Practice

37. **MULTIPLE CHOICE** Which is the best estimate of the value of $\sqrt{54}$?

Ⓐ 6   Ⓑ 7   Ⓒ 8   Ⓓ 27

38. **MULTIPLE CHOICE** If $x^2 = 38$, then a value of x is approximately

Ⓕ 5.   Ⓖ 6.   Ⓗ 7.   Ⓘ 24.

39. **ALGEBRA** Find a number that, when squared, equals 8,100. (Lesson 3-1)

40. **GEOGRAPHY** The Great Lakes cover about 94,000 square miles. Write this number in scientific notation. (Lesson 2-9)

GETTING READY FOR THE NEXT LESSON

PREREQUISITE SKILL Express each decimal as a fraction in simplest form. (Lesson 2-1)

41. 0.15   42. 0.8   43. $0.\overline{3}$ 44. $0.\overline{4}$

msmath3.net/self_check_quiz/fcat

3-3a Problem-Solving Strategy
A Preview of Lesson 3-3

Sunshine State Standards
MA.A.1.3.3-2, MA.A.1.3.3-3, MA.E.1.3.1-1, MA.E.1.3.1-2

What You'll Learn
Solve problems using a Venn diagram.

Use a Venn Diagram

> Of the 12 students who ate lunch with me today, 9 are involved in music activities and 6 play sports. Four are involved in both music and sports.

> How could we organize that information?

Explore	We know how many students are involved in each activity and how many are involved in both activities. We want to organize the information.
Plan	Let's use a **Venn diagram** to organize the information.
Solve	Draw two overlapping circles to represent the two different activities. Since 4 students are involved in both activities, place a 4 in the section that is part of both circles. Use subtraction to determine the number for each other section. 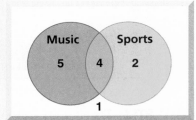 only music: $9 - 4 = 5$ only sports: $6 - 4 = 2$ neither music nor sports: $12 - 5 - 2 - 4 = 1$
Examine	Check each circle to see if the appropriate number of students is represented.

Analyze the Strategy

1. **Tell** what each section of the Venn diagram above represents and the number of students that belong to that category.

2. **Use** the Venn diagram above to determine the number of students who are in either music or sports but not both.

3. **Write** a situation that can be represented by the Venn diagram at the right.

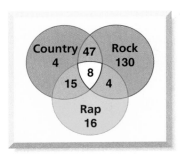

Lesson 3-3a Problem-Solving Strategy: Use a Venn Diagram **123**

Solve. Use a Venn diagram.

4. **MARKETING** A survey showed that 83 customers bought wheat cereal, 83 bought rice cereal, and 20 bought corn cereal. Of those who bought exactly two boxes of cereal, 6 bought corn and wheat, 10 bought rice and corn, and 12 bought rice and wheat. Four customers bought all three. How many customers bought only rice cereal?

5. **FOOD** Napoli's Pizza conducted a survey of 75 customers. The results showed that 35 customers liked mushroom pizza, 41 liked pepperoni, and 11 liked both mushroom and pepperoni pizza. How many liked neither mushroom nor pepperoni pizza?

Mixed Problem Solving

Solve. Use any strategy.

6. **SCIENCE** Emilio created a graph of data he collected for a science project. If the pattern continues, about how far will the marble roll if the end of the tube is raised to an elevation of $3\frac{1}{2}$ feet?

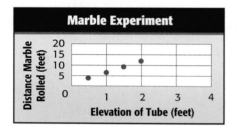

Marble Experiment

7. **MULTI STEP** Three after-school jobs are posted on the job board. The first job pays $5.15 per hour for 15 hours of work each week. The second job pays $10.95 per day for 2 hours of work, 5 days a week. The third job pays $82.50 for 15 hours of work each week. If you want to apply for the best-paying job, which job should you choose? Explain your reasoning.

8. **FACTOR TREE** Copy and complete the factor tree.

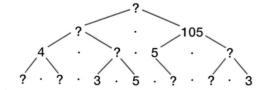

9. **NUMBER THEORY** A subset is a part of a set. The symbol $\subset$ means "is a subset of." Consider the following two statements.

integers $\subset$ rational numbers
rational numbers $\subset$ integers

Are both statements true? Draw a Venn diagram to justify your answer.

HEALTH For Exercises 10 and 11, use the following information.
Dr. Bagenstose is an allergist. Her patients had the following symptoms last week.

Symptom(s)	Number of Patients
runny nose	22
watery eyes	20
sneezing	28
runny nose and watery eyes	8
runny nose and sneezing	15
watery eyes and sneezing	12
runny nose, watery eyes, and sneezing	5

10. Draw a Venn diagram of the data.

11. How many patients had only watery eyes?

12. **STANDARDIZED TEST PRACTICE** FCAT Practice
Which value of x makes $7x - 10 = 9x$ true?
ⓐ -5 ⓑ -4 ⓒ 4 ⓓ 5

> You will use a Venn diagram in the next lesson.

The Real Number System

Sunshine State Standards
MA.A.1.3.2-1, MA.A.1.3.2-2, MA.A.1.3.3-1, MA.A.1.3.3-2, MA.A.1.3.3-3, MA.A.4.3.1-1, MA.A.4.3.1-2

What You'll Learn

Identify and classify numbers in the real number system.

→ NEW Vocabulary

irrational number
real number

↻ REVIEW Vocabulary

rational number: any number that can be expressed in the form $\frac{a}{b}$, where a and b are integers and $b \neq 0$ **(Lesson 2-1)**

WHEN am I ever going to use this?

SPORTS Most sports have rules for the size of the field or court where the sport is played. A diagram of a volleyball court is shown.

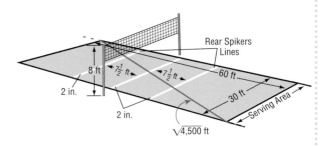

1. The length of the court is 60 feet. Is this number a whole number? Is it a rational number? Explain.

2. The distance from the net to the rear spikers line is $7\frac{1}{2}$ feet. Is this number a whole number? Is it a rational number? Explain.

3. The diagonal across the court is $\sqrt{4,500}$ feet. Can this square root be written as a whole number? a rational number?

Use a calculator to find $\sqrt{4,500}$. $\sqrt{4,500} \approx 67.08203932...$

Although the decimal value of $\sqrt{4,500}$ continues on and on, it does <u>not</u> repeat. Since the decimal does not terminate or repeat, $\sqrt{4,500}$ is *not* a rational number. Numbers that are not rational are called **irrational numbers**. The square root of any number that is not a perfect square is irrational.

Key Concept   Irrational Number

Words   An irrational number is a number that cannot be expressed as $\frac{a}{b}$, where a and b are integers and $b \neq 0$.

Symbols $\sqrt{2} \approx 1.414213562...$ $\sqrt{3} \approx 1.732050808...$

The set of rational numbers and the set of irrational numbers together make up the set of **real numbers**. Study the diagrams below.

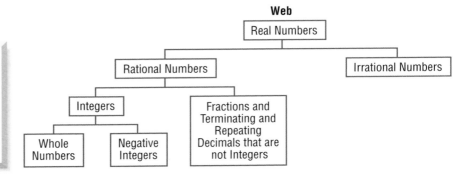

STUDY TIP

Classifying Numbers Always simplify numbers before classifying them.

EXAMPLES Classify Numbers

Name all sets of numbers to which each real number belongs.

① 0.252525. . .   The decimal ends in a repeating pattern. It is a rational number because it is equivalent to $\frac{25}{99}$.

② $\sqrt{36}$ Since $\sqrt{36} = 6$, it is a whole number, an integer, and a rational number.

③ $-\sqrt{7}$ $-\sqrt{7} \approx -2.645751311$. . . Since the decimal does not terminate or repeat, it is an irrational number.

Real numbers follow the number properties that are true for whole numbers, integers, and rational numbers.

Concept Summary   **Real Number Properties**

Property	Arithmetic	Algebra
Commutative	$3.2 + 2.5 = 2.5 + 3.2$ $5.1 \cdot 2.8 = 2.8 \cdot 5.1$	$a + b = b + a$ $a \cdot b = b \cdot a$
Associative	$(2 + 1) + 5 = 2 + (1 + 5)$ $(3 \cdot 4) \cdot 6 = 3 \cdot (4 \cdot 6)$	$(a + b) + c = a + (b + c)$ $(a \cdot b) \cdot c = a \cdot (b \cdot c)$
Distributive	$2(3 + 5) = 2 \cdot 3 + 2 \cdot 5$	$a(b + c) = a \cdot b + a \cdot c$
Identity	$\sqrt{8} + 0 = \sqrt{8}$ $\sqrt{7} \cdot 1 = \sqrt{7}$	$a + 0 = a$ $a \cdot 1 = a$
Additive Inverse	$4 + (-4) = 0$	$a + (-a) = 0$
Multiplicative Inverse	$\frac{2}{3} \cdot \frac{3}{2} = 1$	$\frac{a}{b} \cdot \frac{b}{a} = 1$, where $a, b \neq 0$

The graph of all real numbers is the entire number line without any "holes."

EXAMPLE Graph Real Numbers

④ **Estimate $\sqrt{6}$ and $-\sqrt{3}$ to the nearest tenth. Then graph $\sqrt{6}$ and $-\sqrt{3}$ on a number line.**

Use a calculator to determine the approximate decimal values.

$\sqrt{6} \approx 2.449489743$. . .

$-\sqrt{3} \approx -1.7320508080$. . .

$\sqrt{6} \approx 2.4$ and $-\sqrt{3} \approx -1.7$. Locate these points on the number line.

Your Turn Estimate each square root to the nearest tenth. Then graph the square root on a number line.

a. $\sqrt{5}$ **b.** $-\sqrt{7}$ **c.** $\sqrt{22}$

To compare real numbers, you can use a number line.

EXAMPLES Compare Real Numbers

Replace each ● with <, >, or = to make a true sentence.

5 $\sqrt{7}$ ● $2\frac{2}{3}$

Write each number as a decimal.

$\sqrt{7} \approx 2.645751311\ldots$

$2\frac{2}{3} = 2.666666666\ldots$

Since $2.645751311\ldots$ is less than $2.66666666\ldots$, $\sqrt{7} < 2\frac{2}{3}$.

6 $1.\overline{5}$ ● $\sqrt{2.25}$

Write $\sqrt{2.25}$ as a decimal.

$\sqrt{2.25} = 1.5$

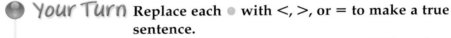

Since $1.\overline{5}$ is greater than 1.5, $1.\overline{5} > \sqrt{2.25}$.

Your Turn Replace each ● with <, >, or = to make a true sentence.

d. $\sqrt{11}$ ● $3\frac{1}{3}$

e. $\sqrt{17}$ ● 4.03

f. $\sqrt{6.25}$ ● $2\frac{1}{2}$

EXAMPLE Use Real Numbers

7 **LIGHTHOUSES** On a clear day, the number of miles a person can see to the horizon is about 1.23 times the square root of his or her distance from the ground, in feet. Suppose Domingo is at the top of the lighthouse at Cape Hatteras and Jewel is at the top of the lighthouse at Cape Lookout. How much farther can Domingo see than Jewel?

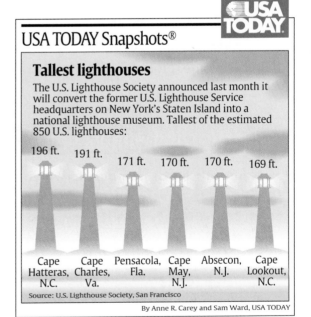

USA TODAY Snapshots®

Tallest lighthouses
The U.S. Lighthouse Society announced last month it will convert the former U.S. Lighthouse Service headquarters on New York's Staten Island into a national lighthouse museum. Tallest of the estimated 850 U.S. lighthouses:

196 ft.   191 ft.   171 ft.   170 ft.   170 ft.   169 ft.

Cape Hatteras, N.C.   Cape Charles, Va.   Pensacola, Fla.   Cape May, N.J.   Absecon, N.J.   Cape Lookout, N.C.

Source: U.S. Lighthouse Society, San Francisco

By Anne R. Carey and Sam Ward, USA TODAY

Use a calculator to approximate the distance each person can see.

Domingo: $1.23\sqrt{196} = 17.22$ Jewel: $1.23\sqrt{169} = 15.99$

Domingo can see about $17.22 - 15.99$ or 1.23 miles farther than Jewel.

Skill and Concept Check

1. **Give a counterexample** for the statement *all square roots are irrational numbers.*

2. **OPEN ENDED** Write an irrational number which would be graphed between 7 and 8 on the number line.

3. **Which One Doesn't Belong?** Identify the number that is not the same type as the other three. Explain your reasoning.

| $\sqrt{7}$ | $\sqrt{11}$ | $\sqrt{25}$ | $\sqrt{35}$ |

GUIDED PRACTICE

Name all sets of numbers to which each real number belongs.

4. $0.050505\ldots$ 5. $-\sqrt{100}$ 6. $\sqrt{17}$ 7. $-3\frac{1}{4}$

Estimate each square root to the nearest tenth. Then graph the square root on a number line.

8. $\sqrt{2}$ 9. $-\sqrt{18}$ 10. $-\sqrt{30}$ 11. $\sqrt{95}$

Replace each ● with <, >, or = to make a true sentence.

12. $\sqrt{15}$ ● 3.5 13. $\sqrt{2.25}$ ● $1\frac{1}{2}$ 14. $2.\overline{21}$ ● $\sqrt{5.2}$

15. Order $5.\overline{5}$, $\sqrt{30}$, $5\frac{1}{2}$, and 5.56 from least to greatest.

16. **GEOMETRY** The area of a triangle with all three sides the same length is $\frac{s^2\sqrt{3}}{4}$, where s is the length of a side. Find the area of the triangle.

6 in.   6 in.

6 in.

Practice and Applications

Name all sets of numbers to which each real number belongs.

17. 14 18. $\frac{2}{3}$ 19. $-\sqrt{16}$ 20. $-\sqrt{20}$

21. 4.83 22. $7.\overline{2}$ 23. $-\sqrt{90}$ 24. $\frac{12}{4}$

25. $-0.\overline{182}$ 26. -13 27. $5\frac{3}{8}$ 28. -108.6

29. Are integers *always*, *sometimes*, or *never* rational numbers? Explain.

30. Are rational numbers *always*, *sometimes*, or *never* integers? Explain.

Estimate each square root to the nearest tenth. Then graph the square root on a number line.

31. $\sqrt{6}$ 32. $\sqrt{8}$ 33. $-\sqrt{22}$ 34. $-\sqrt{27}$

35. $\sqrt{50}$ 36. $-\sqrt{48}$ 37. $-\sqrt{105}$ 38. $\sqrt{150}$

HOMEWORK HELP

For Exericises	See Examples
17–30	1–3
31–38	4
39–48	5, 6
49–50	7

Extra Practice
See pages 622, 650.

Replace each ● with <, >, or = to make a true sentence.

39. $\sqrt{10}$ ● 3.2

40. $\sqrt{12}$ ● 3.5

41. $6\frac{1}{3}$ ● $\sqrt{40}$

42. $2\frac{2}{5}$ ● $\sqrt{5.76}$

43. $5\frac{1}{6}$ ● $5.1\overline{6}$

44. $\sqrt{6.2}$ ● $2.\overline{4}$

45. Order $\sqrt{5}$, $\sqrt{3}$, 2.25, and $2.\overline{2}$ from least to greatest.

46. Order 3.01, $3.\overline{1}$, $3.\overline{01}$, and $\sqrt{9}$ from least to greatest.

47. Order -4.1, $\sqrt{17}$, $-4.\overline{1}$, and 4.01 from greatest to least.

48. Order $-\sqrt{5}$, $\sqrt{6}$, -2.5, and 2.5 from greatest to least.

49. **LAW ENFORCEMENT** Traffic police can use the formula $s = 5.5\sqrt{0.75d}$ to estimate the speed of a vehicle before braking. In this formula, s is the speed of the vehicle in miles per hour, and d is the length of the skid marks in feet. How fast was the vehicle going for the skid marks at the right?

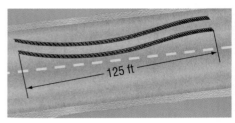

125 ft

50. **WEATHER** Meteorologists use the formula $t^2 = \frac{d^3}{216}$ to estimate the amount of time that a thunderstorm will last. In this formula, t is the time in hours, and d is the distance across the storm in miles. How long will a thunderstorm that is 8.4 miles wide last?

51. **CRITICAL THINKING** Tell whether the following statement is *always*, *sometimes*, or *never* true.

The product of a rational number and an irrational number is an irrational number.

Standardized Test Practice and Mixed Review

FCAT Practice

52. **MULTIPLE CHOICE** To which set of numbers does $-\sqrt{49}$ *not* belong?

Ⓐ whole   Ⓑ rational   Ⓒ integers   Ⓓ real

53. **SHORT RESPONSE** The area of a square playground is 361 square feet. What is the perimeter of the playground?

54. Order 7, $\sqrt{53}$, $\sqrt{32}$, and 6 from least to greatest. (Lesson 3-2)

Solve each equation. (Lesson 3-1)

55. $t^2 = 25$

56. $y^2 = \frac{1}{49}$

57. $0.64 = a^2$

58. **ARCHAEOLOGY** Stone tools found in Ethiopia are estimated to be 2.5 million years old. That is about 700,000 years older than similar tools found in Tanzania. Write and solve an addition equation to find the age of the tools found in Tanzania. (Lesson 1-8)

GETTING READY FOR THE NEXT LESSON

PREREQUISITE SKILL Evaluate each expression. (Lesson 1-2)

59. $3^2 + 5^2$

60. $6^2 + 4^2$

61. $9^2 + 11^2$

62. $4^2 + 7^2$

Vocabulary and Concepts

1. **Graph** $\sqrt{50}$ on a number line. (Lesson 3-2)

2. **Write** an irrational number that would be graphed between 11 and 12 on a number line. (Lesson 3-3)

3. **OPEN ENDED** Give an example of a number that is an integer but not a whole number. (Lesson 3-3)

Skills and Applications

Find each square root. (Lesson 3-1)

4. $\sqrt{1}$

5. $-\sqrt{81}$

6. $\sqrt{36}$

7. $-\sqrt{121}$

8. $-\sqrt{\dfrac{1}{25}}$

9. $\sqrt{0.09}$

10. **GEOMETRY** What is the length of a side of the square? (Lesson 3-1)

11. **ALGEBRA** Estimate the solution of $x^2 = 50$ to the nearest integer. (Lesson 3-2)

> Area = 225 square meters

Estimate to the nearest whole number. (Lesson 3-2)

12. $\sqrt{90}$

13. $\sqrt{28}$

14. $\sqrt{226}$

15. $\sqrt{17}$

16. $\sqrt{21}$

17. $\sqrt{75}$

Name all sets of numbers to which each real number belongs. (Lesson 3-3)

18. $\dfrac{2}{3}$

19. $\sqrt{25}$

20. $-\sqrt{15}$

21. $\sqrt{3}$

22. 10

23. $-\sqrt{4}$

Standardized Test Practice

FCAT Practice

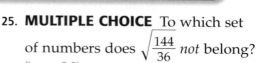

24. **MULTIPLE CHOICE** The area of a square checkerboard is 529 square centimeters. How long is each side of the checkerboard? (Lesson 3-1)

    Ⓐ 21 cm   Ⓑ 22 cm

    Ⓒ 23 cm   Ⓓ 24 cm

25. **MULTIPLE CHOICE** To which set of numbers does $\sqrt{\dfrac{144}{36}}$ *not* belong? (Lesson 3-3)

    Ⓕ integers   Ⓖ rationals

    Ⓗ wholes   Ⓘ irrationals

The Game Zone

A Place To Practice Your Math Skills

Estimate and Eliminate

● **GET READY!**

Players: four

Materials: 40 index cards, 4 markers

● **GET SET!**

- Each player is given 10 index cards.

- Player 1 writes one of each of the whole numbers 1 to 10 on his or her cards. Player 2 writes the square of one of each of the whole numbers 1 to 10. Player 3 writes a different whole number between 11 and 50, that is not a perfect square. Player 4 writes a different whole number between 51 and 99, that is not a perfect square.

● **GO!**

- Mix all 40 cards together. The dealer deals all of the cards.

- In turn, moving clockwise, each player lays down any pair(s) of a perfect square and its square root in his or her hand. The two cards should be laid down as shown at the right. If a player has no perfect square and square root pair, he or she skips a turn.

- After the first round, any player, during his or her turn may:

  (1) lay down a perfect square and square root pair, or

  (2) cover a card that is already on the table. The new card should form a square and *estimated* square root pair with the card next to it. A player makes as many plays as possible during his or her turn.

- After each round, each player passes one card left.

- **Who Wins?** The first person without any cards is the winner.

The Pythagorean Theorem

Sunshine State Standards MA.A.1.3.4-2, MA.B.3.3.1-1, MA.C.1.3.1-4, MA.C.3.3.1-1, MA.C.3.3.1-2, MA.D.1.3.1-2, MA.D.1.3.2-2, MA.D.2.3.1-1, MA.D.2.3.1-6

What You'll Learn
Use the Pythagorean Theorem.

NEW Vocabulary

right triangle
legs
hypotenuse
Pythagorean
　Theorem
converse

LINK To Reading

Everyday Meaning of Leg: limb used to support the body

HANDS-ON Mini Lab

Materials
• grid paper

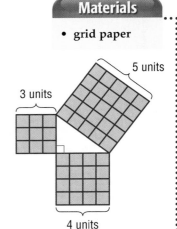

Work with a partner.

Three squares with sides 3, 4, and 5 units are used to form the right triangle shown.

1. Find the area of each square.

2. How are the squares of the sides related to the areas of the squares?

3. Find the sum of the areas of the two smaller squares. How does the sum compare to the area of the larger square?

4. Use grid paper to cut out three squares with sides 5, 12, and 13 units. Form a right triangle with these squares. Compare the sum of the areas of the two smaller squares with the area of the larger square.

A **right triangle** is a triangle with one right angle. A right angle is an angle with a measure of 90°.

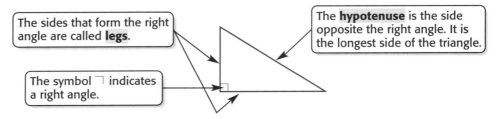

The sides that form the right angle are called **legs**.

The **hypotenuse** is the side opposite the right angle. It is the longest side of the triangle.

The symbol ⌐ indicates a right angle.

The **Pythagorean Theorem** describes the relationship between the lengths of the legs and the hypotenuse for *any* right triangle.

Key Concept

Pythagorean Theorem

Words　　In a right triangle, the square of the length of the hypotenuse is equal to the sum of the squares of the lengths of the legs.

Symbols　　**Arithmetic**　　　**Algebra**　　　**Model**

$5^2 = 3^2 + 4^2$　　$c^2 = a^2 + b^2$

$25 = 9 + 16$

$25 = 25$

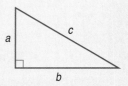

You can use the Pythagorean Theorem to find the length of a side of a right triangle.

EXAMPLE Find the Length of the Hypotenuse

1 KITES Find the length of the kite string.

The kite string forms the hypotenuse of a right triangle. The vertical and horizontal distances form the legs.

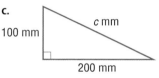

$c^2 = a^2 + b^2$ Pythagorean Theorem

$c^2 = 30^2 + 40^2$ Replace *a* with 30 and *b* with 40.

$c^2 = 900 + 1{,}600$ Evaluate 30^2 and 40^2.

$c^2 = 2{,}500$ Add 900 and 1,600.

$\sqrt{c^2} = \sqrt{2{,}500}$ Take the square root of each side.

$c = 50$ or -50 Simplify.

The equation has two solutions, 50 and -50. However, the length of the kite string must be positive. So, the kite string is 50 feet long.

Your Turn Find the length of each hypotenuse. Round to the nearest tenth if necessary.

a.

b. 16 m, c m, 12 m

c. c mm, 100 mm, 200 mm

EXAMPLE Find the Length of a Leg

2 The hypotenuse of a right triangle is 20 centimeters long and one of its legs is 17 centimeters. Find the length of the other leg.

$c^2 = a^2 + b^2$ Pythagorean Theorem

$20^2 = a^2 + 17^2$ Replace *c* with 20 and *b* with 17.

$400 = a^2 + 289$ Evaluate 20^2 and 17^2.

$400 - 289 = a^2 + 289 - 289$ Subtract 289 from each side.

$111 = a^2$ Simplify.

$\sqrt{111} = \sqrt{a^2}$ Take the square root of each side.

$10.5 \approx a$ Use a calculator.

The length of the other leg is about 10.5 centimeters.

Your Turn Write an equation you could use to find the length of the missing side of each right triangle. Then find the missing length. Round to the nearest tenth if necessary.

d. *b*, 9 ft; *c*, 12 ft   e. *a*, 3 m; *c*, 8 m   f. *a*, 15 in.; *b*, 18 in.

EXAMPLE Use the Pythagorean Theorem

3 **MULTIPLE-CHOICE TEST ITEM** For safety reasons, the base of a 24-foot ladder should be at least 8 feet from the wall. How high can a 24-foot ladder safely reach?

A about 16 feet   B about 22.6 feet

C about 25.3 feet   D about 512 feet

Read the Test Item You know the length of the ladder and the distance from the base of the ladder to the side of the house. Make a drawing of the situation including the right triangle.

Solve the Test Item Use the Pythagorean Theorem.

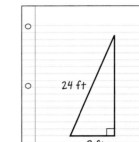

$c^2 = a^2 + b^2$	Pythagorean Theorem
$24^2 = a^2 + 8^2$	Replace c with 24 and b with 8.
$576 = a^2 + 64$	Evaluate 24^2 and 8^2.
$576 - 64 = a^2 + 64 - 64$	Subtract 64 from each side.
$512 = a^2$	Simplify.
$\sqrt{512} = \sqrt{a^2}$	Take the square root of each side.
$22.6 \approx a$	Use a calculator.

The ladder can safely reach a height of 22.6 feet. The answer is B.

If you reverse the parts of the Pythagorean Theorem, you have formed its **converse**. The converse of the Pythagorean Theorem is also true.

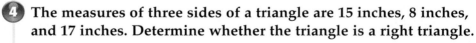

Key Concept   **Converse of the Pythagorean Theorem**

If the sides of a triangle have lengths a, b, and c units such that $c^2 = a^2 + b^2$, then the triangle is a right triangle.

EXAMPLE Identify a Right Triangle

4 The measures of three sides of a triangle are 15 inches, 8 inches, and 17 inches. Determine whether the triangle is a right triangle.

$c^2 = a^2 + b^2$	Pythagorean Theorem
$17^2 \stackrel{?}{=} 15^2 + 8^2$	$c = 17$, $a = 15$, $b = 8$
$289 \stackrel{?}{=} 225 + 64$	Evaluate 17^2, 15^2, and 8^2.
$289 = 289$ ✔	Simplify.

The triangle is a right triangle.

STUDY TIP

Assigning Variables
Remember that the longest side of a right triangle is the hypotenuse. Therefore, c represents the length of the longest side.

Your Turn Determine whether each triangle with sides of given lengths is a right triangle.

g. 18 mi, 24 mi, 30 mi   **h.** 4 ft, 7 ft, 5 ft

Writing Math
Exercise 3

1. **Draw** a right triangle and label all the parts.

2. **OPEN ENDED** State three measures that could be the side measures of a right triangle.

3. **FIND THE ERROR** Catalina and Morgan are writing an equation to find the length of the third side of the triangle. Who is correct? Explain.

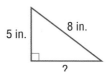

5 in.   8 in.

?

Catalina
$c^2 = 5^2 + 8^2$

Morgan
$8^2 = a^2 + 5^2$

GUIDED PRACTICE

Write an equation you could use to find the length of the missing side of each right triangle. Then find the missing length. Round to the nearest tenth if necessary.

4.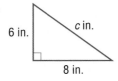
6 in. c in.
8 in.

5.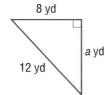
8 yd
a yd
12 yd

6.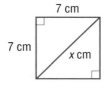
7 cm
7 cm x cm

7. a, 5 ft; c, 6 ft

8. a, 9 m; b, 7 m

9. b, 4 yd; c, 10 yd

Determine whether each triangle with sides of given lengths is a right triangle.

10. 5 in., 10 in., 12 in.

11. 9 m, 40 m, 41 m

Practice and Applications

Write an equation you could use to find the length of the missing side of each right triangle. Then find the missing length. Round to the nearest tenth if necessary.

HOMEWORK HELP

For Exercises	See Examples
12–25, 32	1–3
26–31, 34	4

Extra Practice
See pages 623, 650.

12.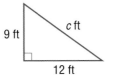
9 ft c ft
12 ft

13.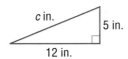
c in.   5 in.
12 in.

14. 10 cm
15 cm a cm

15.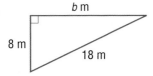
b m
8 m   18 m

16.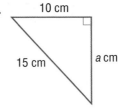
30 cm
x cm   18 cm

17.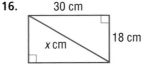
x in.   6 in.
14 in.

18. b, 99 mm; c, 101 mm

19. a, 48 yd; b, 55 yd

20. a, 17 ft; c, 20 ft

21. a, 23 in.; b, 18 in.

22. b, 4.5 m; c, 9.4 m

23. b, 5.1 m; c, 12.3 m

24. The hypotenuse of a right triangle is 12 inches, and one of its legs is 7 inches. Find the length of the other leg.

25. If one leg of a right triangle is 8 feet and its hypotenuse is 14 feet, how long is the other leg?

Determine whether each triangle with sides of given lengths is a right triangle.

26. 28 yd, 195 yd, 197 yd   **27.** 30 cm, 122 cm, 125 cm   **28.** 24 m, 143 m, 145 m

29. 135 in., 140 in., 175 in.   **30.** 56 ft, 65 ft, 16 ft   **31.** 44 cm, 70 cm, 55 cm

32. GEOGRAPHY Wyoming's rectangular shape is about 275 miles by 365 miles. Find the length of the diagonal of the state of Wyoming.

33. RESEARCH Use the Internet or other resource to find the measurements of another state. Then calculate the length of a diagonal of the state.

34. TRAVEL The Research Triangle in North Carolina is formed by Raleigh, Durham, and Chapel Hill. Is this triangle a right triangle? Explain.

35. CRITICAL THINKING About 2000 B.C., Egyptian engineers discovered a way to make a right triangle using a rope with 12 evenly spaced knots tied in it. They attached one end of the rope to a stake in the ground. At what knot locations should the other two stakes be placed in order to form a right triangle? Draw a diagram.

Standardized Test Practice and Mixed Review

  FCAT Practice

36. MULTIPLE CHOICE A hiker walked 22 miles north and then walked 17 miles west. How far is the hiker from the starting point?

     A 374 mi   **B** 112.6 mi   **C** 39 mi   **D** 27.8 mi

37. SHORT RESPONSE What is the perimeter of a right triangle if the lengths of the legs are 10 inches and 24 inches?

Replace each ● with <, >, or = to make each a true sentence. (Lesson 3-3)

38. $\sqrt{12}$ ● 3.5   **39.** $\sqrt{41}$ ● 6.4   **40.** $5.\overline{6}$ ● $\dfrac{17}{3}$ **41.** $\sqrt{55}$ ● $7.\overline{4}$

42. ALGEBRA Estimate the solution of $x^2 = 77$ to the nearest integer. (Lesson 3-2)

GETTING READY FOR THE NEXT LESSON

PREREQUISITE SKILL Solve each equation. Check your solution. (Lesson 1-8)

43. $57 = x + 24$ **44.** $82 = 54 + y$ **45.** $71 = 35 + z$ **46.** $64 = a + 27$

Using the Pythagorean Theorem

Sunshine State Standards MA.A.1.3.4-2, MA.B.3.3.1-1, MA.C.1.3.1-4, MA.C.3.3.1-1, MA.C.3.3.1-2, MA.D.1.3.1-2, MA.D.1.3.2-2, MA.D.2.3.1-1, MA.D.2.3.1-6

What You'll Learn
Solve problems using the Pythagorean Theorem.

⇒ *NEW* Vocabulary
Pythagorean triple

WHEN am I ever going to use this?

GYMNASTICS In the floor exercises of women's gymnastics, athletes cross the diagonal of the mat flipping and twisting as they go. It is important that the gymnast does not step off the mat.

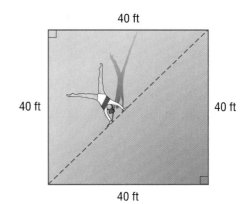

1. What type of triangle is formed by the sides of the mat and the diagonal?

2. Write an equation that can be used to find the length of the diagonal.

The Pythagorean Theorem can be used to solve a variety of problems.

EXAMPLE Use the Pythagorean Theorem

1. **SKATEBOARDING** Find the height of the skateboard ramp.

   Notice the problem involves a right triangle. Use the Pythagorean Theorem.

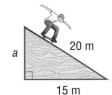

Words	The square of the hypotenuse	equals	the sum of the squares of the legs.
Variables	c^2	$=$	$a^2 + b^2$
Equation	20^2	$=$	$a^2 + 15^2$

$$20^2 = a^2 + 15^2 \qquad \text{Write the equation.}$$
$$400 = a^2 + 225 \qquad \text{Evaluate } 20^2 \text{ and } 15^2.$$
$$400 - 225 = a^2 + 225 - 225 \qquad \text{Subtract 225 from each side.}$$
$$175 = a^2 \qquad \text{Simplify.}$$
$$\sqrt{175} = \sqrt{a^2} \qquad \text{Take the square root of each side.}$$
$$13.2 \approx a \qquad \text{Simplify.}$$

The height of the ramp is about 13.2 meters.

You know that a triangle with sides 3, 4, and 5 units is a right triangle because these numbers satisfy the Pythagorean Theorem. Such whole numbers are called **Pythagorean triples**. By using multiples of a Pythagorean triple, you can create additional triples.

EXAMPLE Write Pythagorean Triples

2 **Multiply the triple 3-4-5 by the numbers 2, 3, 4, and 10 to find more Pythagorean triples.**

You can organize your answers in a table. Multiply each Pythagorean triple entry by the same number and then check the Pythagorean relationship.

	a	b	c	Check: $c^2 = a^2 + b^2$
original	3	4	5	$25 = 9 + 16$ ✓
× 2	6	8	10	$100 = 36 + 64$ ✓
× 3	9	12	15	$225 = 81 + 144$ ✓
× 4	12	16	20	$400 = 144 + 256$ ✓
× 10	30	40	50	$2,500 = 900 + 1,600$ ✓

Skill and Concept Check

Writing Math

Exercises 1 & 3

1. **Explain** why you can use any two sides of a right triangle to find the third side.

2. **OPEN ENDED** Write a problem that can be solved by using the Pythagorean Theorem. Then solve the problem.

3. **Which One Doesn't Belong?** Identify the set of numbers that are not Pythagorean triples. Explain your reasoning.

5-12-13	10-24-26	5-7-9	8-15-17

GUIDED PRACTICE

Write an equation that can be used to answer the question. Then solve. Round to the nearest tenth if necessary.

4. How long is each rafter?

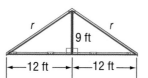

5. How far apart are the planes?

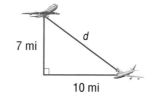

6. How high does the ladder reach?

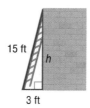

7. **GEOMETRY** An *isosceles* right triangle is a right triangle in which both legs are equal in length. If the leg of an isosceles triangle is 4 inches long, what is the length of the hypotenuse?

Write an equation that can be used to answer the question. Then solve. Round to the nearest tenth if necessary.

HOMEWORK HELP

For Exercises	See Examples
8–19	1, 2

Extra Practice
See pages 623, 650.

8. How long is the kite string?

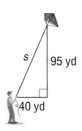

s 95 yd 40 yd

9. How far is the helicopter from the car?

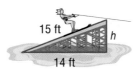
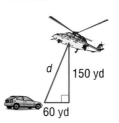

d 150 yd 60 yd

10. How high is the ski ramp?

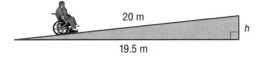

15 ft h 14 ft

11. How long is the lake?

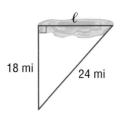

ℓ 18 mi 24 mi

12. How high is the wire attached to the pole?

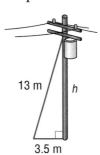

13 m h 3.5 m

13. How high is the wheel chair ramp?

20 m h 19.5 m

14. **VOLLEYBALL** Two ropes and two stakes are needed to support each pole holding the volleyball net. Find the length of each rope.

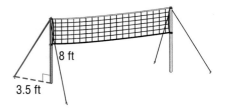

8 ft   3.5 ft

15. **ENTERTAINMENT** Connor loves to watch movies in the letterbox format on his television. He wants to buy a new television with a screen that is at least 25 inches by 13.6 inches. What diagonal size television meets Connor's requirements?

16. **GEOGRAPHY** Suppose Lake City, Gainesville, and Jacksonville, Florida, form a right triangle. What is the distance from Lake City to Jacksonville?

17. **GEOMETRY** A line segment with endpoints on a circle is called a *chord*. Find the distance d from the center of the circle O to the chord $\overline{AB}$ in the circle below.

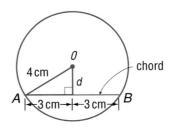

O 4 cm d chord A 3 cm 3 cm B

18. **MULTI STEP** Home builders add corner bracing to give strength to a house frame. How long will the brace need to be for the frame below?

Each board is $1\frac{1}{2}$ in. wide.

16 in. 16 in.

16 in.

8 ft

19. **GEOMETRY** Find the length of the diagonal $\overline{AB}$ in the rectangular prism at the right. (*Hint:* First find the length of $\overline{BC}$.)

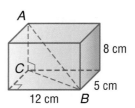

A

8 cm

C

12 cm   B   5 cm

20. **MODELING** Measure the dimensions of a shoebox and use the dimensions to calculate the length of the diagonal of the box. Then use a piece of string and a ruler to check your calculation.

21. **CRITICAL THINKING** Suppose a ladder 100 feet long is placed against a vertical wall 100 feet high. How far would the top of the ladder move down the wall by pulling out the bottom of the ladder 10 feet?

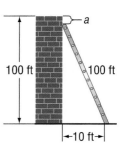

a

100 ft

100 ft

10 ft

Standardized Test Practice and Mixed Review

22. **MULTIPLE CHOICE** What is the height of the tower?

(A) 8 feet   (B) 31.5 feet

(C) 49.9 feet   (D) 992 feet

66 ft

h

58 ft

23. **MULTIPLE CHOICE** Triangle ABC is a right triangle. What is the perimeter of the triangle?

(F) 3 in.   (G) 9 in.

(H) 27 in.   (I) 36 in.

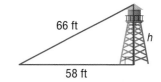

A

15 inches

C   12 inches   B

24. **GEOMETRY** Determine whether a triangle with sides 20 inches, 48 inches, and 52 inches long is a right triangle. (Lesson 3-4)

25. Order $\sqrt{45}$, $6.\overline{6}$, 6.75, and 6.7 from least to greatest. (Lesson 3-3)

Evaluate each expression. (Lesson 2-8)

26. 2^4 27. 3^3 28. $2^3 \cdot 3^2$ 29. $10^5 \cdot 4^2$

GETTING READY FOR THE NEXT LESSON

PREREQUISITE SKILL Graph each point on a coordinate plane. (Page 614)

30. $T(5, 2)$ 31. $A(-1, 3)$ 32. $M(-5, 0)$ 33. $D(-2, -4)$

Sunshine State Standards
MA.A.1.3.3-2, MA.A.1.3.3-3, MA.A.4.3.1-1

What You'll Learn
Graph irrational numbers.

Materials
• grid paper
• compass
• straightedge

Graphing Irrational Numbers

In Lesson 3-3, you found approximate locations for irrational numbers on a number line. You can accurately graph irrational numbers.

ACTIVITY *Work with a partner.*

Graph $\sqrt{34}$ on a number line as accurately as possible.

STEP 1 Find two numbers whose squares have a sum of 34.
$34 = 25 + 9$
$34 = 5^2 + 3^2$

The hypotenuse of a triangle with legs that measure 5 and 3 units will measure $\sqrt{34}$ units.

STEP 2 Draw a number line on grid paper. Then draw a triangle whose legs measure 5 and 3 units.

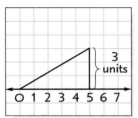 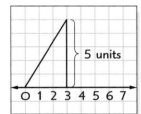

STEP 3 Adjust your compass to the length of the hypotenuse. Place the compass at 0, draw an arc that intersects the number line. The point of intersection is the graph of $\sqrt{34}$.

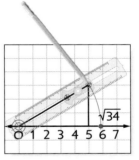 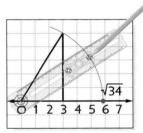

Your Turn Accurately graph each irrational number.

a. $\sqrt{10}$　　b. $\sqrt{13}$　　c. $\sqrt{17}$　　d. $\sqrt{8}$

Writing Math

1. **Explain** how you decide what lengths to make the legs of the right triangle when graphing an irrational number.

2. **Explain** how the graph of $\sqrt{2}$ can be used to graph $\sqrt{3}$.

3. **MAKE A CONJECTURE** Do you think you could graph the square root of any whole number? Explain.

Distance on the Coordinate Plane

Sunshine State Standards
MA.A.1.3.4-2, MA.B.1.3.2-1, MA.B.3.3.1-1, MA.C.3.3.1-2

WHEN am I ever going to use this?

ARCHAEOLOGY Archaeologists keep careful records of the exact locations of objects found at digs. To accomplish this, they set up grids with string. Suppose a ring is found at (1, 3) and a necklace is found at (4, 5). The distance between the locations of these two objects is represented by the blue line.

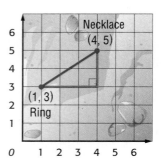

1. What type of triangle is formed by the blue and red lines?

2. What is the length of the two red lines?

3. Write an equation you could use to determine the distance d between the locations where the ring and necklace were found.

4. How far apart were the ring and the necklace?

In mathematics, you can locate a point by using a coordinate system similar to the grid system used by archaeologists. A **coordinate plane** is formed by two number lines that form right angles and intersect at their zero points.

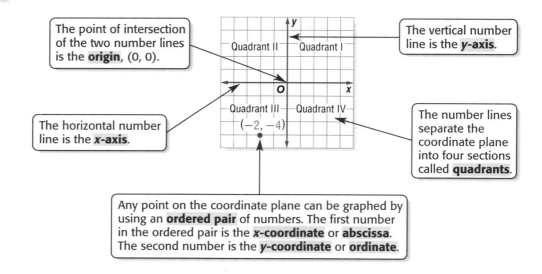

The point of intersection of the two number lines is the **origin**, (0, 0).

The vertical number line is the **y-axis**.

The horizontal number line is the **x-axis**.

The number lines separate the coordinate plane into four sections called **quadrants**.

Any point on the coordinate plane can be graphed by using an **ordered pair** of numbers. The first number in the ordered pair is the **x-coordinate** or **abscissa**. The second number is the **y-coordinate** or **ordinate**.

You can use the Pythagorean Theorem to find the distance between two points on the coordinate plane.

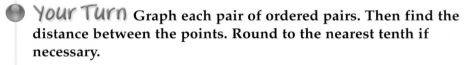

EXAMPLE   Find Distance on the Coordinate Plane

1 Graph the ordered pairs (3, 0) and (7, −5). Then find the distance between the points.

Let c = the distance between the two points, $a = 4$, and $b = 5$.

$c^2 = a^2 + b^2$ Pythagorean Theorem

$c^2 = 4^2 + 5^2$ Replace a with 4 and b with 5.

$c^2 = 16 + 25$ Evaluate 4^2 and 5^2.

$c^2 = 41$ Add 16 and 25.

$\sqrt{c^2} = \sqrt{41}$ Take the square root of each side.

$c \approx 6.4$ Simplify.

The points are about 6.4 units apart.

Your Turn Graph each pair of ordered pairs. Then find the distance between the points. Round to the nearest tenth if necessary.

a. (2, 0), (5, −4)   **b.** (1, 3), (−2, 4)   **c.** (−3, −4), (2, −1)

You can use this technique to find distances on a map.

EXAMPLE   Find Distance on a Map

2 **TRAVEL** The Yeager family is visiting Washington, D.C. A unit on the grid of their map shown at the right is 0.05 mile. Find the distance between the Department of Defense at (−2, 9) and the Madison Building at (3, −3).

Let c = the distance between the Department of Defense and the Madison Building. Then $a = 5$ and $b = 12$.

$c^2 = a^2 + b^2$ Pythagorean Theorem

$c^2 = 5^2 + 12^2$ Replace a with 5 and b with 12.

$c^2 = 25 + 144$ Evaluate 5^2 and 12^2.

$c^2 = 169$ Add 25 and 144.

$\sqrt{c^2} = \sqrt{169}$ Take the square root of each side.

$c = 13$ Simplify.

The distance between the Department of Defense and the Madison Building is 13 units on the map. Since each unit equals 0.05 mile, the distance between the two buildings is $0.05 \cdot 13$ or 0.65 mile.

Skill and Concept Check

1. **Name** the theorem that is used to find the distance between two points on the coordinate plane.

2. **Draw** a triangle that you can use to find the distance between points at $(-3, 2)$ and $(-6, -4)$.

3. **OPEN ENDED** Give the coordinates of a line segment that is neither horizontal nor vertical and has a length of 5 units.

GUIDED PRACTICE

Find the distance between each pair of points whose coordinates are given. Round to the nearest tenth if necessary.

4.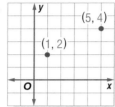

5.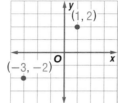

6.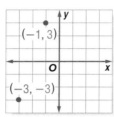

Graph each pair of ordered pairs. Then find the distance between the points. Round to the nearest tenth if necessary.

7. $(1, 5), (3, 1)$

8. $(-1, 0), (2, 7)$

9. $(-5, -2), (2, 3)$

Practice and Applications

Find the distance between each pair of points whose coordinates are given. Round to the nearest tenth if necessary.

10.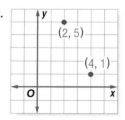

11.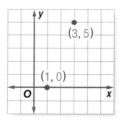

12.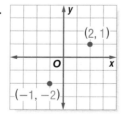

13.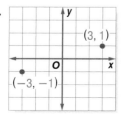

14.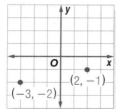

15.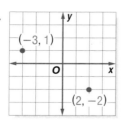

HOMEWORK HELP

For Exercises	See Examples
10–21	1
22–23	2

Extra Practice
See pages 623, 650.

Graph each pair of ordered pairs. Then find the distance between the points. Round to the nearest tenth if necessary.

16. $(4, 5), (2, 2)$

17. $(6, 2), (1, 0)$

18. $(-3, 4), (1, 3)$

19. $(-5, 1), (2, 4)$

20. $(2.5, -1), (-3.5, -5)$

21. $(4, -2.3), (-1, -6.3)$

22. **TECHNOLOGY** A backpacker uses her GPS (Global Positioning System) receiver to find how much farther she needs to travel to get to her stopping point for the day. She is at the red dot on her GPS receiver screen and the blue dot shows her destination. How much farther does she need to travel?

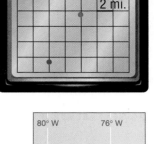

23. **TRAVEL** Corys, North Carolina, has a longitude of 76° W and a latitude of 36° N. Flamingo, Florida, is located at 80° W and 25° N. At this longitude/latitude, each degree is about 72 miles. Find the distance between Corys and Flamingo.

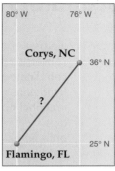

 Data Update What is the distance between where you live and another place of your choice? Visit msmath3.net/data_update to find the longitude and latitude of each city.

24. **CRITICAL THINKING** Find the midpoint of each horizontal or vertical line segment with coordinates of the endpoints given.

   a. (5, 4), (5, 8)   **b.** (3, 2), (3, −4)   **c.** (−2, 5), (−2, −1)   **d.** (a, 5), (b, 5)

25. **CRITICAL THINKING** Study your answers for Exercise 24. Write a rule for finding the midpoint of a horizontal or vertical line.

Standardized Test Practice and Mixed Review

26. **MULTIPLE CHOICE** Find the distance between P and Q.

   Ⓐ 7.8 units   Ⓑ 8.5 units

   Ⓒ 9.5 units   Ⓓ 9.0 units

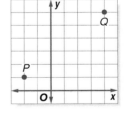

27. **SHORT RESPONSE** Write an equation that can be used to find the distance between $M(-1, 3)$ and $N(3, 5)$.

28. **HIKING** Hunter hikes 3 miles south and then turns and hikes 7 miles east. How far is he from his starting point? (Lesson 3-5)

Find the missing side of each right triangle. Round to the nearest tenth.
(Lesson 3-4)

29. a, 15 cm; b, 18 cm   30. b, 14 in.; c, 17 in.

WebQuest **Interdisciplinary Project**

Bon Voyage!
It's time to complete your project. Use the information and data you have gathered about cruise packages and destination activities to prepare a video or brochure. Be sure to include a diagram and itinerary with your project.
msmath3.net/webquest

Vocabulary and Concept Check

abscissa (p. 142)	origin (p. 142)	right triangle (p. 132)
converse (p. 134)	perfect square (p. 116)	square root (p. 116)
coordinate plane (p. 142)	principal square root (p. 117)	*x*-axis (p. 142)
hypotenuse (p. 132)	Pythagorean Theorem (p. 132)	*x*-coordinate (p. 142)
irrational number (p. 125)	Pythagorean triple (p. 138)	*y*-axis (p. 142)
legs (p. 132)	quadrants (p. 142)	*y*-coordinate (p. 142)
ordered pair (p. 142)	radical sign (p. 116)	
ordinate (p. 142)	real number (p. 125)	

State whether each sentence is *true* or *false*. If *false*, replace the underlined word(s) or number(s) to make a true sentence.

1. An irrational number <u>can</u> be written as a fraction.
2. The <u>hypotenuse</u> is the longest side of a right triangle.
3. The set of numbers <u>{3, 4, 5}</u> is a Pythagorean triple.
4. The number <u>11</u> is a perfect square.
5. The <u>horizontal axis</u> is called the *y*-axis.
6. In an ordered pair, the <u>*y*-coordinate</u> is the second number.
7. The Pythagorean Theorem says that the sum of the squares of the lengths of the <u>legs</u> of a right triangle equals the square of the length of the hypotenuse.
8. The coordinates of the origin are <u>(0, 1)</u>.

Lesson-by-Lesson Exercises and Examples

3-1 **Square Roots** (pp. 116–119)

Find each square root.

9. $\sqrt{81}$
10. $\sqrt{225}$
11. $\sqrt{64}$
12. $-\sqrt{100}$
13. $-\sqrt{\dfrac{4}{9}}$
14. $\sqrt{6.25}$

15. **FARMING** Pecan trees are planted in square patterns to take advantage of land space and for ease in harvesting. For 289 trees, how many rows should be planted and how many trees should be planted in each row?

Example 1 Find $\sqrt{36}$.

$\sqrt{36}$ indicates the *positive* square root of 36.

Since $6^2 = 36$, $\sqrt{36} = 6$.

Example 2 Find $-\sqrt{169}$.

$-\sqrt{169}$ indicates the *negative* square root of 169.

Since $(-13)(-13) = 169$, $-\sqrt{169} = -13$.

 msmath3.net/vocabulary_review

3-2 Estimating Square Roots (pp. 120–122)

Estimate to the nearest whole number.

16. $\sqrt{32}$ 17. $\sqrt{42}$

18. $\sqrt{230}$ 19. $\sqrt{96}$

20. $\sqrt{150}$ 21. $\sqrt{8}$

22. $\sqrt{50.1}$ 23. $\sqrt{19.25}$

24. **ALGEBRA** Estimate the solution of $b^2 = 60$ to the nearest integer.

Example 3 Estimate $\sqrt{135}$ to the nearest whole number.

$121 < 135 < 144$ Write an inequality.

$11^2 < 135 < 12^2$ $121 = 11^2$ and $144 = 12^2$

$11 < \sqrt{135} < 12$ Take the square root of each number.

So, $\sqrt{135}$ is between 11 and 12. Since 135 is closer to 144 than to 121, the best whole number estimate is 12.

3-3 The Real Number System (pp. 125–129)

Name all sets of numbers to which each real number belongs.

25. $-\sqrt{19}$ 26. $0.\overline{3}$

27. 7.43 28. -12

29. $\sqrt{32}$ 30. 101

Example 4 Name all sets of numbers to which $-\sqrt{33}$ belongs.

$-\sqrt{33} \approx -5.744562647$

Since the decimal does not terminate or repeat, it is an irrational number.

3-4 The Pythagorean Theorem (pp. 132–136)

Write an equation you could use to find the length of the missing side of each right triangle. Then find the missing length. Round to the nearest tenth if necessary.

31.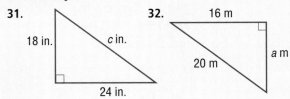

32.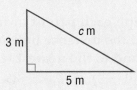

33.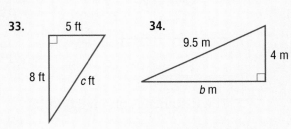

34.

35. a, 5 in.; c, 6 in.   36. a, 6 cm; b, 7 cm

Example 5 Write an equation you could use to find the length of the hypotenuse of the right triangle. Then find the missing length.

$c^2 = a^2 + b^2$ Pythagorean Theorem

$c^2 = 3^2 + 5^2$ Replace a with 3 and b with 5.

$c^2 = 9 + 25$ Evaluate 3^2 and 5^2.

$c^2 = 34$ Simplify.

$c = \sqrt{34}$ Take the square root of each side.

$c \approx 5.8$ Use a calculator.

The hypotenuse is about 5.8 meters long.

Using the Pythagorean Theorem (pp. 137–140)

Write an equation that can be used to answer the question. Then solve. Round to the nearest tenth if necessary.

37. How tall is the light?

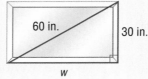

h   25 ft
20 ft

38. How wide is the window?

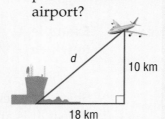

60 in.   30 in.
w

39. How long is the walkway?

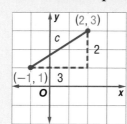

ℓ   5 ft
8 ft

40. How far is the plane from the airport?

d   10 km
18 km

41. GEOMETRY A rectangle is 12 meters by 7 meters. What is the length of one of its diagonals?

Example 6 Write an equation that can be used to find the height of the tree. Then solve.

53 ft   h
25 ft

Use the Pythagorean Theorem to write the equation $53^2 = h^2 + 25^2$. Then solve the equation.

$$53^2 = h^2 + 25^2 \quad \text{Pythagorean Theorem}$$
$$2{,}809 = h^2 + 625 \quad \text{Evaluate } 53^2 \text{ and } 25^2.$$
$$2{,}809 - 625 = h^2 + 625 - 625 \quad \text{Subtract 625 from each side.}$$
$$2{,}184 = h^2 \quad \text{Simplify.}$$
$$\sqrt{2{,}184} = h \quad \text{Take the square root of each side.}$$
$$46.7 \approx h \quad \text{Use a calculator.}$$

The height of the tree is about 47 feet.

Distance on the Coordinate Plane (pp. 142–145)

Graph each pair of ordered pairs. Then find the distance between the points. Round to the nearest tenth if necessary.

42. $(0, -3)$, $(5, 5)$ **43.** $(-1, 2)$, $(4, 8)$

44. $(-2, 1)$, $(2, 3)$ **45.** $(-6, 2)$, $(-4, 5)$

46. $(3, 4)$, $(-2, 0)$ **47.** $(-1, 3)$, $(2, 4)$

48. GEOMETRY The coordinates of points R and S are $(4, 3)$ and $(1, 6)$. What is the distance between the points? Round to the nearest tenth if necessary.

Example 7 Graph the ordered pairs $(2, 3)$ and $(-1, 1)$. Then find the distance between the points.

$$c^2 = a^2 + b^2$$
$$c^2 = 3^2 + 2^2$$
$$c^2 = 9 + 4$$
$$c^2 = 13$$
$$c = \sqrt{13}$$
$$c \approx 3.6$$

The distance is about 3.6 units.

Vocabulary and Concepts

1. **OPEN ENDED** Write an equation that can be solved by taking the square root of a perfect square.

2. **State** the Pythagorean Theorem.

Skills and Applications

Find each square root.

3. $\sqrt{225}$

4. $-\sqrt{25}$

5. $\sqrt{\dfrac{36}{49}}$

Estimate to the nearest whole number.

6. $\sqrt{67}$

7. $\sqrt{108}$

8. $\sqrt{82}$

Name all sets of numbers to which each real number belongs.

9. $-\sqrt{64}$

10. $6.\overline{13}$

11. $\sqrt{14}$

Write an equation you could use to find the length of the missing side of each right triangle. Then find the missing length. Round to the nearest tenth if necessary.

12. a, 5 m; b, 5 m

13. b, 20 ft; c, 35 ft

Determine whether each triangle with sides of given lengths is a right triangle.

14. 12 in., 20 in., 24 in.

15. 34 cm, 30 cm, 16 cm

16. **LANDSCAPING** To make a balanced landscaping plan for a yard, Kelsey needs to know the heights of various plants. How tall is the tree at the right?

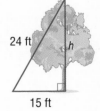

24 ft h

15 ft

17. **GEOMETRY** Find the perimeter of a right triangle with legs of 10 inches and 8 inches.

Graph each pair of ordered pairs. Then find the distance between points. Round to the nearest tenth if necessary.

18. $(-2, -2)$, $(5, 6)$

19. $(1, 3)$, $(-4, 5)$

Standardized Test Practice

Ⓐ   Ⓑ   Ⓒ   Ⓓ

20. **MULTIPLE CHOICE** If the area of a square is 40 square millimeters, what is the approximate length of one side of the square?

Ⓐ 6.3 mm   Ⓑ 7.5 mm   Ⓒ 10 mm   Ⓓ 20 mm

FCAT Practice

PART 1 Multiple Choice

Record your answers on the answer sheet provided by your teacher or on a sheet of paper.

1. Which of the following sets of ordered pairs represents two points on the line below? (Prerequisite Skill, p. 614)

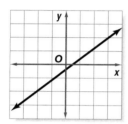

   (A) $\{(3, 1), (2, -1)\}$ (B) $\{(3, 2), (-1, -2)\}$

   (C) $\{(3, 2), (-2, -2)\}$ (D) $\{(3, 3), (-2, -3)\}$

2. The table below shows the income of several baseball teams in 2001. What is the total revenue for all of these teams?
(Lesson 1-4)

Team	Income
Braves	-$14,400,000
Orioles	$1,500,000
Cubs	$4,800,000
Tigers	$500,000
Marlins	-$27,700,000
Yankees	$40,900,000
A's	-$7,100,000
Pirates	-$3,000,000

Source: www.mlb.com

   (F) $99,900,000   (G) $4,500,000

   (H) -$4,500,000   (I) -$99,900,000

3. Which of the following is equivalent to 0.64? (Lesson 2-1)

   (A) $\frac{1}{64}$ (B) $\frac{16}{25}$

   (C) $\frac{100}{64}$ (D) $\frac{64}{10}$

4. Which of the following values are equivalent? (Lesson 2-2)

$$0.08, 0.8, \frac{1}{8}, \frac{4}{5}$$

   (F) 0.08 and $\frac{1}{8}$ (G) 0.8 and $\frac{1}{8}$

   (H) 0.08 and $\frac{4}{5}$ (I) 0.8 and $\frac{4}{5}$

5. Between which two whole numbers is $\sqrt{56}$ located on a number line? (Lesson 3-2)

   (A) 6 and 7   (B) 7 and 8

   (C) 8 and 9   (D) 9 and 10

6. Which of the points on the number line is the *best* representation of $-\sqrt{11}$? (Lesson 3-3)

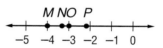

   (F) M (G) N (H) O (I) P

7. What is the value of x? (Lesson 3-4)

   (A) $\sqrt{8 + 11}$

   (B) $\sqrt{8^2 + 11^2}$

   (C) $\frac{8^2 + 11^2}{2}$

   (D) $8^2 + 11^2$

8. Two fences meet in the corner of the yard. The length of one fence is 4 yards, and the other is 6 yards. What is the distance between the far ends of the fences?
(Lesson 3-5)

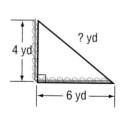

   (F) 6.3 yd   (G) 7.2 yd

   (H) 8.8 yd   (I) 9.5 yd

PART 2   Short Response/Grid In

Record your answers on the answer sheet provided by your teacher or on a sheet of paper.

9. Missy placed a stick near the edge of the water on the beach. If the sum of the distances from the stick is positive, the tide is coming in. If the sum of the distances is negative, the tide is going out. Determine whether the tide is coming in or going out for the readings at the right. (Lesson 1-4)

Wave Distance from Stick (inches)
+3
+5
−4
+2
+8
−6
−3
+7
−5
−4

10. Is the square root of 25 equal to 5, −5, or both? (Lesson 3-1)

11. The value of $\sqrt{134}$ is between what two consecutive whole numbers? (Lesson 3-2)

12. Find the value of x to the nearest tenth. (Lesson 3-4)

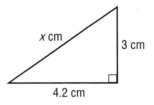

13. Lucas attaches a wire to a young oak tree 4 feet above the ground. The wire is anchored in the ground at an angle from the tree to help the tree stay upright as it grows. If the wire is 5 feet long, what is the distance from the base of the wire to the base of the tree? (Lesson 3-4)

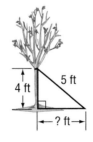

TEST-TAKING TIP

Questions 12 and 13 Remember that the hypotenuse of a right triangle is always opposite the right angle.

14. A signpost casts a shadow that is 6 feet long. The top of the post is 10 feet from the end of the shadow. What is the height of the post? (Lesson 3-5)

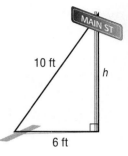

15. Find the distance between the points located on the graph below. Round to the nearest tenth. (Lesson 3-6)

PART 3   Extended Response

Record your answers on a sheet of paper. Show your work.

16. Use the right triangle to answer the following questions. (Lesson 3-4)

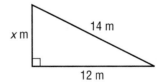

a. Write an equation that can be used to find the length of x.

b. Solve the equation. Justify each step.

c. What is the length of x?

17. Use a grid to graph and answer the following questions. (Lesson 3-6)

a. Graph the ordered pairs (3, 4) and (−2, 1).

b. Describe how to find the distance between the two points.

c. Find the distance between the points.

UNIT 2
Proportional Reasoning

Chapter 4
Proportions, Algebra, and Geometry

Chapter 5
Percent

Although they may seem unrelated, proportions, algebra, and geometry are closely related. In this unit, you will use proportions and algebra to solve problems involving geometry and percents.

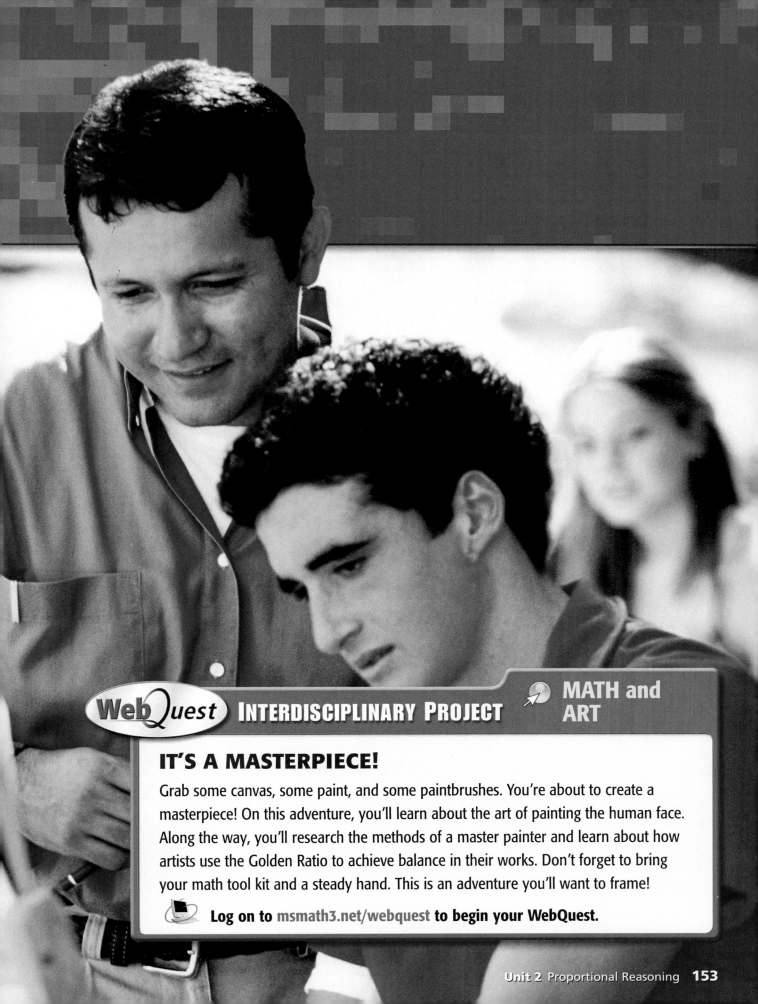

WebQuest INTERDISCIPLINARY PROJECT

MATH and ART

IT'S A MASTERPIECE!

Grab some canvas, some paint, and some paintbrushes. You're about to create a masterpiece! On this adventure, you'll learn about the art of painting the human face. Along the way, you'll research the methods of a master painter and learn about how artists use the Golden Ratio to achieve balance in their works. Don't forget to bring your math tool kit and a steady hand. This is an adventure you'll want to frame!

Log on to msmath3.net/webquest to begin your WebQuest.

CHAPTER 4
Proportions, Algebra, and Geometry

"What do the planets have to do with math?"

The circumference of Earth is about 40,000 kilometers. If you know the circumference of the other planets, you can use **proportions** to make a scale model of our solar system.

You will solve problems involving scale models in Lesson 4-6.

GETTING STARTED

Take this quiz to see whether you are ready to begin Chapter 4. Refer to the lesson or page number in parentheses if you need more review.

▶ Vocabulary Review

Complete each sentence.

1. A ___?___ is a letter used to represent an unknown number. (Lesson 1-2)

2. The coordinate system includes a vertical number line called the ___?___. (Lesson 3-6)

3. An ___?___ names any given point on the coordinate plane with its x-coordinate and y-coordinate. (Lesson 3-6)

▶ Prerequisite Skills

Simplify each fraction. (Page 611)

4. $\dfrac{10}{24}$

5. $\dfrac{88}{104}$

6. $\dfrac{36}{81}$

7. $\dfrac{49}{91}$

Evaluate each expression. (Lesson 1-2)

8. $\dfrac{6-2}{5+5}$

9. $\dfrac{7-4}{8-4}$

10. $\dfrac{3-1}{1+9}$

11. $\dfrac{5+7}{8-6}$

Subtract. (Lesson 1-5)

12. $16 - 7$

13. $5 - 12$

14. $-8 - 10$

15. $4 - (-3)$

16. $-11 - 2$

17. $-8 - (-9)$

Solve each equation. (Lesson 1-9)

18. $5 \cdot 6 = x \cdot 2$

19. $c \cdot 1.5 = 3 \cdot 7$

20. $12 \cdot z = 9 \cdot 4$

21. $7 \cdot 2 = 8 \cdot g$

22. $3 \cdot 11 = 4 \cdot y$

23. $b \cdot 6 = 7 \cdot 9$

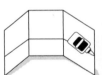

FOLDABLES™ Study Organizer

Using Proportions Make this Foldable to help you organize your notes. Begin with a plain sheet of 11" by 17" paper.

STEP 1 **Fold in thirds**
Fold in thirds widthwise.

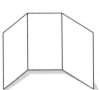

STEP 2 **Open and Fold Again**
Fold the bottom to form a pocket. Glue edges.

STEP 3 **Label**
Label each pocket. Place index cards in each pocket.

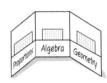

Reading and Writing As you read and study the chapter, write definitions, notes and examples about each topic on index cards and store them in your Foldable.

4-1

Ratios and Rates

Sunshine State Standards
MA.A.1.3.1-1, MA.A.1.3.2-2, MA.A.3.3.2-1, MA.A.3.3.2-2, MA.A.3.3.3-1, MA.B.1.3.2-1

WHEN am I ever going to use this?

TRAIL MIX The diagram shows a batch of trail mix that is made using 3 scoops of raisins and 6 scoops of peanuts.

1. Which combination of ingredients below would you use to make a smaller amount of the same recipe? Explain.

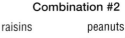

2. In order to make the same recipe of trail mix, how many scoops of peanuts should you use for every scoop of raisins?

Sidebar

What You'll Learn
Express ratios as fractions in simplest form and determine unit rates.

NEW Vocabulary
ratio
rate
unit rate

MATH Symbols
≈ approximately equal to

READING Math

Ratios In Example 1, the ratio 2 out of 7 means that for every 7 cats, 2 are Siamese.

A **ratio** is a comparison of two numbers by division. If a batch of trail mix contains 3 scoops of raisins and 6 scoops of peanuts, then the ratio comparing the raisins to the peanuts can be written as follows.

$$3 \text{ to } 6 \qquad 3:6 \qquad \frac{3}{6}$$

Since a ratio can be written as a fraction, it can be simplified.

EXAMPLES Write Ratios in Simplest Form

1 Express *8 Siamese cats out of 28 cats* in simplest form.

$$\frac{8}{28} = \frac{2}{7}$$ Divide the numerator and denominator by the greatest common factor, 4.

The ratio of Siamese cats to cats is $\frac{2}{7}$ or 2 out of 7.

2 Express *10 ounces to 1 pound* in simplest form.

$$\frac{10 \text{ ounces}}{1 \text{ pound}} = \frac{10 \text{ ounces}}{16 \text{ ounces}}$$ Convert 1 pound to 16 ounces.

$$= \frac{5 \text{ ounces}}{8 \text{ ounces}}$$ Divide the numerator and the denominator by 2.

The ratio in simplest form is $\frac{5}{8}$ or 5:8.

Your Turn Express each ratio in simplest form.

a. 16 pepperoni pizzas out of 24 pizzas
b. 12 minutes to 2 hours

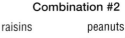

A **rate** is a special kind of ratio. It is a comparison of two quantities with different types of units. Here are two examples of rates.

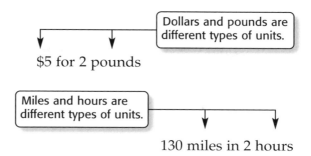

Dollars and pounds are different types of units.

$5 for 2 pounds

Miles and hours are different types of units.

130 miles in 2 hours

When a rate is simplified so it has a denominator of 1, it is called a **unit rate**. An example of a unit rate is *$6.50 per hour*, which means *$6.50 per 1 hour*.

EXAMPLE Find a Unit Rate

3 **TRAVEL** On a trip from Nashville, Tennessee, to Birmingham, Alabama, Darrell drove 187 miles in 3 hours. What was Darrell's average speed in miles per hour?

Write the rate that expresses the comparison of miles to hours. Then find the average speed by finding the unit rate.

$$\frac{187 \text{ miles}}{3 \text{ hours}} \approx \frac{62 \text{ miles}}{1 \text{ hour}}$$ Divide the numerator and denominator by 3 to get a denominator of 1.

Darrell drove an average speed of about 62 miles per hour.

EXAMPLE Compare Unit Rates

4 **CIVICS** For the 2000 census, the population of Texas was about 20,900,000, and the population of Virginia was about 7,000,000. There were 30 members of the U.S. House of Representatives from Texas and 11 from Virginia. In which state did a member represent more people?

For each state, write a rate that compares the state's population to its number of representatives. Then find the unit rates.

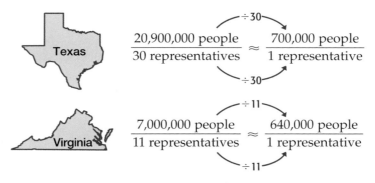

Texas

$$\frac{20,900,000 \text{ people}}{30 \text{ representatives}} \approx \frac{700,000 \text{ people}}{1 \text{ representative}}$$ ÷30   ÷30

Virginia

$$\frac{7,000,000 \text{ people}}{11 \text{ representatives}} \approx \frac{640,000 \text{ people}}{1 \text{ representative}}$$ ÷11   ÷11

Therefore, in Texas, a member of the U.S. House of Representatives represented more people than in Virginia.

1. **OPEN ENDED** Write a ratio about the marbles in the jar. Simplify your ratio, if possible. Then explain the meaning of your ratio.

2. **Explain** how to write a rate as a unit rate.

GUIDED PRACTICE

Express each ratio in simplest form.

3. 12 missed days in 180 school days
4. 12 wins to 18 losses
5. 24 pints:1 quart
6. 8 inches out of 4 feet

Express each rate as a unit rate.

7. $50 for 4 days work
8. 3 feet of snow in 5 hours

9. **SHOPPING** You can buy 4 Granny Smith apples at Ben's Mart for $0.95. SaveMost sells the same quality apples 6 for $1.49. Which store has the better buy? Explain your reasoning.

Practice and Applications

Express each ratio in simplest form.

10. 33 brown eggs to 18 white eggs
11. 56 boys to 64 girls
12. 14 chosen out of 70 who applied
13. 28 out of 100 doctors
14. 400 centimeters to 1 meter
15. 6 feet : 9 yards
16. 2 cups to 1 gallon
17. 153 points in 18 games

HOMEWORK HELP	
For Exercises	See Examples
10–17, 31, 34	1, 2
18–23, 30, 33	3
24–27, 32	4
Extra Practice See pages 624, 651.	

Express each rate as a unit rate.

18. $22 for 5 dozen donuts
19. $73.45 in 13 hours
20. 1,473 people entered the park in 3 hours
21. 11,025 tickets sold at 9 theaters
22. 100 meters in 12.2 seconds
23. 21.5 pounds in 12 weeks

SHOPPING For Exercises 24–27, decide which is the better buy. Explain.

24. a 17-ounce box of cereal for $4.89 or a 21-ounce box for $5.69

25. 6 cans of green beans for $1 or 10 cans for $1.95

26. 1 pound 4 ounces of meat for $4.99 or 2 pounds 6 ounces for $9.75

27. a 2-liter bottle of soda for $1.39 or a 12-pack of 12-ounce cans for $3.49 (*Hint*: 2 liters = 67.63 ounces)

Use ratios to convert the following rates.

28. 60 mi/h = $\underline{\quad ? \quad}$ ft/s
29. 180 gal/h = $\underline{\quad ? \quad}$ oz/min

30. **CARS** Gas mileage is the average number of miles you can drive a car per gallon of gasoline. A test of a new car resulted in 2,250 miles being driven using 125 gallons of gas. Find the car's gas mileage.

SPORTS For Exercises 31 and 32, use the graph at the right.

31. Write a ratio comparing the amount of money Jeff Gordon earned in the Winston Cup Series in 2001 to his number of wins that year.

32. **MULTI STEP** On average, who earned more money per win in their sport in 2001, Jeff Gordon or Tiger Woods? Explain.

33. **ART** At an auction in New York City, a 2.55-square inch portrait of George Washington sold for $1.2 million. About how much did the buyer pay per square inch for the portrait?

34. **WRITE A PROBLEM** Write about a real-life situation that can be represented by the ratio 2:5.

35. **CRITICAL THINKING** Luisa and Rachel have some trading cards. The ratio of their cards is 3:1. If Luisa gives Rachel 2 cards, the ratio will be 2:1. How many cards does Luisa have?

Standardized Test Practice and Mixed Review

  FCAT Practice

36. **MULTIPLE CHOICE** Which of the following cannot be written as a ratio?

　Ⓐ two pages for every one page he reads　　Ⓑ three more chips than she has

　Ⓒ half as many CDs as he has　　Ⓓ twice as many pencils as she has

37. **SHORT RESPONSE** Three people leave at the same time from town A to town B. Sarah averaged 45 miles per hour for the first third of the distance, 55 miles per hour for the second third, and 75 miles per hour for the last third. Darnell averaged 55 miles per hour for the first half of the trip and 70 miles per hour for the second half. Megan drove at a steady speed of 60 miles per hour the entire trip. Who arrived first?

Graph each pair of ordered pairs. Then find the distance between the points. Round to the nearest tenth. (Lesson 3-6)

38. $(1, 4), (6, -3)$　　39. $(-1, 5), (3, -2)$　　40. $(-5, -2), (-1, 0)$　　41. $(-2, -3), (3, 1)$

42. **GYMNASTICS** A gymnast is making a tumbling pass along the diagonal of a square floor exercise mat measuring 40 feet on each side. Find the measure of the diagonal. (Lesson 3-5)

GETTING READY FOR THE NEXT LESSON

PREREQUISITE SKILL Evaluate each expression. (Lesson 1-5)

43. $\dfrac{45 - 33}{10 - 8}$　　44. $\dfrac{85 - 67}{2001 - 1995}$　　45. $\dfrac{29 - 44}{55 - 50}$　　46. $\dfrac{18 - 19}{25 - 30}$

Rate of Change

Sunshine State Standards MA.A.3.3.2-1,
MA.B.1.3.2-1, MA.B.1.3.2-2, MA.D.1.3.2-4, MA.E.1.3.1-1

What You'll Learn
Find rates of change.

→ *NEW* Vocabulary

rate of change

WHEN **am I ever going to use this?**

HOBBIES Alicia likes to collect teddy bears. The graph shows the number of teddy bears in her collection between 1997 and 2002.

1. By how many bears did Alicia's collection increase between 1997 and 1999? Between 1999 and 2002?

2. Between which years did Alicia's collection increase the fastest?

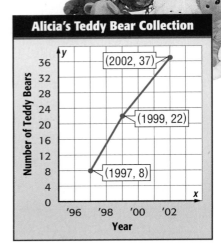

Alicia's Teddy Bear Collection

A **rate of change** is a rate that describes how one quantity changes in relation to another. In the example above, the rate of change in Alicia's teddy bear collection from 1997 to 1999 is shown below.

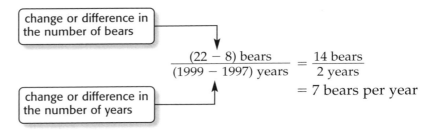

change or difference in the number of bears

change or difference in the number of years

$$\frac{(22 - 8) \text{ bears}}{(1999 - 1997) \text{ years}} = \frac{14 \text{ bears}}{2 \text{ years}}$$
$$= 7 \text{ bears per year}$$

EXAMPLE **Find a Rate of Change**

1 **HEIGHTS** The table at the right shows Ramón's height in inches between the ages of 8 and 13. Find the rate of change in his height between ages 8 and 11.

Age (yr)	8	11	13
Height (in.)	51	58	67

$$\frac{\text{change in height}}{\text{change in age}} = \frac{(58 - 51) \text{ inches}}{(11 - 8) \text{ years}}$$ Ramón grew from 51 to 58 inches tall from age 8 to age 11.

$$= \frac{7 \text{ inches}}{3 \text{ years}}$$ Subtract to find the change in heights and ages.

$$\approx \frac{2.3 \text{ inches}}{1 \text{ year}}$$ Express this rate as a unit rate.

Ramón grew an average of about 2.3 inches per year.

Mental Math
You can also find a unit rate by dividing the numerator by the denominator.

Your Turn

a. Find the rate of change in his height between ages 11 and 13.

A graph of the data in Example 1 is shown at the right. The data points are connected by segments. On a graph, a rate of change measures how fast a segment goes up when the graph is read from left to right.

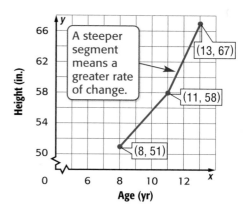

A steeper segment means a greater rate of change.

(13, 67)
(11, 58)
(8, 51)

Height (in.)
Age (yr)

A formula for rate of change using data coordinates is given below.

Key Concept — Rate of Change

Words To find the rate of change, divide the difference in the y-coordinates by the difference in the x-coordinates.

Symbols The rate of change between (x_1, y_1) and (x_2, y_2) is $\dfrac{y_2 - y_1}{x_2 - x_1}$.

Rates of change can be positive or negative. This corresponds to an increase or decrease in the y-value between the two data points.

EXAMPLE — Find a Negative Rate of Change

2 **MUSIC** The graph shows cassette sales from 1995 to 2000. Find the rate of change between 1996 and 2000, and describe how this rate is shown on the graph.

Use the formula for the rate of change.

Let $(x_1, y_1) = (1996, 19.3)$ and $(x_2, y_2) = (2000, 4.9)$.

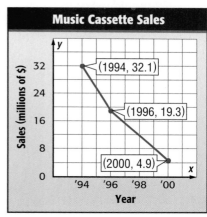

Music Cassette Sales

(1994, 32.1)
(1996, 19.3)
(2000, 4.9)

Sales (millions of $)
Year

Source: Recording Industry Assoc. of America

$$\frac{y_2 - y_1}{x_2 - x_1} = \frac{4.9 - 19.3}{2000 - 1996}$$ Write the formula for rate of change.

$$= \frac{-14.4}{4}$$ Simplify.

$$= \frac{-3.6}{1}$$ Express this rate as a unit rate.

The rate of change is -3.6 million dollars in sales per year. The rate is negative because between 1996 and 2000, the cassette sales *decreased*. This is shown on the graph by a line slanting downward from left to right.

Your Turn

b. In the graph above, find the rate of change between 1994 and 1996.

c. Describe how this rate is shown on the graph.

When a quantity does not change over a period of time, it is said to have a zero rate of change.

EXAMPLES **Zero Rates of Change**

3 **MAIL** The graph shows the cost in cents of mailing a 1-ounce first-class letter. Find a time period in which the cost of a first-class stamp did not change.

Between 1992 and 1994, the cost of a first class stamp did not change. It remained 29¢. This is shown on the graph by a horizontal line segment.

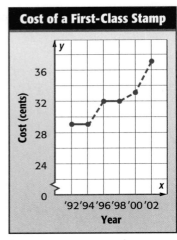

Cost of a First-Class Stamp

Source: www.stamps.org

4 **MAIL** Find the rate of change from 1992 to 1994.

Let $(x_1, y_1) = (1992, 29)$ and $(x_2, y_2) = (1994, 29)$.

$\dfrac{y_2 - y_1}{x_2 - x_1} = \dfrac{29 - 29}{1994 - 1992}$ Write the formula for rate of change.

$\qquad = \dfrac{0}{2}$ or 0 Simplify.

The rate of change in the cost of a first-class stamp between 1992 and 1994 is 0 cents per year.

Your Turn

d. Find another time period in which the cost of a first class stamp did not change. Explain your reasoning.

The table below summarizes the relationship between rates of change and their graphs.

Concept Summary   **Rates of Change**

Rate of Change	positive	zero	negative
Real-Life Meaning	increase	no change	decrease
Graph	slants upward	horizontal line	slants downward

1. **OPEN ENDED** Describe a situation involving a zero rate of change.

2. **NUMBER SENSE** Does the height of a candle as it burns over time show a *positive*, *negative*, or *zero* rate of change? Explain your reasoning.

GUIDED PRACTICE

TEMPERATURE For Exercises 3–6, use the table at the right. It shows the outside air temperature at different times during one day.

Time	Temperature (°F)
6 A.M.	33
8 A.M.	45
12 P.M.	57
3 P.M.	57
4 P.M.	59
8 P.M.	34

3. Find the rate of temperature change in degrees per hour from 6 A.M. to 8 A.M. and from 4 P.M. and 8 P.M.

4. Between which of these two time periods was the rate of change in temperature greater?

5. Make a graph of this data.

6. During which time period(s) was the rate of change in temperature positive? negative? 0° per hour? How can you tell this from your graph?

Practice and Applications

ADVERTISING For Exercises 7–10, use the following information.
Tanisha's job is to neatly fold flyers for the school play. She started folding at 12:55 P.M. The table below shows her progress.

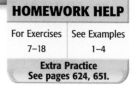

HOMEWORK HELP

For Exercises	See Examples
7–18	1–4

Extra Practice
See pages 624, 651.

Time	12:55	1:00	1:20	1:25	1:30
Flyers Folded	0	21	102	102	125

7. Find the rate of change in flyers per minute between 1:00 and 1:20.

8. Find her rate of change between 1:25 and 1:30.

9. During which time period did her folding rate increase the fastest?

10. Find the rate of change from 1:20 to 1:25 and interpret its meaning.

BIRDS For Exercises 11–14, use the information below and at the right.
The graph shows the approximate number of American Bald Eagle pairs from 1963 to 1998.

11. Find the rate of change in the number of eagle pairs from 1974 to 1984.

12. Find the rate of change in the number of eagle pairs from 1984 to 1994.

13. During which of these two time periods did the eagle population grow faster?

14. Find the rate of change in the population from 1994 to 1998. Then interpret its meaning.

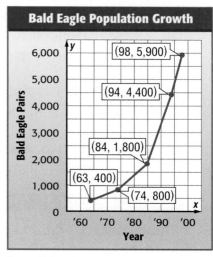

Bald Eagle Population Growth

Source: birding.about.com

FAST FOOD For Exercises 15 and 16, use the graph at the right.

15. During which time period was the rate of change in sales greatest? Explain.

16. Find the rate of change during that period.

CANDY For Exercises 17 and 18, use the following information.

According to the National Confectioners Association, candy sales during the winter holidays in 1995 totaled $1,342 billion. By 2001, this figure had risen to $1,474 billion.

17. Find the rate of change in candy sales during the winter holidays from 1995 to 2001.

18. If this rate of change were to continue, what would the total candy sales during the winter holidays be in 2005?

USA TODAY Snapshots®

Food and drink sales up
Food and beverage sales[1] in the USA keep climbing:

2002[2] **$408 billion**
2000 **$345 billion**
1990 **$239 billion**
1980 **$120 billion**

1 – Includes bars and restaurants, food service contractors, and other retail/vending/recreation businesses. 2 – Projection
Sources: National Restaurant Association, *Tourism Works for America*, 10th annual edition 2001

By Darryl Haralson and Sam Ward, USA TODAY

 Data Update What were candy sales during the winter holidays last year? Visit msmath3.net/data_update to learn more.

19. **CRITICAL THINKING** The rate of change between point A and point B on the graph is 3 meters per day. Find the value of y.

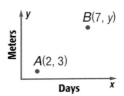

Standardized Test Practice and Mixed Review

Ⓐ Ⓑ Ⓒ Ⓓ   FCAT Practice

20. **SHORT RESPONSE** Nine days ago, the area covered by mold on a piece of bread was 3 square inches. Today the mold covers 9 square inches. Find the rate of change in the mold's area.

21. **MULTIPLE CHOICE** The graph shows the altitude of a falcon over time. Between which two points on the graph was the bird's rate of change in height negative?

Ⓐ A and B   Ⓑ B and C   Ⓒ C and D   Ⓓ D and E

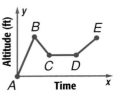

Express each ratio in simplest form. (Lesson 4-1)

22. 42 red cars to 12 black cars

23. 1,500 pounds to 2 tons

24. **GEOMETRY** A triangle has vertices $A(-2, -5)$, $B(-2, 8)$, and $C(1, 4)$. Find the perimeter of triangle ABC. (Lesson 3-6)

GETTING READY FOR THE NEXT LESSON

PREREQUISITE SKILL Evaluate each expression. (Lesson 1-2)

25. $\dfrac{8 - 5}{3 - 1}$

26. $\dfrac{3 - 7}{4 - (-4)}$

27. $\dfrac{-5 - (-2)}{-1 - 8}$

28. $\dfrac{2 - (-4)}{-2 - (-3)}$

What You'll Learn
Find rates of change using a spreadsheet.

Constant Rates of Change

You can calculate rates of change using a spreadsheet.

 ACTIVITY

Andrew earns $18 per hour mowing lawns. Calculate the rate of change in the amount he earns between each consecutive pair of times. Then interpret your results.

Time (h)	Amount ($)
1	18
2	36
3	54
4	72

Set up a spreadsheet like the one shown below.

Lawn Earnings

	A	B	C
1	Time (h)	Amount ($)	Rate of Change
2	x	y	(change in y)/(change in x)
3	1	18	18
4	2	36	18
5	3	54	18
6	4	72	18

Sheet1 / Sheet2

In column A, enter the time *values* in hours.

The spreadsheet evaluates the formula (B5-B4)/(A5-A4).

The spreadsheet evaluates the formula 18*A5.

The rate of change between each consecutive pair of data is the same, or constant—$18 per hour.

 EXERCISES

1. Graph the data given in the activity above. Then describe the figure formed when the points on the graph are connected.

PARKING For Exercises 2–4, use the information in the table. It shows the charges for parking at a football stadium.

Time (h)	Amount ($)
1	5
2	8
3	11
4	14

2. Use a spreadsheet to find the rate of change in the amount charged between each consecutive pair of times.

3. Interpret your results from Exercise 2.

4. Graph the data. Then describe the figure formed when the points on the graph are connected.

Slope

Sunshine State Standards MA.A.3.3.2-1, MA.B.1.3.2-1, MA.B.1.3.2-2, MA.C.3.3.2-2, MA.D.1.3.1-1, MA.D.1.3.2-4, MA.D.2.3.1-4

WHEN am I ever going to use this?

EXERCISE As part of Cameron's fitness program, he tries to run every day. He knows that after he has warmed up, he can maintain a constant running speed of 8 feet per second. This is shown in the table and in the graph.

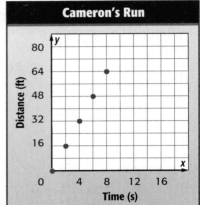

Cameron's Run

Time (s)	0	2	4	6	8
Distance (ft)	0	16	32	48	64

1. Pick several pairs of points from those plotted and find the rate of change between them. Write each rate in simplest form.

2. What is true of these rates?

In the graph above, the rate of change between any two points on a line is always the same. This constant rate of change is called the slope of the line. **Slope** is the ratio of the **rise**, or vertical change, to the **run**, or horizontal change.

$$\text{slope} = \frac{\text{rise}}{\text{run}} \quad \begin{array}{l} \leftarrow \text{ vertical change between any two points} \\ \leftarrow \text{ horizontal change between the same two points} \end{array}$$

EXAMPLE Find Slope Using a Graph

1 Find the slope of the line.

Choose two points on the line. The vertical change is 2 units while the horizontal change is 3 units.

$$\text{slope} = \frac{\text{rise}}{\text{run}} \quad \text{Definition of slope}$$

$$= \frac{2}{3} \quad \text{rise} = 2, \text{run} = 3$$

The slope of the line is $\frac{2}{3}$.

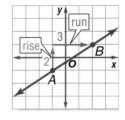

Your Turn Find the slope of each line.

a.
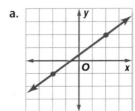

b.
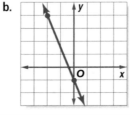

c.
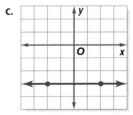

Since slope is a rate of change, it can be positive (slanting upward), negative (slanting downward), or zero (horizontal).

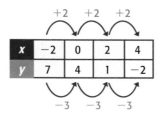

STUDY TIP

Translating Rise and Run

up → positive
down → negative

right → positive
left → negative

EXAMPLE **Find Slope Using a Table**

2) The points given in the table lie on a line. Find the slope of the line. Then graph the line.

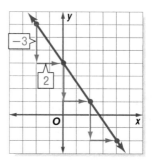

x	−2	0	2	4
y	7	4	1	−2

$$\text{slope} = \frac{\text{rise}}{\text{run}} \quad \leftarrow \text{change in } y \atop \leftarrow \text{change in } x$$

$$= \frac{-3}{2} \text{ or } -\frac{3}{2}$$

Your Turn The points given in each table lie on a line. Find the slope of the line. Then graph the line.

d.

x	−6	−2	2	6
y	−2	−1	0	1

e.

x	−1	0	1	2
y	−4	−4	−4	−4

Since slope is a rate of change, it can have real-life meaning.

EXAMPLES **Use Slope to Solve a Problem**

REAL-LIFE MATH

LIBRARIES With 85 branches, the New York Public Library is the world's largest public library. It has collections totaling 11.6 million items.

Source: www.nypl.org

3) **LIBRARIES** The graph shows the fines charged for overdue books per day at the Eastman Library. Find the slope of the line.

Count the units of vertical and horizontal change between any two points on the line.

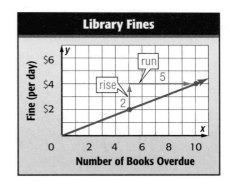

$$\text{slope} = \frac{\text{rise}}{\text{run}} \quad \text{Definition of slope}$$

$$= \frac{2}{5} \quad \text{rise} = 2, \text{run} = 5$$

The slope of the line is $\frac{2}{5}$.

4) **Interpret the meaning of this slope as a rate of change.**

For this graph, a slope of $\frac{2}{5}$ means that the library fine increases $2 for every 5 overdue books. Written as a unit rate, $\frac{\$2}{5}$ is $\frac{\$0.40}{1}$. The fine is $0.40 per overdue book per day.

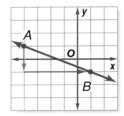

Writing Math
Exercise 2

1. **OPEN ENDED** Graph a line whose slope is 2 and another whose slope is 3. Which line is steeper?

2. **FIND THE ERROR** Juan and Martina are finding the slope of the line graphed at the right. Who is correct? Explain.

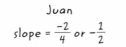

Juan
slope = $\frac{-2}{4}$ or $-\frac{1}{2}$

Martina
slope = $\frac{2}{4}$ or $\frac{1}{2}$

GUIDED PRACTICE

Find the slope of each line.

3.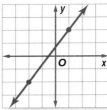

4.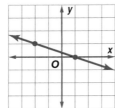

5. The points given in the table at the right lie on a line. Find the slope of the line. Then graph the line.

x	0	1	2	3
y	1	3	5	7

Practice and Applications

Find the slope of each line.

HOMEWORK HELP

For Exercises	See Examples
6–11	1
12–14	2
15–19	3, 4

Extra Practice
See pages 624, 651.

6.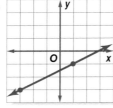

7.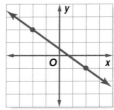

8.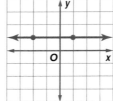

9.

10.

11.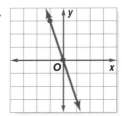

The points given in each table lie on a line. Find the slope of the line. Then graph the line.

12.
x	0	2	4	6
y	9	4	−1	−6

13.
x	−3	3	9	15
y	−3	1	5	9

14.
x	−4	0	4	8
y	7	7	7	7

Find the slope of each line and interpret its meaning as a rate of change.

15.
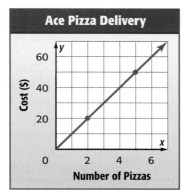

16.
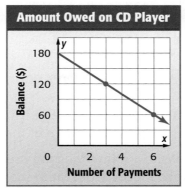

17.
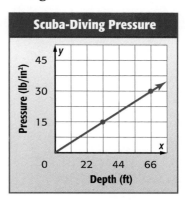

SAVINGS For Exercises 18 and 19, use the following information.

Pedro and Jenna are each saving money to buy the latest video game system. Their savings account balances over 7 weeks are shown in the graph at the right.

18. Find the slope of each person's line.

19. Who is saving more money each week? Explain.

20. **CRITICAL THINKING** According to federal guidelines, wheelchair ramps for access to public buildings are allowed a maximum of one inch of rise for every foot of run. Would a ramp with a slope of $\frac{1}{10}$ comply with this guideline? Explain your reasoning. (*Hint*: Convert feet to inches.)

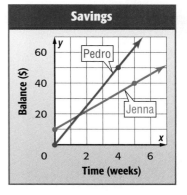

Standardized Test Practice and Mixed Review

FCAT Practice

21. **GRID IN** Find the slope of the roof shown.

22. **MULTIPLE CHOICE** The first major ski slope at a resort rises 8 feet vertically for every 48-foot run. The second rises 12 feet vertically for every 72-foot run. Which statement is true?

Ⓐ The first slope is steeper than the second.

Ⓑ The second slope is steeper than the first.

Ⓒ Both slopes have the same steepness.

Ⓓ This cannot be determined from the information given.

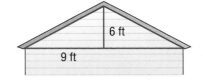

23. **POOL MAINTENANCE** After 15 minutes of filling a pool, the water level is at 2 feet. Twenty minutes later the water level is at 5 feet. Find rate of change in the water level between the first 15 minutes and the last 20 minutes in inches per minute. (Lesson 4-2)

24. Express *$25 for 10 disks* as a unit rate. (Lesson 4-1)

GETTING READY FOR THE NEXT LESSON

PREREQUISITE SKILL Solve each equation. Check your solution. (Lesson 1-9)

25. $5 \cdot x = 6 \cdot 10$ 26. $8 \cdot 3 = 4 \cdot y$ 27. $2 \cdot d = 3 \cdot 5$ 28. $2.1 \cdot 7 = 3 \cdot a$

Solving Proportions

Sunshine State Standards
MA.A.3.3.2-1, MA.A.3.3.2-2, MA.D.1.3.1-2

What You'll Learn
Use proportions to solve problems.

⇒*NEW* Vocabulary
proportion
cross products

WHEN am I ever going to use this?

NUTRITION Part of the nutrition label from a granola bar is shown at the right.

1. Write a ratio that compares the number of Calories from fat to the total number of Calories. Write the ratio as a fraction in simplest form.

2. Suppose you plan to eat two such granola bars. Write a ratio comparing the number of Calories from fat to the total number of Calories.

3. Is the ratio of Calories the same for two granola bars as it is for one granola bar? Why or why not?

Nutrition Facts

Serving Size 1 Bar (28g)
Servings Per Container 10

Amount Per Serving

Calories 110 Calories from Fat 20

	% Daily Value*
Total Fat 2g	3%
Saturated Fat 0.5g	2%
Cholesterol 0mg	0%
Sodium 70mg	3%

In the example above, the ratio $\frac{20}{110}$ simplifies to $\frac{2}{11}$. The equation $\frac{20}{110} = \frac{2}{11}$ indicates that the two ratios are equivalent. This is an example of a **proportion**.

Key Concept		Proportion
Words	A proportion is an equation stating that two ratios are equivalent.	
Symbols	Arithmetic	Algebra
	$\frac{6}{8} = \frac{3}{4}$	$\frac{a}{b} = \frac{c}{d}, b \neq 0, d \neq 0$

In a proportion, the two **cross products** are equal.

$$\begin{matrix} 6 & \times & 3 \\ 8 & = & 4 \end{matrix} \rightarrow 8 \cdot 3 = 24 \\ \rightarrow 6 \cdot 4 = 24$$

The cross products are equal.

STUDY TIP

Mental Math If both ratios simplify to the same fraction, they form a proportion.

$\frac{6}{15} = \frac{2}{5}$ and $\frac{8}{20} = \frac{2}{5}$.

So, $\frac{6}{15} = \frac{8}{20}$.

Key Concept		Property of Proportions
Words	The cross products of a proportion are equal.	
Symbols	If $\frac{a}{b} = \frac{c}{d}$, then $ad = bc$.	

You can use cross products to determine whether a pair of ratios forms a proportion. If the cross products of two ratios are equal, then the ratios form a proportion. If the cross products are *not* equal, the ratios do *not* form a proportion.

EXAMPLE Identify a Proportion

1 Determine whether the ratios $\frac{6}{9}$ and $\frac{8}{12}$ form a proportion.

Find the cross products.

$\begin{matrix} 6 & 8 \\ 9 & 12 \end{matrix} \xrightarrow{} 9 \cdot 8 = 72$
$\xrightarrow{} 6 \cdot 12 = 72$

Since the cross products are equal, the ratios form a proportion.

Your Turn Determine whether the ratios form a proportion.

a. $\frac{2}{5}, \frac{4}{10}$ b. $\frac{6}{16}, \frac{14}{56}$ c. $\frac{30}{35}, \frac{12}{14}$

Proportional Two ratios are said to be *proportional* if they form a proportion.

You can also use cross products to *solve proportions* in which one of the terms is not known.

EXAMPLE Solve a Proportion

2 Solve $\frac{x}{4} = \frac{9}{10}$.

$\frac{x}{4} = \frac{9}{10}$ Write the equation.

$x \cdot 10 = 4 \cdot 9$ Find the cross products.

$10x = 36$ Multiply.

$\frac{10x}{10} = \frac{36}{10}$ Divide each side by 10.

$x = 3.6$ Simplify.

The solution is 3.6. Check the solution by substituting the value of *x* into the original proportion and checking the cross products.

Your Turn Solve each proportion.

d. $\frac{7}{d} = \frac{2}{3}$ e. $\frac{2}{34} = \frac{5}{y}$ f. $\frac{7}{3} = \frac{n}{2.1}$

Proportions can be used to make predictions.

EXAMPLE Use a Proportion to Solve a Problem

3 **LIFE SCIENCE** A microscope slide shows 37 red blood cells out of 60 blood cells. How many red blood cells would be expected in a sample of the same blood that has 925 blood cells?

Write a proportion. Let *r* represent the number of red blood cells.

red blood cells → $\frac{37}{60} = \frac{r}{925}$ ← red blood cells
total blood cells →   ← total blood cells

$37 \cdot 925 = 60 \cdot r$ Find the cross products.

$34{,}225 = 60r$ Multiply.

$\frac{34{,}225}{60} = \frac{r}{60}$ Divide each side by 60.

$570.4 \approx r$ Simplify.

You would expect to find 570 or 571 red blood cells out of 925 blood cells.

Skill and Concept Check

1. **OPEN ENDED** List four different ratios that form a proportion with $\frac{12}{40}$.

2. **NUMBER SENSE** What would be a good estimate of the value of n in the equation $\frac{3}{5} = \frac{n}{11}$? Explain your reasoning.

GUIDED PRACTICE

Determine whether each pair of ratios form a proportion.

3. $\frac{8}{5}, \frac{40}{25}$

4. $\frac{3}{5}, \frac{5}{8}$

5. $\frac{6}{16}, \frac{9}{24}$

Solve each proportion.

6. $\frac{a}{13} = \frac{7}{1}$

7. $\frac{41}{x} = \frac{5}{2}$

8. $\frac{3.2}{9} = \frac{n}{36}$

Write a proportion that could be used to solve for each variable. Then solve.

9. 18 heart beats in 15 seconds
   b times in 60 seconds

10. 483 miles on 14 gallons of gas
    600 miles on g gallons of gas

Practice and Applications

HOMEWORK HELP

For Exercises	See Examples
11–18	1
19–30	2
31–42	3

Extra Practice
See pages 625, 651.

Determine whether each pair of ratios form a proportion.

11. $\frac{8}{7}, \frac{10}{9}$

12. $\frac{12}{14}, \frac{6}{7}$

13. $\frac{16}{12}, \frac{12}{9}$

14. $\frac{3}{11}, \frac{55}{200}$

15. $\frac{42}{56}, \frac{3}{4}$

16. $\frac{5}{18}, \frac{18}{65}$

17. $\frac{0.4}{5}, \frac{0.6}{7.5}$

18. $\frac{1.5}{0.5}, \frac{2.1}{7}$

Solve each proportion.

19. $\frac{k}{7} = \frac{32}{56}$

20. $\frac{44}{p} = \frac{11}{5}$

21. $\frac{45}{y} = \frac{3}{8}$

22. $\frac{x}{13} = \frac{18}{39}$

23. $\frac{6}{25} = \frac{d}{30}$

24. $\frac{48}{9} = \frac{72}{n}$

25. $\frac{15}{2.1} = \frac{12}{c}$

26. $\frac{2.5}{6} = \frac{h}{9}$

27. $\frac{3.5}{8} = \frac{a}{3.2}$

28. $\frac{2}{w} = \frac{0.4}{0.7}$

29. $\frac{2}{3} = \frac{18}{x + 5}$

30. $\frac{m - 4}{10} = \frac{7}{5}$

Write a proportion that could be used to solve for each variable. Then solve.

31. 6 Earth-pounds equals 1 moon-pound
    96 Earth-pounds equals p moon-pounds

32. 2 pages typed in 13 minutes
    25 pages typed in m minutes

33. 3 pounds of seed for 2,000 square feet
    x pounds of seed for 3,500 square feet

34. n cups flour used with $\frac{3}{4}$ cup sugar
    $1\frac{1}{2}$ cups flour used with $\frac{1}{2}$ cup sugar

35. **LIFE SCIENCE** About 4 out of every 5 people are right-handed. If there are 30 students in a class, how many would you expect to be right-handed?

PEOPLE For Exercises 36 and 37, use the following information. Although people vary in size and shape, in general, people do not vary in proportion. The head height to overall height ratio for an adult is given in the diagram at the right.

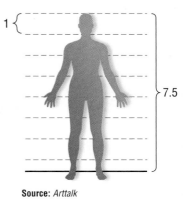

1 {
7.5
Source: *Arttalk*

36. About how tall is an adult with a head height of 9.6 inches?

37. Find the average head height of an adult that is 64 inches tall.

38. RECYCLING The amount of paper recycled is directly proportional to the number of trees that recycling saves. If recycling 2,000 pounds of paper saves 17 trees, how many trees are saved when 5,000 pounds of paper are recycled?

MEASUREMENT For Exercises 39–42, refer to the table. Write and solve a proportion to find each quantity.

Customary System to Metric System
1 inch ≈ 2.54 centimeters
1 mile ≈ 1.61 kilometers
1 gallon ≈ 3.78 liters
1 pound ≈ 0.454 kilogram

39. 12 inches = ■ centimeters

40. 20 miles = ■ kilometers

41. 2 liters = ■ gallons

42. 45 kilograms = ■ pounds

CRITICAL THINKING Classify the following pairs of statements as having a *proportional* or *nonproportional* relationship. Explain.

43. You jump 63 inches and your friend jumps 42 inches. You jump 1.5 times the distance your friend jumps.

44. You jump 63 inches and your friend jumps 42 inches. You jump 21 more inches than your friend jumps.

Standardized Test Practice and Mixed Review

A B C D **FCAT Practice**

45. MULTIPLE CHOICE At Northside Middle School, 30 students were surveyed about their favorite type of music. The results are graphed at the right. If there are 440 students at the middle school, predict how many prefer country music.

Ⓐ 126   Ⓑ 128   Ⓒ 130   Ⓓ 132

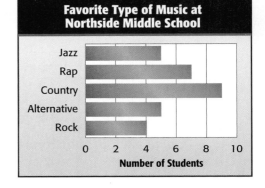

Favorite Type of Music at Northside Middle School

Number of Students

46. SHORT RESPONSE Yutaka can run 3.5 miles in 40 minutes. How many minutes would it take him to run 8 miles at this same rate?

47. The points given in the table lie on a line. Find the slope of the line. Then graph the line. (Lesson 4-3)

x	0	3	6	9
y	−10	−2	6	14

48. GARDENING Three years ago, an oak tree in Emily's back yard was 4 feet 5 inches tall. Today it is 6 feet 3 inches tall. How fast did the tree grow in inches per year? (Lesson 4-2)

GETTING READY FOR THE NEXT LESSON

BASIC SKILL Name the sides of each figure.

49. triangle *ABC*

50. rectangle *DEFG*

51. square *LMNP*

Vocabulary and Concepts

1. **Explain** the meaning of a rate of change of $-2°$ per hour. (Lesson 4-2)
2. **Describe** how to find the slope of a line given two points on the line.
   (Lesson 4-3)

Skills and Applications

Express each ratio in simplest form. (Lesson 4-1)

3. 32 out of 100 dentists
4. 12 chosen out of 60
5. 300 points in 20 games

6. Express *$420 for 15 tickets* as a unit rate. (Lesson 4-1)

TEMPERATURE For Exercises 7 and 8, use the table at the right. (Lesson 4-2)

Time	Temperature (°F)
12 P.M.	88
1 P.M.	86
3 P.M.	60
5 P.M.	66
6 P.M.	64

7. Find the rate of the temperature change in degrees per hour from 1 P.M. to 3 P.M. and from 5 P.M. to 6 P.M.

8. Was the rate of change between 12 P.M. and 3 P.M. positive, negative, or zero?

Find the slope of each line. (Lesson 4-3)

9.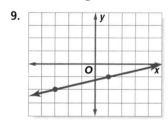

10.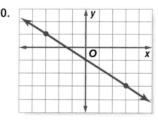

11.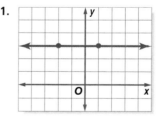

Solve each proportion. (Lesson 4-4)

12. $\dfrac{33}{r} = \dfrac{11}{2}$

13. $\dfrac{x}{36} = \dfrac{15}{24}$

14. $\dfrac{5}{9} = \dfrac{4.5}{a}$

Standardized Test Practice

FCAT Practice

15. **GRID IN** A typical 30-minute TV program in the United States has about 8 minutes of commercials. At that rate, how many commercial minutes are shown during a 2-hour TV movie?
(Lesson 4-4)

16. **MULTIPLE CHOICE** There are 2 cubs for every 3 adults in a certain lion pride. If the pride has 8 cubs, how many adults are there? (Lesson 4-4)

   (A) 12   (B) 16

   (C) 24   (D) 48

The GameZone

A Place To Practice Your Math Skills

Criss Cross

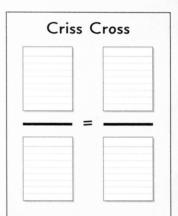

Criss Cross

- **GET READY!**

    Players: two to four
    Materials: paper; scissors; 24 index cards

- **GET SET!**

    - Each player should copy the game board shown onto a piece of paper.
    - Cut each index card in half, making 48 cards.
    - Copy the numbers below, one number onto each card.

1	1	1	1	2	2	2	3	3	3	4	4
4	5	5	5	6	6	6	7	7	7	8	8
8	9	9	9	10	10	11	11	12	12	13	13
14	14	15	15	16	16	18	18	20	22	24	25

    - Deal 8 cards to each player. Place the rest facedown in a pile.

- **GO!**

    - The player to the dealer's right begins by trying to form a proportion using his or her cards. If a proportion is formed, the player says, "Criss cross!" and displays the cards on his or her game board.
    - If the cross products of the proportion are equal, the player forming the proportion is awarded 4 points and those cards are placed in a discard pile. If not that player loses his or her turn.
    - If a player cannot form a proportion, he or she draws a card from the first pile. If the player cannot use the card, play continues to the right.
    - When there are no more cards in the original pile, shuffle the cards in the discard pile and use them.
    - **Who Wins?** The first player to reach 20 points wins the game.

Problem-Solving Strategy
A Preview of Lesson 4-5

Sunshine State Standards
MA.A.3.3.2-1, MA.A.3.3.2-2,
MA.A.3.3.3-1, MA.B.3.3.1-1, MA.D.2.3.1

What You'll Learn
Solve problems by using the draw a diagram strategy.

Draw a Diagram

Cleaning tanks for the city aquarium sure is hard work, and filling them back up seems to take forever. It's been 3 minutes and this 120-gallon tank is only at the 10-gallon mark!

I wonder how much longer it will take? Let's **draw a diagram** to help us picture what's happening.

Explore	The tank holds 120 gallons of water. After 3 minutes, the tank has 10 gallons of water in it. How many more minutes will it take to fill the tank?
Plan	Let's draw a diagram showing the water level after every 3 minutes.
Solve	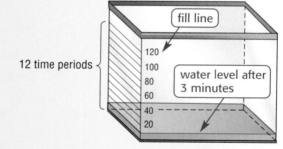 The tank will be filled after twelve 3-minute time periods. This is a total of 12 × 3 or 36 minutes.
Examine	The tank is filling at a rate of 10 gallons every 3 minutes, which is about 3 gallons per minute. So a 120-gallon tank will take about 120 ÷ 3 or 40 minutes to fill. Our answer of 36 minutes seems reasonable.

Analyze the Strategy

1. **Tell** how drawing a diagram helps solve this problem.

2. **Describe** another method the students could have used to find the number of 3-minute time periods it would take to fill the tank.

3. **Write** a problem that can be solved by drawing a diagram. Then draw a diagram and solve the problem.

Solve. Use the draw a diagram strategy.

4. **AQUARIUM** Angelina fills another 120-gallon tank at the same time Kyle is filling the first 120-gallon tank. After 3 minutes, her tank has 12 gallons in it. How much longer will it take Kyle to fill his tank than Angelina?

5. **LOGGING** It takes 20 minutes to cut a log into 5 equally-sized pieces. How long will it take to cut a similar log into 3 equally-sized pieces?

Mixed Problem Solving

Solve. Use any strategy.

6. **STORE DISPLAY** A stock clerk is stacking oranges in the shape of a square-based pyramid, as shown at the right. If the pyramid is to have 5 layers, how many oranges will he need?

FOOD For Exercises 7 and 8, use the following information.
Of the 30 students in a life skills class, 19 like to cook main dishes, 15 prefer baking desserts, and 7 like to do both.

7. How many like to cook main dishes, but not bake desserts?

8. How many do not like either baking desserts or making main dishes?

9. **MOVIES** A section of a theater is arranged so that each row has the same number of seats. You are seated in the 5th row from the front and the 3rd row from the back. If your seat is 6th from the left and 2nd from the right, how many seats are in this section of the theater?

10. **MONEY** Mi-Ling has only nickels in her pocket. Julián has only quarters in his and Aisha has only dimes in hers. Hannah approached all three for a donation for the school fund-raiser. What is the least each person could donate so that each one gives the same amount?

11. **TOURISM** An amusement park in Texas features giant statues of comic strip characters. If you multiply one character's height by 4 and add 1 foot, you will find the height of its statue. If the statue is 65 feet tall, how tall is the character?

TECHNOLOGY For Exercises 12 and 13, use the diagram below and the following information.
Seven closed shapes are used to make the digits 0 to 9 on a digital clock. (The number 1 is made using the line segments on the right side of the figure.)

12. In forming these digits, which line segment is used most often?

13. Which line segment is used the least?

14. **SPORTS** The width of a tennis court is ten more than one-third its length. If the court is 78 feet long, what is its perimeter?

15. **STANDARDIZED TEST PRACTICE** FCAT Practice
Three-inch square tiles that are 2 inches high are being packaged into boxes like the one at the right. If the tiles must be laid flat, how many will fit in one box?

(A) 140   (B) 150   (C) 450   (D) 900

You will use the draw a diagram strategy in the next lesson.

Similar Polygons

Sunshine State Standards
MA.B.2.3.1-1, MA.C.1.3.1-4, MA.C.2.3.1-2, MA.C.3.3.1-1

What You'll Learn

Identify similar polygons and find missing measures of similar polygons.

NEW Vocabulary

polygon
similar
corresponding parts
congruent
scale factor

MATH Symbols

∠ angle

$\overline{AB}$ segment *AB*

~ is similar to

≅ is congruent to

AB measure of $\overline{AB}$

HANDS-ON Mini Lab

Materials
- tracing paper
- centimeter ruler
- scissors

Work with a partner.

Follow the steps below to discover how the triangles at the right are related.

STEP 1 Copy both triangles onto tracing paper.

STEP 2 Measure and record the sides of each triangle.

STEP 3 Cut out both triangles.

1. Compare the angles of the triangles by matching them up. Identify the angle pairs that have equal measure.

2. Express the ratios $\dfrac{DF}{LK}$, $\dfrac{EF}{JK}$, and $\dfrac{DE}{LJ}$ as decimals to the nearest tenth.

3. What do you notice about the ratios of the matching sides of matching triangles?

A simple closed figure in a plane formed by three or more line segments is called a **polygon**. Polygons that have the same shape are called **similar** polygons. In the figure below, polygon *ABCD* is similar to polygon *WXYZ*. This is written as polygon *ABCD* ~ polygon *WXYZ*.

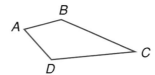
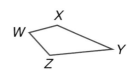

The parts of similar figures that "match" are called **corresponding parts**.

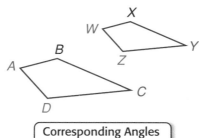

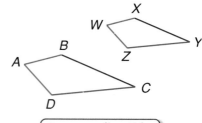

Corresponding Angles
∠A ↔ ∠W, ∠B ↔ ∠X,
∠C ↔ ∠Y, ∠D ↔ ∠Z

Corresponding Sides
$\overline{AB}$ ↔ $\overline{WX}$, $\overline{BC}$ ↔ $\overline{XY}$,
$\overline{CD}$ ↔ $\overline{YZ}$, $\overline{DA}$ ↔ $\overline{ZW}$

The similar triangles in the Mini Lab suggest that the following
properties are true for similar polygons.

Key Concept — **Similar Polygons**

Words	If two polygons are similar, then • their corresponding angles are **congruent**, or have the same measure, and • their corresponding sides are proportional.

Models

$\triangle ABC \sim \triangle XYZ$

Symbols $\quad \angle A \cong \angle X, \angle B \cong \angle Y, \angle C \cong \angle Z$ and $\dfrac{AB}{XY} = \dfrac{BC}{YZ} = \dfrac{AC}{XZ}$

EXAMPLE **Identify Similar Polygons**

1. **Determine whether rectangle *HJKL* is similar to rectangle *MNPQ*. Explain your reasoning.**

   First, check to see if corresponding angles are congruent.

   Since the two polygons are rectangles, all of their angles are right angles. Therefore, all corresponding angles are congruent.

   Next, check to see if corresponding sides are proportional.

   $$\frac{HJ}{MN} = \frac{7}{10} \qquad \frac{JK}{NP} = \frac{3}{6} \text{ or } \frac{1}{2} \qquad \frac{KH}{PM} = \frac{7}{10} \qquad \frac{PQ}{KL} = \frac{3}{6} \text{ or } \frac{1}{2}$$

   Since $\frac{7}{10}$ and $\frac{1}{2}$ are not equivalent ratios, rectangle *HJKL* is *not* similar to rectangle *MNPQ*.

Your Turn

a. Determine whether these polygons are similar. Explain your reasoning.

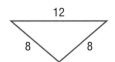

The ratio of the lengths of two corresponding sides of two similar polygons is called the **scale factor**. The squares below are similar.

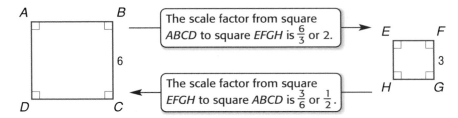

The scale factor from square *ABCD* to square *EFGH* is $\frac{6}{3}$ or 2.

The scale factor from square *EFGH* to square *ABCD* is $\frac{3}{6}$ or $\frac{1}{2}$.

EXAMPLE — Find Missing Measures

2 Given that polygon $ABCD \sim$ polygon $WXYZ$, write a proportion to find the measure of $\overline{XY}$. Then solve.

The scale factor from polygon $ABCD$ to polygon $WXYZ$ is $\dfrac{CD}{YZ}$, which is $\dfrac{10}{15}$ or $\dfrac{2}{3}$. Write a proportion with this scale factor. Let m represent the measure of $\overline{XY}$.

$\dfrac{BC}{XY} = \dfrac{2}{3}$ $\overline{BC}$ corresponds to $\overline{XY}$. The scale factor is $\dfrac{2}{3}$.

$\dfrac{12}{m} = \dfrac{2}{3}$ $BC = 12$ and $XY = m$

$12 \cdot 3 = m \cdot 2$ Find the cross products.

$\dfrac{36}{2} = \dfrac{2m}{2}$ Multiply. Then divide each side by 2.

$18 = m$ Simplify.

Your Turn Write a proportion to find the measure of each side above. Then solve.

   b. $\overline{WZ}$ **c.** $\overline{AB}$

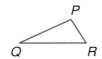

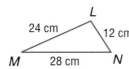

FCAT Practice

A B C D

Standardized Test Practice

EXAMPLE — Scale Factor and Perimeter

3 **MULTIPLE-CHOICE TEST ITEM** Triangle $LMN \sim \triangle PQR$. Each side of $\triangle LMN$ is $1\frac{1}{3}$ times longer than the corresponding sides of $\triangle PQR$.

If the perimeter of $\triangle LMN$ is 64 centimeters, what is the perimeter of $\triangle PQR$?

   A $5\frac{1}{3}$ cm   **B** 16 cm   **C** 48 cm   **D** 61 cm

Read the Test Item Since each side of $\triangle LMN$ is $1\frac{1}{3}$ times longer than the corresponding sides of $\triangle PQR$, the scale factor from $\triangle LMN$ to $\triangle PQR$ is $1\frac{1}{3}$ or $\frac{4}{3}$.

Solve the Test Item Let x represent the perimeter of $\triangle PQR$. The ratio of the perimeters is equal to the ratio of the sides.

ratio of perimeters $\rightarrow$ $\left\{\dfrac{64}{x} = \dfrac{4}{3}\right\}$ $\leftarrow$ ratio of sides

$64(3) = x \cdot 4$ Find the cross products.

$\dfrac{192}{4} = \dfrac{4x}{4}$ Multiply. Then divide each side by 4.

$48 = x$ Simplify.

The answer is C.

Test-Taking Tip
The Princeton Review

Use a Proportion
In similar figures, the ratio of the perimeters is the same as the ratio of corresponding sides. Use a proportion.

1. **Explain** how you can determine whether two polygons are similar.

2. **OPEN ENDED** Draw and label a pair of similar rectangles. Then draw a third rectangle that is not similar to the other two.

GUIDED PRACTICE

Determine whether each pair of polygons is similar. Explain your reasoning.

3.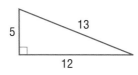

4.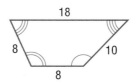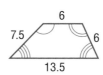

5. In the figure at the right, $\triangle FGH \sim \triangle KLJ$. Write a proportion to find each missing measure. Then solve.

 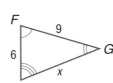

Practice and Applications

Determine whether each pair of polygons is similar. Explain your reasoning.

6.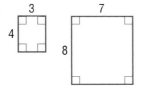

7.

8.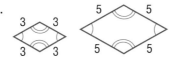

9.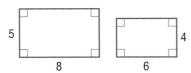

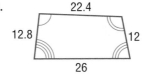

HOMEWORK HELP

For Exercises	See Examples
6–9	1
10–16	2
17	3

Extra Practice
See pages 625, 651.

Each pair of polygons is similar. Write a proportion to find each missing measure. Then solve.

10.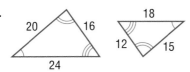

11.

12.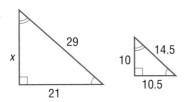

13.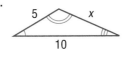

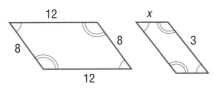

14. YEARBOOK In order to fit 3 pictures across a page, the yearbook staff must reduce their portrait proofs using a scale factor of 8 to 5. Find the dimensions of the pictures as they will appear in the yearbook.

5 in.

4 in.

MOVIES For Exercises 15 and 16, use the following information.
Film labeled 35-millimeter is film that is 35 millimeters wide.

15. When a frame of 35-millimeter movie film is projected onto a movie screen, the image from the film is 9 meters high and 6.75 meters wide. Find the height of the film.

16. If the image from this same film is projected so that it appears 8 meters high, what is the width of the projected image?

17. GEOMETRY Find the ratio of the area of rectangle A to the area of rectangle B for each of the following scale factors of corresponding sides. What can you conclude?

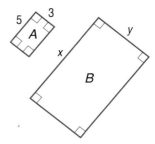

a. $\frac{1}{2}$ **b.** $\frac{1}{3}$ **c.** $\frac{1}{4}$ **d.** $\frac{1}{5}$

CRITICAL THINKING Determine whether each statement is *always*, *sometimes*, or *never* true. Explain your reasoning.

18. Two rectangles are similar. **19.** Two squares are similar.

Standardized Test Practice and Mixed Review

FCAT Practice

20. MULTIPLE CHOICE Which triangle is similar to $\triangle ABC$?

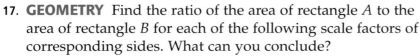

A 4.8 m 12 m 13 m

B 3.4 m 7 m 7.5 m

A 14 m 6.8 m C 15 m B

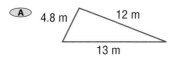

C 3.6 m 9 m 10 m

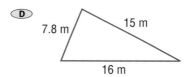

D 7.8 m 15 m 16 m

21. SHORT RESPONSE Polygon $JKLM \sim$ polygon $QRST$. If $JK = 2$ inches and $QR = 2\frac{1}{2}$ inches, find the measure of ST if $LM = 3$ inches.

22. BAKING A recipe calls for 4 cups of flour for 64 cookies. How much flour is needed for 96 cookies? (Lesson 4-4)

Graph each pair of points. Then find the slope of the line that passes through each pair of points. (Lesson 4-3)

23. $(-3, 9), (1, -5)$ **24.** $(2, 4), (-6, 7)$ **25.** $(3, -8), (-1, -8)$

GETTING READY FOR THE NEXT LESSON

PREREQUISITE SKILL Write a proportion and solve for x. (Lesson 4-4)

26. 3 cm is to 5 ft as x cm is to 9 ft **27.** 4 in. is to 5 mi as 5 in. is to x mi

What You'll Learn
Find the value of the golden ratio.

Materials
- grid paper
- scissors
- calculator
- tape measure

Sunshine State Standards
MA.A.3.3.2-1, MA.A.3.3.2-2, MA.C.2.3.1-2, MA.C.3.3.1-1

The Golden Rectangle

INVESTIGATE *Work in groups of three.*

STEP 1 Cut a rectangle out of grid paper that measures 34 units long by 21 units wide. Using your calculator, find the ratio of the length to the width. Express it as a decimal to the nearest hundredth. Record your data in a table like the one below.

length	34	21	?	?	?	?	?
width	21	13	?	?	?	?	?
ratio	?	?	?	?	?	?	?
decimal	?	?	?	?	?	?	?

STEP 2 Cut this rectangle into two parts, in which one part is the largest possible square and the other part is a rectangle. Record the rectangle's length and width. Write the ratio of length to width. Express it as a decimal to the nearest hundredth and record in the table.

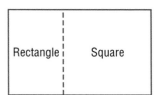

Rectangle · Square

STEP 3 Repeat the procedure described in Step 2 until the remaining rectangle measures 3 units by 5 units.

Writing Math

1. **Describe** the pattern in the ratios you recorded.

2. If the rectangles you cut out are described as *golden rectangles*, **make a conjecture** as to what the value of the *golden ratio* is.

3. **Write** a definition of *golden rectangle*. Use the word *ratio* in your definition. Then describe the shape of a golden rectangle.

4. **Determine** whether all golden rectangles are similar. Explain your reasoning.

5. **RESEARCH** There are many examples of the golden rectangle in architecture. One is shown at the right. Use the Internet or another resource to find three places where the golden rectangle is used in architecture.

Taj Mahal, India

4-6

Scale Drawings and Models

Sunshine State Standards
MA.A.3.3.2-1, MA.A.3.3.2-2, MA.A.3.3.3-1, MA.B.1.3.4-1, MA.B.1.3.4-2, MA.B.2.3.1-1

What You'll Learn
Solve problems involving scale drawings.

⮞NEW Vocabulary
scale drawing
scale model
scale

WHEN am I ever going to use this?

FLOOR PLANS The blueprint for a bedroom is given below.

1. How many units wide is the room?

2. The actual width of the room is 18 feet. Write a ratio comparing the drawing width to the actual width.

3. Simplify the ratio you found and compare it to the scale shown at the bottom of the drawing.

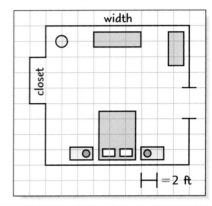

A **scale drawing** or a **scale model** is used to represent an object that is too large or too small to be drawn or built at actual size. Examples are blueprints, maps, models of vehicles, and models of animal anatomy.

The **scale** is determined by the ratio of a given length on a drawing or model to its corresponding actual length. Consider the scales below.

> 1 inch = 4 feet   1 inch represents an actual distance of 4 feet.
>
> 1:30   1 unit represents an actual distance of 30 units.

Distances on a scale drawing are proportional to distances in real-life.

EXAMPLE Find a Missing Measurement

1 **RECREATION** The distance from the roller coaster to the food court on the map is 3.5 centimeters. Find the actual distance to the food court.

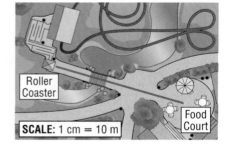

Let x represent the actual distance to the food court. Write and solve a proportion.

STUDY TIP

Scales Scales and scale factors are always written so that the drawing length comes first in the ratio.

Map Scale ⌐⎯⎯⎯⎯⎯⎯⎯⌐⎯⎯⎯⎯⎯⎯ Actual Distance

$$\begin{array}{l} \text{map distance} \rightarrow \\ \text{actual distance} \rightarrow \end{array} \quad \frac{1 \text{ cm}}{10 \text{ m}} = \frac{3.5 \text{ cm}}{x \text{ m}} \quad \begin{array}{l} \leftarrow \text{map distance} \\ \leftarrow \text{actual distance} \end{array}$$

$$1 \cdot x = 10 \cdot 3.5 \qquad \text{Find the cross products.}$$

$$x = 35 \qquad \text{Simplify.}$$

The actual distance to the food court is 35 meters.

184 Chapter 4 Proportions, Algebra, and Geometry

To find the scale factor for scale drawings and models, write the ratio given by the scale in simplest form.

EXAMPLE Find the Scale Factor

2 Find the scale factor for the map in Example 1.

$$\frac{1 \text{ cm}}{10 \text{ m}} = \frac{1 \text{ cm}}{1,000 \text{ cm}}$$ Convert 10 meters to centimeters.

The scale factor is $\frac{1}{1,000}$ or 1:1,000. This means that each distance on the map is $\frac{1}{1,000}$ the actual distance.

EXAMPLE Find the Scale

3 **MODEL TRAINS** A passenger car of a model train is 6 inches long. If the actual car is 80 feet long, what is the scale of the model?

Write a ratio comparing the length of the model to the actual length of the train. Using x to represent the actual length of the train, write and solve a proportion to find the scale of the model.

Length of Train � Model Scale

model length → $\frac{6 \text{ in.}}{80 \text{ ft}} = \frac{1 \text{ in.}}{x \text{ ft}}$ ← model length
actual length →   ← actual length

$6 \cdot x = 80 \cdot 1$ Find the cross products.

$\frac{6x}{6} = \frac{80}{6}$ Multiply. Then divide each side by 6.

$x = 13\frac{1}{3}$ Simplify.

So, the scale is 1 inch = $13\frac{1}{3}$ feet.

To construct a scale drawing of an object, find an appropriate scale.

EXAMPLE Construct a Scale Model

4 **SOCIAL STUDIES** Each column of the Lincoln Memorial is 44 feet tall. Michaela wants the columns of her model to be no more than 12 inches tall. Choose an appropriate scale and use it to determine how tall she should make the model of Lincoln's 19-foot statue.

Try a scale of 1 inch = 4 feet.

$\frac{1 \text{ in.}}{4 \text{ ft}} = \frac{x \text{ in.}}{44 \text{ ft}}$ ← model length
                          ← actual length

$1 \cdot 44 = 4 \cdot x$ Find the cross products.

$44 = 4x$ Multiply.

$11 = x$ Divide each side by 4.

The columns are 11 inches tall.

Use this scale to find the height of the statue.

$\frac{1 \text{ in.}}{4 \text{ ft}} = \frac{y \text{ in.}}{19 \text{ ft}}$

$1 \cdot 19 = 4 \cdot y$

$19 = 4y$

$4\frac{3}{4} = y$

The statue is $4\frac{3}{4}$ inches tall.

Skill and Concept Check

1. **OPEN ENDED** Choose an appropriate scale for a scale drawing of a bedroom 10 feet wide by 12 feet long. Identify the scale factor.

2. **FIND THE ERROR** On a map, 1 inch represents 4 feet. Jacob and Luna are finding the scale factor of the map. Who is correct? Explain.

> Jacob
> scale factor: 1:4

> Luna
> scale factor: 1:48

GUIDED PRACTICE

On a map of the United States, the scale is 1 inch = 120 miles. Find the actual distance for each map distance.

	From	To	Map Distance
3.	South Bend, Indiana	Enid, Oklahoma	6 inches
4.	Atlanta, Georgia	Memphis, Tennessee	$2\frac{3}{4}$ inches

MONUMENTS For Exercises 5 and 6, use the following information.
At 555 feet tall, the Washington Monument is the highest all-masonry tower.

5. A scale model of the monument is 9.25 inches high. What is the model's scale?

6. What is the scale factor?

Practice and Applications

The scale on a set of architectural drawings for a house is 0.5 inch = 3 feet. Find the actual length of each room.

	Room	Drawing Length
7.	Bed Room 2	2 inches
8.	Living Room	3 inches
9.	Kitchen	1.4 inches

	Room	Drawing Length
10.	Dining Room	2.1 inches
11.	Master Bedroom	$2\frac{1}{4}$ inches
12.	Bath	$1\frac{1}{8}$ inches

HOMEWORK HELP

For Exercises	See Examples
7–12, 14	1
13, 21	2
15–16, 18	3
17, 19, 20	4

Extra Practice
See pages 625, 651.

13. Refer to Exercises 7–12. What is the scale factor of these drawings?

14. **MULTI STEP** On the drawings for Exercises 7–12, the area of the living room is 15 square inches. What is the actual area of the living room?

15. **LIFE SCIENCE** In the picture of a paramecium at the right, the length of the single celled organism is 4 centimeters. If the paramecium's actual size is 0.006 millimeter, what is the scale of the drawing?

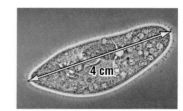
4 cm

16. **MOVIES** One of the models of the gorilla used in the filming of a 1933 movie was only 18 inches tall. In the movie, the gorilla was seen as 24 feet high. What was the scale used?

17. **SPIDERS** A tarantula's body length is 5 centimeters. Choose an appropriate scale for a model of the spider that is to be just over 6 meters long. Use it to determine how long the tarantula's 9-centimeter legs should be.

SPACE For Exercises 18 and 19, use the information in the table.

18. You decide to use a basketball to represent Earth in a scale model of Earth and the moon. A basketball's circumference is about 30 inches. What is the scale of your model?

19. Which of the following should you use to represent the moon in your model? (The number in parentheses is the object's circumference.) Explain your reasoning.

Astrological Body	Approximate Circumference
Earth	40,000 km
moon	11,000 km

   a. a soccer ball (28 in.)   **b.** a tennis ball (8.25 in.)

   c. a golf ball (5.25 in.)   **d.** a marble (4 in.)

20. **CONSTRUCT A SCALE DRAWING** Choose a large rectangular space such as the floor or wall of a room. Find its dimensions and choose an appropriate scale for a scale drawing of the space. Then construct a scale drawing and write a problem that uses your drawing.

21. **NUMBER SENSE** One model of a building is built on a 1:75 scale. Another model of the same building is built on a 1:100 scale. Which model is larger? Explain your reasoning.

22. **CRITICAL THINKING** Describe how you could find the scale of a map that did not have a scale printed on it.

Standardized Test Practice and Mixed Review

A B C D   **FCAT** Practice

23. **MULTIPLE CHOICE** Using which scale would a scale model of a statue appear $\frac{1}{12}$ the size of the actual statue?

   A 4 in. = 8 ft   **B** 3 in. = 36 ft   **C** 3 in. = 4 ft   **D** 4 in. = 4 ft

24. **SHORT RESPONSE** The distance between San Antonio and Houston is $6\frac{3}{4}$ inches on a map with a scale of $\frac{1}{2}$ inch = 15 miles. About how long would it take to drive this distance going 60 miles per hour?

25. Determine whether the polygons at the right are similar. Explain your reasoning. (Lesson 4-5)

Solve each proportion. (Lesson 4-4)

26. $\dfrac{120}{b} = \dfrac{24}{60}$ 27. $\dfrac{0.6}{5} = \dfrac{1.5}{n}$ 28. $\dfrac{10}{6} = \dfrac{p}{26}$

GETTING READY FOR THE NEXT LESSON

PREREQUISITE SKILL In the figure, $\triangle ABC \sim \triangle DFC$. (Lesson 4-5)

29. Identify the corresponding angles in the figure.

30. Identify the corresponding sides in the figure.

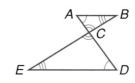

Sunshine State Standards
MA.A.3.3.2-1, MA.A.3.3.2-2, MA.B.2.3.1-1

What You'll Learn
Solve problems involving similar triangles.

NEW Vocabulary

indirect measurement

WHEN am I ever going to use this?

COMICS The caveman is trying to measure the distance to the Sun.

1. How is the caveman measuring the distance to the Sun?

Distances or lengths that are difficult to measure directly can sometimes be found using the properties of similar polygons and proportions. This kind of measurement is called **indirect measurement**.

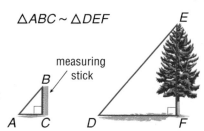

$\triangle ABC \sim \triangle DEF$

One type of indirect measurement is called *shadow reckoning*. Two objects and their shadows form two sides of similar triangles from which a proportion can be written.

stick's shadow → $\dfrac{AC}{DF} = \dfrac{BC}{EF}$ ← stick's height
tree's shadow →   ← tree's height

EXAMPLE Use Shadow Reckoning

1 **FLAGS** One of the tallest flagpoles in the U.S. is in Winsted, Minnesota. At the same time of day that Karen's shadow was about 0.8 meter, the flagpole's shadow was about 33.6 meters. If Karen is 1.5 meters tall, how tall is Winsted's flagpole?

Karen's shadow → $\dfrac{0.8}{33.6} = \dfrac{1.5}{h}$ ← Karen's height
flagpole's shadow →   ← flagpole's height

$0.8h = 33.6 \cdot 1.5$ Find the cross products.

$0.8h = 50.4$ Multiply.

$\dfrac{0.8x}{0.8} = \dfrac{50.4}{0.8}$ Divide each side by 0.8

$x = 63$ Use a calculator.

The flagpole is 63 meters tall.

You can also use similar triangles that do not involve shadows to find missing measurements.

EXAMPLE **Use Indirect Measurement**

2 **SURVEYING** The two triangles shown in the figure are similar. Find the distance d across Coyote Ravine.

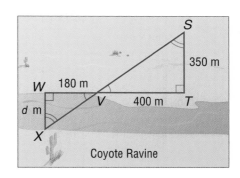

In the figure, $\triangle STV \sim \triangle XWV$. So, $\overline{ST}$ corresponds to $\overline{XW}$, and $\overline{TV}$ corresponds to $\overline{WV}$.

$$\frac{ST}{XW} = \frac{TV}{WV}$$ Write a proportion.

$$\frac{350}{d} = \frac{400}{180}$$ $ST = 350$, $XW = d$, $TV = 400$, and $WV = 180$

$350 \cdot 180 = d \cdot 400$ Find the cross products.

$$\frac{63,000}{400} = \frac{400d}{400}$$ Multiply. Then divide each side by 400.

$157.5 = d$ Use a calculator.

The distance across the ravine is 157.5 meters.

Skill and Concept Check

Writing Math
Exercise 2

1. **Draw and label** similar triangles to illustrate the following problem. Then write an appropriate proportion.
   A building's shadow is 14 feet long, and a street sign's shadow is 5 feet long. If the street sign is 6 feet tall, how tall is the building?

2. **OPEN ENDED** Write a problem that requires shadow reckoning. Explain how to solve the problem.

GUIDED PRACTICE

In Exercises 3 and 4, the triangles are similar. Write a proportion and solve the problem.

3. **ARCHITECTURE** How tall is the pyramid?

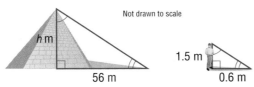

4. **BRIDGES** How far is it across the river?

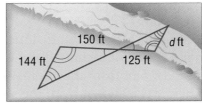

5. A building casts a 18.5-foot shadow. How tall is the building if a 10-foot tall sculpture nearby casts a 7-foot shadow? Draw a diagram of the situation. Then write a proportion and solve the problem.

HOMEWORK HELP

For Exercises	See Examples
6–7, 10–12, 14	1
8–9, 13, 15	2

Extra Practice
See pages 626, 651.

In Exercises 6–9, the triangles are similar. Write a proportion and solve the problem.

6. **REPAIRS** How tall is the telephone pole?

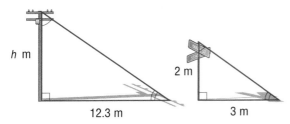

h m

2 m

12.3 m

3 m

7. **LIGHTHOUSE** How tall is the house?

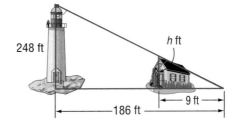

248 ft

h ft

186 ft

9 ft

8. **ZOO** How far are the elephants from the aquarium?

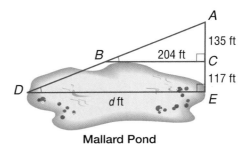

x m

68 m

17 m

20 m

9. **SURVEYING** How far is it across Mallard Pond? (*Hint*: △*ABC* ~ △*ADE*)

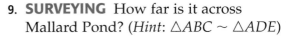

A

135 ft

B   204 ft   *C*

117 ft

D

d ft

E

Mallard Pond

For Exercises 10–15, draw a diagram of the situation. Then write a proportion and solve the problem.

10. **NATIONAL MONUMENT** Devil's Tower in Wyoming was the United States' first national monument. At the same time this natural rock formation casts a 181-foot shadow, a nearby 224-foot tree casts a 32-foot shadow. How tall is the monument?

11. **FAIR** Reaching 212 feet tall, the *Texas Star* at Fair Park in Dallas, Texas, is the tallest Ferris wheel in the United States. A man standing near this Ferris wheel casts an 3-foot shadow. At the same time, the Ferris wheel's shadow is 106 feet long. How tall is the man?

12. **TOWER** The Stratosphere Tower in Las Vegas is the tallest free-standing observation tower in the United States. If the tower cast a 22.5-foot shadow, about how tall is a nearby flagpole that casts a 3-foot shadow? Use the information at the right.

Stratosphere Hotel Facts

Floor 108 Indoor observation deck of 1,149-foot tall tower

Floor 112 World's highest roller coaster, the *High Roller*

Floor 113 *Big Shot* ride shoots riders 160 feet up tower mast in 2.5 seconds, allowing them to free-fall back to the launch pad

13. **LAKES** From the shoreline, the ground slopes down under the water at a constant incline. If the water is 3 feet deep when it is 5 feet from the shore, about how deep will it be when it is 62.5 feet from the shore?

14. **LANDMARKS** The Gateway to the West Arch in St. Louis casts a shadow that is 236 foot 3 inches. At the same time, a 5 foot 4 inch tall tourist casts a 2-foot shadow. How tall is the arch?

15. **SPACE SCIENCE** You cut a square hole $\frac{1}{4}$ inch wide in a piece of cardboard. With the cardboard 30 inches from your face, the moon fits exactly into the square hole. The moon is about 240,000 miles from Earth. estimate the moon's diameter. Draw a diagram of the situation. Then write a proportion and solve the problem.

CRITICAL THINKING For Exercises 16–18, use the following information.

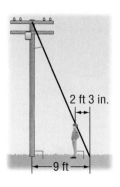

Another method of indirect measurement involves the use of a mirror as shown in the diagram at the right. The two triangles in the diagram are similar.

16. Write a statement of similarity between the two triangles.

17. Write a proportion that could be used to solve for the height h of the light pole.

18. What information would you need to know in order to solve this proportion?

Standardized Test Practice and Mixed Review

19. **MULTIPLE CHOICE** A child $4\frac{1}{2}$ feet tall casts a 6-foot shadow. A nearby statue casts a 12-foot shadow. How tall is the statue?

   Ⓐ $8\frac{1}{4}$ ft   Ⓑ 9 ft   Ⓒ $13\frac{1}{2}$ ft   Ⓓ 24 ft

20. **GRID IN** A guy wire attached to the top of a telephone pole goes to the ground 9 feet from its base. When Jorge stands under the guy wire so that his head touches the wire, he is 2 feet 3 inches from where the wire goes into the ground. If Jorge is 5 feet tall, how tall in feet is the telephone pole in feet?

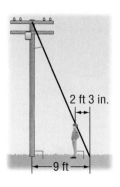

On a city map, the scale is 1 centimeter = 2.5 miles. Find the actual distance for each map distance. (Lesson 4-6)

21. 4 cm   22. 10 cm   23. 13 cm   24. 8.5 cm

25. The triangles at the right are similar. Write a proportion to find the missing measure. Then solve. (Lesson 4-5)

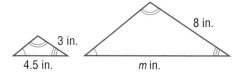

Solve each equation. Check your solution. (Lesson 2-7)

26. $\frac{2}{3}x + 4 = -6$ 27. $a - 2\frac{3}{5} = -6\frac{7}{10}$ 28. $-2.3 = \frac{k}{-8}$ 29. $-4\frac{1}{2}x = 6$

Express each number in scientific notation. (Lesson 2-3)

30. 0.0000236   31. 4,300,000   32. 504,000   33. 0.0000002

GETTING READY FOR THE NEXT LESSON

PREREQUISITE SKILL Graph each pair of ordered pairs. Then find the distance between the points. (Lesson 3-6)

34. (3, 4), (3, 8)   35. (−2, −1), (6, −1)   36. (1, 4), (5, 1)   37. (−1, −2), (4, 10)

Sunshine State Standards
MA.B.3.3.1-1, MA.C.3.3.1-1

Trigonometry

What You'll Learn

Solve problems by using the trigonometric ratios of sine, cosine, and tangent.

Materials

• protractor
• metric rule
• calculator

INVESTIGATE *Work in groups of three.*

Trigonometry is the study of the properties of triangles. The word trigonometry means *triangle measure*. A **trigonometric ratio** is the ratio of the lengths of two sides of a right triangle.

In this Lab you will discover and apply the most common trigonometric ratios: *sine, cosine,* and *tangent.*

In any right triangle, the side **opposite** an angle is the side that is not part of the angle. In the triangle shown,

• side *a* is opposite ∠*A*,

• side *b* is opposite ∠*B*, and

• side *c* is opposite ∠*C*.

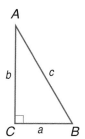

The side that is not opposite an angle and not the hypotenuse is called the **adjacent** side. In △*ABC*,

• side *b* is adjacent to ∠*A*, and

• side *a* is adjacent to ∠*B*.

Each person in the group should complete steps 1–6.

STEP 1 Copy the table shown.

STEP 2 Draw a right triangle *XYZ* so that *m*∠*X* = 30°, *m*∠*Y* = 60°, and *m*∠*Z* = 90°.

STEP 3 Find the length to the nearest millimeter of the leg opposite the angle that measures 30°. Record the length.

	30° angle	60° angle
Length (mm) of opposite leg		
Length (mm) of adjacent leg		
Length (mm) of hypotenuse		
sine		
cosine		
tangent		

STEP 4 Find the length of the leg adjacent to the 30° angle. Record the length.

STEP 5 Find the length of the hypotenuse. Record the length.

STEP 6 Use the measurements and a calculator to find each of the following ratios to the nearest hundredth. Notice that each of these ratios has a special name.

$$\textbf{sine} = \frac{\text{opposite}}{\text{hypotenuse}} \quad \textbf{cosine} = \frac{\text{adjacent}}{\text{hypotenuse}} \quad \textbf{tangent} = \frac{\text{opposite}}{\text{adjacent}}$$

 STEP 7 Compare your ratios with the others in your group.

 STEP 8 Repeat the procedure for the 60° angle. Record the results.

Writing Math

Work with your group.

1. **Make a conjecture** about the ratio of the sides of any 30°-60°-90° triangle.

2. **Repeat the activity** with a triangle whose angles measure 45°, 45°, and 90°.

3. **Make a conjecture** about the ratio of the sides of any 45°-45°-90° triangle.

Use triangle *ABC* to find each of the following ratios to the nearest hundredth.

4. cosine of ∠A

5. sine of ∠A

6. tangent of ∠A

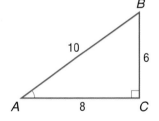

You can use a scientific calculator to find the sine [SIN], cosine [COS], or tangent [TAN] ratio for an angle with a given degree measure. Be sure your calculator is in *degree* mode. Find each value to the nearest thousandth.

7. sin 46°   8. cos 63°   9. tan 82°

10. **SHADOWS** An *angle of elevation* is formed by a horizontal line and a line of sight above it. A flagpole casts a shadow 35 meters long when the angle of elevation of the Sun is 50°. How tall is the flagpole? (*Hint*: Use the tangent ratio.)

11. **Describe** a triangle whose sine and cosine ratios are equal.

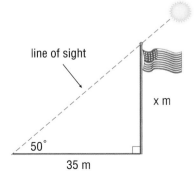

line of sight

x m

50°

35 m

4-8 Dilations

Sunshine State Standards
MA.B.3.3.1-1, MA.C.2.3.1-3

What You'll Learn
Graph dilations on a coordinate plane.

NEW Vocabulary

dilation

∞ LINK To Reading

Everyday Meaning of dilation: the act of enlarging or expanding, as in dilating the pupils of your eyes

HANDS-ON Mini Lab

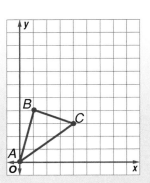

Materials
• graph paper
• ruler

Work with a partner.

Plot $A(0, 0)$, $B(1, 4)$, and $C(4, 3)$ on a coordinate plane. Then draw $\triangle ABC$.

1. Multiply each coordinate by 2 to find the coordinates of points A', B', and C'.

2. On the same coordinate plane, graph points A', B', and C'. Then draw $\triangle A'B'C'$.

3. Determine whether $\triangle ABC \sim \triangle A'B'C'$. Explain your reasoning.

In mathematics, the image produced by enlarging or reducing a figure is called a **dilation**. In the Mini Lab, $\triangle A'B'C'$ has the same shape as $\triangle ABC$, so the two figures are similar. Recall that similar figures are related by a scale factor.

EXAMPLE Graph a Dilation

1 Graph $\triangle JKL$ with vertices $J(3, 8)$, $K(10, 6)$, and $L(8, 2)$. Then graph its image $\triangle J'K'L'$ after a dilation with a scale factor of $\frac{1}{2}$.

To find the vertices of the dilation, multiply each coordinate in the ordered pairs by $\frac{1}{2}$. Then graph both images on the same axes.

$$J(3, 8) \quad \rightarrow \left(3 \cdot \frac{1}{2}, 8 \cdot \frac{1}{2}\right) \rightarrow J'\left(\frac{3}{2}, 4\right)$$

$$K(10, 6) \rightarrow \left(10 \cdot \frac{1}{2}, 6 \cdot \frac{1}{2}\right) \rightarrow K'(5, 3)$$

$$L(8, 2) \quad \rightarrow \left(8 \cdot \frac{1}{2}, 2 \cdot \frac{1}{2}\right) \rightarrow L'(4, 1)$$

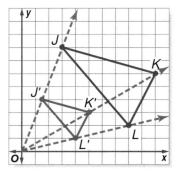

Check Draw lines through the origin and each of the vertices of the original figure. The vertices of the dilation should lie on those same lines.

STUDY TIP

Naming a Dilation
A dilated image is usually named using the same letters as the original figure, but with primes, as in $\triangle JKL \sim \triangle J'K'L'$.

Your Turn Find the coordinates of $\triangle JKL$ after a dilation with each scale factor.

a. scale factor: 2

b. scale factor: $\frac{1}{3}$

Notice that the dilation of △ABC in the Mini Lab is an *enlargement* of the original figure. The dilation of △JKL in Example 1 is a *reduction* of the original figure.

EXAMPLE **Find and Classify a Scale Factor**

2 **Segment V'W' is a dilation of segment VW. Find the scale factor of the dilation, and classify it as an enlargement or as a reduction.**

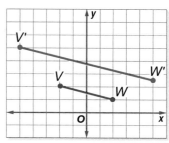

Write a ratio of the *x*- or *y*-coordinate of one vertex of the dilation to the *x*- or *y*-coordinate of the corresponding vertex of the original figure. Use the *y*-coordinates of V(−2, 2) and V'(−5, 5).

$$\frac{y\text{-coordinate of point } V'}{y\text{-coordinate of point } V} = \frac{5}{2}$$

The scale factor is $\frac{5}{2}$. Since the image is larger than the original figure, the dilation is an enlargement.

Your Turn **Segment A'B' is a dilation of segment AB. The endpoints of each segment are given. Find the scale factor of the dilation, and classify it as an *enlargement* or as a *reduction*.**

c. A(4, −8), B(12, −4)
   A'(3, −6), B'(9, −3)

d. A(−5, −7), B(−3, 2)
   A'(−10, −14), B'(−6, 4)

EXAMPLE **Use a Scale Factor**

3 **EYES** **Carleta's optometrist uses medicine to dilate her pupils by a factor of $\frac{5}{3}$. The diagram shows the diameter of Carleta's pupil before dilation. Find the new diameter once her pupil is dilated.**

Before Dilation

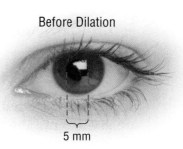

5 mm

Write a proportion using the scale factor.

$$\begin{array}{ll} \text{dilated eye} \rightarrow \\ \text{normal eye} \rightarrow \end{array} \quad \frac{x}{5} = \frac{5}{3} \quad \begin{array}{l} \leftarrow \text{ dilated eye} \\ \leftarrow \text{ normal eye} \end{array}$$

$$x \cdot 3 = 5 \cdot 5 \quad \text{Find the cross products.}$$

$$\frac{3x}{3} = \frac{25}{3} \quad \text{Multiply. Then divide each side by 3.}$$

$$x \approx 8.3 \quad \text{Simplify.}$$

Her pupil will be about 8.3 millimeters in diameter once dilated.

STUDY TIP

Scale Factors
• If the scale factor is between 0 and 1, the dilation is a reduction.
• If the scale factor is greater than 1, the dilation is an enlargement.
• If the scale factor is equal to 1, the dilation is the same size as the original figure.

1. **OPEN ENDED** Draw a triangle on the coordinate plane. Then graph its image after a dilation with a scale factor of 3.

2. **Which One Doesn't Belong?** Identify the pair of points that does not represent a dilation with a factor of 2. Explain your reasoning.

$P(3, -1)$, $P'(5, 1)$	$Q(4, 2)$, $Q'(8, 4)$	$R(-5, 3)$, $R'(-10, 6)$	$S(1, -7)$, $S'(2, -14)$

GUIDED PRACTICE

3. Triangle ABC has vertices $A(-4, 12)$, $B(-2, -4)$, and $C(8, 6)$. Find the coordinates of $\triangle ABC$ after a dilation with a scale factor of $\frac{1}{4}$. Then graph $\triangle ABC$ and its dilation.

4. In the figure at the right, the green rectangle is a dilation of the blue rectangle. Find the scale factor and classify the dilation as an *enlargement* or as a *reduction*.

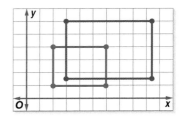

5. Segment $C'D'$ with endpoints $C'(-3, 12)$ and $D'(6, -9)$ is a dilation of segment CD. If segment CD has endpoints $C(-2, 8)$ and $D(4, -6)$, find the scale factor of the dilation. Then classify the dilation as an *enlargement* or as a *reduction*.

Practice and Applications

Find the coordinates of the vertices of polygon $H'J'K'L'$ after polygon $HJKL$ is dilated using the given scale factor. Then graph polygon $HJKL$ and its dilation.

6. $H(-1, 3)$, $J(3, 2)$, $K(2, -3)$, $L(-2, -2)$; scale factor 2

7. $H(0, 2)$, $J(3, 1)$, $K(0, -4)$, $L(-2, -3)$; scale factor 3

8. $H(-6, 2)$, $J(4, 4)$, $K(7, -2)$, $L(-2, -4)$; scale factor $\frac{1}{2}$

9. $H(-8, 4)$, $J(6, 4)$, $K(6, -4)$, $L(-8, -4)$; scale factor $\frac{3}{4}$

10. Write a general rule for finding the new coordinates of any ordered pair (x, y) after a dilation with a scale factor of k.

Segment $P'Q'$ is a dilation of segment PQ. The endpoints of each segment are given. Find the scale factor of the dilation, and classify it as an *enlargement* or as a *reduction*.

11. $P(0, -10)$ and $Q(5, -15)$
    $P'(0, -6)$ and $Q'(3, -9)$

12. $P(-1, 2)$ and $Q(3, -3)$
    $P'(-3, 6)$ and $Q'(9, -9)$

13. $P(-3, -9)$ and $Q(6, -3)$
    $P'(-4, -12)$ and $Q'(8, -4)$

14. $P(-5, 6)$ and $Q(4, 3)$
    $P'(2.5, 3)$ and $Q'(2, 1.5)$

For Exercises 15 and 16, graph each figure on dot paper.

15. a square and its image after a dilation with a scale factor of 4

16. a right triangle and its image after a dilation with a scale factor of 0.5.

HOMEWORK HELP

For Exercises	See Examples
6–10, 15–16	1
11–14, 17–20	2
21–22	3

Extra Practice
See pages 626, 651.

In each figure, the green figure is a dilation of the blue figure. Find the scale factor of each dilation and classify as an _enlargement_ or as a _reduction_.

17.

18.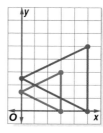

19.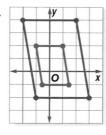

20.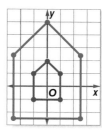

DESIGN For Exercises 21 and 22, use the following information.
Simone designed a logo for her school. The logo, which is 5 inches wide and 8 inches long, will be enlarged and used on a school sweatshirt. On the sweatshirt, the logo will be $12\frac{1}{2}$ inches wide.

21. What is the scale factor for this enlargement?

22. How long will the logo be on the sweatshirt?

ART For Exercises 23 and 24, use the painting at the right and the following information.
Painters use dilations to create the illusion of distance and depth. To create this illusion, the artist establishes a _vanishing point_ on the horizon line. Objects are drawn using intersecting lines that lead to the vanishing point.

23. Find the vanishing point in this painting.

24. **RESEARCH** Use the Internet or other reference to find examples of other paintings that use dilations. Identify the vanishing point in each painting.

Skiffs by Gustave Caillebotte

25. **CRITICAL THINKING** Describe the image of a figure after a dilation with a scale factor of −2.

Standardized Test Practice and Mixed Review

 FCAT Practice

26. **MULTIPLE CHOICE** Square _A_ is a dilation of square _B_. What is the scale factor of the dilation?

   A $\frac{1}{7}$ **B** $\frac{3}{5}$ **C** $\frac{5}{3}$ **D** 7

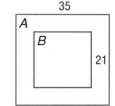

27. **MULTIPLE CHOICE** A photo is 8 inches wide by 10 inches long. You want to make a reduced color copy of the photo that is 5 inches wide for your scrapbook. What scale factor should you choose on the copy machine?

   F $\frac{1}{2}$ or 50%   **G** $\frac{5}{8}$ or 62.5%   **H** $\frac{8}{5}$ or 160%   **I** 2 or 200%

28. **ARCHITECTURE** The Empire State Building casts a shadow 156.25 feet long. At the same time, a nearby building that is 84 feet high casts a shadow 10.5 feet long. How tall is the Empire State Building? (Lesson 4-7)

29. **HOBBIES** A model sports car is 10 inches long. If the actual car is 14 feet, find the scale of the model. (Lesson 4-6)

Vocabulary and Concept Check

congruent (p. 179)	rate (p. 157)	scale drawing (p. 184)
corresponding parts (p. 178)	rate of change (p. 160)	scale factor (p. 179)
cross products (p. 170)	ratio (p. 156)	scale model (p. 184)
dilation (p. 194)	rise (p. 166)	similar (p. 178)
indirect measurement (p. 188)	run (p. 166)	slope (p. 166)
polygon (p. 178)	scale (p. 184)	unit rate (p. 157)
proportion (p. 170)		

Choose the letter of the term that best matches each statement or phrase.

1. polygons that have the same shape
2. a rate with a denominator of one
3. the constant rate of change between two points on a line
4. a comparison of two numbers by division
5. two equivalent ratios
6. ratio of a length on a drawing to its actual length
7. describes how one quantity changes in relation to another
8. the enlarged or reduced image of a figure

a. slope
b. rate of change
c. dilation
d. proportion
e. unit rate
f. similar
g. ratio
h. scale

Lesson-by-Lesson Exercises and Examples

 Ratios and Rates (pp. 156–159)

Express each ratio in simplest form.
9. 7 chaperones for 56 students
10. 12 peaches : 8 pears
11. 5 inches out of 5 feet

Example 1 Express the ratio
10 milliliters to 8 liters in simplest form.

$$\frac{10 \text{ milliliters}}{8 \text{ liters}} = \frac{10 \text{ milliliters}}{8,000 \text{ milliliters}} \text{ or } \frac{1}{800}$$

 Rate of Change (pp. 160–164)

12. **MONEY** The table below shows Victor's weekly allowance for different ages.

Age (yr)	4	6	8	10	12	15
$ per week	0.25	1.00	2.00	2.00	3.00	5.00

Find the rate of change in his allowance between ages 12 and 15.

Example 2 At 5 A.M., it was 54°F. At 11 A.M., it was 78°F. Find the rate of temperature change in degrees per hour.

$$\frac{\text{change in temperature}}{\text{change in hours}} = \frac{(78 - 54)°}{(11 - 5) \text{ hours}}$$

$$= \frac{24°}{6 \text{ hours}} \text{ or } \frac{4°}{1 \text{ hour}}$$

4-3 Slope (pp. 166–169)

Find the slope of each line graphed at the right.

13. $\overline{AB}$

14. $\overline{CD}$

15. The points in the table lie on a line. Find the slope of the line. Then graph the line.

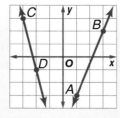

x	−6	−2	2
y	5	2	−1

Example 3 Find the slope of the line.

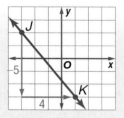

The vertical change from point J to point K is −5 units while the horizontal change is 4 units.

slope $= \dfrac{\text{rise}}{\text{run}}$ Definition of slope

$= \dfrac{-5}{4}$ or $-\dfrac{5}{4}$ rise = −5, run = 4

4-4 Solving Proportions (pp. 170–173)

Solve each proportion.

16. $\dfrac{3}{r} = \dfrac{6}{8}$

17. $\dfrac{7}{4} = \dfrac{n}{2}$

18. $\dfrac{k}{5} = \dfrac{72}{8}$

19. $\dfrac{8}{3.8} = \dfrac{6}{x}$

20. **ANIMALS** A turtle can move 5 inches in 4 minutes. How far will it travel in 10 minutes?

Example 4 Solve $\dfrac{9}{x} = \dfrac{4}{18}$.

$\dfrac{9}{x} = \dfrac{4}{18}$ Write the equation.

$9 \cdot 18 = x \cdot 4$ Find the cross products.

$162 = 4x$ Multiply.

$\dfrac{162}{4} = \dfrac{4x}{4}$ Divide each side by 10.

$40.5 = x$ Simplify.

4-5 Similar Polygons (pp. 178–182)

Each pair of polygons is similar. Write a proportion to find each missing measure. Then solve.

21.

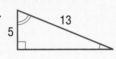

22.

23. **PARTY PLANNING** For your birthday party, you make a map to your house on a 3-inch wide by 5-inch long index card. How long will your map be if you use a copier to enlarge it so that it is 8 inches wide?

Example 5 Rectangle $GHJK$ is similar to rectangle $PQRS$. Find the value of x.

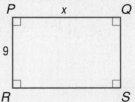

The scale factor from $FGHK$ to $PQRS$ is $\dfrac{GK}{PR}$, which is $\dfrac{3}{9}$ or $\dfrac{1}{3}$.

$\dfrac{GH}{PQ} = \dfrac{1}{3}$ Write a proportion.

$\dfrac{4.5}{x} = \dfrac{1}{3}$ $GH = 4.5$ and $PQ = y$

$13.5 = x$ Find the cross products. Simplify.

4-6 Scale Drawings and Models (pp. 184–187)

The scale on a map is 2 inches = 5 miles. Find the actual distance for each map distance.

24. 12 in.   **25.** 9 in.   **26.** 2.5 in.

27. HOBBIES Mia's sister's dollhouse is a replica of their townhouse. The outside dimensions of the dollhouse are 25 inches by 35 inches. If the actual outside dimensions of the townhouse are 25 feet by 35 feet, what is the scale of the dollhouse?

Example 6 The scale on a model is 3 centimeters = 45 meters. Find the actual length for a model distance of 5 centimeters.

$$\text{model length} \rightarrow \quad \frac{3 \text{ cm}}{45 \text{ m}} = \frac{5 \text{ cm}}{x \text{ m}} \quad \leftarrow \text{model length}$$
$$\text{actual length} \rightarrow \qquad\qquad\qquad \leftarrow \text{actual length}$$
$$3 \cdot x = 45 \cdot 5$$
$$3x = 225$$
$$x = 75$$

The actual length is 75 meters.

4-7 Indirect Measurement (pp. 188–191)

Write a proportion. Then determine the missing measure.

28. MAIL A mailbox casts an 18-inch shadow. A tree casts a 234-inch shadow. If the mailbox is 4 feet tall, how tall is the tree?

29. WATER From the shoreline, the ground slopes down under the water at a constant incline. If the water is $5\frac{1}{2}$ feet deep when it is $2\frac{1}{4}$ feet from the shore, about how deep will it be when it is 6 feet from the shore?

Example 7 A house casts a shadow that is 5 meters long. A tree casts a shadow that is 2.5 meters long. If the house is 20 meters tall, how tall is the tree?

$$\text{house's shadow} \rightarrow \quad \frac{5}{2.5} = \frac{20}{x} \quad \leftarrow \text{house's height}$$
$$\text{tree's shadow} \rightarrow \qquad\qquad\qquad \leftarrow \text{tree's height}$$
$$5 \cdot x = 20 \cdot 2.5$$
$$5x = 50$$
$$x = 10$$

The tree is 10 meters tall.

4-8 Dilations (pp. 194–197)

Segment $C'D'$ is a dilation of segment CD. The endpoints of each segment are given. Find the scale factor of the dilation, and classify it as an *enlargement* or as a *reduction*.

30. $C(-2, 5)$, $D(1, 4)$; $C'(-8, 20)$, $D'(4, 16)$

31. $C(-5, 10)$, $D(0, 5)$; $C'(-2, 4)$, $D'(0, 2)$

Example 8 Segment XY has endpoints $X(-4, 1)$ and $Y(8, -2)$. Find the coordinates of its image for a dilation with a scale factor of $\frac{3}{4}$.

$$X(-4, 1) \rightarrow \left(-4 \cdot \frac{3}{4}, 1 \cdot \frac{3}{4}\right) \rightarrow X'\left(-3, \frac{3}{4}\right)$$

$$Y(8, -2) \rightarrow \left(8 \cdot \frac{3}{4}, -2 \cdot \frac{3}{4}\right) \rightarrow Y'\left(6, -1\frac{1}{2}\right)$$

Vocabulary and Concepts

1. **OPEN ENDED** List four different ratios that form a proportion with $\frac{8}{12}$.

2. **Describe** a reasonable scale for a scale drawing of your classroom.

Skills and Applications

3. Express *15 inches to 1 foot* in simplest form.

4. Express *$1,105 for 26 jerseys* as a unit rate.

BUSINESS For Exercises 5 and 6, use the table at the right.

5. Find the rate of change in new customers per hour between 4 P.M. and 5 P.M.

6. Find the rate of change in new customers per hour between 12 P.M. and 2 P.M. Then interpret its meaning.

Lucky Diner	
Time	New Customers
12 P.M.	30
2 P.M.	6
4 P.M.	15
5 P.M.	32

Find the slope of each line graphed at the right.

7. $\overline{AB}$

8. $\overline{CD}$

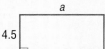

Solve each proportion.

9. $\frac{5}{3} = \frac{20}{y}$

10. $\frac{x}{2} = \frac{5}{8}$

Each pair of polygons is similar. Write a proportion to find each missing measure. Then solve.

11.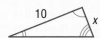

12.

13. **GEOMETRY** Graph triangle *FGH* with vertices $F(-4, -2)$, $G(-1, 2)$, and $H(3, 0)$. Then graph its image after a dilation with a scale factor of $\frac{3}{2}$.

14. On a map, 1 inch = 7.5 miles. How many miles does 2.5 inches represent?

Standardized Test Practice

15. **GRID IN** If it costs an average of $102 to feed a family of three for one week, on average, how much will it cost in dollars to feed a family of five for one week?

16. **MULTIPLE CHOICE** A 36-foot flagpole casts a 9-foot shadow at the same time a building casts a 15-foot shadow. How tall is the building?

    A 21.6 ft   B 60 ft

    C 135 ft   D 375 ft

PART 1 Multiple Choice

Record your answers on the answer sheet provided by your teacher or on a sheet of paper.

1. Which of the numbers below is *not* a prime number? (Prerequisite Skill, p. 609)

   Ⓐ 23　　Ⓑ 49　　Ⓒ 59　　Ⓓ 61

2. One floor of a house is divided into two apartments as shown below.

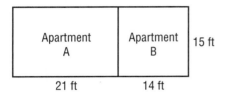

   How much larger is the area of apartment A than the area of apartment B? (Lesson 1-1)

   Ⓕ 90 ft²　　　　Ⓖ 100 ft²

   Ⓗ 105 ft²　　　Ⓘ 115 ft²

3. Which of the following numbers could replace the variable n to make the inequality true? (Lesson 2-2)

   $$\frac{4}{9} < n < 0.72$$

   Ⓐ $\frac{1}{3}$　　Ⓑ $\frac{6}{8}$　　Ⓒ $\frac{3}{2}$　　Ⓓ $\frac{4}{6}$

4. Which of the following could *not* be the side lengths of a right triangle? (Lesson 3-4)

   Ⓕ 2, 3, 5　　　Ⓖ 6, 10, 8

   Ⓗ 8, 15, 17　　Ⓘ 13, 5, 12

TEST-TAKING TIP

Question 3 It will save you time to memorize the decimal equivalents or approximations of some common fractions.

$\frac{3}{4} = 0.75$　$\frac{1}{3} \approx 0.33$　$\frac{2}{3} \approx 0.66$　$\frac{3}{2} = 1.5$

5. Last week, Caleb traveled from home to his grandmother's house. The graph below shows the relationship between his travel time and the distance he traveled.

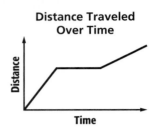

Distance Traveled Over Time

Which best describes his trip? (Lesson 4-2)

   Ⓐ He drove on a high-speed highway, then slowly on a dirt road, and finished his trip on a high-speed highway.

   Ⓑ He drove slowly on a dirt road, stopped for lunch, and then got on a high-speed highway for the rest of his trip.

   Ⓒ He drove slowly on a dirt road, then on a high-speed highway, and finished his trip on a dirt road.

   Ⓓ He started on a high-speed highway, stopped for lunch, and then got on a dirt road for the rest of his trip.

6. You are making a scale model of the car shown below. If your model is to be $\frac{1}{25}$ of car's actual size, which proportion could be used to find the measure ℓ of the model's length? (Lesson 4-6)

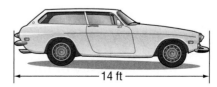

   Ⓕ $\frac{\ell}{14} = \frac{25}{1}$　　　Ⓖ $\frac{14}{\ell} = \frac{1}{25}$

   Ⓗ $\frac{\ell}{14} = \frac{1}{25}$　　　Ⓘ $\frac{14}{25} = \frac{1}{\ell}$

PART 2 Short Response/Grid In

THINK
SOLVE
EXPLAIN

Record your answers on the answer sheet provided by your teacher or on a sheet of paper.

7. The temperature at 9:00 A.M. was $-20°$F. If the temperature rose $15°$ from 9:00 A.M. to 12:00 noon, what was the temperature at noon? (Lesson 1-4)

8. Suppose you made fruit punch for a party using $3\frac{1}{2}$ cups of apple juice, 2 cups of orange juice, and $2\frac{1}{2}$ cups of cranberry juice. How many *quarts* of juice did you make? (Lesson 2-5)

9. Estimate the value of $\sqrt{47}$ to the nearest whole number. (Lesson 3-2)

10. A swan laid 5 eggs. Only 4 of the eggs hatched, and only 3 of these swans grew to become adults. Write the ratio of swans that grew to adulthood to the number of eggs that hatched as a fraction. (Lesson 4-1)

11. Find the slope of the line graphed at the right. (Lesson 4-3)

12. A truck used 6.3 gallons of gasoline to travel 107 miles. How many gallons of gasoline would it need to travel an additional 250 miles? (Lesson 4-4)

13. Triangle *FGH* is similar to triangle *JKL*. The perimeter of triangle *FGH* is 30 centimeters.

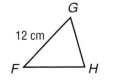

What is the perimeter of triangle *JKL* in centimeters? (Lesson 4-5)

14. The distance between Jasper and Cartersville on a map is 3.8 centimeters. If the actual distance between these two cities is 209 miles, what is the scale for this map? (Lesson 4-7)

15. A 6-foot tall man casts a shadow that is 8 feet long. At the same time, a nearby crane casts a 20-foot long shadow. How tall is the crane? (Lesson 4-7)

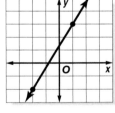
PART 3 Extended Response

Record your answers on a sheet of paper. Show your work.

THINK
SOLVE
EXPLAIN

16. The table below shows how much Susan earns for different amounts of time she works at a fast food restaurant. (Lesson 4-3)

Time (h)	2	4	6	8
Wages ($)	9	18	27	36

a. Graph the data from the table and connect the points with a line.

b. Find the slope of the line.

c. What is Susan's rate of pay?

d. If Susan continues to be paid at this rate, how much money will she make for working 10 hours?

17. Triangle *ABC* has vertices $A(-6, 3)$, $B(3, 6)$, and $C(6, -9)$. (Lesson 4-8)

a. Find the coordinates of the vertices of $\triangle A'B'C'$ after $\triangle ABC$ is dilated using a scale factor of $\frac{2}{3}$.

b. Graph $\triangle ABC$ and its dilation.

c. Name a scale factor that would result in $\triangle ABC$ being enlarged.

d. Find the coordinates of the vertices of $\triangle A'B'C'$ after $\triangle ABC$ after this enlargement.

"What does baseball have to do with math?"

Fans and people involved with baseball often track the ratio of a player's hits to his times at bat. This ratio can be written as a decimal or as a percent.

You will solve problems about baseball and other sports in Lesson 5-1.

GETTING STARTED

Take this quiz to see whether you are ready to begin Chapter 5. Refer to the lesson number in parentheses if you need more review.

▶ **Vocabulary Review**

Choose the correct term to complete each sentence.

1. $2 + 3 = 5$ is an (equation, expression).
   (Lesson 1-7)

2. A (product, ratio) is a comparison of two numbers by division. (Lesson 4-1)

3. Two or more equal ratios can be written to form a (relation, proportion). (Lesson 4-4)

▶ **Prerequisite Skills**

Compute each product mentally.

4. $\frac{1}{3} \cdot 303$ 5. $644 \cdot \frac{1}{2}$

6. $0.1 \cdot 550$ 7. $64 \cdot 0.5$

Write each fraction as a decimal. (Lesson 2-1)

8. $\frac{2}{5}$ 9. $\frac{7}{8}$

10. $\frac{3}{4}$ 11. $\frac{3}{8}$

Solve each equation. (Lesson 2-7)

12. $0.25d = 130$ 13. $48r = 12$

14. $0.4m = 22$ 15. $0.02n = 9$

16. $96 = y \cdot 30 \cdot 4$ 17. $f \cdot 5 \cdot 12 = 21$

18. $33 = 0.5 \cdot c \cdot 6$ 19. $7 = 20 \cdot 0.7 \cdot k$

Solve each proportion. (Lesson 4-4)

20. $\frac{x}{10} = \frac{3}{5}$ 21. $\frac{4}{9} = \frac{14}{b}$

22. $\frac{7}{23} = \frac{s}{46}$ 23. $\frac{6}{z} = \frac{5}{13}$

Percent Make this Foldable to help you organize your notes. Begin with four sheets of $8\frac{1}{2}$" × 11" paper.

STEP 1 **Draw a Circle**
Draw a large circle on one of the sheets of paper.

STEP 2 **Stack and Cut**
Stack the sheets of paper. Place the one with the circle on top. Cut all four sheets in the shape of a circle.

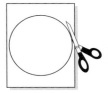

STEP 3 **Staple and Label**
Staple the circles on the left side. Write the first four lesson numbers on each circle.

Lesson 5-1

STEP 4 **Turn and Label**
Turn the circles to the back side so that the staples are still on the left. Write the last four lesson numbers on each circle.

Lesson 5-5

Reading and Writing As you read and study the chapter, write notes and examples for each lesson on the appropriate page.

Ratios and Percents

Sunshine State Standards MA.A.1.3.1-1, MA.A.1.3.2-2, MA.A.1.3.4-1, MA.A.3.3.2-1, MA.A.3.3.2-3

What You'll Learn
Write ratios as percents and vice versa.

NEW Vocabulary
percent

REVIEW Vocabulary
ratio: a comparison of two numbers by division **(Lesson 4-1)**

WHEN am I ever going to use this?

POPULATION The table shows the ratio of people under 18 years of age to the total population for various states.

State	Ratio of People Under 18 to Total Population
Alabama	1 out of 4
Hawaii	6 out of 25
South Dakota	27 out of 100
Utah	8 out of 25

Source: *Time Almanac*

1. Name two states from the table that have ratios in which the second numbers are the same.

2. Of the two states you named in Question 1, which state has a greater ratio of people under 18 to total population? Explain.

3. Describe how to determine which of the four states has the greater ratio of people under 18 to total population.

Ratios such as 27 out of 100 or 8 out of 25 can be written as **percents**.

Key Concept
Percent

A percent is a ratio that compares a number to 100.

Ratio	27 out of 100	**Model**
Symbols	27%	
Words	twenty-seven percent	

EXAMPLES   Write Ratios as Percents

Write each ratio as a percent.

1. **POPULATION** According to the 2000 U.S. Census, 22 out of every 100 people living in West Virginia were younger than 18.

   22 out of 100 = 22%

2. **SPORTS** At a recent triathlon, 180 women competed for every 100 women who competed ten years earlier.

   180 out of 100 = 180%

STUDY TIP

Large Percents
Notice that some percents, such as 180%, are greater than 100%.

Your Turn Write each ratio as a percent.

a. **BASEBALL** During his baseball career, Babe Ruth had a base hit about 34 out of every 100 times he came to bat.

b. **TECHNOLOGY** In a recent year, 41.5 out of 100 households in the United States had access to the Internet.

One way to write a fraction or a ratio as a percent is by finding an equivalent fraction with a denominator of 100.

EXAMPLES Write Ratios and Fractions as Percents

Write each ratio or fraction as a percent.

3 **CARS** About 1 out of 5 luxury cars manufactured in the United States is white.

$$\frac{1}{5} = \frac{20}{100}$$

So, 1 out of 5 equals 20%.

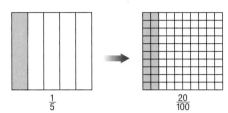

$\frac{1}{5}$ $\frac{20}{100}$

4 **TRAVEL** About $\frac{1}{200}$ of travelers use scheduled buses.

$$\frac{1}{200} = \frac{0.5}{100}$$

So, 1 out of 200 equals 0.5%.

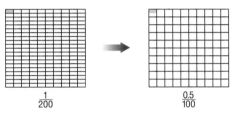

$\frac{1}{200}$ $\frac{0.5}{100}$

Your Turn Write each ratio or fraction as a percent.

c. **TECHNOLOGY** In Finland, almost 3 out of 5 people have cell phones.

d. **ANIMALS** About $\frac{1}{4}$ of the mammals in the world are bats.

STUDY TIP

Small Percents
In Example 4, notice that 0.5% is less than 1%.

You can express a percent as a fraction by writing it as a fraction with a denominator of 100. Then write the fraction in simplest form.

EXAMPLE Write Percents as Fractions

5 **ENVIRONMENT** The circle graph shows an estimate of the percent of each type of trash in landfills. Write the percents for paper and for plastic as a fraction in simplest form.

Paper: $30\% = \frac{30}{100}$ or $\frac{3}{10}$

Plastic: $24\% = \frac{24}{100}$ or $\frac{6}{25}$

Trash in Landfills

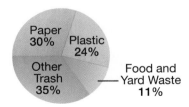

Paper 30%   Plastic 24%
Other Trash 35%   Food and Yard Waste 11%

Source: Franklin Associates, Ltd.

Your Turn Write the percent for each of the following as a fraction in simplest form.

e. food and yard waste   f. other trash

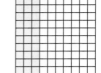

1. **Write** the percent and the fraction in simplest form for the model shown at the right.

2. **OPEN ENDED** Write a percent that is between $\frac{1}{2}$ and $\frac{3}{4}$.

3. **Which One Doesn't Belong?** Identify the number that does not have the same value as the other three. Explain your reasoning.

$\frac{2}{5}$	40%	$\frac{20}{100}$	$\frac{10}{25}$

GUIDED PRACTICE

Write each ratio or fraction as a percent.

4. 17 out of 100   5. 237 out of 100   6. 7:10   7. $\frac{9}{20}$

Write each percent as a fraction in simplest form.

8. 19%   9. 50%   10. 18%   11. 0.4%

12. **TRAVEL** One out of every 50 travelers visiting the United States is from France. Write this ratio as a percent.

Practice and Applications

Write each ratio or fraction as a percent.

13. 23 out of 100   14. 9 out of 100   15. 0.3 out of 100

16. 0.7 out of 100   17. 3:5   18. 9:10

19. 8:25   20. 17:20   21. $\frac{17}{50}$

22. $\frac{7}{20}$ 23. $\frac{39}{20}$ 24. $\frac{47}{25}$

HOMEWORK HELP

For Exercises	See Examples
13–16	1, 2
17–20, 25–26	3
21–24	4
27–41	5

Extra Practice
See pages 626, 652.

25. **PETS** Three out of 25 households in the United States have both a dog and a cat. Write this ratio as a percent.

26. **MUSIC** Eleven out of 25 Americans like rock music. Write this ratio as a percent.

Write each percent as a fraction in simplest form.

27. 29%   28. 43%   29. 40%   30. 70%

31. 45%   32. 28%   33. 64%   34. 65%

35. 125%   36. 240%   37. 0.2%   38. 0.8%

39. **ENERGY** Germany uses about 4% of the world's energy. Write this percent as a fraction.

40. **GEOGRAPHY** About 30% of Minnesota is forested. Write this percent as a fraction.

41. **MUSIC** The influences in the purchases of CDs or cassettes are shown in the graphic at the right. Write each percent as a fraction in simplest form.

42. Which is less, $\frac{1}{4}$ or 30%?

43. Which is less, $\frac{2}{5}$ or 37%?

SCIENCE For Exercises 44–46, use the following information.
In 2000, 5 Tyrannosaurus Rex skeletons were found in Montana. In the previous 100 years, only 15 such skeletons had been found.

44. Write a ratio in simplest form to compare the number of Tyrannosaurus Rex skeletons found in 2000 to the total number of skeletons found during the 101 years.

45. Write the ratio in Exercise 44 as a percent.

46. What percent of the skeletons where found in the previous 100 years?

47. **CRITICAL THINKING** Explain how a student can receive a 86% on a test with 50 questions.

USA TODAY Snapshots®

Radio is strong influence on music buying
What buyers, ages 16-40, of music CDs or cassettes in the last 12 months say most influenced their decision to buy the CD for themselves:

Radio — **45%**
Friend/relative **15%**
Heard/saw in store **10%**
Music video channel **8%**
Live performance **7%**

Source: Edison Media Research

By Cindy Hall and Quin Tian, USA TODAY

Standardized Test Practice and Mixed Review

FCAT Practice

48. **MULTIPLE CHOICE** What percent of the circle at the right is shaded?

   Ⓐ 10%   Ⓑ 20%

   Ⓒ 30%   Ⓓ 40%

49. **MULTIPLE CHOICE** Which value is *not* equal to the other values?

   Ⓕ $\frac{14}{25}$ Ⓖ 56%   Ⓗ 40:75   Ⓘ 28 out of 50

Segment $P'Q'$ is a dilation of segment PQ. The endpoints of each segment are given. Find the scale factor of the dilation, and classify it as an *enlargement* **or as a** *reduction.* (Lesson 4-8)

50. $P(0, 6)$ and $Q(3, -9)$
    $P'(0, 4)$ and $Q'(2, -6)$

51. $P(-1, 2)$ and $Q(-3, 3)$
    $P'(-4, 8)$ and $Q'(-12, 12)$

52. **GEOGRAPHY** The Pyramid of the Sun near Mexico City casts a shadow 13.3 meters long. At the same time, a 1.83-meter tall tourist casts a shadow 0.4 meter long. How tall is the Pyramid of the Sun? (Lesson 4-7)

GETTING READY FOR THE NEXT LESSON

PREREQUISITE SKILL Write each fraction as a decimal. (Lesson 2-1)

53. $\frac{3}{5}$ 54. $\frac{3}{4}$ 55. $\frac{5}{8}$ 56. $\frac{1}{3}$

Fractions, Decimals, and Percents

Sunshine State Standards MA.A.1.3.1-1, MA.A.1.3.2-1, MA.A.1.3.2-2, MA.A.1.3.4-1, MA.A.3.3.1-1, MA.A.3.3.2-1, MA.A.3.3.2-2, MA.A.3.3.2-3, MA.A.3.3.3-1, MA.E.1.3.1-1

What You'll Learn

Write percents as fractions and decimals and vice versa.

WHEN am I ever going to use this?

PETS The table gives the percent of households with various pets.

1. Write each percent as a fraction. Do not simplify the fractions.

2. Write each fraction in Exercise 1 as a decimal.

3. How could you write a percent as a decimal without writing the percent as a fraction first?

Households with Pets	
Pet	**Percent of Households**
dog	39%
cat	34%
freshwater fish	12%
bird	7%
small animal	5%

Source: American Pet Products Manufacturers Association

Fractions, percents, and decimals are all different ways to represent the same number.

Remember that *percent* means *per hundred*. In Lesson 5-1, you wrote percents as fractions with 100 in the denominator. Similarly, you can write percents as decimals by dividing by 100.

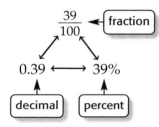

Key Concept · **Decimals and Percents**

- To write a percent as a decimal, divide by 100 and remove the percent symbol.

$$39\% = 39\% = 0.39$$

- To write a decimal as a percent, multiply by 100 and add the percent symbol.

$$0.39 = 0.39 = 39\%$$

EXAMPLES Percents as Decimals

Write each percent as a decimal.

1 **35%**

$35\% = 35\%$ Divide by 100 and remove the percent symbol.

    $= 0.35$

2 **115%**

$115\% = 115\%$ Divide by 100 and remove the percent symbol.

    $= 1.15$

EXAMPLES — Decimals as Percents

Write each decimal as a percent.

3 0.2

$$0.2 = 0.20 \quad \text{Multiply by 100 and add the percent symbol.}$$
$$= 20\%$$

4 1.66

$$1.66 = 1.66 \quad \text{Multiply by 100 and add the percent symbol.}$$
$$= 166\%$$

You have learned to write a fraction as a percent by finding an equivalent fraction with a denominator of 100. This method works well if the denominator is a factor of 100. If the denominator is *not* a factor of 100, you can solve a proportion or you can write the fraction as a decimal and then write the decimal as a percent.

EXAMPLES — Fractions as Percents

STUDY TiP

Look Back You can review **writing fractions as decimals** in Lesson 2-1.

Write each fraction as a percent.

5 $\dfrac{3}{8}$

Method 1 Use a proportion.

$$\frac{3}{8} = \frac{x}{100}$$
$$3 \cdot 100 = 8 \cdot x$$
$$300 = 8x$$
$$\frac{300}{8} = \frac{8x}{8}$$
$$37.5 = x$$

So, $\dfrac{3}{8}$ can be written as 37.5%.

Method 2 Write as a decimal.

$$\frac{3}{8} = 0.375$$
$$= 37.5\%$$

$$\begin{array}{r} 0.375 \\ 8{\overline{\smash{)}3.000}} \\ \underline{2\,4} \\ 60 \\ \underline{56} \\ 40 \\ \underline{40} \\ 0 \end{array}$$

6 $\dfrac{2}{3}$

Method 1 Use a proportion.

$$\frac{2}{3} = \frac{x}{100}$$
$$2 \cdot 100 = 3 \cdot x$$
$$200 = 3x$$
$$\frac{200}{3} = \frac{3x}{3}$$
$$66.\overline{6} = x$$

So, $\dfrac{2}{3}$ can be written as $66.\overline{6}\%$.

Method 2 Write as a decimal.

$$\frac{2}{3} = 0.66\overline{6}$$
$$= 66.\overline{6}\%$$

$$\begin{array}{r} 0.66\ldots \\ 3{\overline{\smash{)}2.0}} \\ \underline{1\,8} \\ 20 \\ \underline{18} \\ 2 \end{array}$$

 Your Turn Write each decimal or fraction as a percent.

a. 0.8 b. 0.564 c. $\dfrac{3}{16}$ d. $\dfrac{1}{9}$

To compare fractions, percents, and decimals, it may be easier to write all of the numbers as percents or decimals.

EXAMPLE Compare Numbers

7 **GEOGRAPHY** About $\frac{3}{20}$ of the land of Earth is covered by desert. North America is about 16% of the total land surface of Earth. Is the area of the deserts on Earth more or less than the area of North America?

Write $\frac{3}{20}$ as a percent.

$\frac{3}{20} = 0.15$ $3 \div 20 = 0.15$

$= 15\%$ Multiply by 100 and add the percent symbol.

Since 15% is less than 16%, the area of the deserts on Earth is just slightly less than the area of North America.

Skill and Concept Check

Writing Math
Exercise 3

1. **Write** a fraction, a percent, and a decimal to represent the shaded part of the rectangle at the right.

2. **OPEN ENDED** Write a fraction that could be easily changed to a percent by using equivalent fractions. Then write a fraction that could *not* be easily changed to a percent by using equivalent fractions. Write each fraction as a percent.

3. **FIND THE ERROR** Kristin and Aislyn are changing 0.7 to a percent. Who is correct? Explain.

Kristin
0.7 = 7%

Aislyn
0.7 = 70%

GUIDED PRACTICE

Write each percent as a decimal.

4. 40%   5. 16%   6. 85%   7. 0.3%

Write each decimal as a percent.

8. 0.68   9. 1.23   10. 0.3   11. 0.725

Write each fraction as a percent.

12. $\frac{11}{25}$ 13. $\frac{7}{8}$ 14. $\frac{13}{40}$ 15. $\frac{5}{6}$

16. **ANIMALS** There are 250 known species of sharks. Of that number, only 27 species have been involved in attacks on humans. What percent of known species of sharks have attacked humans?

Practice and Applications

Write each percent as a decimal.

17. 90%   **18.** 80%   **19.** 15%   **20.** 32%

21. 172%   **22.** 245%   **23.** 27.5%   **24.** 84.2%

25. 7%   **26.** 5%   **27.** 8.2%   **28.** 0.12%

HOMEWORK HELP

For Exercises	See Examples
17–30, 66	1, 2
31–42	3, 4
43–56	5, 6
57–62, 67–68	7

Extra Practice
See pages 627, 652.

29. TELEVISION About 55% of cable TV subscribers decide what program to watch by surfing the channels. Write this number as a decimal.

30. MOVIES In 1936, 85% of movie theaters had double features. Write this number as a decimal.

Write each decimal as a percent.

31. 0.54   **32.** 0.62   **33.** 0.375   **34.** 0.632   **35.** 0.007

36. 0.009   **37.** 0.4   **38.** 0.9   **39.** 2.75   **40.** 1.38

41. CAMPING If 0.21 of adults go camping, what percent of the adults camp?

42. POPULATION In 2010, about 0.25 of the U.S. population will be 55 years old or older. What percent of the population will be 55 or older?

Write each fraction as a percent.

43. $\frac{17}{20}$ **44.** $\frac{12}{25}$ **45.** $\frac{1}{40}$ **46.** $\frac{3}{40}$ **47.** $\frac{8}{5}$

48. $\frac{7}{4}$ **49.** $\frac{1}{200}$ **50.** $\frac{1}{400}$ **51.** $\frac{4}{9}$ **52.** $\frac{2}{3}$

53. TIME Research indicates that $\frac{8}{25}$ of Americans set their watches five minutes ahead. What percent of Americans set their watches five minutes ahead?

54. FOOD About $\frac{3}{20}$ of Americans prefer cold pizza over hot pizza. What percent of Americans prefer cold pizza?

ANIMALS For Exercises 55 and 56, use the information at the right.

55. What percent of a day does a lion spend resting?

56. What percent of a day does a lion spend doing activities?

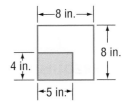

A Day in the Life of a Lion

Activities: 4 hours
Resting: 20 hours

Replace each ● with <, >, or = to make a true sentence.

57. $\frac{5}{9}$ ● 55%   **58.** $\frac{7}{10}$ ● 70%   **59.** 88% ● 8.8

60. 0.03 ● 30%   **61.** 0.5 ● 50%   **62.** 0.09 ● 1%

63. MULTI STEP What percent of the area of the square at the right is shaded?

64. Order $\frac{3}{4}$, 0.8, 8%, and $\frac{7}{10}$ from greatest to least.

65. Order 0.2, $\frac{1}{4}$, 2%, and $\frac{3}{20}$ from least to greatest.

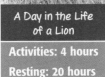

SCHOOL For Exercises 66–68, use the graphic at the right.

66. Write the percent of parents who give themselves an A as a decimal.

67. Did more or less than $\frac{2}{5}$ of the parents give themselves a B?

68. Did more or less than $\frac{1}{5}$ of the parents give themselves a C?

TRAVEL For Exercises 69 and 70, use the following information.
The projected number of household trips in 2010 is 50,000,000. About 14,000,000 of these trips will involve air travel.

69. What fraction of the trips will involve air travel?

70. What percent of the trips will involve air travel?

71. **CRITICAL THINKING** Write $1\frac{3}{5}$ as a percent.

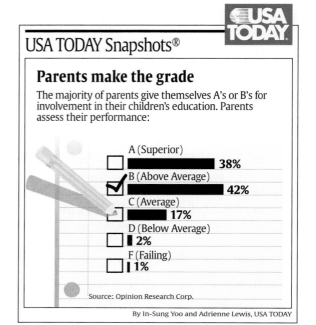

USA TODAY Snapshots®

Parents make the grade
The majority of parents give themselves A's or B's for involvement in their children's education. Parents assess their performance:

A (Superior) 38%
B (Above Average) 42%
C (Average) 17%
D (Below Average) 2%
F (Failing) 1%

Source: Opinion Research Corp.

By In-Sung Yoo and Adrienne Lewis, USA TODAY

Standardized Test Practice and Mixed Review

FCAT Practice

72. **MULTIPLE CHOICE** The graph at the right shows treats Americans prefer during the summer months. Which fraction is *not* equivalent to one of the percents in the graph?

   Ⓐ $\frac{7}{50}$ Ⓑ $\frac{11}{18}$ Ⓒ $\frac{16}{25}$ Ⓓ $\frac{3}{20}$

73. **MULTIPLE CHOICE** Choose the fraction that is less than 35%.

   Ⓕ $\frac{2}{5}$ Ⓖ $\frac{3}{8}$ Ⓗ $\frac{1}{6}$ Ⓘ $\frac{5}{12}$

Write each ratio as a percent. (Lesson 5-1)

74. 27 out of 100

75. 0.6 out of 100

76. 9:20

77. 33:50

Favorite Summer Treat

64%
15% 14%

Ice Cream / Italian Ice / Popsicles

Source: Opinion Research Corporation

78. **GEOMETRY** Graph $\overline{EF}$ with endpoints $E(2, 6)$ and $F(4, -4)$. Then graph its image for a dilation with a scale factor of 2. (Lesson 4-8)

Order the integers in each set from least to greatest. (Lesson 1-3)

79. $\{-12, 5, -5, 13, -1\}$ 80. $\{42, -56, -13, 101, 13\}$ 81. $\{64, -58, -65, 57, -61\}$

GETTING READY FOR THE NEXT LESSON

PREREQUISITE SKILL Solve each proportion. (Lesson 4-4)

82. $\frac{5}{6} = \frac{x}{24}$ 83. $\frac{a}{12} = \frac{2}{15}$ 84. $\frac{2}{7} = \frac{5}{t}$ 85. $\frac{3}{n} = \frac{10}{8}$

Study Skill

HOW TO...

Read Math Problems

Look for words such as *more than*, *times*, or *percent* in problems you are trying to solve. They give you a clue about what operation to use.

The table shows the final standings of the Women's United Soccer Association for the 2002 season.

Women's United Soccer Association					
Team	Games	Wins	Losses	Ties	Points
Carolina	21	12	5	4	40
Philadelphia	21	11	4	6	39
Washington	21	11	5	5	38
Atlanta	21	11	9	1	34
San Jose	21	8	8	5	29
Boston	21	6	8	7	25
San Diego	21	5	11	5	20
New York	21	3	17	1	10

You can compare the data in several ways.

▶**DIFFERENCES** Carolina won 7 more games than San Diego.
$$12 - 5 = 7$$

▶**RATIOS** Boston had 2.5 times more points than New York.
$$25 \div 10 = 2.5$$

▶**PERCENTS** Philadelphia lost about 19% of the games they played.
$$(4 \div 21) \times 100 \approx 19$$

SKILL PRACTICE

Determine whether each problem asks for a *difference*, *ratio*, or *percent*. Write out the key word or words in each problem. Solve each problem.

1. How many times more games did San Jose win than San Diego?

2. How many more games did Washington win than lose or tie?

3. How many fewer points did Atlanta have than Carolina?

4. What percent of the time did Carolina win its games?

5. Write three statements comparing the data in the table. One comparison should be a difference, one should be a ratio, and one should be a percent.

The Percent Proportion

Sunshine State Standards MA.A.1.3.1-1, MA.A.1.3.4-1, MA.A.3.3.1-1, MA.A.3.3.2-1, MA.A.3.3.2-3, MA.D.1.3.2-2, MA.D.2.3.1-1, MA.D.2.3.2-2

What You'll Learn

Solve problems using the percent proportion.

NEW Vocabulary

percent proportion
part
base

∞ LINK To Reading

Everyday Meaning of Base: the bottom of something considered to be its support

---

HANDS-ON Mini Lab

Materials
- grid paper
- markers

Work with a partner.

You can use proportion models to determine the percent represented by 3 out of 5.

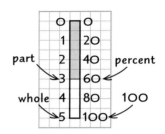

STEP 1 Draw a 10-by-1 rectangle on grid paper. Label the units on the right from 0 to 100.

STEP 2 On the left side, mark equal units from 0 to 5, because 5 represents the whole quantity.

STEP 3 Draw a horizontal line from 3 on the left side of the model. The number on the right side is the percent.

1. Draw a model and find the percent that is represented by each ratio.

   a. 1 out of 2   b. 7 out of 10   c. 2 out of 5

2. Write a proportion that you could use to determine the percent represented by 9 out of 25.

In a **percent proportion**, one of the numbers, called the **part**, is being compared to the whole quantity, called the **base**. The other ratio is the percent, written as a fraction, whose base is 100.

→ Part ←
→ $\dfrac{3}{5} = \dfrac{60}{100}$ ←
→ Base ←

Key Concept   Percent Proportion

Words $\dfrac{\text{part}}{\text{base}} = \dfrac{\text{percent}}{100}$

Symbols $\dfrac{a}{b} = \dfrac{p}{100}$, where a is the part, b is the base, and p is the percent.

EXAMPLE   Find the Percent

1 **22 is what percent of 110?**

Since 22 is being compared to 110, 22 is the part and 110 is the base. You need to find the percent.

$$\frac{a}{b} = \frac{p}{100} \rightarrow \frac{22}{110} = \frac{p}{100}$$ Replace a with 22 and b with 110.

$$22 \cdot 100 = 110 \cdot p$$ Find the cross products.

$$2{,}200 = 110p$$ Multiply.

$$\frac{2{,}200}{110} = \frac{110p}{110}$$ Divide each side by 110.

$$20 = p$$ 22 is 20% of 110.

You can also use the percent proportion to find a missing part or base.

Concept Summary   Type of Percent Problems

Type	Example	Proportion
Find the Percent	7 is what percent of 10? *percent*	$\frac{7}{10} = \frac{P}{100}$
Find the Part	What number is 70% of 10? *part*	$\frac{a}{10} = \frac{70}{100}$
Find the Base	7 is 70% of what number? *base*	$\frac{7}{b} = \frac{70}{100}$

EXAMPLE   **Find the Part**

2 **What number is 80% of 500?**

The percent is 80, and the base is 500. You need to find the part.

$$\frac{a}{b} = \frac{p}{100} \rightarrow \frac{a}{500} = \frac{80}{100}$$ Replace b with 500 and p with 80.

$$a \cdot 100 = 500 \cdot 80$$ Find the cross products.

$$100a = 40{,}000$$ Multiply.

$$\frac{100a}{100} = \frac{40{,}000}{100}$$ Divide each side by 100.

$$a = 400$$ 400 is 80% of 500.

EXAMPLE   **Find the Base**

3 **HISTORY Use the information at the left to determine how many days the Lewis and Clark Expedition spent in Oregon.**

The percent is 89, and the part is 94. You need to find the base.

$$\frac{a}{b} = \frac{p}{100} \rightarrow \frac{94}{b} = \frac{89}{100}$$ Replace a with 94 and p with 89.

$$94 \cdot 100 = b \cdot 89$$ Find the cross products.

$$9{,}400 = 89b$$ Multiply.

$$\frac{9{,}400}{89} = \frac{89b}{89}$$ Divide each side by 89.

$$105.6 \approx b$$ Simplify.

The Lewis and Clark Expedition spent 106 days in Oregon.

REAL-LIFE MATH

HISTORY The members of the Lewis and Clark Expedition spent the winter of 1805–1806 in Oregon. They reported that it rained 94 days, which was about 89% of their days in Oregon.

Source: *Kids Discover*

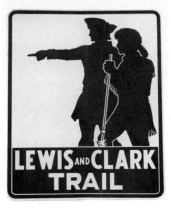

LEWIS AND CLARK TRAIL

1. **Explain** why the value of p in $\frac{a}{b} = \frac{p}{100}$ represents a percent.

2. **OPEN ENDED** Write a real-life problem that could be solved using the proportion $\frac{a}{12} = \frac{25}{100}$.

3. **FIND THE ERROR** Roberto and Jamal are writing percent proportions to solve the following problem. Who is correct? Explain.

95 is 25% of what number?

Roberto
$$\frac{95}{b} = \frac{25}{100}$$

Jamal
$$\frac{a}{95} = \frac{25}{100}$$

GUIDED PRACTICE

Write a percent proportion to solve each problem. Then solve. Round to the nearest tenth if necessary.

4. 70 is what percent of 280?

5. Find 60% of 90.

6. 150 is 60% of what number?

7. What percent of 49 is 7?

8. What is 72.5% of 200?

9. 125 is 30% of what number?

Practice and Applications

Write a percent proportion to solve each problem. Then solve. Round to the nearest tenth if necessary.

10. 3 is what percent of 15?

11. What percent of 64 is 16?

12. What is 15% of 60?

13. Find 35% of 200.

14. 18 is 45% of what number?

15. 95 is 95% of what number?

16. What percent of 56 is 8?

17. 120 is what percent of 360?

18. Find 12.4% of 150.

19. What is 17.2% of 350?

20. 725 is 15% of what number?

21. 225 is 95% of what number?

22. What is 2.5% of 95?

23. Find 5.8% of 42.

24. 17 is what percent of 55?

25. What percent of 27 is 12?

26. 98 is 22.5% of what number?

27. 57 is 13.5% of what number?

HOMEWORK HELP	
For Exercises	See Examples
10–11, 16–17, 24–25, 28–30	1
12–13, 18–19, 22–23, 31–36	2
14–15, 20–21, 26–27	3
Extra Practice See pages 627, 652.	

GAMES For Exercises 28–30, use the following information.
At the start of a game of chess, each player has the pieces listed at the right.

28. What percent of each player's pieces are pawns?

29. What percent of each player's pieces are knights?

30. What percent of each player's pieces are kings?

Chess Pieces
1 king
1 queen
2 bishops
2 knights
2 rooks
8 pawns

ANIMALS For Exercises 31–36, use the graphic at the right.

31. How many of the 4,000,000 households have turtles or tortoises?

32. How many households have snakes?

33. How many households have frogs or toads?

34. How many households have iguanas?

35. How many households have lizards?

36. How many households have newts?

37. **RESEARCH** Use the Internet or another source to find what percent of the total population of the United States is living in your state.

38. **CRITICAL THINKING** Give a counterexample to show the following is *not* true.
10% of a number is added to the number. Then 10% of the sum is subtracted from the sum. The result is the original number.

39. **CRITICAL THINKING** Kwan made 56% of his free throws in the first half of the basketball season. If he makes 7 shots out of the next 13 attempts, will it help or hurt his average? Explain.

USA TODAY Snapshots®

Turtles are right at home
Nearly 4 million households had a reptile or an amphibian as a pet last year. Type they owned:

Turtle/tortoise **46%**
Snake **22%**
Frog/toad **19%**
Iguana **18%**
Lizard **17%**
Newts **5%**

Note: Exceeds 100% due to multiple responses
Source: The NPD Group for American Pet Products Manufacturers Association

By Cindy Hall and Bob Laird, USA TODAY

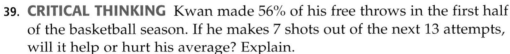

Standardized Test Practice and Mixed Review

FCAT Practice

40. **MULTIPLE CHOICE** The bar graph shows the number of wins for the Chicago Bulls from 1991 to 2001. If they play 82 games in a season, about what percent of games did they win in the 1997–1998 season?

(A) 76%   (B) 82%

(C) 84%   (D) 88%

41. **SHORT RESPONSE** Pure gold is 24-karat gold. In the United States, most jewelry is 18-karat gold. What percent of the 18-karat jewelry is gold?

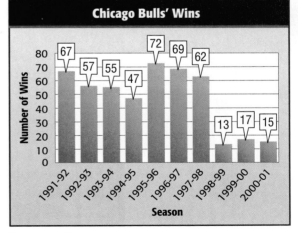

Chicago Bulls' Wins

67, 57, 55, 47, 72, 69, 62, 13, 17, 15

Number of Wins

Seasons: 1991-92, 1992-93, 1993-94, 1994-95, 1995-96, 1996-97, 1997-98, 1998-99, 1999-00, 2000-01

Season

Source: www.nba.com

Write each decimal as a percent. (Lesson 5-2)

42. 0.81   43. 0.12   44. 0.2   45. 1.735

46. Write 48% as a fraction in simplest form. (Lesson 5-1)

GETTING READY FOR THE NEXT LESSON

BASIC SKILL Compute each product mentally.

47. $\frac{1}{2} \cdot 422$ 48. $639 \cdot \frac{1}{3}$ 49. $0.1 \cdot 722$ 50. $0.5 \cdot 680$

Finding Percents Mentally

Sunshine State Standards
MA.A.1.3.4-1, MA.A.3.3.2-1, MA.A.3.3.2-3

What You'll Learn

Compute mentally with percents.

WHEN am I ever going to use this?

SCHOOL The table below lists the enrollment at Roosevelt Middle School by grade level.

1. 50% of the eighth grade class are females. Write 50% as a fraction.

2. Explain how you could find 50% of 104 without using a proportion.

3. Use mental math to find the number of females in the eighth grade class.

4. 25% of the sixth grade class play intramural basketball. Write 25% as a fraction.

5. Use mental math to find the number of students in the sixth grade who play intramural basketball.

Roosevelt Middle School Enrollment	
Grade Level	**Number of Students**
Sixth	84
Seventh	93
Eighth	104

When you compute with common percents like 50% or 25%, it may be easier to use the fraction form of the percent. This number line shows some fraction-percent equivalents.

0%	12.5%	25%	37.5%	50%	62.5%	75%	87.5%	100%
0	$\frac{1}{8}$	$\frac{1}{4}$	$\frac{3}{8}$	$\frac{1}{2}$	$\frac{5}{8}$	$\frac{3}{4}$	$\frac{7}{8}$	1

Some percents are used more frequently than others. So, it is a good idea to be familiar with these percents and their equivalent fractions.

Concept Summary — Percent-Fraction Equivalents

$25\% = \frac{1}{4}$	$20\% = \frac{1}{5}$	$16\frac{2}{3}\% = \frac{1}{6}$	$12\frac{1}{2}\% = \frac{1}{8}$	$10\% = \frac{1}{10}$
$50\% = \frac{1}{2}$	$40\% = \frac{2}{5}$	$33\frac{1}{3}\% = \frac{1}{3}$	$37\frac{1}{2}\% = \frac{3}{8}$	$30\% = \frac{3}{10}$
$75\% = \frac{3}{4}$	$60\% = \frac{3}{5}$	$66\frac{2}{3}\% = \frac{2}{3}$	$62\frac{1}{2}\% = \frac{5}{8}$	$70\% = \frac{7}{10}$
	$80\% = \frac{4}{5}$	$83\frac{1}{3}\% = \frac{5}{6}$	$87\frac{1}{2}\% = \frac{7}{8}$	$90\% = \frac{9}{10}$

STUDY TIP

Look Back
You can review **multiplying fractions** in Lesson 2-3.

EXAMPLES Use Fractions to Compute Mentally

Compute mentally.

1 **20% of 45**

20% of $45 = \frac{1}{5}$ of 45 or 9 Use the fraction form of 20%, which is $\frac{1}{5}$.

2 **$33\frac{1}{3}\%$ of 93**

$33\frac{1}{3}\%$ of $93 = \frac{1}{3}$ of 93 or 31 Use the fraction form of $33\frac{1}{3}\%$, which is $\frac{1}{3}$.

Your Turn Compute mentally.

    a. 25% of 32   **b.** $12\frac{1}{2}\%$ of 160   **c.** 80% of 45

You can also use decimals to find percents mentally. Remember that $10\% = 0.1$ and $1\% = 0.01$.

STUDY TIP

Multiplying Decimals To multiply by 0.1, move the decimal point one place to the left. To multiply by 0.01, move the decimal point two places to the left.

EXAMPLES Use Decimals to Compute Mentally

Compute mentally.

3 **10% of 98**

10% of $98 = 0.1$ of 98 or 9.8

4 **1% of 235**

1% of $235 = 0.01$ of 235 or 2.35

Your Turn Compute mentally.

    d. 10% of 65   **e.** 1% of 450   **f.** 30% of 22

You can use either fractions or decimals to find percents mentally.

EXAMPLE Use Mental Math to Solve a Problem

5 **SCHOOL** At Madison Middle School, 60% of the students voted in an election for student council officers. There are 1,500 students. How many students voted in the election?

Method 1 Use a fraction.

60% of $1,500 = \frac{3}{5}$ of $1,500$ **THINK** $\frac{1}{5}$ of 1,500 is 300.

60% of $1,500$ is 900.   So, $\frac{3}{5}$ of 1,500 is 3 · 300 or 900.

Method 2 Use a decimal.

60% or $1,500 = 0.6$ of $1,500$ **THINK** 0.1 of 1,500 is 150.

60% of $1,500$ is 900.   So, 0.6 of 1,500 is 6 · 150 or 900.

There were 900 students who voted in the election.

1. **Explain** how to find 75% of 40 mentally.

2. **OPEN ENDED** Suppose you wish to find $33\frac{1}{3}\%$ of x. List two values of x for which you could do the computation mentally. Explain.

3. **FIND THE ERROR** Candace and Pablo are finding 10% of 95. Who is correct? Explain.

> Candace
> 10% of 95 = 9.5

> Pablo
> 10% of 95 = 0.95

GUIDED PRACTICE

Compute mentally.

4. 50% of 120

5. $33\frac{1}{3}\%$ of 60

6. $37\frac{1}{2}\%$ of 72

7. 1% of 52

8. 10% of 350

9. 20% of 630

10. **PEOPLE** The average person has about 100,000 hairs on his or her head. However, if people with red hair are taken as a smaller group, they average only 90% of this number. What is the average number of hairs on the head of a person with red hair?

Practice and Applications

Compute mentally.

11. 25% of 44

12. 50% of 62

13. $12\frac{1}{2}\%$ of 64

14. $16\frac{2}{3}\%$ of 54

15. 40% of 35

16. 60% of 15

17. $66\frac{2}{3}\%$ of 120

18. $62\frac{1}{2}\%$ of 160

19. 10% of 57

20. 1% of 81

21. 1% of 28.3

22. 10% of 17.1

23. 3% of 130

24. 7% of 210

25. 150% of 80

26. 125% of 400

27. $66\frac{2}{3}\%$ of 10.8

28. $37\frac{1}{2}\%$ of 41.6

29. Find 1% of $42,200 mentally.

30. Find 10% of $17.40 mentally.

HOMEWORK HELP	
For Exercises	See Examples
11–18, 25–28	1, 2
19–24, 29–30	3, 4
37–39	5

Extra Practice
See pages 627, 652.

Replace each ● with <, >, or = to make a true sentence.

31. 7.5 ● 10% of 80

32. 75% of 80 ● 65

33. 1% of 150 ● 10% of 15

34. $66\frac{2}{3}\%$ of 18 ● 60% of 15

35. Which is greater, 25% of 16 or 5?

36. Which is greater, 75% of 120 or 85?

37. **HEALTH** Many health authorities recommend that a healthy diet contains no more than 30% of its Calories from fat. If Jennie consumes 1,500 Calories each day, what is the maximum number of Calories she should consume from fat?

BASEBALL For Exercises 38 and 39, use the following information.
The graphic shows the results of a survey asking women about their interest in Major League Baseball. Suppose 1,000 women were surveyed.

Women's Interest in Major League Baseball

30% Interested   Not Interested 70%

Source: ESPN

38. How many women said they were interested in Major League Baseball?

39. How many women said they were *not* interested in Major League Baseball?

40. **WRITE A PROBLEM** Write and solve a real-life problem involving percents that uses mental math.

41. **CRITICAL THINKING** Find two numbers, *a* and *b*, such that 10% of *a* is the same as 30% of *b*. Explain.

42. **CRITICAL THINKING** Explain how to determine the 15% tip using mental math.

The waiter brought us just one bill.

That's O.K. I had a hamburger for $2.75 and a cola for $1.25. My part of the bill is $4.

The total bill is for $8.60, so I owe $4.60!

That's great, but how do we determine a 15% tip?

Standardized Test Practice and Mixed Review

FCAT Practice

43. **MULTIPLE CHOICE** Alan and three of his coworkers ate lunch at Old Town Café. They plan to leave a 20% tip for the waiter. Two of his coworkers had turkey sandwiches, one had soup and salad, and Alan had pasta. What information is necessary to determine how much to leave for a tip?

   Ⓐ the cost of the pasta   Ⓑ the cost of the four meals

   Ⓒ what day they had lunch   Ⓓ the soup of the day

44. **GRID IN** Find 10% of 23.

45. **FOOTBALL** Eleven of the 48 members of the football team are on the field. What percent of the team members are playing? (Lesson 5-3)

Write each fraction as a percent. (Lesson 5-2)

46. $\frac{9}{20}$ 47. $\frac{7}{8}$ 48. $\frac{3}{500}$ 49. $\frac{2}{9}$

GETTING READY FOR THE NEXT LESSON

BASIC SKILL Estimate.

50. $\frac{1}{4}$ of 81   51. $\frac{2}{3}$ of 91   52. $\frac{4}{5}$ of 49   53. $\frac{2}{7}$ of 68

Vocabulary and Concepts

1. **Show** two different ways to write $\frac{5}{6}$ as a percent. (Lesson 5-2)

2. **Explain** how to find 75% of 8 mentally. (Lesson 5-4)

Skills and Applications

Write each ratio or fraction as a percent. (Lesson 5-1)

3. 3 out of 20   4. 15.2 out of 100   5. $\frac{13}{25}$ 6. $\frac{7}{20}$

Write each decimal or fraction as a percent. (Lesson 5-2)

7. 0.325   8. $\frac{3}{50}$ 9. $\frac{1}{5}$ 10. 1.02

Write a percent proportion to solve each problem. Then solve. Round to the nearest tenth if necessary. (Lesson 5-3)

11. 63 is what percent of 84?   12. Find 35% of 700.

13. 294 is 35% of what number?

Compute mentally. (Lesson 5-4)

14. 25% of 64   15. 50% of 150   16. 60% of 20   17. 3% of 600

18. **SCHOOL** Santos scored 87% on an English exam. Write this as a decimal. (Lesson 5-2)

FCAT Practice

Standardized Test Practice

Ⓐ Ⓑ Ⓒ Ⓓ

19. **MULTIPLE CHOICE** Fifteen percent of the dogs at a show were Labrador retrievers. Which is *not* true? (Lesson 5-1)

   Ⓐ $\frac{3}{20}$ of the dogs were Labrador retrievers.

   Ⓑ 15 out of every 100 dogs were Labrador retrievers.

   Ⓒ 85% of the dogs were *not* Labrador retrievers.

   Ⓓ 1 out of every 15 dogs were Labrador retrievers.

20. **SHORT RESPONSE** Use the graph below. Does Leah spend more of her day sleeping or at school? Explain. (Lesson 5-2)

Leah's Day

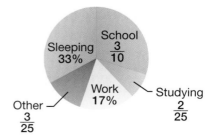

Sleeping 33%

School $\frac{3}{10}$

Work 17%

Studying $\frac{2}{25}$

Other $\frac{3}{25}$

The GameZone

A Place To Practice Your Math Skills

Math Skill
Equivalent Percents,
Fractions,
and Decimals

Per-Fraction

● **GET READY!**

Players: two, three, or four
Materials: 38 index cards, scissors, markers

● **GET SET!**

• Cut each index card in half, making 76 cards.

• Take four cards. On the first card, write a percent from the table on page 220. On the second card, write the corresponding fraction next to the percent. On the third card, write an equivalent fraction. On the fourth card, write the equivalent decimal.

• Repeat the steps until you have used all 19 percents from the table.

● **GO!**

• Deal seven cards to each player. Place the remaining cards facedown on the table. Take the top card and place it faceup next to the deck, forming the discard pile.

• All players check their cards for scoring sets. A scoring set consists of three equivalent numbers.

| 60% | $\frac{3}{5}$ | $\frac{6}{10}$ |

Scoring Set

• The first player draws the top card from the deck or the discard pile. If the player has a scoring set, he or she should place it faceup on the table. The player may also build onto another player's scoring set by placing a card faceup on the table and announcing the set on which the player is building. The player's turn ends when he or she discards a card.

• **Who Wins?** The first person with no cards remaining wins.

Problem-Solving Strategy

A Preview of Lesson 5-5

Sunshine State Standards
MA.A.3.3.1-1, MA.A.3.3.2-1, MA.A.3.3.2-2,
MA.A.3.3.2-3, MA.A.3.3.3-1, MA.A.4.3.1-1,
MA.A.4.3.1-2, MA.B.3.3.1-1

What You'll Learn

Solve problems using the reasonable answer strategy.

Reasonable Answers

Because I work at the Jean Shack, I can buy a $50 jacket there for 60% of its price.

Carla, will you have to pay more or less than $25?

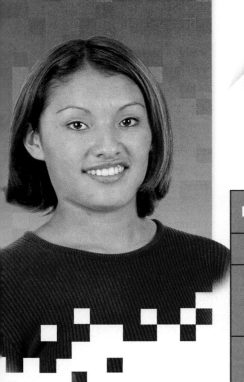

Explore	We know the cost of the jacket. Carla can buy the jacket for 60% of the price. We want to know if the jacket will cost more or less than $25.
Plan	Use mental math to determine a reasonable answer.
Solve	THINK $\frac{25}{50} = \frac{1}{2}$ or 50% Since Carla will pay 60% of the cost, she will have to pay more than $25.
Examine	Find 60% of $50 60% of 50 $= \frac{3}{5}$ of 50 Since $\frac{1}{5}$ of 50 is 10, $\frac{3}{5}$ of 50 is 3 × 10 or 30. Carla will pay $30 which is more than $25.

Analyze the Strategy

1. **Explain** why determining a reasonable answer was an appropriate strategy for solving the above problem.

2. **Explain** why mental math skills are important when using the reasonable answer strategy.

3. **Write** a problem where checking for a reasonable answer is appropriate. Explain how you would solve the problem.

Solve. Use the reasonable answer strategy.

4. **SCHOOL** There are 750 students at Monroe Middle School. If 64% of the students have purchased yearbooks, would the number of yearbooks purchased be about 200, 480, or 700?

5. **MONEY MATTERS** Spencer took $40 to the shopping mall. He has already spent $12.78. He wants to buy two items for $7.25 and $15.78. Does he have enough money with him to make these two purchases?

Mixed Problem Solving

Solve. Use any strategy.

6. **BAKING** Desiree spilled $1\frac{1}{2}$ cups of sugar, which she discarded. She then used half of the remaining sugar to make cookies. If she had $4\frac{1}{2}$ cups left, how much sugar did she have in the beginning?

7. **NUMBER THEORY** Study the pattern.

$$1 \times 1 = 1$$
$$11 \times 11 = 121$$
$$111 \times 111 = 12,321$$
$$1111 \times 1111 = 1,234,321$$

Without doing the multiplication, find 1111111×1111111.

8. **FARMING** An orange grower harvested 1,260 pounds of oranges from one grove, 874 pounds from another, and 602 pounds from a third. The oranges will be placed in crates that hold 14 pounds oranges each. Should the orange grower order 100, 200, or 300 crates for the oranges?

9. **DESIGN** Juanita is designing isosceles triangular tiles for a mosaic. The sides of the larger triangle are $1\frac{1}{2}$ times larger than the sides of the triangle shown. What are the dimensions of the larger triangle?

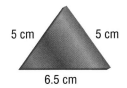

5 cm   5 cm

6.5 cm

10. **GEOMETRY** What percent of the large rectangle is blue?

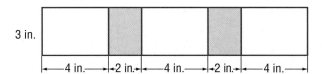

3 in.

|← 4 in. →|←2 in.→|← 4 in. →|←2 in.→|← 4 in. →|

11. **MULTI STEP** Seth is saving for a down payment on a car. He wants to have a down payment of 10% for a car that costs $13,000. So far he has saved $850. If he saves $75 each week for the down payment, how soon can he buy the car?

12. **ECOLOGY** In a survey of 1,413 shoppers, 6% said they would be willing to pay more for environmentally safe products. Is 8.4, 84, or 841 a reasonable estimate for the number of shoppers willing to pay more?

13. **BUILDING** The atrium of a new mega mall will need 2.3×10^5 square feet of ceramic tile. The tiles measure 2 feet by 2 feet and are sold in boxes of 24. How many boxes of tiles will be needed to complete the job?

14. **STANDARDIZED TEST PRACTICE** FCAT Practice

In one month, the Shaffer family spent $121.59, $168.54, $98.67, and $141.78 on groceries. Which amount is a good estimate of the total cost of the groceries for the month?

    Ⓐ $450   Ⓑ $530

    Ⓒ $580   Ⓓ $620

You will determine whether answers are reasonable in the next lesson.

Percent and Estimation

Sunshine State Standards MA.A.1.3.4-1, MA.A.3.3.1-1, MA.A.3.3.2-1, MA.A.3.3.2-3, MA.A.3.3.3-1, MA.A.4.3.1-1, MA.A.4.3.1-2

What You'll Learn

Estimate by using equivalent fractions, decimals, and percents.

NEW Vocabulary

compatible numbers

MATH Symbols

≈ is approximately equal to

WHEN am I ever going to use this?

GEOGRAPHY The total area of Earth is 196,800,000 square miles. The graphic at the right shows the percent of the area of Earth that is land and the percent that is water.

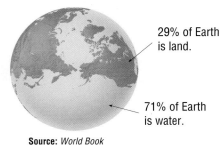

29% of Earth is land.

71% of Earth is water.

Source: *World Book*

1. Round the total area of Earth to the nearest hundred million square miles.

2. Round the percent of Earth that is land to the nearest ten percent.

3. Use what you learned about mental math in Lesson 5-4 to estimate the area of the land on Earth.

Sometimes an exact answer is not needed. One way to estimate a percent of a number is by using compatible numbers. **Compatible numbers** are two numbers that are easy to divide mentally.

EXAMPLES Estimate Percents of Numbers

Estimate.

1 **19% of 30**

19% is about 20% or $\frac{1}{5}$. $\frac{1}{5}$ and 30 are compatible numbers.

$\frac{1}{5}$ of 30 is 6.

So, 19% of 30 is about 6.

2 **25% of 41**

25% is $\frac{1}{4}$, and 41 is about 40. $\frac{1}{4}$ and 40 are compatible numbers.

$\frac{1}{4}$ of 40 is 10.

So, 25% of 41 is about 10.

3 **65% of 76**

65% is about $66\frac{2}{3}$% or $\frac{2}{3}$, and 76 is about 75. $\frac{1}{3}$ and 75 are compatible numbers.

$\frac{2}{3}$ of 75 is 50.

So, 65% or 76 is about 50.

You can use similar techniques to estimate a percent.

EXAMPLES **Estimate Percents**

Estimate each percent.

4 **8 out of 25**

$$\frac{8}{25} \approx \frac{8}{24} \text{ or } \frac{1}{3}$$ 25 is about 24.

$$\frac{1}{3} = 33\frac{1}{3}\%$$

So, 8 out of 25 is about $33\frac{1}{3}\%$.

5 **14 out of 25**

$$\frac{14}{25} \approx \frac{15}{25} \text{ or } \frac{3}{5}$$ 14 is about 15.

$$\frac{3}{5} = 60\%$$

So, 14 out of 25 is about 60%.

6 **89 out of 121**

$$\frac{89}{121} \approx \frac{90}{120} \text{ or } \frac{3}{4}$$ 89 is about 90, and 121 is about 120.

$$\frac{3}{4} = 75\%$$

So, 89 out of 121 is about 75%.

Your Turn Estimate each percent.

a. 7 out of 57   **b.** 9 out of 25   **c.** 7 out of 79

Sometimes estimation is the best answer to a real-life problem.

EXAMPLE **Estimate Percent of an Area**

7 **FIREFIGHTING** Fire fighters use geometry and aerial photography to estimate how much of a forest has been damaged by fire. A grid is superimposed on a photograph of the forest. The gray part of the figure at the right represents the area damaged by a forest fire. Estimate the percent of the forest damaged by the fire.

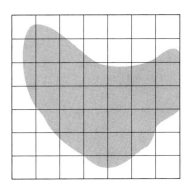

About 24 small squares out of 49 squares are shaded gray.

$$\frac{24}{49} \approx \frac{25}{50} \text{ or } \frac{1}{2}$$ 24 is about 25, and 49 is about 50.

$$\frac{1}{2} = 50\%$$

So, about 50% of the area has been damaged by the fire.

Skill and Concept Check

Writing Math
Exercises 1 & 3

1. **Explain** how you could use fractions and compatible numbers to estimate 26% of $98.98.

2. **OPEN ENDED** Write a percent problem with an estimated answer of 10.

3. **NUMBER SENSE** Use mental math to determine which is greater: 24% of 240 or 51% of 120. Explain.

GUIDED PRACTICE

Estimate.

4. 49% of 160

5. $66\frac{2}{3}\%$ of 20

6. 73% of 65

Estimate each percent.

7. 6 out of 35

8. 8 out of 79

9. 17.5 out of 23

10. **BIOLOGY** The adult skeleton has 206 bones. Sixty of them are in the arms and hands. Estimate the percent of bones that are in the arms and hands.

Practice and Applications

Estimate.

11. 29% of 50

12. 67% of 93

13. 20% of 76

14. 25% of 63

15. 21% of 71

16. 92% of 41

17. 48% of 159

18. 73% of 81

19. 68% of 9.2

20. 26.5% of 123

21. 124% of 41

22. 249% of 119

23. Estimate 34% of 121.

24. Estimate 21% of 348.

HOMEWORK HELP

For Exercises	See Examples
11–24	1–3
25–32, 36–40	4–6
33–35	7

Extra Practice
See pages 628, 652.

Estimate each percent.

25. 7 out of 29

26. 6 out of 59

27. 4 out of 21

28. 6 out of 35

29. 8 out of 13

30. 9 out of 23

31. 150,078 out of 299,000

32. 63,875 out of 245,000

Estimate the percent of the area shaded.

33.

34.

35.

36. **ANIMALS** In the year 2002, 1,053 species of animals were endangered or threatened. Of these species, 340 were mammals. Estimate the percent of endangered or threatened animals that were mammals.

POPULATION For Exercises 37–40, use the following information.

2000 Population		
City	City Population	Entire State Population
New York, New York	8,008,278	18,976,457
Los Angeles, California	3,694,820	33,871,648
Chicago, Illinois	2,896,016	12,419,293

Source: U.S. Bureau of the Census

37. Estimate what percent of the population of the entire state of New York live in New York City.

38. Estimate what percent of the population of the entire state of California live in Los Angeles.

39. Estimate what percent of the population of the entire state of Illinois live in Chicago.

40. Which city has the greatest percent of its state's population?

CRITICAL THINKING Determine whether each statement about estimating percents of numbers is *sometimes*, *always*, or *never* true.

41. If both the percent and the number are rounded up, the estimate will be greater than the actual answer.

42. If both the percent and the number are rounded down, the estimate will be less than the actual answer.

43. If the percent is rounded up and the number is rounded down, the estimate will be greater than the actual answer.

Standardized Test Practice and Mixed Review

  FCAT Practice

44. **MULTIPLE CHOICE** Rick took his father to dinner for his birthday. When the bill came, Rick's father reminded him that it is customary to tip the server 15% of the bill. If the bill was $19.60, a good estimate for the tip is

Ⓐ $6.　　　Ⓑ $5.　　　Ⓒ $4.　　　Ⓓ $3.

45. **MULTIPLE CHOICE** What is the best estimate of the percent represented by 12 out of 35?

Ⓕ 20%　　　Ⓖ $33\frac{1}{3}\%$　　　Ⓗ $37\frac{1}{2}\%$　　　Ⓘ 40%

46. Explain how to find 75% of 84 mentally. (Lesson 5-4)

Write a percent proportion to solve each problem. Then solve. Round to the nearest tenth if necessary. (Lesson 5-3)

47. 7 is what percent of 70?　　48. What is 65% of 200?　　49. 42 is 35% of what number?

GETTING READY FOR THE NEXT LESSON

PREREQUISITE SKILL Solve each equation. (Lesson 2-7)

50. $0.2a = 7$　　　51. $20s = 8$　　　52. $0.35t = 140$　　　53. $30n = 3$

The Percent Equation

Sunshine State Standards MA.A.1.3.4-1, MA.A.3.3.1-1,
MA.A.3.3.2-1, MA.A.3.3.2-2, MA.A.3.3.2-3, MA.A.3.3.3-1,
MA.A.4.3.1-2, MA.D.1.3.2-2, MA.D.2.3.1-1, MA.D.2.3.2-2

What You'll Learn

Solve problems using the percent equation.

NEW Vocabulary

percent equation

REVIEW Vocabulary

equation: a mathematical sentence that contains the equal sign (Lesson 1-8)

WHEN am I ever going to use this?

GEOGRAPHY The following table shows the approximate area of Florida and the percent of the area that is water.

Florida	
Total Area (sq mi)	Percent of Area Occupied by Water
60,000	10%

1. Use the percent proportion to find the area of water in Florida.
2. Express the percent for Florida as a decimal. Multiply the total area of Florida by this decimal.
3. How are the answers for Exercises 1 and 2 related?

The **percent equation** is an equivalent form of the percent proportion in which the percent is written as a decimal.

$$\frac{\text{Part}}{\text{Base}} = \text{Percent}$$ ← The percent is written as a decimal.

$$\frac{\text{Part}}{\text{Base}} \cdot \text{Base} = \text{Percent} \cdot \text{Base}$$ Multiply each side by the base.

$$\text{Part} = \text{Percent} \cdot \text{Base}$$ ← This form is called the percent equation.

Concept Summary — The Percent Equation

Type	Example	Equation
Find the Part	What number is 25% of 60? (part)	$n = 0.25(60)$
Find the Percent	15 is what percent of 60? (percent)	$15 = n(60)$
Find the Base	15 is 25% of what number? (base)	$15 = 0.25n$

EXAMPLE Find the Part

1 **Find 6% of 525.** Estimate 1% of 500 is 5. So, 6% of 500 is 6 · 5 or 30.

The percent is 6%, and the base is 525. Let n represent the part.

$n = 0.06(525)$ Write 6% as the decimal 0.06.

$n = 31.5$ Simplify.

So, 6% of 525 is 31.5. Compare to the estimate.

EXAMPLE Find the Percent

2 **420 is what percent of 600?** Estimate $\frac{420}{600} \approx \frac{400}{600}$ or $66\frac{2}{3}\%$

The part is 420, and the base is 600. Let n represent the percent.

$420 = n(600)$ Write the equation.

$\dfrac{420}{600} = \dfrac{600n}{600}$ Divide each side by 600.

$0.7 = n$ Simplify.

In the percent equation, the percent is written as a decimal. Since $0.7 = 70\%$, 420 is 70% of 600.

Your Turn Solve each problem using the percent equation.

a. What percent of 186 is 62?   **b.** What percent of 90 is 180?

EXAMPLE Find the Base

3 **65 is 52% of what number?** Estimate 65 is 50% of 130.

The part is 65, and the percent is 52%. Let n represent the base.

$65 = 0.52n$ Write 52% as the decimal 0.52.

$\dfrac{65}{0.52} = \dfrac{0.52n}{0.52}$ Divide each side by 0.52.

$125 = n$ Simplify.

So, 65 is 52% of 125.   Compare to the estimate.

Your Turn Solve each problem using the percent equation.

c. 210 is 75% of what number?   **d.** 18% of what number is 54?

e. 0.2% of what number is 25?   **f.** 7 is 2.5% of what number?

EXAMPLE Solve a Real-Life Problem

4 **SALES TAX** A television costs $350. If a 7% tax is added, what is the total cost of the television?

First find the amount of the tax t.

Words	What amount is 7% of $350?		
Symbols	part =	percent ·	base
Equation	t =	0.7 ·	350

$t = 0.07 \cdot 350$ Write the equation.

$t = 24.5$ Simplify.

The amount of the tax is $24.50. The total cost of the television is $350.00 + $24.50 or $374.50.

Writing Math
Exercises 2 & 3

1. **Write** an equation you could use to find the percent of questions answered correctly if 32 out of 40 answers are correct.

2. **OPEN ENDED** Write a percent problem in which you need to find the base. Solve the problem using the percent proportion and using the percent equation. Compare and contrast the two methods of solving the equation.

3. **Which One Doesn't Belong?** Identify the equation that does not have the same solution as the other three. Explain your reasoning.

| $15 = n(20)$ | $3 = n(4)$ | $80 = n(60)$ | $9 = n(12)$ |

GUIDED PRACTICE

Solve each problem using the percent equation.

4. Find 85% of 920.
5. 25 is what percent of 625?
6. 680 is 34% of what number?
7. 2 is what percent of 800?

Practice and Applications

Solve each problem using the percent equation.

8. Find 60% of 30.
9. What is 40% of 90?
10. What percent of 90 is 36?
11. 45 is what percent of 150?
12. 75 is 50% of what number?
13. 15% of what number is 30?
14. 6 is what percent of 3,000?
15. What percent of 5,000 is 6?
16. What number is 130% of 52?
17. Find 240% of 84.
18. 3% of what number is 9?
19. 50 is 10% of what number?
20. 8 is 2.4% of what number?
21. 1.8% of what number is 40?
22. What percent of 675 is 150?
23. 360 is what percent of 270?
24. Find 6.25% of 150.
25. What is 12.5% of 92?

HOMEWORK HELP	
For Exercises	See Examples
8–9, 16–17, 24–25	1
10–11, 14–15, 22–23	2
12–13, 18–21	3
26–29	4
Extra Practice See pages 628, 652.	

26. **REAL ESTATE** A *commission* is a fee paid to a salesperson based on a percent of sales. Suppose a real estate agent earns a 3% commission. How much commission would be earned for the sale of a $150,000 house?

27. **BASKETBALL** In the 2001–2002 National Basketball Association season, Shaquille O'Neal made about 57.74% of his field-goal attempts. If he made 653 field goals, how many attempts did he take?

 Data Update What percent of the field-goal attempts did your favorite player make last season? Visit msmath3.net/data_update to learn more.

28. **MULTI STEP** A sweater costs $45. If a 6.5% sales tax is added, what is the total cost of the sweater?

29. ARCHITECTURE Both the Guggenheim Museum in New York and the Guggenheim Museum in Bilbao, Spain, are known for their interesting architecture. Which museum uses the greater percent of space for exhibits?

Guggenheim Museum in New York
Total area: 79,600 square feet
Exhibition space: 49,600 square feet

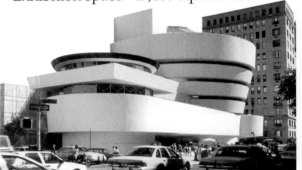

Guggenheim Museum in Bilbao
Total area: 257,000 square feet
Exhibition space: 110,000 square feet

30. CRITICAL THINKING Determine whether $a\%$ of b is *sometimes*, *always*, or *never* equal to $b\%$ of a. Explain.

Standardized Test Practice and Mixed Review

FCAT Practice

31. MULTIPLE CHOICE Fifteen out of the 60 eighth-graders at Seabring Junior High are on the track team. What percent of the eighth-graders are on the track team?

- **A** 15%
- **B** 25%
- **C** 45%
- **D** 60%

32. MULTIPLE CHOICE The graph at the right shows the results of a recent survey asking Americans why we should explore Mars. About how many people were surveyed if 81 of them want to search Mars for a future home for the human race?

- **F** 100
- **G** 125
- **H** 150
- **I** 175

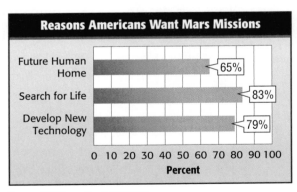

Reasons Americans Want Mars Missions

Source: SPACE.com/Harris Interactive

33. FOOTBALL A quarterback completed 19 out of 30 attempts to pass the football. Estimate his percent of completion. (Lesson 5-5)

Compute mentally. (Lesson 5-4)

34. 15% of $200
35. 62.5% of 96
36. 75% of 84
37. 60% of 150

GETTING READY FOR THE NEXT LESSON

PREREQUISITE SKILL Evaluate each expression. (Lesson 1-3)

38. $|17 - 24|$
39. $|340 - 253|$
40. $|531 - 487|$
41. $|352 - 581|$

Percent of Change

Sunshine State Standards MA.A.1.3.4-1, MA.A.3.3.1-1,
MA.A.3.3.2-1, MA.A.3.3.2-2, MA.A.3.3.2-3, MA.A.3.3.3-1,
MA.D.2.3.1-1

What You'll Learn
Find and use the
percent of increase or
decrease.

WHEN am I ever going to use this?

MONEY MATTERS Over the years, some
prices increase. Study the change in
gasoline prices from 1930 to 1960.

Price of a Gallon of Gasoline	
Year	**Price (¢)**
1930	10
1940	15
1950	20
1960	25

Source: Senior Living

1. How much did the price increase from 1930 to 1940?

2. Write the ratio $\dfrac{\text{amount of increase}}{\text{price in 1930}}$. Then write the ratio
   as a percent.

3. How much did the price increase from 1940 to 1950? Write
   the ratio $\dfrac{\text{amount of increase}}{\text{price in 1940}}$. Then write the ratio as a percent.

4. How much did the price increase from 1950 to 1960? Write
   the ratio $\dfrac{\text{amount of increase}}{\text{price in 1950}}$. The write the ratio as a percent.

5. Compare the amount of increase for each ten-year period.

6. Compare the percents in Exercises 2–4.

7. **Make a conjecture** about why the amounts of increase are the
   same but the percents are different.

In the above application, you expressed the amount of change as a
percent of the original. This percent is called the **percent of change**.

Key Concept

Percent of Change

Words A percent of change is a ratio that compares the change in
quantity to the original amount.

Symbols percent of change $= \dfrac{\text{amount of change}}{\text{original amount}}$

Example original: 12, new: 9

$$\frac{12-9}{12} = \frac{3}{12} = \frac{1}{4} \text{ or } 25\%$$

When the new amount is greater than the original, the percent of change is a **percent of increase**.

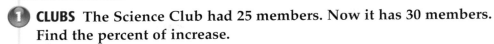 **EXAMPLE** **Find the Percent of Increase**

1 **CLUBS** The Science Club had 25 members. Now it has 30 members. Find the percent of increase.

Step 1 Subtract to find the amount of change. $30 - 25 = 5$

Step 2 Write a ratio that compares the amount of change to the original number of members. Express the ratio as a percent.

$$\text{percent of change} = \frac{\text{amount of change}}{\text{original amount}}$$ Definition of percent of change

$$= \frac{5}{25}$$ The amount of change is 5. The original amount is 25.

$$= 0.2 \text{ or } 20\%$$ Divide. Write as a percent.

The percent of increase is 20%.

Your Turn Find each percent of increase. Round to the nearest tenth if necessary.

a. original: 20
   new: 23

b. original: 50
   new: 67

c. original: 12
   new: 20

When the new amount is less than the original, the percent of change is called a **percent of decrease**.

EXAMPLE **Find the Percent of Change**

2 **COMIC BOOKS** Consuela had 20 comic books. She gave some to her friend. Now she has 13 comic books. Find the percent of change. State whether the percent of change is an *increase* or a *decrease*.

Step 1 Subtract to find the amount of change. $20 - 13 = 7$

Step 2 Write a ratio that compares the amount of change to the original number of comic books. Express the ratio as a percent.

$$\text{percent of change} = \frac{\text{amount of change}}{\text{original amount}}$$ Definition of percent of change

$$= \frac{7}{20}$$ The amount of change is 7. The original amount is 20.

$$= 0.35 \text{ or } 35\%$$ Divide. Write as a percent.

The percent of change is 35%. Since the new amount is less than the original, it is a percent of decrease.

Your Turn Find each percent of change. Round to the nearest tenth if necessary. State whether the percent of change is an *increase* or a *decrease*.

d. original: 10
   new: 6

e. original: 5
   new: 6

f. original: 80
   new: 55

A store sells an item for more than it paid for that item. The extra money is used to cover the expenses and to make a profit. The increase in the price is called the **markup**. The percent of markup is a percent of increase. The amount the customer pays is called the **selling price**.

EXAMPLE Find the Selling Price

3 **MARKETING** Shonny is selling some embroidered jackets on a Web site as shown in the photo. She wants to price the jackets 25% over her cost, which is $35. Find the selling price for a jacket.

Method 1 Find the amount of the markup.

Find 25% of $35. Let m represent the markup.

$m = 0.25(35)$ Write 25% as a decimal. Multiply.

$m = 8.75$

Add the markup to the price Shonny paid for the jacket.

$35 + $8.75 = $43.75

Method 2 Find the total percent.

The customer will pay 100% of the price Shonny paid plus an extra 25% of the price. Find 100% + 25% or 125% of the price Shonny paid for the jacket. Let p represent the price.

$p = 1.25(35)$ Write 125% as a decimal. Multiply.

$p = 43.75$

The selling price of the jacket for the customer is $43.75.

The amount by which a regular price is reduced is called the **discount**. The percent of discount is a percent of decrease.

EXAMPLE Find the Sale Price

4 **SHOPPING** The Sport Chalet is having a sale. Find the sale price of the snowskate at the right.

Method 1 Find the amount of the discount.

Find 35% of $95. Let d represent the amount of the discount.

$d = 0.35(95)$

$d = 33.25$

Subtract the amount of the discount from the original price.

$95 - $33.25 = $61.75

Method 2 Find the percent paid.

If the amount of the discount is 35%, the percent paid is 100% - 35% or 65%. Find 65% of $95. Let s represent the sale price.

$s = 0.65(95)$

$s = 61.75$

The sale price of the snowskate is $61.75.

1. **State** the first step in finding the percent of change.

2. **Explain** how you know whether a percent of change is a percent of increase or a percent of decrease.

3. **OPEN ENDED** Write a percent of increase problem where the percent of increase is greater than 100%.

4. **FIND THE ERROR** Jared and Sydney are solving the following problem. *The price of a movie ticket rose from $5.75 to $6.25. What is the percent of increase for the price of a ticket?* Who is correct? Explain.

Jared

percent of change = $\frac{0.50}{5.75}$

≈ 0.087 or 8.7%

Sydney

percent of change = $\frac{0.50}{6.25}$

$= 0.08$ or 8%

GUIDED PRACTICE

Find each percent of change. Round to the nearest tenth if necessary. State whether the percent of change is an *increase* or a *decrease*.

5. original: 40
   new: 32

6. original: 25
   new: 32

7. original: 325
   new: 400

Find the selling price for each item given the cost to the store and the markup.

8. roller blades: $60, 35% markup

9. coat: $87, $33\frac{1}{3}$% markup

Find the sale price of each item to the nearest cent.

10. CD: $14.50, 10% off

11. sweater: $39.95, 25% off

Practice and Applications

Find each percent of change. Round to the nearest tenth if necessary. State whether the percent of change is an *increase* or a *decrease*.

12. original: 6
    new: 9

13. original: 80
    new: 64

14. original: 560
    new: 420

15. original: 68
    new: 51

16. original: 27
    new: 39

17. original: 98
    new: 150

HOMEWORK HELP

For Exercises	See Examples
12–17, 30, 32	1, 2
18–23	3
24–29	4

Extra Practice
See pages 628, 652.

Find the selling price for each item given the cost to the store and the markup.

18. computer: $700, 30% markup

19. CD player: $120, 20% markup

20. jeans: $25, 45% markup

21. video: $12, 48% markup

22. Find the markup rate on a $60 jacket that sells for $75.

23. What is the markup rate on a $230 game system that sells for $345?

Find the sale price of each item to the nearest cent.

24. video game: $75, 25% off

25. trampoline: $399, 15% off

26. skateboard: $119.95, 30% off

27. television: $675.50, 35% off

28. Find the discount rate on a $24 watch that regularly sells for $32.

29. What is the discount rate on $294 skis that regularly sell for $420?

ANIMALS For Exercises 30 and 31, use the following information.
In 1937, a baby giraffe was born. It was 62 inches tall at birth and grew at the highly unusual rate of 0.5 inch per hour for x hours.

30. By what percent did the height of the giraffe increase in the first day?

31. **MULTI STEP** If the baby giraffe continued to grow at this amazing rate, how long would it take it to reach a height of 18 feet?

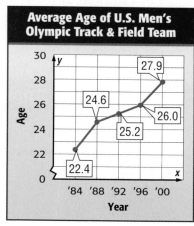

32. **MONEY MATTERS** The table gives the price of milk for various years. During which ten-year period did milk have the greatest percent of increase?

33. **WRITE A PROBLEM** Write and solve a real-life problem involving percent of change.

34. **CRITICAL THINKING** Blake bought a computer listed for $x at a 15% discount. He also paid a 5% sales tax. After 6 months, he decided to sell the computer for $y, which was 55% of what he paid originally. Express y as a function of x.

Price of a Gallon of Milk	
Year	Price ($)
1970	1.23
1980	1.60
1990	2.15
2000	2.78

Source: Senior Living

Standardized Test Practice and Mixed Review

FCAT Practice

35. **SHORT RESPONSE** Use the graph at the right to determine the percent of change in the average age of the U.S. Men's Olympic Track and Field Team from 1984 to 2000. Show your work.

36. **MULTIPLE CHOICE** Find the amount of discount for a pair of $89 shoes that are on sale at a discount of 30%.

   Ⓐ $17.80   Ⓑ $26.70

   Ⓒ $35.60   Ⓓ none of the above

37. **TAXES** An average of 40% of the cost of gasoline goes to state and federal taxes. If gasoline sells for $1.35 per gallon, how much goes to taxes? (Lesson 5-6)

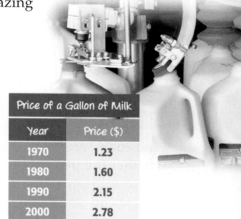

Average Age of U.S. Men's Olympic Track & Field Team

27.9, 24.6, 26.0, 25.2, 22.4

Source: Sports Illustrated

Estimate. (Lesson 5-5)

38. 21% of 60

39. 25% of 83

40. 12% of 31

41. 34% of 95

GETTING READY FOR THE NEXT LESSON

PREREQUISITE SKILL Solve each equation. (Lesson 2-7)

42. $45 = 300 \cdot a \cdot 3$

43. $24 = 200 \cdot 0.04 \cdot y$

44. $21 = 60 \cdot m \cdot 5$

45. $18 = 90 \cdot b \cdot 5$

Simple Interest

Sunshine State Standards MA.A.1.3.4-1, MA.A.3.3.1-1, MA.A.3.3.2-1, MA.A.3.3.2-2, MA.A.3.3.2-3, MA.D.2.3.1-1, MA.D.2.3.1-2

What You'll Learn
Solve problems involving simple interest.

⇒ NEW Vocabulary

interest
principal

WHEN am I ever going to use this?

COLLEGE SAVINGS
Hector received $1,000 from his grandparents. He plans to save it for college expenses. The graph shows rates for various investments for one year.

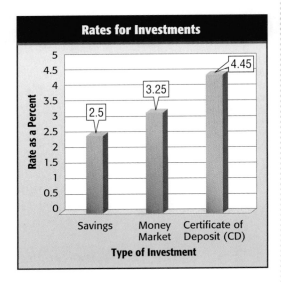

Rates for Investments

1. If Hector puts his money in a savings account, he will receive 2.5% of $1,000 in interest for one year. Find the interest Hector will receive.

2. Find the interest Hector will receive if he puts his money in a money market for one year.

3. Find the interest Hector will receive if he puts his money in a certificate of deposit for one year.

Interest is the amount of money paid or earned for the use of money. For a savings account, you earn interest from the bank. For a credit card, you pay interest to the bank. To solve problems involving simple interest, use the following formula.

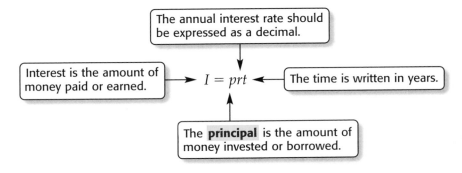

The annual interest rate should be expressed as a decimal.

Interest is the amount of money paid or earned.

$I = prt$

The time is written in years.

The **principal** is the amount of money invested or borrowed.

STUDY TIP

Reading Math
$I = prt$ is read *interest equals principal times rate times time.*

EXAMPLE Find Simple Interest

1. **Find the simple interest for $500 invested at 6.25% for 3 years.**

$I = prt$ Write the simple interest formula.

$I = 500 \cdot 0.0625 \cdot 3$ Replace *p* with 500, *r* with 0.0625, and *t* with 3.

$I = 93.75$ The simple interest is $93.75.

EXAMPLE Find the Total Amount

2 **GRID-IN TEST ITEM** Find the total amount of money in an account where $95 is invested at 7.5% for 8 months.

Read the Test Item

You need to find the total amount in an account. Notice that the time is given in months. Eight months is $\frac{8}{12}$ or $\frac{2}{3}$ year.

Solve the Test Item

$I = prt$

$I = 95 \cdot 0.075 \cdot \frac{2}{3}$

$I = 4.75$

The amount in the account is $95 + $4.75 or $99.75.

Fill in the Grid

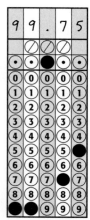

You can use the simple interest formula and what you know about solving equations to find the principal, the interest rate, or the amount of time.

EXAMPLE Find the Interest Rate

3 **CAR SALES** Tonya borrowed $3,600 to buy a used car. She will be paying $131.50 each month for the next 36 months. Find the simple interest rate for her loan.

First find the total amount of money Tonya will pay.
$131.50 \cdot 36 = $4,734
Tonya will pay a total of $4,734.

She will pay $4,734 − $3,600 or $1,134 in interest.

The loan will be for 36 months or 3 years.

Use the simple interest formula to find the interest rate.

Words	Interest equals principal times rate times time.
Variables	$I = p \cdot r \cdot t$
Equation	$1{,}134 = 3{,}600 \cdot r \cdot 3$

$1{,}134 = 3{,}600 \cdot r \cdot 3$ Write the equation.

$1{,}134 = 10{,}800r$ Simplify.

$\dfrac{1{,}134}{10{,}800} = \dfrac{10{,}800r}{10{,}800}$ Divide each side by 10,800.

$0.105 = r$ Simplify.

The simple interest rate is 10.5%.

REAL-LIFE CAREERS

How Does a Car Salesperson Use Math?
A car salesperson must calculate the price of a car including any discounts, dealer preparation costs, and state taxes. They may also help customers by determining the amount of their car payments.

Online Research
For information about a career as a car salesperson, visit:
msmath3.net/careers

1. **Explain** what each variable in the simple interest formula represents.

2. **OPEN ENDED** Give a principal and interest rate where the amount of simple interest earned in two years would be $50.

3. **NUMBER SENSE** Yoshiko needs to find the simple interest on a savings account of $600 at 7% interest for one-half year. She writes $I = 600 \cdot 0.035$. Will this equation give her the correct answer? Explain. (*Hint*: Use the Associative Property of Multiplication.)

GUIDED PRACTICE

Find the simple interest to the nearest cent.

4. $300 at 7.5% for 5 years

5. $230 at 12% for 8 months

Find the total amount in each account to the nearest cent.

6. $660 at 5.25% for 2 years

7. $385 at 12.6% for 9 months

8. **HOUSING** After World War II, William Levitt and his family members began to develop suburbs priced for the middle class. The prices of a ranch in Levittown, New York, are given at the right. Determine the simple interest rate for the investment of a ranch in Levittown from 1947 to 2000.

Price of Ranch in Levittown	
Year	Price
1947	$9,500
2000	$200,000

Source: *Century of Change*

Practice and Applications

Find the simple interest to the nearest cent.

9. $250 at 6.5% for 3 years

10. $725 at 4.5% for 4 years

11. $834 at 7.25% for $1\frac{1}{2}$ years

12. $3,070 at 8.65% for $2\frac{1}{4}$ years

13. $1,000 at 7.5% for 30 months

14. $5,200 at 13.5% for 18 months

HOMEWORK HELP	
For Exercises	See Examples
9–16	1
17–25	2
26–27	3
Extra Practice See pages 629, 652.	

15. Suppose $1,250 is placed in a savings account for 2 years. Find the interest if the simple interest rate is 4.5%.

16. Suppose $580 is placed in a savings account at a simple interest rate of 5.5%. How much interest will the account earn in 3 years?

Find the total amount in each account to the nearest cent.

17. $2,250 at 5.5% for 3 years

18. $5,060 at 7.2% for 5 years

19. $575 at 4.25% for $2\frac{1}{2}$ years

20. $950 at 7.85% for $3\frac{2}{3}$ years

21. $12,000 at 7.5% for 39 months

22. $2,600 at 5.8% for 54 months

23. A savings account starts with $980. If the simple interest rate is 5%, find the total amount after 9 months.

24. Suppose $800 is deposited into a savings account with a simple interest rate of 6.5%. Find the total amount of the account after 15 months.

25. **MONEY MATTERS** Generation X (Gen Xers) are people who were born in the late 1960s or the early 1970s. According to the graphic at the right, most Gen Xers would invest an unexpected $50,000. Suppose someone invested this money at a simple interest rate of 4.5%. How much money would they have at the end of 10 years?

26. **INVESTMENTS** Booker earned $1,200 over the summer. He invested the money in stocks. To his surprise, the stocks increased in value to $1,335 in only 9 months. Find the simple interest rate for the investment.

27. **CRITICAL THINKING** Ethan's bank account listed a balance of $328.80. He originally opened the account with a $200 deposit and a simple interest rate of 4.6%. If there were no deposits or withdrawals, how long ago was the account opened?

USA TODAY Snapshots®

Generation X wants to invest
How Gen Xers would spend an unexpected $50,000:

Invest in future	55%
Make down payment for a home	11%
Buy a new car	7%
Take a vacation	7%
Pay credit card bills	6%
Go on shopping spree	6%
Other	8%

Source: Greenfield Online for MainStay Mutual Funds

By Shannon Reilly and Adrienne Lewis, USA TODAY

Standardized Test Practice and Mixed Review

FCAT Practice

28. **MULTIPLE CHOICE** Suppose Mr. and Mrs. Owens placed $1,500 in a college savings account with a simple interest rate of 4% when Lauren was born. How much will be in the account in 18 years when Lauren is ready to go to college? Assume no more deposits or no withdrawals were made.

　Ⓐ $1,080　　　Ⓑ $2,580　　　Ⓒ $10,800　　　Ⓓ $12,300

29. **SHORT RESPONSE** A $750 investment earned $540 in 6 years. Write an equation you can use to find the simple interest rate. Then find the simple interest rate.

30. **SALES** What is the sale price of a $250 bicycle on sale at 10% off the regular price? (Lesson 5-7)

Solve each equation using the percent equation. (Lesson 5-6)

31. What percent of 70 is 17.5?

32. 18 is 30% of what number?

Web Quest **Interdisciplinary Project**

It's a Masterpiece
It's time to complete your project. Use the information and data you have gathered about your artist and the Golden Ratio to prepare a Web page or poster. Be sure to include your reports and calculations with your project.
msmath3.net/webquest

Spreadsheet Investigation

A Follow-Up of Lesson 5-8

Sunshine State Standards
MA.A.1.3.4-1, MA.A.3.3.1-1, MA.A.3.3.2-1, MA.A.3.3.2-2, MA.A.3.3.3-1

What You'll Learn
Find compound interest.

Compound Interest

Simple interest, which you studied in Lesson 5-8, is paid only on the initial principal of a savings account or a loan. *Compound interest* is paid on the initial principal and on interest earned in the past. You can use a spreadsheet to investigate the growth of compound interest.

ACTIVITY

SAVINGS Find the value of a $2,000 savings account after four years if the account pays 8% interest compounded semiannually.

8% interest compounded semiannually means that the interest is paid twice a year, or every 6 months. The interest rate is 8% ÷ 2 or 4%.

The interest rate is entered as a decimal.

The spreadsheet evaluates the formula A4 × B1.

The interest is added to the principal every 6 months. The spreadsheet evaluates the formula A4 + B4.

	A	B	C	D
			Compound Interest	
1	RATE	0.04		
2				
3	PRINCIPAL	INTEREST	NEW PRINCIPAL	TIME (YR)
4	$2000.00	$80.00	$2080.00	0.5
5	$2080.00	$83.20	$2163.20	1
6	$2163.20	$86.53	$2249.73	1.5
7	$2249.73	$89.99	$2339.72	2
8	$2339.72	$93.59	$2433.31	2.5
9	$2433.31	$97.33	$2530.64	3
10	$2530.64	$101.23	$2631.86	3.5
11	$2631.86	$105.27	$2737.14	4
12	$2737.14	$109.49	$2846.62	4.5
13	$2846.62	$113.86	$2960.49	5
14	$2960.49	$118.42	$3078.91	5.5
15	$3078.91	$123.16	$3202.06	6

Sheet1 / Sheet2

The value of the savings account after four years is $2,737.14.

EXERCISES

1. **Use a spreadsheet** to find the amount of money in a savings account if $2,000 is invested for four years at 8% interest compounded quarterly.

2. Suppose you leave $1,000 in each of three bank accounts paying 6% interest per year. One account pays simple interest, one pays interest compounded semiannually, and one pays interest compounded quarterly. Use a spreadsheet to find the amount of money in each account after three years.

3. **MAKE A CONJECTURE** How does the amount of interest change if the compounding occurs more frequently?

Vocabulary and Concept Check

base (p. 216)
compatible numbers (p. 228)
discount (p. 238)
interest (p. 241)
markup (p. 238)

part (p. 216)
percent (p. 206)
percent equation (p. 232)
percent of change (p. 236)
percent of decrease (p. 237)

percent of increase (p. 237)
percent proportion (p. 216)
principal (p. 241)
selling price (p. 238)

Choose the correct term or number to complete each sentence.

1. A (proportion, percent) is a ratio that compares a number to 100.

2. In a percent proportion, the whole quantity is called a (part, base).

3. The proportion $\frac{1}{10} = \frac{p}{100}$ is an example of a (percent proportion, discount).

4. (Percents, Compatible numbers) are numbers that are easy to divide mentally.

5. A (markup, discount) is an increase in price.

6. A (markup, discount) is a decrease in price.

7. The (interest, principal) is the amount borrowed.

8. The (interest, principal) is the money paid for the use of money.

9. 25% of 16 is (4, 40).

10. The interest formula is ($I = prt$, $p = Irt$).

Lesson-by-Lesson Exercises and Examples

5-1 **Ratios and Percents** (pp. 206–209)

Write each ratio or fraction as a percent.

11. $\frac{4}{5}$

12. $\frac{7}{5}$

13. 16.5 out of 100

14. 0.8 out of 100

15. **WEATHER** There is a 1 in 5 chance of rain tomorrow. Write this as a percent.

Write each percent as a fraction in simplest form.

16. 90%

17. 120%

18. **GAMES** 80% of students at Monroe Middle School play video games. Write this as a fraction in simplest form.

Example 1 Write $\frac{1}{4}$ as a percent.

$$\frac{1}{4} = \frac{25}{100} \quad \text{So, } \frac{1}{4} = 25\%.$$

with $\times 25$ arrows.

Example 2 Write 35% as a fraction in simplest form.

$$35\% = \frac{35}{100} \quad \text{Definition of percent}$$

$$= \frac{7}{20} \quad \text{Simplify.}$$

So, $35\% = \frac{7}{20}$.

 msmath3.net/vocabulary_review

5-2 Fractions, Decimals, and Percents (pp. 210–214)

Write each percent as a decimal.

19. 4.3%
20. 90%
21. 13%
22. 33.2%
23. 147%
24. 0.7%

Write each decimal as a percent.

25. 0.655
26. 0.35
27. 0.7
28. 0.38
29. 0.015
30. 2.55

Write each fraction as a percent.

31. $\frac{7}{8}$
32. $\frac{3}{40}$
33. $\frac{24}{25}$
34. $\frac{1}{6}$

Example 3 Write 24% as a decimal.

$24\% = 24\%$ Divide by 100 and remove the percent symbol.
$= 0.24$

Example 4 Write 0.04 as a percent.

$0.04 = 0.04$ Multiply by 100 and add the percent symbol.
$= 4\%$

Example 5 Write $\frac{9}{25}$ as a percent.

$\frac{9}{25} = 0.36$ Write as a decimal.
$= 36\%$ Change the decimal to a percent.

5-3 The Percent Proportion (pp. 216–219)

Write a percent proportion to solve each problem. Then solve. Round to the nearest tenth if necessary.

35. 15 is 30% of what number?
36. Find 45% of 18.
37. 75 is what percent of 250?

38. **SCHOOL** Hernando hired a band to play at the school dance. The band charges $3,000 and requires a 20% deposit. How much money does Hernando need for the deposit?

Example 6 18 is what percent of 27? Round to the nearest tenth.

$\dfrac{a}{b} = \dfrac{p}{100} \rightarrow \dfrac{18}{27} = \dfrac{p}{100}$ Percent proportion

$18 \cdot 100 = 27 \cdot p$ Find the cross products.

$1,800 = 27p$ Multiply.

$\dfrac{1,800}{27} = \dfrac{27p}{27}$ Divide each side by 27.

$66.7 \approx p$ Simplify.

So, 18 is 66.7% of 27.

5-4 Finding Percents Mentally (pp. 220–223)

Compute mentally.

39. 90% of 100
40. 10% of 18.3
41. $66\frac{2}{3}\%$ of 24
42. 20% of 60
43. 1% of 243
44. 6% of 200

Example 7 Compute 50% of 42 mentally.

50% of $42 = \frac{1}{2}$ of 42 or 21 $50\% = \frac{1}{2}$

5-5 Percent and Estimation (pp. 228–231)

Estimate.

45. 12.5% of 83   **46.** 67% of 60

47. 41% of 39   **48.** 34% of 61

Estimate each percent.

49. 33 out of 98   **50.** 19 out of 52

Example 8   Estimate 8% of 104.

104 is about 100.

8% of 100 is 8.

So, 8% of 104 is about 8.

5-6 The Percent Equation (pp. 232–235)

Solve each problem using the percent equation.

51. What is 66% of 7,000?

52. 60 is what percent of 500?

53. Find 15% of 82.

54. 25 is what percent of 125?

Example 9   70 is 25% of what number?

$70 = 0.25n$ Write 25% as the decimal 0.25.

$\dfrac{70}{0.25} = \dfrac{0.25n}{0.25}$ Divide each side by 0.25

$280 = n$ Simplify.

So, 70 is 25% of 280.

5-7 Percent of Change (pp. 236–240)

Find each percent of change. Round to the nearest tenth if necessary. State whether the percent of change is an *increase* or a *decrease*.

55. original: 10   **56.** original: 8
  new: 15   new: 10

57. original: 37.5   **58.** original: 18
  new: 30   new: 12

59. HOBBIES Mariah collects comic books. Last year she had 50 comic books. If she now has 74 comic books, what is the percent of increase?

Example 10   Find the percent of change if the original amount is 900 and the new amount is 725. Round to the nearest tenth.

$900 - 725 = 175$ The amount of change is 175.

$$\text{percent of change} = \frac{\text{amount of change}}{\text{original amount}}$$

$$= \frac{175}{900}$$

$$\approx 0.194 \text{ or } 19.4\%$$

5-8 Simple Interest (pp. 241–244)

Find the simple interest to the nearest cent.

60. $780 at 6% for 8 months

61. $100 at 8.5% for 2 years

62. $350 at 5% for 3 years

63. $260 at 17.5% for 18 months

Example 11   Find the simple interest for $250 invested at 5.5% for 2 years.

$I = prt$ Simple interest formula

$I = 250 \cdot 0.055 \cdot 2$ Write 5.5% as 0.055.

$I = 27.50$ Simplify.

The simple interest is $27.50.

Vocabulary and Concepts

1. **Write** a percent, a decimal, and a fraction in simplest form for the model shown.

2. **Write** the percent proportion and the percent equation. Use a for the part, b for the base, and p for the percent.

Skills and Applications

Write each ratio or fraction as a percent. Round to the nearest tenth.

3. 7 out of 10

4. 2:40

5. $\frac{1}{6}$

Express each percent as a decimal.

6. 135%

7. 14.6%

Compute mentally.

8. 30% of 60

9. $33\frac{1}{3}\%$ of 90

Estimate.

10. 23% of 16

11. 9% of 81

Solve each problem. Round to the nearest tenth.

12. What is 2% of 3,600?

13. 62 is 90% of what number?

14. Find 45% of 600.

15. 75 is what percent of 30?

Find each percent of change. Round to the nearest tenth if necessary. State whether the percent of change is an *increase* or a *decrease*.

16. original: 15
    new: 12

17. original: 40
    new: 55

18. **BUSINESS** A store prices items 30% over the price paid by the store. If the store purchases a tennis racket for $65, find the selling price of the racket.

19. **MONEY MATTERS** Find the simple interest if $300 is invested at 8% for 3 years.

FCAT Practice

Standardized Test Practice

20. **MULTIPLE CHOICE** Kevin opened a savings account with $125. The account earns 5.2% interest annually. If he does not deposit or withdraw any money for 18 months, how much will he have in his account?

　ⒶＭ $9.75　　　ⒷＮ $117　　　Ⓒ $134.75　　　Ⓓ $242

🔶 FCAT Practice

Record your answers on the answer sheet provided by your teacher or on a sheet of paper.

1. Which of these is the least number?
   (Basic Skill)
   Ⓐ three-thousandths
   Ⓑ three and one-thousandth
   Ⓒ three-hundredths
   Ⓓ three and one-hundredth

2. When evaluating the following expression, which operation should be performed first?
   (Lesson 1-2)

   $42 - 12 + [10(26 - 9)] \div 3$

   Ⓕ Divide 9 by 3.
   Ⓖ Subtract 12 from 42.
   Ⓗ Multiply 10 time 26.
   Ⓘ Subtract 9 from 26.

3. Charles made 8 cups of lemonade. He poured himself $1\frac{1}{2}$ cups, his sister $1\frac{1}{3}$ cups, his mother $2\frac{1}{4}$ cups, and his father $2\frac{5}{8}$ cups. How much did he have left? (Lesson 2-6)
   Ⓐ $\frac{7}{24}$ c
   Ⓑ $\frac{5}{12}$ c
   Ⓒ $\frac{7}{8}$ c
   Ⓓ $1\frac{1}{12}$ c

4. What is the distance between the points in the graph below? (Lesson 3-6)

   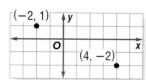

   Ⓕ 3 units
   Ⓖ 4.5 units
   Ⓗ 6 units
   Ⓘ 6.7 units

5. Which of the following percents is more than 2 out of 6 but less than 3 out of 5?
   (Lesson 5-1)
   Ⓐ 33%   Ⓑ 50%   Ⓒ 75%   Ⓓ 85%

6. Which of the following is the *best* estimate for the shaded portion of the rectangle below? (Lesson 5-2)

   Ⓕ 10%   Ⓖ $\frac{1}{3}$ Ⓗ 0.60   Ⓘ 80%

7. Jesse got a 88% on his science test. How many questions did Jesse answer correctly?
   (Lesson 5-3)
   Ⓐ 88 questions
   Ⓑ less than 88 questions
   Ⓒ more than 88 questions
   Ⓓ cannot be determined from the information

8. The pair of jeans is on sale for 25% off the regular price of $47. How much money is discounted off the regular price? (Lesson 5-7)
   Ⓕ $6.25   Ⓖ $8.55
   Ⓗ $11.75   Ⓘ $35.25

9. Ms. Katz took out a loan for $1,200. The loan had an simple interest rate of 8.5%. If she paid off the loan in 6 months, which of the following expressions gives the total amount of interest she had to pay?
   (Lesson 5-8)
   Ⓐ $1,200 - 0.085 \times 0.6$
   Ⓑ $1,200 \times 0.085 \div 6$
   Ⓒ $1,200 \times 0.085 \times 0.5$
   Ⓓ $1,200 \times 0.085 + 5$

FCAT Practice

PART 2   Short Response/Grid In

Record your answers on the answer sheet provided by your teacher or on a sheet of paper.

10. The perimeter of the rectangular rug below is 42 feet. What is its length? (Lesson 1-1)

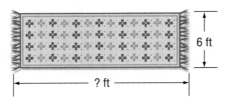

6 ft

? ft

11. Find the length of the third side of the triangle. (Lesson 3-5)

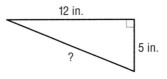

12 in.

5 in.

?

12. Which is less, 0.02% or $\frac{1}{7}$%? (Lesson 5-1)

13. Which is the greatest number? (Lesson 5-2)

0.3%,   2%, $\frac{4}{26}$, $\frac{1}{4}$

14. Hakeem enlarged a photograph to 250% of its original size. If the length of the original photograph is indicated below, what is length of the copy of the photograph? (Lesson 5-4)

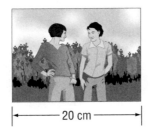

← 20 cm →

TEST-TAKING TIP

Question 13 Use your ability to convert percents, decimals, and fractions to your advantage. For example, you may find Question 13 easiest to answer if you convert all of the answer choices to fractions.

FCAT Practice

PART 3   Extended Response

Record your answers on a sheet of paper. Show your work.

15. A local health clinic found that 1,497 blood donors had a positive Rh factor. The following is the blood-type breakdown of these donors with a positive Rh factor. (Lessons 5-5 and 5-6)

Blood Type	Percent of Donors
O	45%
A	40%
B	10%
AB	5%

a. Explain how you could estimate the number of people with type A blood. Find the actual number of people with type A blood. Compare the actual number with your estimate.

b. Explain how you could estimate the number of people with type AB blood. Find the actual number of people with type AB blood. Compare the actual number with your estimate.

16. The Dow Jones Average is used to measure changes in stock values on the New York Stock Exchange. Three major drops in the Dow Jones Average for one day are listed below. (Lesson 5-7)

Date	Opening	Closing
October 29, 1929	261.07	230.07
October 19, 1987	2,246.74	1,738.74
September 17, 2001	9,605.51	8,920.70

Source: www.mdleasing.com

a. Which day had the greatest decrease in amount?

b. Did this decrease represent the biggest percent of decrease of the three drops? Explain your reasoning.

UNIT 3
Geometry and Measurement

Chapter 6

Geometry

Chapter 7

Geometry: Measuring Area and Volume

Our world is made up of lines, angles, and shapes, both two- and three-dimensional. In this unit, you will learn about the properties and measures of geometric figures.

WebQuest INTERDISCIPLINARY PROJECT

 MATH and
ARCHITECTURE

UNDER CONSTRUCTION

Can you build it? Yes, you can! You've been selected to head the architectural and construction teams on a house of your own design. You'll create the uniquely shaped floor plan, research different floor coverings for the rooms in your house, and finally research different loans to cover the cost of purchasing these floor coverings. So grab a hammer and some nails, and don't forget your geometry and measurement tool kits. You're about to construct a cool adventure!

Log on to msmath3.net/webquest to begin your WebQuest.

6 Geometry

"How is geometry used in the game of pool?"

A billiard ball is struck so that it bounces off the cushion of a pool table and heads for a corner pocket. **The three angles created by the path of the ball and the cushion together form a straight angle that measures 180°.** Pool players use such angle relationships and the properties of reflections to make their shots.

You will solve problems about angle relationships in Lesson 6-1.

GETTING STARTED

Take this quiz to see whether you are ready to begin Chapter 6. Refer to the lesson number in parentheses if you need more review.

▶ Vocabulary Review

State whether each sentence is *true* or *false*. If *false*, replace the underlined word or number to make a true sentence.

1. For the right triangle shown, the Pythagorean Theorem states that <u>$a^2 + c^2 = b^2$</u>. (Lesson 3-4)

2. A rectangle is also a <u>polygon</u>. (Lesson 4-5)

▶ Prerequisite Skills

Solve each equation. (Lesson 1-8)

3. $49 + b + 45 = 180$

4. $t + 98 + 55 = 180$

5. $15 + 67 + k = 180$

Find the missing side length of each right triangle. Round to the nearest tenth, if necessary. (Lesson 3-4)

6. a, 8 m; b, 6 m

7. b, 9 ft; a, 7 ft

8. a, 4 in.; c, 5 in.

9. c, 10 yd; a, 3 yd

Decide whether the figures are congruent. Write *yes* or *no* and explain your reasoning. (Lesson 4-5)

10.

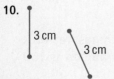

11.

FOLDABLES™ Study Organizer

Geometry Make this Foldable to help you organize your notes. Begin with a plain piece of 11" × 17" paper.

STEP 1 **Fold**
Fold the paper in fifths lengthwise.

STEP 2 **Open and Fold**
Fold a $2\frac{1}{2}$" tab along the short side. Then fold the rest in half.

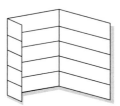

STEP 3 **Label**
Draw lines along folds and label each section as shown.

	words	model
lines		
polygons		
symmetry		
trans-formations		

Reading and Writing As you read and study the chapter, complete the table by writing down important definitions and key concepts for each heading.

6-1 Line and Angle Relationships

Sunshine State Standards
MA.C.1.3.1-1, MA.C.2.3.1-1

What You'll Learn

Identify special pairs of angles and relationships of angles formed by two parallel lines cut by a transversal.

→NEW Vocabulary

acute angle
right angle
obtuse angle
straight angle
vertical angles
adjacent angles
complementary
   angles
supplementary
   angles
perpendicular lines
parallel lines
transversal

HANDS-ON Mini Lab

Work with a partner.

STEP 1 Draw two different pairs of intersecting lines and label the angles formed as shown.

STEP 2 Find and record the measure of each angle.

STEP 3 Color angles that have the same measure.

1. For each set of intersecting lines, identify the pairs of angles that have the same measure.

2. What is true about the sum of the measures of the angles sharing a side?

Materials

- straightedge
- protractor
- colored pencils
- notebook paper

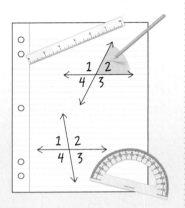

Angles can be classified by their measures.

- **Acute angles** have measures less than 90°.
- **Right angles** have measures equal to 90°.
- **Obtuse angles** have measures between 90° and 180°.
- **Straight angles** have measures equal to 180°.

Pairs of angles can be classified by their relationship to each other. Recall that angles with the same measure are congruent.

Key Concept — Special Pairs of Angles

Vertical angles are opposite angles formed by intersecting lines. Vertical angles are congruent.

$\angle 1$ and $\angle 2$ are vertical angles.
$\angle 1 \cong \angle 2$

Adjacent angles have the same vertex, share a common side, and do not overlap.

$\angle 1$ and $\angle 2$ are adjacent angles.
$m\angle ABC = m\angle 1 + m\angle 2$

The sum of the measures of **complementary angles** is 90°.

$\angle ABD$ and $\angle DBC$ are complementary angles.
$m\angle ABD + m\angle DBC = 90°$

The sum of the measures of **supplementary angles** is 180°.

$\angle C$ and $\angle D$ are supplementary angles.
$m\angle C + m\angle D = 180°$

READING Math

Angle Measure
Read $m\angle 1$ as *the measure of angle 1*.

EXAMPLES **Classify Angles and Angle Pairs**

Classify each angle or angle pair using all names that apply.

1  $m\angle 1$ is greater than $90°$
So, $\angle 1$ is an obtuse angle.

2 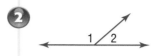 $\angle 1$ and $\angle 2$ are adjacent angles since they have the same vertex, share a common side, and do not overlap. Together, they form a straight angle measuring $180°$. So, $\angle 1$ and $\angle 2$ are also supplementary angles.

Your Turn Classify each angle or angle pair using all names that apply.

a.    b.    c.

You can use the relationships between pairs of angles to find missing measures.

EXAMPLE **Find a Missing Angle Measure**

3 In the figure, $m\angle ABC = 90°$. Find the value of x.

$m\angle ABD + m\angle DBC =$	90	Write an equation.
$x + 65 =$	90	$m\angle ABD = x$ and $m\angle DBC = 65$
$-65 = -65$		Subtract 65 from each side.
$x =$	25	Simplify.

Your Turn Find the value of x in each figure.

d. 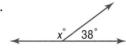   e.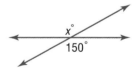

READING Math

Parallel and Perpendicular Lines
Read $m \perp n$ as *m is perpendicular to n*.
Read $p \parallel q$ as *p is parallel to q*.

Lines that intersect at right angles are called **perpendicular lines**. Two lines in a plane that never intersect or cross are called **parallel lines**.

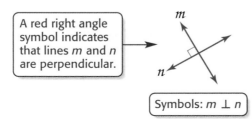

A red right angle symbol indicates that lines m and n are perpendicular.

Red arrowheads indicate that lines p and q are parallel.

Symbols: $m \perp n$ Symbols: $p \parallel q$

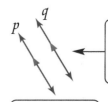

A line that intersects two or more other lines is called a **transversal**. When a transversal intersects two lines, eight angles are formed that have special names.

If the two lines cut by a transversal are *parallel*, then these special pairs of angles are congruent.

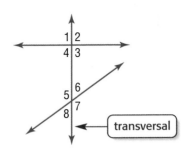

transversal

Key Concept

Parallel Lines

If two parallel lines are cut by a transversal, then the following statements are true.

- **Alternate interior angles**, those on opposite sides of the transversal and inside the other two lines, are congruent.
  Example: ∠2 ≅ ∠8

- **Alternate exterior angles**, those on opposite sides of the transversal and outside the other two lines, are congruent.
  Example: ∠4 ≅ ∠6

- **Corresponding angles**, those in the same position on the two lines in relation to the transversal, are congruent.
  Example: ∠3 ≅ ∠7

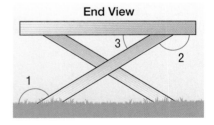

You can use congruent angle relationships to solve real-life problems.

EXAMPLE **Find an Angle Measure**

4 **CARPENTRY** You are building a bench for a picnic table. The top of the bench will be parallel to the ground. If $m\angle 1 = 148°$, find $m\angle 2$ and $m\angle 3$.

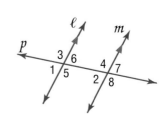

End View

Since ∠1 and ∠2 are alternate interior angles, they are congruent. So, $m\angle 2 = 148°$.

Since ∠2 and ∠3 are supplementary, the sum of their measures is 180°. Therefore, $m\angle 3 = 180° - 148°$ or 32°.

Your Turn For Exercises f–h, use the figure at the right.

f. Find $m\angle 2$ if $m\angle 1 = 63°$.

g. Find $m\angle 3$ if $m\angle 8 = 100°$.

h. Find $m\angle 4$ if $m\angle 7 = 82°$.

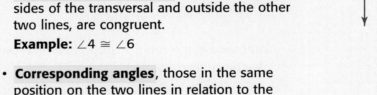

Skill and Concept Check

1. **OPEN ENDED** Draw a pair of complementary angles.

2. **Draw** a pair of parallel lines and a third line intersecting them. Choose one angle and mark it with a ✔. Then mark all other angles that are congruent to that angle with a ✔. Explain.

Writing Math
Exercise 2

GUIDED PRACTICE

Classify each angle or angle pair using all names that apply.

3.

4.  117° 63°

5. 3 4

Find the value of x in each figure.

6. 60° $x°$

7. $x°$ 27°

8. 37° $x°$

For Exercises 9–12, use the figure at the right.

9. Find $m\angle 4$ if $m\angle 5 = 43°$.

10. Find $m\angle 1$ if $m\angle 3 = 135°$.

11. Find $m\angle 6$ if $m\angle 8 = 126°$.

12. Find $m\angle A$ if $m\angle B = 15°$ and $\angle A$ and $\angle B$ are supplementary.

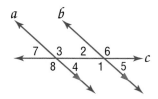

a b 7 3 2 6 8 4 1 5 c

Practice and Applications

Classify each angle or angle pair using all names that apply.

13.

14.

15. 2 1

16. 3 4

17. 5 6

18. 7 8

HOMEWORK HELP	
For Exercises	See Examples
13–18	1, 2
19–29	3
30–38	4
Extra Practice	
See pages 629, 653.	

Find the value of x in each figure.

19. 140° $x°$

20. 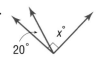 87° $x°$

21. $x°$ 144°

22. 24° $x°$

23. $x°$ 45°

24. $x°$ 20°

25. 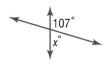 107° $x°$

26. 80° $x°$

27. **ALGEBRA** Angles P and Q are vertical angles. If $m\angle P = 45°$ and $m\angle Q = (x + 25)°$, find the value of x.

28. **ALGEBRA** Angles A and B are supplementary. If $m\angle A = 2x°$ and $m\angle B = 80°$, find the value of x.

29. **POOL** Aaron is trying a complicated pool shot. He wants to hit the number 8 ball into the corner pocket. If Aaron knows the angle measures shown in the diagram, what angle x must the path of the ball take to go into the corner pocket?

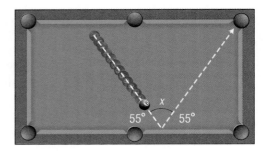

For Exercises 30–37, use the figure at the right.

30. Find $m\angle 2$ if $m\angle 3 = 108°$.

31. Find $m\angle 6$ if $m\angle 7 = 111°$.

32. Find $m\angle 5$ if $m\angle 8 = 85°$.

33. Find $m\angle 8$ if $m\angle 1 = 63°$.

34. Find $m\angle 8$ if $m\angle 2 = 50°$.

35. Find $m\angle 4$ if $m\angle 1 = 59°$.

36. Find $m\angle 5$ if $m\angle 4 = 72°$.

37. Find $m\angle 5$ if $m\angle 7 = 98°$.

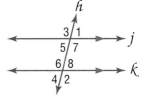

38. **PARKING** Engineers angled the parking spaces along a downtown street so that cars could park and back out easily. All of the lines marking the parking spaces are parallel. If $m\angle 1 = 55°$, find $m\angle 2$. Explain your reasoning.

39. **CRITICAL THINKING** Suppose two parallel lines are cut by a transversal. How are the interior angles on the same side of the transversal related? Use a diagram to explain your reasoning.

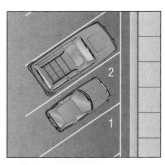

Standardized Test Practice and Mixed Review

40. **SHORT RESPONSE** If $m\angle A = 81°$ and $\angle A$ and $\angle B$ are complementary, what is $m\angle B$?

41. **MULTIPLE CHOICE** Find the value of x in the figure at the right.

   (A) 30   (B) 40   (C) 116   (D) 124

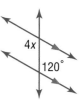

42. A savings account starts with $560. If the simple interest rate is 3%, find the total amount after 18 months. (Lesson 5-8)

Find each percent of change. Round to the nearest tenth if necessary. State whether the percent of change is an *increase* or a *decrease*. (Lesson 5-7)

43. original: 20   44. original: 45   45. original: 620   46. original: 260
     new: 27   new: 18   new: 31   new: 299

GETTING READY FOR THE NEXT LESSON

PREREQUISITE SKILL Solve each equation. Check your solution. (Lesson 1-8)

47. $n + 32 + 67 = 180$ 48. $45 + 89 + x = 180$ 49. $180 = 120 + a + 15$

What You'll Learn

Construct a line parallel to a given line.

Materials
- compass
- straightedge
- paper

Constructing Parallel Lines

In this lab, you will construct a line parallel to a given line.

ACTIVITY

STEP 1 Draw a line and label it *p*. Then draw and label a point *A* not on line *p*.

STEP 2 Draw a line through point *A* so that it intersects line *p*. Label the point of intersection point *B*.

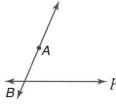

Steps 1–2

STEP 3 Place the compass at point *B* and draw a large arc. Label the point where the arc crosses line *p* as point *C*, and label where it crosses line *AB* as point *D*.

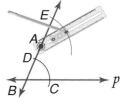

Steps 3–4

STEP 4 With the same compass opening, place the compass at point *A* and draw a large arc. Label the point of intersection with line *AB* as point *E*.

STEP 5 Use your compass to measure the distance between points *D* and *C*.

STEP 6 With the compass opened the same amount, place the compass at point *E* and draw an arc to intersect the arc already drawn. Label this point *F*.

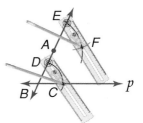

Steps 5–6

STEP 7 Draw a line through points *A* and *F*. Label this line *q*. You have drawn *q* ∥ *p*.

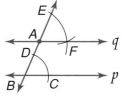

Step 7

Your Turn Draw a line. Then construct a line parallel to it.

Writing Math

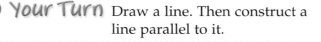

Work with a partner. Use the information in the activity above.

1. **Classify** ∠*DBC* and ∠*FAE* in relationship to lines *p*, *q*, and transversal *AB*.

2. **Explain** why you should expect ∠*ABC* to be congruent to ∠*FAE*.

Triangles and Angles

Sunshine State Standards
MA.C.1.3.1-1, MA.C.1.3.1-4

What You'll Learn

Find missing angle measures in triangles and classify triangles by their angles and sides.

→ NEW Vocabulary

triangle
acute triangle
obtuse triangle
right triangle
scalene triangle
isosceles triangle
equilateral triangle

HANDS-ON Mini Lab

Work with a partner.

Investigate the relationship among the measures of the angles of a triangle.

STEP 1 Use a straightedge to draw a triangle on your paper. Then shade each angle of the triangle using a different color and cut out the triangle.

STEP 2 Cut off each angle and arrange the pieces as shown so that the three angles are adjacent.

Repeat the steps above with several other triangles.

1. What do you think is the sum of the measures of the three angles of any triangle? Explain your reasoning.

Materials

• paper
• straightedge
• colored pencils
• scissors

A **triangle** is a figure formed by three line segments that intersect only at their endpoints. Recall that triangles are named by the letters at their vertices.

Triangle *LMN* is written △*LMN*.

Key Concept — Angles of a Triangle

Words	The sum of the measures of the angles of a triangle is 180°.	Model
Symbols	$x + y + z = 180$	

READING Math

Naming Triangles
Read △*RST* as *triangle RST*.

EXAMPLE — Find a Missing Angle Measure

① Find the value of x in △*RST*.

$$m\angle R + m\angle S + m\angle T = 180 \quad \text{The sum of the measures is 180.}$$

$$x + 72 + 74 = 180 \quad \begin{array}{l}\text{Replace } m\angle R \text{ with } x, m\angle S \\ \text{with 72, and } m\angle T \text{ with 74.}\end{array}$$

$$x + 146 = 180 \quad \text{Simplify.}$$

$$\underline{-146 = -146} \quad \text{Subtract 146 from each side.}$$

$$x = 34 \quad \text{The value of } x \text{ is 34.}$$

All triangles have at least two acute angles. Triangles can be classified by the measure of the third angle.

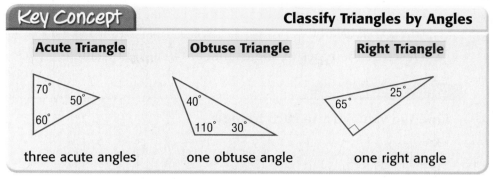

Key Concept — Classify Triangles by Angles

| Acute Triangle | Obtuse Triangle | Right Triangle |

70° 50° 60° — three acute angles

40° 110° 30° — one obtuse angle

25° 65° — one right angle

In an *equiangular* triangle, all angles have the same measure, 60°.

Triangles can also be classified by the number of congruent sides. Congruent sides are often marked with tick marks.

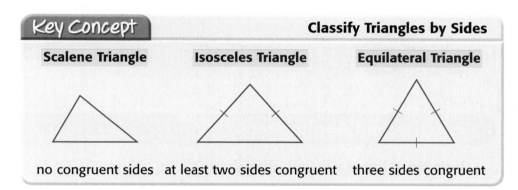

Key Concept — Classify Triangles by Sides

| Scalene Triangle | Isosceles Triangle | Equilateral Triangle |

no congruent sides   at least two sides congruent   three sides congruent

EXAMPLES Classify Triangles

Classify each triangle by its angles and by its sides.

2 A 71° 38° C 71° B

Angles △ABC has all acute angles.

Sides △ABC has two congruent sides.

So, △ABC is an acute isosceles triangle.

3 X 60° 30° Z Y

Angles △XYZ has one right angle.

Sides △XYZ has no congruent sides.

So, △XYZ is a right scalene triangle.

Your Turn Classify each triangle by its angles and by its sides.

a.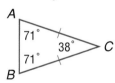
78° 57° 45°

b.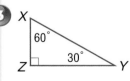
35° 110° 35°

c.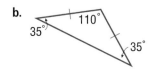
60° 60° 60°

Skill and Concept Check

Writing Math
Exercises 1 & 2

1. **OPEN ENDED** Name a real-life object that is shaped like an isosceles triangle. Explain.

2. **Describe** the types of angles that are in a right triangle.

GUIDED PRACTICE

Find the value of *x* in each triangle.

3.
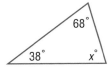
68°
38°
x°

4.
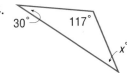
30°
117°
x°

5.
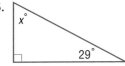
x°
29°

Classify each triangle by its angles and by its sides.

6.

75°
75°
30°

7.

Stillwater
48°
OKLAHOMA
Albus
Lawton
55° 77°

8.
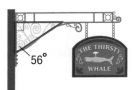
56°

9.
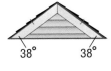
38°   38°

Practice and Applications

Find the value of *x* in each triangle.

HOMEWORK HELP

For Exercises	See Examples
10–15	1
16–32	2, 3

Extra Practice
See pages 629, 653.

10.

x°
64°   60°

11.

x°
50°

12.

122°   *x*°
25°

13.
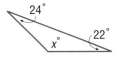
24°
22°
x°

14.
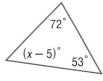
72°
(*x* − 5)°
53°

15.
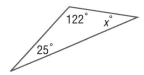
36°
2*x*°

Classify each triangle by its angles and by its sides.

16.
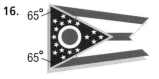
65°
65°

17.
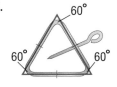
60°
60°   60°

18.

Huntsville
100°
Tuscaloosa
ALABAMA
Dothan

19.

15°
75°

20.

YIELD

21.

45°
45°

22.

80°

23.
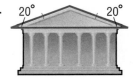
20°   20°

24. **BRIDGE BUILDING** At a Science Olympiad tournament, your team is to design and construct a bridge that will hold the most weight for a given span. Your team knows that triangles add stability to bridges. Below is a side view of your team's design. Name and classify three differently-shaped triangles in their design.

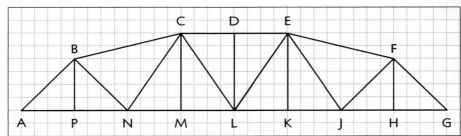

Draw each triangle. If it is not possible to draw the triangle, write *not possible.*

25. three acute angles

26. two obtuse angles

27. obtuse isosceles with two acute angles

28. obtuse equilateral

29. right equilateral

30. right scalene

Determine whether each statement is *sometimes,* *always,* **or** *never* **true.**

31. Isosceles triangles are equilateral.   32. Equilateral triangles are isosceles.

33. **CRITICAL THINKING** Explain why all triangles have at least two acute angles.

Standardized Test Practice and Mixed Review

FCAT Practice

34. **SHORT RESPONSE** Triangle *ABC* is isosceles. What is the value of *x*?

35. **MULTIPLE CHOICE** Which term describes the relationship between the two acute angles of a right triangle?

   A adjacent   **B** complementary

   C vertical   **D** supplementary

Find the measure of each angle in the figure if $m \parallel n$ **and** $m\angle 7 = 95°$.
(Lesson 6-1)

36. $\angle 4$ 37. $\angle 3$ 38. $\angle 1$ 39. $\angle 2$

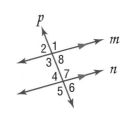

40. **SAVINGS** Shala's savings account earned $4.56 in 6 months at a simple interest rate of 4.75%. How much was in her account at the beginning of that 6-month period? (Lesson 5-8)

GETTING READY FOR THE NEXT LESSON

PREREQUISITE SKILL Find the missing side length of each right triangle. Round to the nearest tenth if necessary. (Lesson 3-4)

41. *a*, 5 ft; *b*, 8 ft   42. *b*, 10 m; *c*, 12 m   43. *a*, 6 in.; *c*, 13 in.   44. *a*, 7 yd; *b*, 7 yd

What You'll Learn
Bisect an angle.

Materials
• compass
• straightedge
• paper

∞ *LINK* **To Reading**

Everyday Meaning of Bisect: to divide into two equal parts

Bisecting Angles

In this lab, you will learn to bisect an angle.

ACTIVITY

STEP 1 Draw ∠JKL.

STEP 2 Place the compass at point *K* and draw an arc that intersects both sides of the angle. Label the intersections *X* and *Y*.

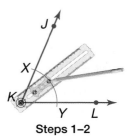
Steps 1–2

STEP 3 With the compass at point *X*, draw an arc in the interior of ∠JKL.

STEP 4 Using this setting, place the compass at point *Y*. Draw another arc.

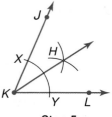
Steps 3–4

STEP 5 Label the intersection of these arcs *H*. Then draw $\overrightarrow{KH}$. $\overrightarrow{KH}$ is the **bisector** of ∠JKL.

Your Turn Draw each kind of angle. Then bisect it.

a. acute   b. obtuse

Step 5

Writing Math

Work with a partner. Use the information in the activity above.

1. **Describe** what is true about the measures of ∠JKH and ∠HKL.

2. **Explain** why we say that $\overrightarrow{KH}$ is the bisector of ∠JKL.

3. The point where the bisectors of all three angles of a triangle meet is called the *incenter*. Draw a triangle. Then locate its incenter using only a compass and straightedge.

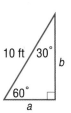

6-3

Special Right Triangles

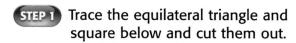

Sunshine State Standards
MA.C.3.3.1-2

What You'll Learn

Find missing measures in 30°-60° right triangles and 45°-45° right triangles.

○ REVIEW Vocabulary

Pythagorean Theorem: in a right triangle, the square of the length of the hypotenuse is equal to the sum of the squares of the lengths of the legs **(Lesson 3-4)**

HANDS-ON **Mini Lab**

Materials
• pencil
• paper
• scissors
• ruler

Work with a partner.

STEP 1 Trace the equilateral triangle and square below and cut them out.

STEP 2 Measure each angle.

STEP 3 Fold the triangle so that one half matches the other. Fold the square in half along a diagonal.

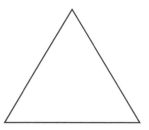

1. What type of triangles have you formed?

2. What are the measures of the angles of the folded triangle?

3. Measure and describe the relationship between the shortest and longest sides of this triangle.

4. What are the measures of the angles of the triangle formed by folding the square?

5. Measure and describe the relationship between the legs of this triangle.

The sides of a triangle whose angles measure 30°, 60°, and 90° have a special relationship. The hypotenuse is always twice as long as the side opposite the 30° angle.

side opposite 30° angle

$c = 2a$ or $a = \frac{1}{2}c$

EXAMPLE **Find Lengths of a 30°-60° Right Triangle**

1 Find each missing length.

Step 1 Find a.

$a = \frac{1}{2}c$ Write the equation.

$a = \frac{1}{2}(10)$ or 5   Replace c with 10.

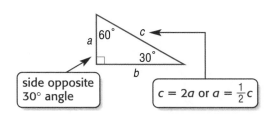

(continued on the next page)

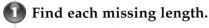

msmath3.net/extra_examples/fcat

Step 2 Find b.

$$c^2 = a^2 + b^2$$ Pythagorean Theorem

$$10^2 = 5^2 + b^2$$ Replace c with 10 and a with 5.

$$100 = 25 + b^2$$ Evaluate 10^2 and 5^2.

$$100 - 25 = 25 + b^2 - 25$$ Subtract 25 from each side.

$$75 = b^2$$ Simplify.

$$\sqrt{75} = \sqrt{b^2}$$ Take the square root of each side.

$$8.7 \approx b$$ Use a calculator.

The length of a is 5 feet, and the length of b is about 8.7 feet.

Your Turn Find each missing length. Round to the nearest tenth if necessary.

a.

b.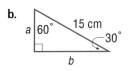

c.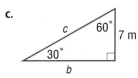

A 45°-45° right triangle is also an isosceles triangle because two angle measures are the same. Thus, the legs are always congruent.

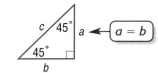

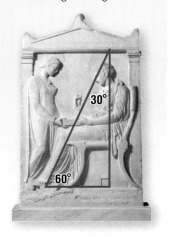
EXAMPLE **Find Lengths of a 45°-45° Right Triangle**

2 **ART** The ancient Greeks sometimes used 45°-45° right triangles in their art. The sculpture on the right is based on such a triangle. Suppose the base of a reproduction of the sculpture shown is 15 feet long. Find each missing length.

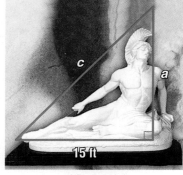

Achilles Wounded

Step 1 Find a.

Sides a and b are the same length. Since $b = 15$ feet, $a = 15$ feet.

Step 2 Find c.

$$c^2 = a^2 + b^2$$ Pythagorean Theorem

$$c^2 = 15^2 + 15^2$$ Replace a with 15 and b with 15.

$$c^2 = 225 + 225$$ Evaluate 15^2.

$$c^2 = 450$$ Add 225 and 225.

$$\sqrt{c^2} = \sqrt{450}$$ Take the square root of each side.

$$c \approx 21.2$$ Simplify.

Skill and Concept Check

Writing Math
Exercise 1

1. **Write** a sentence describing the relationship between the hypotenuse of a 30°-60° right triangle and the leg opposite the 30° angle.

2. **OPEN ENDED** Give a real-life example of a 45°-45° right triangle.

GUIDED PRACTICE

Find each missing length. Round to the nearest tenth if necessary.

3.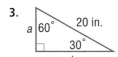

4.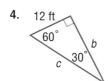

5.

Practice and Applications

Find each missing length. Round to the nearest tenth if necessary.

6.

7.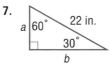

8.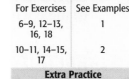

9.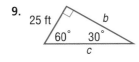

10.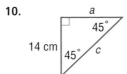

11.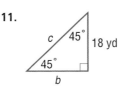

HOMEWORK HELP

For Exercises	See Examples
6–9, 12–13, 16, 18	1
10–11, 14–15, 17	2

Extra Practice
See pages 630, 653.

12. The length of the hypotenuse of a 30°-60° right triangle is 7.5 meters. Find the length of the side opposite the 30° angle.

13. In a 30°-60° right triangle, the length of the side opposite the 30° angle is 5.8 centimeters. What is the length of the hypotenuse?

14. The length of one of the legs of a 45°-45° right triangle is 6.5 inches. Find the lengths of the other sides.

15. In a 45°-45° right triangle, the length of one leg is 7.5 feet. What are the lengths of the other sides?

16. **HISTORY** Redan lines such as the one below were used at many battlefields in the Civil War. A *redan* is a triangular shape that goes out from the main line of defense. What is the distance *h* from the base of each redan to its farthest point?

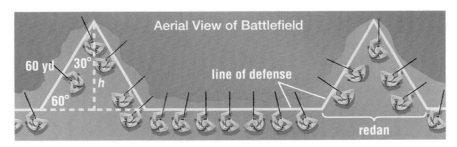

17. **QUILTING** Refer to the photograph at the right. Each triangle in the Flying Geese pattern is a 45°-45° right triangle. If the length of a leg is $2\frac{1}{2}$ inches, find the length of each hypotenuse.

18. **SKIING** A ski jump is constructed so that the length of the board necessary for the surface of the ramp is twice as long as the ramp is high *h*. If the ramp forms a right triangle, what is the measure of ∠1? Explain your reasoning.

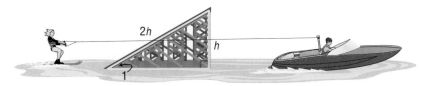

19. **WRITE A PROBLEM** Write a real-life problem involving a 30°-60° right triangle or a 45°-45° right triangle. Then solve the problem.

20. **CRITICAL THINKING** Find the length of each leg of a 45°-45° right triangle whose hypotenuse measures $\sqrt{162}$ centimeters.

Standardized Test Practice and Mixed Review

FCAT Practice

21. **MULTIPLE CHOICE** The midpoints of the sides of the square at the right are joined to form a smaller square. What is the area of the smaller square?

    (A) 196 in²   (B) 98 in²   (C) 49 in²   (D) 9.9 in²

22. **MULTIPLE CHOICE** Which values represent the sides of a 30°-60° right triangle?

    (F) 3 cm, 4 cm, 5 cm   (G) 8 cm, 8 cm, $\sqrt{128}$ cm

    (H) 4 cm, 8 cm, $\sqrt{48}$ cm   (I) 5 cm, 12 cm, 13 cm

23. Two sides of a triangle are congruent. The angles opposite these sides each measure 70°. Classify the triangle by its angles and by its sides. (Lesson 6-2)

Classify each angle or angle pair using all names that apply. (Lesson 6-1)

24.   25.   26.   27.

                    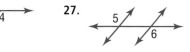

Multiply. Write in simplest form. (Lesson 2-3)

28. $\frac{2}{3} \cdot \frac{5}{8}$ 29. $-\frac{2}{5} \cdot \frac{3}{4}$ 30. $-1\frac{1}{2}\left(-2\frac{1}{3}\right)$ 31. $2\frac{2}{3}\left(-2\frac{1}{4}\right)$

GETTING READY FOR THE NEXT LESSON

PREREQUISITE SKILL Solve each equation. Check your solution. (Lesson 1-8)

32. $x + 90 + 50 + 100 = 360$ 33. $45 + 150 + x + 85 = 360$

What You'll Learn

Construct a perpendicular bisector of a segment.

Materials

• compass
• straightedge
• protractor
• paper

Constructing Perpendicular Bisectors

In this lab, you will learn to construct a line perpendicular to a segment so that it bisects that segment.

ACTIVITY

STEP 1 Draw $\overline{AB}$. Then place the compass at point A. Using a setting greater than one half the length of $\overline{AB}$, draw an arc above and below $\overline{AB}$.

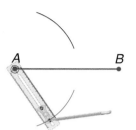

STEP 2 Using this setting, place the compass at point B. Draw another set of arcs above and below $\overline{AB}$ as shown.

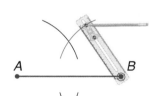

STEP 3 Label the intersection of these arcs X and Y as shown. Then draw $\overline{XY}$. $\overline{XY}$ is the **perpendicular bisector** of $\overline{AB}$. Label the intersection of $\overline{AB}$ and this new line segment M.

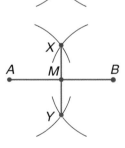

Your Turn

Draw a line segment. Then construct the perpendicular bisector of the segment.

Writing Math

Work with a partner. Use the information in the activity above.

1. **Describe** what is true about the measures of $\overline{AM}$ and $\overline{MB}$.

2. **Find** $m\angle XMB$. Then describe the relationship between $\overline{AB}$ and $\overline{XY}$.

3. **Explain** how to construct a 45°-45° right triangle with legs half as long as the segment below. Then construct the triangle.

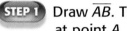

C   D

Classifying Quadrilaterals

Sunshine State Standards
MA.C.1.3.1-4

What You'll Learn

Find missing angle measures in quadrilaterals and classify quadrilaterals.

NEW Vocabulary

quadrilateral
trapezoid
parallelogram
rectangle
rhombus
square

∞LINK To Reading

Everyday meaning of prefix quadri-: four

HANDS-ON Mini Lab

Materials
- paper
- straightedge
- protractor

Work with a partner.

The polygon at the right is a **quadrilateral**, since it has four sides and four angles.

STEP 1 Draw a quadrilateral.

STEP 2 Pick one vertex and draw the diagonal to the opposite vertex.

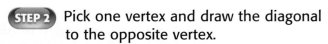

1. Name the shape of the figures formed when you drew the diagonal. How many figures were formed?

2. You know that the sum of the angle measures of a triangle is 180°. Use this fact to find the sum of the angle measures in a quadrilateral. Explain your reasoning.

3. Find the measure of each angle of your quadrilateral. Compare the sum of these measures to the sum you found in Exercise 2.

The angles of a quadrilateral have a special relationship.

Key Concept   **Angles of a Quadrilateral**

Words   The sum of the measures of the angles of a quadrilateral is 360°.

Model   **Symbols** $w + x + y + z = 360$

EXAMPLE   **Find a Missing Angle Measure**

1. Find the value of w in quadrilateral WXYZ.

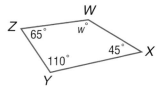

$$m\angle W + m\angle X + m\angle Y + m\angle Z = 360 \quad \text{The sum of the measures is 360.}$$

$$w + 45 + 110 + 65 = 360 \quad \text{Replace } m\angle W \text{ with } w, m\angle X \text{ with 45,}$$
$$\text{} m\angle Y \text{ with 110, and } m\angle Z \text{ with 65.}$$

$$w + 220 = 360 \quad \text{Simplify.}$$

$$\underline{-220 = -220} \quad \text{Subtract 220 from each side.}$$

$$w = 140 \quad \text{Simplify.}$$

READING Math

Isosceles Trapezoid
A trapezoid with one pair of opposite congruent sides is classified as an *isosceles* trapezoid.

The concept map below shows how quadrilaterals are classified. Notice that the diagram goes from the most general type of quadrilateral to the most specific.

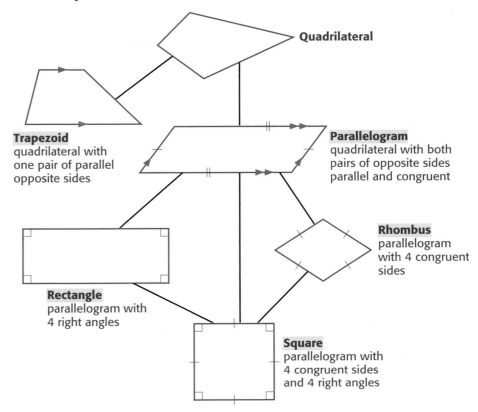

Quadrilateral

Trapezoid
quadrilateral with one pair of parallel opposite sides

Parallelogram
quadrilateral with both pairs of opposite sides parallel and congruent

Rhombus
parallelogram with 4 congruent sides

Rectangle
parallelogram with 4 right angles

Square
parallelogram with 4 congruent sides and 4 right angles

The best description of a quadrilateral is the one that is the most specific.

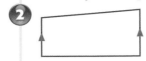

STUDY TIP

Classifying Quadrilaterals
When classifying a quadrilateral, begin by counting the number of parallel lines. Then count the number of right angles and the number of congruent sides.

EXAMPLES Classify Quadrilaterals

Classify each quadrilateral using the name that *best* describes it.

2  The quadrilateral has one pair of parallel sides. It is a trapezoid.

3 The quadrilateral is a parallelogram with four congruent sides. It is a rhombus.

Your Turn Classify each quadrilateral using the name that *best* describes it.

a.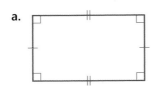

b.

Skill and Concept Check

1. **Explain** why a square is a type of rhombus.

2. **OPEN ENDED** Give a real-life example of a parallelogram.

3. **Which One Doesn't Belong?** Identify the quadrilateral that does not belong with the other three. Explain your reasoning.

| rhombus | rectangle | square | trapezoid |

GUIDED PRACTICE

Find the value of *x* in each quadrilateral.

4.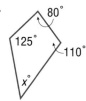

5.

6.

Classify each quadrilateral using the name that *best* describes it.

7.

8.

9.

Practice and Applications

Find the value of *x* in each quadrilateral.

10.

11.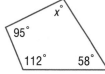

12.

HOMEWORK HELP

For Exercises	See Examples
10–15, 25–26	1
16–24, 27–30	2, 3

Extra Practice
See pages 630, 653.

13.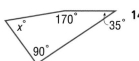

14.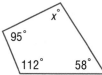

15.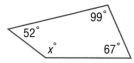

Classify each quadrilateral using the name that *best* describes it.

16.

17.

18.

19.

20.

21.

22.

23.

24. **INTERIOR DESIGN** The stained glass window shown is an example of how geometric figures can be used in decorating. Identify all of the quadrilaterals within the print.

25. **ALGEBRA** In parallelogram $WXYZ$, $m\angle W = 45°$, $m\angle X = 135°$, $m\angle Y = 45°$, and $m\angle Z = (x + 15)°$. Find the value of x.

26. **ALGEBRA** In trapezoid $ABCD$, $m\angle A = 2a°$, $m\angle B = 40°$, $m\angle C = 110°$, and $m\angle D = 70°$. Find the value of a.

Name all quadrilaterals with the given characteristic.

27. only one pair of parallel sides

28. opposite sides congruent

29. all sides congruent

30. all angles are right angles

CRITICAL THINKING Determine whether each statement is *true* or *false*. If *false*, draw a counterexample.

31. All trapezoids are quadrilaterals.

32. All squares are rectangles.

33. All rhombi (plural of rhombus) are squares.

34. A trapezoid can have only one right angle.

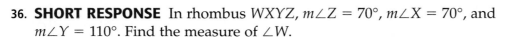

Standardized Test Practice and Mixed Review

FCAT Practice

35. **MULTIPLE CHOICE** Which of the following does *not* describe the quadrilateral at the right?

    A parallelogram   **B** square

    C trapezoid   **D** rhombus

36. **SHORT RESPONSE** In rhombus $WXYZ$, $m\angle Z = 70°$, $m\angle X = 70°$, and $m\angle Y = 110°$. Find the measure of $\angle W$.

37. The length of the hypotenuse of a 30°-60° right triangle is 16 feet. Find the length of the side opposite the 60° angle. Round to the nearest tenth. (Lesson 6-3)

38. The length of one of the legs of a 45°-45° right triangle is 8 meters. Find the length of the hypotenuse. Round to the nearest tenth. (Lesson 6-3)

Classify each triangle by its angles and by its sides. (Lesson 6-2)

39.

40.

41.
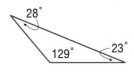

GETTING READY FOR THE NEXT LESSON

PREREQUISITE SKILL Decide whether the figures are congruent. Write *yes* or *no* and explain your reasoning. (Lesson 4-5)

42.
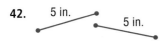

43.
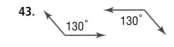

44.

6-4b Problem-Solving Strategy
A Follow-Up of Lesson 6-4

What You'll Learn
Solve problems using the logical reasoning strategy.

Use Logical Reasoning

Jacy, how can we be sure this playing field we've marked out is a rectangle? We don't have anything we can use to measure its angles.

Someone told me that there is something special about the diagonals of a rectangle. Zach, let's see if we can **use logical reasoning** to figure out what that is.

Explore	The playing field is a parallelogram because its opposite sides are the same length. Our math teacher said that means they are also parallel. We need to see what the relationship is between the diagonals of a rectangle.
Plan	Let's draw several different rectangles, measure the diagonals, and see if there is a pattern.
Solve	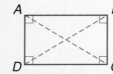 It appears that the diagonals of a rectangle are congruent. If the diagonals of our field are congruent, then we can reason that it is a rectangle.
Examine	Do all parallelograms, not just rectangles, have congruent diagonals? The counterexample at the right suggests that this statement is false.

Analyze the Strategy

1. *Deductive reasoning* uses an existing rule to make a decision. **Determine** where Zach and Jacy used deductive reasoning. Explain.

2. *Inductive reasoning* is the process of making a rule after observing several examples and using that rule to make a decision. **Determine** where Zach and Jacy used inductive reasoning. Explain.

3. **Write** about a situation in which you use inductive reasoning to solve a problem. Then solve the problem.

Solve. Use logical reasoning.

4. **GEOMETRY** Draw several parallelograms and measure their angles. What can you conclude about opposite angles of parallelograms? Did you use deductive or inductive reasoning?

opposite angles

5. **SPORTS** Noah, Brianna, Mackenzie, Antoine, and Bianca were the first five finishers of a race. From the given clues, give the order in which they finished.

• Noah passed Mackenzie just before the finish line.
• Bianca finished 5 seconds ahead of Noah.
• Brianna crossed the finish line after Mackenzie.
• Antoine was fifth at the finish line.

Solve. Use any strategy.

6. **GEOMETRY** If the sides of the pentagons shown are 1 unit long, find the perimeter of 8 pentagons arranged according to the pattern below.

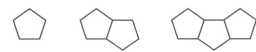

7. **MONEY** After a trip to the mall, Alex and Marcus counted their money to see how much they had left. Alex said, "If I had $4 more, I would have as much as you." Marcus replied, "If I had $4 more, I would have twice as much as you." How much does each boy have?

8. **WEATHER** Based on the data shown, what is a reasonable estimate for the difference in the July high and low temperatures in Statesboro?

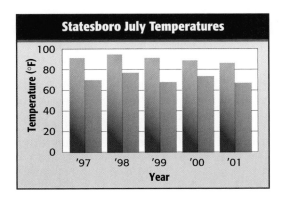

9. **MEASUREMENT** You have a large container of pineapple juice, an empty 4-pint container, and an empty 5-pint container. Explain how you can use these containers to measure 2 pints of juice for a punch recipe.

10. **LAUNDRY** You need two clothespins to hang one towel on a clothesline. One clothespin can be used on a corner of one towel and a corner of the towel next to it. What is the least number of clothespins you need to hang 8 towels?

11. **STANDARDIZED TEST PRACTICE** FCAT Practice
Vanessa and Ashley varied the length of a pendulum and measured the time it took for the pendulum to complete one swing back and forth. Based on their data, how long do you think a pendulum with a swing of 5 seconds is?

Time (s)	1	2	3	4
Length (ft)	1	4	9	16

Ⓐ 21 ft   Ⓑ 23 ft
Ⓒ 24 ft   Ⓓ 25 ft

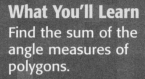
What You'll Learn
Find the sum of the angle measures of polygons.

Materials
- paper
- straightedge

⟳ *REVIEW* Vocabulary

polygon: a simple closed figure in a plane formed by three or more line segments **(Lesson 4-5)**

Sunshine State Standards
MA.C.1.3.1-2

Angles of Polygons

In this lab, you will use the fact that the sum of the angle measures of a triangle is 180° to find the sum of the angle measures of any polygon.

INVESTIGATE *Work with a partner.*

Copy and complete the table below using the procedure described at the top of page 272.

Number of Sides	Sketch of Figure	Number of Triangles	Sum of Angle Measures
3		1	1(180°) = 180°
4		2	2(180°) = 360°
5			
6			
7			

Writing Math

1. **Predict** the number of triangles in an octagon and the sum of its angle measures. Check your prediction by drawing a figure.

2. **Write** an algebraic expression that tells the number of triangles in an n-sided polygon. Then write an expression for the sum of the angle measures in an n-sided polygon.

REGULAR POLYGONS A **regular polygon** is one that is *equilateral* (all sides congruent) and *equiangular* (all angles congruent).

equilateral triangle

square

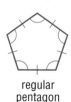
regular pentagon

regular hexagon

3. Use your results from Exercise 2 to find the measure of each angle in the four regular polygons shown above. Check your results by using a protractor to measure one angle of each polygon.

4. **Write** an algebraic expression that tells the measure of each angle in an n-sided regular polygon. Use it to predict the measure of each angle in a regular octagon.

Congruent Polygons

Sunshine State Standards
MA.C.2.3.1-2

What You'll Learn
Identify congruent polygons.

NEW Vocabulary

congruent polygons

WHEN am I ever going to use this?

QUILTING A template, or pattern, for a quilt block contains the minimum number of shapes needed to create the pattern.

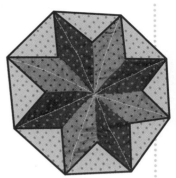

1. How many different triangles are shown in the *Winter Stars* quilt at the right? Explain your reasoning and draw each triangle.

2. Copy the quilt and label all matching triangles with the same number, starting with 1.

Polygons that have the same size and shape are called **congruent polygons**. Recall that the parts of polygons that "match" are called *corresponding* parts.

Key Concept   Congruent Polygons

Words   If two polygons are congruent, their corresponding sides are congruent and their corresponding angles are congruent.

Model

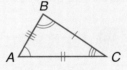
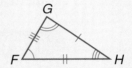

Symbols   Congruent angles: $\angle A \cong \angle F$, $\angle B \cong \angle G$, $\angle C \cong \angle H$

Congruent sides: $\overline{BC} \cong \overline{GH}$, $\overline{AC} \cong \overline{FH}$, $\overline{AB} \cong \overline{FG}$

In a congruence statement, the letters identifying each polygon are written so that corresponding vertices appear in the same order. For example, for the diagram below, write $\triangle CBD \cong \triangle PQR$.

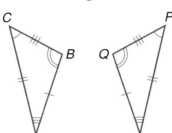

$$\triangle CBD \cong \triangle PQR$$

Vertex *C* corresponds to vertex *P*.
Vertex *B* corresponds to vertex *Q*.
Vertex *D* corresponds to vertex *R*.

Two polygons are congruent if all pairs of corresponding angles are congruent and all pairs of corresponding sides are congruent.

Identify Congruent Polygons

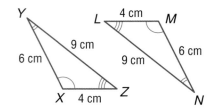

1. **Determine whether the triangles shown are congruent. If so, name the corresponding parts and write a congruence statement.**

   Angles The arcs indicate that $\angle X \cong \angle M$, $\angle Y \cong \angle N$, and $\angle Z \cong \angle L$.

   Sides The side measures indicate that $\overline{XY} \cong \overline{MN}$, $\overline{YZ} \cong \overline{NL}$, and $\overline{XZ} \cong \overline{ML}$.

   Since all pairs of corresponding angles and sides are congruent, the two triangles are congruent. One congruence statement is $\triangle XYZ \cong \triangle MNL$.

STUDY TIP

Congruence Statements Other possible congruence statements for Example 1 are $\triangle YZX \cong \triangle NLM$, $\triangle ZXY \cong \triangle LMN$, $\triangle YXZ \cong \triangle NML$, $\triangle XZY \cong \triangle MLN$, and $\triangle ZYX \cong \triangle LNM$.

Your Turn Determine whether the polygons shown are congruent. If so, name the corresponding parts and write a congruence statement.

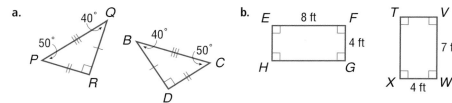

a.

b.

You can use corresponding parts to find the measures of an angle or side in a figure that is congruent to a figure with known measures.

Find Missing Measures

In the figure, $\triangle AFH \cong \triangle QRN$.

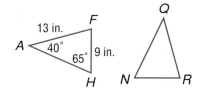

2. **Find $m\angle Q$.**

   According to the congruence statement, $\angle A$ and $\angle Q$ are corresponding angles. So, $\angle A \cong \angle Q$. Since $m\angle A = 40°$, $m\angle Q = 40°$.

READING Math

Recall that symbols like *NR* refer to the measure of the segment with those endpoints.

3. **Find NR.**

   $\overline{FH}$ corresponds to $\overline{NR}$. So, $\overline{FH} \cong \overline{NR}$. Since $FH = 9$ inches, $NR = 9$ inches.

Your Turn In the figure, quadrilateral $ABCD$ is congruent to quadrilateral $WXYZ$. Find each measure.

c. $m\angle X$

d. YX

e. $m\angle Y$

 msmath3.net/extra_examples/fcat

Skill and Concept Check

1. **OPEN ENDED** Draw and label a pair of congruent polygons. Be sure to indicate congruent angles and sides on your drawing.

2. **FIND THE ERROR** Justin and Amanda are writing a congruence statement for the triangles at the right. Who is correct? Explain.

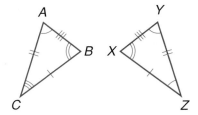

Justin
△ABC ≅ △XYZ

Amanda
△ABC ≅ △YXZ

GUIDED PRACTICE

Determine whether the polygons shown are congruent. If so, name the corresponding parts and write a congruence statement.

3.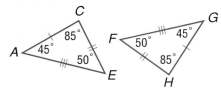

4.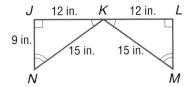

In the figure, △PQR ≅ △YWX. Find each measure.

5. $m\angle X$

6. YW

7. XY

8. $m\angle W$

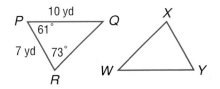

Practice and Applications

Determine whether the polygons shown are congruent. If so, name the corresponding parts and write a congruence statement.

9.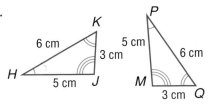

10.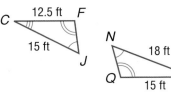

HOMEWORK HELP	
For Exercises	See Examples
9–13	1
14–23	2, 3
Extra Practice See pages 630, 653.	

11.

12.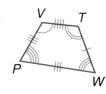

13. **BIRDS** The wings of a hummingbird are shaped like triangles. Determine whether these triangles are congruent. If so, name the corresponding parts and write a congruence statement.

In the figure, △JKL ≅ △PNM. Find each measure.

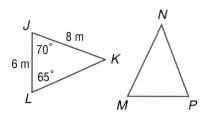

14. *PN* **15.** *PM*

16. *m∠P* **17.** *m∠N*

In the figure, quadrilateral *ABCD* ≅ quadrilateral *HEFG*. Find each measure.

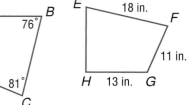

18. *AD* **19.** *m∠H*

20. *m∠G* **21.** *CD*

22. ALGEBRA Find the value of *x* in the two congruent triangles.

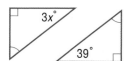

23. TRAVEL An overhead sign on an interstate highway is shown at the right. In the scaffolding, △ABC ≅ △DCB, AC = 2.5 meters, BC = 1 meter, and AB = 2.7 meters. What is the length of $\overline{BD}$?

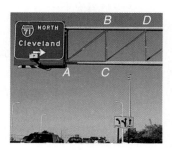

24. CRITICAL THINKING Tell whether the following statement is *sometimes*, *always* or *never* true. Explain your reasoning.

If the perimeters of two triangles are equal, then the triangles are congruent.

Standardized Test Practice and Mixed Review

FCAT Practice

25. SHORT RESPONSE Which of the following polygons appear congruent?

a.    b.   c.   d.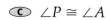

26. MULTIPLE CHOICE If △AFG ≅ △PQR, which statement is *not* true?

   A ∠G ≅ ∠R   **B** $\overline{AG} ≅ \overline{PQ}$ **C** ∠P ≅ ∠A   **D** $\overline{AG} ≅ \overline{QR}$

Classify each quadrilateral using the name that *best* describes it. (Lesson 6-4)

27.   **28.**   **29.**

30. The length of each leg of a 45°-45° right triangle is 14 feet. Find the length of the hypotenuse. (Lesson 6-3)

GETTING READY FOR THE NEXT LESSON

BASIC SKILL Which figure *cannot* be folded so one half matches the other?

31.

32.

6-5b HANDS-ON LAB

A Follow-Up of Lesson 6-5

What You'll Learn
Construct congruent triangles.

Materials
- compass
- straightedge
- protractor
- paper

Constructing Congruent Triangles

ACTIVITY

STEP 1 Use a straightedge to draw a line. Put a point on it labeled *X*.

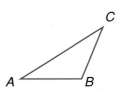

STEP 2 Open your compass to the same width as the length of $\overline{AB}$. Put the compass point at *X*. Draw an arc that intersects the line. Label this point of intersection *Y*.

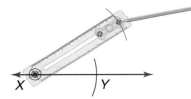

STEP 3 Open your compass to the same width as the length of $\overline{AC}$. Place your compass point at *X* and draw an arc above the line.

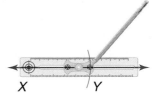

STEP 4 Open your compass to the same width as the length of $\overline{BC}$. Place the compass point at *Y* and draw an arc above the line so that it intersects the arc drawn in Step 3. Label this point *Z*.

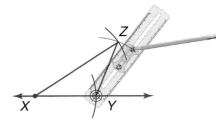

STEP 5 Draw $\overline{YZ}$ and $\overline{XZ}$. $\triangle ABC \cong \triangle XYZ$.

Writing Math

1. **Explain** why the corresponding sides of $\triangle ABC$ and $\triangle XYZ$ are congruent.

2. **Draw** three different triangles. Then construct a triangle that is congruent to each one.

Vocabulary and Concepts

1. **Describe** three ways to classify triangles by their sides. (Lesson 6-2)
2. **List and define** five types of quadrilaterals. (Lesson 6-4)

Skills and Applications

For Exercises 3–5, use the figure at the right. (Lesson 6-1)

3. Find $m\angle 6$ if $m\angle 7 = 84°$.
4. Find $m\angle 5$ if $m\angle 1 = 35°$.

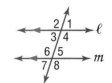

Find the value of x in each figure. (Lessons 6-2 and 6-4)

5.

6.

7.

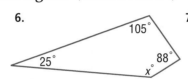

8. **FLAGS** The "Union Jack", a common name for the flag of the United Kingdom, is shown at the right. The blue portions of the flag are triangular. Determine whether the triangles indicated are congruent. If so, write a congruence statement. (Lesson 6-5)

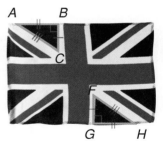

Standardized Test Practice

FCAT Practice

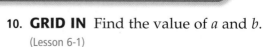

9. **MULTIPLE CHOICE** How many pairs of congruent triangles are formed by the diagonals of a rectangle? (Lesson 6-5)

    (A) 2   (B) 3

    (C) 4   (D) 5

10. **GRID IN** Find the value of a and b. (Lesson 6-1)

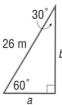

The GameZone

A Place To Practice Your Math Skills

Polygon Bingo

GET READY!

Players: two
Materials: 10 counters, 1 number cube, marker, 1 large red cube, 1 large blue cube, 2 square sheets of paper

GET SET!

- Write *quadrilateral*, *trapezoid*, *parallelogram*, *rectangle*, *rhombus*, and *square* on different faces of the red cube.

- In the same manner, write *scalene*, *isosceles*, *equilateral*, *acute*, *right*, and *obtuse* on different faces of the blue cube.

- Create two boards like the one shown by drawing a different polygon in each square. Use no shape more than once.

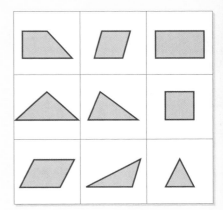

GO!

- The starting player rolls the number cube. If an even number is rolled, the player rolls the red cube. If an odd number is rolled, the player rolls the blue cube.

- The player covers with a counter any shape that matches the information on the top face of the cube. If a player cannot find a figure matching the information, he or she loses a turn.

- **Who Wins?** The first player to get three counters in a row wins.

Symmetry

Sunshine State Standards
MA.C.2.3.1-1

HANDS-ON Mini Lab

What You'll Learn
Identify line symmetry and rotational symmetry.

NEW Vocabulary
line symmetry
line of symmetry
rotational
    symmetry
angle of rotation

Work with a partner.

Trace the outline of the starfish shown onto both a piece of tracing paper and a transparency.

1. Draw a line down the center of your starfish outline. Then fold your paper across this line. What do you notice about the two halves?

2. Are there other lines you can draw on your outline that will produce the same result? If so, how many?

3. Place the transparency over the outline on your tracing paper. Use your pencil point at the centers of the starfish to hold the transparency in place. How many times can you rotate the transparency from its original position so that the two figures match? Do not count the original position.

4. Find the first angle of rotation by dividing 360° by the number of times the figures matched.

5. List the other angles of rotation by adding the first angle of rotation to the previous angle. Stop when you reach 360°.

Materials
- tracing paper
- transparency
- pencil
- overhead markers

A figure has **line symmetry** if it can be folded over a line so that one half of the figure matches the other half. This fold line is called the **line of symmetry**.

vertical line
of symmetry

horizontal line
of symmetry

no line
of symmetry

Some figures, such as the starfish in the Mini Lab above, have more than one line of symmetry. The figure at the right has one vertical, one horizontal, and two diagonal lines of symmetry.

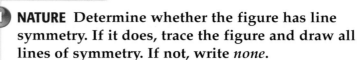

EXAMPLE **Identify Line Symmetry**

1 **NATURE** Determine whether the figure has line symmetry. If it does, trace the figure and draw all lines of symmetry. If not, write *none*.

This figure has one vertical line of symmetry.

Your Turn Determine whether each figure has line symmetry. If it does, trace the figure and draw all lines of symmetry. If not, write *none*.

a. 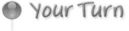   b.   c.

A figure has **rotational symmetry** if it can be rotated or turned less than 360° about its center so that the figure looks exactly as it does in its original position. The degree measure of the angle through which the figure is rotated is called the **angle of rotation**. Some figures have just one angle of rotation, while others, like the starfish, have several.

EXAMPLES **Identify Rotational Symmetry**

LOGOS Determine whether each figure has rotational symmetry. Write *yes* or *no*. If *yes*, name its angle(s) of rotation.

2

Yes, this figure has rotational symmetry. It will match itself after being rotated 180°.

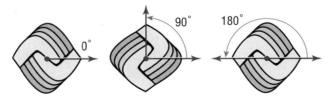

3

Yes, this figure has rotational symmetry. It will match itself after being rotated 120° and 240°.

Skill and Concept Check

1. **OPEN ENDED** Draw a figure that has rotational symmetry.

2. **Which One Doesn't Belong?** Identify the capital letter that does not have the type of symmetry as the other three. Explain your reasoning.

| A | B | M | S |

GUIDED PRACTICE

SPORTS For Exercises 3–6, complete parts a and b for each figure.

a. Determine whether the logo has line symmetry. If it does, trace the figure and draw all lines of symmetry. If not, write *none*.

b. Determine whether the logo has rotational symmetry. Write *yes* or *no*. If yes, name its angle(s) of rotation.

3.

4.

5.

6.

Practice and Applications

JAPANESE FAMILY CRESTS For Exercises 7–14, complete parts a and b for each figure.

a. Determine whether the figure has line symmetry. If it does, trace the figure and draw all lines of symmetry. If not, write *none*.

b. Determine whether the figure has rotational symmetry. Write *yes* or *no*. If *yes*, name its angle(s) of rotation.

HOMEWORK HELP	
For Exercises	See Examples
7–15, 17, 21	1
7–15, 16, 18	2, 3
Extra Practice See pages 631, 653.	

7.

8.

9.

10.

11.

12.

13.

14.

15. **TRIANGLES** Which types of triangles—*scalene, isosceles, equilateral*—have line symmetry? Which have rotational symmetry?

16. **ALPHABET** What capital letters of the alphabet produce the same letter after being rotated 180°?

ROAD SIGNS For Exercises 17 and 18, use the diagrams below.

a.   b.   c.   d.

17. Determine whether each sign has line symmetry. If it does, trace the sign and draw all lines of symmetry. If not, write *none*.

18. Which of the signs above could be rotated and still look the same?

19. **RESEARCH** Use the Internet or other resource to find other examples of road signs that have line and/or rotational symmetry.

20. **ART** Artist Scott Kim uses reflections of words or names as part of his art. Patricia's reflected name is at the right. Create a reflection design for your name using tracing paper.

CRITICAL THINKING Determine whether each statement is *true* or *false*. If *false*, give a counterexample.

21. If a figure has one horizontal and one vertical line of symmetry, then it also has rotational symmetry.

22. If a figure has rotational symmetry, it also has line symmetry.

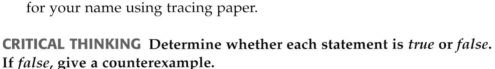

Standardized Test Practice and Mixed Review

 **FCAT Practice**

23. **MULTIPLE CHOICE** Which shape has *only* two lines of symmetry?

 A 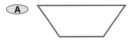   B   C   D

24. **SHORT RESPONSE** Copy the figure at the right. Then shade two squares so that the figure has rotational symmetry.

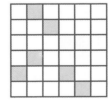

25. **DESIGN** The former symbol for the National Council of Teachers of Mathematics is shown at the right. Which triangles in the symbol appear to be congruent? (Lesson 6-5)

26. In parallelogram $ABCD$, $m\angle A = 55°$, $m\angle B = 125°$, $m\angle C = x°$, and $m\angle D = 125°$. Find the value of x. (Lesson 6-4)

GETTING READY FOR THE NEXT LESSON

PREREQUISITE SKILL Graph each point on a coordinate plane. (Page 614)

27. $A(3, 2)$ 28. $B(-1, 4)$ 29. $C(-2, -1)$ 30. $D(0, 3)$

Reflections

Sunshine State Standards
MA.C.2.3.1-3

What You'll Learn
Graph reflections on a coordinate plane.

NEW Vocabulary

reflection
line of reflection
transformation

∞ LINK To Reading

Everyday Meaning of reflection: the production of an image by or as if by a mirror

WHEN am I ever going to use this?

PHOTOGRAPHY The undisturbed surface of a pond acts like a mirror and can provide the subject for beautiful photographs.

1. Compare the shape and size of the bird to its image in the water.

2. Compare the perpendicular distance from the water line to each of the points shown. What do you observe?

3. The points *A*, *B*, and *C* appear *counterclockwise* on the bird. How are these points oriented on the bird's image?

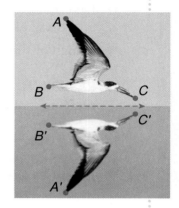

The mirror image produced by flipping a figure over a line is called a **reflection**. This line is called the **line of reflection**. A reflection is one type of **transformation** or mapping of a geometric figure.

Key Concept — Properties of Reflections

1. Every point on a reflection is the same distance from the line of reflection as the corresponding point on the original figure.

2. The image is congruent to the original figure, but the orientation of the image is *different* from that of the original figure.

Model

EXAMPLE Draw a Reflection

1 **Copy △JKL at the right on graph paper. Then draw the image of the figure after a reflection over the given line.**

Step 1 Count the number of units between each vertex and the line of reflection.

Step 2 Plot a point for each vertex the same distance away from the line on the other side.

Step 3 Connect the new vertices to form the image of △JKL, △J'K'L'.

READING Math

Notation Read *P'* as *P prime*. It is the image of point *P*.

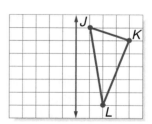

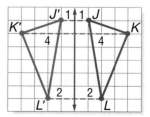

EXAMPLE **Reflect a Figure over the x-axis**

2 Graph △PQR with vertices P(−3, 4), Q(4, 2), and R(−1, 1). Then graph the image of △PQR after a reflection over the x-axis, and write the coordinates of its vertices.

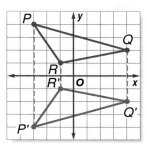

The coordinates of the vertices of the image are P′(−3, −4), Q′(4, −2), and R′(−1, 1). Examine the relationship between the coordinates of each figure.

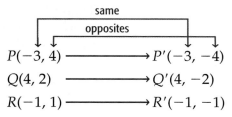

Notice that the y-coordinate of a point reflected over the x-axis is the opposite of the y-coordinate of the original point.

EXAMPLE **Reflect a Figure over the y-axis**

3 Graph quadrilateral ABCD with vertices A(−4, 1), B(−2, 3), C(0, −3), and D(−3, −2). Then graph the image of ABCD after a reflection over the y-axis, and write the coordinates of its vertices.

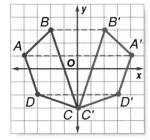

The coordinates of the vertices of the image are A′(4, 1), B′(2, 3), C′(0, −3), and D′(3, −2). Examine the relationship between the coordinates of each figure.

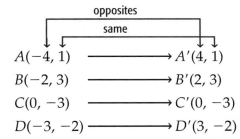

Notice that the x-coordinate of a point reflected over the y-axis is the opposite of the x-coordinate of the original point.

STUDY TIP

Points on Line of Reflection Notice that if a point lies on the line of reflection, the image of that point has the same coordinates as those of the point on the original figure.

Your Turn Graph △FGH with vertices F(1, −1), G(5, −3), and H(2, −4). Then graph the image of △FGH after a reflection over the given axis, and write the coordinates of its vertices.

a. x-axis   **b.** y-axis

If a figure touches the line of reflection as it does in Example 3, then the figure and its image form a new figure that has line symmetry. The line of reflection is a line of symmetry.

EXAMPLE   Use a Reflection

4 **MASKS** Copy and complete the mask shown so that the completed figure has a vertical line of symmetry.

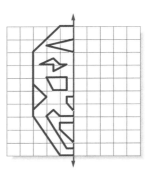

You can reflect the half of the mask shown over the indicated vertical line.

Find the distance from each vertex on the figure to the line of reflection.

Then plot a point that same distance away on the opposite side of the line. Connect vertices as appropriate.

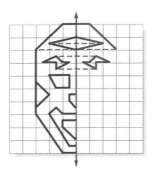

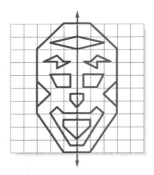

Skill and Concept Check

Writing Math
Exercises 2 & 3

1. **OPEN ENDED** Draw a triangle on grid paper. Then draw a horizontal line below the triangle. Finally, draw the image of the triangle after it is reflected over the horizontal line.

2. **Explain** how a reflection and line symmetry are related.

3. **Which One Doesn't Belong?** Identify the transformation that is not the same as the other three. Explain your reasoning.

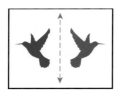 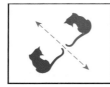 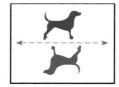 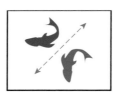

GUIDED PRACTICE

4. Copy the figure at the right on graph paper. Then draw the image of the figure after a reflection over the given line.

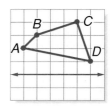

Graph the figure with the given vertices. Then graph its image after a reflection over the given axis, and write the coordinates of its vertices.

5. parallelogram *QRST* with vertices *Q*(−3, 3), *R*(2, 4), *S*(3, 2), and *T*(−2, 1); *x*-axis

6. triangle *JKL* with vertices *J*(−2, 3), *K*(−1, −4), and *L*(−4, −2); *y*-axis

HOMEWORK HELP

For Exercises	See Examples
7–16, 27–28	1
17–24	2, 3
25–26	4

Extra Practice
See pages 631, 653.

Copy each figure onto graph paper. Then draw the image of the figure after a reflection over the given line.

7.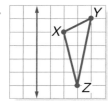

8.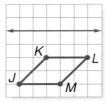

9.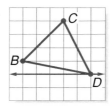

10.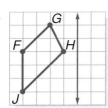

11.

12.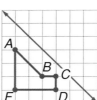

For Exercises 13–16, determine whether the figure in green is a reflection of the figure in blue over the line *n*. Write *yes* or *no*. Explain.

13.

14.

15.

16.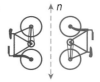

Graph the figure with the given vertices. Then graph its image after a reflection over the given axis, and write the coordinates of its vertices.

17. triangle *ABC* with vertices *A*(−1, −1), *B*(−2, −4), and *C*(−4, −1); *x*-axis

18. triangle *FGH* with vertices *F*(3, 3), *G*(4, −3), and *H*(2, 1); *y*-axis

19. square *JKLM* with vertices *J*(−2, 0), *K*(−1, −2), *L*(−3, −3), and *M*(−4, −1); *y*-axis

20. quadrilateral *PQRS* with vertices *P*(1, 3), *Q*(3, 5), *R*(5, 2), and *S*(3, 1); *x*-axis

Name the line of reflection for each pair of figures.

21.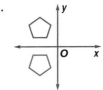

22.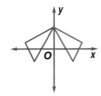

23.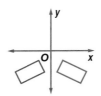

24.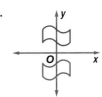

25. **DESIGN** Does the rug below have line symmetry? If so, sketch the rug and draw the line(s) of symmetry.

26. **DESIGN** Copy and complete the rug pattern shown so that the completed figure has line symmetry.

ALPHABET For Exercises 27 and 28, use the figure at the right. It shows that the capital letter A looks the same after a reflection over a vertical line. It does not look the same after a reflection over a horizontal line.

27. What other capital letters look the same after a reflection over a vertical line?

28. Which capital letters look the same after a reflection over a horizontal line?

29. **CRITICAL THINKING** Suppose a point P with coordinates $(-4, 5)$ is reflected so that the coordinates of its image are $(-4, -5)$. Without graphing, which axis was this point reflected over? Explain.

Standardized Test Practice and Mixed Review

FCAT Practice

SHORT RESPONSE For Exercises 30 and 31, use the drawing at the right.

30. The drawing shows the pattern for the left half of the front of the shirt. Copy the pattern onto grid paper. Then draw the outline of the pattern after it has been flipped over a vertical line. Label it "Right Front."

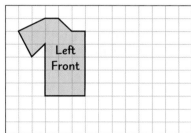

31. Use two geometric terms to explain the relationship between the left and right fronts of the shirt.

32. **MULTIPLE CHOICE** Which of the following is the reflection of $\triangle ABC$ with vertices $A(1, -1)$, $B(4, -1)$, and $C(2, -4)$ over the x-axis?

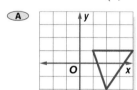 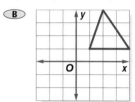 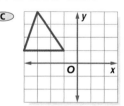 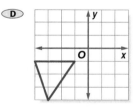

CARDS Determine whether each card has rotational symmetry. Write *yes* or *no*. If *yes*, name its angle(s) of rotation. (Lesson 6-6)

33.

34.

35.

36.

37. Find the value of x if the triangles at the right are congruent. (Lesson 6-5)

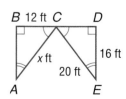

GETTING READY FOR THE NEXT LESSON

PREREQUISITE SKILL Add. (Lesson 1-4)

38. $-4 + (-1)$ 39. $-5 + 3$ 40. $-1 + 4$ 41. $2 + (-2)$

Study Skill
HOW TO...
Study Math Vocabulary

DEFINITION MAP

Understanding a math term requires more than just memorizing a definition. Try completing a definition map to expand your understanding of a geometry vocabulary word.

A definition map can help you visualize the parts of a good definition. Ask yourself these questions about the vocabulary terms.

- What is it? (Category)
- What can it be compared to? (Comparisons)
- What is it like? (Properties)
- What are some examples? (Illustrations)

Here's a definition map for *reflection*.

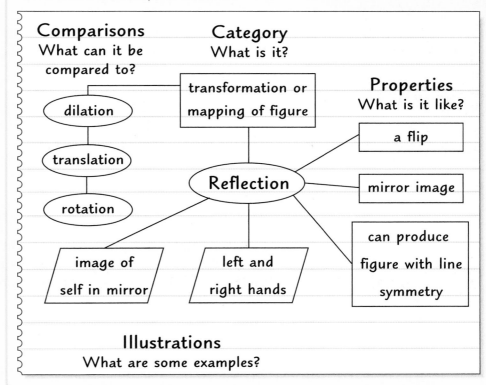

Comparisons
What can it be compared to?
- dilation
- translation
- rotation

Category
What is it?
transformation or mapping of figure

Reflection

Properties
What is it like?
- a flip
- mirror image
- can produce figure with line symmetry

Illustrations
What are some examples?
- image of self in mirror
- left and right hands

SKILL PRACTICE
Make a definition map for each term.

1. complementary angles (Page 256)
2. perpendicular lines (Page 257)
3. isosceles triangle (Page 263)
4. square (Page 273)

Translations

Sunshine State Standards
MA.C.2.3.1-3

What You'll Learn

Graph translations on a coordinate plane.

NEW Vocabulary

translation

WHEN am I ever going to use this?

CHESS In chess, there are rules governing how many spaces and in what direction each game piece can be moved during a player's turn. The diagram at the right shows one legal move of a knight.

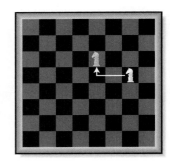

1. Describe the motion involved in moving the knight.

2. Compare the shape, size, and orientation of the knight in its original position to that of the knight in its new position.

A **translation** (sometimes called a *slide*) is the movement of a figure from one position to another without turning it.

Key Concept   Properties of Translations

1. Every point on the original figure is moved the same distance and in the same direction.

2. The image is congruent to the original figure, and the orientation of the image is *the same* as that of the original figure.

Model

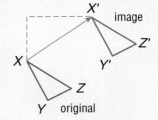

EXAMPLE   Draw a Translation

1. Copy parallelogram *WXYZ* at the right on graph paper. Then draw the image of the figure after a translation 4 units left and 2 units down.

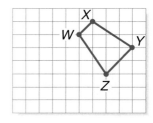

Step 1 Move each vertex of the trapezoid 4 units left and 2 units down.

Step 2 Connect the new vertices to form the image.

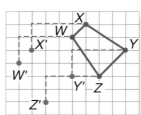   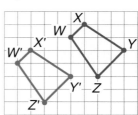

2 Graph △*JKL* with vertices *J*(−3, 4), *K*(1, 3), and *L*(−4, 1). Then graph the image of △*JKL* after a translation 2 units right and 5 units down. Write the coordinates of its vertices.

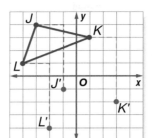   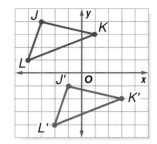

The coordinates of the vertices of the image are *J*′(−1, −1), *K*′(3, −2), and *L*′(−2, −4). Notice that these vertices can also be found by adding 2 to the *x*-coordinates and −5 to the *y*-coordinates, or (2, −5).

Original		Add (2, −5).		Image
$J(-3, 4)$	→	$(-3 + 2, 4 + (-5))$	→	$J'(-1, -1)$
$K(1, 3)$	→	$(1 + 2, 3 + (-5))$	→	$K'(3, -2)$
$L(-4, 1)$	→	$(-4 + 2, 1 + (-5))$	→	$L'(-2, -4)$

Your Turn Graph △*ABC* with vertices *A*(4, −3), *B*(0, 2), and *C*(5, 1). Then graph its image after each translation, and write the coordinates of its vertices.

a. 2 units down   **b.** 4 units left and 3 units up

EXAMPLE   Use a Translation

3 **MULTIPLE-CHOICE TEST ITEM** Point *N* is moved to a new location, *N*′. Which white shape shows where the shaded figure would be if it was translated in the same way?

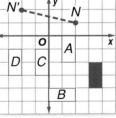

Ⓐ A   Ⓑ B   Ⓒ C   Ⓓ D

Read the Test Item

You are asked to determine which figure has been moved according to the same translation as Point *N*.

Solve the Test Item

Point *N* is translated 4 units left and 1 unit up. Identify the figure that is a translation of the shaded figure 4 units left and 1 unit up.

Figure A: 2 units left and 2 units up

Figure B: represents a turn, not a translation

Figure C: 4 units left and 1 unit up

The answer is C.

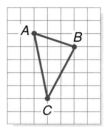

1. **Which One Doesn't Belong?** Identify the transformation that is not the same as the other three. Explain your reasoning.

2. **OPEN ENDED** Draw a rectangle on grid paper. Then draw the image of the rectangle after a translation 2 units right and 3 units down.

GUIDED PRACTICE

3. Copy the figure at the right on graph paper. Then draw the image of the figure after a translation 4 units left and 1 unit up.

Graph the figure with the given vertices. Then graph the image of the figure after the indicated translation, and write the coordinates of its vertices.

4. triangle XYZ with vertices $X(-4, -4)$, $Y(-3, -1)$, and $Z(2, -2)$ translated 3 units right and 4 units up

5. trapezoid $EFGH$ with vertices $E(0, 3)$, $F(3, 3)$, $G(4, 1)$, and $H(-2, 1)$ translated 2 units left and 3 units down

Practice and Applications

Copy each figure onto graph paper. Then draw the image of the figure after the indicated translation.

6. 5 units right and 3 units up   7. 3 units right and 4 units down

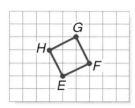   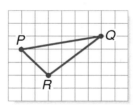

HOMEWORK HELP

For Exercises	See Examples
6–7	1
8–11	2
13–14	3

Extra Practice
See pages 631, 653.

Graph the figure with the given vertices. Then graph the image of the figure after the indicated translation, and write the coordinates of its vertices.

8. $\triangle ABC$ with vertices $A(1, 2)$, $B(3, 1)$, and $C(3, 4)$ translated 2 units left and 1 unit up

9. $\triangle RST$ with vertices $R(-5, -2)$, $S(-2, 3)$, and $T(2, -3)$ translated 1 unit left and 3 units down

10. rectangle $JKLM$ with vertices $J(-3, 2)$, $K(3, 5)$, $L(4, 3)$, and $M(-2, 0)$ translated by 1 unit right and 4 units down

11. parallelogram $ABCD$ with vertices $A(6, 3)$, $B(4, 0)$, $C(6, -2)$, and $D(8, 1)$ translated 3 units left and 2 units up

12. **ART** Explain why Andy Warhol's 1962 *Self Portrait*, shown at the right, is an example of an artist's use of translations.

MUSIC For Exercises 13 and 14, use the following information.
The sound wave of a tuning fork is given below.

13. Look for a pattern in the sound wave. Then copy the sound wave and indicate where this pattern repeats or is translated.

14. How many translations of the original pattern are shown?

15. **CRITICAL THINKING** Triangle *RST* has vertices *R*(4, 2), *S*(−8, 0), and *T*(6, 7). When translated, *R'* has coordinates (−2, 4). Find the coordinates of *S'* and *T'*.

Standardized Test Practice and Mixed Review

  FCAT Practice

16. **MULTIPLE CHOICE** Which of the following is a vertex of the figure shown at the right after a translation 4 units down?

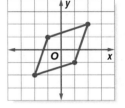

   A (1, −5)   **B** (−6, −2)   **C** (1, −3)   **D** (−2, −2)

17. **SHORT RESPONSE** What are the coordinates of *W*(−6, 3) after it is translated 2 units right and 1 unit down?

18. Graph polygon *ABCDE* with vertices *A*(−5, −3), *B*(−2, 1), *C*(−3, 4), *D*(0, 2), and *E*(0, −3). Then graph the image of the figure after a reflection over the *y*-axis, and write the coordinates of its vertices. (Lesson 6-7)

LIFE SCIENCE For Exercises 19 and 20, use the diagram of the diatom at the right. (Lesson 6-6)

19. Does the diatom have line symmetry? If so, trace the figure and draw any lines of symmetry. If not, write *none*.

20. Does the diatom have rotational symmetry? Write *yes* or *no*. If *yes*, name its angle(s) of rotation.

GETTING READY FOR THE NEXT LESSON

PREREQUISITE SKILL Determine whether each figure has rotational symmetry. Write *yes* or *no*. If *yes*, name its angles of rotation. (Lesson 6-6)

21.   22.   23.   24.

Rotations

Sunshine State Standards
MA.C.2.3.1-3

What You'll Learn
Graph rotations on a coordinate plane.

→**NEW Vocabulary**

rotation
center of rotation

↺**REVIEW Vocabulary**

angle of rotation:
the degree measure of the angle through which a figure is rotated (Lesson 6-6)

HANDS-ON Mini Lab

Materials
• tracing paper
• straightedge
• tape
• protractor

A **rotation** is a transformation involving the turning or spinning of a figure around a fixed point called the **center of rotation**.

STEP 1 Draw a polygon, placing a dot at one vertex. Place a second dot, the center of rotation, in a nearby corner.

STEP 2 Form an angle of rotation by connecting the first dot, the center of rotation, and a point on the edge of the paper.

STEP 3 Place a second paper over the first and trace the figure, the dots, and the ray passing through the figure.

STEP 4 With your pencil on the center of rotation, turn the top paper until its ray lines up with the ray passing through the edge of the first paper. Tape the papers together.

Step 1
center of rotation

Step 2
angle of rotation

Step 3

Step 4

1. Measure the distances from points on the original figure and corresponding points on the image to the center of rotation. What do you observe?

2. Measure the angles formed by connecting the center of rotation to pairs of corresponding points. What do you observe?

The Mini Lab suggests the following properties of rotations.

Key Concept

Properties of Rotations

1. Corresponding points are the same distance from *R*. The angles formed by connecting *R* to corresponding points are congruent.

2. The image is congruent to the original figure, and their orientations are *the same*.

Model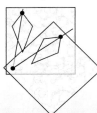

$m\angle XRX' = m\angle YRY'$

EXAMPLE **Rotations in the Coordinate Plane**

① Graph △*XYZ* with vertices *X*(2, 2), *Y*(4, 3), and *Z*(3, 0). Then graph the image of △*XYZ* after a rotation 90° counterclockwise about the origin, and write the coordinates of its vertices.

Step 1 Lightly draw a line connecting point *X* to the origin.

Step 2 Lightly draw $\overline{OX}$ so that *m∠X'OX* = 90° and *OX'* = *OX*.

Step 3 Repeat steps 1–3 for points *Y* and *Z*. Then erase all lightly drawn lines and connect the vertices to form △*X'Y'Z'*.

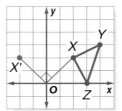 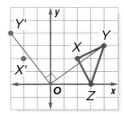 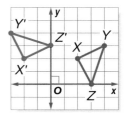

Triangle *X'Y'Z'* has vertices *X'*(−2, 2), *Y'*(−3, 4), and *Z'*(0, 3).

Your Turn Graph △*ABC* with vertices *A*(1, −2), *B*(4, 1), and *C*(3, −4). Then graph the image of △*ABC* after the indicated rotation about the origin, and write the coordinates of its vertices.

a. 90° counterclockwise   **b.** 180° counterclockwise

If a figure touches its center of rotation, then one or more rotations of the figure can be used to create a new figure that has rotational symmetry.

EXAMPLE **Use a Rotation**

② **FOLK ART** Copy and complete the barn sign shown so that the completed figure has rotational symmetry with 90°, 180°, and 270° as its angles of rotation.

Use the procedure described above and the points indicated to rotate the figure 90°, 180°, and 270° counterclockwise. Use a 90° rotation *clockwise* to produce the same rotation as a 270° rotation counterclockwise.

90° counterclockwise   180° counterclockwise   90° clockwise

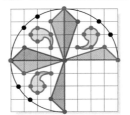 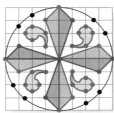

1. **OPEN ENDED** Give three examples of rotating objects you see every day.

2. **FIND THE ERROR** Anita and Manuel are graphing △MNP with vertices M(−3, 2), N(−1, −1), and P(−4, −2) and its image after a rotation 90° counterclockwise about the origin. Who is correct? Explain.

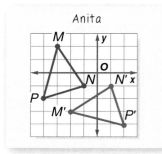

Anita

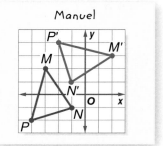

Manuel

GUIDED PRACTICE

Graph the figure with the given vertices. Then graph the image of the figure after the indicated rotation about the origin, and write the coordinates of its vertices.

3. triangle ABC with vertices A(−2, −4), B(2, −1), and C(4, −3); 90° counterclockwise

4. quadrilateral DFGH with vertices D(−3, 2), F(−1, 0), G(−3, −4), and H(−4, −2); 180°

Practice and Applications

Graph the figure with the given vertices. Then graph the image of the figure after the indicated rotation about the origin, and write the coordinates of its vertices.

5. triangle VWX with vertices V(−4, 2), W(−2, 4), and X(2, 1); 180°

6. triangle BCD with vertices B(−5, 3), C(−2, 5), and D(−3, 2); 90° counterclockwise

7. trapezoid LMNP with vertices L(0, 3), M(4, 3), N(1, −3), and P(−1, 1); 90° counterclockwise

8. quadrilateral FGHJ with vertices F(−5, 4), G(−3, 4), H(0, −1), and J(−5, 2); 180°

HOMEWORK HELP

For Exercises	See Examples
5–12	1
13	2

Extra Practice
See pages 632, 653.

Determine whether the figure in green is a rotation of the figure in blue about the origin. Write *yes* or *no*. Explain.

9.

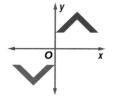

10.

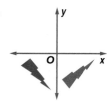

11.

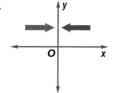

12.

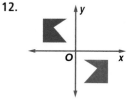

13. **FABRIC DESIGN** Copy and complete the handkerchief design at the right so that it has rotational symmetry. Rotate the figure 90°, 180°, and 270° counterclockwise about point C.

14. **CRITICAL THINKING** What are the new coordinates of a point at (x, y) after the point is rotated 90° counterclockwise? 180°?

C

Standardized Test Practice and Mixed Review

A B C D

15. **SHORT RESPONSE** Draw a rectangle. Then draw the image of the rectangle after it has been translated 1.5 inches to the right and then rotated 90° counterclockwise about the bottom left vertex. Label this rectangle I.

16. **MULTIPLE CHOICE** Which illustration shows the figure at the right rotated 180°?

A   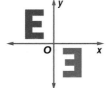   B   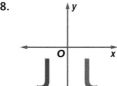   C   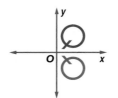   D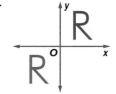

Identify each transformation as a *reflection*, a *translation*, or a *rotation*.
(Lessons 6-7, 6-8, and 6-9)

17.   18.   19.   20.

For Exercises 21–25, use the graphic at the right.

21. Describe a translation used in this graphic. (Lesson 6-8)

22. Trace at least two examples of figures or parts of figures in the graphic that appear to have line symmetry. Then draw all lines of symmetry. (Lesson 6-6)

23. Trace a figure or part of a figure used in the graphic that appears to have rotational symmetry. (Lesson 6-6)

24. Trace and then classify the quadrilateral that makes up the top portion of the collar on the shirt. (Lesson 6-4)

25. Trace the two triangles that make up the collar on the shirt. Classify each triangle by its angles and by its sides and then determine whether the two triangles are congruent. (Lessons 6-2 and 6-5)

USA TODAY Snapshots®

Crafts are tops on this gift list

If they were a father, what kids say they would like to receive for Father's Day:

51% Something your child made

19% A night of peace and quiet

15% A night out on the town

7% Offer to help around the house

Source: WGBH in conjunction with Applied Research & Consulting LLC for ZOOM

By Cindy Hall and Suzy Parker, USA TODAY

What You'll Learn
Create Escher-like drawings using translations and rotations.

Materials
- index cards
- scissors
- tape
- paper

Sunshine State Standards
MA.C.2.3.2-1, MA.C.2.3.2-2

Tessellations

Maurits Cornelis Escher (1898–1972) was a Dutch artist whose work used tessellations. A **tessellation** is a tiling made up of copies of the same shape or shapes that fit together without gaps and without overlapping. The sum of the angle measures where vertices meet in a tessellation must equal 360°. For this reason, equilateral triangles and squares will tessellate a plane.

Symmetry drawing E70 by M.C. Escher. © 2002 Cordon Art-Baarn-Holland. All rights reserved.

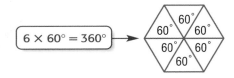

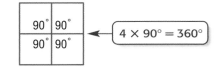

ACTIVITY

1 Create a tessellation using a translation.

STEP 1 Draw a square on the back of an index card. Then draw a triangle inside the top of the square as shown below.

STEP 2 Cut out the square. Then cut out the triangle and translate it from the top to the bottom of the square.

STEP 3 Tape the triangle and square together to form a pattern.

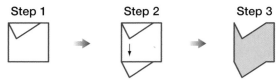

Step 1   Step 2   Step 3

STEP 4 Trace this pattern onto a sheet of paper as shown to create a tessellation.

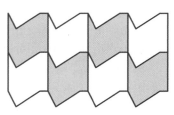

Your Turn Make an Escher-like drawing using each pattern.

a.   b.   c.

2 Create a tessellation using a rotation.

STEP 1 Draw an equilateral triangle on the back of an index card. Then draw a right triangle inside the left side of the triangle as shown below.

STEP 2 Cut out the equilateral triangle. Then cut out the right triangle and rotate it so that the right triangle is on the right side as indicated.

STEP 3 Tape the right triangle and equilateral triangle together to form a pattern unit.

Step 1   Step 2   Step 3

STEP 4 Trace this pattern onto a sheet of paper as shown to create a tessellation.

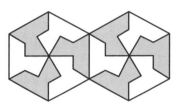

Your Turn Make an Escher-like drawing using each pattern.

d.   e.   f.

Writing Math

1. **Design and draw** a pattern for an Escher-like drawing.

2. **Describe** how to use your pattern to create a pattern unit for your tessellation. Then create a tessellation using your pattern.

3. **Name** another regular polygon other than an equilateral triangle or square that will tessellate a plane. Explain your reasoning.

Determine whether each of the following figures will tessellate a plane. Explain your reasoning.

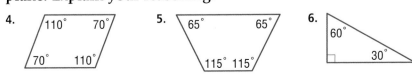

4. 110° 70° / 70° 110°

5. 65° 65° / 115° 115°

6. 60° 30°

Vocabulary and Concept Check

acute angle (p. 256)	line of symmetry (p. 286)	rotation (p. 300)
acute triangle (p. 263)	line symmetry (p. 286)	rotational symmetry (p. 287)
adjacent angles (p. 256)	obtuse angle (p. 256)	scalene triangle (p. 263)
alternate exterior angles (p. 258)	obtuse triangle (p. 263)	square (p. 273)
alternate interior angles (p. 258)	parallel lines (p. 257)	straight angle (p. 256)
angle of rotation (p. 287)	parallelogram (p. 273)	supplementary angles (p. 256)
center of rotation (p. 300)	perpendicular lines (p. 257)	transformation (p. 290)
complementary angles (p. 256)	quadrilateral (p. 272)	translation (p. 296)
congruent polygons (p. 279)	rectangle (p. 273)	transversal (p. 258)
corresponding angles (p. 258)	reflection (p. 290)	trapezoid (p. 273)
equilateral triangle (p. 263)	rhombus (p. 273)	triangle (p. 262)
isosceles triangle (p. 263)	right angle (p. 256)	vertical angles (p. 256)
line of reflection (p. 290)	right triangle (p. 263)	

State whether each sentence is *true* or *false*. If *false*, replace the underlined word to make a true sentence.

1. A(n) <u>acute</u> angle has a measure greater than 90° and less than 180°.

2. The sum of the measures of <u>supplementary</u> angles is 180°.

3. <u>Parallel</u> lines intersect at a right angle.

4. In a(n) <u>scalene</u> triangle, all three sides are congruent.

5. A(n) <u>rhombus</u> is a parallelogram with four congruent sides.

6. An isosceles trapezoid has <u>rotational</u> symmetry.

7. The orientations of a figure and its reflected image are <u>different</u>.

Lesson-by-Lesson Exercises and Examples

 Line and Angle Relationships (pp. 256–260)

Find the value of *x* in each figure.

8. 125° / x°

9. 43° x°

For Exercises 10 and 11, use the figure at the right.

10. Find $m\angle 8$ if $m\angle 4 = 118°$.

11. Find $m\angle 6$ if $m\angle 2 = 135°$.

Example 1
If $m\angle 1 = 105°$, find $m\angle 3$, $m\angle 5$, and $m\angle 8$.

Since $\angle 1$ and $\angle 3$ are vertical angles $\angle 1 \cong \angle 3$. So, $m\angle 3 = 105°$.

Since $\angle 1$ and $\angle 5$ are corresponding angles, $\angle 1 \cong \angle 5$. Therefore, $m\angle 5 = 105°$.

Since $\angle 5$ and $\angle 8$ are supplementary, $m\angle 8 = 180° - 105°$ or 75°.

msmath3.net/vocabulary_review

6-2

Triangles and Angles (pp. 262–265)

Find the value of *x* in each triangle.

12.

13.

14. Classify the triangle in Exercise 14 by its angles and by its sides.

Example 2
Find the value of *x* in △*JKL*.

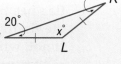

$x + 20 + 20 = 180$

$\qquad x = 180 - (20 + 20)$ or 140

6-3

Special Right Triangles (pp. 267–270)

Find each missing length. Round to the nearest tenth if necessary.

15.

16.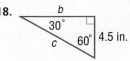

17.

18.

Example 3 Find each missing length.

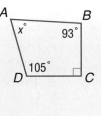

$a = \dfrac{1}{2}(6)$ or 3 m

To find *b*, use the Pythagorean Theorem.

$6^2 = 3^2 + b^2 \qquad c^2 = a^2 + b^2$

$36 = 9 + b^2 \qquad$ Evaluate 6^2 and 9^2.

$27 = b^2 \qquad$ Subtract 9 from each side.

$5.2 \approx b \qquad$ Take the square root of each side.

6-4

Classifying Quadrilaterals (pp. 272–275)

19. In quadrilateral *JKLM*, $m\angle J = 123°$, $m\angle K = 90°$, and $m\angle M = 45°$. Find $m\angle L$.

20. Classify the quadrilateral shown using the name that *best* describes it.

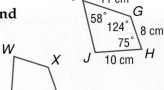

Example 4 Find the value of *x* in quadrilateral *ABCD*.

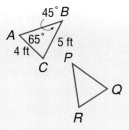

$x + 93 + 90 + 105 = 360$

$\qquad x + 228 = \quad 360$

$\underline{\qquad -228 = -228}$

$\qquad\qquad x = 72$

6-5

Congruent Polygons (pp. 279–282)

In the figure, *FGHJ* ≅ *YXWZ*. Find each measure.

21. $m\angle X$

22. WZ

23. YX

24. $m\angle Z$

Example 5
In the figure, △*ABC* ≅ △*RPQ*. Find *PQ*.

PQ corresponds to *BC*. Since *BC* = 5 feet, *PQ* = 5 feet.

Symmetry (pp. 286–289)

BOATING Determine whether each signal flag has line symmetry. If it does, trace the figure and draw all lines of symmetry. If not, write *none*.

25.   26.   27.

28. Which of the figures above has rotational symmetry? Name the angle(s) of rotation.

Example 6 Determine whether the logo at the right has rotational symmetry. If it does, name its angles of rotation.

The logo has rotational symmetry. Its angles of rotation are 90°, 180°, and 270°.

Reflections (pp. 290–294)

Graph parallelogram *QRST* with vertices *Q*(2, 5), *R*(4, 5), *S*(3, 1), and *T*(1, 1). Then graph its image after a reflection over the given axis, and write the coordinates of its vertices.

29. *x*-axis   30. *y*-axis

Example 7 Graph △*FGH* with vertices *F*(1, −1), *G*(3, 1), and *H*(2, −3) and its image after a reflection over the *y*-axis.

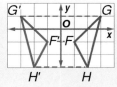

Translations (pp. 296–299)

Graph △*ABC* with vertices *A*(2, 2), *B*(3, 5), and *C*(5, 3). Then graph its image after the indicated translation, and write the coordinates of its vertices.

31. 6 units down

32. 2 units left and 4 units down

Example 8 Graph △*XYZ* with vertices *X*(−3, −1), *Y*(−1, 0), and *Z*(−2, −3) and its image after a translation 4 units right and 1 unit up.

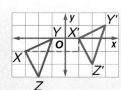

Rotations (pp. 300–303)

Graph △*JKL* with vertices *J*(−1, 3), *K*(1, 1), and *L*(3, 4). Then graph its image after the indicated rotation about the origin, and write the coordinates of its vertices.

33. 90° counterclockwise

34. 180° counterclockwise

Example 9 Graph △*PQR* with vertices *P*(1, 3), *Q*(2, 1), and *R*(4, 2) and its image after a rotation of 90° counterclockwise about the origin.

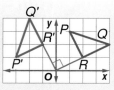

Vocabulary and Concepts

1. **Draw** a pair of complementary angles. Label the angles $\angle 1$ and $\angle 2$.

2. **OPEN ENDED** Draw an obtuse isosceles triangle.

Skills and Applications

For Exercises 3–5, use the figure at the right.

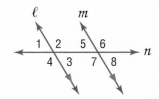

3. Find $m\angle 6$ if $m\angle 5 = 60°$.   4. Find $m\angle 8$ if $m\angle 1 = 82°$.

5. Name a pair of corresponding angles.

Find each missing measure. Round to the nearest tenth.

6.
7.

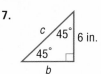

DESIGN **Identify each quadrilateral in the stained glass window using the name that best describes it.**

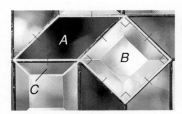

8. A   9. B   10. C

In the figure at the right, $\triangle MNP \cong \triangle ZYX$. Find each measure.

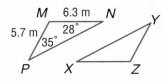

11. ZY 12. $\angle Z$

MUSIC **Determine whether each figure has line symmetry. If it does, trace the figure and draw all lines of symmetry. If not, write *none*.**

13.    14.   15.

16. Which of the figures in Exercises 13–15 has rotational symmetry?

Graph $\triangle JKL$ with vertices $J(2, 3)$, $K(-1, 4)$, and $L(-3, -5)$. Then graph its image, and write the coordinates of its vertices after each transformation.

17. reflection over the x-axis   18. translation by $(-2, 5)$ 19. rotation 180°

Standardized Test Practice

20. **MULTIPLE CHOICE** $\overline{WY}$ is a diagonal of rectangle $WXYZ$. Which angle is congruent to $\angle WYZ$?

    (A) $\angle WXY$ (B) $\angle WYX$ (C) $\angle ZWY$ (D) $\angle XWY$

🔶 **FCAT Practice**

PART 1 Multiple Choice

Record your answers on the answer sheet provided by your teacher or on a sheet of paper.

1. One week Alexandria ran 500 meters, 600 meters, 800 meters, and 1,100 meters. How many kilometers did she run that week? (Prerequisite Skill, pp. 606–607)

   A 1   **B** 2   **C** 3   **D** 4

2. The graph shows the winning times in seconds of the women's 4 × 100-meter freestyle relay for several Olympic games.

   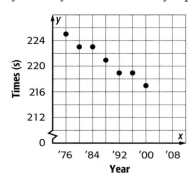

   What is a reasonable prediction for the winning time in 2008? (Lesson 1-1)

   F 212 s   **G** 215 s   **H** 218 s   **I** 221 s

3. Which expression is equivalent to xy^2z^{-1}? (Lesson 2-2)

   A $\dfrac{1}{x \cdot y \cdot y \cdot z}$ **B** $x + y + y - z$

   C $x \cdot y \cdot y - z$ **D** $\dfrac{x \cdot y \cdot y}{z}$

   ### TEST-TAKING TIP

   Question 3 Answer every question when there is no penalty for guessing. If you must guess, eliminate answers you know are incorrect. For Question 3, eliminate Choice B since $xy^2 = x \cdot y \cdot y$.

4. Aleta went to the grocery store and paid $19.71 for her purchases. A portion of her receipt is shown below.

   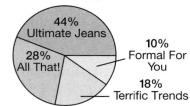

   About how much did the beef cost per pound? (Lesson 4-1)

   F $2.60   **G** $3.34   **H** $3.80   **I** $4.25

5. Which of the stores represented in the circle graph is preferred by $\dfrac{7}{25}$ of the students? (Lesson 5-1)

   Preferred Clothing Stores

   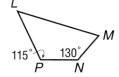

   A Ultimate Jeans   **B** All That!

   C Terrific Trends   **D** Formal For You

6. Which of the following could *not* be the measure of $\angle M$? (Lesson 6-4)

   F 35°   **G** 50°   **H** 45°   **I** 116°

7. Which of the following figures is *not* a rotation of the figure at the right? (Lesson 6-9)

   A   **B**

   C   **D**

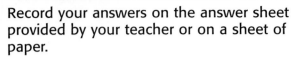

FCAT Practice

PART 2 Short Response/Grid In

THINK
SOLVE
EXPLAIN

Record your answers on the answer sheet provided by your teacher or on a sheet of paper.

8. Ms. Neville has 26 students in her homeroom. All of her students take at least one foreign language. Thirteen students take Spanish, 11 students take French, and 5 students take Japanese. How many students take more than one language? (Lesson 1-1)

9. Write $\sqrt{8}$, $\sqrt{6}$, 3.2, and $\frac{1}{3}$ in order from least to greatest. (Lesson 3-3)

10. An area of 2,500 square feet of grass produces enough oxygen for a family of 4. What is the area of grass needed to supply a family of 5 with oxygen? (Lesson 4-4)

11. You buy a sweater on sale for $29.96. You paid 25% less than the original price. What was the original price of the sweater? (Lesson 5-8)

12. If $a \parallel b$, find the value of x. (Lesson 6-1)

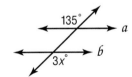

13. Name a quadrilateral with one pair of parallel sides and one pair of sides that are not parallel. (Lesson 6-4)

14. How many lines of symmetry does the figure at the right have? (Lesson 6-6)

15. If $\triangle JKL \cong \triangle MNP$, name the segment in $\triangle MNP$ that is congruent to $\overline{LJ}$. (Lesson 6-5)

16. If $\triangle ABC$ is reflected about the y-axis, what are the coordinates of point A'? (Lesson 6-7)

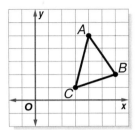

Record your answers on a sheet of paper. Show your work.

THINK
SOLVE
EXPLAIN

17. The graph below shows Arm 1 of the design for a company logo. (Lesson 6-9)

Arm 1

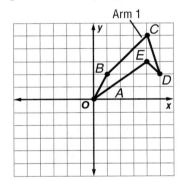

a. To create Arm 2 of the logo, graph the image of figure $ABCDE$ after a rotation 90° counterclockwise about the origin. Write the coordinates of the vertices of Arm 2.

b. To create Arm 3 of the logo, graph the coordinates of figure $ABCDE$ after a rotation 180° about the origin. Write the coordinates of the vertices of Arm 3.

c. To create Arm 4 of the logo, graph the coordinates of Arm 2 after a rotation 180° about the origin. Write the coordinates of the vertices of Arm 4.

d. Does the completed logo have rotational symmetry? If so, name its angle(s) of rotation.

CHAPTER 7

Geometry: Measuring Area and Volume

"How is math used in packaging candy?"

When marketing a product such as candy, how the product is packaged can be as important as how it tastes. A marketer must decide what shape container is best, how much candy the container should hold, and how much material it will take to make the chosen container. **To make these decisions, you must be able to identify three-dimensional objects, calculate their volumes, and calculate their surface areas.**

You will solve problems about packaging in Lesson 7-5.

GETTING STARTED

Take this quiz to see whether you are ready to begin Chapter 7. Refer to the lesson or page number in parentheses if you need more review.

▶ Vocabulary Review

Choose the correct term to complete each sentence.

1. A quadrilateral with exactly one pair of parallel opposite sides is called a (parallelogram, trapezoid). (Lesson 6-4)

2. Polygons that have the same size and shape are called (congruent, similar) polygons. (Lesson 6-5)

▶ Prerequisite Skills

Multiply. (Lesson 6-4)

3. $\frac{1}{3} \cdot 8 \cdot 12$

4. $\frac{1}{3} \cdot 4 \cdot 9^2$

Find the value of each expression to the nearest tenth.

5. $8.3 \cdot 4.1$

6. $9 \cdot 5.2$

7. $7.36 \div 4$

8. $12 \div 0.06$

Use the π key on a calculator to find the value of each expression. Round to the nearest tenth.

9. $\pi \cdot 15$

10. $2 \cdot \pi \cdot 3.2$

11. $\pi \cdot 7^2$

12. $\pi \cdot (19 \div 2)^2$

Classify each polygon according to its number of sides.

13.

14.

15.

16.

Area and Volume Make this Foldable to help you organize your notes. Begin with a plain piece of $8\frac{1}{2}$" × 11" paper.

STEP 1 Fold
Fold in half widthwise.

STEP 2 Open and Fold Again
Fold the bottom to form a pocket. Glue edges.

STEP 3 Label
Label each pocket. Place several index cards in each pocket.

Reading and Writing As you read and study the chapter, write down important definitions, formulas, and key concepts under each heading.

Area of Parallelograms, Triangles, and Trapezoids

Sunshine State Standards
MA.B.1.3.3-1, MA.B.1.3.3-2, MA.C.1.3.1-4

What You'll Learn
Find the areas of parallelograms, triangles, and trapezoids.

NEW Vocabulary

base
altitude

HANDS-ON Mini Lab

Materials
• grid paper

Work with a partner.

STEP 1 Draw a rectangle on grid paper.

STEP 2 Shift the top line 3 units right and draw a parallelogram.

STEP 3 Draw a line connecting two opposite vertices of the parallelogram and form two triangles.

1. What dimensions are the same in each figure?
2. Compare the areas of the three figures. What do you notice?

The area of a parallelogram can be found by multiplying the measures of its base and its height.

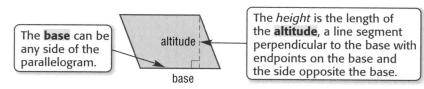

The **base** can be any side of the parallelogram.

The *height* is the length of the **altitude**, a line segment perpendicular to the base with endpoints on the base and the side opposite the base.

Key Concept — Area of a Parallelogram

Words The area A of a parallelogram is the product of any base b and its height h.

Model

Symbols $A = bh$

EXAMPLE — Find the Area of a Parallelogram

1 Find the area of the parallelogram.

The base is 5 feet. The height is 7 feet.

$A = bh$ Area of a parallelogram

$A = 5 \cdot 7$ Replace b with 5 and h with 7.

$A = 35$ Multiply.

The area is 35 square feet.

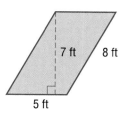

A diagonal of a parallelogram separates the parallelogram into two congruent triangles.

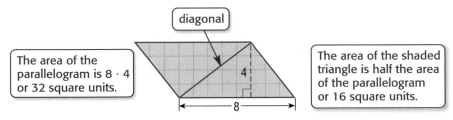

The area of the parallelogram is 8 · 4 or 32 square units.

diagonal

The area of the shaded triangle is half the area of the parallelogram or 16 square units.

Using the formula for the area of a parallelogram, you can find the formula for the area of a triangle.

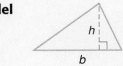
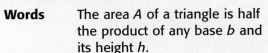

EXAMPLE   **Find the Area of a Triangle**

2 **Find the area of the triangle.**

The base is 12 meters. The height is 8 meters.

12 m

8 m

$A = \frac{1}{2}bh$ Area of a triangle

$A = \frac{1}{2}(12)(8)$ Replace b with 12 and h with 8.

$A = \frac{1}{2}(96)$ Multiply. $12 \times 8 = 96$

$A = 48$ Multiply. $\frac{1}{2} \times 96 = 48$.

The area is 48 square meters.

In Chapter 6, you learned that a trapezoid is a quadrilateral with exactly one pair of parallel sides. These parallel sides are its bases. A trapezoid can be separated into two triangles. Consider trapezoid $EFGH$.

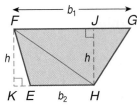

The triangles are $\triangle FGH$ and $\triangle EFH$.
The measure of a base of $\triangle FGH$ is b_1 units.
The measure of a base of $\triangle EFH$ is b_2 units.
The altitudes of the triangles, $\overline{FK}$ and $\overline{JH}$, are congruent. Both are h units long.

area of trapezoid $EFGH$ = area of $\triangle FGH$ + area of $\triangle EFH$

$= \frac{1}{2}b_1h + \frac{1}{2}b_2h$

$= \frac{1}{2}h(b_1 + b_2)$ Distributive Property

Key Concept — Area of a Trapezoid

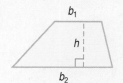

Words The area A of a trapezoid is half the product of the height h and the sum of the bases, b_1 and b_2.

Model

Symbols $A = \frac{1}{2}h(b_1 + b_2)$

EXAMPLE Find the Area of a Trapezoid

3 **Find the area of the trapezoid.**

The height is 4 yards. The lengths of the bases are 7 yards and 3 yards.

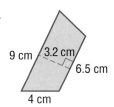

$A = \frac{1}{2}h(b_1 + b_2)$ Area of a trapezoid

$A = \frac{1}{2}(4)(7 + 3)$ Replace h with 4, b_1 with 7, and b_2 with 3.

$A = \frac{1}{2}(4)(10)$ or 20 Simplify.

The area of the trapezoid is 20 square yards.

Your Turn Find the area of each figure.

a.

5.6 cm
3.2 cm
6 cm

b.

15 in.
$4\frac{1}{2}$ in.
10 in.

c.

9 cm 3.2 cm
6.5 cm
4 cm

EXAMPLE Use Area to Solve a Real-Life Problem

4 **LANDSCAPING** You are buying grass seed for the lawn surrounding three sides of an office building. If one bag covers 2,000 square feet, how many bags should you buy?

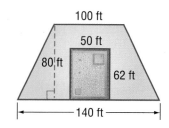

To find the area to be seeded, subtract the area of the rectangle from the area of the trapezoid.

Area of trapezoid

$A = \frac{1}{2}h(b_1 + b_2)$

$A = \frac{1}{2}(80)(100 + 140)$

$A = 9,600$

Area of rectangle

$A = \ell w$

$A = (50)(62)$

$A = 3,100$

The area to be seeded is $9,600 - 3,100$ or 6,500 square feet. If one bag seeds 2,000 square feet, then you will need $6,500 \div 2,000$ or 3.25 bags. Since you cannot buy a fraction of a bag, you should buy 4 bags.

Skill and Concept Check

Writing Math
Exercises 1 & 3

1. **Compare** the formulas for the area of a rectangle and the area of a parallelogram.

2. **OPEN ENDED** Draw and label two different triangles that have the same area.

3. **FIND THE ERROR** Anthony and Malik are finding the area of the trapezoid at the right. Who is correct? Explain.

Anthony
$A = \frac{1}{2}(14.2)(8.5)$
$A = 60.35 \text{ mm}^2$

Malik
$A = \frac{1}{2}(8.5)(14.2 + 11.8)$
$A = 110.5 \text{ mm}^2$

GUIDED PRACTICE

Find the area of each figure.

4.

5.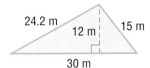

6.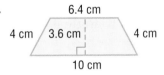

Practice and Applications

Find the area of each figure.

7.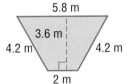

8.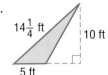

9.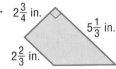

HOMEWORK HELP	
For Exercises	See Examples
7, 10, 13–14	1
8, 11, 15–16	2
9, 12, 17–18	3
21–26	4
Extra Practice See pages 632, 654.	

10.

11.

12.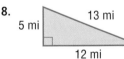

13. parallelogram: base, $4\frac{2}{3}$ in.; height, 6 in.

14. parallelogram: base, 3.8 m; height, 4.2 m

15. triangle: base, 12 cm; height, 5.4 cm

16. triangle: base, $15\frac{3}{4}$ ft; height, $5\frac{1}{2}$ ft

17. trapezoid: height, 3.6 cm; bases, 2.2 cm and 5.8 cm

18. trapezoid: height, 8 yd; bases, $10\frac{1}{2}$ yd and $15\frac{1}{3}$ yd

19. **ALGEBRA** Find the height of a triangle with a base of 6.4 centimeters and an area of 22.4 square centimeters.

20. **ALGEBRA** A trapezoid has an area of 108 square feet. If the lengths of the bases are 10 feet and 14 feet, find the height.

GEOGRAPHY For Exercises 21–24, estimate the area of each state using the scale given.

21.
Tennessee
1 cm = 200 km

22.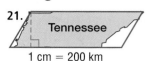
Arkansas
1 cm = 250 km

23.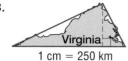
Virginia
1 cm = 250 km

24.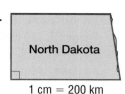
North Dakota
1 cm = 200 km

25. **RESEARCH** Use the Internet or another reference to find the actual area of each state listed above. Compare to your estimate.

26. **MULTI STEP** A deck shown is constructed in the shape of a trapezoid, with a triangular area cut out for an existing oak tree. You want to waterproof the deck with a sealant. One can of sealant covers 400 square feet. Find the area of the deck. Then determine how many cans of sealant you should buy.

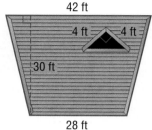
42 ft
4 ft   4 ft
30 ft
28 ft

CRITICAL THINKING For Exercises 27 and 28, decide how the area of each figure is affected.

27. The height of a triangle is doubled, but the length of the base remains the same.

28. The length of each base of a trapezoid is doubled and its height is also doubled.

Standardized Test Practice and Mixed Review

A B C D

FCAT Practice

29. **MULTIPLE CHOICE** Which figure does *not* have an area of 120 square feet?

A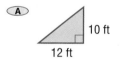
10 ft
12 ft

B
16 ft
7.5 ft

C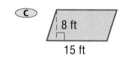
8 ft
15 ft

D
5.3 ft
12 ft
14.7 ft

30. **MULTIPLE CHOICE** Which of the following is the best estimate of the area of the shaded region?

F 20 in²   G 40 in²   H 60 in²   I 80 in²

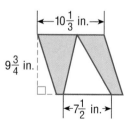
←10⅓ in.→
9¾ in.
←7½ in.→

For Exercises 31–33, use the following information.
Triangle *XYZ* has vertices *X*(−4, 1), *Y*(−1, 4), and *Z*(−3, −3). Then graph the image of △*XYZ* after the indicated transformation and write the coordinates of its vertices. (Lessons 6-7, 6-8, and 6-9)

31. translated by (3, −2)   32. reflected over the *x*-axis   33. rotated 180°

GETTING READY FOR THE NEXT LESSON

BASIC SKILL Use the π key on a calculator to find the value of each expression. Round to the nearest tenth.

34. $\pi \cdot 27$ 35. $2 \cdot \pi \cdot 9.3$ 36. $\pi \cdot 5^2$ 37. $\pi \cdot (15 \div 2)^2$

Circumference and Area of Circles

Sunshine State Standards
MA.B.1.3.3-1, MA.B.1.3.3-2, MA.C.1.3.1-4

HANDS-ON Mini Lab

Work with a partner.

STEP 1 Measure and record the distance d across the circular part of the object, through its center.

STEP 2 Place the object on a piece of paper. Mark the point where the object touches the paper on both the object and on the paper.

STEP 3 Carefully roll the object so that it makes one complete rotation. Then mark the paper again.

STEP 4 Finally, measure the distance C between the marks.

1. What distance does C represent?

2. Find the ratio $\frac{C}{d}$ for this object.

3. Repeat the steps above for at least two other circular objects and compare the ratios of C to d. What do you observe?

4. Plot the data you collected as ordered pairs, (d, C). Then find the slope of a best-fit line through these points.

Materials

- several different cylindrical objects like a can or battery
- ruler
- marker

What You'll Learn

Find the circumference and area of circles.

NEW Vocabulary

circle
center
radius
diameter
circumference
pi

MATH Symbols

π pi
$\approx$ approximately equal to

STUDY TIP

Pi The numbers 3.14 and $\frac{22}{7}$ are often used as approximations for π.

A **circle** is a set of points in a plane that are the same distance from a given point in the plane, called the **center**. The distance from the center to any point on the circle is called the **radius**. The distance across the circle through the center is its **diameter**. The distance around the circle is called the **circumference**.

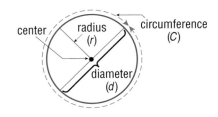

The diameter of a circle is twice its radius or $d = 2r$.

The relationship you discovered in the Mini Lab is true for all circles. The ratio of the circumference of a circle to its diameter is always 3.1415926 …. The Greek letter π **(pi)** represents this number.

Words The circumference C of a circle is equal to its diameter d times π, or 2 times its radius r times π.

Model

Symbols $C = \pi d$ or $C = 2\pi r$

STUDY TIP

Calculating with Pi Unless otherwise specified, use the π key on a calculator to evaluate expressions involving π.

EXAMPLES Find the Circumferences of Circles

Find the circumference of each circle.

①

$C = \pi d$ Circumference of a circle

$C = \pi \cdot 9$ Replace d with 9.

$C = 9\pi$ This is the *exact* circumference.

Use a calculator to find 9π.   9 ☒ π 🄴 28.27433388

The circumference is about 28.3 inches.

②

$C = 2\pi r$ Circumference of a circle

$C = 2 \cdot \pi \cdot 7.2$ Replace r with 7.2.

$C \approx 45.2$ Use a calculator.

The circumference is about 45.2 centimeters.

Finding the area of a circle can be related to finding the area of a parallelogram. A circle can be separated into congruent wedge-like pieces. Then the pieces can be rearranged to form the figure below.

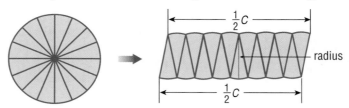

Since the circle has an area that is relatively close to the area of the parallelogram-shaped figure, you can use the formula for the area of a parallelogram to find the area of a circle.

$A = bh$ Area of a parallelogram

$A = \left(\frac{1}{2} \cdot C\right)r$ The base of the parallelogram is one-half the circumference and the height is the radius.

$A = \left(\frac{1}{2} \cdot 2\pi r\right)r$ Replace C with $2\pi r$.

$A = \pi \cdot r \cdot r$ or πr^2 Simplify.

Key Concept   Area of a Circle

Words The area A of a circle is equal to π times the square of the radius r.

Model

Symbols $A = \pi r^2$

EXAMPLES Find the Areas of Circles

Find the area of each circle.

3
8 km

$A = \pi r^2$ Area of a circle

$A = \pi \cdot 8^2$ Replace r with 8.

$A = \pi \cdot 64$ Evaluate 8^2.

$A \approx 201.1$ Use a calculator.

The area is about 201.1 square kilometers.

4
15 ft

$A = \pi r^2$ Area of a circle

$A = \pi(7.5)^2$ Replace r with half of 15 or 7.5.

$A = \pi \cdot 56.25$ Evaluate 7.5^2.

$A \approx 176.7$ Use a calculator.

The area is about 176.7 square feet.

Your Turn Find the circumference and area of each circle. Round to the nearest tenth.

a.
11 cm

b.
5 mi

c.
$2\frac{3}{4}$ in.

EXAMPLE Use Circumference and Area

5 **TREES** During a construction project, barriers are placed around trees. For each inch of trunk diameter, the protected zone should have a radius of $1\frac{1}{2}$ feet. Find the area of this zone for a tree with a trunk circumference of 63 inches.

d in. $1\frac{1}{2}d$ ft

First find the diameter of the tree.

$C = \pi d$ Circumference of a circle

$63 = \pi \cdot d$ Replace C with 63.

$\dfrac{63}{\pi} = d$ Divide each side by π.

$20.1 \approx d$ Use a calculator.

The diameter d of the tree is about 20.1 inches. The radius r of the protected zone should be $1\frac{1}{2}d$ feet. That is, $r = 1\frac{1}{2}(20.1)$ or 30.15 feet. Use this radius to find the area of the protected zone.

$A = \pi r^2$ Area of a circle

$A = \pi(30.15)^2$ or about 2,855.8   Replace r with 30.15 and use a calculator.

The area of the protected zone is about 2,855.8 square feet.

Skill and Concept Check

Writing Math

Exercise 2

1. **OPEN ENDED** Draw and label a circle that has a circumference between 10 and 20 centimeters.

2. **NUMBER SENSE** If the radius of a circle is doubled, how will this affect its circumference? its area? Explain your reasoning.

GUIDED PRACTICE

Find the circumference and area of each circle. Round to the nearest tenth.

3.
12 yd

4.
18 cm

5.
21 ft

6.
14.5 m

7. The diameter is 5.3 miles.

8. The radius is $4\frac{3}{4}$ inches.

Practice and Applications

Find the circumference and area of each circle. Round to the nearest tenth.

HOMEWORK HELP

For Exercises	See Examples
9–18	1–4
19–24	5

Extra Practice
See pages 632, 654.

9.
10 in.

10.
24 mm

11.
38 mi

12.
17 km

13.
19.4 m

14.
$7\frac{1}{4}$ ft

15. The radius is 3.5 centimeters.

16. The diameter is 8.6 kilometers.

17. The diameter is $10\frac{3}{8}$ feet.

18. The radius is $6\frac{2}{5}$ inches.

19. **CARS** If the tires on a car each have a diameter of 25 inches, how far will the car travel in 100 rotations of its tires?

20. **SPORTS** Three tennis balls are packaged one on top of the other in a can. Which measure is greater, the can's height or circumference? Explain.

21. **ANIMALS** A California ground squirrel usually stays within 150 yards of its burrow. Find the area of a California ground squirrel's world.

22. **LAWN CARE** The pattern of water distribution from a sprinkler is commonly a circle or part of a circle. A certain sprinkler is set to cover part of a circle measuring 270°. Find the area of the grass watered if the sprinkler reaches a distance of 15 feet.

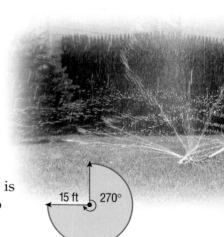
15 ft   270°

PIZZA For Exercises 23 and 24, use the diagram at the right.

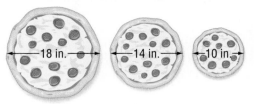
HOLIDAY PIZZERIA

23. Find the area of each size pizza.

24. **MULTI STEP** The pizzeria has a special that offers one large, two medium, or three small pizzas for $12. Which offer is the best buy? Explain your reasoning.

Large   Medium   Small

ALGEBRA For Exercises 25 and 26, round to the nearest tenth.

25. What is the diameter of a circle if its circumference is 41.8 feet?

26. Find the radius of a circle if its area is 706.9 square millimeters.

CRITICAL THINKING Find the area of each shaded region.

27.

28.

29.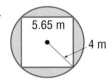

30.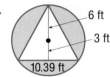

EXTENDING THE LESSON A *central angle* is an angle that intersects a circle in two points and has its vertex at the center of the circle. A *chord* is a line segment joining two points on a circle.

central angle *ABC*

31. Draw and label a circle with a central angle *JKL* measuring 120°.

32. *True* or *False*? One side of a central angle can be a chord of the circle. Explain your reasoning.

chord $\overline{DE}$

Standardized Test Practice and Mixed Review

FCAT Practice

33. **MULTIPLE CHOICE** One lap around the outside of a circular track is 352 yards. If you jog from one side of the track to the other through the center, about how far do you travel?

Ⓐ 11 yd   Ⓑ 56 yd   Ⓒ 112 yd   Ⓓ 176 yd

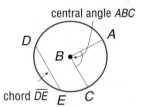

34. **SHORT RESPONSE** The circumference of a circle is 16.5 feet. What is its area to the nearest tenth of a square foot?

Find the area of each figure described. (Lesson 7-1)

35. triangle: base, 4 cm
       height, 8.7 cm

36. trapezoid: height, 4 in.
       bases, 2.5 in. and 5 in.

37. Graph △*WXY* with vertices *W*(1, −3), *X*(4, −2), and *Y*(4, −5). Then graph its image after a rotation of 90° counterclockwise about the origin and write the coordinates of its vertices. (Lesson 6-9)

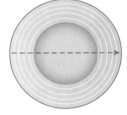

GETTING READY FOR THE NEXT LESSON

BASIC SKILL Add.

38. 450 + 210.5   39. 16.4 + 8.7   40. 25.9 + 134.8   41. 213.25 + 86.9

Solve a Simpler Problem

What You'll Learn
Solve problems by solving a simpler problem.

Mr. Lewis wants to know the largest number of pieces of pizza that can be made by using 8 straight cuts. That's a hard problem!

Well, maybe we can make it easier by **solving a simpler problem**, or maybe even a few simpler problems.

Explore	Mr. Lewis said that a "cut" does not have to be along a diameter, but it must be from edge to edge. Also, the pieces do not have to be the same size.
Plan	Let's draw diagrams to find the largest number of pieces formed by 1, 2, 3, and 4 cuts and then look for a pattern.
Solve	

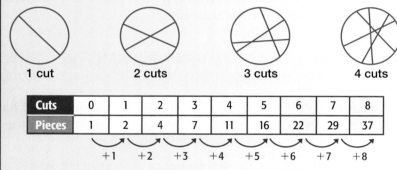

1 cut 2 cuts 3 cuts 4 cuts

Cuts	0	1	2	3	4	5	6	7	8
Pieces	1	2	4	7	11	16	22	29	37

+1 +2 +3 +4 +5 +6 +7 +8

So the largest number of pieces formed by 8 cuts is 37.

Examine	Two cuts formed 2 · 2 or 4 pieces, and 4 cuts formed about 3 · 4 or 12 pieces. It is reasonable to assume that 6 cuts would form about 4 · 6 or 24 pieces and 8 cuts about 5 · 8 or 40 pieces. Our answer is reasonable.

Analyze the Strategy

1. **Explain** why it was helpful for Kimi and Paige to solve a simpler problem to answer Mr. Lewis' question.

2. **Explain** how you could use the solve a simpler problem strategy to find the thickness of one page in this book.

3. **Write** about a situation in which you might need to solve a simpler problem in order to find the solution to a more complicated problem. Then solve the problem.

Solve. Use the solve a simpler problem strategy.

4. **GEOMETRY** How many squares of any size are in the figure at the right?

5. **TABLES** A restaurant has 25 square tables that can be pushed together to form one long table for a banquet. Each square table can seat only one person on each side. How many people can be seated at the banquet table?

Mixed Problem Solving

Solve. Use any strategy.

6. **PARTY SUPPLIES** Paper cups come in packages of 40 or 75. Monica needs 350 paper cups for the school party. How many packages of each size should she buy?

7. **SOFT DRINKS** The graph below represents a survey of 400 students. Determine the difference in the number of students who preferred cola to lemon-lime soda.

Soft Drink Preferences

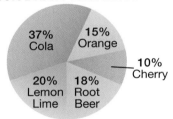

8. **GIFT WRAPPING** During the holidays, Tyler and Abigail earn extra money by wrapping gifts at a department store. Tyler wraps 8 packages an hour while Abigail wraps 10 packages an hour. Working together, about how long will it take them to wrap 40 packages?

READING For Exercises 9 and 10, use the following information.
Carter Middle School has 487 fiction books and 675 nonfiction books. Of the nonfiction books, 84 are biographies.

9. Draw a Venn diagram of this situation.

10. How many books are *not* biographies?

11. **MONEY** Mario has $12 to spend at the movies. After he pays the $6.50 admission, he estimates that he can buy a tub of popcorn that costs $4.25 and a medium drink that is $2.50. Is this reasonable? Explain.

12. **HEALTH** A human heart beats an average of 72 times in one minute. Estimate the number of times a human heart beats in one year.

13. **TRAVEL** When Mrs. Lopez started her trip from Jacksonville, Florida, to Atlanta, Georgia, her odometer read 35,400 miles. When she reached Atlanta, her odometer read 35,742 miles. If the trip took $5\frac{1}{2}$ hours, what was her average speed?

14. **NUMBER SENSE** Find the sum of all the whole numbers from 1 to 40, inclusive.

15. **STANDARDIZED TEST PRACTICE**
Three different views of a cube are shown. If the fish is currently faceup, what figure is facedown?

　(A) heart   (B) lightning bolt
　(C) question mark   (D) tree

You will use the solve a simpler problem strategy in the next lesson.

Area of Complex Figures

WHEN am I ever going to use this?

CARPETING When carpeting, you must calculate the amount of carpet needed for the floor space you wish to cover. Sometimes the space is made up of several shapes.

1. Identify some of the polygons that make up the family room, nook, and foyer area shown in this floor plan.

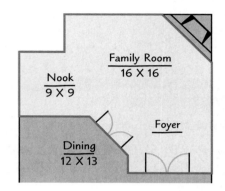

We have discussed the following area formulas.

Parallelogram	Triangle	Trapezoid	Circle
$A = bh$	$A = \frac{1}{2}bh$	$A = \frac{1}{2}h(b_1 + b_2)$	$A = \pi r^2$

You can use these formulas to help you find the area of complex figures. A **complex figure** is made up of two or more shapes.

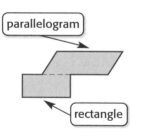

parallelogram

rectangle

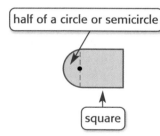

half of a circle or semicircle

square

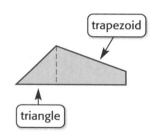

trapezoid

triangle

To find the area of a complex figure, separate the figure into shapes whose areas you know how to find. Then find the sum of these areas.

EXAMPLE Find the Area of a Complex Figure

1. **Find the area of the complex figure.**

   The figure can be separated into a rectangle and a triangle.

Area of rectangle	**Area of triangle**
$A = \ell w$	$A = \frac{1}{2}bh$
$A = 15 \cdot 12$	$A = \frac{1}{2} \cdot 15 \cdot 4$
$A = 180$	$A = 30$

   The area of the figure is $180 + 30$ or 210 square feet.

EXAMPLE Find the Area of a Complex Figure

2 Find the area of the complex figure.

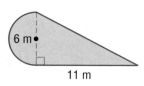
6 m•
11 m

The figure can be separated into a semicircle and a triangle.

Area of semicircle	**Area of triangle**

$A = \frac{1}{2}\pi r^2$ $A = \frac{1}{2}bh$

$A = \frac{1}{2} \cdot \pi \cdot 3^2$ $A = \frac{1}{2} \cdot 6 \cdot 11$

$A \approx 14.1$ $A = 33$

The area of the figure is about $14.1 + 33$ or 47.1 square meters.

Your Turn Find the area of each figure. Round to the nearest tenth if necessary.

a.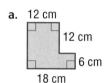
12 cm
12 cm
6 cm
18 cm

b.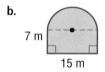
7 m
15 m

c.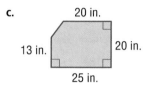
20 in.
13 in.   20 in.
25 in.

EXAMPLE Use the Area of a Complex Figure

3 **SHORT-RESPONSE TEST ITEM** The plans for one hole of a miniature golf course are shown. How many square feet of turf will be needed to cover the putting green if one square represents 1.5 square foot?

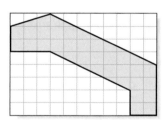

Read the Test Item You need to find the area of the putting green in square units and then multiply this result by 1.5 to find the area of the green in square feet.

Solve the Test Item Find the area of the green by dividing it into smaller areas.

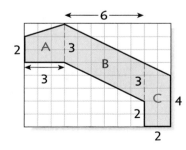
6
2 A 3
B
3   3
C 4
2
2

Region A Trapezoid

$A = \frac{1}{2}h(b_1 + b_2)$

$A = \frac{1}{2}(3)(2 + 3)$ or 7.5

Region B Parallelogram

$A = bh$

$A = 6 \cdot 3$ or 18

Region C Trapezoid

$A = \frac{1}{2}h(b_1 + b_2)$

$A = \frac{1}{2}(2)(4 + 5)$ or 9

The total area is $7.5 + 18 + 9$ or 34.5 square units. So, $1.5(34.5)$ or 51.75 square feet of turf is needed to cover the green.

1. **OPEN ENDED** Draw an example of a complex figure that can be separated into at least two different shapes whose area you know how to find. Then show how you would separate this figure to find its area.

2. **Explain** at least two different ways of finding the area of the figure at the right.

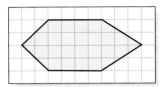

GUIDED PRACTICE

Find the area of each figure. Round to the nearest tenth if necessary.

3.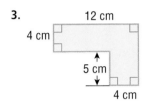

4.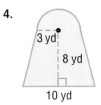

5.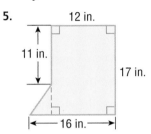

Practice and Applications

Find the area of each figure. Round to the nearest tenth if necessary.

HOMEWORK HELP

For Exercises	See Examples
6–13	1, 2
14–19	3

Extra Practice
See pages 633, 654.

6.

7.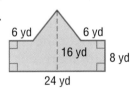

8.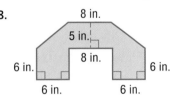

9.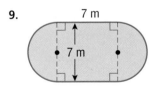

10.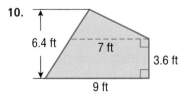

11.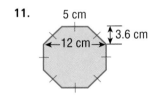

12. What is the area of a figure that is formed using a square with sides 15 yards and a triangle with a base of 8 yards and a height of 12 yards?

13. What is the area of a figure that is formed using a trapezoid with one base of 9 meters, one base of 15 meters, and a height of 6 meters and a semicircle with a diameter of 9 meters?

FLAGS For Exercises 14–16, use the diagram of Ohio's state flag at the right.

14. Find the area of the flag. Describe your method.

15. Find the area of the triangular region of the flag.

16. What percent of the total area of the flag is the triangular region?

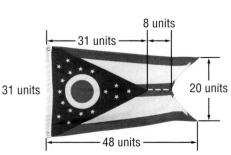

HOME IMPROVEMENT For Exercises 17 and 18, use the diagram of one side of a house and the following information.

Suppose you are painting one side of your house. One gallon of paint covers 350 square feet and costs $21.95.

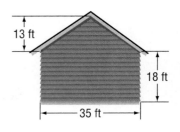

17. If you are only planning to apply one coat of paint, how many cans should you buy? Explain your reasoning.

18. Find the total cost of the paint, not including tax.

19. **MULTI STEP** A school's field, shown at the right, must be mowed before 10:00 A.M. on Monday. The maintenance crew says they can mow at a rate of 1,750 square feet of grass per minute. If the crew begins mowing at 9:30 that morning, will the field be mowed in time? Explain your reasoning.

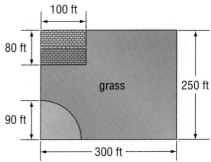

20. **CRITICAL THINKING** In the diagram at the right, a 3-foot wide wooden walkway surrounds a garden. What is the area of the walkway?

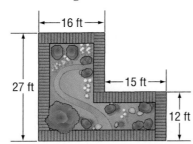

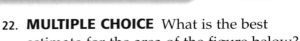

21. **MULTIPLE CHOICE** What is the area of the figure below?

   Ⓐ 17.5 m²   Ⓑ 25.5 m²

   Ⓒ 437.5 m²   Ⓓ 637.5 m²

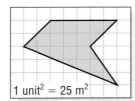

1 unit² = 25 m²

22. **MULTIPLE CHOICE** What is the best estimate for the area of the figure below?

   Ⓕ 36 units²   Ⓖ 48 units²

   Ⓗ 54 units²   Ⓘ 56 units²

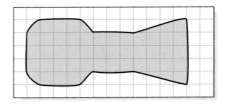

23. **MONUMENTS** Stonehenge is a circular array of giant stones in England. The diameter of Stonehenge is 30.5 meters. Find the approximate distance around Stonehenge. (Lesson 7-2)

Find the area of each figure. (Lesson 7-1)

24. triangle: base, 4 mm
             height, 3.5 mm

25. trapezoid: height, 11 ft
               bases, 17 ft and 23 ft

GETTING READY FOR THE NEXT LESSON

BASIC SKILL Classify each polygon according to its number of sides.

26.

27.

28.

29.

What You'll Learn
Build and draw three-dimensional figures.

Materials
• cubes
• isometric dot paper

∞ **LINK** To Reading

Everyday Meaning of Perspective: the ability to view things in their true relationship or importance to one another.

Sunshine State Standards
MA.C.1.3.1-3

Building Three-Dimensional Figures

Different views of a stack of cubes are shown in the activity below. A point of view is called a **perspective**. You can build or draw three-dimensional figures using different perspectives.

ACTIVITY *Work with a partner.*

The top, side, and front views of a three-dimensional figure are shown. Use cubes to build the figure. Then, draw your model on isometric dot paper.

top   side   front

Build Base Using Top View

Complete Figure Using Side View

Check Figure Using Front View

The base is a 2 by 3 rectangle.

The 1st and 2nd rows are 1 unit high.

The 3rd row is 2 units high.

The overall width is 2 units.

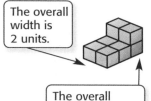

The overall height is 2 units.

Now draw your model on isometric dot paper as shown at the right. Label the front and the side of your figure.

front   side

Your Turn The top, side, and front views of three-dimensional figures are shown. Use cubes to build each figure. Then draw your model on isometric dot paper, labeling its front and side.

a.   top   side   front   b.   top   side   front

Writing Math

1. **Determine** which view, *top*, *side*, or *front*, would show that a building has multiple heights.

2. **Build** your own figure using up to 20 cubes and draw it on isometric dot paper. Then **draw** the figure's top, side, and front views. **Explain** your reasoning.

Three-Dimensional Figures

Sunshine State Standards
MA.C.1.3.1-2, MA.C.1.3.1-3, MA.C.1.3.1-4

Amethyst

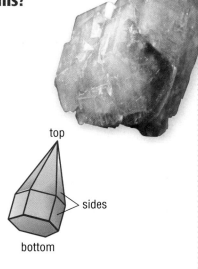

What You'll Learn
Identify and draw three-dimensional figures.

NEW Vocabulary

plane
solid
polyhedron
edge
face
vertex
prism
base
pyramid

WHEN am I ever going to use this?

CRYSTALS A two-dimensional figure has two dimensions, length and width. A three-dimensional figure, like the Amethyst crystal shown at the right, has three dimensions, length, width, and depth (or height).

1. Name the two-dimensional shapes that make up the sides of this crystal.

2. If you observed the crystal from directly above, what two-dimensional figure would you see?

3. How are two- and three-dimensional figures related?

A **plane** is a two-dimensional flat surface that extends in all directions. There are different ways that planes may be related in space.

Intersect in a Line	Intersect at a Point	No Intersection

These are called *parallel planes*.

Intersecting planes can also form three-dimensional figures or **solids**. A **polyhedron** is a solid with flat surfaces that are polygons.

An **edge** is where two planes intersect in a line.

A **face** is a flat surface.

A **vertex** is where three or more planes intersect at a point.

A **prism** is a polyhedron with two parallel, congruent faces called **bases**. A **pyramid** is a polyhedron with one base that is a polygon and faces that are triangles.

Prisms and pyramids are named by the shape of their bases.

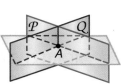
prism

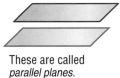
pyramid

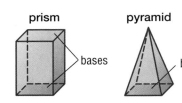

Common Polyhedrons

triangular prism   rectangular prism   triangular pyramid   rectangular pyramid

EXAMPLES **Identify Prisms and Pyramids**

Identify each solid. Name the number and shapes of the faces. Then name the number of edges and vertices.

1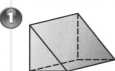

The figure has two parallel congruent bases that are triangles, so it is a triangular prism. The other three faces are rectangles. It has a total of 5 faces, 9 edges, and 6 vertices.

2

The figure has one base that is a pentagon, so it is a pentagonal pyramid. The other faces are triangles. It has a total of 6 faces, 10 edges, and 6 vertices.

Your Turn Identify each solid. Name the number and shapes of the faces. Then name the number of edges and vertices.

a.    b.   c.

EXAMPLES **Analyze Real-Life Drawings**

3 **ARCHITECTURE** An artist's drawing shows the plans for a new office building. Each unit on the drawing represents 50 feet. Draw and label the top, front, and side views.

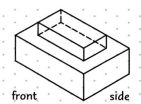

front   side

top view   front view   side view

4 **ARCHITECTURE** Find the area of the top floor.

You can see from the front and side views that the top floor is a rectangle that is 2 units wide by 4 units long. The actual dimensions are 4(50) feet by 2(50) feet or 200 feet by 100 feet.

$A = 200 \cdot 100$ $A = \ell \cdot w$

$A = 20{,}000$ Simplify.

The area of the top floor is 20,000 square feet.

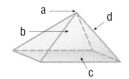

1. **Identify** the indicated parts of the polyhedron at the right.

2. **OPEN ENDED** Give a real-life example of three intersecting planes and describe their intersection.

GUIDED PRACTICE

Identify each solid. Name the number and shapes of the faces. Then name the number of edges and vertices.

3.

4.

5.

6. **PETS** Your pet lizard lives in an aquarium with a hexagonal base and a height of 5 units. Draw the aquarium using isometric dot paper.

Practice and Applications

Identify each solid. Name the number and shapes of the faces. Then name the number of edges and vertices.

7.

8.

9.

10.

HOMEWORK HELP

For Exercises	See Examples
7–10	1, 2
11–12, 16–18	3

Extra Practice
See pages 633, 654.

ARCHITECTURE For Exercises 11 and 12, complete parts a–c for each architectural drawing.

a. Draw and label the top, front, and side views.

b. Find the overall height of the solid in feet.

c. Find the area of the shaded region.

11. Sculpture Pedestal

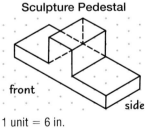

front side

1 unit = 6 in.

12. Porch Steps

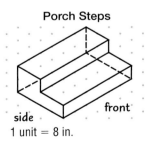

side front

1 unit = 8 in.

Determine whether each statement is *sometimes*, *always*, or *never* true. Explain your reasoning.

13. Three planes do not intersect in a point.

14. A prism has two congruent bases.

15. A pyramid has five vertices.

CRYSTALS For Exercises 16–18, complete parts a and b for each crystal.

a. **Identify the solid or solids that form the crystal.**

b. **Draw and label the top and one side view of the crystal.**

16.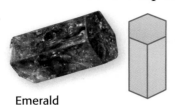

Emerald

17.

Fluorite

18.

Quartz

19. **CRITICAL THINKING** A pyramid with a triangular base has 6 edges and a pyramid with a rectangular base has 8 edges. Write a formula that gives the number of edges *E* for a pyramid with an *n*-sided base.

EXTENDING THE LESSON *Skew lines* do not intersect, but are also not parallel. They lie in different planes. In the figure at the right, the lines containing $\overline{AD}$ and $\overline{CG}$ are skew. $\overline{BH}$ is a *diagonal* of this prism because it joins two vertices that have no faces in common.

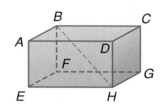

For Exercises 20–22, use the rectangular prism above.

20. Identify three other diagonals that could have been drawn.

21. Name two other segments that are skew to $\overline{BH}$.

22. State whether $\overline{DH}$ and $\overline{CG}$ are *parallel*, *skew*, or *intersecting*.

Standardized Test Practice and Mixed Review

FCAT Practice

For Exercises 23 and 24, use the figure at the right.

23. **SHORT RESPONSE** Identify the two polyhedrons that make up the figure.

24. **MULTIPLE CHOICE** Identify the shaded part of the figure.

   Ⓐ edge   Ⓑ face   Ⓒ vertex   Ⓓ prism

Find the area of each figure. Round to the nearest tenth. (Lesson 7-3)

25.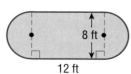

26.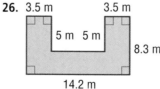

27.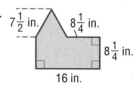

28. **MANUFACTURING** The label that goes around a jar of peanut butter overlaps itself by $\frac{3}{8}$ inch. If the diameter of the jar is 2 inches, what is the length of the label? (Lesson 7-2)

GETTING READY FOR THE NEXT LESSON

PREREQUISITE SKILL Find the area of each triangle described. (Lesson 7-1)

29. base, 3 in.; height, 10 in.   30. base, 8 ft; height, 7 feet   31. base, 5 cm; height, 11 cm

7-5 Volume of Prisms and Cylinders

Sunshine State Standards
MA.B.1.3.1-1, MA.B.1.3.1-2, MA.B.1.3.3-1, MA.B.1.3.3-2, MA.C.1.3.1-4

What You'll Learn
Find the volumes of prisms and cylinders.

NEW **Vocabulary**

volume
cylinder
complex solid

HANDS-ON Mini Lab

Materials
• 12 cubes

The rectangular prism at the right has a volume of 12 cubic units.

STEP 1 Model three other rectangular prisms with a volume of 12 cubic units.

STEP 2 Copy and complete the following table.

Prism	Length (units)	Width (units)	Height (units)	Area of Base (units²)
A	4	1	3	4
B				
C				
D				

1. Describe how the volume V of each prism is related to its length ℓ, width w, and height h.

2. Describe how the area of the base B and the height h of each prism is related to its volume V.

Volume is the measure of the space occupied by a solid. Standard measures of volume are cubic units such as cubic inches (in^3) or cubic feet (ft^3).

Key Concept — Volume of a Prism

Words The volume V of a prism is the area of the base B times the height h.

Models

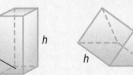

Symbols $V = Bh$

EXAMPLE — Find the Volume of a Rectangular Prism

1 **Find the volume of the prism.**

$V = Bh$ Volume of a prism

$V = (\ell \cdot w)h$ The base is a rectangle, so $B = \ell \cdot w$.

$V = (9 \cdot 5)6.5$ $\ell = 9, w = 5, h = 6.5$

$V = 292.5$ Simplify.

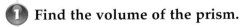

The volume is 292.5 cubic centimeters.

EXAMPLE Find the Volume of a Triangular Prism

2 **Find the volume of the prism.**

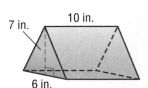

7 in.   10 in.

6 in.

$V = Bh$ Volume of a prism

$V = \left(\frac{1}{2} \cdot 6 \cdot 7\right)h$ The base is a triangle, so $B = \frac{1}{2} \cdot 6 \cdot 7$.

$V = \left(\frac{1}{2} \cdot 6 \cdot 7\right)10$ The height of the prism is 10.

$V = 210$ Simplify.

The volume is 210 cubic inches.

A **cylinder** is a solid whose bases are congruent, parallel circles, connected with a curved side. You can use the formula $V = Bh$ to find the volume of a cylinder, where the base is a circle.

Key Concept   **Volume of a Cylinder**

Words   The volume V of a cylinder with radius r is the area of the base B times the height h.

Model

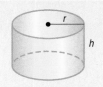

Symbols $V = Bh$ or $V = \pi r^2 h$, where $B = \pi r^2$

EXAMPLES Find the Volumes of Cylinders

Find the volume of each cylinder.

3

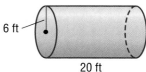
6 ft

20 ft

$V = \pi r^2 h$ Volume of a cylinder

$V = \pi \cdot 6^2 \cdot 20$ Replace r with 6 and h with 20.

$V \approx 2{,}261.9$ Simplify.

The volume is about 2,261.9 cubic feet.

4 **diameter of base, 13 m; height, 15.2 m**

Since the diameter is 13 meters, the radius is 6.5 meters.

$V = \pi r^2 h$ Volume of a cylinder

$V = \pi \cdot 6.5^2 \cdot 15.2$ Replace r with 6.5 and h with 15.2.

$V \approx 2{,}017.5$ Simplify.

The volume is about 2,017.5 cubic meters.

Your Turn **Find the volume of each prism. Round to the nearest tenth if necessary.**

a.

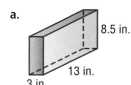
8.5 in.
13 in.
3 in.

b.

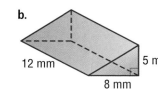
12 mm
8 mm
5 mm

c.

2 in.
7 in.

Many objects in real-life are made up of more than one type of solid. Such figures are called **complex solids**. To find the volume of a complex solid, separate the figure into solids whose volumes you know how to find.

EXAMPLE Find the Volume of a Complex Solid

5 **DISPENSERS** Find the volume of the soap dispenser at the right.

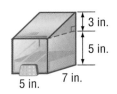

The dispenser is made of one rectangular prism and one triangular prism. Find the volume of each prism.

Rectangular Prism

5 in.   7 in.

$V = Bh$

$V = (5 \cdot 7)5$ or 175

Triangular Prism

5 in.   7 in.

$V = Bh$

$V = \left(\frac{1}{2} \cdot 7 \cdot 3\right)5$ or 52.5

The volume of the dispenser is $175 + 52.5$ or 227.5 cubic inches.

Skill and Concept Check

Writing Math

Exercises 1–3

1. **Write** another formula for the volume of a rectangular prism and explain how it is related to the formula $V = Bh$.

2. **FIND THE ERROR** Erin and Dulce are finding the volume of the prism shown at the right. Who is correct? Explain.

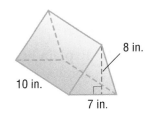

8 in.

10 in.

7 in.

Erin
$A = Bh$
$A = (10 \cdot 7) \cdot 8$
$A = 560 \text{ in}^3$

Dulce
$A = Bh$
$A = \left(\frac{1}{2} \cdot 7 \cdot 8\right) \cdot 10$
$A = 280 \text{ in}^3$

3. **OPEN ENDED** Find the volume of a can or other cylindrical object, being sure to include appropriate units. Explain your method.

GUIDED PRACTICE

Find the volume of each solid. Round to the nearest tenth if necessary.

4.

6 ft
2 ft   3 ft

5.
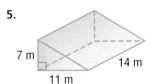
7 m   14 m
11 m

6.
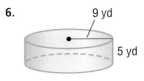
9 yd
5 yd

7.
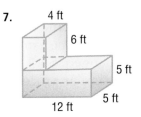
4 ft
6 ft
5 ft
5 ft
12 ft

Practice and Applications

Find the volume of each solid. Round to the nearest tenth if necessary.

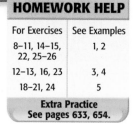

For Exercises	See Examples
8–11, 14–15, 22, 25–26	1, 2
12–13, 16, 23	3, 4
18–21, 24	5

HOMEWORK HELP

Extra Practice
See pages 633, 654.

8. 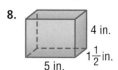 4 in. 5 in. $1\frac{1}{2}$ in.

9. 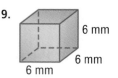 6 mm 6 mm 6 mm

10. 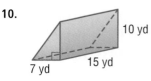 10 yd 15 yd 7 yd

11. 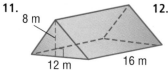 8 m 12 m 16 m

12. 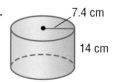 7.4 cm 14 cm

13. 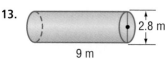 2.8 m 9 m

14. rectangular prism: length, 4 in.; width, 6 in.; height, 17 in.

15. triangular prism: base of triangle, 5 ft; altitude, 14 ft; height of prism, $8\frac{1}{2}$ ft

16. cylinder: diameter, 7.2 cm; height, 5.8 cm

17. hexagonal prism: base area 48 mm²; height, 12 mm

18. 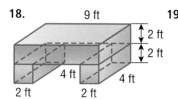 9 ft 2 ft 2 ft 4 ft 4 ft 2 ft 2 ft

19. 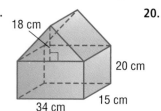 18 cm 20 cm 34 cm 15 cm

20. 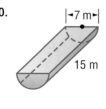 |◄7 m►| 15 m

21. 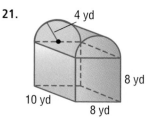 4 yd 8 yd 10 yd 8 yd

22. ALGEBRA Find the height of a rectangular prism with a length of 6.8 meters, a width of 1.5 meters, and a volume of 91.8 cubic meters.

23. ALGEBRA Find the height of a cylinder with a radius of 4 inches and a volume of 301.6 cubic inches.

24. Explain how you would find the volume of the hexagonal prism shown at the right. Then find its volume.

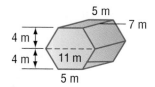
5 m 7 m 4 m 4 m 11 m 5 m

POOLS For Exercises 25 and 26, use the following information.
A wading pool is to be 20 feet long, 11 feet wide, and 1.5 feet deep.

25. Approximately how much water will the pool hold?

26. The excavated dirt is to be hauled away by wheelbarrow. If the wheelbarrow holds 9 cubic feet of dirt, how many wheelbarrows of dirt must be hauled away from the site?

CONVERTING UNITS OF MEASURE For Exercises 27–29, use the cubes at the right.
The volume of the left cube is 1 cubic yard. The right cube is the same size, but the unit of measure has been changed. So, 1 cubic yard = (3)(3)(3) or 27 cubic feet. Use a similar process to convert each measurement.

1 yd 1 yd 1 yd   3 ft 3 ft 3 ft

27. 1 ft³ = ▦ in³

28. 1 cm³ = ▦ mm³

29. 1 m³ = ▦ cm³

30. PACKAGING The Cooking Club is selling their own special blends of rice mixes. They can choose from the two containers at the right to package their product. Which container will hold more rice? Explain your reasoning.

A
9 cm

8 cm

B

16 cm

10 cm   3 cm

31. FARMING When filled to capacity, a silo can hold 8,042 cubic feet of grain. The circumference C of the silo is approximately 50.3 feet. Find the height h of the silo to the nearest foot.

32. WRITE A PROBLEM Write about a real-life problem that can be solved by finding the volume of a rectangular prism or a cylinder. Explain how you solved the problem.

h

C

CRITICAL THINKING For Exercises 33–36, describe how the volume of each solid is affected after the indicated change in its dimension(s).

33. You double one dimension of a rectangular prism.

34. You double two dimensions of a rectangular prism.

35. You double all three dimensions of a rectangular prism.

36. You double the radius of a cylinder.

Standardized Test Practice and Mixed Review

A B C D

FCAT Practice

37. MULTIPLE CHOICE A bar of soap in the shape of a rectangular prism has a volume of 16 cubic inches. After several uses, it measures $2\frac{1}{4}$ inches by 2 inches by $1\frac{1}{2}$ inches. How much soap was used?

  A $6\frac{3}{4}$ in³   **B** $9\frac{1}{4}$ in³   **C** $10\frac{1}{4}$ in³   **D** 108 in³

38. MULTIPLE CHOICE Which is the best estimate of the volume of a cylinder that is 20 meters tall and whose diameter is 10 meters?

  F 200 m³   **G** 500 m³   **H** 600 m³   **I** 1500 m³

39. PAINTING You are painting a wall of this room red. Find the area of the red wall to the nearest square foot. (Lesson 7-3)

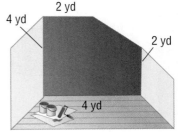

2 yd

4 yd

2 yd

4 yd

40. How many edges does an octagonal pyramid have? (Lesson 7-4)

Write each percent as a fraction or mixed number in simplest form. (Lesson 5-4)

41. 0.12%   **42.** 225%   **43.** 135%   **44.** $\frac{3}{8}$%

GETTING READY FOR THE NEXT LESSON

PREREQUISITE SKILL Multiply. (Lesson 2-5)

45. $\frac{1}{3} \cdot 6 \cdot 10$ **46.** $\frac{1}{3} \cdot 7 \cdot 15$ **47.** $\frac{1}{3} \cdot 4^2 \cdot 9$ **48.** $\frac{1}{3} \cdot 6^2 \cdot 20$

Vocabulary and Concepts

1. Draw and label a trapezoid with an area of 20 square inches. (Lesson 7-1)

2. **Compare and contrast** the characteristics of prisms and pyramids. (Lesson 7-4)

Skills and Applications

3. Find the area of a triangle with a 30-meter base and 12-meter height. (Lesson 7-1)

4. **SPORTS** A shot-putter must stay inside a circle with a radius of 7 feet. What is the circumference and area of the region in which the athlete is able to move in this competition? Round to the nearest tenth. (Lesson 7-2)

Find the area of each figure. Round to the nearest tenth. (Lesson 7-3)

5.
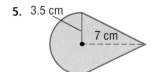
3.5 cm
7 cm

6.
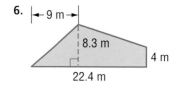
9 m
8.3 m
4 m
22.4 m

STORAGE For Exercises 7 and 8, use the diagram of the storage shed at the right.

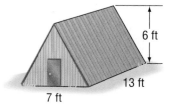
6 ft
13 ft
7 ft

7. Identify the solid. Name the number and shapes of the faces. Then name the number of edges and vertices. (Lesson 7-4)

8. Find the volume of this storage shed. (Lesson 7-5)

Find the volume of each solid. Round to the nearest tenth. (Lesson 7-5)

9.
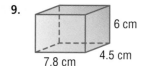
6 cm
7.8 cm
4.5 cm

10.
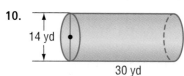
14 yd
30 yd

Standardized Test Practice

FCAT Practice
Ⓐ Ⓑ Ⓒ Ⓓ

11. **MULTIPLE CHOICE** Which of the following solids is *not* a polyhedron? (Lesson 7-4)

   Ⓐ prism   Ⓑ cylinder

   Ⓒ pyramid   Ⓓ cube

12. **MULTIPLE CHOICE** Find the volume of a cube-shaped box with edges 15 inches long. (Lesson 7-5)

   Ⓕ 225 in³   Ⓖ 900 in³

   Ⓗ 1,350 in³   Ⓘ 3,375 in³

The Game Zone

A Place To Practice Your Math Skills

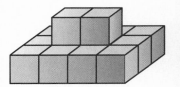

Math Skill
3-Dimensional Figures

Archi-test

● GET READY!

Players: two
Materials: cubes, manila folders, index cards cut in half

● GET SET!

- Players each receive 15 cubes and a manila folder.

- Each player designs a structure with some of his or her cubes, using the manila folder to hide the structure from the other player's view. The player then draws the top, front, back, and side views of the structure on separate index cards. The player also computes the structure's volume in cubic units, writing this on a fourth index card.

● GO!

- Player A tries to guess Player B's structure. Player A does this by asking Player B for one of the index cards that shows one of the views of the structure. Player A tries to build Player B's structure.

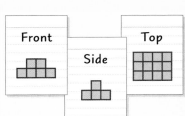

- Player A receives 4 points for correctly building Players B's structure after receiving only one piece of information, 3 points for correctly building after only two pieces of information, and so on.

- If Player A cannot build Player B's structure after receiving all 4 pieces of information, then Player B receives 2 points.

- Player B now tries to build Player A's structure.

- **Who Wins?** Play continues for an agreed-upon number of structures. The player with the most points at the end of the game wins.

Volume of Pyramids and Cones

Sunshine State Standards
MA.B.1.3.1-1, MA.B.1.3.1-2, MA.B.1.3.3-1, MA.B.1.3.3-2, MA.C.1.3.1-4

What You'll Learn

Find the volumes of pyramids and cones.

NEW Vocabulary

cone

REVIEW Vocabulary

pyramid: a polyhedron with one base that is a polygon and faces that are triangles (Lesson 7-4)

HANDS-ON Mini Lab

Materials
- construction paper
- ruler
- scissors
- tape
- rice

Work with a partner.

In this Mini Lab, you will investigate the relationship between the volume of a pyramid and the volume of a prism with the same base area and height.

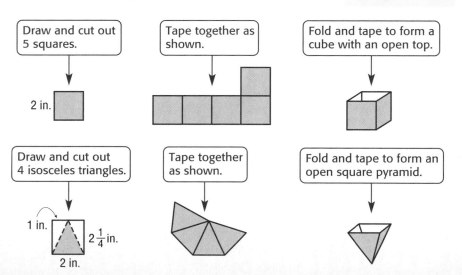

Draw and cut out 5 squares.

2 in.

Tape together as shown.

Fold and tape to form a cube with an open top.

Draw and cut out 4 isosceles triangles.

1 in. $2\frac{1}{4}$ in.   2 in.

Tape together as shown.

Fold and tape to form an open square pyramid.

1. Compare the base areas and the heights of the two solids.

2. Fill the pyramid with rice, sliding a ruler across the top to level the amount. Pour the rice into the cube. Repeat until the prism is filled. How many times did you fill the pyramid in order to fill the cube?

3. What fraction of the cube's volume does one pyramid fill?

The volume of a pyramid is one third the volume of a prism with the same base area and height.

Key Concept — Volume of a Pyramid

Words The volume V of a pyramid is one third the area of the base B times the height h.

Symbols $V = \frac{1}{3}Bh$

Model

h

B

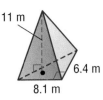

EXAMPLE Find the Volume of a Pyramid

1 Find the volume of the pyramid.

$V = \frac{1}{3}Bh$ Volume of a pyramid

$V = \frac{1}{3}\left(\frac{1}{2} \cdot 8.1 \cdot 6.4\right)11$ $B = \frac{1}{2} \cdot 8.1 \cdot 6.4, h = 11$

$V = 95.04$ Simplify.

The volume is about 95.0 cubic meters.

11 m

6.4 m

8.1 m

EXAMPLE Use Volume to Solve a Problem

2 **ARCHITECTURE** The area of the base of the Pyramid Arena in Memphis, Tennessee, is 360,000 square feet. If its volume is 38,520,000 cubic feet, find the height of the structure.

$V = \frac{1}{3}Bh$ Volume of a pyramid

$38{,}520{,}000 = \frac{1}{3} \cdot 360{,}000 \cdot h$ Replace V with 38,520,000 and B with 360,000.

$38{,}520{,}000 = 120{,}000 \cdot h$ Simplify.

$321 = h$ Divide each side by 120,000.

The height of the Pyramid Arena is 321 feet.

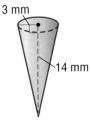
A **cone** is a three-dimensional figure with one circular base. A curved surface connects the base and the vertex. The volumes of a cone and a cylinder are related in the same way as those of a pyramid and prism.

Key Concept   Volume of a Cone

Words The volume V of a cone with radius r is one-third the area of the base B times the height h.

Symbols $V = \frac{1}{3}Bh$ or $V = \frac{1}{3}\pi r^2 h$

Model

EXAMPLE Find the Volume of a Cone

3 Find the volume of the cone.

$V = \frac{1}{3}\pi r^2 h$ Volume of a cone

$V = \frac{1}{3} \cdot \pi \cdot 3^2 \cdot 14$ Replace r with 3 and h with 14.

$V \approx 131.9$ Simplify.

The volume is about 131.9 cubic millimeters.

3 mm

14 mm

Skill and Concept Check

1. **NUMBER SENSE** Which would have a greater effect on the volume of a cone, doubling its radius or doubling its height? Explain your reasoning.

2. **OPEN ENDED** Draw and label a rectangular pyramid with a volume of 48 cubic centimeters.

GUIDED PRACTICE

Find the volume of each solid. Round to the nearest tenth if necessary.

3.
7 m 5 m

4.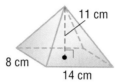
11 cm 8 cm 14 cm

5.
7 ft 4 ft 3 ft

Practice and Applications

Find the volume of each solid. Round to the nearest tenth if necessary.

HOMEWORK HELP

For Exercises	See Examples
8–11, 14–15	1
20–22	2
6–7, 12–13	3

Extra Practice
See pages 634, 654.

6.
22 ft 9 ft

7.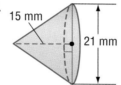
15 mm 21 mm

8.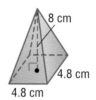
8 cm 4.8 cm 4.8 cm

9.
5 in. 4 in. $6\frac{1}{2}$ in.

10.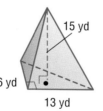
15 yd 6 yd 13 yd

11.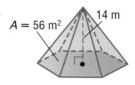
$A = 56$ m² 14 m

12. cone: diameter, 12 mm; height, 5 mm

13. cone: radius, $3\frac{1}{2}$ in.; height, 18 in.

14. octagonal pyramid: base area, 120 ft²; height, 19 ft

15. triangular pyramid: triangle base, 10 cm; triangle height, 7 cm; prism height, 15 cm

16.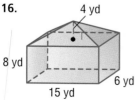
4 yd 8 yd 6 yd 15 yd

17.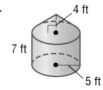
4 ft 7 ft 5 ft

18.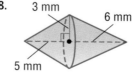
3 mm 6 mm 5 mm

19.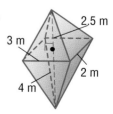
2.5 m 3 m 2 m 4 m

20. **VOLCANO** A model of a volcano constructed for a science project is cone-shaped with a diameter of 10 inches. If the volume of the model is about 287 cubic inches, how tall is the model?

ICE CREAM For Exercises 21 and 22, use the diagram at the right and the following information.
You are filling cone-shaped glasses with frozen custard. Each glass is 8 centimeters wide and 15 centimeters tall.

8 cm
15 cm

21. Estimate the volume of custard each glass will hold assuming you fill each one level with the top of the glass.

22. One gallon is equivalent to 4,000 cubic centimeters. Estimate how many glasses you can fill with one gallon of custard.

23. **WRITE A PROBLEM** Write about a real-life situation that can be solved by finding volume of a cone. Then solve the problem.

24. **CRITICAL THINKING** How could you change the height of a cone so that its volume would remain the same when its radius was tripled?

EXTENDING THE LESSON A *sphere* is the set of all points in space that are a given distance from a given point, called the center. The volume V of a sphere with radius r is given by the formula $V = \frac{4}{3}\pi r^3$.

r

Find the volume of each sphere described. Round to the nearest tenth.

25. radius, 3 in.   26. radius, 6 in.   27. diameter, 10 m   28. diameter, 9 ft

29. How does doubling a sphere's radius affect its volume? Explain.

Standardized Test Practice and Mixed Review

FCAT Practice

30. **MULTIPLE CHOICE** If each of the following solids has a height of 8 centimeters, which has the greatest volume?

 A
←10 cm→

 B
←10 cm→

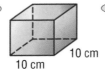 **C**
10 cm
10 cm

 D
10 cm
10 cm

31. **SHORT RESPONSE** A triangular prism has a volume of 135 cubic inches. Find the volume in cubic centimeters of a triangular pyramid with the same base area and height as this prism.

32. **PETS** Find the volume of a doghouse with a rectangular space that is 3 feet wide, 4 feet deep, and 5 feet high and has a triangular roof $1\frac{1}{2}$ feet higher than the walls of the house. (Lesson 7-5)

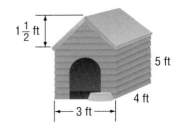
$1\frac{1}{2}$ ft
5 ft
4 ft
←3 ft→

33. Name the number and shapes of the faces of a trapezoidal prism. Then name the number of edges and vertices. (Lesson 7-4)

GETTING READY FOR THE NEXT LESSON

PREREQUISITE SKILL Find the circumference of each circle. Round to the nearest tenth. (Lesson 7-2)

34. diameter, 9 in.   35. diameter, $5\frac{1}{2}$ ft   36. radius, 2 m   37. radius, 3.8 cm

What You'll Learn
Represent three-dimensional objects as nets.

Materials
- empty box with tuck-in lid
- scissors

Sunshine State Standards
MA.C.1.3.1-3

Nets

Work with a partner.

Open the lid of a box and make 5 cuts as shown. Then open the box up and lay it flat. The result is a net. **Nets** are two-dimensional patterns of three-dimensional figures. You can use a net to build a three-dimensional figure.

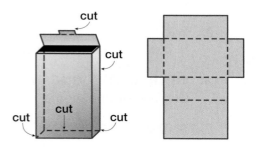

ACTIVITY

STEP 1 Copy the net onto a piece of paper, shading the base as shown.

STEP 2 Use scissors to cut out the net.

STEP 3 Fold on the dashed lines and tape the sides together.

STEP 4 Sketch the figure and draw its top, side, and front views.

top   side   front

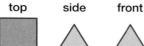

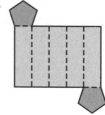

Your Turn Use each net to build a figure. Then sketch the figure, and draw and label its top, side, and front views.

a.
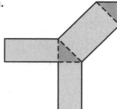

b.
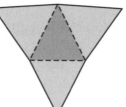

c.

Writing Math

1. **Describe** each shape that makes up the three nets above.

2. **Identify** each of the solids formed by the three nets above.

Surface Area of Prisms and Cylinders

Sunshine State Standards
MA.B.1.3.1-1, MA.B.1.3.1-2, MA.B.1.3.3-1, MA.B.1.3.3-2

What You'll Learn
Find the surface areas of prisms and cylinders.

⇒NEW Vocabulary

surface area

HANDS-ON Mini Lab

Materials
• 3 different-sized boxes
• centimeter ruler

The **surface area** of a solid is the sum of the areas of all its surfaces, or faces. In this lab, you will find the surface areas of rectangular prisms.

1. Estimate the area in square centimeters of each face for one of your boxes. Then find the sum of these six areas.

2. Now use your ruler to measure the sides of each face. Then find the area of each face to the nearest square centimeter. Find the sum of these areas and compare to your estimate.

3. Estimate and then find the surfaces areas of your other boxes.

One way to easily visualize all of the surfaces of a prism is to sketch a two-dimensional pattern of the solid, called a *net*, and label all its dimensions.

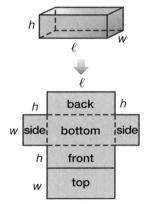

Faces	Area
top and bottom	$(\ell \cdot w) + (\ell \cdot w) = 2\ell w$
front and back	$(\ell \cdot h) + (\ell \cdot h) = 2\ell h$
two sides	$(w \cdot h) + (w \cdot h) = 2wh$
Sum of areas	→ $2\ell w + 2\ell h + 2wh$

Key Concept — Surface Area of a Rectangular Prism

Words The surface area S of a rectangular prism with length ℓ, width w, and height h is the sum of the areas of the faces.

Model

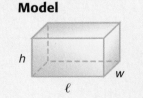

Symbols $S = 2\ell w + 2\ell h + 2wh$

EXAMPLE — Surface Area of a Rectangular Prism

1. **Find the surface area of the rectangular prism.**

   $S = 2\ell w + 2\ell h + 2wh$ Write the formula.

   $S = 2(7)(3) + 2(7)(12) + 2(3)(12)$ Substitution.

   $S = 282$ Simplify.

   The surface area is 282 square meters.

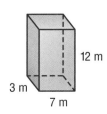

2 **SKATEBOARDING** A skateboarding ramp called a *wedge* is built in the shape of a triangular prism. You plan to paint all surfaces of the ramp. Find the surface area to be painted.

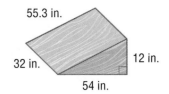

A triangular prism consists of two congruent triangular faces and three rectangular faces.

Draw and label a net of this prism. Find the area of each face.

bottom	$54 \cdot 32 = 1{,}728$
left side	$55.3 \cdot 32 = 1{,}769.6$
right side	$12 \cdot 32 = 384$
two bases	$2\left(\frac{1}{2} \cdot 54 \cdot 12\right) = 648$

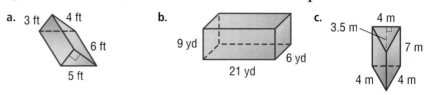

Add to find the total surface area.

$1{,}728 + 1{,}769.6 + 384 + 648 = 4{,}529.6$

The surface area of the ramp is 4,529.6 square inches.

Your Turn Find the surface area of each prism.

a. 3 ft 4 ft 6 ft 5 ft

b. 9 yd 21 yd 6 yd

c. 3.5 m 4 m 7 m 4 m 4 m

You can find the surface area of a cylinder by finding the area of its two bases and adding the area of its curved side. If you unroll a cylinder, its net is two circles and a rectangle.

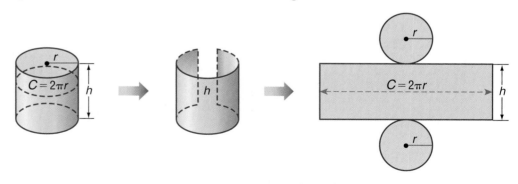

Model	Net	Area
2 circular bases	2 congruent circles with radius r	$2(\pi r^2)$ or $2\pi r^2$
1 curved surface	1 rectangle with width h and length $2\pi r$	$2\pi r \cdot h$ or $2\pi rh$

So, the surface area S of a cylinder is $2\pi r^2 + 2\pi rh$.

Words   The surface area *S* of a cylinder with height *h* and radius *r* is the area of the two bases plus the area of the curved surface.

Model

Symbols $S = 2\pi r^2 + 2\pi rh$

EXAMPLE   **Surface Area of a Cylinder**

3 **Find the surface area of the cylinder. Round to the nearest tenth.**

$S = 2\pi r^2 + 2\pi rh$ Surface area of a cylinder

$S = 2\pi(2)^2 + 2\pi(2)(3)$ Replace *r* with 2 and *h* with 3.

$S \approx 62.8$ Simplify.

The surface area is 62.8 square feet.

Your Turn Find the surface area of each cylinder. Round to the nearest tenth.

d. 5 mm, 10 mm

e.  6.5 in., 4 in.

f. 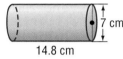 7 cm, 14.8 cm

Skill and Concept Check

Writing Math   Exercises 1 & 2

1. **Determine** whether the following statement is *true* or *false*. If *false*, give a counterexample.

   If two rectangular prisms have the same volume, then they also have the same surface area.

2. **NUMBER SENSE** If you double the edge length of a cube, explain how this affects the surface area of the prism.

3. **OPEN ENDED** The surface area of a rectangular prism is 96 square feet. Name one possible set of dimensions for this prism.

GUIDED PRACTICE

Find the surface area of each solid. Round to the nearest tenth if necessary.

4. 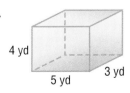 4 yd, 5 yd, 3 yd

5. 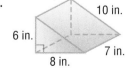 10 in., 6 in., 8 in., 7 in.

6. 8 m, 9.4 m

7. rectangular prism: length, 12.2 cm; width, 4.8 cm; height, 10.3 cm

8. cylinder: radius, 16 yd; height, 25 yd

Practice and Applications

Find the surface area of each solid. Round to the nearest tenth if necessary.

9.
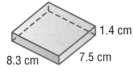
1.4 cm
8.3 cm 7.5 cm

10.
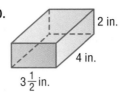
2 in.
4 in.
$3\frac{1}{2}$ in.

11.
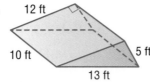
12 ft
10 ft
5 ft
13 ft

12.
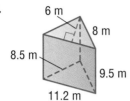
6 m
8 m
8.5 m
9.5 m
11.2 m

13.
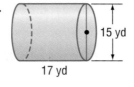
15 yd
17 yd

14.
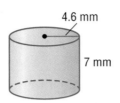
4.6 mm
7 mm

15. cube: edge length, 12 m

16. cylinder: diameter, 18 yd, height, 21 yd

17. cylinder: radius, 7 in.; height, $9\frac{1}{2}$ in.

18. rectangular prism: length, $1\frac{1}{2}$ cm; width, $5\frac{3}{4}$ cm; height, $3\frac{1}{4}$ cm

19. **POOL** A vinyl liner covers the inside walls and bottom of the swimming pool shown below. Find the area of this liner to the nearest square foot.

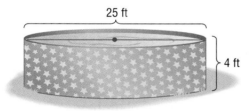
25 ft
4 ft

20. **GARDENING** The door of the greenhouse shown below has an area of 4.5 square feet. How many square feet of plastic are needed to cover the roof and sides of the greenhouse?

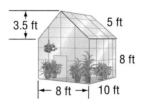
3.5 ft
5 ft
8 ft
8 ft
10 ft

21. **MULTI STEP** An airport has changed the carrels used for public telephones. The old carrels consisted of four sides of a rectangular prism. The new carrels are half of a cylinder with an open top. How much less material is needed to construct a new carrel than an old carrel?

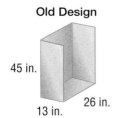
Old Design   New Design
45 in.   45 in.
13 in.
26 in.   26 in.

22. **CAMPING** A camping club has designed a tent with canvas sides and floor as shown below. About how much canvas will the club members need to construct the tent? (*Hint:* Use the Pythagorean Theorem to find the height of the triangular base.)

2 yd   2 yd
1 yd 1 yd   3 yd

HOMEWORK HELP

For Exercises	See Examples
9–10, 15, 18	1
11–12	2
13–14, 16–17	3

Extra Practice
See pages 634, 654.

350 Chapter 7 Geometry: Measuring Area and Volume

23. CRITICAL THINKING Will the surface area of a cylinder increase more if you double the height or double the radius? Explain your reasoning.

CRITICAL THINKING The length of each edge of a cube is 3 inches. Suppose the cube is painted and then cut into 27 smaller cubes that are 1 inch on each side.

24. How many of the smaller cubes will have paint on exactly three faces?

25. How many of the smaller cubes will have paint on exactly two faces?

26. How many of the smaller cubes will have paint on only one face?

27. How many of the smaller cubes will have no paint on them at all?

28. Find the answers to Exercises 24–27 if the cube is 10 inches on a side and cut into 1,000 smaller cubes.

EXTENDING THE LESSON If you make cuts in a solid, different two-dimensional cross sections result, as shown at the right.

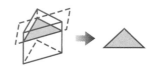

Describe the cross section of each figure cut below.

29.   30. 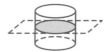   31.   32.

Standardized Test Practice and Mixed Review

FCAT Practice

33. **MULTIPLE CHOICE** The greater the surface area of a piece of ice the faster it will melt. Which block of ice described will be the *last* to melt?

   Ⓐ 1 in. by 2 in. by 32 in. block   Ⓑ 4 in. by 8 in. by 2 in. block

   Ⓒ 16 in. by 4 in. by 1 in. block   Ⓓ 4 in. by 4 in. by 4 in. block

34. **SHORT RESPONSE** Find the amount of metal needed to construct the mailbox at the right to the nearest tenth of a square inch.

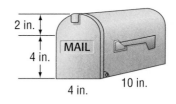

Find the volume of each solid described. Round to the nearest tenth if necessary. (Lesson 7-6)

35. rectangular pyramid: length, 14 m; width, 12 m; height, 7 m

36. cone: diameter 22 cm; height, 24 cm

37. **HEALTH** The inside of a refrigerator in a medical laboratory measures 17 inches by 18 inches by 42 inches. You need at least 8 cubic feet to refrigerate some samples from the lab. Is the refrigerator large enough for the samples? Explain. (Lesson 7-5)

GETTING READY FOR THE NEXT LESSON

PREREQUISITE SKILL Multiply. (Lesson 2-5)

38. $\frac{1}{2} \cdot 2.8$

39. $\frac{1}{2} \cdot 10 \cdot 23$

40. $\frac{1}{2} \cdot 2.5 \cdot 16$

41. $\frac{1}{2}\left(3\frac{1}{2}\right)(20)$

Surface Area of Pyramids and Cones

Sunshine State Standards
MA.B.1.3.1-1, MA.B.1.3.1-2, MA.B.1.3.3-1, MA.B.1.3.3-2

WHEN am I ever going to use this?

HISTORY In 1485, Leonardo Da Vinci sketched a pyramid-shaped parachute in the margin of his notebook. In June 2000, using a parachute created with tools and materials available in medieval times, Adrian Nicholas proved Da Vinci's design worked by descending 7,000 feet.

1. How many cloth faces does this pyramid have? What shape are they?

2. How could you find the total area of the material used for the parachute?

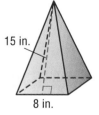

What You'll Learn
Find the surface areas of pyramids and cones.

NEW Vocabulary
lateral face
slant height
lateral area

LINK To Reading
Everyday Meaning of lateral: situated on the side

The triangular sides of a pyramid are called **lateral faces**. The triangles intersect at the vertex. The altitude or height of each lateral face is called the **slant height**.

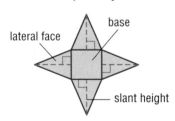

Model of Square Pyramid   Net of Square Pyramid

The sum of the areas of the lateral faces is the **lateral area**. The surface area of a pyramid is the lateral area plus the area of the base.

EXAMPLE Surface Area of a Pyramid

1 **Find the surface area of the square pyramid.**

Find the lateral area and the area of the base.

Area of each lateral face

$A = \frac{1}{2}bh$ Area of a triangle

$A = \frac{1}{2}(8)(15)$ or 60   Replace *b* with 8 and *h* with 15.

15 in.

8 in.

There are 4 faces, so the lateral area is 4(60) or 240 square inches.

Area of base

$A = s^2$ Area of a square

$A = 8^2$ or 64 Replace s with 8.

The surface area of the pyramid is the sum of the lateral area and the area of the base, $240 + 64$ or 304 square inches.

You can find the surface area of a cone with radius r and slant height ℓ by finding the area of its bases and adding the area of its curved side. If you unroll a cone, its net is a circle and a portion of a larger circle.

Model of Cone   Net of Cone

Model	Net	Area
lateral area	portion of circle with radius ℓ	$\pi r \ell$
circular base	circle with radius r	πr^2

So, the surface area S of a cone is $\pi r\ell + \pi r^2$.

Key Concept   **Surface Area of a Cone**

Words	The surface area S of a cone with slant height ℓ and radius r is the lateral area plus the area of the base.	**Model**
Symbols	$S = \pi r \ell + \pi r^2$	

STUDY TIP

Slant Height Be careful not to use the height of a pyramid or cone in place of its slant height. Remember that a slant height lies along a cone or pyramid's lateral surface.

EXAMPLE   **Surface Area of a Cone**

2   Find the surface area of the cone.

$S = \pi r \ell + \pi r^2$ Surface area of a cone

$S = \pi(7)(13) + \pi(7)^2$ Replace r with 7 and ℓ with 13.

$S \approx 439.8$ Simplify.

The surface area of the cone is about 439.8 square centimeters.

Your Turn Find the surface area of each solid. Round to the nearest tenth if necessary.

a.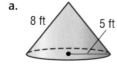
    8 ft   5 ft

b.
    18 mm   11 mm   11 mm

c.
    $3\frac{1}{2}$ in.   10 in.

Writing Math

Exercise 1

1. **Explain** how the slant height and the height of a pyramid are different.

2. **OPEN ENDED** Draw a square pyramid, giving measures for its slant height and base side length. Then find its lateral area.

GUIDED PRACTICE

Find the surface area of each solid. Round to the nearest tenth if necessary.

3.

4.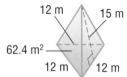

5.

Practice and Applications

Find the surface area of each solid. Round to the nearest tenth if necessary.

HOMEWORK HELP	
For Exercises	See Examples
6–9, 13, 16	1
10–12, 15, 17	2
Extra Practice See pages 634, 654.	

6.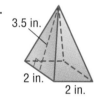

7.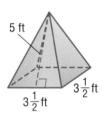

8.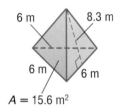

9.

10.

11.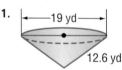

12. cone: diameter, 11.4 ft; slant height, 25 ft

13. square pyramid: base side length, $6\frac{1}{2}$ cm; slant height $8\frac{1}{4}$ cm

14. Find the surface area of the complex solid at the right. Round to the nearest tenth.

15. **ROOFS** A cone-shaped roof has a diameter of 20 feet and a slant height of 16 feet. If roofing material comes in 120 square-foot rolls, how many rolls will be needed to cover this roof? Explain your reasoning.

16. **GLASS** The Luxor Hotel in Las Vegas, Nevada, is a pyramid-shaped building standing 350 feet tall and covered with glass. Its base is a square with each side 646 feet long. Find the surface area of the glass on the Luxor. (*Hint*: Use the Pythagorean Theorem to find the pyramid's slant height ℓ.)

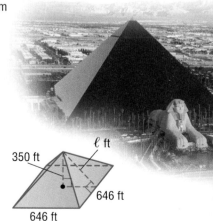

17. GEOMETRY A *frustum* is the part of a solid that remains after the top portion of the solid has been cut off by a plane parallel to the base. The lampshade at the right is a frustum of a cone. Find the surface area of the lampshade.

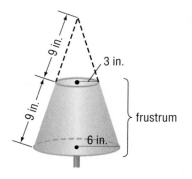

CRITICAL THINKING For Exercises 18–20, use the drawings of the pyramid below, whose lateral faces are equilateral triangles.

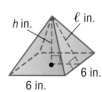

Side View

18. Find the exact measure of the slant height ℓ.

19. Use the slant height to find the exact height h of the pyramid.

20. Find the exact volume and surface area of the pyramid.

EXTENDING THE LESSON The surface area S of a sphere with radius r is given by the formula $S = 4\pi r^2$.

Find the surface area of each sphere to the nearest tenth.

21.

3 m

22.

10 in.

23.
16 ft
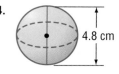

24.

4.8 cm

Standardized Test Practice and Mixed Review

25. **MULTIPLE CHOICE** Which is the best estimate for the surface area of a cone with a radius of 3 inches and a slant height of 5 inches?

   Ⓐ 45 in²   Ⓑ 72 in²   Ⓒ 117 in²   Ⓓ 135 in²

26. **MULTIPLE CHOICE** What is the lateral area of the pentagonal pyramid at the right if the slant height is 9 centimeters?

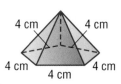

   Ⓕ 18 cm²   Ⓖ 72 cm²   Ⓗ 90 cm²   Ⓘ 180 cm²

27. **GEOMETRY** Find the surface area of a cylinder whose diameter is 22 feet and whose height is 7.5 feet.

28. **MULTI STEP** The cylindrical air duct of a large furnace has a diameter of 30 inches and a height of 120 feet. If it takes 15 minutes for the contents of the duct to be expelled into the air, what is the volume of the substances being expelled each hour? (Lesson 7-6)

GETTING READY FOR THE NEXT LESSON

BASIC SKILL Find the value of each expression to the nearest tenth.

29. $8.35 + 54.2$ 30. $7 - 2.89$ 31. $4.2 \cdot 6.13$ 32. $9.31 \div 5$

A Follow-Up of Lesson 7-8

Sunshine State Standards
MA.B.1.3.3-1, MA.B.1.3.3-2, MA.B.1.3.3-3,
MA.B.2.3.1-1

What You'll Learn
Investigate the volume and surface area of similar solids.

⚙ REVIEW Vocabulary

proportion: an equation stating that two ratios are equivalent (Lesson 4-4)

Similar Solids

The pyramids are **similar solids**, because they have the same shape and their corresponding linear measures are proportional.

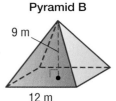

Pyramid A   Pyramid B

6 m   9 m   8 m   12 m

The number of times you increase or decrease the linear dimensions of a solid is called the **scale factor**. The heights of pyramid A and pyramid B are 6 meters and 9 meters, respectively. So the scale factor from pyramid A to pyramid B is $\frac{6}{9}$ or $\frac{2}{3}$.

ACTIVITY

Find the surface area and volume of the prism at the right. Then find the surface areas and volumes of similar prisms with scale factors of 2, 3, and 4.

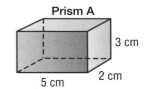

Prism A
3 cm
2 cm
5 cm

Similar Prisms

	A	B	C	D	E	F	G
1	Prism	Scale Factor	Length	Width	Height	Surface Area	Volume
2	A	1	5	2	3	62	30
3	B	2	10	4	6	248	240
4	C	3	15	6	9	558	810
5	D	4	20	8	12	992	1920

The spreadsheet evaluates the formula 2*C3*D3+2*C3*E3+2*D3*E3.

The spreadsheet evaluates the formula C5*D5*E5.

EXERCISES

1. How many times greater than the surface area of prism A is the surface area of prism B? prism C? prism D?

2. How are the answers to Exercise 1 related to the scale factors?

3. How many times greater than the volume of prism A is the volume of prism B? prism C? prism D?

4. How are the answers to Exercise 3 related to the scale factors?

5. Considering the rectangular prism in the activity above, write expressions for the surface area and volume of a similar prism with scale factor x.

ACTIVITY

Find the surface area and volume of the cylinder at the right. Then find the surface areas and volumes of similar cylinders with scale factors of 2, 3, and 4.

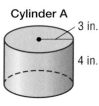

Cylinder A

3 in.

4 in.

Similar Cylinders

	A	B	C	D	E	F	G
1	Cylinder	Scale Factor	Radius	Height	Surface Area	Volume	
2	A	1	3	4	103.7	113.1	
3	B	2	6	8	414.7	904.78	
4	C	3	9	12	933.1	3053.6	
5	D	4	12	16	1658.8	7238.2	

The spreadsheet evaluates the formula PI()*C3^2+2*PI()*C3*D3.

The spreadsheet evaluates the formula PI()*C5^2*D5.

EXERCISES

6. How many times greater than the surface area of cylinder A is the surface area of cylinder B? cylinder C? cylinder D?

7. How are the answers to Exercise 6 related to the scale factors of each cylinder?

8. How many times greater than the volume of cylinder A is the volume of cylinder B? cylinder C? cylinder D?

9. How are the answers to Exercise 8 related to the scale factors of each cylinder?

10. Considering the cylinder in the activity above, write expressions for the surface area and volume of a similar cylinder with scale factor x.

11. **Make a conjecture** about how the volume and surface area of a pyramid are affected when all edges of this solid are multiplied by a scale factor of x.

For Exercises 12 and 13, use the diagram of the two similar prisms at the right.

Prism A   Prism B

4 ft   6 ft

12. If the surface area of prism A is 52 square feet, find the surface area of prism B.

13. If the volume of prism A is 24 cubic feet, find the volume of prism B.

Precision and Significant Digits

Sunshine State Standards
MA.B.4.3.1-4, MA.B.4.3.1-2, MA.B.4.3.1-3, MA.B.4.3.1-4, MA.B.4.3.1-5

What You'll Learn
Analyze measurements.

NEW Vocabulary
precision
significant digits

WHEN am I ever going to use this?

CARTOONS Consider the cartoon below.

Hi & Lois

1. How precisely has the daughter, Dot, measured each piece?
2. Give an example of a situation where this degree of accuracy might be appropriate.

The **precision** of a measurement is the exactness to which a measurement is made. Precision depends upon the smallest unit of measure being used, or the *precision unit*. A measurement is accurate to the nearest precision unit.

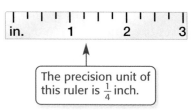

The precision unit of this ruler is $\frac{1}{4}$ inch.

EXAMPLE Identify Precision Units

1. **Identify the precision unit of the flask.**

   There are two spaces between each 50 milliliter-mark, so the precision unit is $\frac{1}{2}$ of 50 milliliters or 25 milliliters.

Your Turn Identify the precision unit of each measuring instrument.

a.    b.

One way to record a measure is to estimate to the nearest precision unit. A more precise method is to include all of the digits that are actually measured, plus one estimated digit. The digits you record when you measure this way are called significant digits. **Significant digits** indicate the precision of the measurement.

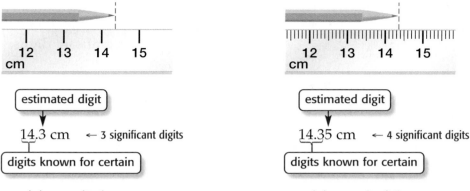

estimated digit

14.3 cm ← 3 significant digits

digits known for certain

precision unit: 1 cm
actual measure: 14–15 cm
estimated measure: 14.3 cm

estimated digit

14.35 cm ← 4 significant digits

digits known for certain

precision unit: 0.1 cm
actual measure: 14.3–14.4 cm
estimated measure: 14.35

There are special rules for determining significant digits in a given measurement. Numbers are analyzed for significant digits by counting

Number	Significant Digits	Rule
2.45	3	All nonzero digits are significant.
140.06	5	Zeros between two significant digits are significant.
0.013	2	Zeros used to show place value of the decimal are not significant.
120.0	4	In a number with a decimal point, all zeros to the right of a nonzero digit are significant.
350	2	In a number without a decimal point, any zeros to the right of the last nonzero digit are *not* significant.

EXAMPLES **Identify Significant Digits**

Determine the number of significant digits in each measure.

2 **10.25 g**
4 significant digits

3 **0.003 L**
1 significant digit

When adding or subtracting measurements, the sum or difference should have the *same precision* as the least precise measurement.

EXAMPLE **Add Measurements**

4 **LIFTING** You are attempting to lift three packages that weigh 5.125 pounds, 6.75 pounds, and 4.6 pounds. Write the combined weight of the packages using the correct precision.

$$
\begin{array}{rl}
6.75 & \leftarrow \text{2 decimal places} \\
5.125 & \leftarrow \text{3 decimal places} \\
+\ 4.6 & \leftarrow \text{1 decimal place} \\
\hline
16.475 &
\end{array}
$$

The least precise measurement has 1 decimal place, so round the sum to 1 decimal place.

The combined weight of the packages is about 16.5 pounds.

When multiplying or dividing measurements, the product or quotient should have the *same number of significant digits* as the measurement with the least number of significant digits.

EXAMPLE **Multiply Measurements**

5 **GEOMETRY** Use the correct number of significant digits to find the area of the parallelogram.

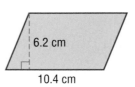
6.2 cm
10.4 cm

10.4 ← 3 significant digits

× 6.2 ← 2 significant digits

64.48

> This measurement has the least number of significant digits, 2.

Round the product, 64.48, so that it has 2 significant digits. The area of the parallelogram is about 64 square centimeters.

Your Turn

c. Find 3.48 liters − 0.2 liters using the correct precision.

d. Use the correct number of significant digits to calculate 0.45 meter ÷ 0.8 meter.

Skill and Concept Check

Writing Math
Exercises 1 & 3

1. Determine which measurement of a bag of dog food would be the most precise: 5 pounds, 74 ounces, or 74.8 ounces. Explain.

2. OPEN ENDED Write a 5-digit number with 3 significant digits.

3. Which One Doesn't Belong? Identify the number that does not have the same number of significant digits as the other three. Explain.

| 20.6 | 0.0815 | 4,260 | 375.0 |

GUIDED PRACTICE

Identify the precision of the unit of each measuring instrument.

4.
in.   1   2

5.
50° 60° 70° 80° 90° 100° 110°

Determine the number of significant digits in each measure.

6. 138.0 g   **7.** 0.0037 mm   **8.** 50 min   **9.** 206.04 cm

Find each sum or difference using the correct precision.

10. 45 in. + 12.7 in.   **11.** 7.38 m − 5.9 m

Find each product or quotient using the correct number of significant digits.

12. 8.2 yd · 4.5 yd   **13.** 7.31 s ÷ 5.4 s

Identify the precision unit of each measuring instrument.

14.

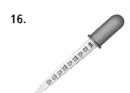

15.

16.

17.

Determine the number of significant digits in each measure.

18. 0.025 mL   19. 3,450 km   20. 40.03 in.   21. 7.0 kg

22. 104.30 mi   23. 3.06 s   24. 0.009 mm   25. 380 g

Find each sum or difference using the correct precision.

26. 12.85 cm + 5.4 cm   27. 14.003 L − 4.61 L   28. 34 g − 15.2 g

29. 150 m + 44.7 m   30. 100 mi + 63.7 mi   31. 14.37 s − 9.2 s

Find each product or quotient using the correct number of significant digits.

32. 0.8 cm · 9.4 cm   33. 3.82 ft · 3.5 ft   34. 10 mi · 1.2 mi

35. 200 g ÷ 2.6 g   36. 88.5 lb ÷ 0.05 lb   37. 7.50 mL ÷ 0.2 mL

38. **GEOMETRY** A triangle's sides measure 17.04 meters, 8.2 meters, and 7.375 meters. Write the perimeter using the correct precision.

39. **SURVEYING** A surveyor measures the dimensions of a field and finds that the length is 122.5 meters and the width is 86.4 meters. What is the area of the field? Round to the correct number of significant digits.

SCHOOL For Exercises 40–42, refer to the graphic at the right.

40. Are the numbers exact? Explain.

41. How many significant digits are used to describe the number of children enrolled in public school?

42. Find the difference between public and private school enrollment using the correct precision.

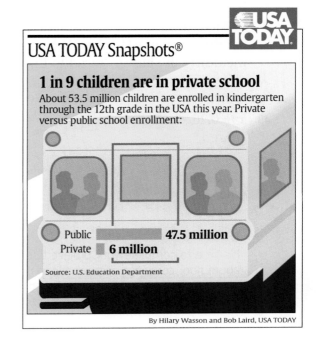

USA TODAY Snapshots®

1 in 9 children are in private school

About 53.5 million children are enrolled in kindergarten through the 12th grade in the USA this year. Private versus public school enrollment:

Public — 47.5 million
Private — 6 million

Source: U.S. Education Department

By Hilary Wasson and Bob Laird, USA TODAY

43. **CRITICAL THINKING** Find the surface area of the square pyramid at the right. Use the correct precision or number of significant digits as appropriate.

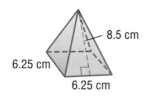

EXTENDING THE LESSON The *greatest possible error* is one-half the precision unit. It can be used to describe the actual measure. The cotton swab below appears to be about 7.8 centimeters long.

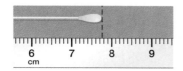

greatest possible error = $\frac{1}{2}$ · precision unit

$= \frac{1}{2}$ · 0.1 cm or 0.05 cm

The possible actual length of the cotton swab is 0.05 centimeter less than or 0.05 centimeter more than 7.8 centimeters. So, it is between 7.75 and 7.85 centimeters long.

44. **SPORTS** An Olympic swimmer won the gold medal in the 100-meter backstroke with a time of 61.19 seconds. Find the greatest possible error of the measurement and use it to determine between which two values is the swimmer's actual time.

Standardized Test Practice and Mixed Review

 FCAT Practice

45. **MULTIPLE CHOICE** Choose the measurement that is most precise.
   Ⓐ 54 kg   Ⓑ 5.4 kg   Ⓒ 54 g   Ⓓ 54 mg

46. **GRID IN** Use the correct number of significant digits to find the volume of a cylinder in cubic feet whose radius is 4.0 feet and height is 10.2 feet.

47. **DESSERT** Find the surface area of the waffle cone at the right. (Lesson 7-8)

5 cm

9.5 cm

48. **HISTORY** The great pyramid of Khufu in Egypt was originally 481 feet high, had a square base 756 feet on a side, and slant height of about 611.8 feet. What was its surface area, not including the base? Round to the nearest tenth. (Lesson 7-7)

Solve each equation. Check your solution. (Lesson 2-9)

49. $x + 0.26 = -3.05$ 50. $\frac{3}{5} = a - \frac{1}{2}$ 51. $-\frac{1}{6} = -\frac{1}{4}n$ 52. $\frac{y}{2.4} = -6.5$

WebQuest **Interdisciplinary Project**

Under Construction

It's time to complete your project. Use the information and data you have gathered about floor coverings costs and loan rates to prepare a Web page or brochure. Be sure to include a labeled scale drawing with your project.

msmath3.net/webquest

Vocabulary and Concept Check

altitude (p. 314)	diameter (p. 319)	prism (p. 331)
base (pp. 314, 331)	edge (p. 331)	pyramid (p. 331)
center (p. 319)	face (p. 331)	radius (p. 319)
circle (p. 319)	lateral area (p. 352)	significant digits (p. 358)
circumference (p. 319)	lateral face (p. 352)	slant height (p. 352)
complex figure (p. 326)	pi (π) (p. 319)	solid (p. 331)
complex solid (p. 337)	plane (p. 331)	surface area (p. 347)
cone (p. 343)	polyhedron (p. 331)	vertex (p. 331)
cylinder (p. 336)	precision (p. 358)	volume (p. 335)

Choose the letter of the term that best matches each phrase.

1. a flat surface of a prism
2. the measure of the space occupied by a solid
3. a figure that has two parallel, congruent circular bases
4. any three-dimensional figure
5. the sides of a pyramid
6. the distance around a circle
7. the exactness to which a measurement is made
8. any side of a parallelogram
9. a solid figure with flat surfaces that are polygons

a. volume
b. face
c. precision
d. cylinder
e. base
f. solid
g. polyhedron
h. circumference
i. lateral face

Lesson-by-Lesson Exercises and Examples

 7-1 **Area of Parallelograms, Triangles, and Trapezoids** (pp. 314–318)

Find the area of each figure.

10.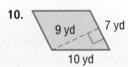
9 yd, 7 yd, 10 yd

11.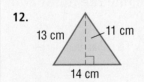
$16\frac{1}{2}$ in., 20 in., 17 in.

12.
13 cm, 11 cm, 14 cm

13.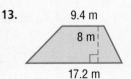
9.4 m, 8 m, 17.2 m

Example 1
Find the area of the trapezoid.

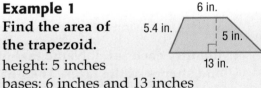
6 in., 5.4 in., 5 in., 13 in.

height: 5 inches
bases: 6 inches and 13 inches

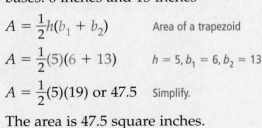

$A = \frac{1}{2}h(b_1 + b_2)$ Area of a trapezoid

$A = \frac{1}{2}(5)(6 + 13)$ $h = 5, b_1 = 6, b_2 = 13$

$A = \frac{1}{2}(5)(19)$ or 47.5 Simplify.

The area is 47.5 square inches.

7-2 Circumference and Area of Circles (pp. 319–323)

Find the circumference and area of each circle. Round to the nearest tenth.

14.
18 in.

15.
6 cm

16. The diameter is $4\frac{1}{3}$ feet.

17. The radius is 2.6 meters.

Example 2 Find the circumference and area of the circle.

5 yd

The radius of the circle is 5 yards.

$C = 2\pi r$ $A = \pi r^2$

$C = 2 \cdot \pi \cdot 5$ $A = \pi \cdot 5^2$

$C \approx 31.4$ yd $A \approx 78.5$ yd^2

7-3 Area of Complex Figures (pp. 326–329)

Find the area of each figure. Round to the nearest tenth if necessary.

18.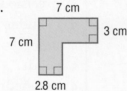
7 cm
7 cm
3 cm
2.8 cm

19.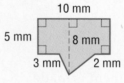
10 mm
5 mm
8 mm
3 mm
2 mm

20.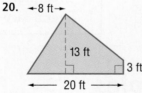
8 ft
13 ft
3 ft
20 ft

21.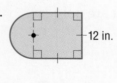
12 in.

Example 3 Find the area of the complex figure.

4 m
6 m
10 m

Area of semicircle   **Area of trapezoid**

$A = \frac{1}{2} \cdot \pi \cdot 2^2$ $A = \frac{1}{2}(6)(4 + 10)$

$A \approx 6.3$ $A = 42$

The area is about $6.3 + 42$ or 48.3 square meters.

7-4 Three-Dimensional Figures (pp. 331–334)

Identify each solid. Name the number and shapes of the faces. Then name the number of edges and vertices.

22.

23.

Example 4 Name the number and shapes of the faces of a rectangular prism. Then name the number of edges and vertices.

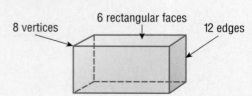
8 vertices
6 rectangular faces
12 edges

7-5 Volume of Prisms and Cylinders (pp. 335–339)

Find the volume of each solid.

24.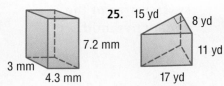

25. 15 yd, 8 yd, 7.2 mm, 11 yd, 3 mm, 4.3 mm, 17 yd

26. **FOOD** A can of green beans has a diameter of 10.5 centimeters and a height of 13 centimeters. Find its volume.

Example 5 Find the volume of the solid.

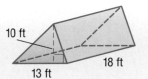

The base of this prism is a triangle.

$V = Bh$ B = area of base, h = height of prism

$V = \left(\frac{1}{2} \cdot 13 \cdot 10\right)18$

$V = 1{,}170 \text{ ft}^3$

7-6 Volume of Pyramids and Cones (pp. 342–345)

Find the volume of each solid. Round to the nearest tenth if necessary.

27.
10 ft, 7 ft, 7 ft

28.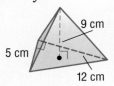
9 cm, 5 cm, 12 cm

29. cone: diameter, 9 yd; height, 21 yd

Example 6 Find the volume of the pyramid.

8 in., 6 in., 12 in.

The base B of the pyramid is a rectangle.

$V = \frac{1}{3}Bh$ Volume of pyramid or cone

$V = \frac{1}{3}(12 \cdot 6)8$

$V = 192 \text{ in}^3$

7-7 Surface Area of Prisms and Cylinders (pp. 347–351)

Find the surface area of each solid. Round to the nearest tenth if necessary.

30.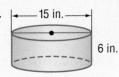
15 in., 6 in.

31.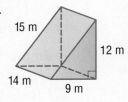
15 m, 12 m, 14 m, 9 m

32. **SET DESIGN** All but the bottom of a platform 15 feet long, 8 feet wide, and 3 feet high is to be painted for use in a play. Find the area of the surface to be painted.

Example 7 Find the surface area of the cylinder.

8 mm, 11 mm

Find the area of the two circular bases and add the area of the curved surface.

$S = 2\pi r^2 + 2\pi rh$ Surface area of a cylinder

$S = 2\pi(8)^2 + 2\pi(8)(11)$ $r = 8$ and $h = 11$

$S \approx 955.0$ Use a calculator.

The surface area is about 955.0 square millimeters.

7-8 Surface Area of Pyramids and Cones (pp. 352–355)

Find the surface area of each solid. Round to the nearest tenth if necessary.

33.
7 ft
5 ft   5 ft

34.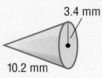
3.4 mm
10.2 mm

35.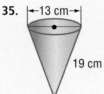
|←13 cm→|
19 cm

36.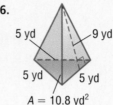
5 yd   9 yd
5 yd   5 yd
$A = 10.8 \text{ yd}^2$

37. **DECORATING** All but the underside of a 10-foot tall conical-shaped tree is to be covered with fake snow. The base of the tree has a radius of 5 feet, and its slant height is about 11.2 feet. How much area is to be covered with fake snow?

Example 8 Find the surface area of the square pyramid.

7 m
3 m   3 m

$A = \frac{1}{2}bh$ Area of triangle

$A = \frac{1}{2}(3)(7)$ or 10.5

The total lateral area is 4(10.5) or 42 square meters. The area of the base is 3(3) or 9 square meters. So the total surface area of the pyramid is 42 + 9 or 51 square meters.

Example 9 Find the surface area of the cone.

13 in.   4 in.

$S = \pi r \ell + \pi r^2$ Surface area of a cone
$S = \pi(4)(13) + \pi(4)^2$ $r = 4$ and $\ell = 13$
$S \approx 213.6$ Use a calculator.

The surface area is about 213.6 square inches.

7-9 Precision and Significant Digits (pp. 358–362)

38. **MEASUREMENT** Order the following measures from least precise to most precise.
0.50 cm, 0.005 cm, 0.5 cm, 50 cm

Determine the number of significant digits in each measure.

39. 0.14 ft   40. 7.0 L   41. 9.04 s

Find each sum or difference using the correct precision.

42. 40 g + 15.7 g   43. 45.3 lb − 0.02 lb

Find each product or quotient using the correct number of significant digits.

44. 6.4 yd · 2 yd   45. 200.8 m ÷ 12.0 m

Example 10 Determine the number of significant digits in a measure of 180 miles.

In a number without a decimal point, any zeros to the right of the last nonzero digit are *not* significant. Therefore, 180 miles has 2 significant digits, 1 and 8.

Example 11 Use the correct number of significant digits to find 701 feet · 0.04 feet.

701 ← 2 significant digits
× 0.04 ← 1 significant digit ← [least number]
28.04

The product, rounded to 1 significant digit, is 30 square feet.

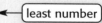

Vocabulary and Concepts

1. **Explain** how to find the volume of any prism.
2. **Explain** how to find the surface area of any prism.

Skills and Applications

Find the area of each figure. Round to the nearest tenth if necessary.

3.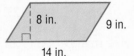
8 in. 9 in.
14 in.

4.
3 ft 5 ft
4 ft

5.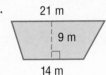
21 m
9 m
14 m

6.
9.4 cm

7. **CIRCUS** The elephants at a circus are paraded around the edge of the center ring two times. If the ring has a radius of 25 yards, about how far do the elephants walk during this part of the show?

8. **CAKE DECORATION** Mrs. Chávez designed the flashlight birthday cake shown at the right. If one container of frosting covers 250 square inches of cake, how many containers will she need to frost the top of this cake? Explain your reasoning.

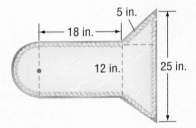
5 in.
18 in.
12 in.
25 in.

Find the volume and surface area of each solid. Round to the nearest tenth if necessary.

9.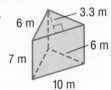
6 m 3.3 m
6 m
7 m
10 m

10.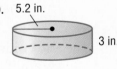
5.2 in.
3 in.

11.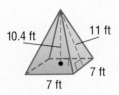
10.4 ft 11 ft
7 ft
7 ft

12.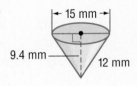
15 mm
9.4 mm
12 mm

13. Determine the number of significant digits in 0.089 milliliters.
14. Find 18.2 milligrams − 7.34 milligrams using the correct precision.
15. Find 0.5 yards · 18.3 yards using the correct number of significant digits.

FCAT Practice

Standardized Test Practice

Ⓐ Ⓑ Ⓒ Ⓓ

16. **MULTIPLE CHOICE** Find the volume of the solid at the right.

Ⓐ 2,160 ft³   Ⓑ 2,520 ft³
Ⓒ 3,600 ft³   Ⓓ 7,200 ft³

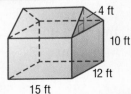

4 ft
10 ft
12 ft
15 ft

🔺 **FCAT Practice**

Record your answers on the answer sheet provided by your teacher or on a sheet of paper.

1. Unleaded gasoline costs 1.49\frac{9}{10}$ per gallon. What is the best estimate of the cost of 8.131 gallons of unleaded gasoline? (Prerequisite Skill, pp. 600–601)

   Ⓐ $8   Ⓑ $9   Ⓒ $12   Ⓓ $16

2. Which equation is equivalent to $n + 7 = -4$? (Lesson 1-8)

   Ⓕ $n = 3$

   Ⓖ $n + 7 - 7 = -4 + 7$

   Ⓗ $n + 14 = -8$

   Ⓘ $n + 7 - 7 = -4 - 7$

3. Jamie started at point F and drove 28 miles due north to point G. He then drove due west to point H. He was then 35 miles from his starting point. What was the distance from point G to point H? (Lesson 3-5)

   Ⓐ 7 mi   Ⓑ 14 mi

   Ⓒ 21 mi   Ⓓ 31.5 mi

4. In 1990, the population of Tampa, Florida, was about 281,000. In 2000, the population was about 303,000. What was the approximate percent of increase in population over this ten-year period? (Lesson 5-7)

   Ⓕ 7%   Ⓖ 8%   Ⓗ 22%   Ⓘ 93%

5. In the diagram, $\angle A \cong \angle B$. Find the measure of $\angle A$. (Lessons 6-1, 6-2)

   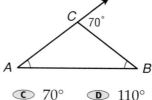

   Ⓐ 35°   Ⓑ 55°   Ⓒ 70°   Ⓓ 110°

6. Keisha needed to paint a triangular wall that was 19 feet long and 8 feet tall. When she stopped to rest, she still had 25 square feet of wall unpainted. How many square feet of wall did she paint before she stopped to rest? (Lesson 7-1)

   Ⓕ 51 ft²   Ⓖ 76 ft²

   Ⓗ 101 ft²   Ⓘ 127 ft²

7. If a circle's circumference is 28 yards, what is the *best* estimate of its diameter? (Lesson 7-2)

   Ⓐ 9 yd   Ⓑ 14 yd   Ⓒ 21 yd   Ⓓ 84 yd

8. The drawing shows a solid figure built with cubes. Which drawing represents a view of this solid from directly above? (Lesson 7-4)

   Front

   Ⓕ

   Front

   Ⓖ

   Front

   Ⓗ

   Front

   Ⓘ

   Front

9. The volume of the pyramid at the right is 54 cubic meters. Find the height of the prism. (Lesson 7-6)

   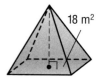

   18 m²

   Ⓐ 3 m   Ⓑ 9 m   Ⓒ 18 m   Ⓓ 36 m

TEST-TAKING TIP

Question 9 Most standardized tests will include any commonly used formulas at the front of the test booklet, but it will save you time to memorize many of these formulas. For example, you should memorize that the volume of a pyramid is one-third the area of the base times the height of the pyramid.

FCAT Practice

PART 2   Short Response/Grid In

Record your answers on the answer sheet provided by your teacher or on a sheet of paper.

10. You need $1\frac{2}{3}$ cups of chocolate chips to make one batch of chocolate chip cookies. How many $\frac{1}{3}$-cups of chocolate chips is this? (Lesson 2-4)

11. Four days ago, Evan had completed 5 pages of his term paper. Today he has completed a total of 15 pages. Find the rate of change in his progress in pages per day. (Lesson 4-2)

12. A boy who is $5\frac{1}{2}$ feet tall casts a shadow 4 feet long. A nearby tree casts a shadow 10 feet long.

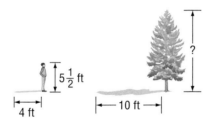

What is the height of the tree in feet? (Lesson 4-7)

13. Find the area of the top of a compact disc if its diameter is 12 centimeters and the diameter of the hole is 1.5 centimeters. (Lesson 7-2)

14. Mr. Brauen plans to carpet the part of his house shown on the floor plan below. How many square feet of carpet does he need? (Lesson 7-3)

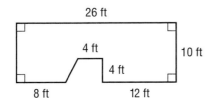

15. The curved part of a can will be covered by a label. What is the area of the label to the nearest tenth of a square centimeter? (Lesson 7-7)

FCAT Practice

PART 3   Extended Response

Record your answers on a sheet of paper. Show your work.

16. A prism with a triangular base has 9 edges, and a prism with a rectangular base has 12 edges.

9 edges   **12 edges**

Explain in words or symbols how to determine the number of edges for a prism with a 9-sided base. Be sure to include the number of edges in your explanation. (Lesson 7-4)

17. The diagrams show the design of the trashcans in the school cafeteria. (Lessons 7-5 and 7-7)

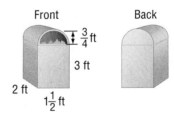

a. Find the volume of trash each can is designed to hold to the nearest tenth.

b. The top and sides of the cans need to be painted. Find the surface area of each can to the nearest tenth.

c. The paint used by the school covers 200 square feet per gallon. How many trashcans can be covered with 1 gallon of paint?

UNIT 4
Probability and Statistics

People often base their decisions about the future on data they've collected. In this unit, you will learn how to make such predictions using probability and statistics.

WebQuest INTERDISCIPLINARY PROJECT ⟶ MATH and SCIENCE

IT'S ALL IN THE GENES

Mirror, mirror on the wall... why do I look like my parents at all? You've been selected to join a team of genetic researchers to find an answer to this very question. On this adventure, you'll research basic genetic lingo and learn how to use a Punnett square. Then you'll gather information about the genetic traits of your classmates. You'll also make predictions based on an analysis of your findings. So grab your lab coat and your probability and statistics tool kits. This is one adventure you don't want to miss.

 Log on to msmath3.net/webquest to begin your WebQuest.

8 Probability

"How are math and bicycles related?"

Bicycles come in many styles, colors, and sizes. **To find how many different types of bicycles a manufacturer makes, you can use a tree diagram or the Fundamental Counting Principle.**

You will solve problems about different types of bicycles in Lesson 8-2.

GETTING STARTED

Take this quiz to see whether you are ready to begin Chapter 8. Refer to the lesson or page number in parentheses if you need more review.

▶ Vocabulary Review

Complete each sentence.

1. The equation $\dfrac{6}{15} = \dfrac{2}{5}$ is a ___?___ because it contains two equivalent ratios. (Lesson 4-4)

2. Percent is a ratio that compares a number to ___?___. (Lesson 5-1)

▶ Prerequisite Skills

Write each fraction in simplest form.

(Page 611)

3. $\dfrac{48}{72}$
4. $\dfrac{35}{60}$
5. $\dfrac{21}{99}$
6. $\dfrac{30}{82}$

Evaluate $x(x - 1)(x - 2)(x - 3)$ for each value of x. (Lesson 1-2)

7. $x = 11$
8. $x = 6$
9. $x = 9$
10. $x = 7$

Evaluate each expression. (Lesson 1-2)

11. $\dfrac{7 \cdot 6 \cdot 5}{3 \cdot 2 \cdot 1}$
12. $\dfrac{12 \cdot 11}{2 \cdot 1}$
13. $\dfrac{8 \cdot 7 \cdot 6 \cdot 5}{4 \cdot 3 \cdot 2 \cdot 1}$
14. $\dfrac{5 \cdot 4 \cdot 3}{3 \cdot 2 \cdot 1}$

Multiply. Write in simplest form. (Lesson 2-3)

15. $\dfrac{2}{3} \cdot \dfrac{3}{4}$
16. $\dfrac{4}{15} \cdot \dfrac{5}{7}$
17. $\dfrac{7}{8} \cdot \dfrac{4}{9}$
18. $\dfrac{3}{5} \cdot \dfrac{1}{6}$

Solve each problem. (Lessons 5-3 and 5-6)

19. Find 28% of 80.
20. Find 55% of 34.

FOLDABLES™ Study Organizer

Probability Make this Foldable to help you organize your notes. Begin with two sheets of $8\frac{1}{2}"\times 11"$ unlined paper.

STEP 1 Fold in Quarters
Fold each sheet in quarters along the width.

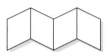

STEP 2 Tape
Unfold each sheet and tape to form one long piece.

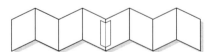

STEP 3 Label
Label each page with the lesson number as shown. Refold to form a booklet.

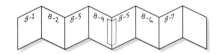

Reading and Writing As you read and study the chapter, write notes and examples for each lesson on each page of the journal.

Probability of Simple Events

Sunshine State Standards
MA.A.1.3.4-1, MA.E.2.3.2-1

WHEN am I ever going to use this?

GAMES The game of double-six dominoes is played with 28 tiles. Seven of the tiles are called doubles.

1. Write the ratio that compares the number of double tiles to the total number of tiles.
2. What percent of the tiles are doubles?
3. Write a fraction in simplest form that represents the part of the tiles that are doubles.
4. Write a decimal that represents the part of the tiles that are doubles.
5. Suppose you pick a domino without looking at the spots. Would you be more likely to pick a tile that is a double or one that is not a double? Explain.

Double

In the game of double-six dominoes, there are 28 tiles that can be picked. These tiles are called the **outcomes**. A list of all the tiles is called the **sample space**. If all outcomes occur by chance, the outcomes happen at **random**.

A **simple event** is a specific outcome or type of outcome. When picking dominoes, one event is picking a double. **Probability** is the chance that an event will happen.

Key Concept   **Probability**

Words   The probability of an event is a ratio that compares the number of favorable outcomes to the number of possible outcomes.

Symbols $P(\text{event}) = \dfrac{\text{number of favorable outcomes}}{\text{number of possible outcomes}}$

Example $P(\text{doubles}) = \dfrac{7}{28}$ or $\dfrac{1}{4}$

The probability that an event will happen is between 0 and 1 inclusive. A probability can be expressed as a fraction, a decimal, or a percent.

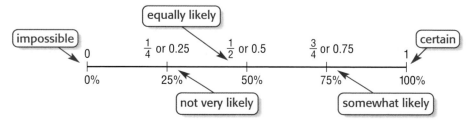

A box contains 5 green pens, 3 blue pens, 8 black pens, and 4 red pens. A pen is picked at random.

1 **What is the probability the pen is green?**

There are $5 + 3 + 8 + 4$ or 20 pens in the box.

$P(\text{green}) = \dfrac{\text{green pens}}{\text{total number of pens}}$ Definition of probability

$\quad\quad\quad = \dfrac{5}{20}$ or $\dfrac{1}{4}$ There are 5 green pens out of 20 pens.

The probability the pen is green is $\dfrac{1}{4}$. The probability can also be written as 0.25 or 25%.

2 **What is the probability the pen is blue or red?**

$P(\text{blue or red}) = \dfrac{\text{blue pens} + \text{red pens}}{\text{total number of pens}}$ Definition of probability

$\quad\quad\quad\quad\quad = \dfrac{3 + 4}{20}$ or $\dfrac{7}{20}$ There are 3 blue pens and 4 red pens.

The probability the pen is blue or red is $\dfrac{7}{20}$. The probability can also be written as 0.35 or 35%.

3 **What is the probability the pen is gold?**

Since there are no gold pens, the probability is 0.

Your Turn **The spinner is used for a game. Write each probability as a fraction, a decimal, or a percent.**

a. $P(6)$ **b.** $P(\text{odd})$

c. $P(5 \text{ or even})$ **d.** $P(\text{a number less than } 7)$

Suppose you roll a number cube. The events of rolling a 6 and of *not* rolling a 6 are **complementary events**. The sum of the probabilities of complementary events is 1.

EXAMPLE Probability of a Complementary Event

4 **PURCHASES** **A computer company manufactures 2,500 computers each day. An average of 100 of these computers are returned with defects. What is the probability that the computer you purchased is *not* defective?**

$2,500 - 100$ or 2,400 computers were not defective.

$P(\text{not defective}) = \dfrac{\text{nondefective computers}}{\text{total number of computers}}$ Definition of probability

$\quad\quad\quad\quad\quad\quad = \dfrac{2,400}{2,500}$ or $\dfrac{24}{25}$ There are 2,400 nondefective computers.

The probability that your computer is *not* defective is $\dfrac{24}{25}$.

Writing Math
Exercises 2 & 3

1. **Draw** a spinner where the probability of an outcome of white is $\frac{3}{8}$.

2. **OPEN ENDED** Give an example of an event with a probability of 1.

3. **FIND THE ERROR** Masao and Brian are finding the probability of getting a 2 when a number cube is rolled. Masao says it is $\frac{1}{6}$, and Brian says it is $\frac{2}{6}$. Who is correct? Explain.

GUIDED PRACTICE

The spinner is used for a game. Write each probability as a fraction, a decimal, and a percent.

4. $P(5)$

5. $P(\text{even})$

6. $P(\text{greater than 5})$

7. $P(\text{not 2})$

8. $P(\text{an integer})$

9. $P(\text{less than 7})$

10. **GAMES** A card game has 25 red cards, 25 green cards, 25 yellow cards, 25 blue cards, and 8 wild cards. What is the probability that the first card dealt is a wild card?

Practice and Applications

A beanbag is tossed on the square at the right. It lands at random in a small square. Write each probability as a fraction, a decimal, and a percent.

HOMEWORK HELP	
For Exercises	See Examples
11–20, 24–25	1–3
21	4
Extra Practice See pages 635, 655.	

11. $P(\text{red})$

12. $P(\text{blue})$

13. $P(\text{white or yellow})$

14. $P(\text{blue or red})$

15. $P(\text{not green})$

16. $P(\text{brown})$

17. What is the probability that a month picked at random starts with J?

18. What is the probability that a day picked at random is a Saturday?

19. A number cube is tossed. Are the events of rolling a number greater than 3 and a number less than 3 complementary events? Explain.

20. A coin is tossed twice and shows heads both times. What is the probability that the coin will show a tail on the next toss? Explain.

21. **WEATHER** A weather reporter says that there is a 40% chance of rain. What is the probability of *no* rain?

22. **WRITE A PROBLEM** Write a real-life problem with a probability of $\frac{1}{6}$.

23. **RESEARCH** Use the Internet or other resource to find the probability that a person from your state picked at random will be from your city or community.

HISTORY For Exercises 24–26, use the table at the right and the information below.

The U.S. Census Bureau divides the United States into four regions: Northeast, Midwest, South, and West.

U.S. Population (thousands)		
Region	1890	2000
Northeast	17,407	53,594
Midwest	22,410	64,393
South	20,028	100,237
West	3,134	63,198

Source: U.S. Census Bureau

24. Suppose a person living in the United States in 1890 was picked at random. What is the probability that the person lived in the West? Write as a decimal to the nearest thousandth.

25. Suppose a person living in the United States in 2000 was picked at random. What is the probability that the person lived in the West? Write as a decimal to the nearest thousandth.

26. How has the population of the West changed?

27. **CRITICAL THINKING** A box contains 5 red, 6 blue, 3 green, and 2 yellow crayons. How many red crayons must be added to the box so that the probability of randomly picking a red crayon is $\frac{2}{3}$?

EXTENDING THE LESSON The *odds* of an event occurring is a ratio that compares the number of favorable outcomes to the number of unfavorable outcomes. Suppose a number cube is rolled.

Find the odds of each outcome.

28. a 6

29. not a 6

30. an even number

Standardized Test Practice and Mixed Review

FCAT Practice

For Exercises 31 and 32, the following cards are put into a box.

| 2 | 6 | 7 | 5 | 8 | 9 | 8 | 4 |

31. **MULTIPLE CHOICE** Emma picks a card at random. The number on the card will *most likely* be

Ⓐ a number greater than 6.   Ⓑ a number less than 6.

Ⓒ an even number.   Ⓓ an odd number.

32. **MULTIPLE CHOICE** What is the probability of *not* getting an 8?

Ⓕ 25%   Ⓖ 30%   Ⓗ 50%   Ⓘ 75%

Analyze each measurement. Give the precision, significant digits if appropriate, greatest possible error, and relative error to two significant digits. (Lesson 7-9)

33. 8 cm   34. 0.36 kg   35. 4.83 m   36. 410 cm

37. **GEOMETRY** Find the surface area of a cone with radius of 5 inches and slant height of 12 inches. (Lesson 7-8)

GETTING READY FOR THE NEXT LESSON

BASIC SKILL Multiply.

38. $5 \cdot 6 \cdot 2$ 39. $5 \cdot 5 \cdot 8$ 40. $12 \cdot 5 \cdot 3$ 41. $7 \cdot 8 \cdot 2$

Problem-Solving Strategy
A Preview of Lesson 8-2

What You'll Learn
Solve problems by making an organized list.

Make an Organized List

We have all the orders for the Valentine's Day bouquets. Each student could choose any combination of red, pink, white, or yellow carnations for their bouquets.

How many different bouquets do you think there are?

Explore	We want to know how many different bouquets can be made from four different colors of carnations.
Plan	Let's make an organized list.
Solve	Four-color bouquets: red, pink, white, yellow
	Three-color bouquets: red, pink, white red, pink, yellow red, white, yellow pink, white, yellow
	Two-color bouquets: red, pink red, white red, yellow pink, white pink, yellow white, yellow
	One-color bouquets: red pink white yellow
	There is 1 four-color bouquet, 4 three-color bouquets, 6 two-color bouquets, and 4 one-color bouquets. There are $1 + 4 + 6 + 4$ or 15 bouquets.
Examine	Check the list. Make sure that every color combination is listed and that no color combination is listed more than once.

Analyze the Strategy

1. **Explain** why the list of possible bouquets was divided into four-color, three-color, two-color, and one-color bouquets.

2. **Explain** why a red and white bouquet is the same as a white and red bouquet.

3. **Write** a problem that can be solved by making an organized list. Include the organized list you would use to solve the problem.

Solve. Make an organized list.

4. **MONEY MATTERS** Destiny wants to buy a cookie from a vending machine. The cookie costs 45¢. If Destiny uses exact change, how many different combinations of nickels, dimes, and quarters can she use?

5. **READING** Rosa checked out three books from the library. While she was at the library, she visited the fiction, nonfiction, and biography sections. What are the possible combinations of book types she could have checked out?

Mixed Problem Solving

Solve. Use any strategy.

6. **GAMES** Steven and Derek are playing a guessing game. Steven says he is thinking of two integers between -10 and 10 that have a product of -12. If Derek has one guess, what is the probability that he will guess the pair of numbers?

7. **COOKING** The graph shows the number of types of outdoor grills sold. How does the number of charcoal grills compare to the number of gas grills?

Source: Barbecue Industry Association

BASEBALL For Exercises 8–10, use the following information.
In the World Series, two teams play each other until one team wins 4 games.

8. What is the least number of games needed to determine a winner of the World Series?

9. What is the greatest number of games needed to determine a winner?

10. How many different ways can a team win the World Series in six games or less? (*Hint*: The team that wins the series must win the last game.)

11. **SLEEP** What is the probability that a person between the ages of 35 and 49 talks in his or her sleep? Write the probability as a fraction and as a decimal.

Source: The Better Sleep Council

12. **MULTI STEP** At 2:00 P.M., Cody began writing the final draft of a report. At 3:30 P.M., he had written 5 pages. If he works at the same pace, when should he complete 8 pages?

13. **MONEY MATTERS** Rebecca is shopping for fishing equipment. She has $135 and has already selected items that total $98.50. If the sales tax is 8%, will she have enough to purchase a fishing net that costs $23?

14. **STANDARDIZED TEST PRACTICE** FCAT Practice
Which equation best identifies the pattern in the table?

x	y
-2	2
-1	0.5
0	0
1	0.5
2	2

Ⓐ $y = x^2$

Ⓑ $y = 2x^2$

Ⓒ $y = 0.5x^2$

Ⓓ $y = -x^2$

You will use the make an organized list strategy in the next lesson.

Counting Outcomes

WHEN am I ever going to use this?

BICYCLES Antonio wants to buy a Dynamo bicycle.

1. How many different styles are available?

2. How many different colors are available?

3. How many different sizes are available?

4. Make an organized list to determine how many different bicycles are available.

Dynamo Bicycles are the Best!

Choose your Dynamo Today!

Styles: Mountain or 10-Speed
Colors: Red, Black, or Green
Sizes: 26-inch or 28-inch

An organized list can help you determine the number of possible combinations or outcomes. One type of organized list is a **tree diagram**.

EXAMPLE Use a Tree Diagram

1 **BICYCLES** Draw a tree diagram to determine the number of different bicycles described in the real-life example above.

List each style of bicycle.

Each color is paired with each style of bicycle.

Each size is paired with each style and color of bicycle.

List of all the outcomes when choosing a bicycle.

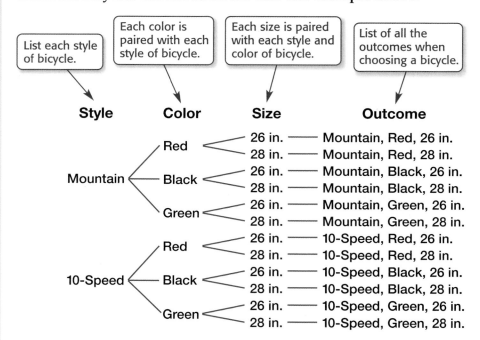

Style	Color	Size	Outcome
Mountain	Red	26 in.	Mountain, Red, 26 in.
		28 in.	Mountain, Red, 28 in.
	Black	26 in.	Mountain, Black, 26 in.
		28 in.	Mountain, Black, 28 in.
	Green	26 in.	Mountain, Green, 26 in.
		28 in.	Mountain, Green, 28 in.
10-Speed	Red	26 in.	10-Speed, Red, 26 in.
		28 in.	10-Speed, Red, 28 in.
	Black	26 in.	10-Speed, Black, 26 in.
		28 in.	10-Speed, Black, 28 in.
	Green	26 in.	10-Speed, Green, 26 in.
		28 in.	10-Speed, Green, 28 in.

There are 12 different Dynamo bicycles.

You can also find the total number of outcomes by multiplying. This principle is known as the **Fundamental Counting Principle**.

Key Concept   **Fundamental Counting Principle**

Words   If event M can occur in m ways and is followed by event N that can occur in n ways, then the event M followed by the event N can occur in $m \cdot n$ ways.

Example   If a number cube is rolled and a coin is tossed, there are $6 \cdot 2$ or 12 possible outcomes.

You can also use the Fundamental Counting Principle when there are more than two events.

REAL-LIFE MATH

COMMUNICATIONS On October 27, 1920, KDKA in Pittsburgh, Pennsylvania, became the first licensed radio station.

Source: *Time Almanac*

EXAMPLE   **Use the Fundamental Counting Principle**

2   **COMMUNICATIONS** In the United States, radio and television stations use call letters that start with K or W. How many different call letters with 4 letters are possible?

Use the Fundamental Counting Principle.

number of possible letters for the first letter	×	number of possible letters for the second letter	×	number of possible letters for the third letter	×	number of possible letters for the fourth letter	=	total number of possible call letters
2	×	26	×	26	×	26	=	35,152

There 35,152 possible call letters.

Your Turn Use the Fundamental Counting Principle to find the number of possible outcomes.

a. A hair dryer has 3 settings for heat and 2 settings for fan speed.

b. A restaurant offers a choice of 3 types of pasta with 5 types of sauce. Each pasta entrée comes with or without a meatball.

EXAMPLE   **Find Probability**

3   **GAMES** What is the probability of winning a lottery game where the winning number is made up of three digits from 0 to 9 chosen at random?

First, find the number of possible outcomes. Use the Fundamental Counting Principle.

choices for the first digit	×	choices for the second digit	×	choices for the third digit	=	total number of outcomes
10	×	10	×	10	=	1,000

There are 1,000 possible outcomes. There is 1 winning number. So, the probability of winning with one ticket is $\frac{1}{1,000}$. This can also be written as a decimal, 0.001, or a percent, 0.1%.

1. **Describe** a possible advantage for using a tree diagram rather than the Fundamental Counting Principle.

2. **OPEN ENDED** Give an example of a situation that has 15 outcomes.

3. **NUMBER SENSE** Whitney has a choice of a floral, plaid, or striped blouse to wear with a choice of a tan, black, navy, or white skirt. How many more outfits can she make if she buys a print blouse?

GUIDED PRACTICE

The spinner at the right is spun two times.

4. Draw a tree diagram to determine the number of outcomes.

5. What is the probability that both spins will land on red?

6. What is the probability that the two spins will land on different colors?

7. **FOOD** A pizza parlor has regular, deep-dish, and thin crust, 2 different cheeses, and 4 toppings. How many different one-cheese and one-topping pizzas can be ordered?

8. **GOVERNMENT** The first three digits of a social security number are a geographic code. The next two digits are determined by the year and the state where the number is issued. The final four digits are random numbers. How many possible ways can the last four digits be assigned?

Practice and Applications

Draw a tree diagram to determine the number of outcomes.

9. A penny, a nickel, and a dime are tossed.

10. A number cube is rolled and a penny is tossed.

11. A sweatshirt comes in small, medium, large, and extra large. It comes in white or red.

12. The Sweet Treats Shoppe has three flavors of ice cream: chocolate, vanilla, and strawberry; and two types of cones, regular and sugar.

Use the Fundamental Counting Principle to find the number of possible outcomes.

13. The day of the week is picked at random and a number cube is rolled.

14. A number cube is rolled 3 times.

15. There are 5 true-false questions on a history quiz.

16. There are 4 choices for each of 5 multiple-choice questions on a science quiz.

HOMEWORK HELP

For Exercises	See Examples
9–12, 17	1
13–16, 22–23	2
18–21	3

Extra Practice
See pages 635, 655.

For Exercises 17–20, each of the spinners at the right is spun once.

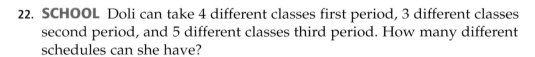

17. Draw a tree diagram to determine the number of outcomes.

18. What is the probability that both spinners land on the same color?

19. What is the probability that at least one spinner lands on blue?

20. What is the probability that at least one spinner lands on yellow?

21. **PROBABILITY** What is the probability of winning a lottery game where the winning number is made up of five digits from 0 to 9 chosen at random?

22. **SCHOOL** Doli can take 4 different classes first period, 3 different classes second period, and 5 different classes third period. How many different schedules can she have?

23. **STATES** In 2003, Ohio celebrated its bicentennial. The state issued bicentennial license plates with 2 letters, followed by 2 numbers and then 2 more letters. How many bicentennial license plates could the state issue?

24. **CRITICAL THINKING** If x coins are tossed, write an algebraic expression for the number of possible outcomes.

Standardized Test Practice and Mixed Review

 FCAT Practice

25. **MULTIPLE CHOICE** At the café, Dion can order one of the flavors of tea listed at the right. He can order the tea in a small, medium, or large cup. How many different ways can Dion order tea?

Flavors of Tea
mint
orange
peach
raspberry
strawberry

    Ⓐ 5   Ⓑ 8   Ⓒ 12   Ⓓ 15

26. **GRID IN** Felisa has a red and a white sweatshirt. Courtney has a black, a green, a red, and a white sweatshirt. Each girl picks a sweatshirt at random to wear to the picnic. What is the probability the girls will wear the same color sweatshirt?

Each letter of the word *associative* is written on 11 identical slips of paper. A piece of paper is chosen at random. Find each probability. (Lesson 8-1)

27. $P(s)$ 28. $P(vowel)$ 29. $P(not\ r)$ 30. $P(d)$

31. **MEASUREMENT** How many significant digits are in the measurement 14.4 centimeters? (Lesson 7-9)

GETTING READY FOR THE NEXT LESSON

PREREQUISITE SKILL Evaluate $n(n-1)(n-2)(n-3)$ for each value of n.
(Lesson 1-2)

32. $n = 5$ 33. $n = 10$ 34. $n = 12$ 35. $n = 8$

Permutations

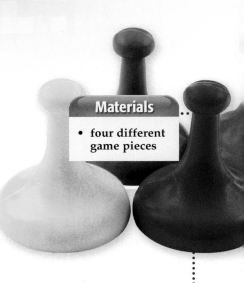

HANDS-ON Mini Lab

Materials
• four different game pieces

Work with a partner.

Suppose you are playing a game with 4 different game pieces. Show all of the ways the game pieces can be chosen first and second. Record each arrangement.

1. How many different arrangements did you make?

2. How many different game pieces could you pick for the first place?

3. Once you picked the first-place game piece, how many game pieces could you pick for the second place?

4. Use the Fundamental Counting Principle to determine the number of arrangements for first and second places.

5. How do the numbers in Exercises 1 and 4 compare?

When deciding who goes first and who goes second, order is important. An arrangement or listing in which order is important is called a **permutation**.

EXAMPLE Find a Permutation

1. **FOOD** An ice cream shop has 31 flavors. Carlos wants to buy a three-scoop cone with three different flavors. How many cones can he buy if order is important?

number of possible flavors for the first scoop	×	number of possible flavors for the second scoop	×	number of possible flavors for the third scoop	=	total number of possible cones
31	×	30	×	29	=	26,970

There are 26,970 different cones Carlos can order.

The symbol $P(31, 3)$ represents the number of permutations of 31 things taken 3 at a time.

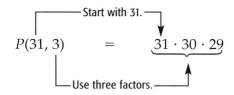

Start with 31.

$$P(31, 3) \quad = \quad 31 \cdot 30 \cdot 29$$

Use three factors.

EXAMPLES Use Permutation Notation

Find each value.

2 $P(8, 3)$

$P(8, 3) = 8 \cdot 7 \cdot 6$ or 336   8 things taken 3 at a time.

3 $P(6, 6)$

$P(6, 6) = 6 \cdot 5 \cdot 4 \cdot 3 \cdot 2 \cdot 1$ or 720   6 things taken 6 at a time.

Your Turn Find each value.

a. $P(12, 2)$ b. $P(4, 4)$ c. $P(10, 5)$

In Example 3, $P(6, 6) = 6 \cdot 5 \cdot 4 \cdot 3 \cdot 2 \cdot 1$. The mathematical notation 6! also means $6 \cdot 5 \cdot 4 \cdot 3 \cdot 2 \cdot 1$. The symbol 6! is read *six* **factorial**. $n!$ means the product of all counting numbers beginning with n and counting backward to 1. We define 0! as 1.

Standardized Test Practice

EXAMPLE Find Probability

4 **MULTIPLE-CHOICE TEST ITEM** Consider all of the four-digit numbers that can be formed using the digits 1, 2, 3, and 4 where no digit is used twice. Find the probability that one of these numbers picked at random is between 1,000 and 2,000.

Ⓐ $33\frac{1}{3}\%$ Ⓑ 25%   Ⓒ 20%   Ⓓ 10%

Read the Test Item

You are considering all of the permutations of 4 digits taken 4 at a time. You wish to find the probability that one of these numbers picked at random is greater than 1,000, but less than 2,000.

Solve the Test Item

Find the number of possible four-digit numbers. $P(4, 4) = 4!$

In order for a number to be between 1,000 and 2,000, the thousands digit must be 1.

number of ways to pick the first digit	×	number of ways to pick the last three digits	=	number of permutations between 1,000 and 2,000
1	×	$P(3, 3)$	=	$P(3, 3)$ or 3!

$P(\text{between } 1{,}000 \text{ and } 2{,}000)$

$= \dfrac{\text{number of permutations between 1,000 and 2,000}}{\text{total number of permutations}}$

$= \dfrac{3!}{4!}$ Substitute.

$= \dfrac{\cancel{3} \cdot \cancel{2} \cdot 1}{4 \cdot \cancel{3} \cdot \cancel{2} \cdot 1}$ Definition of factorial

$= \dfrac{1}{4}$ or 25%   The probability is 25%, which is B.

msmath3.net/extra_examples/fcat

1. **Tell** the difference between 9! and $P(9, 5)$.

2. **OPEN ENDED** Write a problem that can be solved by finding the value of $P(7, 3)$.

3. **FIND THE ERROR** Daniel and Bailey are evaluating $P(7, 3)$. Who is correct? Explain.

Daniel
$P(7, 3) = 7 \cdot 6 \cdot 5 \cdot 4 \cdot 3$
$= 2,520$

Bailey
$P(7, 3) = 7 \cdot 6 \cdot 5$
$= 210$

GUIDED PRACTICE

Find each value.

4. $P(5, 3)$ 5. $P(7, 4)$ 6. 3!   7. 8!

8. In a race with 7 runners, how many ways can the runners end up in first, second, and third place?

9. How many ways can you arrange the letters in the word *equals*?

10. **SPORTS** There are 9 players on a baseball team. How many ways can the coach pick the first 4 batters?

Practice and Applications

Find each value.

11. $P(6, 3)$ 12. $P(9, 2)$ 13. $P(5, 5)$ 14. $P(7, 7)$

15. $P(14, 5)$ 16. $P(12, 4)$ 17. $P(25, 4)$ 18. $P(100, 3)$

19. 2!   20. 5!   21. 11!   22. 12!

HOMEWORK HELP	
For Exercises	See Examples
11–22	2, 3
23–26, 29–32	1
27–28	4
Extra Practice See pages 636, 655.	

23. How many ways can the 4 runners on a relay team be arranged?

24. **FLAGS** The flag of Mexico is shown at the right. How many ways could the Mexican government have chosen to arrange the three bar colors (green, white, and red) on the flag?

25. A security system has a pad with 9 digits. How many four-number "passwords" are available if no digit is repeated?

26. Of the 10 games at the theater's arcade, Tyrone plans to play 3 different games. In how many orders can he play the 3 games?

27. **MULTI STEP** Each arrangement of the letters in the word *quilt* is written on a piece of paper. One paper is drawn at random. What is the probability that the word begins with *q*?

28. **MULTI STEP** Each arrangement of the letters in the word *math* is written on a piece of paper. One paper is drawn at random. What is the probability that the word ends with *th*?

29. **SOCCER** The teams of the Eastern Conference of Major League Soccer are listed at the right. If there are no ties for placement in the conference, how many ways can the teams finish the season from first to last place?

Eastern Conference
Chicago Fire
Columbus Crew
D.C. United
MetroStars
New England Revolution

ENTERTAINMENT **For Exercises 30–32, use the following information.**
In the 2002 Tournament of Roses Parade, there were 54 floats, 23 bands, and 26 equestrian groups.

30. In how many ways could the first 3 bands be chosen?

31. In how many ways could the first 3 equestrian groups be chosen?

32. Two of the 54 floats were entered by the football teams competing in the Rose Bowl. If they cannot be first or second, how many ways can the first 3 floats be chosen?

 Data Update How many floats, bands, and equestrian groups were in the last Tournament of Roses Parade? Visit **msmath3.net/data_update** to learn more.

33. **CRITICAL THINKING** If 9! = 362,880, use mental math to find 10! Explain.

34. **CRITICAL THINKING** Compare $P(n, n)$ and $P(n, n - 1)$, where n is any whole number greater than one. Explain.

Standardized Test Practice and Mixed Review

FCAT Practice

35. **MULTIPLE CHOICE** How many seven-digit phone numbers are available if a digit can only be used once and the first number cannot be 0 or 1?
    Ⓐ 5,040   Ⓑ 483,840   Ⓒ 544,320   Ⓓ 10,000,000

36. **MULTIPLE CHOICE** The school talent show is featuring 13 acts. In how many ways can the talent show coordinator order the first 5 acts?
    Ⓕ 6,227,020,800   Ⓖ 371,293   Ⓗ 154,440   Ⓘ 1,287

37. **SPORTS** The Silvercreek Ski Resort has 4 ski lifts up the mountain and 11 trails down the mountain. How many different ways can a skier take a ski lift up the mountain and then ski down? (Lesson 8-2)

A number cube is rolled. Find each probability. (Lesson 8-1)

38. $P(5 \text{ or } 6)$ 39. $P(\text{odd})$ 40. $P(\text{less than } 10)$ 41. $P(1 \text{ or even})$

42. Write an equation you could use to find the length of the missing side of the triangle at the right. Then find the missing length. (Lesson 3-4)

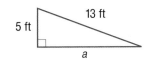

13 ft
5 ft
a

GETTING READY FOR THE NEXT LESSON

PREREQUISITE SKILL Evaluate each expression. (Lesson 1-2)

43. $\dfrac{6 \cdot 5 \cdot 4}{3 \cdot 2 \cdot 1}$ 44. $\dfrac{10 \cdot 9 \cdot 8 \cdot 7}{4 \cdot 3 \cdot 2 \cdot 1}$ 45. $\dfrac{20 \cdot 19}{2 \cdot 1}$ 46. $\dfrac{6 \cdot 5 \cdot 4 \cdot 3 \cdot 2}{5 \cdot 4 \cdot 3 \cdot 2 \cdot 1}$

Combellations

What You'll Learn
Find the number of combinations of objects.

NEW Vocabulary
combination

MATH Symbols
$C(a, b)$ the number of combinations of a things taken b at a time

HANDS-ON Mini Lab

Work in a group of 6.

Each member of the group should shake hands with every other member of the group. Make a list of each handshake.

1. How many different handshakes did you record?
2. Find $P(6, 2)$.
3. Is the number of handshakes equal to $P(6, 2)$? Explain.

In the Mini Lab, it did not matter whether you shook hands with your friend, or your friend shook hands with you. Order is not important. An arrangement or listing where order is not important is called a **combination**. Let's look at a simpler form of the handshake problem.

EXAMPLE   Find a Combination

1. **GEOMETRY** Four points are located on a circle. How many line segments can be drawn with these points as endpoints?

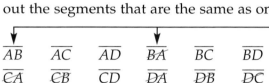

Method 1

First list all of the possible permutations of A, B, C, and D taken two at a time. Then cross out the segments that are the same as one another.

| $\overline{AB}$ | $\overline{AC}$ | $\overline{AD}$ | $\overline{BA}$ | $\overline{BC}$ | $\overline{BD}$ |
| $\overline{CA}$ | $\overline{CB}$ | $\overline{CD}$ | $\overline{DA}$ | $\overline{DB}$ | $\overline{DC}$ |

$\overline{AB}$ is the same as $\overline{BA}$, so cross off one of them.

There are only 6 different segments.

Method 2

Find the number of permutations of 4 points taken 2 at a time.

$P(4, 2) = 4 \cdot 3$ or 12

Since order is not important, divide the number of permutations by the number of ways 2 things can be arranged.

$$\frac{12}{2!} = \frac{12}{2 \cdot 1} \text{ or } 6$$

There are 6 segments that can be drawn.

Your Turn

a. If there are 8 people in a room, how many handshakes will occur if each person shakes hands with every other person?

The symbol $C(4, 2)$ represents the number of combinations of 4 things taken 2 at a time.

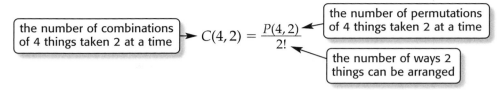

the number of combinations of 4 things taken 2 at a time ⟶ $C(4, 2) = \dfrac{P(4, 2)}{2!}$ ⟵ the number of permutations of 4 things taken 2 at a time

⟵ the number of ways 2 things can be arranged

EXAMPLE · Use Combination Notation

2 **Find $C(7, 4)$.**

$C(7, 4) = \dfrac{P(7, 4)}{4!}$ Definition of $C(7, 4)$

$= \dfrac{7 \cdot \overset{1}{\cancel{6}} \cdot 5 \cdot \overset{1}{\cancel{4}}}{\underset{1}{\cancel{4}} \cdot \underset{1}{\cancel{3}} \cdot \underset{1}{\cancel{2}} \cdot 1}$ or 35 $P(7, 4) = 7 \cdot 6 \cdot 5 \cdot 4$ and $4! = 4 \cdot 3 \cdot 2 \cdot 1$

EXAMPLES · Combinations and Permutations

MUSIC The makeup of a symphony is shown in the table at the right.

3 **A group of 3 musicians from the strings section will talk to students at Madison Middle School. Does this represent a combination or a permutation? How many possible groups could talk to the students?**

Makeup of the Symphony	
Instrument	**Number**
Strings	45
Woodwinds	8
Brass	8
Percussion	3
Harps	2

This is a combination problem since the order is not important.

$C(45, 3) = \dfrac{P(45, 3)}{3!}$ 45 musicians taken 3 at a time

$= \dfrac{\overset{15}{\cancel{45}} \cdot \overset{22}{\cancel{44}} \cdot 43}{\underset{1}{\cancel{3}} \cdot \underset{1}{\cancel{2}} \cdot 1}$ or 14,190 $P(45, 3) = 45 \cdot 44 \cdot 43$ and $3! = 3 \cdot 2 \cdot 1$

There are 14,190 different groups that could talk to the students.

4 **One member from the strings section will talk to students at Brown Middle School, another to students at Oak Avenue Middle School, and another to students at Jefferson Junior High. Does this represent a combination or a permutation? How many possible ways can the strings members talk to the students?**

Since it makes a difference which member goes to which school, order is important. This is a permutation.

$P(45, 3) = 45 \cdot 44 \cdot 43$ or 85,140   Definition of $P(45, 3)$

There are 85,140 ways for the members to talk to the students.

1. **OPEN ENDED** Give an example of a combination and an example of a permutation.

2. **Which One Doesn't Belong?** Identify the situation that is not the same as the other three. Explain your reasoning.

choosing 3 toppings for the pizzas to be served at the party	choosing 3 members for the decorating committee	choosing 3 people to chair 3 different committees	choosing 3 desserts to serve at the party

GUIDED PRACTICE

Find each value.

3. $C(6, 2)$ 4. $C(10, 5)$ 5. $C(7, 6)$ 6. $C(8, 4)$

Determine whether each situation is a *permutation* or a *combination*.

7. writing a four-digit number using no digit more than once

8. choosing 3 shirts to pack for vacation

9. How many different starting squads of 6 players can be picked from 10 volleyball players?

10. How many different combinations of 2 colors can be chosen as school colors from a possible list of 8 colors?

Practice and Applications

Find each value.

11. $C(9, 2)$ 12. $C(6, 3)$ 13. $C(9, 8)$ 14. $C(8, 7)$

15. $C(9, 5)$ 16. $C(10, 4)$ 17. $C(18, 4)$ 18. $C(20, 3)$

Determine whether each situation is a *permutation* or a *combination*.

19. choosing a committee of 5 from the members of a class

20. choosing 2 co-captains of the basketball team

21. choosing the placement of 9 model cars in a line

22. choosing 3 desserts from a dessert tray

23. choosing a chairperson and an assistant chairperson for a committee

24. choosing 4 paintings to display at different locations

25. How many three-topping pizzas can be ordered from a list of toppings at the right?

26. **GEOMETRY** Eight points are located on a circle. How many line segments can be drawn with these points as endpoints?

HOMEWORK HELP

For Exercises	See Examples
11–18	2
19–24, 27–32	3, 4
25–26	1

Extra Practice
See pages 636, 655.

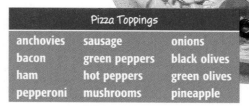

Pizza Toppings		
anchovies	sausage	onions
bacon	green peppers	black olives
ham	hot peppers	green olives
pepperoni	mushrooms	pineapple

27. There are 20 runners in a race. In how many ways can the runners take first, second, and third place?

28. How many ways can 7 people be arranged in a row for a photograph?

29. How many five-card hands can be dealt from a standard deck of 52 cards?

30. **GAMES** In the game of cribbage, a player gets 2 points for each combination of cards that totals 15. How many points for totals of 15 are in the hand at the right?

ENTERTAINMENT For Exercises 31 and 32, use the following information.
An amusement park has 15 roller coasters. Suppose you only have time to ride 8 of the coasters.

31. How many ways are there to ride 8 coasters if order is important?

32. How many ways are there to ride 8 coasters if order is not important?

33. **CRITICAL THINKING** Is the value of $P(x, y)$ *sometimes*, *always*, or *never* greater than the value of $C(x, y)$? Explain. Assume x and y are positive integers and $x \geq y$.

Standardized Test Practice and Mixed Review

FCAT Practice

34. **MULTIPLE CHOICE** Which situation is represented by $C(8, 3)$?
   Ⓐ the number of arrangements of 8 people in a line
   Ⓑ the number of ways to pick 3 out of 8 vegetables to add to a salad
   Ⓒ the number of ways to pick 3 out of 8 students to be the first, second, and third contestant in a spelling bee
   Ⓓ the number of ways 8 people can sit in a row of 3 chairs

35. **SHORT RESPONSE** The enrollment for Centerville Middle School is given at the right. How many different four-person committees could be formed from the students in the 8th grade?

Centerville Middle School		
Class	**Boys**	**Girls**
6th grade	42	47
7th grade	55	49
8th grade	49	53

Find each value. (Lesson 8-3)

36. $P(7, 2)$ 37. $P(15, 4)$ 38. $10!$ 39. $7!$

40. **SCHOOL** At the school cafeteria, students can choose from 4 entrees and 3 beverages. How many different lunches of one entree and one beverage can be purchased at the cafeteria? (Lesson 8-2)

GETTING READY FOR THE NEXT LESSON

PREREQUISITE SKILL Multiply. Write in simplest form. (Lesson 2-3)

41. $\dfrac{4}{5} \cdot \dfrac{3}{8}$ 42. $\dfrac{3}{10} \cdot \dfrac{5}{6}$ 43. $\dfrac{7}{12} \cdot \dfrac{3}{14}$ 44. $\dfrac{2}{3} \cdot \dfrac{9}{10}$

A Follow-Up of Lesson 8-4

Sunshine State Standards
MA.E.2.3.1-2

Combinations and Pascal's Triangle

For many years, mathematicians have been interested in a pattern called *Pascal's Triangle*.

What You'll Learn
Identify patterns in Pascal's Triangle.

Materials
• paper
• pencil

Row									Sum
0				1					$1 = 2^0$
1			1		1				$2 = 2^1$
2		1		2		1			$4 = 2^2$
3	1		3		3		1		$8 = 2^3$
4	1	4		6		4		1	$16 = 2^4$

ACTIVITY *Work with a partner.*

1 Find all possible outcomes if you toss a penny and a dime.

STEP 1 Copy and complete the tree diagram shown below.

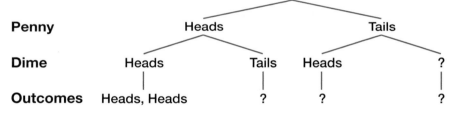

STEP 2 In the tree diagram above, how many outcomes have exactly no heads? one head? two heads?

STEP 3 Use a tree diagram to determine the outcomes of tossing a penny, a nickel, and a dime. How many outcomes have exactly no head, one head, two heads, three heads?

Writing Math

1. **Describe** the pattern in the numbers in Pascal's Triangle. Use the pattern to write the numbers in Rows 5, 6, and 7.

2. **Explain** how your tree diagrams are related to Pascal's Triangle.

3. Suppose you toss a penny, nickel, dime, and quarter. **Make a conjecture** about how many outcomes have exactly no head, one head, two heads, and so on. **Test your conjecture**.

Pascal's Triangle can also be used to find probabilities of events for which there are only two possible outcomes, such as heads-tails, boy-girl, and true-false.

ACTIVITY   *Work with a partner.*

2 **In a five-item true-false quiz, what is the probability of getting exactly three right answers by guessing?**

STEP 1 Since there are five items, look at Row 5.

Number Right	0	1	2	3	4	5
Row 5	1	5	10	10	5	1

There are 10 ways to get exactly three right answers.

STEP 2 Find the total possible outcomes.

$$1 + 5 + 10 + 10 + 5 + 1 = 32$$

STEP 3 Find the probability.

$$\frac{\text{number of ways to guess 3 right answers}}{\text{number of outcomes}} = \frac{10}{32} \text{ or } \frac{5}{16}$$

So, the probability of guessing exactly three right answers is $\frac{5}{16}$.

Writing Math

4. Suppose you guess on a five-item true-false test. What is the probability of getting all of the right answers?

5. There are ten true-false questions on a quiz. What is the probability of guessing at least six correct answers and passing the quiz?

6. If you toss eight coins, you would expect there to be four heads and four tails. What is the probability this will happen?

For Exercises 7–9, use the following information.
The Band Boosters are selling pizzas. You can choose to add onions, pepperoni, mushrooms, and/or green pepper to the basic cheese pizza.

7. Find each number of combinations of toppings.

   a. $C(4, 0)$ **b.** $C(4, 1)$ **c.** $C(4, 2)$ **d.** $C(4, 3)$ **e.** $C(4, 4)$

8. How many different combinations are there in all?

9. Suppose the Boosters decide to offer hot peppers as an additional choice. How many combinations of pizzas are available?

Mid-Chapter Practice Test

Vocabulary and Concepts

1. **Draw** a spinner where P(green) is $\frac{1}{4}$. (Lesson 8-1)

2. **Write** a problem that is solved by finding the value of $P(8, 3)$. (Lesson 8-3)

Skills and Applications

There are 6 purple, 5 blue, 3 yellow, 2 green, and 4 brown marbles in a bag. One marble is selected at random. Write each probability as a fraction, a decimal, and a percent. (Lesson 8-1)

3. P(purple)

4. P(blue)

5. P(not brown)

6. P(purple or blue)

7. P(not green)

8. P(blue or green)

For Exercises 9–11, a penny is tossed, and a number cube is rolled. (Lesson 8-2)

9. Draw a tree diagram to determine the number of outcomes.

10. What is the probability that the penny shows heads and the number cube shows a six?

11. What is the probability that the penny shows heads and the number cube shows an even number?

Find each value. (Lessons 8-3 and 8-4)

12. $P(5, 3)$

13. $P(6, 2)$

14. $P(5, 5)$

15. $C(5, 3)$

16. $C(6, 2)$

17. $C(5, 5)$

18. **SCHOOL** How many ways can 2 student council members be elected from 7 candidates? (Lesson 8-4)

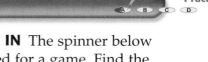

FCAT
Practice

Standardized Test Practice

19. **MULTIPLE CHOICE** A pizza shop advertises that it has 3 different crusts, 3 different meat toppings, and 5 different vegetables. If Carlotta wants a pizza with one meat and one vegetable, how many different pizzas can she order? (Lesson 8-2)

   Ⓐ 11   Ⓑ 15

   Ⓒ 45   Ⓓ 90

20. **GRID IN** The spinner below is used for a game. Find the probability that the spinner will *not* land on yellow. (Lesson 8-1)

The Game Zone

A Place To Practice Your Math Skills

Winning Numbers

● **GET READY!**

Players: three
Materials: 15 index cards, scissors, markers, 3 paper bags

● **GET SET!**

- Cut each index card in half, making 30 cards.
- Give each player 10 cards.
- Each player writes one number from 0 to 9 on each card.
- Each player takes a different bag and places his or her cards in the bag.

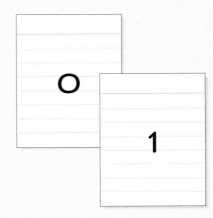

● **GO!**

- Each player writes down three numbers each between 0 and 9. Repeat numbers are allowed.
- Each player draws a card from his or her paper bag without looking. These are the winning numbers.
- Each player scores 2 points if one number matches, 16 points if two numbers match, and 32 points if all three numbers match. Order is not important.
- Replace the cards in the paper bags. Repeat the process.
- **Who Wins?** The first person to get a total of 100 points is the winner.

Probability of Compound Events

Sunshine State Standards
MA.A.3.3.2-1, MA.E.2.3.1-2

What You'll Learn
Find the probability of independent and dependent events.

⇒ NEW Vocabulary

compound event
independent events
dependent events

WHEN am I ever going to use this?

GAMES A game uses a number cube and the spinner shown at the right.

1. A player rolls the number cube. What is P(odd number)?

2. The player spins the spinner. What is P(red)?

3. What is the product of the probabilities in Exercises 1 and 2?

4. Draw a tree diagram to determine the probability that the player will get an odd number and red.

5. Compare your answers for Exercises 3 and 4.

The combined action of rolling a number cube and spinning a spinner is a compound event. In general, a **compound event** consists of two or more simple events.

The outcome of the spinner does not depend on the outcome of the number cube. These events are independent. For **independent events**, the outcome of one event does not affect the other event.

Key Concept — Probability of Two Independent Events

Words The probability of two independent events can be found by multiplying the probability of the first event by the probability of the second event.

Symbols $P(A \text{ and } B) = P(A) \cdot P(B)$

EXAMPLE Probability of Independent Events

1 The two spinners are spun. What is the probability that both spinners will show an even number?

$P(\text{first spinner is even}) = \dfrac{3}{7}$

$P(\text{second spinner is even}) = \dfrac{1}{2}$

$P(\text{both spinners are even}) = \dfrac{3}{7} \cdot \dfrac{1}{2} \text{ or } \dfrac{3}{14}$

EXAMPLE — Use Probability to Solve a Problem

2 **POPULATION** Use the information in the table. In the United States, what is the probability that a person picked at random will be under the age of 18 and live in an urban area?

$P(\text{younger than } 18) = \frac{1}{4}$

$P(\text{urban area}) = \frac{4}{5}$

$P(\text{younger than } 18 \text{ and urban area})$

$\quad = \frac{1}{4} \cdot \frac{4}{5}$ or $\frac{1}{5}$

The probability that the two events will occur is $\frac{1}{5}$.

United States	
Demographic Group	**Fraction of the Population**
Under age 18	$\frac{1}{4}$
18 to 64 years old	$\frac{5}{8}$
65 years or older	$\frac{1}{8}$
Urban	$\frac{4}{5}$
Rural	$\frac{1}{5}$

Source: U.S. Census Bureau

If the outcome of one event affects the outcome of another event, the compound events are called **dependent events**.

Key Concept — Probability of Two Dependent Events

Words If two events, A and B, are dependent, then the probability of both events occurring is the product of the probability of A and the probability of B after A occurs.

Symbols $P(A \text{ and } B) = P(A) \cdot P(B \text{ following } A)$

EXAMPLE — Probability of Dependent Events

3 There are 2 white, 8 red, and 5 blue marbles in a bag. Once a marble is selected, it is not replaced. Find the probability that two red marbles are chosen.

Since the first marble is not replaced, the first event affects the second event. These are dependent events.

$P(\text{first marble is red}) = \frac{8}{15}$ ← number of red marbles ← total number of marbles

$P(\text{second marble is red}) = \frac{7}{14}$ ← number of red marbles after one red marble is removed ← total number of marbles after one red marble is removed

$P(\text{two red marbles}) = \frac{\overset{4}{\cancel{8}}}{15} \cdot \frac{\overset{1}{\cancel{7}}}{\underset{1}{\cancel{\underset{7}{14}}}}$ or $\frac{4}{15}$

Your Turn Find each probability.

a. $P(\text{two blue marbles})$

b. $P(\text{a white marble and then a blue marble})$

Skill and Concept Check

1. **Compare and contrast** independent events and dependent events.

2. **OPEN ENDED** Give an example of dependent events.

3. **FIND THE ERROR** The spinner at the right is spun twice. Evita and Tia are finding the probability that both spins will result in an odd number. Who is correct? Explain.

Evita	Tia
$\frac{3}{5} \cdot \frac{3}{5} = \frac{9}{25}$	$\frac{3}{5} \cdot \frac{2}{4} = \frac{6}{20}$ or $\frac{3}{10}$

GUIDED PRACTICE

A penny is tossed, and a number cube is rolled. Find each probability.

4. P(tails and 3)

5. P(heads and odd)

Two cards are drawn from a deck of ten cards numbered 1 to 10. Once a card is selected, it is not returned. Find each probability.

6. P(two even cards)

7. P(a 6 and then an odd number)

8. **MARKETING** A discount supermarket has found that 60% of their customers spend more than $75 each visit. What is the probability that the next two customers will spend more than $75?

Practice and Applications

A number cube is rolled, and the spinner at the right is spun. Find each probability.

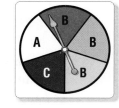

9. P(1 and A)

10. P(3 and B)

11. P(even and C)

12. P(odd and B)

13. P(greater than 2 and A)

14. P(less than 3 and B)

15. What is the probability of tossing a coin 3 times and getting heads each time?

16. What is the probability of rolling a number cube 3 times and getting numbers greater than 4 each time?

There are 3 yellow, 5 red, 4 blue, and 8 green candies in a bag. Once a candy is selected, it is not replaced. Find each probability.

17. P(two red candies)

18. P(two blue candies)

19. P(a yellow candy and then a blue candy)

20. P(a green candy and then a red candy)

21. P(two candies that are not green)

22. P(two candies that are neither blue nor green)

HOMEWORK HELP	
For Exercises	See Examples
9–16	1
17–22	3
23–24	2
Extra Practice See pages 636, 655.	

KITCHENS For Exercises 23 and 24, use the table at the right. Round to the nearest tenth of a percent.

23. What is the probability that a household picked at random will have both an electric frying pan and a toaster?

24. What is the probability that a household picked at random will use both a mixer and a drip coffee maker?

EXTENDING THE LESSON If two events cannot happen at the same time, they are said to be *mutually exclusive*. For example, suppose you randomly select a card from a standard deck of 52 cards. Getting a 5 or getting a 6 are mutually exclusive events. To find the probability of two mutually exclusive events, add the probabilities.

$$P(5 \text{ or } 6) = P(5) + P(6)$$
$$= \frac{1}{13} + \frac{1}{13} \text{ or } \frac{2}{13}$$

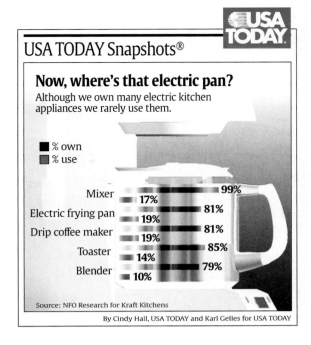

USA TODAY Snapshots®

Now, where's that electric pan?
Although we own many electric kitchen appliances we rarely use them.

■ % own
■ % use

	% use	% own
Mixer	17%	99%
Electric frying pan	19%	81%
Drip coffee maker	19%	81%
Toaster	14%	85%
Blender	10%	79%

Source: NFO Research for Kraft Kitchens

By Cindy Hall, USA TODAY and Karl Gelles for USA TODAY

Consider a standard deck of 52 cards. Find each probability.

25. *P*(face card or an ace)

26. *P*(club or a red card)

27. **CRITICAL THINKING** There are 9 marbles in a bag having 3 colors of marbles. The probability of picking 2 red marbles at random and without replacement is $\frac{1}{6}$. How many red marbles are in the bag?

Standardized Test Practice and Mixed Review

FCAT Practice

28. **MULTIPLE CHOICE** Jeremy tossed a coin and rolled a number cube. What is the probability that he will get tails and roll a multiple of 3?

  Ⓐ $\frac{1}{2}$ Ⓑ $\frac{1}{3}$ Ⓒ $\frac{1}{4}$ Ⓓ $\frac{1}{6}$

29. **GRID IN** Suppose you pick 3 cards from a standard deck of 52 cards without replacement. What is the probability all of the cards will be red?

Find each value. (Lesson 8-4)

30. $C(8, 5)$ 31. $C(7, 2)$ 32. $C(6, 5)$ 33. $C(9, 3)$

34. **SPORTS** There are 10 players on a softball team. How many ways can a coach pick the first 3 batters? (Lesson 8-3)

GETTING READY FOR THE NEXT LESSON

PREREQUISITE SKILL Write each fraction in simplest form. (Page 611)

35. $\frac{52}{120}$ 36. $\frac{33}{90}$ 37. $\frac{49}{70}$ 38. $\frac{24}{88}$

8-6 Experimental Probability

Sunshine State Standards
MA.A.3.3.2-1, MA.A.4.3.1-1, MA.A.4.3.1-2, MA.E.2.3.1-1, MA.E.3.3.1-2

What You'll Learn

Find experimental probability.

NEW Vocabulary

experimental probability
theoretical probability

REVIEW Vocabulary

proportion: a statement of equality of two or more ratios, $\frac{a}{b} = \frac{c}{d}$, $b \neq 0$, $d \neq 0$
(Lesson 4-4)

HANDS-ON Mini Lab

Materials
• paper bag containing 10 colored marbles

Work with a partner.

Draw one marble from the bag, record its color, and replace it in the bag. Repeat this 50 times.

1. Compute the ratio $\frac{\text{number of times color was drawn}}{\text{total number of draws}}$ for each color of marble.

2. Is it possible to have a certain color marble in the bag and never draw that color?

3. Open the bag and count the marbles. Find the ratio $\frac{\text{number of each color marble}}{\text{total number of marbles}}$ for each color of marble.

4. Are the ratios in Exercises 1 and 3 the same? Explain why or why not.

In the Mini Lab above, you determined a probability by conducting an experiment. Probabilities that are based on frequencies obtained by conducting an experiment are called **experimental probabilities**. Experimental probabilities usually vary when the experiment is repeated.

Probabilities based on known characteristics or facts are called **theoretical probabilities**. For example, you can compute the theoretical probability of picking a certain color marble from a bag. Theoretical probability tells you what *should* happen in an experiment.

EXAMPLES Experimental Probability

Michelle is conducting an experiment to find the probability of getting various sums when two number cubes are rolled. The results of her experiment are given at the right.

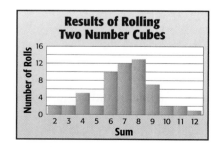

Results of Rolling Two Number Cubes

1. According to the experimental probability, is Michelle likely to get a sum of 12 on the next roll?

Based on the results of the rolls so far, a sum of 12 is not very likely.

2. How many possible outcomes are there for a pair of number cubes?

There are 6 · 6 or 36 possible outcomes.

EXAMPLE Theoretical Probability

3 What is the theoretical probability of rolling a double six?

The theoretical probability is $\frac{1}{6} \cdot \frac{1}{6}$ or $\frac{1}{36}$.

The experimental probability and the theoretical probability seem to be consistent.

EXAMPLE Experimental Probability

4 **MARKETING** Two hundred teenagers were asked whether they purchased certain household items in the past year. The table gives the results of the survey. What is the experimental probability that a teenager bought a photo frame in the last year?

Item	Number Who Purchased the Item
candles	110
photo frames	95

There were 200 teenagers surveyed and 95 purchased a photo frame in the last year. The experimental probability is $\frac{95}{200}$ or $\frac{19}{40}$.

Your Turn

a. What is the experimental probability that a teenager bought a candle in the last year?

You can use past performance to predict future events.

EXAMPLES Use Probability to Predict

FARMING Over the last 8 years, the probability that corn seeds planted by Ms. Diaz produced corn is $\frac{5}{6}$.

5 Is this probability experimental or theoretical? Explain.

This is an experimental probability since it is based on what happened in the past.

6 If Ms. Diaz wants to have 10,000 corn-bearing plants, how many seeds should she plant?

This problem can be solved using a proportion.

| 5 out of 6 seeds should produce corn. | → $\frac{5}{6} = \frac{10,000}{x}$ ← | 10,000 out of x seeds should produce corn. |

Solve the proportion.

$\frac{5}{6} = \frac{10,000}{x}$ Write the proportion.

$5 \cdot x = 6 \cdot 10,000$ Find the cross products.

$5x = 60,000$ Multiply.

$\frac{5x}{5} = \frac{60,000}{5}$ Divide each side by 5.

$x = 12,000$ Ms. Diaz should plant 12,000 seeds.

STUDY TIP

Mental Math
For every 5 corn bearing plants, Ms. Diaz must plant an extra seed. Think: $10,000 \div 5 = 2,000$ Ms. Diaz must plant 2,000 extra seeds. She must plant a total of $10,000 + 2,000$ or $12,000$ seeds. The answer is correct.

Skill and Concept Check

Writing Math
Exercises 1 & 2

1. **Explain** why you would *not* expect the theoretical probability and the experimental probability of an event to always be the same.

2. **OPEN ENDED** Two hundred fifty people are surveyed about their favorite color. Make a possible table of results if the experimental probability that the favorite color is blue is $\frac{2}{5}$.

GUIDED PRACTICE

For Exercises 3–7, use the table that shows the results of tossing a coin.

3. Based on your results, what is the probability of getting heads?

4. Based on the results, how many heads would you expect to occur in 400 tries?

5. What is the theoretical probability of getting heads?

6. Based on the theoretical probability, how many heads would you expect to occur in 400 tries?

7. Compare the theoretical probability to your experimental probability.

Result	Number of Times
heads	26
tails	24

For Exercises 8 and 9, use the table at the right showing the results of a survey of cars that passed the school.

8. What is the probability that the next car will be white?

9. Out of the next 180 cars, how many would you expect to be white?

Cars Passing the School	
Color	**Number of Cars**
white	35
red	23
green	12
other	20

Practice and Applications

SCHOOL For Exercises 10 and 11, use the following information. In keyboarding class, Cleveland made 4 typing errors in 60 words.

10. What is the probability that his next word will have an error?

11. In a 1,000-word essay, how many errors would you expect Cleveland to make?

12. **SCHOOL** In the last 40 school days, Esteban's bus has been late 8 times. What is the experimental probability the bus will be late tomorrow?

FOOD For Exercises 13 and 14, use the survey results at the right.

13. What is the probability that a person's favorite snack while watching television is corn chips?

14. Out of 450 people, how many would you expect to have corn chips as their favorite snack with television?

15. **SPORTS** In practice, Crystal made 80 out of 100 free throws. What is the experimental probability that she will make a free throw?

HOMEWORK HELP

For Exercises	See Examples
10, 12–13, 15–16, 18	1, 4, 5
11, 14, 17, 19	6
20–21	2, 3

Extra Practice
See pages 637, 655.

Favorite Snack While Watching Television	
Snack	**Number**
potato chips	55
corn chips	40
popcorn	35
pretzels	15
other	5

SPORTS For Exercises 16 and 17, use the results of a survey of 90 teens shown at the right.

Sports Participation by Teens	
Sport	Number of Participants
basketball	42
volleyball	26
soccer	24
football	16

16. What is the probability that a teen plays soccer?

17. Out of 300 teens, how many would you expect to play soccer?

For Exercises 18–22, toss two coins 50 times and record the results.

18. What is the experimental probability of tossing two heads?

19. Based on your results, how many times would you expect to get two heads in 800 tries?

20. What is the theoretical probability of tossing two heads?

21. Based on the theoretical probability, how many times would you expect to get two heads in 800 tries?

22. Compare the theoretical and experimental probability.

23. **CRITICAL THINKING** An inspector found that 15 out of 250 cars had a loose front door and that 10 out of 500 cars had headlight problems. What is the probability that a car has both problems?

Standardized Test Practice and Mixed Review

FCAT Practice

24. **MULTIPLE CHOICE** Kylie and Tonya are playing a game where the difference of two rolled number cubes determines the outcome of each play. The graph shows the results of rolls of the number cubes so far in the game. Kylie needs a difference of 2 on her next roll to win the game. Based on past results, what is the probability that Kylie will win on her next roll?

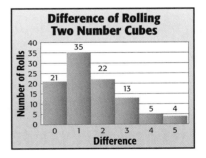

Difference of Rolling Two Number Cubes

Ⓐ $\frac{7}{20}$ Ⓑ $\frac{11}{50}$ Ⓒ $\frac{1}{20}$ Ⓓ $\frac{1}{25}$

25. **SHORT RESPONSE** A local video store has advertised that one out of every four customers will receive a free box of popcorn with their video rental. So far, 15 out of 75 customers have won popcorn. Compare the experimental and theoretical probability of getting popcorn.

There are 3 red marbles, 4 green marbles, and 5 blue marbles in a bag. Once a marble is selected, it is not replaced. Find the probability of each outcome. (Lesson 8-5)

26. 2 green marbles

27. a blue marble and then a red marble

28. **FOOD** Pepperoni, mushrooms, onions, and green peppers can be added to a basic cheese pizza. How many 2-item pizzas can be prepared? (Lesson 8-4)

GETTING READY FOR THE NEXT LESSON

PREREQUISITE SKILL Solve each problem. (Lessons 5-3 and 5-6)

29. Find 35% of 90.

30. Find 42% of 340.

31. What is 18% of 90?

8-6b Graphing Calculator Investigation

A Follow-Up of Lesson 8-6

Sunshine State Standards
MA.E.2.3.1-1, MA.E.3.3.1-2

What You'll Learn

Use a graphing calculator to simulate probability experiments.

Materials
• graphing calculator
• paper
• pencil

Simulations

A **simulation** is an experiment that is designed to act out a given situation. You can use items such as a number cube, a coin, a spinner, or a random number generator on a graphing calculator. From the simulation, you can calculate experimental probabilities.

 **ACTIVITY 1** *Work with a partner.*

Simulate rolling a number cube 50 times.

Use the random number generator on a TI-83 Plus graphing calculator. Enter 1 as the lower bound and 6 as the upper bound for 50 trials.

Keystrokes: [MATH] [◀] 5 1 [,] 6 [,] 50 [)] [ENTER]

A set of 50 numbers ranging from 1 to 6 appears. Use the right arrow key to see the next number in the set. Record all 50 numbers on a separate sheet of paper.

● Your Turn

a. Use the simulation to determine the experimental probability of each number showing on the number cube.

b. Compare the experimental probabilities found in Step 2 to the theoretical probabilities.

STUDY TIP

Simulations
Repeating a simulation may result in different probabilities since the numbers generated are different each time.

ACTIVITY 2 *Work with a partner.*

A company is placing one of 8 different cards of action heroes in its boxes of cereal. If each card is equally likely to appear, what is the experimental probability that a person who buys 12 boxes of cereal will get all 8 cards?

Let the numbers 1 through 8 represent the cards. Use the random number generator on a graphing calculator. Enter 1 as the lower bound and 8 as the upper bound for 12 trials.

Keystrokes: [MATH] [◀] 5 1 [,] 8 [,] 12 [)] [ENTER]

Record whether all of the numbers are represented.

msmath3.net/other_calculator_keystrokes

c. Repeat the simulation thirty times.

d. Use the simulation to find the experimental probability that a person who buys 12 boxes of cereal will get all 8 cards.

EXERCISES

1. **Describe** what you would expect if you repeated the simulation in Activity 1 more than 50 times.

2. **Explain** how you could use a graphing calculator to simulate tossing a coin 40 times.

3. **CLOTHING** Rodolfo must wear a tie when he works at the mall on Friday, Saturday, and Sunday. Each day, he picks one of his 6 ties at random. Create a simulation to find the experimental probability that he wears a different tie each day of the weekend.

4. **TOYS** A fast food restaurant is putting 3 different toys in their children's meals. If the toys are placed in the meals at random, create a simulation to determine the experimental probability that a child will have all 3 toys after buying 5 meals.

5. **SCIENCE** Suppose a mouse is placed in the maze at the right. If each decision about direction is made at random, create a simulation to determine the probability that the mouse will find its way out before coming to a dead end or going out the In opening.

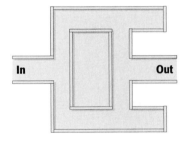

6. **WRITE A PROBLEM** Write a real-life problem that could be answered by using a simulation.

For Exercises 7–9, use the following information.
Suppose you play a game where there are three containers, each with 10 balls numbered 0 to 9. One number is randomly picked from each container. Pick three numbers each between 0 and 9. Then use the random number generator to simulate the game. Score 2 points if one number matches, 16 points if two numbers match, and 32 points if all three numbers match. Notice that numbers can appear more than once.

7. Play the game if the order of the numbers does *not* matter. Total your score for 10 simulations.

8. Now play the game if order of the numbers does matter. Total your score for 10 simulations.

9. With which game rules did you score more points?

Using Sampling to Predict

Sunshine State Standards
MA.E.3.3.2-1, MA.E.3.3.2-2, MA.E.3.3.2-3, MA.E.3.3.2-4

What You'll Learn
Predict the actions of a larger group by using a sample.

→ NEW Vocabulary

sample
population
unbiased sample
simple random
   sample
stratified random
   sample
systematic random
   sample
biased sample
convenience
   sample
voluntary response
   sample

WHEN **am I ever going to use this?**

ENTERTAINMENT The manager of a radio station wants to conduct a survey to determine what type of music people like.

1. Suppose she decides to survey a group of people at a rock concert. Do you think the results would represent all of the people in the listening area? Explain.

2. Suppose she decides to survey students at your middle school. Do you think the results would represent all of the people in the listening area? Explain.

3. Suppose she decides to call every 100th household in the telephone book. Do you think the results would represent all of the people in the listening area? Explain.

What Type of Music Do You Like?

Country
Alternative
Rock
Oldies
Top 40
Urban
Adult Contemporary

The manager of the radio station cannot survey everyone in the listening area. A smaller group called a **sample** is chosen. A sample is representative of a larger group called a **population**.

For valid results, a sample must be chosen very carefully. An **unbiased sample** is selected so that it is representative of the entire population. Three ways to pick an unbiased sample are listed below.

Concept Summary — Unbiased Samples

Type	Definition	Example
Simple Random Sample	A simple random sample is a sample where each item or person in the population is as likely to be chosen as any other.	The name of each student attending a school is written on a piece of paper. The names are placed in a bowl, and names are picked without looking.
Stratified Random Sample	In a stratified random sample, the population is divided into similar, non-overlapping groups. A simple random sample is then selected from each group.	Students are picked at random from each grade level at a school.
Systematic Random Sample	In a systematic random sample, the items or people are selected according to a specific time or item interval.	From an alphabetical list of all students attending a school, every 20th person is chosen.

In a **biased sample**, one or more parts of the population are favored over others. Two ways to pick a biased sample are listed below.

Concept Summary		Biased Samples
Type	**Definition**	**Example**
Convenience Sample	A convenience sample includes members of a population that are easily accessed.	To represent all the students attending a school, the principal surveys the students in one math class.
Voluntary Response Sample	A voluntary response sample involves only those who want to participate in the sampling.	Students at a school who wish to express their opinion are asked to come to the office after school.

EXAMPLES Describe Samples

Describe each sample.

1 To determine what videos their customers like, every tenth person to walk into the video store is surveyed.

Since the population is the customers of the video store, the sample is a systematic random sample. It is an unbiased sample.

2 To determine what people like to do in their leisure time, the customers of a video store are surveyed.

The customers of a video store probably like to watch videos in their leisure time. This is a biased sample. The sample is a convenience sample since all of the people surveyed are in one location.

EXAMPLES Using Sampling to Predict

SCHOOL The school bookstore sells 3-ring binders in 4 different colors; red, green, blue, and yellow. The students who run the store survey 50 students at random. The colors they prefer are indicated at the right.

Color	Number
red	25
green	10
blue	13
yellow	2

3 What percent of the students prefer blue binders?

13 out of 50 students prefer blue binders.

$13 \div 50 = 0.26$ 26% of the students prefer blue binders.

4 If 450 binders are to be ordered to sell in the store, how many should be blue?

Find 26% of 450.

$0.26 \times 450 = 117$ About 117 binders should be blue.

STUDY TIP

Mental Math
26% is about $\frac{1}{4}$, and 450 is about 440. The answer should be about $\frac{1}{4}$ of 440 or 110. Therefore, 117 is reasonable.

Skill and Concept Check

Writing Math
Exercises 1 & 2

1. **Compare** taking a survey and finding an experimental probability.

2. **OPEN ENDED** Give a counterexample to the following statement.
The results of a survey are always valid.

GUIDED PRACTICE

Describe each sample.

3. To determine how much money the average family in the United States spends to heat their home, a survey of 100 households from Arizona are picked at random.

4. To determine what benefits employees consider most important, one person from each department of the company is chosen at random.

ELECTIONS For Exercises 5 and 6, use the following information.
Three students are running for class president. Jonathan randomly surveyed some of his classmates and recorded the results at the right.

5. What percent said they were voting for Della?

6. If there are 180 students in the class, how many do you think will vote for Della?

Candidate	Number
Luke	7
Della	12
Ryan	6

Practice and Applications

Describe each sample.

7. To evaluate the quality of their product, a manufacturer of cell phones pulls every 50th phone off the assembly line to check for defects.

8. To determine whether the students will attend a spring music concert at the school, Rico surveys her friends in the chorale.

9. To determine the most popular television stars, a magazine asks its readers to complete a questionnaire and send it back to the magazine.

10. To determine what people in Texas think about a proposed law, 2 people from each county in the state are picked at random.

11. To pick 2 students to represent the 28 students in a science class, the teacher uses the computer program to randomly pick 2 numbers from 1 to 28. The students whose names are next to those numbers in his grade book will represent the class.

12. To determine if the oranges in 20 crates are fresh, the produce manager at a grocery store takes 5 oranges from the top of the first crate off the delivery truck.

13. **SCHOOL** Suppose you are writing an article for the school newspaper about the proposed changes to the cafeteria. Describe an unbiased way to conduct a survey of students.

HOMEWORK HELP

For Exercises	See Examples
7–12, 19–20	1, 2
14–18	3, 4

Extra Practice
See pages 637, 655.

SALES For Exercises 14 and 15, use the following information.
A random survey of shoppers shows that 19 prefer whole milk, 44 prefer low-fat milk, and 27 prefer skim milk.

14. What percent prefer skim milk?

15. If 800 containers of milk are ordered, how many should be skim milk?

16. **MARKETING** A grocery store is considering adding a world foods area. They survey 500 random customers, and 350 customers agree the world foods area is a good idea. Should the store add this area? Explain.

FOOD For Exercises 17–20, conduct a survey of the students in your math class to determine whether they prefer hamburgers or pizza.

17. What percent prefer hamburgers?

18. Use your survey to predict how many students in your school prefer hamburgers.

19. Is your survey a good way to determine the preferences of the students in your school? Explain.

20. How could you improve your survey?

21. **CRITICAL THINKING** How could the wording of a question or the tone of voice of the interviewer affect a survey? Give at least two examples.

Standardized Test Practice and Mixed Review

FCAT Practice

22. **MULTIPLE CHOICE** The Star Theater records the number of food items sold at its concessions. If the manager orders 5,000 food items for next week, approximately how many trays of nachos should she order?

    Ⓐ 1,025   Ⓑ 850   Ⓒ 800   Ⓓ 400

Food Items Sold at Movie Concessions During the Past Week	
Item	**Number**
popcorn	620
nachos	401
candy	597
slices of pizza	336

23. **MULTIPLE CHOICE** Brett wants to conduct a survey about who stays for after-school activities at his school. Who should he ask?

    Ⓕ his friends on the bus   Ⓖ members of the football team

    Ⓗ community leaders   Ⓘ every 10th student entering school

24. **MANUFACTURING** An inspector finds that 3 out of the 250 DVD players he checks are defective. What is the experimental probability that a DVD player is defective? (Lesson 8-6)

Each spinner at the right is spun once. Find each probability. (Lesson 8-5)

25. P(3 and B)   26. P(even and consonant)

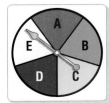

Vocabulary and Concept Check

biased sample (p. 407)	independent events (p. 396)	stratified random sample (p. 406)
combination (p. 388)	outcome (p. 374)	systematic random sample
complementary events (p. 375)	permutation (p. 384)	(p. 406)
compound events (p. 396)	population (p. 406)	theoretical probability (p. 400)
convenience sample (p. 407)	probability (p. 374)	tree diagram (p. 380)
dependent events (p. 397)	random (p. 374)	unbiased sample (p. 406)
experimental probability (p. 400)	sample (p. 406)	voluntary response sample
factorial (p. 385)	sample space (p. 374)	(p. 407)
Fundamental Counting	simple event (p. 374)	
Principle (p. 381)	simple random sample (p. 406)	

Choose the correct term to complete each sentence.

1. A list of all the possible outcomes is called the (sample space, event).

2. (Outcome, Probability) is the chance that an event will happen.

3. The Fundamental Counting Principle says that you can find the total number of outcomes by (multiplying, dividing).

4. A (combination, permutation) is an arrangement where order matters.

5. A (combination, compound event) consists of two or more simple events.

6. For (independent events, dependent events), the outcome of one does not affect the other.

7. (Theoretical probability, Experimental probability) is based on known characteristics or facts.

8. A (simple random sample, convenience sample) is a biased sample.

Lesson-by-Lesson Exercises and Examples

 8-1 **Probability of Simple Events** (pp. 374–377)

A bag contains 6 white, 7 blue, 11 red, and 1 black marbles. A marble is picked at random. Write each probability as a fraction, a decimal, and a percent.

9. P(white)   10. P(blue)

11. P(not blue)   12. P(white or blue)

13. P(red or blue)   14. P(yellow)

15. If a month is picked at random, what is the probability that the month will start with M?

Example 1   A box contains 4 green, 7 blue, and 9 red pens. Write the probability that a pen picked at random is green.

There are $4 + 7 + 9$ or 20 pens in the box.

$$P(\text{green}) = \frac{\text{green pens}}{\text{total number of pens}}$$

$$= \frac{4}{20} \text{ or } \frac{1}{5} \quad \begin{array}{l}\text{There are 4 green} \\ \text{pens out of 20 pens.}\end{array}$$

The probability the pen is green is $\frac{1}{5}$.

Counting Outcomes (pp. 380–383)

A penny is tossed and a 4 sided number cube with sides of 1, 2, 3, and 4 is rolled.

16. Draw a tree diagram to show the possible outcomes.

17. Find the probability of getting a head and a 3.

18. Find the probability of getting a tail and an odd number.

19. Find the probability of getting a head and a number less than 4.

20. **FOOD** A restaurant offers 15 main menu items, 5 salads, and 8 desserts. How many meals of a main menu item, a salad, and a dessert are there?

Example 2   BUSINESS A car manufacturer makes 8 different models in 12 different colors. They also offer standard or automatic transmission. How many choices does a customer have?

number of models	×	number of colors	×	number of transmissions	=	total number of cars
8	×	12	×	2	=	192

The customer can choose from 192 cars.

Permutations (pp. 384–387)

Find each value.

21. $P(6, 1)$
22. $P(4, 4)$
23. $P(5, 3)$
24. $P(7, 2)$
25. $P(10, 3)$
26. $P(4, 1)$

27. **NUMBER THEORY** How many 3-digit whole numbers can you write using the digits 1, 2, 3, 4, 5, and 6 if no digit can be used twice?

Example 3   Find $P(4, 2)$.

$P(4, 2)$ represents the number of permutations of 4 things taken 2 at a time.

$P(4, 2) = 4 \cdot 3$ or 12

Combinations (pp. 388–391)

Find each value.

28. $C(5, 5)$
29. $C(4, 3)$
30. $C(12, 2)$
31. $C(9, 5)$
32. $C(3, 1)$
33. $C(7, 2)$

34. **PETS** How many different pairs of puppies can be selected from a litter of 8?

Example 4   Find $C(4, 2)$.

$C(4, 2)$ represents the number of combinations of 4 things taken 2 at a time.

$$C(4, 2) = \frac{P(4, 2)}{2!} \quad \text{Definition of } C(4, 2)$$

$$= \frac{\overset{2}{\cancel{4}} \cdot 3}{\underset{1}{\cancel{2}} \cdot 1} \text{ or } 6 \quad \begin{array}{l} P(4, 2) = 4 \cdot 3 \text{ and} \\ 2! = 2 \cdot 1 \end{array}$$

8-5 **Probability of Compound Events** (pp. 396–399)

A number cube is rolled, and a penny is tossed. Find each probability.

35. P(2 and heads)   36. P(even and heads)

37. P(1 or 2 and tails) 38. P(odd and tails)

39. P(divisible by 3 and tails)

40. P(less than 7 and heads)

41. **GAMES** A card is picked from a standard deck of 52 cards and is not replaced. A second card is picked. What is the probability that both cards are red?

Example 5 A bag of marbles contains 7 white and 3 blue marbles. Once selected, the marble is not replaced. What is the probability of choosing 2 blue marbles?

P(first marble is blue) $= \dfrac{3}{10}$

P(second marble is blue) $= \dfrac{2}{9}$

P(two blue marbles) $= \dfrac{3}{10} \cdot \dfrac{2}{9}$

$\phantom{P\text{(two blue marbles) }} = \dfrac{6}{90}$ or $\dfrac{1}{15}$

8-6 **Experimental Probability** (pp. 400–403)

A spinner has four sections. Each section is a different color. In the last 30 spins, the pointer landed on red 5 times, blue 10 times, green 8 times, and yellow 7 times. Find each experimental probability.

42. P(red)   43. P(green)

44. P(red or blue)   45. P(not yellow)

Example 6 In an experiment, 3 coins are tossed 50 times. Five times no tails were showing. Find the experimental probability of no tails.

Since no tails were showing 5 out of the 50 tries, the experimental probability is $\dfrac{5}{50}$ or $\dfrac{1}{10}$.

8-7 **Using Sampling to Predict** (pp. 406–409)

Station WXYZ is taking a survey to determine how many people would attend a rock festival.

46. Describe the sample if the station asks listeners to call the station.

47. Describe the sample if the station asks people coming out of a rock concert.

48. If 12 out of 80 people surveyed said they would attend the festival, what percent said they would attend?

49. Use the result in Exercise 48 to determine how many out of 800 people would be expected to attend the festival.

Example 7 In a survey, 25 out of 40 students in the school cafeteria preferred chocolate to white milk.

a. **What percent preferred chocolate milk?**

$25 \div 40 = 0.625$

62.5% of the students prefer chocolate milk.

b. **How much chocolate milk should the school buy for 400 students?**

Find 62.5% of 400.

$0.625 \times 400 = 250$

About 250 cartons of chocolate milk should be ordered.

Practice Test

Vocabulary and Concepts

1. **Write** a probability problem that involves dependent events.
2. **Describe** the difference between biased and unbiased samples.

Skills and Applications

In a bag, there are 12 red, 3 blue, and 5 green candies. One is picked at random. Write each probability as a fraction, a decimal, and a percent.

3. $P(\text{red})$
4. $P(\text{no green})$
5. $P(\text{red or green})$

Find each value.

6. $C(10, 5)$
7. $P(6, 3)$
8. $P(5, 2)$
9. $C(7, 4)$

10. In how many ways can 6 students stand in a line?

11. How many teams of 5 players can be chosen from 15 players?

There are 4 blue, 3 red, and 2 white marbles in a bag. Once selected, it is not replaced. Find each probability.

12. $P(\text{2 blue})$
13. $P(\text{red, then white})$
14. $P(\text{white, then blue})$

15. Are these events in Exercises 12–14 dependent or independent?

16. **FOOD** Students at West Middle School can purchase a box lunch to take on their field trip. They choose one item from each category. How many lunches can be ordered?

Sandwich	Fruit	Cookie
ham	apple	chocolate
roast beef	banana	oatmeal
tuna	orange	sugar
turkey		

Two coins are tossed 20 times. No tails were tossed 4 times, one tail was tossed 11 times, and 2 tails were tossed 5 times.

17. What is the experimental probability of no tails?

18. Draw a tree diagram to show the outcomes of tossing two coins.

19. Use the tree diagram in Exercise 18 to find the theoretical probability of getting no tails when two coins are tossed.

Standardized Test Practice

20. **MULTIPLE CHOICE** A school board wants to know if it has community support for a new school. How should they conduct a valid survey?

  Ⓐ Ask parents at a school open house.

  Ⓑ Ask people at the Senior Center.

  Ⓒ Call every 50th number in the phone book.

  Ⓓ Ask people to call with their opinions.

 msmath3.net/chapter_test/fcat

PART 1 Multiple Choice

Record your answers on the answer sheet provided by your teacher or on a sheet of paper.

1. Which of these would be the next number in the following pattern? (Lesson 1-1)

   4, 12, 22, 34, …

   ⒶＡ 40   ⒷＢ 44

   ⒸＣ 46   ⒹＤ 48

2. Ms. Yeager asked the students in math class to tell one thing they did during the summer.

Activity	Number of Students
traveled with family	12
went to camp	6
worked on a summer job	10
other	2

   What fraction of the class said they went to camp or worked a summer job? (Lesson 2-1)

   Ⓕ $\frac{2}{5}$ Ⓖ $\frac{8}{15}$

   Ⓗ $\frac{11}{15}$ Ⓘ $\frac{6}{5}$

3. Find the length of side *FH*. (Lesson 3-4)

   Ⓐ 14 m

   Ⓑ 16 m

   Ⓒ 17 m

   Ⓓ 18 m

   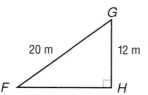

4. What is the area of the circle? (Lesson 7-2)

   Ⓕ 540 in^2

   Ⓖ 907.5 in^2

   Ⓗ 1,017.9 in^2

   Ⓘ 1,105.1 in^2

5. In the spinner below, what color should the blank portion of the spinner be so that the probability of landing on this color is $\frac{3}{8}$? (Lesson 8-1)

   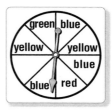

   Ⓐ red   Ⓑ blue

   Ⓒ yellow   Ⓓ green

6. Ed, Lauren, Sancho, James, Sofia, Tamara, and Haloke are running for president, vice-president, secretary, and recorder of the student council. Each of them would be happy to take any of the 4 positions, and none of them can take more than one position. How many ways can the offices be filled? (Lesson 8-3)

   Ⓕ 28   Ⓖ 210

   Ⓗ 840   Ⓘ 2,520

7. Alonso surveyed people leaving a pizza parlor to determine whether people in his area like pizza. Explain why this might *not* have been a valid survey. (Lesson 8-7)

   Ⓐ The survey is biased because Alonso should have asked people coming out of an ice cream parlor.

   Ⓑ Alonso should have mailed survey questionnaires to people.

   Ⓒ The survey is biased because Alonso was asking only people who had chosen to eat pizza.

   Ⓓ Alonso should have conducted the survey on a weekend.

PART 2 Short Response/Grid In

THINK
SOLVE
EXPLAIN

Record your answers on the answer sheet provided by your teacher or on a sheet of paper.

8. The first super computer, the Cray-1, was installed in 1976. It was able to perform 160 million different operations in a second. Use scientific notation to represent the number of operations the Cray-1 could perform in one day. (Lesson 2-9)

9. What is the value of x if x is a whole number? (Lesson 5-5)

$$34\tfrac{1}{3}\% \text{ of } 27 < x < 75\% \text{ of } 16$$

10. Find the coordinates of the fourth vertex of the parallelogram. (Lesson 6-4)

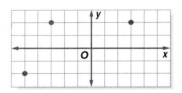

11. Ling knows the circumference of a circle and wants to find its radius. After she divides the circumference by π, what should she do next? (Lesson 7-2)

12. The eighth-grade graduation party is being catered. The caterers offer 4 appetizers, 3 salads, and 2 main courses for each eighth-grade student to choose for dinner. If the caterers would like 48 different combinations of dinners, how many desserts should they offer? (Lesson 8-2)

13. There are 15 glass containers of different flavored jellybeans in the candy store. If Jordan wants to try 4 different flavors, how many different combinations of flavors can he try? (Lesson 8-4)

PART 3 Extended Response

Record your answers on a sheet of paper. Show your work.

THINK
SOLVE
EXPLAIN

14. A red number cube and a blue number cube are tossed. (Lesson 8-2)

   a. Make a tree diagram to show the outcomes.

   b. Use the Fundamental Counting Principle to determine the number of outcomes. What are the advantages of using the Fundamental Counting Principle? of using a tree diagram?

   c. What is the probability that the sum of the two number cubes is 8?

15. Tiffany has a bag of 10 yellow, 10 red, and 10 green marbles. Tiffany picks two marbles at random and gives them to her sister. (Lesson 8-5)

   a. What is the probability of choosing 2 yellow marbles?

   b. Of the marbles left, what is the probability of choosing a green marble next?

   c. Of the marbles left, what color has a probability of $\tfrac{1}{3}$ of being picked? Explain how you determined your answer.

TEST-TAKING TIP

Question 15 Extended response questions often involve several parts. When one part of the question involves the answer to a previous part of the question, make sure you check your answer to the first part before moving on. Also, remember to show all of your work. You may be able to get partial credit for your answers, even if they are not entirely correct.

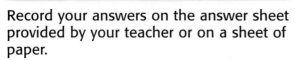

❝What does football have to do with math?❞

Many numbers, or data, are recorded during football games. These numbers include passing yards, running yards, interceptions, points scored, and distance of punts. These data can be represented by different types of graphs or by different measures of central tendency.

You will solve problems about football in Lesson 9-1.

GETTING STARTED

Take this quiz to see whether you are ready to begin Chapter 9. Refer to the lesson or page number in parentheses if you need more review.

▶ Vocabulary Review

State whether each sentence is *true* or *false*. If *false*, replace the underlined word or number to make a true sentence.

1. If one or more parts of a population are favored over others in a sample, then the sample is <u>unbiased</u>. (Lesson 8-7)

2. A <u>line plot</u> is a graph that uses an X above a number line to represent each number in a set of data. (Page 602)

▶ Prerequisite Skills

Graph each set of points on a number line. (Lesson 1-3)

3. {7, 8, 10, 15, 16}   4. {15, 20, 21, 25, 30}

5. {1, 4, 6, 10, 13}   6. {5, 7, 9, 13, 17}

Add or subtract. (Lessons 1-4 and 1-5)

7. $-4 + (-8)$ 8. $-5 + 2$

9. $7 + (-3)$ 10. $1 - (-5)$

11. $-9 - 3$ 12. $-2 - (-7)$

Order each set of rational numbers from least to greatest. (Lesson 2-2)

13. 0.23, 2.03, 0.32

14. 5.4, 5.64, 5.46, 5.6

15. 0.01, 1.01, 0.10, 1.10

16. 0.7, 0.17, 0.07

Solve each problem. (Lessons 5-3 and 5-6)

17. Find 52% of 360.   18. What is 36% of 360?

19. Find 14% of 360.   20. What is 8% of 360?

Statistics and Matrices
Make this Foldable to help you organize your notes. Begin with four pieces of $8\frac{1}{2}$" by 11" paper.

STEP 1 **Stack Pages**
Place 4 sheets of paper $\frac{3}{4}$ inch apart.

STEP 2 **Roll Up Bottom Edges**
All tabs should be the same size.

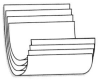

STEP 3 **Crease and Staple**
Staple along the fold.

STEP 4 **Label**
Label the tabs with topics from the chapter.

9-1 Histograms
9-2 Circle Graphs
9-3 Appropriate Display
9-4 Central Tendency
9-5 Measures of Variation
9-6 Box-and-Whisker
9-7 Misleading Statistics
9-8 Matrices

Reading and Writing As you read and study each lesson, write notes and examples under the appropriate tab.

Problem-Solving Strategy
A Preview of Lesson 9-1

Sunshine State Standards
MA.E.1.3.1-1, MA.E.1.3.1-2

What You'll Learn
Solve problems using the make a table strategy.

Make a Table

In science class, we used a pH meter to determine whether various substances were acids or bases. I listed the pH values in a table.

Substances with numbers less than 7 are acids, and substances with numbers greater than 7 are bases. Substances with the number 7 are neutral. How many acids, bases, and neutral substances did we test?

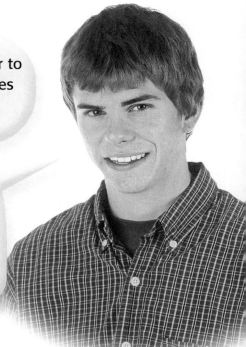

Explore	We have a list of the numbers shown on the pH meter. We need to know how many substances have a pH number of less than 7, greater than 7, and equal to 7.
Plan	Let's make a frequency table.

pH number	Tally	Frequency
Less than 7	JHT IIII	9
7	IIII	4
Greater than 7	JHT II	7

Solve

We tested 9 acids, 7 bases, and 4 neutral substances.

Examine	The students tested 9 + 4 + 7 or 20 substances. Since there are 20 numbers listed, the table seems reasonable.

7	8	4	3
8	7	9	7
5	2	3	7
4	6	8	5
9	9	8	6

Analyze the Strategy

1. **Tell** an advantage and disadvantage of listing the values in a table.

2. **Describe** two types of information you have seen recorded in a table.

3. **Write** a problem that can be answered using a table.

Solve. Use the make a table strategy.

4. **FORESTS** What percent of the tree diameters below are from 4 to 9.9 inches?

Sample Tree Diameters from Cumberland National Forest		
Diameter (in.)	Tally	Frequency
2.0–3.9	ＨＴ I	6
4.0–5.9	ＨＴ ＨＴ ＨＴ ＨＴ ＨＴ ＨＴ	30
6.0–7.9	ＨＴ ＨＴ ＨＴ ＨＴ ＨＴ III	28
8.0–9.9	ＨＴ ＨＴ ＨＴ ＨＴ IIII	24
10.0–11.9	ＨＴ ＨＴ ＨＴ IIII	19
12.0–13.9	IIII	4

5. **ALLOWANCES** The list shows weekly allowances.

$2.50	$3.00	$3.75	$4.25	$4.25
$4.50	$4.75	$4.75	$5.00	$5.00
$5.00	$5.00	$5.50	$5.50	$5.75
$5.80	$6.00	$6.00	$6.00	$6.50
$6.75	$7.00	$8.50	$10.00	$10.00

a. Organize the data in a table using intervals $2.00–$2.99, $3.00–$3.99, $4.00–$4.99, and so on.

b. What is the most common interval of allowance amounts?

Solve. Use any strategy.

6. **MULTI STEP** The oldest magazine in the United States was first published in 1845. If 12 issues were published each year, how many issues would be published through 2005?

7. **SPORTS** In a recent survey of 120 students, 50 students said they play baseball, and 60 said they play soccer. If 20 play both sports, how many students do not play either baseball or soccer?

8. **GEOGRAPHY** Name three countries that have a combined area of forests approximately equal to the area of forest in Russia.

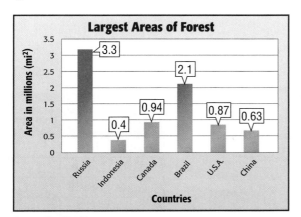

Largest Areas of Forest

Russia 3.3, Indonesia 0.4, Canada 0.94, Brazil 2.1, U.S.A. 0.87, China 0.63

Area in millions (mi^2)

Countries

Source: *Top Ten Things*

9. **CARS** Dexter's brother wants to buy a used car. The list shows the model year of the cars listed in the classified ads. Which year is listed most frequently?

1998	2000	1999	1999	2001	2001
2002	1998	2000	2000	1997	2001
1998	1999	2001	2001	1999	2000
2000	1997	1999	1998	2002	1997
2000	1999	2000	2001	1999	1999

10. **BASKETBALL** The average salary of an NBA player is $4.5 million per season. The average salary of a WNBA player is $43,000 per season. About what percent of the NBA player's salary is the WNBA player's salary?

11. **STANDARDIZED TEST PRACTICE** FCAT Practice

What are the dimensions of the rectangle?

Area = 24 m^2

Perimeter = 22 m

Ⓐ 8 m by 3 m Ⓑ 6 m by 4 m

Ⓒ 12 m by 2 m Ⓓ 24 m by 1 m

You will solve problems by making tables in the next lesson.

Histograms

Sunshine State Standards
MA.E.1.3.1-1, MA.E.1.3.1-2

What You'll Learn
Construct and interpret histograms.

NEW Vocabulary

histogram

REVIEW Vocabulary

bar graph: a graphic form using bars to make comparisons of statistics (page 602)

WHEN am I ever going to use this?

CONCERTS The table shows the number of concerts with an average ticket price in each price range.

1. What do you notice about the price intervals?

2. What does each tally mark represent?

3. How is the frequency for each price range determined?

Average Ticket Prices of Top 20 Money Earning Concerts for 2001		
Price	**Tally**	**Frequency**
$25.00–$49.99	ＨＨ IIII	9
$50.00–$74.99	ＨＨ II	7
$75.00–$99.99	I	1
$100.00–$124.99	II	2
$125.00–$149.99		0
$150.00–$174.99	I	1

Source: Pollstar

Data from a frequency table, such as the one above, can be displayed as a histogram. A **histogram** is a type of bar graph used to display numerical data that have organized into equal intervals.

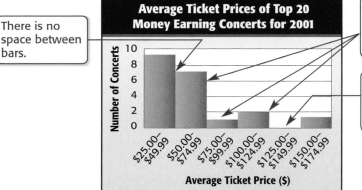

There is no space between bars.

Because all of the intervals are equal, all of the bars have the same width.

Intervals with a frequency of 0 have a bar height of 0.

EXAMPLE Draw a Histogram

1. **FOOD** The frequency table at the right shows the number of Calories in certain soup-in-a-cup products. Draw a histogram to represent the data.

Step 1 Draw and label a horizontal and vertical axis. Include a title.

Calories of Soup-in-a-Cup		
Calories	**Tally**	**Frequency**
100–149	II	2
150–199	ＨＨ II	7
200–249	ＨＨ III	8
250–299	I	1
300–349	I	1
350–399	I	1

Step 2 Show the intervals from the frequency table on the horizontal axis.

Step 3 For each Calorie interval, draw a bar whose height is given by the frequencies.

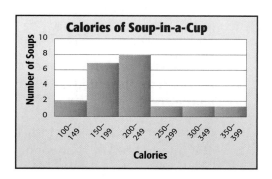

EXAMPLE Interpret Data

2 HISTORY How many presidents were younger than 50 when they were first inaugurated?

Two presidents were 40–44 years old, and six presidents were 45–49 years old. Therefore, 2 + 6 or 8 presidents were younger than 50 when they were first inaugurated.

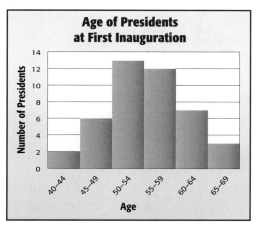

Source: *The World Almanac*

READING Math

At Least Recall that at least means *greater than or equal to.*

EXAMPLE Compare Two Sets of Data

3 FOOTBALL Determine which bowl game below has had a winning team score of at least 40 points more often.

Scores of Winning Teams through 2002

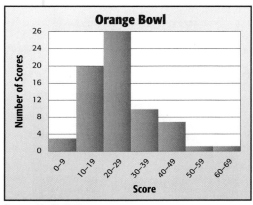

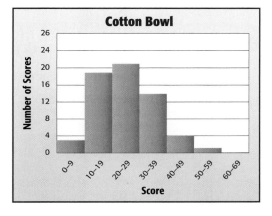

Source: *The World Almanac*

In the Orange Bowl, 7 + 1 + 1 or 9 winning teams scored at least 40 points. In the Cotton Bowl, 4 + 1 + 0 or 5 winning teams scored at least 40 points. The winning teams in the Orange Bowl scored 40 points more often than the winning teams in the Cotton Bowl.

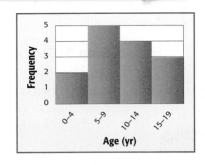

Writing Math
Exercise 2

1. **OPEN ENDED** Give a set of data that could be represented by the histogram at the right.

2. **Which One Doesn't Belong?** Identify the interval below that is not equal to the other three. Explain your reasoning.

15–19	30–34
40–45	45–49

GUIDED PRACTICE

3. **WEATHER** Draw a histogram to represent the data below.

Record High Temperatures for Each State		
Temperature (°F)	Tally	Frequency
100–104	III	3
105–109	JHT III	8
110–114	JHT JHT JHT II	17
115–119	JHT JHT II	12
120–124	JHT II	7
125–129	II	2
130–134	I	1

Source: National Climatic Data Center

4. **AUTO RACING** How many races had winning average speeds that were at least 150 miles per hour?

Source: *The World Almanac*

Practice and Applications

Draw a histogram to represent each set of data.

5.
New Broadway Productions for Each Year from 1960 to 2001		
Number of Shows	Tally	Frequency
20–29	II	2
30–39	JHT JHT IIII	14
40–49	JHT II	7
50–59	JHT JHT	10
60–69	JHT III	8
70–79	I	1

Source: The League of American Theatres and Producers

6.
National League's Greatest Number of Individual Strikeouts from 1960 to 2001		
Strikeouts	Tally	Frequency
150–199	II	2
200–249	JHT JHT II	12
250–299	JHT JHT JHT I	16
300–349	JHT IIII	9
350–399	III	3

Source: *The World Almanac*

HOMEWORK HELP

For Exercises	See Examples
5–8	1
9–18	2
19–21	3

Extra Practice
See pages 637, 656.

7. Calories of various types of frozen bars

   25, 35, 200, 280, 80, 80, 90, 40, 45, 50, 50, 60, 90, 100, 120, 40, 45, 60, 70, 350

8. maximum height in feet of various species of trees in the United States

   278, 272, 366, 302, 163, 161, 147, 223, 219, 216, 177

LIBRARIES **For Exercises 9–13, use the histogram at the right.**

9. Which interval represents the most number of states?

10. Which state has the most public libraries?

11. How many states have at least 600 public libraries?

12. How many states have between 400 and 800 public libraries?

13. How many states have less than 400 public libraries?

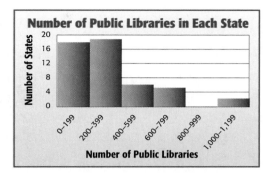

Source: Public Libraries Survey

BASKETBALL **For Exercises 14–18, use the histogram at the right.**

14. Which interval represents the most number of courts?

15. How many courts have less than 19,000 seats?

16. Which court has the least number of seats?

17. How many courts have between 18,000 and 19,999 seats?

18. How many courts have at least 20,000 seats?

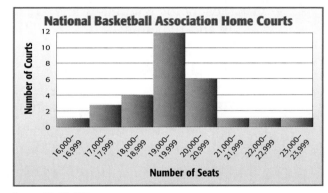

Source: *The World Almanac*

GEOGRAPHY **For Exercises 19–21, use the histograms.**

Land Area of Counties

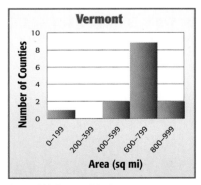 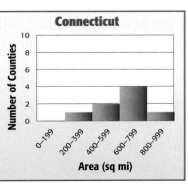

Source: U.S. Bureau of the Census

19. Which state has the smallest county by area?

20. Which state has more counties?

21. How many counties in the two states have less than 600 square miles?

22. CRITICAL THINKING Describe what is wrong with the histogram at the right.

23. RESEARCH Use the Internet or other resource to find the populations of each county, census division, or parish in your state. Make a histogram using your data. How does your county, census division, or parish compare with others in your state?

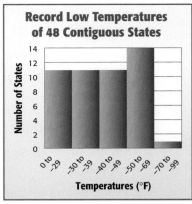

Record Low Temperatures of 48 Contiguous States

Source: National Climatic Data Center

24. MULTIPLE CHOICE Which statement can be concluded for the histogram at the right?

   Ⓐ The lowest winning score was 10.

   Ⓑ The highest winning score was 59.

   Ⓒ Most of the winning teams scored between 10 and 29 points.

   Ⓓ Most of the winning teams scored between 20 and 39 points.

Winning Scores at the First 36 Super Bowls

Source: *The World Almanac*

25. MULTIPLE CHOICE Use the histogram to determine how many winning teams scored less than 30 points.

   Ⓕ 31 teams   Ⓖ 17 teams

   Ⓗ 14 teams   Ⓘ 13 teams

26. ELECTIONS Would a survey of your neighborhood be a good indication of who will be elected governor of your state? Explain. (Lesson 8-7)

27. GOLF Tamika is practicing her putting from a certain place on the green. If she made 24 out of her last 32 attempts, what is the experimental probability that she will make her next putt? (Lesson 8-6)

Write each percent as a fraction in simplest form. (Lesson 5-1)

28. 24%   **29.** 55%   **30.** 29%   **31.** 66%

Solve each proportion. (Lesson 4-4)

32. $\dfrac{t}{7} = \dfrac{12}{42}$ **33.** $\dfrac{8}{m} = \dfrac{96}{60}$ **34.** $\dfrac{3}{7} = \dfrac{36}{x}$ **35.** $\dfrac{9}{5} = \dfrac{a}{7}$

GETTING READY FOR THE NEXT LESSON

PREREQUISITE SKILL Solve each problem. (Lessons 5-3 and 5-6)

36. Find 26% of 360.   **37.** What is 53% of 360?   **38.** Find 73% of 360.

Graphing Calculator Investigation

9-1b

A Follow-Up of Lesson 9-1

Sunshine State Standards
MA.E.1.3.1-2, MA.E.1.3.3-2

Histograms

You can make a histogram using a TI-83 Plus graphing calculator.

 ACTIVITY

Mr. Yamaguchi's second period class has listed the distance each student lives from the school. Make a histogram.

Distance from School (miles)											
4	2	6	1	10	3	19	5	20	1	1	9
22	15	2	4	12	8	1	4	16	3	6	7

STEP 1 **Enter the data.**
Clear any existing data in list L1.

Keystrokes: [STAT] [ENTER] [▲] [CLEAR] [ENTER]

Then enter the data into L1. Input each number and press [ENTER].

STEP 2 **Format the graph.**
Turn on the statistical plot.

Keystrokes: [2nd] [STAT PLOT] [ENTER] [ENTER]

Select the histogram and L1 as the Xlist.

Keystrokes: [▼] [▶] [▶] [ENTER] [▼] [2nd] L1 [ENTER]

STEP 3 **Graph the histogram.**
Set the viewing window to be [0, 25] scl: 5 by [0, 12] scl: 1. Then graph.

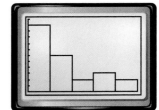

Keystrokes:
[WINDOW] 0 [ENTER] 25
[ENTER] 5 [ENTER] 0 [ENTER]
12 [ENTER] 1 [ENTER] [GRAPH]

 EXERCISES

1. Press [TRACE]. Find the frequency of each interval using the right arrow keys.

2. Discuss why the domain is from 0 to 25 for this data set.

3. Make a histogram on the graphing calculator of your classmates' heights in inches.

Circle Graphs

Sunshine State Standards
MA.E.1.3.1-1, MA.E.1.3.1-2

What You'll Learn
Construct and interpret circle graphs.

NEW Vocabulary

circle graph

REVIEW Vocabulary

line plot: a graph that uses an X above a number on a number line each time that number occurs in a set of data (page 602)

WHEN am I ever going to use this?

ROADS The graphic shows who owns the public roads in the United States.

1. What percent of the public roads are owned by the counties?

2. What government owns 19.6% of the public roads?

3. How do you know that all types of government have been accounted for?

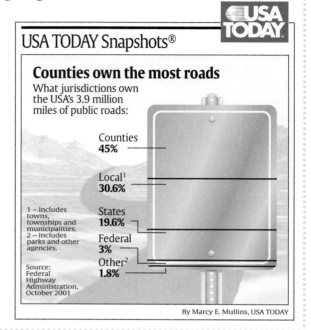

USA TODAY Snapshots®

Counties own the most roads

What jurisdictions own the USA's 3.9 million miles of public roads:

Counties **45%**
Local[1] **30.6%**
States **19.6%**
Federal **3%**
Other[2] **1.8%**

1 – Includes towns, townships and municipalities.
2 – Includes parks and other agencies.

Source: Federal Highway Administration, October 2001

By Marcy E. Mullins, USA TODAY

The graphic above compares parts of a set of data to the whole set. A **circle graph** also compares parts to the whole.

EXAMPLE Draw a Circle Graph

1 **ROADS** Make a circle graph using the information above.

Step 1 There are 360° in a circle. So, multiply each percent by 360 to find the number of degrees for each section of the graph.

Counties: 45% of 360 = 0.45 · 360 or 162
Local: 30.6% of 360 = 0.306 · 360 or about 110
States: 19.6% of 360 = 0.196 · 360 or about 71
Federal: 3% of 360 = 0.03 · 360 or about 11
Other: 1.8% of 360 = 0.018 · 360 or about 6

Step 2 Use a compass to draw a circle and a radius. Then use a protractor to draw a 162° angle. This section represents county roads. From the new radius, draw the next angle. Repeat for each of the remaining angles. Label each section. Then give the graph a title.

Who Owns Public Roads?

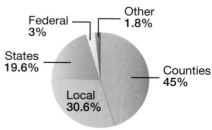

Federal **3%**
Other **1.8%**
States **19.6%**
Local **30.6%**
Counties **45%**

When percents are not known, you must first determine what part of the whole each item represents.

EXAMPLES Use Circle Graphs to Interpret Data

2 **HISTORY** Make a circle graph using the information in the histogram at the right.

Step 1 Find the total number of signers of the Declaration of Independence.

$3 + 17 + 19 + 10 + 6 + 1 = 56$

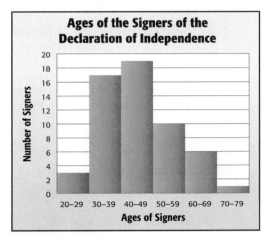

Ages of the Signers of the Declaration of Independence

Source: *The World Almanac*

Step 2 Find the ratio that compares the number in each age group to the total number of signers. Round to the nearest hundredth.

20 to 29: $3 ÷ 56 ≈ 0.05$ 50 to 59: $10 ÷ 56 ≈ 0.18$

30 to 39: $17 ÷ 56 ≈ 0.30$ 60 to 69: $6 ÷ 56 ≈ 0.11$

40 to 49: $19 ÷ 56 ≈ 0.34$ 70 to 79: $1 ÷ 56 ≈ 0.02$

Step 3 Use these ratios to find the number of degrees of each section. Round to the nearest degree if necessary.

20 to 29: $0.05 · 360 = 18$

30 to 39: $0.30 · 360 = 108$

40 to 49: $0.34 · 360 = 122.4$ or about 122

50 to 59: $0.18 · 360 = 64.8$ or about 65

60 to 69: $0.11 · 360 = 39.6$ or about 40

70 to 79: $0.02 · 360 = 7.2$ or about 7

Step 4 Use a compass and a protractor to draw a circle and the appropriate sections. Label each section and give the graph a title. Write the ratios as percents.

Ages of the Signers of the Declaration of Independence

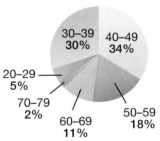

3 Use the circle graph to describe the makeup of the ages of the signers of the Declaration of Independence.

More signers of the Declaration of Independence were in their 40s than any other age group. Over $\frac{3}{4}$ of the signers were between 30 and 59.

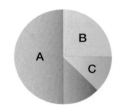

1. **Compare and contrast** the histogram and the circle graph in Example 2 on page 427.

2. **NUMBER SENSE** What percent of the circle graph is represented by Section A? by Section B? by Section C?

3. **OPEN ENDED** Make a circle graph with five categories showing how you spend 24 hours for a typical weekday.

GUIDED PRACTICE

Make a circle graph for each set of data.

4.

How Often Teens Borrow a CD from Their Parents	
frequently	11%
occasionally	34%
never/rarely	55%

Source: *USA WEEKEND*

5.

Area (square miles) of the Five Counties of Hawaii	
Hawaii	4,028
Honolulu	600
Kalawao	13
Kauai	623
Maui	1,159

Source: U.S. Department of Commerce

Practice and Applications

Make a circle graph for each set of data.

6.

Major Influences for Teens on Music Choices	
radio	43%
friends	30%
television	16%
parents	7%
concerts	3%
magazines	1%

Source: *USA WEEKEND*

7.

Types of Flowers and Plants Purchased for Mother's Day	
garden plants	37%
cut flowers	36%
flowering plants	18%
green plants	9%

Source: California Cut Flower Commission

HOMEWORK HELP

For Exercises	See Examples
6–7	1
8–10	2
11	3

Extra Practice
See pages 638, 657.

8.

Acres (millions) Planted in Cotton	
Texas	6.2
Georgia	1.5
Mississippi	1.2
Arkansas	1.0
North Carolina	0.9
Other	4.1

Source: U.S. Department of Agriculture

9.

U.S. Population (millions) by Age	
0–19 years	78.8
20–39 years	78.1
40–59 years	75.2
60–79 years	36.5
80+ years	9.5

Source: U.S. Census Bureau

10. **HISTORY** The table shows the birthplaces of the signers of the Declaration of Independence. Make a circle graph of the data.

Location	Signers	Location	Signers	Location	Signers
Connecticut	5	Massachusetts	9	Rhode Island	2
Delaware	2	New York	3	South Carolina	4
Maine	1	New Jersey	3	United Kingdom	8
Maryland	5	Pennsylvania	5	Virginia	9

Source: *The World Almanac*

11. **ENERGY** Use the circle graph to describe how we heat our homes.

Type of Fuel Used to Heat Homes

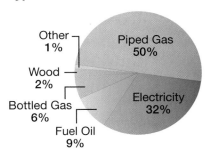

Other 1%
Wood 2%
Bottled Gas 6%
Fuel Oil 9%
Piped Gas 50%
Electricity 32%

Source: U.S. Census Bureau

12. **CRITICAL THINKING** Make a circle graph using the data in the table.

Favorite NBA Team	
Los Angeles Lakers	12%
Chicago Bulls	6.3%
Philadelphia 76ers	3.7%
New York Knicks	3.3%
Boston Celtics	2.1%
None	56%

Source: ESPN

Standardized Test Practice and Mixed Review

FCAT Practice

13. **MULTIPLE CHOICE** Which statement *cannot* be determined from the graph at the right?

   Ⓐ Most adults want to live to 100.

   Ⓑ Nearly one third of adults do not want to live to 100.

   Ⓒ Five people who were surveyed "don't know."

   Ⓓ One twentieth of the adults "don't know."

Do You Want to See Your 100th Birthday?

Don't Know 5%
No 32%
Yes 63%

Source: Alliance for Aging

14. **GRID IN** Find the measure in degrees of the angle of the "no" section of the circle graph.

15. **FOOD** The number of Calories in single serving, frozen pizzas are listed below. Make a histogram of the data. (Lesson 9-1)

200, 270, 290, 300, 310, 320, 330, 350, 360, 380, 380, 390, 390, 420, 440, 450

16. **RADIO LISTENING** A radio station asks listeners to call in and state their favorite band. Explain why this is a biased sample. (Lesson 8-7)

GETTING READY FOR THE NEXT LESSON

PREREQUISITE SKILL Make a line plot for each set of data. (Page 602)

17. 2, 5, 9, 8, 2, 6, 2, 5, 8, 10

18. 14, 12, 9, 7, 12, 10, 14, 7, 8, 12

What You'll Learn
Choose an appropriate display for a set of data.

⟳ **REVIEW Vocabulary**

line graph: a type of statistical graph using lines to show how values change over a period of time **(page 602)**

9-3 Choosing an Appropriate Display

Sunshine State Standards
MA.E.1.3.1-2

WHEN am I ever going to use this?

SCHOOL The following are four different ways a teacher can display the grades on a test.

Stem-and-Leaf Plot

Stem	Leaf
6	4 8
7	0 2 2 4 6 6 6 8 8 8
8	0 2 2 2 2 2 6 6 8 8 8
9	2 2 6

6 | 4 = 64%

Line Plot

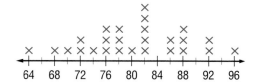

Histogram

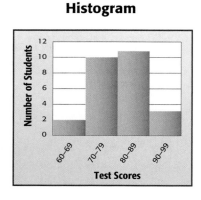

Circle Graph

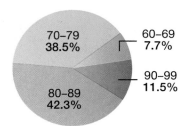

1. Which display(s) show all of the individual test scores?

2. Do any of the displays allow you to find the test score of a certain student? If not, what type of display would show this type of information?

Some of the ways to display data and their uses are listed below.

Concept Summary — **Statistical Displays**

Display	Use
Bar Graph	shows the number of items in specific categories
Circle Graph	compares parts of the data to the whole
Histogram	shows the frequency of data that has been organized into equal intervals
Line Graph	shows change over a period of time
Line Plot	shows how many times each number occurs in the data
Pictograph	shows the number of items in specific categories
Stem-and-Leaf Plot	lists all individual numerical data in a condensed form
Table	may list all the data individually or by groups

As you decide what type of display to use, ask the following questions.
- What type of information is this?
- What do I want my graph or display to show?

EXAMPLES **Choose an Appropriate Display**

Choose an appropriate type of display for each situation. Then make a display.

1 **CELLULAR PHONES** The table shows cellular phone subscribers.

Cellular Phones					
Year	Subscribers (millions)	Year	Subscribers (millions)	Year	Subscribers (millions)
1993	34	1996	145	1999	471
1994	55	1997	214	2000	650
1995	91	1998	319	2001	900

Source: International Telecommunication Union

This data deals with change over time. A line graph would be a good way to show the change over time.

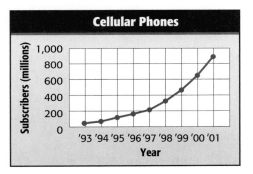

2 **BICYCLES** The results of a survey of a group of students asked to give their favorite bicycle color are given at the right.

In this case, there are specific categories. If you want to show the specific number, use a bar graph or a pictograph. If you want to show how each part is related to the whole, use a circle graph.

Red   ЖHT ЖHT ЖHT I
Blue   ЖHT III
Silver   ЖHT ЖHT ЖHT ЖHT
Black   ЖHT ЖHT III

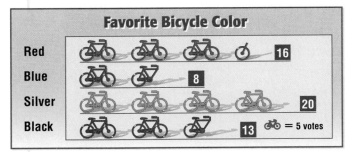

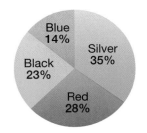

Favorite Bicycle Color

1. **Compare and contrast** bar graphs and histograms.

2. **OPEN ENDED** Give an example of data that could be represented using a line graph.

GUIDED PRACTICE

Choose an appropriate type of display for each situation.

3. the parts of a landfill used for various types of trash

4. plant height measurements made every 2 days in a science fair report

5. **FOOD** Choose an appropriate type of display for the following situation. Then make a display.

Grams of Carbohydrates in a Serving of Various Vegetables											
3	8	10	4	7	6	1	5	19	6	1	2
3	12	23	34	17	37	10	28	7	28	11	

Practice and Applications

Choose an appropriate type of display for each situation.

6. points scored by individual members of a basketball team compared to the team total

7. numbers of Americans whose first language is Spanish, Mandarin, or Hindi

8. the profits of a company every year for the last ten years

9. the populations of the states arranged by intervals

10. the number of students who wish to order each size of T-shirt

11. the price of an average computer for the last twenty years

Choose an appropriate type of display for each situation. Then make a display.

HOMEWORK HELP

For Exercises	See Examples
6–13	1, 2

Extra Practice
See pages 638, 656.

12.

Americans Studying in Selected Countries	
Country	Number
United Kingdom	27,720
Spain	12,292
Italy	11,281
France	10,479

Source: Open Doors 2000

13.

Average Height of Girls	
Age (years)	Height (inches)
2	35
3	39
4	42
5	44
6	46
7	48
8	51
9	53
10	56

Source: www.babybag.com

14. **CRITICAL THINKING** Display the data from the bar graph at the right using another type of display. Compare the displays.

15. **RESEARCH** Find a display of data in a newspaper or on the Internet. Do you think the most appropriate type of display was used?

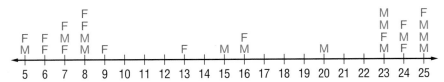

Standardized Test Practice and Mixed Review

FCAT Practice

16. **MULTIPLE CHOICE** All of the students in Mrs. Gomez's first period class walk to school. The line plot shows the time students take to walk to school. The data is labeled with "M" for male and "F" for female. Which statement is supported by the information in the graph?

Time it Takes to Walk to School (min)

```
                                              M    F
                F                             M    M
                F                             M    F
        F       F                             F    M
    F   F   M   M                    F        M    M
    M   F   F   M   F         F   M           M    M
   ─┼───┼───┼───┼───┼───┼───┼───┼───┼───┼───┼───┼───┼──▶
    5   6   7   8   9  10  11  12  13  14  15  16  17  18  19  20  21  22  23  24  25
```

  (A) The majority of females live more than 15 minutes away.

  (B) Most of the students live more than 18 minutes away.

  (C) Most of the students live less than 10 minutes or more than 22 minutes away.

  (D) There are 25 students in the class.

17. **SHORT RESPONSE** Make a histogram of the data in the above line plot.

18. **NATIONAL PARKS** Yellowstone National Park has 3,159 square miles in Wyoming, 264 square miles in Montana, and 49 square miles in Idaho. Make a circle graph to show what part of Yellowstone National Park is in each state. (Lesson 9-2)

GETTING READY FOR THE NEXT LESSON

PREREQUISITE SKILL Evaluate each expression. (Lesson 1-2)

19. $\dfrac{14 + 22 + 18 + 28}{4}$

20. $\dfrac{23 + 19 + 2 + 8 + 18}{5}$

21. $\dfrac{7 + 9 + 2 + 1 + 14 + 6}{6}$

What You'll Learn

Show statistics on maps.

Materials

• outline of map of the United States
• markers

Sunshine State Standards
MA.E.1.3.1-1, MA.E.1.3.1-2

MAPS AND STATISTICS

Follow the steps below to make a map using data from the table.

Average Number of Tornadoes Each Year									
State	**No.**	**State**	**No.**	**State**	**No.**	**State**	**No.**	**State**	**No.**
AL	22	HI	1	MA	3	NM	9	SD	29
AK	0	ID	3	MI	19	NY	6	TN	12
AZ	4	IL	27	MN	20	NC	15	TX	139
AR	20	IN	20	MS	26	ND	21	UT	2
CA	5	IA	36	MO	26	OH	15	VT	1
CO	26	KS	40	MT	6	OK	47	VA	6
CT	1	KY	10	NE	37	OR	1	WA	2
DE	1	LA	28	NV	1	PA	10	WV	2
FL	53	ME	2	NH	2	RI	0	WI	21
GA	21	MD	3	NJ	3	SC	10	WY	12

Source: National Severe Storm Forecast Center

STEP 1 Make a line plot of the data using the state abbreviations instead of ×s.

STEP 2 Find natural breaks in the data and organize the data into fewer than 7 categories.

STEP 3 Color each state according to its category. Include a key.

Writing Math

1. **Explain** how you could change the categories to show that a greater number of states have many tornadoes. How could you change the categories to show that only a few states have many tornadoes?

2. What information is obvious in the map that would not be found in a table?

3. **RESEARCH** Use the Internet or another source to find data about the 50 states. Make two different maps of the data showing two different points of view.

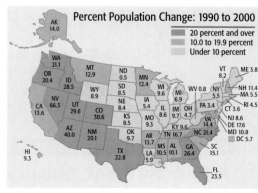

Percent Population Change: 1990 to 2000
- 20 percent and over
- 10.0 to 19.9 percent
- Under 10 percent

Source: U.S. Bureau of the Census

Measures of Central Tendency

Sunshine State Standards
MA.A.1.3.2-2, MA.E.1.3.2-1, MA.E.1.3.2-2, MA.E.1.3.2-3

WHEN am I ever going to use this?

VACATION DAYS Use the table to answer each question.

1. What is the *average* number of days for these nine countries?

2. Order the numbers from least to greatest. What is the middle number in your list?

3. What number(s) appear more than once?

4. Which of the number or numbers from Exercises 1–3 might be representative of the set of data? Explain.

Average Number of Vacation Days Per Year for Selected Countries	
Country	**Vacation Days**
Brazil	34
Canada	26
France	37
Germany	35
Italy	42
Japan	25
Korea	25
United Kingdom	28
United States	13

Source: World Tourism Organization

Measures of central tendency are numbers that describe a set of data. The most common measures are **mean**, **median**, and **mode**.

Concept Summary   **Measures of Central Tendency**

Measure	Description
mean	the sum of the data divided by the number of items in the data set
median	the middle number of the data ordered from least to greatest, or the mean of the middle two numbers
mode	the number or numbers that occur most often

EXAMPLE Find Measures of Central Tendency

① Find the mean, median, and mode of the set of data.
22, 18, 24, 32, 24, 18

Mean $\dfrac{22 + 18 + 24 + 32 + 24 + 18}{6} = \dfrac{138}{6}$

$= 23$ The mean is 23.

Median Arrange the numbers in order from least to greatest.

18   18 $\underbrace{22 \qquad 24}$ 24   32

$\dfrac{22 + 24}{2} = 23$ The median is 23.

Mode The data has two modes, 18 and 24.

Sometimes one or two measures of central tendency are more representative of the data than the other measure(s).

EXAMPLES **Using Appropriate Measures**

REAL-LIFE MATH

GEOGRAPHY Although no one actually resides on Antarctica, about 1,000 scientists live at over 30 scientific stations during the summer. Some scientists even stay through the winter, when the temperatures can drop to −94°F. Icy winds make the temperature seem even colder.

Source: *World Book*

GEOGRAPHY Use the table to answer each question.

Population of the Seven Continents	
Content	**Population (millions)**
North America	481
South America	347
Europe	729
Asia	3,688
Africa	805
Australia and Oceania	31
Antarctica	0

Source: *The World Almanac for Kids*

2 **What is the mean, median, and mode of the data?**

Mean $\dfrac{481 + 347 + 729 + 3,688 + 805 + 31 + 0}{7} = \dfrac{6,081}{7}$
≈ 868.7

The mean is about 868.7 million.

Median Arrange the numbers from least to greatest.

0, 31, 347, 481, 729, 805, 3,688

The median is the middle number or 481 million.

Mode Since each number only occurs once, there is no mode.

3 **Which measure of central tendency is most representative of the data?**

Since there is no mode, you must decide whether the mean, 868.7 million, or the median, 481 million, is more representative of the data.

Notice that the extremely large population of Asia greatly affected the mean. In fact, the only continent with a population greater than the mean is Asia.

The best representation of the data is the median, 481 million.

Different circumstances determine which of the measures of central tendency are most useful.

Concept Summary **Using Mean, Median, and Mode**

Measure	Most Useful When . . .
mean	• the data has no extreme values
median	• the data has extreme values • there are no big gaps in the middle of the data
mode	• data has many identical numbers

msmath3.net/extra_examples/fcat

1. **Determine** whether all measures of central tendency must be members of the set of data. Explain.

2. **OPEN ENDED** Construct a set of data that has a mode of 4 and a median of 3.

3. **FIND THE ERROR** Tobias and Erica are finding the median of 93, 90, 94, 99, 92, 93, and 100. Who is correct? Explain.

> Tobias
> 93, 90, 94, <u>99</u>, 92, 93, 100
> The median is 99.

> Erica
> 90, 92, 93, <u>93</u>, 94, 99, 100
> The median is 93.

GUIDED PRACTICE

Find the mean, median, and mode of each set of data. Round to the nearest tenth if necessary.

4. 19, 21, 18, 17, 18, 22, 46

5. 10, 3, 17, 1, 8, 6, 12, 15

FOOTBALL For Exercises 6 and 7, use the graphic.

6. Find the mean, median, and mode of the data.

7. Which measure of central tendency is most representative of the data? Explain.

Touchdown Passes Completed on Monday Night Football Through 2001	
Quarterback	**Number of Touchdown Passes**
Dan Marino	74
Steve Young	42
Joe Montana	36
Jim Kelly	31
Brett Favre	27
Ken Stabler	27
Danny White	27

Source: NFL

Practice and Applications

Find the mean, median, and mode of each set of data. Round to the nearest tenth if necessary.

8. 9, 8, 15, 8, 20

9. 23, 16, 5, 6, 14

10. 78, 80, 75, 73, 84, 81, 84, 79

11. 36, 38, 33, 34, 32, 30, 34, 35

12. 8.5, 8.7, 6.9, 7.5, 7, 9.8, 5.4, 8.9, 6.5, 8.2, 8, 9.4

13. 1.2, 1.78, 1.73, 1.9, 1.19, 1.8, 1.24, 1.92, 1.54, 1.7, 1.42, 1

14.

15.

HOMEWORK HELP	
For Exercises	See Examples
8–16	1, 2
17	3
Extra Practice See pages 638, 656.	

CIVICS For Exercises 16 and 17, use the stem-and-leaf plot. It shows the number of members in the House of Representatives for each state.

Stem	Leaf
0	1 1 1 1 1 1 1 2 2 2 2 2 3 3 3 3 3 4 4
	4 5 5 5 5 6 6 7 7 7 8 8 8 8 9 9 9 9
1	0 1 3 3 3 5 8 9 9
2	5 9
3	2
4	
5	3

5 | 3 = 53 members

Source: *The World Almanac*

16. Find the mean, median, and mode of the data.

17. Which measure of central tendency is most representative of the data? Explain.

18. **WRITE A PROBLEM** Write a problem that asks for the measures of central tendency. Use data from a newspaper or magazine. Tell which measure is most representative of the data.

19. **CRITICAL THINKING** Give a counterexample to show that the following statement is false.
The median is always representative of the data.

Standardized Test Practice and Mixed Review

FCAT Practice

20. **MULTIPLE CHOICE** A consumer group tested several brands of headphones and compared their ratings (G-good, P-poor) with their price. Which statement is *not* supported by the information in the graph?

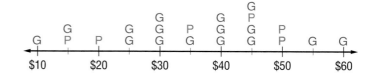

   A The mean price for a pair of headphones is $40.

   B There are 16 headphones that are rated good.

   C There is 1 headphone that is rated good and one that is rated poor for $35.

   D $45 is the mode for the data set.

21. **MULTIPLE CHOICE** In the following list of data, which number is the median? 27, 13, 26, 26, 17, 14, 15, 26, 16

   F 16   **G** 17   **H** 20   **I** 26

Choose an appropriate type of display for each situation. (Lesson 9-3)

22. the amount of each flavor of ice cream sold relative to the total sales

23. the intervals of ages of the people attending the fair

24. **TENNIS** Of the Americans who play tennis, 63% play at public parks, 26% play at private clubs, 6% play at apartment or condo complexes, and 5% play at other places. Make a circle graph of the data. (Lesson 9-2)

GETTING READY FOR THE NEXT LESSON

PREREQUISITE SKILL Order each set of rational numbers from least to greatest. (Lesson 2-2)

25. 3.1, 3.25, 3.2, 2.9, 2.89   26. 91.3, 93.1, 94.7, 93.11, 93   27. 17.4, 16.8, 16.79, 15.01, 15.1

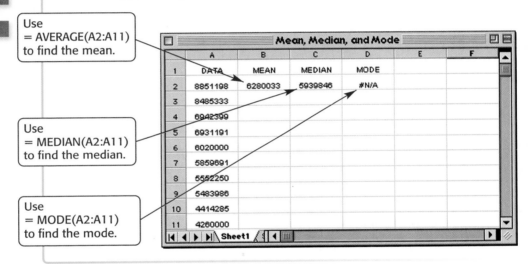

9-4b Spreadsheet Investigation

A Follow-Up of Lesson 9-4

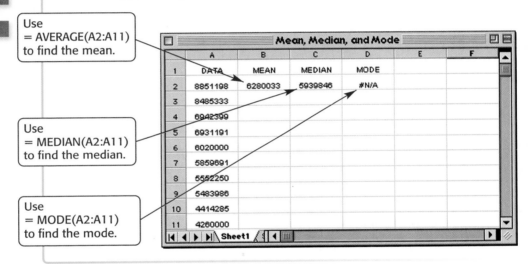Sunshine State Standards
MA.E.1.3.2-1, MA.E.1.3.3-1, MA.E.1.3.3-2

What You'll Learn
Use a spreadsheet to find mean, median, and mode.

Mean, Median, and Mode

You can use a spreadsheet to find the mean, median, and mode of data.

ACTIVITY

The following is a list of the top ten salaries of quarterbacks in the NFL in 2001. Make a spreadsheet for the data.

Top Ten Salaries of Quarterbacks in the NFL in 2001				
$8,851,198	$6,942,399	$6,020,000	$5,552,250	$4,414,285
$8,485,333	$6,931,191	$5,859,691	$5,483,986	$4,260,000

Source: NFL Players Association

Use = AVERAGE(A2:A11) to find the mean.

Use = MEDIAN(A2:A11) to find the median.

Use = MODE(A2:A11) to find the mode.

	A	B	C	D	E	F
1	DATA	MEAN	MEDIAN	MODE		
2	8851198	6280033	5939846	#N/A		
3	8485333					
4	6942399					
5	6931191					
6	6020000					
7	5859691					
8	5552250					
9	5483986					
10	4414285					
11	4260000					

Mean, Median, and Mode — Sheet1

EXERCISES

For Exercises 1–3, use the following tables.

Top Ten Salaries of Running Backs in the NFL in 2001	
$8,455,125	$4,400,000
$5,000,000	$4,300,000
$4,962,703	$4,066,666
$4,800,000	$3,334,718
$4,783,600	$2,928,571

Top Ten Salaries of Defensive Ends in the NFL in 2001	
$8,750,000	$4,535,500
$5,249,411	$4,445,833
$5,050,000	$4,259,166
$4,843,666	$4,163,674
$4,600,000	$3,850,000

1. Use spreadsheets to find the mean, median, and mode of the top ten salaries for each position.

2. Compare the highest salary for the three positions.

3. Compare the mean and median of the three positions.

Vocabulary and Concepts

1. **Compare and contrast** a bar graph and a histogram. (Lesson 9-1)

2. **OPEN ENDED** Give an example of data that could be displayed using a pictograph. (Lesson 9-3)

Skills and Applications

FOOD The frequency table shows the grams of sugar per serving in 28 cereals made for adults.

3. Use the intervals 0–2, 3–5, 6–8, and 9–11 to make a histogram of the data. (Lesson 9-1)

4. Make a circle graph of the data. (Lesson 9-2)

Choose an appropriate type of display for each situation. (Lesson 9-3)

5. percent of students in each grade level in a school

6. prices of different brands of ice cream by intervals

Find the mean, median, and mode of each set of data. Round to the nearest tenth if necessary. (Lesson 9-4)

7. 7, 3, 8, 6, 2

8. 73, 78, 71, 95, 86, 88, 86

Sugar in Cereal		
Grams	Tally	Frequency
0	‖‖‖	5
1		0
2	‖‖	3
3	‖‖‖ ‖	6
4	‖	1
5	‖‖‖	5
6	‖‖‖‖	4
7	‖	1
8		0
9	‖	1
10	‖	1
11	‖	1

Standardized Test Practice

FCAT Practice
Ⓐ Ⓑ Ⓒ Ⓓ

For Exercises 9 and 10, use the graph.

Nevada's Budget

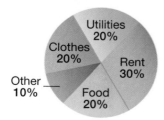

Utilities 20%
Clothes 20%
Rent 30%
Other 10%
Food 20%

9. **GRID IN** If Nevada makes $1,200 per month, how much does she budget in dollars for rent? (Lesson 9-2)

10. **MULTIPLE CHOICE** Which statement cannot be determined from the graph? (Lesson 9-2)

Ⓐ Nevada budgets half of her money for rent and food.

Ⓑ Nevada budgets the same amount of money for clothes as food.

Ⓒ Nevada budgets more money for food and clothes than rent.

Ⓓ Nevada does not spend any money on going to the movies.

The Game Zone

A Place To Practice Your Math Skills

What's the Average?

● GET READY!

Players: four
Materials: 4 index cards, 2 spinners

● GET SET!

- Each player should write five whole numbers on an index card. The numbers should be from 1 through 10.

- Label a spinner that has two equal regions with the words *mean* and *median*.

- Label a spinner that has four equal regions with the words *add/increase*, *add/decrease*, *remove/increase*, and *remove/decrease*.

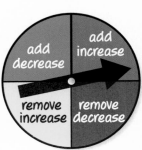

● GO!

- Mix the index cards and turn them facedown.

- The first player randomly selects a card and spins each spinner once. Then the player adjusts the data set as instructed. For example, if the player gets *mean* and *add/decrease*, the player must add a piece of data to the data set so the mean decreases. If the player gets *median* and *remove/increase*, the player must remove a piece of data from the data set so the median of the set increases.

- The other players then check his or her work.

- A player scores two points for each correct solution and loses one point for each incorrect solution.

- **Who Wins?** The first player to get 10 points is the winner.

Measures of Variation

Sunshine State Standards
MA.A.1.3.2-2, MA.E.1.3.2-2

WHEN am I ever going to use this?

What You'll Learn
Find the range and quartiles of a set of data.

NEW Vocabulary

measures of variation
range
quartiles
lower quartile
upper quartile
interquartile range
outlier

∞ LINK To Reading

Everyday Meaning of Quart: one fourth of a gallon

ONLINE TIME The average number of hours that teens in various cities spend online is given in the table.

1. What is the greatest number of hours spent online?

2. What is the least number of hours spent online?

3. Find the difference between the greatest number and the least number of hours spent online.

4. Write a sentence explaining what the answer to Exercise 3 says about the data.

Average Number of Hours Teens Spend Online Each Week	
City	**Hours Online**
Pittsburgh	15.8
New York	14.9
Cleveland	14.9
San Diego	14.4
Miami	14.2
Hartford	13.4
Los Angeles	13.3
Detroit	13.1
Philadelphia	12.9
Milwaukee	12.9

Source: Digital Marketing Services

Measures of variation are used to describe the distribution of the data. One measure of variation is the range. The **range** indicates how "spread out" the data are.

Key Concept — Range

The range of a set of data is the difference between the greatest and the least numbers in the set.

Quartiles are the values that divide the data into four equal parts. Recall that the median separates the data in two equal parts.

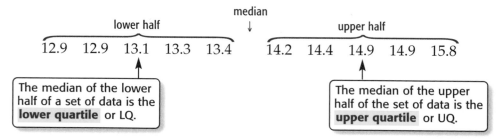

lower half					median ↓	upper half				
12.9	12.9	13.1	13.3	13.4		14.2	14.4	14.9	14.9	15.8

The median of the lower half of a set of data is the **lower quartile** or LQ.

The median of the upper half of the set of data is the **upper quartile** or UQ.

So, one half of the data lie between the lower quartile and the upper quartile. Another measure of variation is the **interquartile range**.

Key Concept — Interquartile Range

The interquartile range is the range of the middle half of the data. It is the difference between the upper quartile and the lower quartile.

FOOD Use the table at the right.

1 Find the range of the Calories.

The greatest number of Calories is 180.
The least number of Calories is 35. The
range is 180 − 35 or 145 Calories.

2 Find the median and the upper and
lower quartiles.

Arrange the numbers in order from least
to greatest.

Calories in a Serving of Juice	
Juice	**Calories**
Apple	120
Carrot	80
Grape	170
Grapefruit	100
Orange	120
Pineapple	110
Prune	180
Tomato	35

Source: Center for Science in the Public Interest

lower quartile   median   upper quartile
↓   ↓   ↓

35   80   100   110   120   120   170   180

$\frac{80 + 100}{2} = 90$ $\frac{110 + 120}{2} = 115$ $\frac{120 + 170}{2} = 145$

The median is 115, the lower quartile is 90, and the upper
quartile is 145.

3 Find the interquartile range.

Interquartile Range = 145 − 90 or 55

Data that are more than 1.5 times the value of the interquartile range
beyond the quartiles are called **outliers**.

EXAMPLE Find Outliers

4 **CHOCOLATE** Find any
outliers for the data
in the table.

Annual Chocolate Sales	
Country	**Sales (billion dollars)**
United States	16.6
United Kingdom	6.5
Germany	5.1
Russia	4.9
Japan	3.2
France	2.1
Brazil	2.0

upper quartile → United Kingdom

median → Russia

lower quartile → France

Source: Euromonitor

Interquartile Range = 6.5 − 2.1 or 4.4

Multiply the interquartile range, 4.4, by 1.5.   4.4 × 1.5 = 6.6

Find the limits for the outliers.

Subtract 6.6 from the lower quartile.   2.1 − 6.6 = −4.5

Add 6.6 to the upper quartile.   6.5 + 6.6 = 13.1

The limits for the outliers are −4.5 and 13.1. The only outlier is 16.6.

1. **OPEN ENDED** Write a list of data with at least eight numbers that has an interquartile range of 20 and one outlier.

2. **Which One Doesn't Belong?** Identify the statistical value that is not the same as the other three. Explain your reasoning.

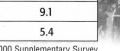

| mean | median | range | mode |

GUIDED PRACTICE

Find the range, median, upper and lower quartiles, interquartile range, and any outliers for each set of data.

3. 54, 58, 58, 59, 60, 62, 63

4. 9, 0, 2, 8, 19, 5, 3, 2

POPULATION For Exercises 5–10, use the graphic at the right.

5. Find the range of the data.

6. Find the median of the data.

7. Find the upper and lower quartile of the data.

8. Find the interquartile range of the data.

9. Find any outliers of the data.

10. Use the information in Exercises 5–9 to describe the data.

Top Ancestral Origins of Americans	
Country	**Number (millions)**
Germany	46.5
Ireland	33.0
England	28.3
Italy	15.9
France	9.8
Poland	9.1
Scotland	5.4

Source: Census 2000 Supplementary Survey

Practice and Applications

Find the range, median, upper and lower quartiles, interquartile range, and any outliers for each set of data.

11. 43, 55, 49, 49, 53, 48, 57, 60, 57, 60, 47, 51, 59, 22

12. 55, 76, 104, 65, 62, 79, 63, 57, 52, 72, 57, 73, 55, 60, 80, 53

13. 19.8, 16.6, 19, 15.5, 14.6, 18.4, 13.5, 18, 14.5

14. 2.3, 2.3, 3.8, 2.6, 3.7, 2.9, 6.1, 2.3, 2.9, 2.5, 3.5

HOMEWORK HELP

For Exercises	See Examples
11–23	1–4

Extra Practice
See pages 639, 656.

15.

```
                    ×
              × × ×
        × ×   × × × ×
        × × × × × × × ×
   +--+--+--+--+--+--+--+--+--+--+
   0        0.5        1.0
```

16.

```
                    × ×
              × × ×
        ×     × × × ×
        × × × × × × × ×       ×
   +--+--+--+--+--+--+--+--+--+--+
   2.0        2.5        3.0
```

MOVIES For Exercises 17 and 18, use the stem-and-leaf plot at the right showing the ages of the Best Actress Academy Award winners from 1976 to 2002.

17. Find the median and upper and lower quartiles of the data.

18. Between what two ages were the middle half of the actresses when they won the award?

Stem	Leaf
2	1 6 7 8 9
3	0 1 3 3 4 5 8 8
4	1 1 2 2 3 4 9
5	2 5
6	1
7	4
8	0 2	1 = 21 years old

Source: The World Almanac

WEATHER For Exercises 19–23, use the table at the right.

19. Which city has a greater range of temperatures?

20. Find the median and upper and lower quartile ranges of the average temperatures for San Francisco.

21. Find the median and upper and lower quartile ranges of the average temperatures for Philadelphia.

22. Compare the medians of the average temperatures.

23. Compare the interquartile ranges of the average temperatures.

24. **WRITE A PROBLEM** Write a real-life problem that asks for the interquartile range.

25. **CRITICAL THINKING** Create two different sets of data that meet the following conditions.

    a. the same range, different interquartile ranges

    b. the same median and quartiles, but different ranges

Average Temperatures (°F)		
Month	San Francisco	Philadelphia
January	49	30
February	52	33
March	53	42
April	56	52
May	58	63
June	62	72
July	63	77
August	64	76
September	65	68
October	61	56
November	55	46
December	49	36

Source: *The World Almanac*

Standardized Test Practice and Mixed Review

FCAT Practice

26. **MULTIPLE CHOICE** High temperatures (°F) of twelve cities on March 20 were 40, 72, 74, 35, 58, 64, 40, 67, 40, 75, 68, and 51. What is the range of this set of data?

    Ⓐ 75°F   Ⓑ 51°F   Ⓒ 40°F   Ⓓ 11°F

27. **GRID IN** Find the interquartile range of the data in the stem-and-leaf plot.

Stem	Leaf
4	2 3 3 7
5	0 1 1 5 8 9
6	
7	2 3 4|2 = 4.2 meters

Find the mean, median, and mode for each set of data. Round to the nearest tenth if necessary. (Lesson 9-4)

28. 6, 4, 6, 12, 10, 8, 7, 12, 11, 9   29. 14, 3, 6, 8, 11, 9, 3, 2, 7

30. **RADIO LISTENING** Choose an appropriate display for the data below. Then make a display. (Lesson 9-3)

Adult Audience of Oldies Radio				
Age	18 to 24	25 to 34	35 to 44	45 to 54	55 or older
Percent of Audience	10%	14%	29%	33%	14%

Source: Interep Research Division

GETTING READY FOR THE NEXT LESSON

PREREQUISITE SKILL Graph each set of points on a number line. (Lesson 1-3)

31. {3, 5, 8, 9, 10}   32. {13, 15, 20, 27, 31}   33. {9, 13, 16, 17, 21}   34. {3, 9, 10, 15, 19}

Box-and-Whisker Plots

Sunshine State Standards
MA.E.1.3.2-2

What You'll Learn

Display and interpret data in a box-and-whisker plot.

NEW Vocabulary

box-and-whisker plot

WHEN am I ever going to use this?

WILDFIRES The table gives the number of wildfires for various states.

1. What is the least value in the data?

2. What is the lower quartile of the data?

3. What is the median of the data?

4. What is the upper quartile of the data?

5. What is the greatest value in the data?

6. Name any outliers.

Wildfires in 2001	
State	Number of Fires
Alaska	343
Nevada	1,098
Washington	1,209
Colorado	1,391
Montana	1,412
Idaho	1,566
Utah	1,646
Oregon	3,011
Florida	4,899
California	7,668

Source: National Interagency Fire Center

A **box-and-whisker plot** uses a number line to show the distribution of a set of data. The *box* is drawn around the quartile values, and the *whiskers* extend from each quartile to the extreme data points that are not outliers.

EXAMPLE Draw a Box-and-Whisker Plot

1. **WILDFIRES** Use the data in the table above to draw a box-and-whisker plot.

Step 1 Draw a number line that includes the least and greatest number in the data.

Step 2 Mark the extremes, the median, and the upper and lower quartile above the number line. Since the data have an outlier, mark the greatest value that is not an outlier.

Step 3 Draw the box and the whiskers.

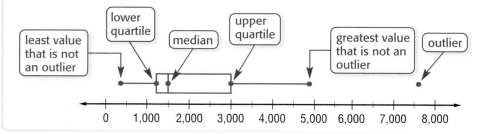

Box-and-whisker plots separate data into four parts. Although the parts usually differ in length, each part contains one fourth of the data.

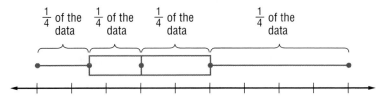

A long whisker or box indicates that the data in that quartile or quartiles have a greater range. A short whisker or box indicates the data in that quartile or quartiles have a lesser range.

EXAMPLE Interpret Data

2 **DIET** What does the length of the box-and-whisker plot tell you about the data?

Calories in Fast Food Sandwiches

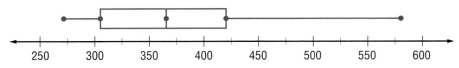

The median line seems to divide the box into two approximately equal parts, so data in the second and third quartiles are similarly spread out. The whisker at the right is longer than the other parts of the plot, so the data in the fourth quartile are more spread out.

A double box-and-whisker plot can be used to compare data.

EXAMPLE Compare Data

3 **MULTIPLE-CHOICE TEST ITEM** Use the box-and-whisker plots below to determine which statement is *not* true.

Ages of the U.S.A. 2002 Olympic Hockey Players

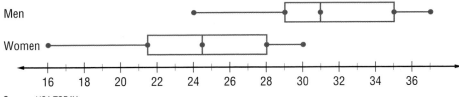

Men

Women

16 18 20 22 24 26 28 30 32 34 36

Source: *USA TODAY*

Ⓐ The women's ages have a greater range than the men's ages.

Ⓑ The women's ages were all less than the men's median age.

Ⓒ The men's ages were all greater than the women's median age.

Ⓓ Most of the men were 29 or older.

Read the Test Item You need to study the box-and-whisker plot.

Solve the Test Item The ages of the men were not all greater than the median age of the women. The answer is C.

Check to make sure A, B, and D are true.

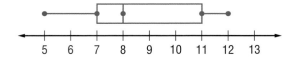
1. **Describe** the meaning of the box in a box-and-whisker plot.

2. **OPEN ENDED** Write a set of data that could be represented by the box-and-whisker at the right.

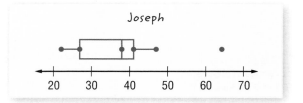

3. **FIND THE ERROR** Chapa and Joseph are making a box-and-whisker plot for the following set of data. Who is correct? Explain.

22, 23, 27, 30, 34, 38, 39, 40, 41, 47, 64

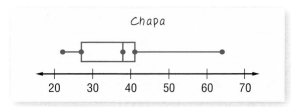

GUIDED PRACTICE

Draw a box-and-whisker plot for each set of data.

4. 38, 43, 36, 37, 32, 37, 29, 51

5. 100, 70, 70, 90, 50, 90, 50, 90, 100, 50, 90, 100, 90, 50, 25, 80

FOOD For Exercises 6–8, use the following box-and-whisker plot.

Calories in Fast Food Muffins

6. What is the interquartile range of the data?

7. Three fourths of the muffins have at least how many Calories?

Practice and Applications

Draw a box-and-whisker plot for each set of data.

8. 49, 45, 55, 32, 28, 53, 26, 38, 35, 35, 51

9. 77, 85, 72, 76, 95, 90, 73, 82, 82, 80, 73

10. 540, 460, 520, 350, 500, 480, 475, 525, 450, 515

11. 225, 245, 220, 270, 350, 280, 230, 240, 225, 270

12. 42, 38, 42, 45, 43, 80, 55, 50, 34, 36, 40, 35

13. 52, 58, 67, 63, 47, 44, 52, 15, 49, 65, 52, 59

14. **HISTORY** The population in thousands of the American colonies in 1770 are listed below. Make a box-and-whisker plot of the data.

31.3, 62.4, 10.0, 235.3, 58.2, 183.9, 162.9, 117.4, 240.1, 35.5, 202.6, 447.0, 197.2, 124.2, 23.4, 15.7, 1.0

HOMEWORK HELP

For Exercises	See Examples
8–14	1
15–18	2, 3

Extra Practice
See pages 639, 656.

GAS MILEAGE For Exercises 15–18, use the following box-and-whisker plot.

Highway Gas Mileage for 2002 Two-Wheel-Drive Sports Utility Vehicles (SUV)

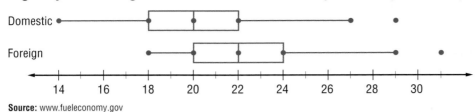

Source: www.fueleconomy.gov

15. Which set of data has a greater range?

16. What percent of these domestic SUVs get at least 20 miles per gallon?

17. What percent of these foreign SUVs get at least 20 miles per gallon?

18. In general, do domestic two-wheel-drive SUVs get more or less gas mileage than the foreign ones? Explain.

 Data Update What is the gas mileage of current SUVs? Visit msmath3.net/data_update to learn more.

19. **CRITICAL THINKING** Write a set of data that could be represented by the box-and-whisker plot at the right.

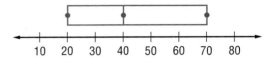

Standardized Test Practice and Mixed Review   Ⓐ Ⓑ Ⓒ Ⓓ   **FCAT Practice**

For Exercises 20 and 21, use the box-and-whisker plot.

20. **MULTIPLE CHOICE** Twenty-five percent of the data are found between what two values?

   Ⓐ 55 and 75   Ⓑ 60 and 80   Ⓒ 75 and 95   Ⓓ 60 and 75

21. **SHORT RESPONSE** What is the range of the data?

Find the range, median, upper and lower quartiles, interquartile range, and any outliers for each set of data. (Lesson 9-5)

22. 73, 52, 31, 54, 46, 28, 47, 49, 58

23. 87, 63, 84, 94, 89, 74, 50, 85, 91, 78, 99, 81, 77, 86, 65, 81, 74

24. **LIFE SCIENCE** Find the mean, median, and mode of the plant heights 22, 4, 1, 12, 5, 22, 5, 25, 25, 19, 23, 24, 11, 16, 3, and 22 inches. (Lesson 9-4)

GETTING READY FOR THE NEXT LESSON

PREREQUISITE SKILL Describe each sample as *biased* or *unbiased.* **Explain.** (Lesson 8-7)

25. To determine how the neighborhood park should be improved, a survey is taken of every other house in the neighborhood.

26. To determine who will be elected governor, a survey is taken of every other house in one neighborhood.

Misleading Graphs and Statistics

Sunshine State Standards
MA.E.1.3.1-1, MA.E.1.3.1-2

What You'll Learn
Recognize when graphs and statistics are misleading.

↻ REVIEW Vocabulary

biased sample: a sample where one or more parts of the population are favored over others (Lesson 8-7)

WHEN am I ever going to use this?

MOVIES Study the graphs.

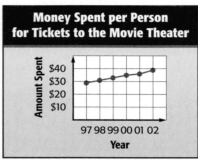

Graph A

Money Spent per Person for Tickets to the Movie Theater

Graph B

Money Spent per Person for Tickets to the Movie Theater

Source: Veronis, Schler and Associates, Inc.

1. Do both graphs show the same data?

2. Which graph seems to show a greater increase in spending? Why?

When dealing with statistics, you must interpret the information carefully. The scale in Graph B above may make people think the increase is greater than the actual increase.

EXAMPLE Identify a Misleading Graph

① **CARS** Which graph could be used to indicate a greater increase in sales of automotive equipment? Explain.

Graph A

Sales of Speciality Automotive Equipment (billions)

$12.2 $24.9
1990   2000

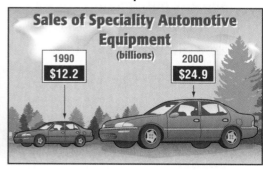

Graph B

Sales of Speciality Automotive Equipment (billions)

1990 $12.2   2000 $24.9

Source: Speciality Equipment Market Association

STUDY TIP

Statistics A graph should have a title and labels on both scales.

Both graphs show the amount of sales of specialty automotive equipment has about doubled. The ratio of the areas of the bars in Graph A is about 1 : 2. The ratio of the areas of the cars in Graph B is about 1 : 4. Graph B seems to show a greater increase in sales.

ROLLER COASTERS The world's longest roller coaster is the Steel Dragon in Mie, Japan. It is 8,133 feet long.

Source: *The World Almanac for Kids*

Recall that there are three different measures of central tendency or types of averages. They are mean, median, and mode. These different values can be used to show different points of view.

EXAMPLES **Identify Different Uses of Statistics**

ROLLER COASTERS The ride times of the roller coasters at an amusement park are 220, 150, 150, 150, 120, 90, 90, and 52 seconds.

2 Find the mean, median, and mode of the ride times.

Mean $\dfrac{\text{sum of values}}{\text{number of values}} = \dfrac{1{,}022}{8}$ or 127.75

Median $\dfrac{150 + 120}{2} = \dfrac{270}{2}$ or 135

Mode 150

The mean is 127.75 seconds, the median is 135 seconds, and the mode is 150 seconds.

3 Which average would the amusement park use to encourage people who like roller coasters to come to the park? Explain.

People who like roller coasters would probably enjoy longer rides. Therefore, the amusement park would use the mode since it is the greatest of the averages.

Skill and Concept Check

Writing Math

Exercise 1

1. **List** two ways a graph can be misleading.

2. **OPEN ENDED** Write a set of data in which the mean is not representative.

GUIDED PRACTICE

3. Which graph would you use to indicate that Norway had many more medals than the Soviet Union? Explain.

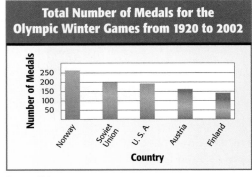

Graph A

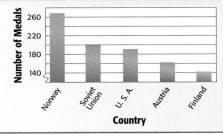

Graph B

Source: IOC

4. **SCHOOL** Drew received a 75% on his history test. If the mean of the test scores is 80%, the median is 78%, and the mode is 70%. Which average might Drew use when describing his score to his parents? Explain.

HOMEWORK HELP

For Exercises	See Examples
5–6	1
8–12	2, 3

Extra Practice
See pages 639, 656.

5. Which graph would you use to indicate that a male householder has a much greater median income than a female householder? Explain.

Graph A

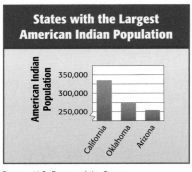

Graph B

Source: U.S. Bureau of the Census

6. Which graph would you use to indicate that there are many more American Indians living in California than Oklahoma? Explain.

Graph A

Graph B

Source: U.S. Bureau of the Census

7. **ADVERTISING** Study the situation below. Is the advertisement false? Is it misleading? Explain.

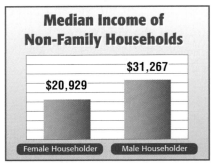

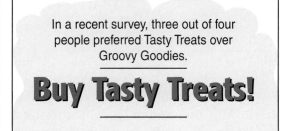

SALARIES For Exercises 8–12, use the table. It compares the salaries of the employees of two small manufacturing companies.

Salaries	
Company A	Company B
$69,800	$25,100
$21,500	$23,650
$18,000	$23,600
$17,600	$23,100
$17,400	$21,750
$17,300	$21,600
$16,150	$21,500
$16,050	$20,680
$15,100	$19,670
$14,900	$19,450

8. What is the mean, median, and mode of the salaries of the employees of Company A?

9. What average would Company A use to try to encourage someone to work for them? Explain.

10. What is the mean, median, and mode of the salaries of the employees of Company B?

11. What average would Company B use to encourage someone to work for them? Explain.

12. If you were choosing to work for Company A or Company B, which might you choose? Explain.

13. **CRITICAL THINKING** The number of admissions to movie theaters went from 1.14 billion in 1991 to 1.49 billion in 2001.

   a. Make a graph that shows a small change in admissions.

   b. Make a graph that shows a large change in admissions.

Standardized Test Practice and Mixed Review

For Exercises 14 and 15, use the data in the table.

14. **MULTIPLE CHOICE** Which value is the most misleading average of the data?

   Ⓐ mean   Ⓑ median

   Ⓒ mode   Ⓓ none of the averages

Borders of the United States			
Border	Miles	Border	Miles
Mainland/Canada	3,987	Gulf of Mexico Coast	1,631
Alaska/Canada	1,538	Pacific Coast	7,623
Mexican	1,933	Arctic Coast	1,060
Atlantic Coast	2,069		

Source: *The World Almanac for Kids*

15. **MULTIPLE CHOICE** Which value is the most representative of the data?

   Ⓕ mean   Ⓖ median   Ⓗ mode   Ⓘ none of the averages

Draw a box-and-whisker plot for each set of data. (Lesson 9-6)

16. 55, 63, 72, 52, 55, 68, 64, 61, 58

17. 53, 49, 43, 5, 28, 38, 34, 45, 51, 45

18. **MUSIC** The numbers of pages in a magazine in the last nine issues were 196, 188, 184, 200, 168, 176, 192, 160, and 180. Find the median, upper quartile, and lower quartile of this data. (Lesson 9-5)

Write each number in scientific notation. (Lesson 2-9)

19. 70,200   20. 0.000081   21. 0.000456   22. 620,000,000

GETTING READY FOR THE NEXT LESSON

PREREQUISITE SKILL Add or subtract. (Lessons 1-4 and 1-5)

23. $8 + (-5)$ 24. $-7 + (-4)$ 25. $-6 - (-4)$ 26. $-7 - 8$

9-8 Matrices

What You'll Learn
Use matrices to organize data.

NEW Vocabulary

matrix
row
column
element
dimensions

WHEN am I ever going to use this?

WEATHER The record temperatures for each continent are listed in the table below.

Road Temperatures		
Continent	**Highest Temperature (°F)**	**Lowest Temperature (°F)**
Africa	136	−11
Antarctica	59	−129
Asia	129	−90
Australia	128	−9
Europe	122	−67
North America	134	−81
South America	120	−27

Source: *The World Almanac for Kids*

1. How many continents are listed?
2. How many temperatures are given for each continent?
3. What does the number −129 represent?
4. What does the number 136 represent?

The table above has rows and columns of data. A rectangular arrangement of numerical data is called a **matrix**.

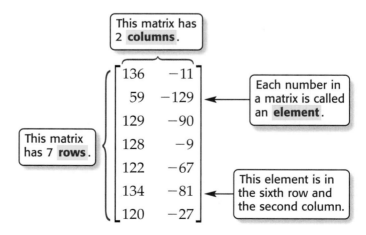

This matrix has 2 **columns**.

Each number in a matrix is called an **element**.

This matrix has 7 **rows**.

This element is in the sixth row and the second column.

$$\begin{bmatrix} 136 & -11 \\ 59 & -129 \\ 129 & -90 \\ 128 & -9 \\ 122 & -67 \\ 134 & -81 \\ 120 & -27 \end{bmatrix}$$

A matrix is described by its **dimensions**, or the number of rows and columns, with the number of rows stated first. The dimensions of this matrix are 7 by 2.

1 State the dimensions of $\begin{bmatrix} -3 & 6 & ⑨ & 3 \\ -5 & 8 & 4 & 0 \end{bmatrix}$. Then identify the position of the circled element.

The matrix has 2 rows and 4 columns. The dimensions of the matrix are 2 by 4.

The circled element is in the first row and the third column.

Your Turn State the dimensions of each matrix. Then identify the position of the circled element.

a. $[7 \quad -8 \quad ⑤]$

b. $\begin{bmatrix} 14 & 17 \\ ㉗ & 8 \end{bmatrix}$

c. $\begin{bmatrix} -5 & -7 & 3 & 2 \\ 9 & -3 & -10 & -1 \\ 2 & ④ & 9 & -4 \end{bmatrix}$

READING Math

Matrices The plural of matrix is *matrices*. It is pronounced MAY tra cees.

If two matrices have the same dimensions, you can add or subtract them. To do this, add or subtract corresponding elements of the two matrices.

EXAMPLES Add and Subtract Matrices

Add or subtract. If there is no sum or difference, write *impossible*.

2 $\begin{bmatrix} 4 & 3 & 2 \\ 7 & -1 & 5 \end{bmatrix} + \begin{bmatrix} 0 & 2 & -1 \\ 12 & 5 & 6 \end{bmatrix}$

$\begin{bmatrix} 4 & 3 & 2 \\ 7 & -1 & 5 \end{bmatrix} + \begin{bmatrix} 0 & 2 & -1 \\ 12 & 5 & 6 \end{bmatrix} = \begin{bmatrix} 4+0 & 3+2 & 2+(-1) \\ 7+12 & -1+5 & 5+6 \end{bmatrix}$

$= \begin{bmatrix} 4 & 5 & 1 \\ 19 & 4 & 11 \end{bmatrix}$

3 $[-7 \quad 8 \quad 4 \quad -9] - [-5 \quad -6 \quad 3 \quad 10]$

$[-7 \quad 8 \quad 4 \quad -9] - [-5 \quad -6 \quad 3 \quad 10]$

$= [-7-(-5) \quad 8-(-6) \quad 4-3 \quad -9-10]$

$= [-2 \quad 14 \quad 1 \quad -19]$

4 $\begin{bmatrix} 7 & -8 \\ 4 & -6 \end{bmatrix} - \begin{bmatrix} 8 & -19 \\ 22 & 4 \\ -8 & 12 \end{bmatrix}$

The first matrix has 2 rows and 2 columns. The second matrix has 3 rows and 2 columns. Since the matrices do not have the same dimensions, it is impossible to subtract them.

Your Turn Add or subtract. If there is no sum or difference, write *impossible*.

d. $\begin{bmatrix} 6 & -8 \\ 5 & 4 \end{bmatrix} - \begin{bmatrix} 5 & 8 \\ -10 & -3 \end{bmatrix}$

e. $\begin{bmatrix} 5 & -9 \\ 6 & 7 \\ -8 & -2 \end{bmatrix} + \begin{bmatrix} -1 & 5 & -6 \\ 4 & 7 & -3 \end{bmatrix}$

Skill and Concept Check

1. **Describe** the difference between a 3-by-2 matrix and a 2-by-3 matrix.

2. **OPEN ENDED** Write two matrices whose sum is $\begin{bmatrix} 3 & 8 & -2 \\ -1 & 0 & 6 \end{bmatrix}$.

GUIDED PRACTICE

State the dimensions of each matrix. Then identify the position of the circled element.

3. $\begin{bmatrix} 5 \\ -7 \\ ④ \end{bmatrix}$

4. $\begin{bmatrix} 8 & 5 & ⑥ \\ 10 & 18 & -5 \\ -10 & 14 & 7 \end{bmatrix}$

5. $\begin{bmatrix} 12 & 6 & 18 & 14 & 0 \\ 9 & 5 & 22 & ⑯ & 25 \end{bmatrix}$

Add or subtract. If there is no sum or difference, write *impossible*.

6. $\begin{bmatrix} 5 & -8 & 4 \\ -2 & 6 & 3 \end{bmatrix} + \begin{bmatrix} -1 & -3 & 7 \\ 4 & -9 & -2 \end{bmatrix}$

7. $\begin{bmatrix} 5 & 9 & -6 \\ 4 & -2 & -8 \\ 7 & -1 & 0 \end{bmatrix} - \begin{bmatrix} 7 & -4 & 2 \\ -7 & -5 & 10 \\ 12 & 7 & -8 \end{bmatrix}$

Practice and Applications

State the dimensions of each matrix. Then identify the position of the circled element.

HOMEWORK HELP

For Exercises	See Examples
8–13	1
14–17, 19	2–4

Extra Practice
See pages 640, 656.

8. $\begin{bmatrix} 5 \\ ⑥ \\ 2 \end{bmatrix}$

9. $[7 \quad -9 \quad ⑤]$

10. $\begin{bmatrix} 9 & -7 \\ 4 & ③ \end{bmatrix}$

11. $\begin{bmatrix} 7 & -9 & 3 \\ -4 & 2 & 0 \\ ⑱ & 12 & -6 \end{bmatrix}$

12. $\begin{bmatrix} 6 & -8 & 5 & ⑨ \\ -3 & -2 & 0 & 0 \end{bmatrix}$

13. $\begin{bmatrix} 12 & -4 & 3 & 7 \\ -5 & -2 & 0 & 1 \\ 14 & ⑪ & 25 & -9 \end{bmatrix}$

Add or subtract. If there is no sum or difference, write *impossible*.

14. $\begin{bmatrix} 2 & 1 \\ 3 & 0 \\ 5 & -4 \end{bmatrix} + \begin{bmatrix} 6 & 0 \\ 2 & -3 \\ 5 & 8 \end{bmatrix}$

15. $\begin{bmatrix} 8 & 1 & 9 \\ 2 & 0 & 7 \\ 3 & -2 & 4 \end{bmatrix} - \begin{bmatrix} 6 & 0 & 7 \\ -1 & 5 & 8 \\ 2 & 6 & -4 \end{bmatrix}$

16. $\begin{bmatrix} 4 & -2 \\ 3 & 1 \end{bmatrix} - \begin{bmatrix} 1 & 5 & 7 \\ 3 & -2 & 1 \\ 0 & 4 & 11 \end{bmatrix}$

17. $\begin{bmatrix} 1 \\ 2 \\ 9 \end{bmatrix} + [-5 \quad -2 \quad -7]$

SPORTS For Exercises 18 and 19, use the following information.

1998 Winter Olympics Medals			
Country	Gold	Silver	Bronze
Canada	6	5	4
Norway	10	10	5
U.S.A.	6	3	4

Source: *The World Almanac*

2002 Winter Olympics Medals			
Country	Gold	Silver	Bronze
Canada	6	3	8
Norway	11	7	6
U.S.A.	10	13	11

Source: *USA TODAY*

18. Make a matrix for the information on each of the Olympics.

19. Use addition of matrices to find the total number of each type of medals won by the countries in the two Olympics. Write as a matrix.

20. **CRITICAL THINKING** Find the values of a, b, c, and d if
$$\begin{bmatrix} 1 & 4 \\ 3 & -2 \end{bmatrix} + \begin{bmatrix} a & b \\ c & d \end{bmatrix} = \begin{bmatrix} -1 & 7 \\ 5 & -2 \end{bmatrix}.$$

For Exercises 21 and 22, use the table below. It shows the attendance for four school concerts.

	Fall	Holiday	Winter	Spring
Friday Night	112	100	95	99
Saturday Night	101	103	75	60
Sunday Matinee	89	88	90	86

21. **MULTIPLE CHOICE** Choose the matrix that correctly displays the attendance.

A $\begin{bmatrix} 89 & 101 & 112 \\ 88 & 103 & 100 \\ 90 & 75 & 95 \\ 86 & 60 & 99 \end{bmatrix}$

B $\begin{bmatrix} 112 & 101 & 89 & 100 \\ 103 & 88 & 95 & 75 \\ 90 & 99 & 60 & 86 \end{bmatrix}$

C $\begin{bmatrix} 112 & 100 & 95 & 99 \\ 101 & 103 & 75 & 60 \\ 89 & 88 & 90 & 86 \end{bmatrix}$

D $\begin{bmatrix} 112 & 101 & 89 \\ 100 & 103 & 88 \\ 95 & 75 & 90 \\ 99 & 60 & 86 \end{bmatrix}$

22. **MULTIPLE CHOICE** Which concert had the greatest attendance?

F Fall   G Holiday   H Winter   I Spring

23. **EXERCISE** A person working fairly hard on a treadmill will burn about 700 Calories per hour. A person working fairly hard on a stairstepper will burn about 625 Calories per hour. Make a graph of the data showing the treadmill is much better than the stairstepper in burning Calories. (Lesson 9-7)

Draw a box-and-whisker plot for each set of data. (Lesson 9-6)

24. 43, 47, 48, 50, 53, 54, 56, 56, 59   25. 37, 40, 56, 57, 57, 64, 68, 72

26. What is the probability that a month picked at random ends in *–ber*? (Lesson 8-1)

WebQuest **Interdisciplinary Project**

It's all in the Genes
It's time to complete your project. Use the information and data you have gathered about genetics and the traits of your classmates to prepare a Web page or poster. Be sure to include a chart displaying your data with your project.
msmath3.net/webquest

Vocabulary and Concept Check

box-and-whisker plot (p. 446)	lower quartile (p. 442)	mode (p. 435)
circle graph (p. 426)	matrix (p. 454)	outlier (p. 443)
column (p. 454)	mean (p. 435)	quartiles (p. 442)
dimensions (p. 454)	measures of central	range (p. 442)
element (p. 454)	tendency (p. 435)	row (p. 454)
histogram (p. 420)	measures of variation (p. 442)	upper quartile (p. 442)
interquartile range (p. 442)	median (p. 435)	

State whether each sentence is *true* or *false*. If *false*, replace the underlined word or number to make a true sentence.

1. A <u>histogram</u> is a bar graph that shows the frequency of data in intervals.

2. The range is one of the <u>measures of central tendency</u>.

3. The <u>mean</u> is the sum of the data divided by the number of pieces of data.

4. If you want to show how the parts compare to the whole, use a <u>circle graph</u>.

5. The <u>mode</u> is the middle number of a set of data.

6. A <u>matrix</u> is a rectangular arrangement of numbers.

7. A matrix is described by its <u>rows</u>.

8. Each number in a matrix is called an <u>element</u>.

Lesson-by-Lesson Exercises and Examples

 Histograms (pp. 420–424)

For Exercises 9–10, use the histogram at the right.

9. How many students received a score of at least 80?

10. How many students received a score less than 70?

11. **ANIMALS** The following is a list of years various types of animals are expected to live. Draw a histogram to represent the data.

1, 3, 5, 5, 6, 7, 8, 8, 10, 10, 10, 12, 12, 12, 12, 15, 15, 15, 15, 16, 18, 20, 20, 25, 35

Example 1
Make a histogram to represent the following English test scores.

56, 87, 87, 74, 87, 84, 94, 80, 72, 58, 87, 90, 68, 90, 70, 73, 74, 82, 68, 64

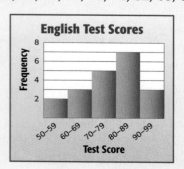

9-2 Circle Graphs (pp. 426–429)

12. GEOGRAPHY Lake Erie is 9,910 square miles, Lake Huron is 23,010 square miles, Lake Michigan is 22,300 square miles, Lake Ontario is 7,540 square miles, and Lake Superior is 31,700 square miles. Make a circle graph showing what percent of the total area of the Great Lakes is represented by each lake.

Example 2

CARS Which country made the same amount of motor vehicles as Canada and Japan?

Motor Vehicle Production

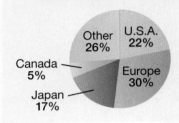

The answer is the U.S.A.

9-3 Choosing an Appropriate Display (pp. 430-433)

Choose an appropriate type of display for each situation.

13. percent of income people spend on different expenses each month

14. populations of counties in Pennsylvania arranged by intervals

Example 3

Choose an appropriate display for the number of students who prefer each color.

An appropriate display would be a table, bar graph, or pictograph.

9-4 Measures of Central Tendency (pp. 435–438)

Find the mean, median, and mode for each set of data. Round to the nearest tenth if necessary.

15. 13, 15, 15, 15, 18, 19, 20

16. 5.6, 6.5, 6.8, 9.6, 10.1

17. 5, 6, 7, 7, 8, 8, 8, 9, 9, 10

Example 4

Find the mean, median, and mode for the data 8, 8, 9, 9, 9, 13, 14.

$$\text{mean} = \frac{8 + 8 + 9 + 9 + 9 + 13 + 14}{7} \text{ or } 10$$

median = 9   mode = 9

9-5 Measures of Variation (pp. 442–445)

Find the range, median, upper and lower quartiles, interquartile range, and any outliers for each set of data.

18. 0, 5, 7, 11, 13, 13, 13, 14, 15

19. 1, 2, 2, 3, 3, 3, 4, 4, 5, 6, 12

20. 3, 5, 7, 7, 7, 8, 8, 9

21. 8, 9, 5, 10, 7, 6, 2, 4

Example 5

Find the range, median, upper and lower quartiles, and interquartile range.

2, 3, 4, 5, 6, 9, 9, 9, 9, 9, 10

Range = 10 − 2 or 8   Median = 9

Lower Quartile = 4   Upper Quartile = 9

Interquartile Range = 9 − 4 or 5

Box-and-Whisker Plots (pp. 446–449)

Draw a box-and-whisker plot for each set of data.

22. 0, 5, 7, 11, 13, 13, 13, 14, 15
23. 1, 2, 2, 3, 3, 3, 4, 4, 5, 6, 6
24. 2, 5, 7, 7, 7, 8, 8, 9
25. 8, 9, 5, 10, 7, 6, 2, 4

Example 6
Draw a box-and-whisker plot for the set of data.

2, 3, 4, 5, 6, 7, 9, 9, 9, 9, 10

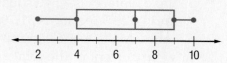

Misleading Graphs and Statistics (pp. 450–453)

SCHOOL For Exercises 26–30, use the list of test grades for two group of students.
Group A: 100, 95, 90, 89, 88, 45, 42, 40, 40
Group B: 99, 98, 93, 89, 88, 85, 85, 75, 72

26. What is the mean, median, and mode of the grades of Group A?
27. What average is most favorable for Group A?
28. What is the mean, median, and mode of the grades for Group B?
29. What average is most favorable for Group B?
30. Since both groups have the same median, did both groups do as well on the test? Explain.

Example 7
Which graph makes Program E appear to be much more popular than Program A? Explain.

Graph X makes Program E appear to be much more popular than Program A, because the scale does not start with 0.

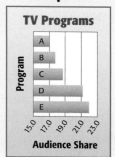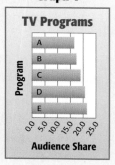

Matrices (pp. 454–457)

Add or subtract. If there is no sum or difference, write *impossible.*

31. $\begin{bmatrix} 6 & 5 \\ -2 & 1 \end{bmatrix} - \begin{bmatrix} -3 & -2 \\ 0 & -1 \end{bmatrix}$

32. $\begin{bmatrix} 2 & -1 \\ -3 & 9 \end{bmatrix} + \begin{bmatrix} 10 & -3 \\ 8 & -3 \end{bmatrix}$

33. $\begin{bmatrix} 3 & -2 \\ 4 & -1 \end{bmatrix} + \begin{bmatrix} -3 & 8 \\ 0 & 6 \\ -4 & 2 \end{bmatrix}$

Example 8
Add. If there is no sum, write *impossible.*

$\begin{bmatrix} 3 & 5 \\ -2 & 1 \end{bmatrix} + \begin{bmatrix} -2 & -1 \\ -3 & 5 \end{bmatrix}$

$= \begin{bmatrix} 3 + (-2) & 5 + (-1) \\ -2 + (-3) & 1 + 5 \end{bmatrix}$

$= \begin{bmatrix} 1 & 4 \\ -5 & 6 \end{bmatrix}$

Vocabulary and Concepts

1. **Explain** how to draw a box-and-whisker plot.
2. **Describe** the difference between a 4 by 2 matrix and a 2 by 4 matrix.

Skills and Applications

EXERCISE For Exercises 3–5, use the histogram.

3. How many people were surveyed?
4. How many people spend more than 8 hours per week exercising?
5. Make a circle graph of the data.

6. **SPORTS** Toni made a survey about students' favorite sports. What type of graph should she use to show the percent of students who picked each sport?

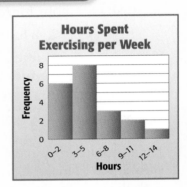

For Exercises 7–12, use the following list of the ages of 13 people.

45, 36, 27, 16, 19, 46, 40, 38, 22, 23, 25, 40, 17.

7. Find the mean.
8. What is the median?
9. Find the range.
10. Find the upper and lower quartile.
11. What is the interquartile range?
12. Draw a box-and-whisker plot.

13. **SCHOOL** Which graph indicates that there are many more eighth grade students that have a B average or better than students in the sixth grade with the same average? Explain.

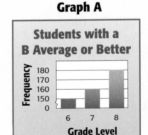

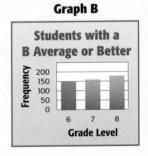

Add or subtract. If no sum or difference exists, write *impossible*.

14. $\begin{bmatrix} 3 & 2 \\ -2 & 5 \end{bmatrix} - \begin{bmatrix} -2 & -1 \\ 0 & 5 \end{bmatrix}$

15. $\begin{bmatrix} 2 & 5 \\ 6 & 1 \end{bmatrix} + \begin{bmatrix} 2 & 5 & 0 \\ 3 & 1 & 2 \end{bmatrix}$

Standardized Test Practice

Ⓐ Ⓑ Ⓒ Ⓓ

16. **MULTIPLE CHOICE** Alma's monthly earnings were $540, $450, $800, $560, $350, $400, $350, $380, $500, $450, $600, and $200. How much more would Alma have needed to earn to have an average monthly income of $500?

 Ⓐ 0   Ⓑ $220   Ⓒ $420   Ⓓ $500

🔴 **FCAT Practice**

PART 1 Multiple Choice

Record your answers on the answer sheet provided by your teacher or on a sheet of paper.

1. Federico counted the number of bacteria in one dish each hour for four hours.

    2, 4, 8, 16

    If the pattern continues, how many bacteria will there be in the fifth hour? (Lesson 2-8)

    Ⓐ 20 bacteria   Ⓑ 24 bacteria
    Ⓒ 32 bacteria   Ⓓ 64 bacteria

2. Trina is planning a rectangular garden. She would like to make a diagonal walking path through the garden with stone tiles. If each tile is a 6-inch square, how many tiles will she need? (Lesson 3-5)

    12 ft
    5 ft

    Ⓕ 12 tiles   Ⓖ 13 tiles
    Ⓗ 26 tiles   Ⓘ 78 tiles

3. Describe the triangles.
    (Lesson 4-5)

    Ⓐ right   Ⓑ similar
    Ⓒ congruent   Ⓓ corresponding

4. In the figure below, lines M and N are parallel, and lines O and P are parallel. Find the measure of ∠1. (Lesson 6-1)

    Ⓕ 85°   Ⓖ 95°   Ⓗ 105°   Ⓘ 125°

5. Samuel is setting up his tent for the night. What is the area of the canvas needed to form the front of his tent? (Lesson 7-1)

    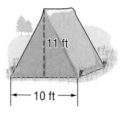
    11 ft
    10 ft

    Ⓐ 25 ft²   Ⓑ $27\frac{1}{2}$ ft²
    Ⓒ 55 ft²   Ⓓ 110 ft²

6. Alyssa is making a beaded bracelet. She has 12 red beads, 18 blue beads, 8 yellow beads, and 10 green beads. If she randomly chooses a bead from the bag, what is the probability that she will select a blue bead? (Lesson 8-1)

    Ⓕ $\frac{1}{4}$ Ⓖ $\frac{3}{8}$
    Ⓗ $\frac{5}{12}$ Ⓘ $\frac{9}{19}$

7. Which of the following would best display the information in a frequency table that has been divided into intervals? (Lesson 9-1)

    Ⓐ circle graph
    Ⓑ stem-and-leaf plot
    Ⓒ double bar graph
    Ⓓ histogram

8. Which of the following would best display data about different types of activities offered by a summer camp and the percent of time spent on each activity? (Lesson 9-3)

    Ⓕ histogram   Ⓖ line plot
    Ⓗ bar graph   Ⓘ circle graph

9. The following is a list of the number of minutes Melanie spent on math homework each night. What is the lower quartile of the set of data? (Lesson 9-5)

    26, 19, 45, 32, 40, 15,
    34, 12, 37, 25, 43, 21

    Ⓐ 19   Ⓑ 20   Ⓒ 21   Ⓓ 22

FCAT Practice

PART 2 Short Response/Grid In

THINK
SOLVE
EXPLAIN

Record your answers on the answer sheet provided by your teacher or on a sheet of paper.

10. Hunter deposited $450 in a savings account that receives 4.5% simple interest. Marina deposited $550 in a savings account that receives 2.8% simple interest. Who will earn the greater amount of interest after 18 months? (Lesson 5-8)

11. The triangle at the right is translated 2 units up and 2 units to the left. What are the vertices of the translated triangle? (Lesson 6-8)

12. Curtis wants to cover the box at the right with decorative paper. How much paper will he need to cover the box? (Lesson 7-7)

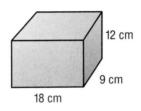

13. In Springwood Middle School, there are 50 students in the eighth grade. The class will elect 4 students as different class officers. If no student can hold more than one of these offices, how many different ways can the positions be filled? (Lesson 8-3)

14. Mr. Francis has told his students that he will remove the lowest exam score for each student at the end of the grading period. Seki received grades of 43, 78, 84, 85, 88, and 90 on her exams. What will be the difference between the mean of her original grades and the mean of her five grades after Mr. Francis removes one grade? (Lesson 9-4)

15. The number of students who attended science club for each meeting last semester are listed.

    21, 33, 38, 12, 47, 18, 42, 51, 17, 35, 46

    Is the number 39 the upper quartile, the lower quartile, or the range of the set of numbers? (Lesson 9-5)

16. Alfonso's bowling scores are listed below. What is the interquartile range? (Lesson 9-5)

    125, 142, 167, 138, 176, 102, 156, 130, 142

FCAT Practice

PART 3 Extended Response

Record your answers on a sheet of paper. Show your work.

THINK
SOLVE
EXPLAIN

17. A pet store has 8 black dogs, 10 brown dogs, 2 white dogs, 6 spotted dogs, and 5 multicolored dogs. (Lesson 9-3)

    a. Make a graph that shows the number of each type of coloring the pet store has.

    b. Make a graph that shows what part of the total number of dogs is represented by each type of coloring.

    c. Describe an advantage of each type of graph you drew.

18. The price of a CD increased from $12 to $14. (Lesson 9-7)

    a. Make a graph to indicate that the increase in price was not too much.

    b. Make a graph to indicate that the increase in price is too much.

TEST-TAKING TIP

Question 16 Review any terms that you have learned before you take a test. For example, for a test on data and statistics, be sure that you understand such terms as *mean, median, mode, range, upper quartile, lower quartile,* and *interquartile range.*

UNIT 5

Algebra: Linear and Nonlinear Functions

In this unit, you will build on your understanding of algebra to solve problems involving linear and nonlinear functions.

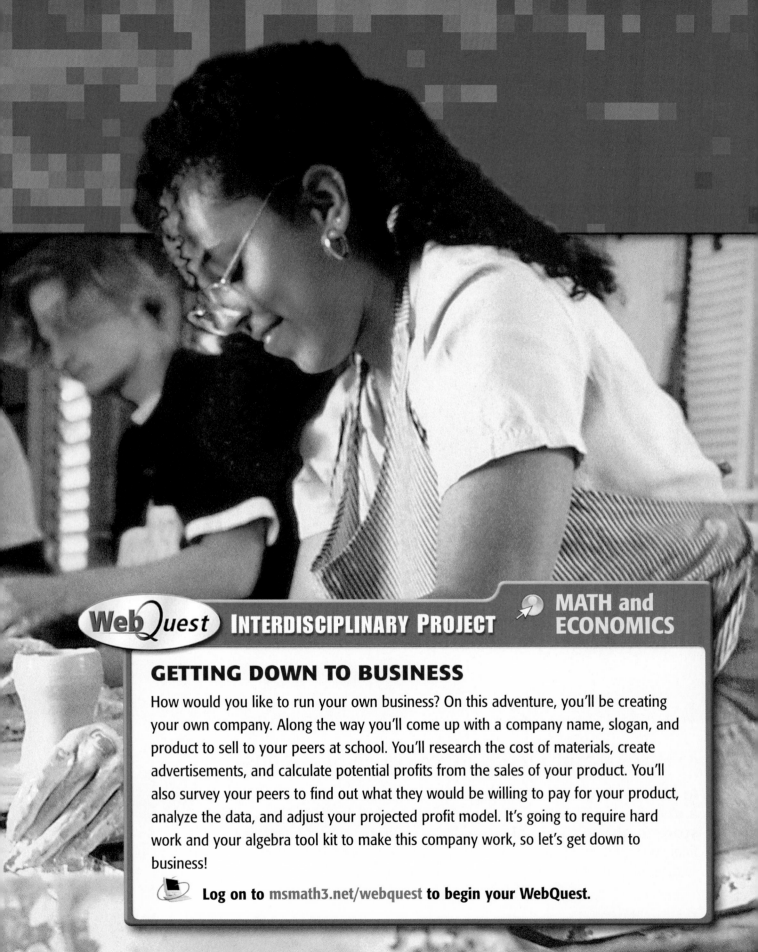

GETTING DOWN TO BUSINESS

How would you like to run your own business? On this adventure, you'll be creating your own company. Along the way you'll come up with a company name, slogan, and product to sell to your peers at school. You'll research the cost of materials, create advertisements, and calculate potential profits from the sales of your product. You'll also survey your peers to find out what they would be willing to pay for your product, analyze the data, and adjust your projected profit model. It's going to require hard work and your algebra tool kit to make this company work, so let's get down to business!

Log on to msmath3.net/webquest to begin your WebQuest.

10 Algebra: More Equations and Inequalities

""How is math used in skiing competitions?""

In aerial skiing competitions, the total judges score is multiplied by a *degree of difficulty* factor and then added to the skier's current score to obtain the final score. If you know your current score, the leader's final score, and your jump's degree of difficulty, you can solve a two-step equation to determine what score you need to win a competition.

You will solve a problem about aerial skiing in Lesson 10-3.

GETTING STARTED

Take this quiz to see whether you are ready to begin Chapter 10. Refer to the lesson or page number in parentheses if you need more review.

▶ Vocabulary Review

Complete each sentence.

1. A(n) __?__ expression contains a variable, a number, and at least one operation symbol. (Lesson 1-2)

2. A sentence that compares two numbers or quantities is called a(n) __?__. (Lesson 1-3)

▶ Prerequisite Skills

Determine whether each statement is *true* or *false*. (Lesson 1-3)

3. $10 > 4$

4. $3 < -3$

5. $-7 < -8$

6. $-1 < 0$

Write an algebraic equation for each verbal sentence. (Lesson 1-7)

7. Ten increased by a number is -8.

8. The difference of -5 and $3x$ equals 32.

9. Twice a number decreased by 4 is 26.

10. The sum of 9 and a number is 14.

Solve each equation. Check your solution. (Lessons 1-8 and 1-9)

11. $n + 8 = -9$

12. $4 = m + 19$

13. $-4 + c = 15$

14. $z - 6 = -10$

15. $p - 12 = 2$

16. $21 = y - (-3)$

17. $3c = -18$

18. $-2x = 18$

19. $-42 = -6b$

20. $\dfrac{w}{4} = -8$

21. $12 = \dfrac{r}{-7}$

22. $\dfrac{a}{-3} = -5$

Equations and Inequalities Make this Foldable to help you organize your notes. Begin with a plain sheet of 11" × 17" paper.

STEP 1 Fold
Fold in half lengthwise.

STEP 2 Fold Again
Fold the top to the bottom.

STEP 3 Cut
Open and cut along the second fold to make two tabs.

STEP 4 Label
Label each tab as shown.

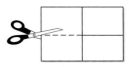

Reading and Writing As you read and study the chapter, write notes and examples for each topic under the appropriate tab.

10-1a HANDS-ON LAB

A Preview of Lesson 10-1

What You'll Learn
Model and solve equations using algebra tiles.

Materials
• algebra tiles

Algebra Tiles

In Chapter 1, you used cups and counters to model equations. In this lab and throughout the rest of this book, you will use **algebra tiles**. The table below shows how these two types of models are related.

Type of Model	Variable x	Integer 1	Integer -1
Cups and Counters	(cup)	$\oplus$	$\ominus$
Algebra Tiles	x	1	-1

You will use an equation mat to model and solve equations using algebra tiles in the same way as you did with cups and counters.

ACTIVITY *Work with a partner.*

Use algebra tiles to model and solve $x + 3 = -2$.

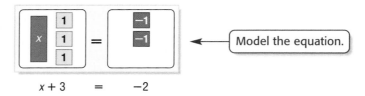

$x + 3 = -2$

Model the equation.

$x + 3 + (-3) = -2 + (-3)$

Add three -1-tiles to each side of the mat. The left side now contains zero pairs.

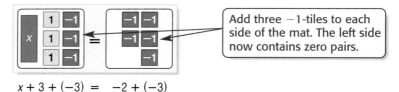

$x = -5$

Remove the zero pairs from the left side. The x-tile is now isolated. There are 5 negative tiles on the right side of the mat.

Therefore, $x = -5$. Since $-5 + 3 = -2$, the solution is correct.

Your Turn Use algebra tiles to model and solve each equation.

a. $x + 2 = 3$ b. $4 + x = 6$ c. $x + 2 = -1$ d. $-4 = x + 3$

e. $x - 3 = 2$ f. $x - 1 = -3$ g. $2x = -4$ h. $3 = 3x$

Simplifying Algebraic Expressions

Sunshine State Standards
MA.A.3.3.1-3, MA.D.2.3.1-7, MA.D.2.3.2-1

What You'll Learn
Use the Distributive Property to simplify algebraic expressions.

NEW Vocabulary
equivalent expressions
term
coefficient
like terms
constant
simplest form
simplifying the expression

∞ LINK To Reading
Everyday Meaning of Constant: unchanging

HANDS-ON Mini Lab

Materials
• algebra tiles

You can use algebra tiles to rewrite the algebraic expression $2(x + 3)$.

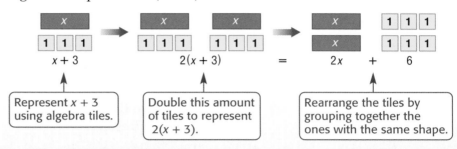

Represent $x + 3$ using algebra tiles.

Double this amount of tiles to represent $2(x + 3)$.

Rearrange the tiles by grouping together the ones with the same shape.

1. Choose two positive and one negative value for x. Then evaluate $2(x + 3)$ and $2x + 6$ for each of these values. What do you notice?

2. Use algebra tiles to rewrite the expression $3(x - 2)$. (*Hint:* Use one green x-tile and 2 red -1-tiles to represent $x - 2$.)

In Chapter 1, you learned that expressions like $2(4 + 3)$ can be rewritten using the Distributive Property and then simplified.

$$2(4 + 3) = 2(4) + 2(3) \quad \text{Distributive Property}$$
$$= 8 + 6 \text{ or } 14 \quad \text{Multiply. Then add.}$$

The Distributive Property can also be used to simplify an algebraic expression like $2(x + 3)$.

$$2(x + 3) = 2(x) + 2(3) \quad \text{Distributive Property}$$
$$= 2x + 6 \quad \text{Multiply.}$$

The expressions $2(x + 3)$ and $2x + 6$ are **equivalent expressions**, because no matter what x is, these expressions have the same value.

EXAMPLES Write Equivalent Expressions

Use the Distributive Property to rewrite each expression.

1 $4(x + 7)$

$$4(x + 7) = 4(x) + 4(7)$$
$$= 4x + 28 \quad \text{Simplify.}$$

2 $(y + 2)5$

$$(y + 2)5 = y \cdot 5 + 2 \cdot 5$$
$$= 5y + 10 \quad \text{Simplify.}$$

Your Turn Use the Distributive Property to rewrite each expression.

a. $6(a + 4)$　　　**b.** $(n + 3)8$　　　**c.** $-2(x + 1)$

EXAMPLES Write Expressions with Subtraction

Use the Distributive Property to rewrite each expression.

3 $6(p - 5)$

$$\begin{aligned}
6(p - 5) &= 6[p + (-5)] && \text{Rewrite } p - 5 \text{ as } p + (-5).\\
&= 6(p) + 6(-5) && \text{Distributive Property}\\
&= 6p + (-30) && \text{Simplify.}\\
&= 6p - 30 && \text{Definition of subtraction}
\end{aligned}$$

4 $-2(x - 8)$

$$\begin{aligned}
-2(x - 8) &= -2[x + (-8)] && \text{Rewrite } x - 8 \text{ as } x + (-8).\\
&= -2(x) + (-2)(-8) && \text{Distributive Property}\\
&= -2x + 16 && \text{Simplify.}
\end{aligned}$$

Your Turn Use the Distributive Property to rewrite each expression.

d. $3(y - 10)$ **e.** $-7(w - 4)$ **f.** $(n - 2)(-9)$

When a plus sign separates an algebraic expression into parts, each part is called a **term**. The numerical factor of a term that contains a variable is called the **coefficient** of the variable.

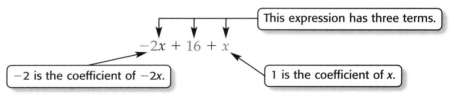

This expression has three terms.

$-2x + 16 + x$

−2 is the coefficient of −2x.   1 is the coefficient of x.

Like terms contain the same variables, such as $2x$ and x. A term without a variable is called a **constant**. Constant terms are also like terms.

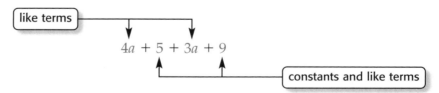

like terms

$4a + 5 + 3a + 9$

constants and like terms

Rewriting a subtraction expression using addition will help you identify the like terms of an expression that contains subtraction.

EXAMPLE Identify Parts of an Expression

5 Identify the terms, like terms, coefficients, and constants in the expression $6n - 7n - 4 + n$.

$$\begin{aligned}
6n - 7n - 4 + n &= 6n + (-7n) + (-4) + n && \text{Definition of subtraction}\\
&= 6n + (-7n) + (-4) + 1n && \text{Identity Property; } n = 1n
\end{aligned}$$

The terms are $6n$, $-7n$, -4, and n. The like terms are $6n$, $-7n$, and n. The coefficients are 6, −7, and 1. The constant is −4.

An algebraic expression is in **simplest form** if it has no like terms and no parentheses. You can use the Distributive Property to combine like terms. This is called **simplifying the expression**.

EXAMPLES Simplify Algebraic Expressions

Simplify each expression.

6 $3y + y$

$3y$ and y are like terms.

$$3y + y = 3y + 1y \quad \text{Identity Property; } y = 1y$$
$$= (3 + 1)y \quad \text{Distributive Property}$$
$$= 4y \quad \text{Simplify.}$$

7 $-9k + 4 + 9k$

$-9k$ and $9k$ are like terms.

$$-9k + 4 + 9k = -9k + 9k + 4 \quad \text{Commutative Property}$$
$$= (-9 + 9)k + 4 \quad \text{Distributive Property}$$
$$= 0k + 4 \quad -9 + 9 = 0$$
$$= 0 + 4 \text{ or } 4 \quad 0k = 0 \cdot k \text{ or } 0$$

8 $5x - 2 - 7x + 6$

$5x$ and $-7x$ are like terms. -2 and 6 are also like terms.

$$5x - 2 - 7x + 6 = 5x + (-2) + (-7x) + 6 \quad \text{Definition of subtraction}$$
$$= 5x + (-7x) + (-2) + 6 \quad \text{Commutative Property}$$
$$= [5 + (-7)]x + (-2) + 6 \quad \text{Distributive Property}$$
$$= -2x + 4 \quad \text{Simplify.}$$

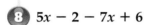 Simplify each expression.

g. $4z - z$ h. $6 + 3n - 8n$ i. $2g - 3 + 11 - 2g$

EXAMPLE Translate Phrases into Expressions

9 **FOOD** At a baseball game, you buy some hot dogs that cost $3 each and the same number of soft drinks for $2.50 each. Write an expression in simplest form that represents the total amount of money spent on food and drinks.

If x represents the number of hot dogs you buy, then x also represents the number of drinks you buy. To find the total amount spent, multiply the cost of each item by the number of items purchased. Then add the expressions.

$$3x + 2.50x = (3 + 2.50)x \quad \text{Distributive Property}$$
$$= 5.50x \quad \text{Simplify.}$$

The expression $\$5.50x$ represents the total amount of money spent on food and drink, where x is the number of hot dogs or drinks.

Equivalent Expressions To check whether $3y + y$ and $4y$ are equivalent expressions, substitute any value for y and see whether the expressions have the same value.

1. **Define** *like terms*.

2. **OPEN ENDED** Write an expression that has four terms and simplifies to $3n + 2$. Identify the coefficient(s) and constant(s) in your expression.

3. **Which One Doesn't Belong?** Identify the expression that is not equivalent to the other three. Explain your reasoning.

| $x - 3 + 4x$ | $5(x - 3)$ | $6 + 5x - 9$ | $5x - 3$ |

GUIDED PRACTICE

Use the Distributive Property to rewrite each expression.

4. $5(x + 4)$ 5. $-3(a + 9)$ 6. $-6(g - 2)$

Identify the terms, like terms, coefficients, and constants in each expression.

7. $8a + 4 - 6a$ 8. $7 - 3d - 8 + d$ 9. $5n - n + 3 - 2n$

Simplify each expression.

10. $5x + 2x$ 11. $8n + n$ 12. $10y - 17y$

13. $12c - c$ 14. $4p - 7 + 6p$ 15. $11x - 12 - 6x + 9$

Practice and Applications

Use the Distributive Property to rewrite each expression.

16. $3(x + 8)$ 17. $7(m + 6)$ 18. $-8(b + 5)$

19. $-7(n + 2)$ 20. $-4(k + 8)$ 21. $(c - 8)(-8)$

22. $-5(a - 9)$ 23. $(x - 6)(-4)$ 24. $2(a + b)$

25. $4(x - y)$ 26. $3(2y + 1)$ 27. $-4(3x + 5)$

HOMEWORK HELP

For Exercises	See Examples
16–31	1–4
32–37	5
38–49	6–8
50–53	9

Extra Practice
See pages 640, 657.

GEOMETRY Write two equivalent expressions for the area of each figure.

28.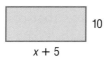
$x + 5$; 10

29.
12 ; $x - 7$

30.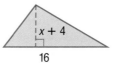
$x + 4$; 16

31.
18 ; $x - 3$

Identify the terms, like terms, coefficients, and constants in each expression.

32. $2 + 3a + 9a$ 33. $7 - 5x + 1$ 34. $4 + 5y - 6y + y$

35. $n + 4n - 7n - 1$ 36. $-3d + 8 - d - 2$ 37. $9 - z + 3 - 2z$

Simplify each expression.

38. $4y + 7y$ 39. $n + 5n$ 40. $12x - 5x$ 41. $4k - 7k$

42. $10k - k$ 43. $5x + 4 + 9x$ 44. $2 + 3d + d$ 45. $6 - 4c + c$

46. $2m + 5 - 8m$ 47. $3r + 7 - 3r$ 48. $9y - 4 - 11y + 7$ 49. $3x + 2 - 10 - 3x$

For Exercises 50–53, write an expression in simplest form that represents the total amount in each situation.

50. **MOVIES** You buy 2 drinks that each cost x dollars, a large bag of popcorn for $3.50, and a chocolate bar for $1.50.

51. **PHYSICAL EDUCATION** Each lap around the school track is a distance of y yards. You ran 2 laps on Monday, $3\frac{1}{2}$ laps on Wednesday, and 100 yards on Friday.

52. **SHOPPING** You buy x shirts that each cost $15.99, the same number of jeans for $34.99 each, and a pair of sneakers for $58.99.

53. **FUND-RAISING** You have sold t tickets for a school fund-raiser. Your friend has sold 24 more than you.

54. **CRITICAL THINKING** Is $2(x - 1) + 3(x - 1) = 5(x - 1)$ a true statement? If so, explain your reasoning. If not, give a counterexample.

Standardized Test Practice and Mixed Review

 FCAT Practice

55. **SHORT RESPONSE** Write an expression in simplest form for the perimeter of the figure.

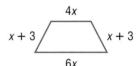

56. **MULTIPLE CHOICE** Dustin is 3 years younger than his older sister. If his older sister is y years old, which expression represents the sum of their ages?

  Ⓐ $2y - 3$ Ⓑ $y - 3$ Ⓒ $y^2 - 3$ Ⓓ $2y + 3$

State the dimensions of each matrix. Then identify the position of the circled element. (Lesson 9-8)

57. $[\boxed{3} \quad -2]$

58. $\begin{bmatrix} -4 & ⑤ \\ 0 & 2 \end{bmatrix}$

59. $\begin{bmatrix} 4 \\ -2 \\ ⑦ \end{bmatrix}$

60. $\begin{bmatrix} 9 & 3 & 5 \\ -4 & 7 & ① \end{bmatrix}$

TECHNOLOGY For Exercises 61 and 62, refer to the graphs at the right. (Lesson 9-7)

61. Which graph gives the impression that the number of DVD players sold in 2001 was more than 5 times the amount sold in 1999?

62. About how many times more DVD's were sold in 2001 than in 1999?

GETTING READY FOR THE NEXT LESSON

PREREQUISITE SKILL Solve each equation. Check your solution.

(Lessons 1-8 and 1-9)

63. $x + 8 = 2$ 64. $y - 5 = -9$ 65. $32 = -4n$ 66. $\frac{a}{3} = -15$

10-2 Solving Two-Step Equations

Sunshine State Standards MA.D.1.3.1-2, MA.D.1.3.2-2, MA.D.2.3.1-3, MA.D.2.3.1-6, MA.D.2.3.2-2

What You'll Learn
Solve two-step equations.

NEW Vocabulary
two-step equation

WHEN am I ever going to use this?

BOOK SALE Linda bought four books at a book sale benefiting a local charity. The handwritten receipt she received was missing the cost for the hardback books she purchased.

1. Explain how you could use the work backward strategy to find the cost of each hardback book. Then find the cost.

Thank You for Your Support!

3 hardback(s) $ ___

1 paperback(s) $ 1

Total paid $ 7

The solution to this problem can also be found by solving the equation $3x + 1 = 7$, where x is the cost per hardback book. This equation can be modeled using algebra tiles.

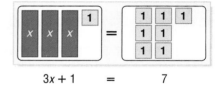

$3x + 1$ =   7

A **two-step equation** contains two operations. In the equation $3x + 1 = 7$, x is multiplied by 3 and then 1 is added. To solve two-step equations, undo each operation in reverse order.

EXAMPLE   Solve a Two-Step Equation

1   Solve $3x + 1 = 7$.

Method 1 Use a model.

Remove one 1-tile from the mat.

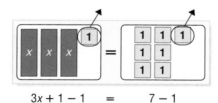

$3x + 1 - 1$ = $7 - 1$

Separate the remaining tiles into 3 equal groups.

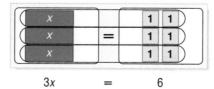

$3x$ =   6

There are 2 tiles in each group.

The solution is 2.

Method 2 Use symbols.

Use the Subtraction Property of Equality.

$$3x + 1 = 7 \quad \text{Write the equation.}$$
$$\underline{ - 1 = -1} \quad \text{Subtract 1}$$
$$3x = 6 \quad \text{from each side.}$$

Use the Division Property of Equality.

$$3x = 6$$
$$\frac{3x}{3} = \frac{6}{3} \quad \text{Divide each side by 3.}$$
$$x = 2 \quad \text{Simplify.}$$

EXAMPLES Solve Two-Step Equations

Solve each equation. Check your solution.

2 $4x - 3 = 25$

Method 1 Vertical Method

$$4x - 3 = 25 \quad \text{Write the equation.}$$

$$\begin{aligned} 4x - 3 &= 25 \\ \underline{+ 3 = + 3} \quad &\text{Add 3 to each side.} \\ 4x &= 28 \quad \text{Simplify.} \\ \frac{4x}{4} &= \frac{28}{4} \quad \text{Divide each side by 4.} \\ x &= 7 \end{aligned}$$

Method 2 Horizontal Method

$$\begin{aligned} 4x - 3 &= 25 \\ 4x - 3 + 3 &= 25 + 3 \\ 4x &= 28 \\ \frac{4x}{4} &= \frac{28}{4} \\ x &= 7 \end{aligned}$$

Check $\quad 4x - 3 = 25 \quad$ Write the equation.

$\quad\quad\quad 4(7) - 3 \overset{?}{=} 25 \quad$ Replace x with 7 and check to see if the sentence is true.

$\quad\quad\quad\quad 25 = 25 \checkmark \quad$ The sentence is true.

The solution is 7.

3 $-1 = \frac{1}{2}m + 9$

$$-1 = \frac{1}{2}m + 9 \quad \text{Write the equation.}$$

$$-1 - 9 = \frac{1}{2}m + 9 - 9 \quad \text{Subtract 9 from each side.}$$

$$-10 = \frac{1}{2}m \quad \text{Simplify.}$$

$$2(-10) = 2 \cdot \frac{1}{2}m \quad \text{Multiply each side by 2.}$$

$$-20 = m \quad \text{Simplify.}$$

The solution is -20. $\quad$ Check this solution.

Some two-step equations have a term with a negative coefficient.

STUDY TIP

Common Error A common mistake when solving the equation in Example 4 is to divide each side by 3 instead of -3. Remember that you are dividing by the coefficient of the variable, which in this instance is a negative number.

EXAMPLE Equations with Negative Coefficients

4 Solve $6 - 3x = 21$.

$$6 - 3x = 21 \quad \text{Write the equation.}$$

$$6 + (-3x) = 21 \quad \text{Definition of subtraction}$$

$$6 - 6 + (-3x) = 21 - 6 \quad \text{Subtract 6 from each side.}$$

$$-3x = 15 \quad \text{Simplify.}$$

$$\frac{-3x}{-3} = \frac{15}{-3} \quad \text{Divide each side by } -3.$$

$$x = -5 \quad \text{Simplify.}$$

The solution is -5. $\quad$ Check this solution.

Your Turn Solve each equation. Check your solution.

a. $\frac{n}{-3} - 2 = -18$ $\quad\quad$ **b.** $19 = 3x - 2$ $\quad\quad$ **c.** $5 - 2n = -1$

Sometimes it is necessary to combine like terms before solving an equation.

EXAMPLE   Combine Like Terms Before Solving

5 Solve $-2y + y - 5 = 11$. Check your solution.

$-2y + y - 5 = 11$	Write the equation.
$-2y + 1y - 5 = 11$	Identity Property; $y = 1y$
$-y - 5 = 11$	Combine like terms; $-2y + 1y = (-2 + 1)y$ or $-y$.
$-y - 5 + 5 = 11 + 5$	Add 5 to each side.
$-y = 16$	Simplify.
$\dfrac{-1y}{-1} = \dfrac{16}{-1}$	$-y = -1y$; divide each side by -1.
$y = -16$	Simplify.

Check	$-2y + y - 5 = 11$	Write the equation.
	$-2(-16) + (-16) - 5 \overset{?}{=} 11$	Replace y with -16.
	$32 + (-16) - 5 \overset{?}{=} 11$	Multiply.
	$11 = 11$ ✔	The statement is true.

The solution is -16.

Your Turn Solve each equation. Check your solution.

d. $x + 4x = 45$ e. $10 = 2a + 13 - a$ f. $-3 = 6 - 5w + 2w$

Skill and Concept Check

Writing Math
Exercises 1 & 3

1. **Explain** how you can use the work backward problem-solving strategy to solve a two-step equation.

2. **OPEN ENDED** Write a two-step equation that can be solved by using the Addition and Division Properties of Equality.

3. **FIND THE ERROR** Alexis and Tomás are solving the equation $2x + 7 = 16$. Who is correct? Explain.

Alexis
$2x + 7 = 16$
$\dfrac{2x}{2} + 7 = \dfrac{16}{2}$
$x + 7 = 8$
$x + 7 - 7 = 8 - 7$
$x = 1$

Tomás
$2x + 7 = 16$
$2x + 7 - 7 = 16 - 7$
$2x = 9$
$\dfrac{2x}{2} = \dfrac{9}{2}$
$x = 4.5$

GUIDED PRACTICE

Solve each equation. Check your solution.

4. $6x + 5 = 29$ 5. $9m - 11 = -2$ 6. $1 = 2p + 13$

7. $10 = \dfrac{a}{4} + 3$ 8. $\dfrac{c}{-2} - 4 = 3$ 9. $3 - 5y = -37$

10. $4 - d = 11$ 11. $7 = -2n + 1$ 12. $6k - 10k = 16$

Solve each equation. Check your solution.

13. $2h + 9 = 21$ **14.** $11 = 2b + 17$ **15.** $5 = 4a - 7$

16. $6p - 5 = -17$ **17.** $2g - 3 = -19$ **18.** $16 = 5x - 9$

19. $3 + 8c = 35$ **20.** $13 + 3d = -8$ **21.** $13 = \frac{g}{3} + 4$

22. $5 + \frac{y}{8} = -3$ **23.** $-\frac{1}{2}x - 7 = -11$ **24.** $-\frac{1}{4}w + 15 = 28$

25. SCHOOL TRIP At a theme park, each student is given $19. This covers the cost of 2 meals at x dollars each plus $7 worth of snacks. Solve $2x + 7 = 19$ to find the amount each student can spend per meal.

26. SHOPPING You receive a $75 online gift to a music site. You want to purchase CDs that cost $14 each. There is a $5 shipping and handling fee. Solve $14n + 5 = 75$ to find the number of CDs you can purchase.

Solve each equation. Check your solution.

27. $5 - 3c = 14$ **28.** $9 - 5y = 19$ **29.** $-6 = 4 - 2x$

30. $2 = 18 - 4d$ **31.** $8 - k = 17$ **32.** $-7 - p = -15$

33. $12 = 6 - x$ **34.** $-2 = 4 - t$ **35.** $5w - 8w = -12$

36. $28 = 3m - 7m$ **37.** $y + 5y + 11 = 35$ **38.** $3 - 6x + 8x = 9$

39. $-21 = 9a - 15 - 3a$ **40.** $26 = g + 10 - 3g$ **41.** $8x + 5 - x = -2$

42. $6h + 5 + h = -30$ **43.** $-n + 9 - 2n - 1 = -13$ **44.** $10 = 6a + 4 - 9a + a$

45. CRITICAL THINKING Work backward to write a two-step equation whose solution is -5.

Standardized Test Practice and Mixed Review

46. MULTIPLE CHOICE If $3x + 10 = 4$, what is the value of $2 + 5x$?

   A -14 **B** -8 **C** -2 **D** 12

47. SHORT RESPONSE Write an equation for the given diagram. Then find the value of x.

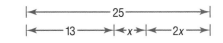

Use the Distributive Property to rewrite each expression. (Lesson 10-1)

48. $6(a + 6)$ **49.** $-3(x + 5)$ **50.** $(y - 8)4$ **51.** $-8(p - 7)$

52. Find $[6 \quad 3 \quad -1] - [2 \quad 8 \quad 9]$. If there is no difference, write *impossible*.
(Lesson 9-8)

GETTING READY FOR THE NEXT LESSON

PREREQUISITE SKILL Write an algebraic equation for each verbal sentence. (Lesson 1-7)

53. Four times a number increased by 5 is 17. **54.** 8 less than twice a number equals 10.

Writing Two-Step Equations

Sunshine State Standards MA.D.1.3.2-2, MA.D.2.3.1-1,
MA.D.2.3.1-2, MA.D.2.3.1-3, MA.D.2.3.1-6, MA.D.2.3.2-2

What You'll Learn

Write two-step equations that represent real-life situations.

WHEN am I ever going to use this?

HOME ENTERTAINMENT Your parents offer to loan you the money to buy a $600 sound system. You give them $125 as a down payment and agree to make monthly payments of $25 until you have repaid the loan.

Payments	Amount Paid
0	125 + 25(0) = $125
1	125 + 25(1) = $150
2	125 + 25(2) = $175
3	125 + 25(3) = $200
⋮	⋮

1. Let n represent the number of payments. Write an expression that represents the amount of the loan paid after n payments.

2. Write and solve an equation to find the number of payments you will have to make in order to pay off your loan.

3. What type of equation did you write for Exercise 2? Explain your reasoning.

In Chapter 1, you learned how to write verbal sentences as one-step equations. Some verbal sentences translate to two-step equations.

Words	The sum of 125 and 25 times a number is 600.
Variable	Let n = the number.
Equation	The sum of 125 and 25 times a number is 600. $125 + 25n = 600$

EXAMPLES Translate Sentences into Equations

Translate each sentence into an equation.

Sentence	**Equation**
① Eight less than three times a number is −23.	$3n - 8 = -23$
② Thirteen is 7 more than twice a number.	$13 = 2n + 7$
③ The quotient of a number and 4, decreased by 1, is equal to 5.	$\dfrac{n}{4} - 1 = 5$

Your Turn Translate each sentence into an equation.

a. Fifteen equals three more than six times a number.

b. If 10 is increased by the quotient of a number and 6, the result is 5.

c. The difference between 12 and a twice a number is 18.

EXAMPLE Translate and Solve an Equation

④ **Nine more than four times a number is 21. Find the number.**

Words	Nine more than four times a number is 21.
Variable	Let n = the number.
Equation	$4n + 9 = 21$

$$4n + 9 = 21 \qquad \text{Write the equation.}$$
$$4n + 9 - 9 = 21 - 9 \qquad \text{Subtract 9 from each side.}$$
$$4n = 12 \qquad \text{Simplify.}$$
$$n = 3 \qquad \text{Mentally divide each side by 4.}$$

Therefore, the number is 3.

In many real-life situations, you start with a given amount and then increase it at a certain rate. These situations can be represented by two-step equations.

EXAMPLE Write and Solve a Two-Step Equation

⑤ **FUND-RAISING** Your Class Council needs $600 for the Spring Dance. With only $210 in their treasury, the Council decides to raise the rest by selling donuts for a profit of $1.50 per dozen. How many dozen donuts will they need to sell?

The council already has $210 and will sell donuts for a profit of $1.50 per dozen until they have $600. Organize the data for the first few dozen donuts sold into a table and look for a pattern.

Dozens	Amount
0	$210 + 1.50(0) = 210.00$
1	$210 + 1.50(1) = 211.50$
2	$210 + 1.50(2) = 213.00$
3	$210 + 1.50(3) = 214.50$

Write an equation to represent the situation. Let d represent the number of dozens.

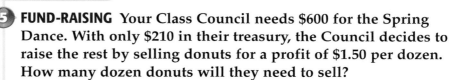

amount already in treasury — plus — d dozen sold at a profit of $1.50 per dozen — equals — $600

$$210 \qquad + \qquad 1.50d \qquad = \qquad 600$$

$$210 + 1.50d = 600 \qquad \text{Write the equation.}$$
$$210 - 210 + 1.50d = 600 - 210 \qquad \text{Subtract 210 from each side.}$$
$$1.50d = 390 \qquad \text{Simplify.}$$
$$\frac{1.50d}{1.50} = \frac{390}{1.50} \qquad \text{Divide each side by 1.50.}$$
$$d = 260 \qquad \text{Simplify.}$$

They need to sell 260 dozen donuts.

1. **NUMBER SENSE** Identify the operation indicated by the word *twice*.

2. **OPEN ENDED** Write two different statements that translate into the same two-step equation.

GUIDED PRACTICE

Translate each sentence into an equation. Then find each number.

3. One more than three times a number is 7.

4. Seven less than twice a number is -1.

5. The quotient of a number and 5, less 10, is 3.

6. **FINES** You return a book that is 5 days overdue. Including a previous unpaid overdue balance of $1.30, your new balance is $2.05. Write and solve an equation to find the fine for a book that is one day overdue.

Practice and Applications

Translate each sentence into an equation. Then find each number.

7. Four less than five times a number is equal to 11.

8. Fifteen more than twice a number is 9.

9. Eight more than four times a number is -12.

10. Six less than seven times a number is equal to -20.

11. Nine more than the quotient of a number and 3 is 14.

12. The quotient of a number and -7, less 4, is -11.

13. The difference between three times a number and 10 is 17.

14. The difference between twice a number and 1 is -21.

HOMEWORK HELP

For Exercises	See Examples
7–14	1–4
15–19	5

Extra Practice
See pages 641, 657.

Solve each problem by writing and solving an equation.

15. **VACATION** While on vacation, you purchase 4 identical T-shirts for some friends and a watch for yourself, all for $75. You know that the watch cost $25. How much did each T-shirt cost?

16. **PERSONAL FITNESS** Angelica joins a local gym called Fitness Solutions. If she sets aside $1,000 in her annual budget for gym costs, use the ad at the right to determine how many hours she can spend with a personal trainer.

Fitness Solutions

Annual Membership: $720

Personal Trainers Available
($35/h)

17. **PHONE SERVICE** A telephone company advertises long distance service for 7¢ per minute plus a monthly fee of $3.95. If your bill one month was $12.63, find the number of minutes you used making long distance calls.

18. **GAMES** You and two friends share the cost of renting a video game system for 5 nights. Each person also rents one video game for $6.33. If each person pays $11.33, what is the cost of renting the system?

19. SKIING In aerial skiing competitions, the total judges score is multiplied by the jump's *degree of difficulty* and then added to the skier's current score to obtain their final score. The table shows the first-round scores of a competition. After her second jump, Toshiro's final score is 216.59. The degree of difficulty for Martin's second jump is 4.45. Write and solve an equation to find what the judge's score for Martin's jump must be in order for her to tie Toshiro for first.

Skier	Score
Martin, S.	100.23
Toshiro, M.	105.34
Moseley, K.	93.99
Long, A.	87.50
Cruz, P.	80.63
Thompson, L.	75.23

20. WRITE A PROBLEM Write about a real-life situation that can be solved using a two-step equation. Then write the equation and solve the problem.

21. CRITICAL THINKING Student Council has a total of $200 to divide among the top class finishers in a used toy drive. Second place will receive twice as much as third place. First place will receive $15 more than second place. Write and solve an equation to find how much each winning class will receive.

Standardized Test Practice and Mixed Review

Ⓐ Ⓑ Ⓒ Ⓓ

FCAT Practice

22. MULTIPLE CHOICE Ms. Anderson receives a weekly base salary of $325 plus 7% of her weekly sales. At the end of one week, she earned $500. Which equation can be used to find her sales s for that week?

Ⓐ $325s + 0.07 = 500$ Ⓑ $325 + 7s = 500$

Ⓒ $325 + 0.7s = 500$ Ⓓ $325 + 0.07s = 500$

23. GRID IN Find the value of x in the parallelogram at the right.

SHORT RESPONSE For Exercises 24 and 25, use the following information.

In a basketball game, 2 points are awarded for making a regular basket, and 1 point is awarded for making a foul shot. Emeril scored 21 points during one game. Three of those points were for foul shots. The rest were for regular goals.

24. Write an equation to find the number of regular baskets b Emeril made during the game.

25. Solve the equation to find the number of regular baskets he made.

Solve each equation. Check your solution. (Lesson 10-2)

26. $5x + 2 = 17$ **27.** $-7b + 13 = 27$ **28.** $\frac{n}{8} + 1 = -6$ **29.** $-15 = -4p + 9$

Determine the number of significant digits in each measure. (Lesson 7-9)

30. 140 ft **31.** 7.0 L **32.** 9.04 s **33.** 1,000.2 mi

GETTING READY FOR THE NEXT LESSON

PREREQUISITE SKILL Simplify each expression. (Lesson 10-1)

34. $5x + 6 - x$ **35.** $8 - 3n + 3n$ **36.** $7a - 7a - 9$ **37.** $3 - 4y + 9y$

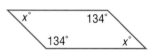

Sunshine State Standards
MA.D.2.3.1-3

Equations with Variables on Each Side

What You'll Learn
Solve equations with variables on each side using algebra tiles.

Materials
• algebra tiles

You can also use algebra tiles to solve equations that have variables on each side of the equation.

ACTIVITY *Work with a partner.*

① Use algebra tiles to model and solve $3x + 1 = x + 5$.

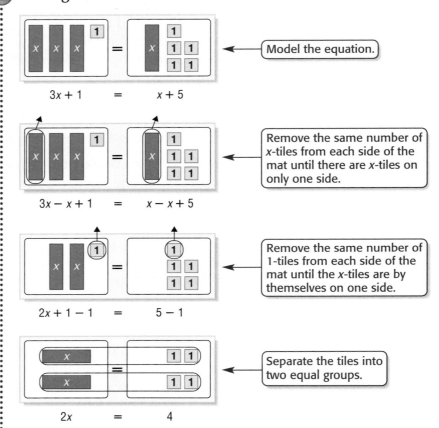

$3x + 1 = x + 5$ — Model the equation.

$3x - x + 1 = x - x + 5$ — Remove the same number of *x*-tiles from each side of the mat until there are *x*-tiles on only one side.

$2x + 1 - 1 = 5 - 1$ — Remove the same number of 1-tiles from each side of the mat until the *x*-tiles are by themselves on one side.

$2x = 4$ — Separate the tiles into two equal groups.

Therefore, $x = 2$. Since $3(2) + 1 = 2 + 5$, the solution is correct.

Your Turn Use algebra tiles to model and solve each equation.

a. $x + 2 = 2x + 1$ **b.** $2x + 7 = 3x + 4$ **c.** $2x - 5 = x - 7$

d. $8 + x = 3x$ **e.** $4x = x - 6$ **f.** $2x - 8 = 4x - 2$

Writing Math

1. **Identify** the property of equality that allows you to remove a 1-tile or −1-tile from each side of an equation mat.

2. **Explain** why you can remove an *x*-tile from each side of the mat.

ACTIVITY *Work with a partner.*

2 Use algebra tiles to model and solve $x - 4 = 2x + 2$.

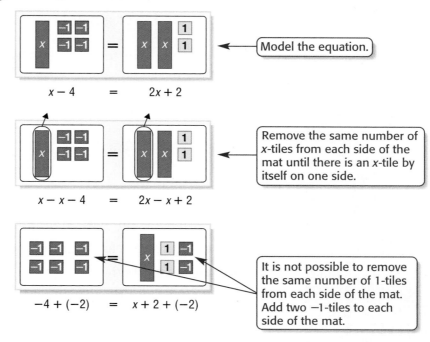

Therefore, $x = -6$. Since $-6 - 4 = 2(-6) + 2$, the solution is correct.

Your Turn Use algebra tiles to model and solve each equation.

g. $x + 6 = 3x - 2$ **h.** $x - 3 = 3x + 5$ **i.** $2x + 1 = x - 7$

j. $x - 4 = 2x + 5$ **k.** $3x - 2 = 2x + 3$ **l.** $2x + 5 = 4x - 1$

Writing Math

3. Solve $x + 4 = 3x - 4$ by removing 1-tiles first. Then solve the equation by removing x-tiles first. Does it matter whether you remove x-tiles or 1-tiles first? Is one way more convenient? Explain.

4. In the set of algebra tiles, $-x$ is represented by ▮$-x$▮. **Make a conjecture** and explain how you could use $-x$-tiles and other algebra tiles to solve $-3x + 4 = -2x - 1$.

 **Solving Equations with Variables on Each Side**

Sunshine State Standards
MA.D.2.3.1-3, MA.D.2.3.1-6, MA.D.2.3.2-2

What You'll Learn
Solve equations with variables on each side.

WHEN am I ever going to use this?

SPORTS You and your friend are having a race. You give your friend a 15-meter head start. During the race, you average 6 meters per second and your friend averages 5 meters per second.

Time (s)	Friend's Distance (m)	Your Distance (m)
0	$15 + 5(0) = 15$	$6(0) = 0$
1	$15 + 5(1) = 20$	$6(1) = 6$
2	$15 + 5(2) = 25$	$6(2) = 12$
3	$15 + 5(3) = 30$	$6(3) = 18$
⋮	⋮	⋮

1. Copy the table. Continue filling in rows to find how long it will take you to catch up to your friend.

2. Write an expression for your distance after x seconds.

3. Write an expression for your friend's distance after x seconds.

4. What is true about the distances you and your friend have gone when you catch up to your friend?

5. Write an equation that could be used to find how long it will take for you to catch up to your friend.

Some equations, like $15 + 5x = 6x$, have variables on each side of the equals sign. To solve these equations, use the Addition or Subtraction Property of Equality to write an equivalent equation with the variables on one side of the equals sign. Then solve the equation.

EXAMPLE Equations with Variables on Each Side

 Solve $15 + 5x = 6x$. Check your solution.

$$15 + 5x = 6x \qquad \text{Write the equation.}$$
$$15 + 5x - 5x = 6x - 5x \qquad \text{Subtract } 5x \text{ from each side.}$$
$$15 = x \qquad \text{Simplify by combining like terms.}$$

Subtract $5x$ from the left side of the equation to isolate the variable.

Subtract $5x$ from the right side of the equation to keep it balanced.

To check your solution, replace x with 15 in the original equation.

Check
$$5x + 15 = 6x \qquad \text{Write the equation.}$$
$$5(15) + 15 \stackrel{?}{=} 6(15) \qquad \text{Replace } x \text{ with 15.}$$
$$90 = 90 ✓ \qquad \text{The sentence is true.}$$

The solution is 15.

The problem in Example 2 could have also been solved by first subtracting $6n$ from each side.

$$6n - 1 = 4n - 5$$
$$\underline{-6n = -6n}$$
$$-1 = -2n - 5$$

Notice that this method results in a term with a negative coefficient, but the solution is the same.

EXAMPLE Equations with Variables on Each Side

2 Solve $6n - 1 = 4n - 5$.

$6n - 1 = 4n - 5$	Write the equation.
$6n - 4n - 1 = 4n - 4n - 5$	Subtract $4n$ from each side.
$2n - 1 = -5$	Simplify.
$2n - 1 + 1 = -5 + 1$	Add 1 to each side.
$2n = -4$	Simplify.
$n = -2$	Mentally divide each side by 2.

The solution is -2.   Check this solution.

Your Turn Solve each equation. Check your solution.

a. $8a = 5a + 21$ b. $3x - 7 = 8x + 23$ c. $7g - 12 = 3 + 2g$

EXAMPLE Use an Equation to Solve a Problem

3 **GRID-IN TEST ITEM** Find the value of x so that polygons have the same perimeter.

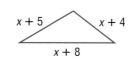

Read the Test Item You need to find the value of x that will make the perimeter of the triangle equal to the perimeter of the rectangle.

Test-Taking Tip
The Princeton Review

Fractions
To grid in a fraction, grid in the numerator, then a slash, then the denominator. To grid in an answer that is a mixed number, you must first rewrite your answer as an improper fraction.

Solve the Test Item Write expressions for the perimeter of each figure. Then set the two expressions equal to each other and solve for x.

Triangle
$$(x + 5) + (x + 4) + (x + 8) = 3x + 17$$

Rectangle
$$(2x + 3) + (2x + 3) + (x + 5) + (x + 5) = 6x + 16$$

$$\underbrace{3x + 17}_{\text{Perimeter of Triangle}} = \underbrace{6x + 16}_{\text{Perimeter of Rectangle}}$$

$$3x - 3x + 17 = 6x - 3x + 16$$
$$17 = 3x + 16$$
$$17 - 16 = 3x + 16 - 16$$
$$1 = 3x$$
$$\frac{1}{3} = \frac{3x}{3}$$
$$\frac{1}{3} = x$$

Fill in the Grid

1	/	3	

Writing Math

Exercise 2

1. **Name** the property of equality that allows you to add $3x$ to each side of the equation $1 - 3x = 5x - 7$.

2. **OPEN ENDED** Write an equation that has variables on each side. Then list the steps you would use to isolate the variable.

GUIDED PRACTICE

Solve each equation. Check your solution.

3. $5n + 9 = 2n$

4. $3k + 14 = k$

5. $10x = 3x - 28$

6. $7y - 8 = 6y + 1$

7. $2a + 21 = 8a - 9$

8. $-4p - 3 = 2 + p$

9. Eighteen less than three times a number is twice the number. Define a variable, write an equation, and solve to find the number.

Practice and Applications

Solve each equation. Check your solution.

10. $7a + 10 = 2a$

11. $11x = 24 + 8x$

12. $9g - 14 = 2g$

13. $m - 18 = 3m$

14. $5p + 2 = 4p - 1$

15. $8y - 3 = 6y + 17$

16. $15 - 3n = n - 1$

17. $3 - 10b = 2b - 9$

18. $-6f + 13 = 2f - 11$

19. $2z - 31 = -9z + 24$

20. $2.5h - 15 = 4h$

21. $21.6 - d = 5d$

22. $1 - 3c = 9c + 7$

23. $7k + 12 = 8 - 9k$

24. $13.4w + 17 = 5w - 4$

25. $8.1a + 2.3 = 5.1a - 3.1$

26. $\frac{2}{3}x + 5 = \frac{1}{3}x + 14$

27. $\frac{1}{2}a - 3 = 7 - \frac{3}{4}a$

HOMEWORK HELP

For Exercises	See Examples
10–13, 20–21	1
14–19, 22–27	2
30–31	3

Extra Practice
See pages 641, 657.

Define a variable and write an equation to find each number. Then solve.

28. Twice a number is 42 less than five times a number. What is the number?

29. Two more than 4 times a number is the number less 7. What is the number?

Write an equation to find the value of x so that each pair of polygons has the same perimeter. Then solve.

30.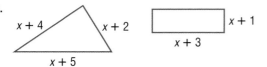

31.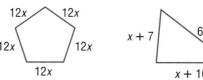

32. **MOVIES** For an annual fee of $30, you can join a movie club that will allow you to purchase tickets for $5.50 each at your local theater. If the theater charges $8 for movie tickets, write and solve an equation to determine how many movie tickets you will have to buy through the movie club for the cost to equal that of buying regularly priced tickets.

33. DISCOUNTS Band members are selling the coupons shown at the right for $14 each. Write and solve an equation to determine how much money you would have to spend on food and drinks for the cost to equal that of buying the concessions without the discount.

34. FOOD DRIVES The seventh graders at your school have collected 345 cans for the canned food drive and are averaging 115 cans per day. The eighth graders have collected 255 cans, but vow to win the contest by collecting an average of 130 cans per day. If both grades continue collecting at these rates, after how many days will the number of cans they have collected be equal?

35. CRAFT FAIRS The Art Club is selling handcrafted mugs at a local craft fair. Vendors at the fair must pay $5 for a booth plus 10% of their sales. It costs $8 in materials to make each mug. If the club sells each mug for $10, write and solve an equation to find how many mugs they must sell to break even. (*Hint:* Total cost must equal total income.)

36. CRITICAL THINKING Find the area of the parallelogram at the right.

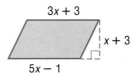

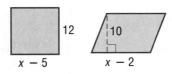

Band Booster
Plus Card

20% off on all food
and drink at the Lions'
Concession Stand

Good this season only

Standardized Test Practice and Mixed Review

FCAT Practice

37. MULTIPLE CHOICE Phone company A charges $28.25 a month plus 18¢ per minute for local calls. Company B charges $19.85 per month plus 32¢ per minute for local calls. Which equation can be used to find the number of minutes for which the companies' plans cost the same?

Ⓐ $28.25x + 0.18 = 19.85x + 0.32$ Ⓑ $28.25 + 0.32x = 19.85 + 0.18x$

Ⓒ $28.25 + 0.18x = 19.85 + 0.32x$ Ⓓ $(28.25 + 0.18)x = (19.85 + 0.32)x$

38. SHORT RESPONSE Find the value of x so that the two figures at the right have the same area.

Translate each sentence into an equation. Then find each number. (Lesson 10-3)

39. Eight more than four times a number is 60.

40. Five less than the quotient of a number and 3 equals -9.

Solve each equation. Check your solution. (Lesson 10-2)

41. $7r + 10 = -11$ **42.** $3g - 7 = 8$ **43.** $8 - p = -10$ **44.** $2 + \dfrac{a}{-5} = 6$

GETTING READY FOR THE NEXT LESSON

PREREQUISITE SKILL Determine whether each statement is *true* or *false*. (Lesson 1-3)

45. $8 > 11$ **46.** $3 \geq -6$ **47.** $-5 \geq -5$ **48.** $-2 < -9$

10-4b Problem-Solving Strategy
A Follow-Up of Lesson 10-4

Guess and Check

What You'll Learn
Solve problems using the guess and check strategy.

Wow! The Fall Carnival was really a success! We collected 150 tickets at the Balloon Pop and Bean-Bag Toss booths alone.

But how many came from each booth? They are all mixed together. We can **guess and check** to figure this out.

Explore	The Bean-Bag Toss was 3 tickets, and the Balloon Pop was 2 tickets. The person running the Bean-Bag Toss said 10 more games were played at her booth than at the Balloon Pop.
Plan	Let's make a guess and check to see if it is correct. Remember, the number we guess for the Bean-Bag Toss must be 10 more than the number we guess for the Balloon Pop.
Solve	We need to find the combination that gives us 150 total tickets. In our list, p is the number of Balloon Pop games, and t is the number of Bean-Bag Toss games.

p	t	$3p + 2t$	Check
12	22	$3(12) + 2(22) = 80$	too low
30	40	$3(30) + 2(40) = 170$	too high
28	38	$3(28) + 2(38) = 160$	still too high
26	36	$3(26) + 2(36) = 150$	correct

So 3(26) or 78 tickets were from the Balloon Pop and 2(36) or 72 tickets were from the Bean-Bag Toss.

Examine	36 Balloon Pop games is 10 more than 26 Bean-Bag Toss games. Since 78 tickets plus 72 tickets is 150 tickets, the guess is correct.

Analyze the Strategy

1. **Describe** how to solve a problem using the guess and check strategy.

2. **Explain** why it is important to make an organized list of your guesses and their results when using the guess and check strategy.

3. **Write** a problem that could be solved by guessing and checking. Then write the steps you would take to find the solution to your problem.

488 Chapter 10 Algebra: More Equations and Inequalities

Solve. Use the guess and check strategy.

4. **NUMBER THEORY** The product of a number and its next two consecutive whole numbers is 60. Find the number.

5. **MONEY MATTERS** Adam has exactly $2 in quarters, dimes, and nickels. If he has 12 coins, how many of each coin does he have?

Mixed Problem Solving

Solve. Use any strategy.

6. **DESIGN** Edu-Toys is designing a new package to hold a set of 30 alphabet blocks like the one shown. Give two possible dimensions for the box.

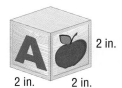
2 in.
2 in.   2 in.

7. **RECREATION** During a routine, ballet dancers are evenly spaced in a circle. If the sixth person is directly opposite the sixteenth person, how many people are in the circle?

8. **TECHNOLOGY** The average Internet user spends $6\frac{1}{2}$ hours online each week. What percent of the week does the average user spend online?

9. **READING** Terrence is reading a 255-page book. He needs to read twice as many pages as he has already read to finish the book. How many pages has he read so far?

10. **DINING** The cost of your meal comes to $8.25. If you want to leave a 15% tip, would it be more reasonable to expect the tip to be about $1.25 or about $1.50?

11. **GEOMETRY** The length ℓ of the rectangle below is longer than its width w. List the possible whole number dimensions for the rectangle, and identify the possibility that gives the smallest perimeter.

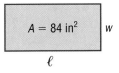
$A = 84$ in^2 w
ℓ

FOOD For Exercises 12 and 13, use the following information.
The school cafeteria surveyed 36 students about their dessert preference. The results are listed below.

Number of Students	Preference of Students
25	cake
20	ice cream
15	pie
2	all three
1	no desserts
15	cake or ice cream
8	pie or cake
3	ice cream only

12. How many students prefer only pie?

13. How many prefer either pie or ice cream?

14. **NUMBER SENSE** Find the product of $1 - \frac{1}{2}$, $1 - \frac{1}{3}$, $1 - \frac{1}{4}$, ..., $1 - \frac{1}{48}$, $1 - \frac{1}{49}$, and $1 - \frac{1}{50}$.

15. **STANDARDIZED TEST PRACTICE**   **FCAT Practice**
At a souvenir shop, a mug costs $3, and a pin costs $2. Chase bought either a mug or a pin for each of his 11 friends. If he spent $30 on these gifts and bought at least one of each type of souvenir, how many of each did he buy?
ⓐ 7 mugs, 4 pins
ⓑ 8 mugs, 3 pins
ⓒ 9 mugs, 2 pins
ⓓ 10 mugs, 1 pin

Vocabulary and Concepts

1. **Explain** what is meant by *like terms*. Then give an example of two terms that are considered like terms and two that are not. (Lesson 10-1)

2. **OPEN ENDED** Write a two-step equation whose solution is 12. (Lesson 10-2)

Skills and Applications

Use the Distributive Property to rewrite each expression. (Lesson 10-1)

3. $3(x + 2)$

4. $-2(a - 3)$

5. $5(3c - 7)$

6. Identify the terms, like terms, coefficients, and constants in the expression $5 - 4x + x - 3$. (Lesson 10-1)

Simplify each expression. (Lesson 10-1)

7. $2a - 13a$

8. $6b + 5 - 6b$

9. $7x + 2 - 8x + 5$

Solve each equation. Check your solution. (Lessons 10-2 and 10-4)

10. $3m + 5 = 14$

11. $11 = \frac{1}{3}a + 2$

12. $-2k + 7 = -3$

13. $3x + 7 = 2x$

14. $7p - 6 = 4p$

15. $3y - 5 = 5y + 7$

16. Two less than 5 times a number is 23. Write and solve an equation to find the number. (Lesson 10-3)

17. **CAR RENTALS** A rental car company charges $52 per day and $0.32 per mile to rent a car. Ms. Misel was charged $202.40 for a 3-day rental. Write and solve an equation to determine how many miles she drove. (Lesson 10-3)

18. **GEOMETRY** Write and solve an equation to find the value of x so that the polygons have the same perimeter. (Lesson 10-4)

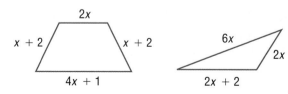

Standardized Test Practice

FCAT Practice

19. **MULTIPLE CHOICE** Which expression is equivalent to $2(3x + 1 - x - 5)$? (Lesson 10-1)

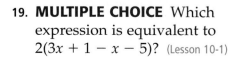

   Ⓐ $6x - 8$ Ⓑ $4x - 8$

   Ⓒ $4x - 4$ Ⓓ $7x - 4$

20. **SHORT RESPONSE** The length of a rectangular room is 3 feet more than twice its width. If the perimeter of the room is 78 feet, find its width. (Lesson 10-2)

The Game Zone

A Place To Practice Your Math Skills

Math-O

● **GET READY!**

Players: two, three, or four
Materials: 52 index cards and 4 different colored markers

● **GET SET!**

- Make a set of four cards by using the markers to put a different-colored stripe at the top of each card.

- Then write a different two-step equation on each card. The solution of each equation should be 1.

- Continue to make sets of four cards having equations with solutions of 2, 3, 4, 5, 6, −1, −2, −3, −4, −5 and −6.

- Mark the remaining set of four cards "Wild".

$2x - 3 = 5$

$-5x + 8 = 3$

● **GO!**

- The dealer shuffles the cards and deals five to each person. The remaining cards are placed in a pile facedown in the middle of the table. The dealer turns the top card faceup.

- The player to the left of the dealer plays a card with the same color or solution as the faceup card. Wild cards can be played any time. If the player cannot play a card, he or she takes a card from the pile and plays it, if possible. If it is not possible to play, the player places the card in his or her hand, and it is the next player's turn.

- **Who Wins?** The first person to play all cards in his or her hand is the winner.

Inequalities

Sunshine State Standards MA.D.1.3.2-2, MA.D.2.3.1-1, MA.D.2.3.1-2, MA.D.2.3.1-3, MA.D.2.3.1-6, MA.D.2.3.2-2

What You'll Learn
Write and graph inequalities.

MATH Symbols

≤ less than or equal to
≥ greater than or equal to

WHEN **am I ever going to use this?**

SIGNS The first highway sign at the right indicates that trucks *more than* 10 feet 6 inches tall cannot pass. The second sign indicates that a speed of 45 miles per hour *or less* is legal.

1. Name three truck heights that can safely pass on a road where the first sign is posted. Can a truck that is 10 feet 6 inches tall pass? Explain.

2. Name three speeds that are legal according to the second sign. Is a car traveling at 45 miles per hour driving at a legal speed? Explain.

In Chapter 1, you learned that a mathematical sentence that contains > or < is called an inequality. When used to compare a variable and a number, inequalities can describe a range of values.

EXAMPLES Write Inequalities with < or >

Write an inequality for each sentence.

1 **SAFETY** A package must weigh less than 80 pounds.

Let w = package's weight.

$w < 80$

2 **AGE** You must be over 55 years old to join.

Let a = person's age.

$a > 55$

Some inequalities use the symbols ≤ or ≥. They are combinations of the symbol < or > with part of the equals sign. The symbol ≤ is read *is less than or equal to*, while the symbol ≥ is read *is greater than or equal to*.

EXAMPLES Write Inequalities with ≤ or ≥

Write an inequality for each sentence.

3 **VOTING** You must be 18 years of age or older to vote.

Let a = person's age.

$a \geq 18$

4 **DRIVING** Your speed must be 65 miles per hour or less.

Let s = car's speed.

$s \leq 65$

Inequalities				
Words	• is less than • is fewer than	• is greater than • is more than • exceeds	• is less than or equal to • is no more than • is at most	• is greater than or equal to • is no less than • is at least
Symbols	$<$	$>$	$\leq$	$\geq$

Inequalities with variables are open sentences. When the variable in an open sentence is replaced with a number, the inequality may be true or false.

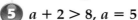 **Determine the Truth of an Inequality**

For the given value, state whether each inequality is *true* or *false*.

 $a + 2 > 8,\ a = 5$

$a + 2 > 8$ Write the inequality.

$5 + 2 \overset{?}{>} 8$ Replace a with 5.

$7 \not> 8$ Simplify.

Since 7 is not greater than 8, $7 > 8$ is false.

$10 \leq 7 - x,\ x = -3$

$10 \leq 7 - x$ Write the inequality.

$10 \overset{?}{\leq} 7 - (-3)$ Replace x with -3.

$10 \leq 10$ Simplify.

While $10 < 10$ is false, $10 = 10$ is true, so $10 \leq 10$ is true.

Your Turn **For the given value, state whether each inequality is *true* or *false*.**

a. $n - 6 < 15,\ n = 18$ b. $-3p \geq 24,\ p = 8$ c. $-2 > 5y - 7,\ y = 1$

Inequalities can be graphed on a number line. Since it is impossible to show all the values that make an inequality true, an open or closed circle is used to indicate where these values begin, and an arrow to the left or to the right is used to indicate that they continue in the indicated direction.

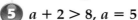 **Graph an Inequality**

Graph each inequality on a number line.

 $n < 3$

Place an open circle at 3. Then draw a line and an arrow to the left.

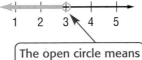

> The open circle means the number 3 is *not* included in the graph.

 $n \geq 3$

Place a closed circle at 3. Then draw a line and an arrow to the right.

> The closed circle means the number 3 *is* included in the graph.

Your Turn **Graph each inequality on a number line.**

d. $x > 2$ e. $x < 1$ f. $x \leq 5$ g. $x \geq -4$

STUDY TIP

Symbols Read $7 \not> 8$ as *7 is not greater than 8.*

Skill and Concept Check

1. **OPEN ENDED** Write an inequality using $\leq$ or $\geq$. Then give a situation that can be represented by the inequality.

2. **NUMBER SENSE** Integers that are greater than or equal to zero are classified as what types of numbers? Represent this classification of numbers using an inequality.

GUIDED PRACTICE

Write an inequality for each sentence.

3. **RESTAURANTS** Children under the age of 6 eat free.

4. **TESTING** You are allowed a maximum of 45 minutes to complete one section of a standardized test.

For the given value, state whether each inequality is *true* or *false*.

5. $x - 11 < 9, x = 20$
6. $42 \geq 6a, a = 8$
7. $\frac{n}{3} + 1 \leq 6; n = 15$

Graph each inequality on a number line.

8. $n > 4$
9. $p \leq 2$
10. $x \geq 0$
11. $a < 7$

Practice and Applications

Write an inequality for each sentence.

12. **MOVIES** Children under 13 are not permitted without an adult.

13. **SHOPPING** You must spend more than $100 to receive a discount.

14. **ELEVATORS** An elevator's maximum load is 3,400 pounds.

15. **FITNESS** You must run at least 4 laps around the track.

16. **GRADES** A grade of no less than 70 is considered passing.

17. **MONEY** The cost can be no more than $25.

HOMEWORK HELP

For Exercises	See Examples
12–17	1–4
18–23	5, 6
24–33	7. 8

Extra Practice
See pages 641, 657.

For the given value, state whether each inequality is *true* or *false*.

18. $12 + a < 20, a = 9$
19. $15 - k > 6, k = 8$
20. $-3y < 21; y = 8$
21. $32 \leq 2x, x = 16$
22. $\frac{n}{4} \geq 5, n = 12$
23. $\frac{-18}{x} > 9, x = -2$

Graph each inequality on a number line.

24. $x > 6$
25. $a > 0$
26. $y < 8$
27. $h < 2$
28. $w \leq 3$
29. $p \geq 7$
30. $n \geq 1$
31. $d \leq 4$
32. $-5 > b$
33. $-3 \leq y$

Write an inequality for each sentence.

34. A number increased by 5 is at most 15.

35. Eight times a number is no less than 24.

36. Sixteen is more than the quotient of number and 2.

37. Four less than a number is less than 12.

TELEVISION For Exercises 38 and 39, use the information in the graphic.

38. Rashid decides that he spends at least 100 more hours than the average time spent by kids watching television each year. Write an inequality for Rashid's TV viewing time.

39. Gabriela determines that she spends at most the same amount of time watching TV each year as the average amount of time kids spend attending school. Write an inequality to represent Gabriela's TV viewing time.

EQUIVALENT INEQUALITIES The inequality $3 < x$ is equivalent to $x > 3$. Write an equivalent inequality for each of the following.

40. $14 \leq a$ 41. $-2 > n$ 42. $-5 \geq y$

43. **RESEARCH** Use the Internet or another resource to find who first used the symbols $<$ for *less than* and $>$ for *greater than*.

44. **CRITICAL THINKING** Determine whether the following statement is *sometimes*, *always*, or *never* true. Explain your reasoning. *If x is a real number, then $x \geq x$.*

Standardized Test Practice and Mixed Review

FCAT Practice

45. **MULTIPLE CHOICE** What inequality is graphed below?

$-6\ -5\ -4\ -3\ -2\ -1\ 0\ 1\ 2\ 3\ 4\ 5\ 6$

Ⓐ $x < -3$ Ⓑ $x \leq -3$ Ⓒ $x > -3$ Ⓓ $x \geq -3$

46. **MULTIPLE CHOICE** Which inequality represents *a number is at least 24*?

Ⓕ $n \geq 24$ Ⓖ $n < 24$ Ⓗ $n \leq 24$ Ⓘ $n > 24$

Solve each equation. Check your solution. (Lesson 10-4)

47. $2x + 16 = 6x$ 48. $5y - 1 = 3y + 11$ 49. $4a - 9 = 7a + 6$ 50. $n + 0.8 = -n + 1$

51. **WEATHER** The temperature is $-3°$F. It is expected to rise $6°$ each hour for the next several hours. Write and solve an equation to find in how many hours the temperature will be $21°$F. (Lesson 10-3)

GETTING READY FOR THE NEXT LESSON

PREREQUISITE SKILL Solve each equation. (Lesson 1-8)

52. $y + 15 = 31$ 53. $n + 4 = -7$ 54. $a - 8 = 25$ 55. $-12 = x - 3$

Solving Inequalities by Adding or Subtracting

Sunshine State Standards
MA.D.2.3.1-3, MA.D.2.3.2-2

What You'll Learn

Solve inequalities by using the Addition or Subtraction Properties of Inequality.

WHEN am I ever going to use this?

FAMILY The table shows the age of each member of Victoria's family. Notice that Victoria is younger than her brother, since $13 < 16$. Will this be true 10 years from now?

Family Member	Age
Dad	43
Mom	41
Brother	16
Victoria	13

1. Add 10 to each side of the inequality $13 < 16$. Write the resulting inequality and decide whether it is *true* or *false*.

2. Was Victoria's dad younger or older than Victoria's mom 13 years ago? Explain your reasoning using an inequality.

The examples above demonstrate properties of inequality.

Key Concept — Addition and Subtraction Properties of Inequality

Words   When you add or subtract the same number from each side of an inequality, the inequality remains true.

Symbols   For all numbers a, b, and c,

1. if $a > b$, then $a + c > b + c$ and $a - c > b - c$.
2. if $a < b$, then $a + c < b + c$ and $a - c < b - c$.

Examples

$$2 > -3 \qquad\qquad 3 < 8$$
$$2 + 5 > -3 + 5 \qquad\qquad 3 - 4 < 8 - 4$$
$$7 > 2 \checkmark \qquad\qquad -1 < 4 \checkmark$$

These properties are also true for $a \geq b$ and $a \leq b$.

Solving an inequality means finding values for the variable that make the inequality true.

EXAMPLE   Solve an Inequality Using Addition

1 Solve $n - 8 < 15$. Check your solution.

$n - 8 < 15$	Write the inequality.
$n - 8 + 8 < 15 + 8$	Add 8 to each side.
$n < 23$	Simplify.
Check $n - 8 < 15$	Write the inequality.
$22 - 8 \overset{?}{<} 15$	Replace n with a number less than 23, such as 22.
$14 < 15 \checkmark$	This statement is true.

Any number less than 23 will make the statement true, so the solution is $n < 23$.

EXAMPLE Solve an Inequality Using Subtraction

2 Solve $-4 \geq a + 7$. Check your solution.

$$-4 \geq a + 7 \qquad \text{Write the inequality.}$$

$$-4 - 7 \geq a + 7 - 7 \qquad \text{Subtract 7 from each side.}$$

$$-11 \geq a \text{ or } a \leq -11 \qquad \text{Simplify.}$$

Check Replace a in the original inequality with −11 and then with a number less than −11.

The solution is $a \leq -11$.

EXAMPLE Graph the Solutions of an Inequality

3 Solve $y - \dfrac{1}{3} \leq 5$. Then graph the solution on a number line.

$$y - \frac{1}{3} \leq 5$$

$$y - \frac{1}{3} + \frac{1}{3} \leq 5 + \frac{1}{3}$$

$$y \leq 5\frac{1}{3}$$

The solution is $y \leq 5\frac{1}{3}$.

Graph the solution.

Place a closed circle at $5\frac{1}{3}$. Draw a line and arrow to the left.

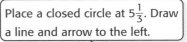

Your Turn Solve each inequality and check your solution. Then graph the solution on a number line.

a. $t + 3 > 12$ **b.** $2 > p - 5$ **c.** $n + \dfrac{1}{2} \geq 4$

EXAMPLE Use an Inequality to Solve a Problem

4 **ANIMALS** Suppose a South American manatee weighs 968 pounds. Use the information at the left to determine how much more weight this manatee might gain.

Words	The phrase *up to* means *less than or equal to*. So, the manatee's current weight plus any weight gained must be less than or equal to 1,300 pounds.
Variable	Let w = weight gained by the manatee.

manatee's current weight	plus	weight gained	must be less than or equal to	1,300 pounds
968	+	w	≤	1,300

$$968 + w \leq 1,300 \qquad \text{Write the inequality.}$$

$$968 - 968 + w \leq 1,300 - 968 \qquad \text{Subtract 968 from each side.}$$

$$w \leq 332 \qquad \text{Simplify.}$$

The manatee might gain up to 332 more pounds.

1. **Explain** how solving an inequality by using subtraction is similar to solving an equation by using subtraction.

2. **OPEN ENDED** Write an inequality whose solution is $n > 5$ that can be solved by using the Addition or Subtraction Property of Equality.

GUIDED PRACTICE

Solve each inequality. Check your solution.

3. $b + 5 > 9$ 4. $12 + n \leq 4$ 5. $-6 \leq 7 + g$

6. $x - 4 < 10$ 7. $k - 9 \geq -2$ 8. $8 > y - 8$

Solve each inequality and check your solution. Then graph the solution on a number line.

9. $c + 9 < 7$ 10. $m - 1 \geq 3$ 11. $a - \dfrac{1}{2} > 3$

Practice and Applications

Solve each inequality. Check your solution.

12. $a + 7 < 21$ 13. $5 + x \leq 18$ 14. $10 + n \geq -2$

15. $-4 < k + 6$ 16. $3 < y + 8$ 17. $c + 10 < 9$

18. $r - 9 \leq 7$ 19. $g - 4 \geq 13$ 20. $-2 < b - 6$

21. $s - 12 \leq -5$ 22. $t - 3 < -9$ 23. $-17 \leq w - 15$

24. $2 + m \geq 3.5$ 25. $q + 0.8 \leq -0.5$ 26. $v - 6 > 2.7$

27. $p - 4.8 > -6$ 28. $d - \dfrac{2}{3} \leq \dfrac{1}{2}$ 29. $5 > f + 1\dfrac{1}{4}$

HOMEWORK HELP	
For Exercises	See Examples
12–33	1, 2
34–45	3
46–49	4

Extra Practice
See pages 642, 657.

Write an inequality and solve each problem.

30. Five more than a number is at least 13.

31. The difference between a number and 11 is less than 8.

32. Nine less than a number is more than 4.

33. The sum of a number and 17 is no more than 6.

Solve each inequality and check your solution. Then graph the solution on a number line.

34. $c + 1 < 4$ 35. $n + 8 > 12$ 36. $2 \leq 7 + p$

37. $-10 \geq x + 6$ 38. $a - 3 \leq 5$ 39. $-11 > g - 4$

40. $-12 < k - 9$ 41. $h - 6 \geq -4$ 42. $y - 1.5 < 2$

43. $b - 0.75 \leq 7$ 44. $t + \dfrac{2}{3} > 8$ 45. $w + 5\dfrac{1}{3} < 10$

46. **INSECTS** There are more than 250,000 species of beetles. A science museum has a collection representing 320 of these species. Write and solve an inequality to find how many beetle species are not represented.

HEALTH For Exercises 47 and 48, use the diagram at the right.

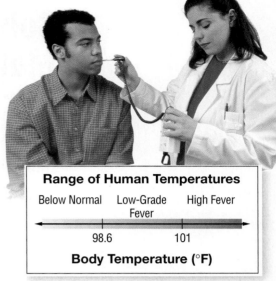

47. An adult is considered to have a high fever if his or her temperature goes above 101°F. Suppose Mr. Herr has a temperature of 99.2°F. Write and solve an inequality to find how much his temperature must increase before he is considered to have a high fever.

48. *Hypothermia* occurs when a person's body temperature falls below 95°F. Write and solve an inequality that describes how much lower the body temperature of a person with hypothermia will be than a person with a normal body temperature of 98.6°F.

Range of Human Temperatures

Below Normal   Low-Grade   High Fever
                Fever

← 98.6   101 →

Body Temperature (°F)

49. **GEOMETRY** The base of the rectangle shown is greater than its height. Write and solve an inequality to find the possible values of x.

$x - 3$ cm

15 cm

50. **WRITE A PROBLEM** Write about a real-life situation that can be solved by using an addition inequality. Then write an inequality and solve the problem.

51. **CRITICAL THINKING** Is it *sometimes*, *always*, or *never* true that $x > x + 1$? Explain your reasoning.

Standardized Test Practice and Mixed Review

FCAT Practice

52. **MULTIPLE CHOICE** Adriana has $30 to spend on food and rides at a carnival. She has already spent $12 on food. Which inequality represents how much money she can spend on rides?

    A $m < 18$ **B** $m \leq 18$ **C** $m > 18$ **D** $m \geq 18$

53. **MULTIPLE CHOICE** If $x - 6 > 17$, then x could be which of the following values?

    F 11   **G** 22   **H** 23   **I** 24

For the given value, state whether each inequality is *true* or *false*. (Lesson 10-5)

54. $18 - n > 4, n = 11$ 55. $13 + x < 21, x = 8$ 56. $34 \leq 5p, p = 7$ 57. $\dfrac{a}{-4} \geq 3, a = -12$

58. **CAR RENTAL** Suppose you can rent a car for either $35 a day plus $0.40 a mile or for $20 a day plus $0.55 per mile. Write and solve an equation to find the number of miles that results in the same cost for one day. (Lesson 10-4)

59. If $\angle F$ and $\angle G$ are supplementary and $m\angle G = 47°$, find $m\angle F$. (Lesson 6-1)

GETTING READY FOR THE NEXT LESSON

PREREQUISITE SKILL Solve each equation. (Lesson 1-9)

60. $3y = -15$ 61. $-18 = -2a$ 62. $\dfrac{w}{4} = 12$ 63. $20 = \dfrac{x}{-5}$

Solving Inequalities by Multiplying or Dividing

Sunshine State Standards
MA.D.2.3.1-3, MA.D.2.3.2-2

WHEN am I ever going to use this?

SHOPPING The table shows the prices of the same brand name of shoes at a sports apparel store. Notice that walking shoes cost less than cross-training shoes, since $80 < 150$. Will this inequality be true if the store sells both pairs of shoes at half price?

Shoe Style	Regular Price ($)
athletic sandal	60
walking	80
running	100
basketball	120
cross training	150

1. Divide each side of the inequality $80 < 150$ by 2. Write the resulting inequality and decide whether it is *true* or *false*.

2. Would the cost of three pairs of basketball shoes be greater or less than the cost of three pairs of running shoes all sold at the regular price? Explain your reasoning using an inequality.

The examples above demonstrate additional properties of inequality.

Key Concept | **Multiplication and Division By a Positive Number**

Words When you multiply or divide each side of an inequality by a positive number, the inequality remains true.

Symbols For all numbers a, b, and c, where $c > 0$,

1. if $a > b$, then $ac > bc$ and $\dfrac{a}{c} > \dfrac{b}{c}$.

2. if $a < b$, then $ac < bc$ and $\dfrac{a}{c} < \dfrac{b}{c}$.

Examples

$$5 < 8 \qquad\qquad 2 > -10$$
$$4(5) < 4(8) \qquad \dfrac{2}{2} > \dfrac{-10}{2}$$
$$20 < 32 \qquad\qquad 1 > -5$$

These properties also hold true for $a \geq b$ and $a \leq b$.

EXAMPLE **Divide by a Positive Number**

1 **Solve $7y > -42$. Check your solution.**

$7y > -42$ Write the inequality.

$\dfrac{7y}{7} > \dfrac{-42}{7}$ Divide each side by 7.

$y > -6$ Simplify.

The solution is $y > -6$. You can check this solution by substituting numbers greater than -6 into the inequality.

EXAMPLE **Multiply by a Positive Number**

2 Solve $\frac{1}{3}x \le 8$ and check your solution. Then graph the solution on a number line.

$\frac{1}{3}x \le 8$ Write the inequality.

$3\left(\frac{1}{3}x\right) \le 3(8)$ Multiply each side by 3.

$x \le 24$ Simplify.

The solution is $x \le 24$. You can check this solution by substituting 24 and a number less than 24 into the inequality.

Graph the solution, $x \le 24$.

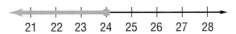

Your Turn Solve each inequality and check your solution. Then graph the solution on a number line.

a. $3a \ge 45$ **b.** $\frac{n}{4} < -16$ **c.** $81 \le 9p$

What happens when each side of an inequality is multiplied or divided by a negative number?

Graph 3 and 5 on a number line.

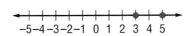

Multiply each number by -1.

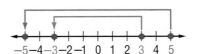

Since 3 is to the left of 5, $3 < 5$.

Since -3 is to the right of -5, $-3 > -5$.

Notice that the numbers being compared switched positions as a result of being multiplied by a negative number. In other words, their order reversed. These and other examples suggest the following properties.

STUDY TIP

Common Error
Do not reverse the inequality symbol just because there is a negative sign in the inequality, as in Example 1. Only reverse the inequality symbol when you *multiply* or *divide* each side by a negative number.

Key Concept **Multiplication and Division By a Negative Number**

Words   When you multiply or divide each side of an inequality by a negative number, the direction of the inequality symbol must be reversed for the inequality to remain true.

Symbols   For all numbers a, b, and c, where $c < 0$,

1. if $a > b$, then $ac < bc$ and $\frac{a}{c} < \frac{b}{c}$.

2. if $a < b$, then $ac > bc$ and $\frac{a}{c} > \frac{b}{c}$.

Examples

$8 > 5$ $-3 < 9$

$-1(8) < -1(5)$ Reverse the inequality symbols. $\frac{-3}{-3} > \frac{9}{-3}$

$-8 < -5$ $1 > -3$

These properties also hold true for $a \ge b$ and $a \le b$.

EXAMPLES Multiply or Divide by a Negative Number

3 Solve $\frac{a}{-2} \geq 8$. Check your solution.

$$\frac{a}{-2} \geq 8 \qquad \text{Write the inequality.}$$

$$-2\left(\frac{a}{-2}\right) \leq -2(8) \qquad \text{Multiply each side by } -2 \text{ and reverse the inequality symbol.}$$

$$a \leq -16 \qquad \text{Simplify.}$$

The solution is $a \leq -16$. You can check this solution by replacing a in the original inequality with -16 and a number less than -16.

4 Solve $-6n < -24$. Then graph the solution on a number line.

$$-6n < -24 \qquad \text{Write the inequality.}$$

$$\frac{-6n}{-6} > \frac{-24}{-6} \qquad \text{Divide each side by } -6 \text{ and reverse the inequality symbol.}$$

$$n > 4 \qquad \text{Check this result.}$$

Graph the solution, $n > 4$.

-2 -1 0 1 2 3 4 5 6 7 8 9

Your Turn Solve each inequality and check your solution. Then graph the solution on a number line.

d. $\frac{c}{-7} < -14$ **e.** $-5d \geq 30$ **f.** $-3 \leq \frac{w}{-8}$

REAL-LIFE MATH

WORK If you are 14 or 15 and have a part-time job, you can work no more than 3 hours on a school day, 18 hours in a school week, 8 hours on a nonschool day, or 40 hours in a nonschool week.

Source: www.youthrules.dol.gov

Some inequalities involve more than one operation. To solve, work backward to undo the operations as you did in solving two-step equations.

EXAMPLE Solve a Two-Step Inequality

5 **WORK** Jason wants to earn at least $30 this week to go to the state fair. His dad will pay him $12 to mow the lawn. For washing their cars, his neighbors will pay him $8 per car. If Jason mows the lawn, write and solve an inequality to find how many cars he needs to wash to earn at least $30.

The phrase *at least* means *greater than or equal to*. Let $c =$ the number of cars he needs to wash. Then write an inequality.

$12	plus	$8 per car	is greater than or equal to	$30.
12	+	8c	≥	30

$$12 + 8c \geq 30 \qquad \text{Write the inequality.}$$

$$12 - 12 + 8c \geq 30 - 12 \qquad \text{Subtract 12 from each side.}$$

$$8c \geq 18 \qquad \text{Simplify}$$

$$\frac{8c}{8} \geq \frac{18}{8} \qquad \text{Divide each side by 8.}$$

$$c \geq 2.25 \qquad \text{Simplify.}$$

Since he will not get paid for washing a fourth of a car, Jason must wash at least 3 cars.

Writing Math
Exercise 2

1. **OPEN ENDED** Write an inequality that can be solved using the Multiplication Property of Equality where the inequality symbol needs to be reversed.

2. **FIND THE ERROR** Olivia and Lakita each solved $8a \leq -56$. Who is correct? Explain.

Olivia
$8a \leq -56$
$\dfrac{8a}{8} \geq \dfrac{-56}{8}$
$a \geq -7$

Lakita
$8a \leq -56$
$\dfrac{8a}{8} \leq \dfrac{-56}{8}$
$a \leq -7$

GUIDED PRACTICE

Solve each inequality and check your solution. Then graph the solution on a number line.

3. $8x \leq -72$

4. $-4y > 32$

5. $-56 \leq -7p$

6. $\dfrac{h}{4} \geq -6$

7. $\dfrac{g}{-2} < -7$

8. $\dfrac{d}{-3} \geq -3$

Solve each inequality. Check your solution.

9. $2a - 8 < -24$

10. $-4k + 3 > -13$

11. $\dfrac{m}{-3} + 7 \leq -2$

Practice and Applications

Solve each inequality and check your solution. Then graph the solution on a number line.

HOMEWORK HELP	
For Exercises	See Examples
12–33	1–4
34–45	5

Extra Practice
See pages 642, 657.

12. $5x < 15$

13. $9n \leq 45$

14. $14k \geq -84$

15. $-12 > 3g$

16. $-100 \leq 50p$

17. $2y < -22$

18. $-4w \geq 20$

19. $-3r > 9$

20. $-72 < -12h$

21. $-6c \geq -6$

22. $\dfrac{v}{-4} > 4$

23. $\dfrac{a}{-3} \geq 5$

24. $\dfrac{x}{9} \leq -3$

25. $\dfrac{n}{7} < -14$

26. $\dfrac{m}{-2} < -7$

27. $\dfrac{t}{-5} \leq -2$

28. $-8 \leq \dfrac{y}{0.2}$

29. $-\dfrac{1}{2}k > -10$

30. **BUS TRAVEL** A city bus company charges $2.50 per trip. They also offer a monthly pass for $85.00. Write and solve an inequality to find how many times a person should use the bus so that the pass is less expensive than buying individual tickets.

31. **BABY-SITTING** You want to buy a pair of $42 inline skates with the money you make baby-sitting. If you charge $5.25 an hour, write and solve an inequality to find how many whole hours you must baby-sit to buy the skates.

ROADS For Exercises 32 and 33, use the information in the graphic at the right.

32. Write and solve an inequality to find the approximate circumference of Earth.

33. Write and solve an inequality to find the approximate distance from Earth to the moon and back.

 Data Update What is the circumference of Earth? the distance from Earth to the moon and back? Visit msmath3.net/data_update to learn more.

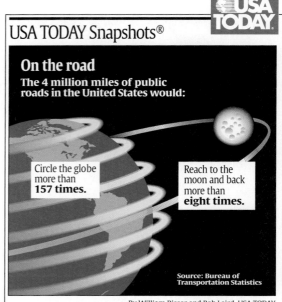

USA TODAY Snapshots®

On the road
The 4 million miles of public roads in the United States would:

Circle the globe more than **157 times.**

Reach to the moon and back more than **eight times.**

Source: Bureau of Transportation Statistics

By William Risser and Bob Laird, USA TODAY

Solve each inequality. Check your solution.

34. $5y - 2 > 13$

35. $8k + 3 \le -5$

36. $-3g + 8 \ge -4$

37. $7 + \frac{n}{3} < 4$

38. $\frac{w}{8} - 4 \le -5$

39. $\frac{c}{-4} + 8 < 1$

40. $3a - 8 < 5a$

41. $10 - 3x \ge 25 + 2x$

Write an inequality for each sentence. Then solve the inequality.

42. Three times a number is less than -60.

43. The quotient of a number and -5 is at most 7.

44. The quotient of a number and 3 is at least -12.

45. The product of -2 and a number is greater than -18.

46. **CRITICAL THINKING** You have scores of 88, 92, 85, and 87 on four tests. What number of points must you get on your fifth test to have a test average of at least 90?

Standardized Test Practice and Mixed Review

  FCAT Practice

47. **MULTIPLE CHOICE** Which number is a possible base length of the triangle if its area is greater than 45 square yards?

  Ⓐ 3   Ⓑ 4   Ⓒ 5   Ⓓ 6

18 yd

x yd

48. **MULTIPLE CHOICE** As a salesperson, you are paid $60 per week plus $5 per sale. This week you want your pay to be at least $120. Which inequality can be used to find the number of sales you must make this week?

  Ⓕ $60 + 5x \ge 120$ Ⓖ $60x + 5 \ge 120$

  Ⓗ $60 + 5x \le 120$ Ⓘ $60x + 5 \le 120$

Solve each inequality. Check your solution. (Lesson 10-6)

49. $y + 7 < 9$ 50. $a - 5 \le 2$ 51. $j - 8 \ge -12$ 52. $-14 > 8 + n$

Write an inequality for each sentence. (Lesson 10-5)

53. **HEALTH** Your heart beats over 100,000 times a day.

54. **BIRDS** A peregrine falcon can spot a pigeon up to 8 kilometers away.

Vocabulary and Concept Check

coefficient (p. 470)	simplest form (p. 471)
constant (p. 470)	simplifying the expression (p. 471)
equivalent expressions (p. 469)	term (p. 470)
like terms (p. 470)	two-step equation (p. 474)

Choose the letter of the term that best matches each statement or phrase.

1. terms that contain the same variables
2. an equation that contains two operations
3. a term without a variable
4. the parts of an algebraic expression separated by a plus sign
5. an algebraic expression that has no like terms and no parentheses
6. the numerical part of a term that contains a variable

a. constant
b. coefficient
c. two-step equation
d. like terms
e. term
f. simplest form

Lesson-by-Lesson Exercises and Examples

10-1 Simplifying Algebraic Expressions (pp. 469–473)

Use the Distributive Property to rewrite each expression.

7. $4(a + 3)$
8. $-6(x + 7)$
9. $(n - 5)(-7)$
10. $-2(6x - 3)$

Simplify each expression.

11. $p + 6p$
12. $6b + 7b - 3 + 5$

Example 1 Use the Distributive Property to rewrite $-8(x - 9)$.

$-8(x - 9)$ Write the expression.

$= -8[x + (-9)]$ $x - 9 = x + (-9)$

$= -8(x) + (-8)(-9)$ Distributive Property

$= -8x + 72$ Simplify.

10-2 Solving Two-Step Equations (pp. 474–477)

Solve each equation. Check your solution.

13. $2x + 5 = 17$
14. $3d + 20 = 2$
15. $10 = 3 - g$
16. $4 = -3y - 2$
17. $\frac{c}{5} + 2 = 9$
18. $a + 6a + 11 = 39$

Example 2 Solve $5h + 8 = -12$.

$5h + 8 = -12$ Write the equation.

$5h + 8 - 8 = -12 - 8$ Subtract 8 from each side.

$5h = -20$ $-12 - 8 = -12 + (-8)$

$\frac{5h}{5} = \frac{-20}{5}$ Divide each side by 5.

$h = -4$ Simplify.

The solution is -4.   Check this solution.

 msmath3.net/vocabulary_review

10-3 Writing Two-Step Equations (pp. 478–481)

Translate each sentence into an equation. Then find the number.

19. Six more than twice a number is −4.
20. Three less than 2 times a number equals 11.
21. The quotient of a number and 8, less 2, is 5.

Example 3 Translate the following sentence into an equation.

6 less than 4 times a number is 10.

6 less than   4 times a number   is   10.

$$4n - 6 = 10$$

10-4 Solving Equations with Variables on Each Side (pp. 484–487)

Solve each equation. Check your solution.

22. $11x = 20x + 18$
23. $4n + 13 = n - 8$
24. $3a + 5 = 2a + 7$
25. $7b - 3 = -2b + 24$
26. $9 - 2y = 8y - 6$

Example 4 Solve $7x + 5 = 6x - 19$.

$$7x + 5 = 6x - 19$$
$$7x - 6x + 5 = 6x - 6x - 19$$
$$x + 5 = -19$$
$$x + 5 - 5 = -19 - 5$$
$$x = -24$$

10-5 Inequalities (pp. 492–495)

Write an inequality. Then graph the inequality on a number line.

27. **GRADES** a grade of 92 or better
28. **SPORTS** qualifying time must be less than 2 minutes

Example 5 Graph $a < -4$.

Place an open circle at −4. Then draw a line and an arrow to the left.

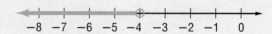

10-6 Solving Inequalities by Adding or Subtracting (pp. 496–499)

Solve each inequality. Check your solution.

29. $y + 7 \leq 5$ 30. $x - 2 < 7$
31. $18 < 4 + d$ 32. $a - 6 > -2$

Example 6 Solve $k + 2 > -5$.

$$k + 2 > -5 \qquad \text{Write the inequality.}$$
$$k + 2 - 2 > -5 - 2 \qquad \text{Subtract 2 from each side.}$$
$$k > -7 \qquad \text{Simplify. Check this result.}$$

10-7 Solving Inequalities by Multiplying or Dividing (pp. 500–504)

Solve each inequality. Check your solution.

33. $13c \leq -26$ 34. $-2a \geq -10$
35. $-6m > 18$ 36. $22 \geq -3x - 2$

Example 7 Solve $-9n < 54$.

$$-9n < 54 \qquad \text{Write the inequality.}$$
$$\frac{-9n}{-9} > \frac{54}{-9} \qquad \text{Divide each side by −9 and reverse the inequality symbol.}$$
$$n > -6 \qquad \text{Simplify. Check this result.}$$

Vocabulary and Concepts

1. **Explain** how you determine whether or not an expression is in simplest form.

2. **Give** three examples of phrases that indicate the inequality symbol $\leq$.

Skills and Applications

3. Use the Distributive Property to rewrite the expression $-7(x - 10)$.

4. Simplify the expression $9a - a + 15 - 10a - 6$.

Solve each equation. Check your solution.

5. $3n + 18 = 6$

6. $\dfrac{k}{2} - 11 = 5$

7. $-23 = 3p + 5 + p$

8. $4x - 6 = 5x$

9. $-3a - 2 = 2a + 3$

10. $-2y + 5 = y - 1$

11. Translate *the quotient of a number and 6, plus 3, is 11* into an equation. Then find the number.

12. **FUND-RAISER** The band buys coupon books for a one-time fee of $60 plus $5 per book. If they sell the books for $10 each, write and solve an equation to find how many books they must sell to break even.

13. **COMPUTERS** A disk can hold at most 1.38 megabytes of data. Write an inequality. Then graph the inequality on a number line.

Solve each inequality and check the solution. Then graph the solution on a number line.

14. $x + 5 \geq 3$

15. $5 > a - 2$

16. $-3d \leq 18$

17. $-4 > \dfrac{c}{9}$

18. $-2g + 15 > 45$

19. $\dfrac{m}{-5} + 4 \geq 1$

Standardized Test Practice

20. **MULTIPLE CHOICE** The perimeter of the parallelogram at the right is no more than 44 inches. Which of the following inequalities represents all possible values for x?

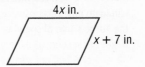

4x in.

$x + 7$ in.

Ⓐ $x \leq 3$

Ⓑ $x \geq 3$

Ⓒ $x \leq 7.4$

Ⓓ $x \geq 7.4$

PART 1 Multiple Choice

Record your answers on the answer sheet provided by your teacher or on a sheet of paper.

1. On Saturday Jennifer rode her bike to Robert's house. They then biked together to the library. Finally, Jennifer rode home alone from the library. If each unit on the grid represents 1 mile, what was the total distance that Jennifer biked on Saturday? (Lesson 3-6)

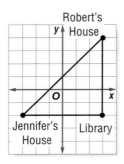

   Ⓐ about 3.5 mi   Ⓑ about 8.5 mi
   Ⓒ about 15.5 mi   Ⓓ about 20.5 mi

2. The data below was collected from four different remote-controlled car tests.

Car	Distance Traveled (ft)	Time (s)
Speedster	45	9
Turbo	31	5
Cruiser	33	6
Hurricane	51	$8\frac{1}{2}$

Which car traveled at the fastest rate?
(Lesson 4-1)

   Ⓕ Speedster   Ⓖ Turbo
   Ⓗ Cruiser   Ⓘ Hurricane

TEST-TAKING TIP

Question 2 You can often use estimation to eliminate incorrect answers. In Question 2, the Turbo's rate of speed is about 30 ÷ 5 or 6 feet per second, and the Cruiser's is about 30 ÷ 6 or 5 feet per second. Thus, the Cruiser can be eliminated since the Turbo's speed is faster.

3. In 1990, the number of students attending a school was 865. In 2000, the number was 680. By what percent did the number decrease from 1990 to 2000? (Lesson 5-7)

   Ⓐ 20%   Ⓑ 21%   Ⓒ 43%   Ⓓ 79%

4. What is the volume of paint in a can that has a diameter of 10 inches and a height of 12 inches? (Lesson 7-5)

   Ⓕ 188.4 in³   Ⓖ 376.8 in³
   Ⓗ 942 in³   Ⓘ 1,884 in³

5. The graph shows the results of the election for club president. Which statement is supported by the information on the graph? (Lesson 9-2)

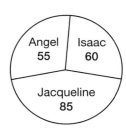

   Ⓐ The total number of votes was 190.
   Ⓑ Isaac received 30% of the votes.
   Ⓒ The ratio of Angel's votes to Isaac's votes was 5 to 6.
   Ⓓ Jacqueline received half the votes.

6. Find the value of x so that the isosceles trapezoid at the right has a perimeter of 200 inches. (Lesson 10-3)

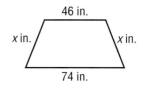

   Ⓕ 35   Ⓖ 40   Ⓗ 55   Ⓘ 80

7. The number line below is the graph of which inequality? (Lesson 10-7)

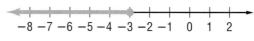

   Ⓐ $4y \leq -12$ Ⓑ $-5y \leq 15$
   Ⓒ $5y > -15$ Ⓓ $-4y > 12$

PART 2   Short Response/Grid In

THINK
SOLVE
EXPLAIN

Record your answers on the answer sheet provided by your teacher or on a sheet of paper.

8. You are serving soup to 7 people. Each serving is $\frac{3}{4}$ of a cup. If each can of soup contains $2\frac{1}{2}$ cups, how many cans do you need in order to give every person a full serving? (Lessons 2-3 and 2-4)

9. Draw the figure that results from rotating the figure at the right 90° counterclockwise about its center and then reflecting it over the indicated vertical axis. (Lessons 6-7 and 6-9)

10. Santiago is running the duck pond at the youth carnival. Each duck has a number on the bottom indicating the level of the prize awarded. The table shows how many ducks of each prize level are currently in the pond.

Prize Level	1	2	3	4
Number of Ducks	4	6	8	?

How many prize-level-4 ducks should Santiago place in the pond so that the chances of randomly selecting one of these ducks is $\frac{1}{2}$? (Lesson 8-1)

11. The statistics below were listed on the board at the end of the grading period for a class of 9 students.

mean: 87
median: 88
range: 15

List a possible set of grades for the students in this class. (Lesson 9-4)

12. Write an expression with four terms, two of which are constants. The other terms should be like terms, one with a coefficient of −2 and the other with a coefficient of 6. Then simplify your expression. (Lesson 10-1)

13. If $8 + 5w = 11$, find the value of $2w$. (Lesson 10-2)

14. A restaurant has s small tables that will seat 4 people each. They also have ℓ large tables that will seat 10 people each. Write an inequality representing the maximum number of people p that can be seated at this restaurant. (Lesson 10-5)

FCAT Practice

PART 3   Extended Response

Record your answers on a sheet of paper. Show your work.

THINK
SOLVE
EXPLAIN

The table below gives prices for two different bowling alleys in your area. (Lessons 10-4 and 10-5)

Bowling Alley	Shoe Rental	Cost per Game
X	$2.50	$4.00
Y	$3.50	$3.75

15. Write an equation to find the number of games g for which the total cost to bowl at each alley would be equal.

16. Explain how you would solve the equation you wrote in Question 15.

17. How many games would you have to play for the cost to bowl at each alley to be equal?

18. Write an inequality giving the number of games g for which Bowling Alley X would be cheaper.

19. Write an inequality giving the number of games g for which Bowling Alley Y would be cheaper.

"What does mountain climbing have to do with math?"

As mountain climbers ascend the mountain, the temperature becomes colder. So, the temperature *depends* on the altitude. **In mathematics, you say that the temperature is a function of the altitude.**

You will solve problems about temperature changes and climbing mountains in Lesson 11-3.

Take this quiz to see whether you are ready to begin Chapter 11. Refer to the lesson number in parentheses if you need more review.

▶ Vocabulary Review

State whether each sentence is *true* or *false*. If *false*, replace the underlined word or number to make a true sentence.

1. In terms of slope, the rise is the <u>horizontal</u> change. (Lesson 4-3)

2. $y \geq x + 4$ is an example of an <u>inequality</u>. (Lesson 4-3)

▶ Prerequisite Skills

Graph each point on the same coordinate plane. (Page 614)

3. $A(-3, -4)$
4. $B(2, -1)$
5. $C(0, -2)$
6. $D(-4, 3)$

Evaluate each expression if $x = 6$. (Lesson 1-2)

7. $3x$
8. $4x - 9$
9. $2x + 8$
10. $5 + x$
11. $18 - x$
12. $7x$

Solve each equation. (Lesson 1-8)

13. $14 = n + 9$
14. $z - 3 = 8$
15. $-17 = b - 21$
16. $23 + r = 16$
17. $a + 7 = 2$
18. $-18 = 16 + y$

Find the slope of each line. (Lesson 4-3)

19.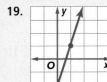

20.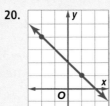

Linear Functions Make this Foldable to help you organize your notes. Begin with a plain piece of notebook paper.

STEP 1 **Fold in Half**
Fold the paper lengthwise to the holes.

STEP 2 **Fold**
Fold the paper in fourths.

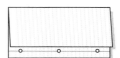

STEP 3 **Cut**
Open. Cut one side along the folds to make four tabs.

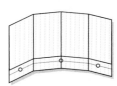

STEP 4 **Label**
Label each tab with the main topics as shown.

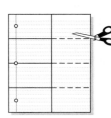

Sequences and Functions

Graphing Linear Functions

Systems of Equations

Graphing Linear Inequalities

Reading and Writing As you read and study the chapter, write notes and examples under each tab.

Sequences

Sunshine State Standards
MA.A.5.3.1-2, MA.B.1.3.3-1

HANDS-ON Mini Lab

What You'll Learn
Recognize and extend arithmetic and geometric sequences.

NEW Vocabulary

sequence
term
arithmetic sequence
common difference
geometric sequence
common ratio

Work with a partner.

Consider the following pattern.

Number of Triangles	1 triangle	2 triangles	3 triangles
Number of Toothpicks	3 toothpicks	5 toothpicks	7 toothpicks

1. Continue the pattern for 4, 5, and 6 triangles. How many toothpicks are needed for each case?

2. Study the pattern of numbers. How many toothpicks will you need for 7 triangles?

Now, consider another pattern.

Number of Squares	1 square	2 squares	3 squares
Number of Toothpicks	4 toothpicks	7 toothpicks	10 toothpicks

3. Continue the pattern for 4, 5, and 6 squares. How many toothpicks are needed for each case?

4. How many toothpicks will you need for 7 squares?

The numbers of toothpicks needed for each pattern form a sequence. A **sequence** is an ordered list of numbers. Each number is called a **term**.

An **arithmetic sequence** is a sequence in which the difference between any two consecutive terms is the same.

$$3, \quad 5, \quad 7, \quad 9, \quad 11, \ldots$$
$$+2 \quad +2 \quad +2 \quad +2$$

The difference is called the **common difference**.

To find the next number in an arithmetic sequence, add the common difference to the last term.

EXAMPLE Identify Arithmetic Sequences

1 State whether the sequence 17, 12, 7, 2, −3, ... is arithmetic. If it is, state the common difference. Write the next three terms of the sequence.

$$17, \quad 12, \quad 7, \quad 2, \quad -3$$
$$-5 \quad -5 \quad -5 \quad -5$$

Notice that 12 − 17 = −5, 7 − 12 = −5, and so on.

The terms have a common difference of -5, so the sequence is arithmetic. Continue the pattern to find the next three terms.

$$-3 \quad -8 \quad -13 \quad -18$$
$$\underset{-5}{\frown} \quad \underset{-5}{\frown} \quad \underset{-5}{\frown}$$

The next three terms are -8, -13, and -18.

READING Math

And So On The three dots following a list of numbers are read as *and so on*.

A **geometric sequence** is a sequence in which the quotient between any two consecutive terms is the same.

$$2, \quad 6, \quad 18, \quad 54, \quad 162, \ldots$$
$$\underset{\times 3}{\frown} \quad \underset{\times 3}{\frown} \quad \underset{\times 3}{\frown} \quad \underset{\times 3}{\frown}$$

The quotient is called the **common ratio**.

To find the next number in a geometric sequence, multiply the last term by the common ratio.

EXAMPLES Identify Geometric Sequences

State whether each sequence is geometric. If it is, state the common ratio. Write the next three terms of each sequence.

2 **96, −48, 24, −12, 6, …**

$$96, \quad -48, \quad 24, \quad -12, \quad 6$$
$$\underset{\times\left(-\frac{1}{2}\right)}{\frown} \quad \underset{\times\left(-\frac{1}{2}\right)}{\frown} \quad \underset{\times\left(-\frac{1}{2}\right)}{\frown} \quad \underset{\times\left(-\frac{1}{2}\right)}{\frown}$$

Notice that $-48 \div 96 = -\frac{1}{2}$, $24 \div (-48) = -\frac{1}{2}$, and so on.

The terms have a common ratio of $-\frac{1}{2}$, so the sequence is geometric.

Continue the pattern to find the next three terms.

$$6, \quad -3, \quad \frac{3}{2}, \quad -\frac{3}{4}$$
$$\underset{\times\left(-\frac{1}{2}\right)}{\frown} \quad \underset{\times\left(-\frac{1}{2}\right)}{\frown} \quad \underset{\times\left(-\frac{1}{2}\right)}{\frown}$$

The next three terms are -3, $\frac{3}{2}$, and $-\frac{3}{4}$.

3 **2, 4, 12, 48, 240, …**

$$2, \quad 4, \quad 12, \quad 48, \quad 240$$
$$\underset{\times 2}{\frown} \quad \underset{\times 3}{\frown} \quad \underset{\times 4}{\frown} \quad \underset{\times 5}{\frown}$$

Since there is no common ratio, the sequence is not geometric. However, the sequence does have a pattern. Multiply the last term by 6, the next term by 7, and the following term by 8.

$$240, \quad 1{,}440, \quad 10{,}080, \quad 80{,}640$$
$$\underset{\times 6}{\frown} \quad \underset{\times 7}{\frown} \quad \underset{\times 8}{\frown}$$

The next three terms are 1,440, 10,080, and 80,640.

Your Turn State whether each sequence is *arithmetic*, *geometric*, or *neither*. If it is arithmetic or geometric, state the common difference or common ratio. Write the next three terms of each sequence.

a. 8, 12, 16, 20, 24, …

b. -5, 25, -125, 625, $-3{,}125$, …

c. 1, 2, 4, 7, 11, …

d. 243, 81, 27, 9, 3, …

1. **Explain** how to determine whether a sequence is geometric.

2. **OPEN ENDED** Give a counterexample to the following statement.
   All sequences are either arithmetic or geometric.

3. **Which One Doesn't Belong?** Identify the sequence that is not the same type as the others. Explain your reasoning.

1, 2, 4, 8, 16, ...	125, 25, 5, 1, $\frac{1}{5}$, ...
5, 10, 15, 20, 25, ...	−2, 6, −18, 54, −162, ...

GUIDED PRACTICE

State whether each sequence is *arithmetic*, *geometric*, or *neither*. If it is arithmetic or geometric, state the common difference or common ratio. Write the next three terms of each sequence.

4. 2, 4, 6, 8, 10, ...   5. 11, 4, −2, −7, −11, ... 6. 3, −6, 12, −24, 48, ...

FOOD For Exercises 7–9, use the figure at the right.

7. Make a list of the number of cans in each level starting with the top.

8. State whether the sequence is *arithmetic*, *geometric*, or *neither*.

9. If the stack of cans had six levels, how many cans would be in the bottom level?

Practice and Applications

State whether each sequence is *arithmetic*, *geometric*, or *neither*. If it is arithmetic or geometric, state the common difference or common ratio. Write the next three terms of each sequence.

HOMEWORK HELP

For Exercises	See Examples
10–31	1–3

Extra Practice
See pages 642, 658.

10. 20, 24, 28, 32, 36, ...   11. 1, 10, 100, 1,000, 10,000, ...

12. 486, 162, 54, 18, 6, ...   13. 88, 85, 82, 79, 76, ...

14. 1, 1, 2, 6, 24, ...   15. 1, 2, 5, 10, 17, ...

16. −6, −4, −2, 0, 2, ...   17. 5, −15, 45, −135, 405, ...

18. 189, 63, 21, 7, $2\frac{1}{3}$, ...   19. 4, $6\frac{1}{2}$, 9, $11\frac{1}{2}$, 14, ...   20. 1, $\frac{1}{2}$, $\frac{1}{6}$, $\frac{1}{24}$, $\frac{1}{120}$, ...

21. 16, −4, 1, $-\frac{1}{4}$, $\frac{1}{16}$, ...   22. −1, 1, −1, 1, −1, ...   23. $4\frac{1}{2}$, $4\frac{1}{6}$, $3\frac{5}{6}$, $3\frac{1}{2}$, $3\frac{1}{6}$, ...

24. What are the first four terms of an arithmetic sequence with a common difference of $3\frac{1}{3}$ if the first term is 4?

25. What are the first four terms of a geometric sequence with a common ratio of −6 if the first term is 100?

GEOMETRY For Exercises 26 and 27, use the sequence of squares.

1 unit   2 units   3 units   4 units

26. Write a sequence for the areas of the squares. Is the sequence *arithmetic, geometric,* or *neither*?

27. Write a sequence for the perimeters of the squares. Is the sequence *arithmetic, geometric,* or *neither*?

SKIING For Exercises 28–32, use the following information.
A ski resort advertises a one-day lift pass for $40 and a yearly lift pass for $400.

28. Copy and complete the table.

29. Is the sequence formed by the row for the total cost with one day passes *arithmetic, geometric,* or *neither*?

Number of Times At the Ski Resort	1	2	3	4	5
Total Cost with One Day Passes	$40	$80			
Total Cost with Yearly Pass	$400	$400			

30. Can the sequence formed by the total cost with a yearly pass be considered arithmetic? Explain.

31. Can the sequence formed by the total cost with a yearly pass be considered geometric? Explain.

32. Extend each sequence to determine how many times a person would have to go skiing to make the yearly pass a better buy.

33. **CRITICAL THINKING** What are the first six terms of an arithmetic sequence where the second term is 5 and the fourth term is 15?

Standardized Test Practice and Mixed Review

34. **MULTIPLE CHOICE** At the beginning of each week, Lina increases the time of her daily jog. If she continues the pattern shown in the table, how many minutes will she spend jogging each day during her fifth week of jogging?

  Ⓐ 32 min   Ⓑ 40 min   Ⓒ 48 min   Ⓓ 56 min

Week	Time Jogging (minutes)
1	8
2	16
3	24
4	32
5	?

35. **MULTIPLE CHOICE** Which sequence is geometric?

  Ⓕ 5, 5, 10, 30, 120, …   Ⓖ $-12, -8, -4, 0, 4, \ldots$

  Ⓗ 1, 1, 2, 3, 5, …   Ⓘ $-1{,}280, 320, -80, 20, -5, \ldots$

36. **GEOMETRY** The length of a rectangle is 6 inches. Its area is greater than 30 square inches. Write an inequality for the situation. (Lesson 10-7)

Solve each inequality. (Lesson 10-6)

37. $b + 15 > 32$ 38. $y - 24 \le 12$ 39. $9 \le 16 + t$ 40. $18 \ge a - 6$

GETTING READY FOR THE NEXT LESSON

PREREQUISITE SKILL Evaluate each expression if $x = 9$. (Lesson 1-2)

41. $2x$ 42. $x - 12$ 43. $17 + x$ 44. $3x - 5$

Sunshine State Standards
MA.A.5.3.1-2

The Fibonacci Sequence

What You'll Learn
Determine numbers that make up the Fibonacci Sequence.

Materials
• grid or dot paper
• colored pencils

INVESTIGATE *Work in groups of three.*

Leonardo, also known as Fibonacci, created story problems based on a series of numbers that became known as the *Fibonacci sequence*. In this lab, you will investigate this sequence.

STEP 1 Using grid paper or dot paper, draw a "brick" that is 2 units long and 1 unit wide. If you build a "road" of grid paper bricks, there is only one way to build a road that is 1 unit wide.

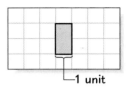

└ 1 unit

STEP 2 Using two bricks, draw all of the different roads that are 2 units wide. There are two ways to build the road.

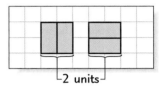

└ 2 units ┘

STEP 3 Using three bricks, draw all of the different roads that are 3 units wide. There are three ways to build the road.

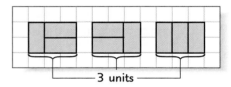

└─── 3 units ───┘

STEP 4 Draw all of the different roads that are 4 units, 5 units, and 6 units long using the brick.

STEP 5 Copy and complete the table.

Length of Road	0	1	2	3	4	5	6
Number of Ways to Build the Road	1	1	2	3			

Writing Math

Work with a partner.

1. **Explain** how each number is related to the previous numbers in the pattern.

2. **Tell** the number of ways there are to build a road that is 8 units long. Do not draw a model.

3. **MAKE A CONJECTURE** Write a rule describing how you generate numbers in the Fibonacci sequence.

4. **RESEARCH** Use the Internet or other resource to find how the Fibonacci sequence is related to nature, music, or art.

Functions

Sunshine State Standards
MA.D.1.3.1-4, MA.D.1.3.1-5, MA.D.1.3.2-1

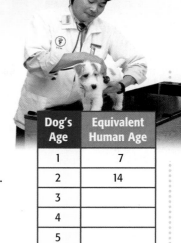

What You'll Learn
Complete function tables.

NEW Vocabulary

function
function table
independent variable
dependent variable
domain
range

MATH Symbols

$f(x)$ the function of x

WHEN am I ever going to use this?

ANIMALS Veterinarians have used the rule that one year of a dog's life is equivalent to seven years of human life.

1. Copy and complete the table at the right.

2. If a dog is 6 years old, what is its equivalent human age?

3. Explain how to find the equivalent human age of a dog that is 10 years old.

Dog's Age	Equivalent Human Age
1	7
2	14
3	
4	
5	

The equivalent human age of a dog depends on, or is a function of, its age in years. A relationship where one thing depends upon another is called a **function**. In a function, one or more operations are performed on one number to get another.

Functions are often written as equations.

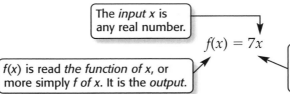

The *input x* is any real number.

$$f(x) = 7x$$

The operations performed in the function are sometimes called the *rule*.

$f(x)$ is read *the function of x*, or more simply *f of x*. It is the *output*.

To find the value of a function for a certain number, substitute the number into the function value.

EXAMPLES Find a Function Value

Find each function value.

1. $f(9)$ if $f(x) = x - 5$

$f(x) = x - 5$

$f(9) = 9 - 5$ or 4   Substitute 9 for x into the function rule.

So, $f(9) = 4$.

2. $f(-3)$ if $f(x) = 2x + 1$

$f(x) = 2x + 1$

$f(-3) = 2(-3) + 1$ Substitute -3 for x into the function rule.

$f(-3) = -6 + 1$ or -5 Simplify.

So, $f(-3) = -5$.

Your Turn Find each function value.

a. $f(2)$ if $f(x) = x - 4$ b. $f(6)$ if $f(x) = 2x - 8$

Input and Output
The variable for
the input is called
the **independent**
variable because it
can be any number.
The variable for the
output is called
the **dependent**
variable because
it *depends* on the
input value.

You can organize the input, rule, and output of a function into a
function table.

EXAMPLE Make a Function Table

3 Complete the function table
for $f(x) = x + 5$.

Substitute each value of x,
or input, into the function
rule. Then simplify to find
the output.

$f(x) = x + 5$

$f(-2) = -2 + 5$ or 3

$f(-1) = -1 + 5$ or 4

$f(0) = 0 + 5$ or 5

$f(1) = 1 + 5$ or 6

$f(2) = 2 + 5$ or 7

Input	Rule	Output
x	x + 5	f(x)
−2		
−1		
0		
1		
2		

Input	Rule	Output
x	x + 5	f(x)
−2	−2 + 5	3
−1	−1 + 5	4
0	0 + 5	5
1	1 + 5	6
2	2 + 5	7

The set of input values in a function is called the **domain**. The set
of output values is called the **range**. In Example 3, the domain is
$\{-2, -1, 0, 1, 2\}$. The range is $\{3, 4, 5, 6, 7\}$.

Sometimes functions do not use the $f(x)$ notation. Instead they use two
variables. One variable, usually x, represents the input and the other,
usually y, represents the output. The function in Example 3 can also be
written as $y = x + 5$.

EXAMPLES Functions with Two Variables

ZOOKEEPER The zoo needs 1.5 tons of specially mixed elephant
chow to feed its elephants each week.

4 Write a function using two variables to represent the amount of
elephant chow needed for w weeks.

Words   Amount of chow equals 1.5 times the number of weeks.

Function c =   1.5   · w

The function $c = 1.5w$ represents the situation.

5 How much elephant chow will the zoo need to feed its elephants
for 12 weeks?

Substitute 12 for w into the function rule.

$c = 1.5w$

$c = 1.5(12)$ or 18   The zoo needs 18 tons of elephant chow.

1. **State** the mathematical names for the input values and the output values.

2. **OPEN ENDED** If $f(x) = 2x - 4$, find a value of x that will make the function value a negative number.

3. **FIND THE ERROR** Mitchell and Tomi are finding the function value of $f(x) = 5x$ if the input is 10. Who is correct? Explain.

Mitchell
$f(x) = 5x$
$10 = 5x$
$\dfrac{10}{5} = \dfrac{5x}{5}$
$2 = x$

Tomi
$f(x) = 5x$
$f(10) = 5(10)$ or 50

GUIDED PRACTICE

Find each function value.

4. $f(4)$ if $f(x) = x - 6$

5. $f(-2)$ if $f(x) = 4x + 1$

Copy and complete each function table.

6. $f(x) = 8 - x$

x	8 − x	f(x)
−3		
−1		
2		
4		

7. $f(x) = 5x + 1$

x	5x + 1	f(x)
−2		
0		
1		
3		

8. $f(x) = 3x - 2$

x	3x − 2	y
−5		
−2		
2		
5		

Practice and Applications

Find each function value.

9. $f(7)$ if $f(x) = 5x$

10. $f(9)$ if $f(x) = x + 13$

11. $f(4)$ if $f(x) = 3x - 1$

12. $f(5)$ if $f(x) = 2x + 5$

13. $f(-6)$ if $f(x) = -3x + 1$

14. $f(-8)$ if $f(x) = 3x + 24$

15. $f\left(\dfrac{5}{6}\right)$ if $f(x) = 2x + \dfrac{1}{3}$

16. $f\left(\dfrac{5}{8}\right)$ if $f(x) = 4x - \dfrac{1}{4}$

HOMEWORK HELP

For Exercises	See Examples
9–16	1, 2
17–20	3
21–24	4, 5

Extra Practice
See pages 643, 658.

Copy and complete each function table.

17. $f(x) = 6x - 4$

x	6x − 4	f(x)
−5		
−1		
2		
7		

18. $f(x) = 5 - 2x$

x	5 − 2x	y
−2		
0		
3		
5		

19. $f(x) = 7 + 3x$

x	7 + 3x	y
−3		
−2		
1		
6		

20. Make a function table for $y = 3x + 5$ using any four values for x.

GEOMETRY For Exercises 21 and 22, use the following information.
The perimeter of a square equals 4 times the length of a side.

21. Write a function using two variables to represent the situation.

22. What is the perimeter of a square with a side 14 inches long?

PARTY PLANNING For Exercises 23 and 24, use the following information.
Sherry is having a birthday party at the Swim Center. The cost of renting the pool is $45 plus $3.50 for each person.

23. Write a function using two variables to represent the situation.

24. What is the total cost if 20 people attend the party?

25. WRITE A PROBLEM Write a real-life problem involving a function.

26. CRITICAL THINKING Write the function rule for each function table.

a.
x	f(x)
−3	−30
−1	−10
2	20
6	60

b.
x	f(x)
−5	−9
−1	−5
3	−1
7	3

c.
x	y
−2	−3
1	3
3	7
5	11

d.
x	y
−2	−5
1	1
3	5
5	9

Standardized Test Practice and Mixed Review

FCAT Practice

27. MULTIPLE CHOICE Which function matches the function table at the right?

x	y
−5	−1
−2	0.2
1	1.4
3	2.2

Ⓐ $y = 0.4x + 1$

Ⓑ $y = 4x - 0.4$

Ⓒ $y = \frac{1}{4}x + 1$

Ⓓ $y = \frac{1}{4}x - 1$

28. SHORT RESPONSE A nautical mile is a measure of distance frequently used in sea travel. One nautical mile equals about 6,076 feet. Write a function to represent the number of feet in x nautical miles.

State whether each sequence is *arithmetic*, *geometric*, or *neither*. If it is arithmetic or geometric, state the common difference or common ratio. Write the next three terms of each sequence. (Lesson 11-1)

29. 3, −6, 12, −24, 48, …

30. 74, 71, 68, 65, 62, …

31. 2, 3, 5, 8, 12, …

32. BAND The school band makes $0.50 for every candy bar they sell. They want to make at least $500 on the candy sale. Write and solve an inequality to find how many candy bars they must sell. (Lesson 10-7)

33. ALGEBRA Solve $\frac{n}{2} + 31 = 45$. (Lesson 10-2)

GETTING READY FOR THE NEXT LESSON

PREREQUISITE SKILL Graph each point on the same coordinate plane. (Page 614)

34. $A(-4, 2)$

35. $B(3, -1)$

36. $C(0, -3)$

37. $D(1, 4)$

Sunshine State Standards
MA.E.3.3.1-2

Graphing Relationships

What You'll Learn
Graph relationships.

Materials
• pencil
• paper cup
• 2 paper clips
• large rubber band
• tape
• ruler
• 10 pennies
• grid paper

INVESTIGATE *Work in groups of four.*

In this lab, you will investigate a relationship between the number of pennies in a cup and how far the cup will stretch a rubber band.

STEP 1 Using a pencil, punch a small hole in the bottom of the paper cup. Place one paper clip onto the rubber band. Push the other end of the rubber band through the hole in the cup. Attach the second paper clip to the other end of the rubber band. Place it horizontally across the bottom of the cup to keep it from coming through the hole.

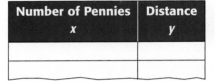

STEP 2 Copy the table at the right.

Number of Pennies x	Distance y

STEP 3 Tape the top paper clip to the edge of a desk. Measure and record the distance from the bottom of the desk to the bottom of the cup. Drop one penny into the cup. Measure and record the new distance from the bottom of the desk to the bottom of the cup.

STEP 4 Continue adding one penny at a time. Measure and record the distance after each addition.

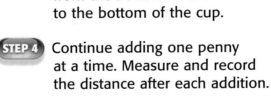

Work with a partner.

1. **Examine** the data. Do you think the number of pennies affects the distance? Explain.

2. **Graph** the ordered pairs formed by your data. Do the points resemble a straight line?

3. **Predict** the distance of the bottom of the cup from the bottom of the desk if 15 pennies are placed in the cup.

4. **Find** the ratio of each distance to the number of pennies. What do you notice about these ratios?

11-3 Graphing Linear Functions

Sunshine State Standards MA.C.3.3.2-1, MA.C.3.3.2-3, MA.D.1.3.1-1, MA.D.1.3.1-5, MA.D.1.3.2-1, MA.D.1.3.2-3, MA.D.2.3.1-4, MA.D.2.3.1-5

What You'll Learn

Graph linear functions by using function tables and plotting points.

→ NEW Vocabulary

linear function
x-intercept
y-intercept

↻ REVIEW Vocabulary

ordered pair: a pair of numbers used to locate a point on a coordinate plane
(Lesson 3-6)

WHEN am I ever going to use this?

ROLLER COASTERS The *Millennium Force* roller coaster has a maximum speed of 1.5 miles per minute. If x represents the minutes traveled at this maximum speed, the function rule for the distance traveled is $y = 1.5x$.

1. Copy and complete the following function table.

Input	Rule	Output	(Input, Output)
x	$1.5x$	y	(x, y)
1	1.5(1)	1.5	(1, 1.5)
2	1.5(2)		
3			
4			

2. Graph the ordered pairs on a coordinate plane.

3. What do you notice about the points on your graph?

Ordered pairs of the form (input, output), or (x, y), can represent a function. These ordered pairs can then be graphed on a coordinate plane as part of the graph of the function.

EXAMPLE Graph a Function

1 Graph $y = x + 2$.

Step 1 Choose some values for x. Make a function table. Include a column of ordered pairs of the form (x, y).

Step 2 Graph each ordered pair. Draw a line that passes through each point. Note that the ordered pair for any point on this line is a solution of $y = x + 2$. The line is the complete graph of the function.

x	$x + 2$	y	(x, y)
0	0 + 2	2	(0, 2)
1	1 + 2	3	(1, 3)
2	2 + 2	4	(2, 4)
3	3 + 2	5	(3, 5)

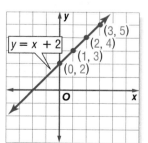

Check It appears from the graph that $(-2, 0)$ is also a solution. Check this by substitution.

$y = x + 2$ Write the function.

$0 \stackrel{?}{=} -2 + 2$ Replace x with -2 and y with 0.

$0 = 0$ ✔   Simplify.

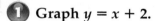

A function in which the graph of the solutions forms a line is called a **linear function**. Therefore, $y = x + 2$ is a *linear equation*.

Concept Summary — **Representing Functions**

Words	The value of y is one less than the corresponding value of x.
Equation	$y = x - 1$
Ordered Pairs	$(0, -1), (1, 0), (2, 1), (3, 2)$

Table

x	y
0	-1
1	0
2	1
3	2

Graph

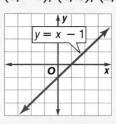

The value of x where the graph crosses the x-axis is called the **x-intercept**. The value of y where the graph crosses the y-axis is called the **y-intercept**.

FCAT Practice

Standardized Test Practice

EXAMPLE Use x- and y-intercepts

2 MULTIPLE-CHOICE TEST ITEM Which graph represents $y = 3x - 6$?

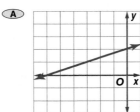

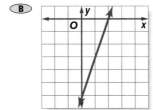

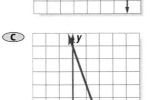

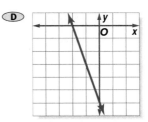

Test-Taking Tip
The Princeton Review

Use Different Methods
Always work each problem using the method that is easiest for you. You could solve the problem at the right in several ways.

- You could test the coordinates of several points on each graph.
- You could graph the function and see which graph matched your graph.
- You could determine the intercepts and see which graph had those intercepts.

Which method is easiest for you?

Read the Test Item You need to decide which of the four graphs represents $y = 3x - 6$.

Solve the Test Item

The graph will cross the x-axis when $y = 0$.

$0 = 3x - 6$	Replace y with 0.
$0 + 6 = 3x - 6 + 6$	Add 6.
$6 = 3x$	Simplify.
$\dfrac{6}{3} = \dfrac{3x}{3}$	Divide by 3.
$2 = x$	Simplify.

The graph will cross the y-axis when $x = 0$.

$y = 3(0) - 6$	Replace x with 0.
$y = 0 - 6$	Simplify.
$y = -6$	Simplify.

The x-intercept is 2, and the y-intercept is -6. Graph B is the only graph with both of these intercepts. The answer is B.

msmath3.net/extra_examples/fcat

Writing Math
Exercises 1 & 3

1. **Explain** how a function table can be used to graph a function.

2. **OPEN ENDED** Draw a graph of a linear function. Name the coordinates of three points on the graph.

3. **Which One Doesn't Belong?** Identify the ordered pair that is not a solution of $y = 2x - 3$. Explain your reasoning.

| (1, –1) | (2, 1) | (O, 3) | (–2, –7) |

GUIDED PRACTICE

4. Copy and complete the function table at the right. Then graph $y = x + 5$.

x	x + 5 ·	y	(x, y)
−3			
−1			
0			
2			

Graph each function.

5. $y = 3x$

6. $y = 3x + 1$

7. $y = \dfrac{x}{2} - 1$

FOOD For Exercises 8 and 9, use the following information.
The function $y = 40x$ describes the relationship between the number of gallons of sap y used to make x gallons of maple syrup.

8. Graph the function.

9. Use your graph to determine the amount of sap needed to make $2\frac{1}{2}$ gallons of syrup.

Practice and Applications

Copy and complete each function table. Then graph the function.

10. $y = x - 4$

11. $y = 2x$

x	x − 4	y	(x, y)
−1			
1			
3			
5			

x	2x	y	(x, y)
−2			
0			
1			
2			

HOMEWORK HELP

For Exercises	See Examples
10–23	1
27	2

Extra Practice
See pages 643, 658.

Graph each function.

12. $y = 4x$

13. $y = -3x$

14. $y = x - 3$

15. $y = x + 1$

16. $y = 3x - 7$

17. $y = 2x + 3$

18. $y = \dfrac{x}{3} + 1$

19. $y = \dfrac{x}{2} - 3$

20. Draw the graph of $y = 5 - x$.

21. Graph the function $y = -\dfrac{1}{2}x + 5$.

22. **GEOMETRY** The equation $s = 180(n - 2)$ relates the sum of the measures of angles s formed by the sides of a polygon to the number of sides n. Find four ordered pairs (n, s) that are solutions of the equation.

MOUNTAIN CLIMBING For Exercises 23 and 24, use the following information.
If the temperature is 80°F at sea level, the function $t = 80 - 3.6h$ describes the temperature t at a height of h thousand feet above sea level.

23. Graph the temperature function.

24. The top of Mount Everest is about 29 thousand feet above sea level. What is the temperature at its peak on a day that is 80°F at sea level?

25. CRITICAL THINKING Name the coordinates of four points that satisfy each function. Then give the function rule.

a.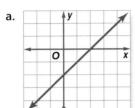

b.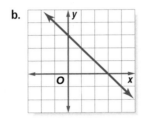

26. CRITICAL THINKING The vertices of a triangle are at $(-1, -1)$, $(1, -2)$, and $(5, 1)$. The triangle is translated 1 unit left and 2 units up and then reflected across the graph of $y = x - 1$. What are the coordinates of the image? (*Hint:* Use a ruler.)

Standardized Test Practice and Mixed Review   A B C D   **FCAT Practice**

27. MULTIPLE CHOICE Which function is graphed at the right?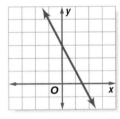

Ⓐ $y = -2x + 3$ Ⓑ $y = -2x - 3$

Ⓒ $y = 2x + 3$ Ⓓ $y = 2x - 3$

28. SHORT RESPONSE An African elephant eats 500 pounds of vegetation a day. Write a function for the amount of vegetation y it eats in x days. Graph the function.

Find each function value. (Lesson 11-2)

29. $f(6)$ if $f(x) = 7x - 3$ **30.** $f(-5)$ if $f(x) = 3x + 15$ **31.** $f(3)$ if $f(x) = 2x - 7$

32. SCIENCE Each time a certain ball hits the ground, it bounces up $\frac{2}{3}$ of its previous height. If the ball is dropped from 27 inches off the ground, write a sequence showing the height of the ball after each of the first three bounces. (Lesson 11-1)

GETTING READY FOR THE NEXT LESSON

PREREQUISITE SKILL Find the slope of each line. (Lesson 4-3)

33.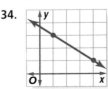

34.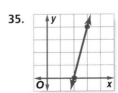

35.

The Slope Formula

Sunshine State Standards
MA.B.1.3.2-2, MA.C.3.3.2-3, MA.D.1.3.2-4

What You'll Learn

Find the slope of a line using the slope formula.

→ NEW Vocabulary

slope formula

↺ REVIEW Vocabulary

slope: the ratio of the rise, or vertical change, to the run, or horizontal change (Lesson 4-3)

MATH Symbols

x_2 x sub 2

HANDS-ON Mini Lab

Materials
• grid paper

Work with a partner.

On a coordinate plane, graph $A(2, 1)$ and $B(4, 4)$. Draw the line through points A and B as shown.

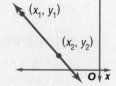

1. Find the slope of the line by counting units of vertical and horizontal change.

2. Subtract the y-coordinate of A from the y-coordinate of B. Call this value t.

3. Subtract the x-coordinate of A from the x-coordinate of B. Call this value s.

4. Write the ratio $\frac{t}{s}$. Compare the slope of the line with $\frac{t}{s}$.

You can find the slope of a line by using the coordinates of any two points on the line. One point can be represented by (x_1, y_1) and the other by (x_2, y_2). The small numbers slightly below x and y are called *subscripts*.

Key Concept   Slope Formula

Words The slope m of a line passing through points (x_1, y_1) and (x_2, y_2) is the ratio of the difference in the y-coordinates to the corresponding difference in the x-coordinates.

Model

(x_1, y_1)

(x_2, y_2)

Symbols $m = \dfrac{y_2 - y_1}{x_2 - x_1}$, where $x_2 \neq x_1$

EXAMPLE Positive Slope

① Find the slope of the line that passes through $C(-1, -4)$ and $D(2, 2)$.

$m = \dfrac{y_2 - y_1}{x_2 - x_1}$ Definition of slope

$m = \dfrac{2 - (-4)}{2 - (-1)}$ $(x_1, y_1) = (-1, -4)$, $(x_2, y_2) = (2, 2)$

$m = \dfrac{6}{3}$ or 2 Simplify.

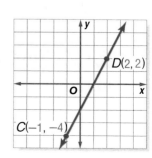

Check When going from left to right, the graph of the line slants upward. This is consistent with a positive slope.

EXAMPLE Negative Slope

2 Find the slope of the line that passes through $R(1, 2)$ and $S(-4, 3)$.

$$m = \frac{y_2 - y_1}{x_2 - x_1}$$ Definition of slope

$$m = \frac{3 - 2}{-4 - 1}$$ $(x_1, y_1) = (1, 2),$
              $(x_2, y_2) = (-4, 3)$

$$m = \frac{1}{-5} \text{ or } -\frac{1}{5}$$ Simplify.

Check   When going from left to right, the graph of the line slants downward. This is consistent with a negative slope.

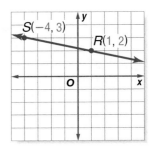

EXAMPLE Zero Slope

3 Find the slope of the line that passes through $V(-5, -1)$ and $W(2, -1)$.

$$m = \frac{y_2 - y_1}{x_2 - x_1}$$ Definition of slope

$$m = \frac{-1 - (-1)}{2 - (-5)}$$ $(x_1, y_1) = (-5, -1),$
              $(x_2, y_2) = (2, -1)$

$$m = \frac{0}{7} \text{ or } 0$$ Simplify.

The slope is 0. The slope of any horizontal line is 0.

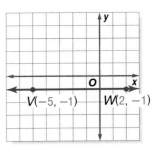

EXAMPLE Undefined Slope

4 Find the slope of the line that passes through $X(4, 3)$ and $Y(4, -1)$

$$m = \frac{y_2 - y_1}{x_2 - x_1}$$ Definition of slope

$$m = \frac{-1 - 3}{4 - 4}$$ $(x_1, y_1) = (4, 3),$
              $(x_2, y_2) = (4, -1)$

$$m = \frac{-4}{0}$$ Simplify.

Division by 0 is not defined. So, the slope is undefined. The slope of any vertical line is undefined.

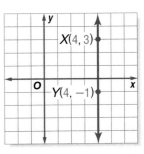

Your Turn Find the slope of the line that passes through each pair of points.

a. $M(2, 2), N(5, 3)$ b. $A(-2, 1), B(0, -3)$

c. $C(-5, 6), D(-5, 0)$ d. $E(-1, 1), F(3, 1)$

Writing
Math
Exercises 1 & 3

1. **Explain** why the slope formula, which states $m = \dfrac{y_2 - y_1}{x_2 - x_1}$, says that x_2 cannot equal x_1.

2. **OPEN ENDED** Write the coordinates of two points. Show that you can define either point as (x_1, y_1) and the slope of the line containing the points will be the same.

3. **FIND THE ERROR** Martin and Dylan are finding the slope of the line that passes through $X(0, 2)$ and $Y(2, 3)$. Who is correct? Explain.

Martin
$$m = \frac{3 - 2}{0 - 2}$$
$$m = \frac{1}{-2} \text{ or } -\frac{1}{2}$$

Dylan
$$m = \frac{3 - 2}{2 - 0}$$
$$m = \frac{1}{2}$$

GUIDED PRACTICE

Find the slope of the line that passes through each pair of points.

4. $A(-3, -2)$, $B(5, 4)$ 5. $C(-4, 2)$, $D(1, 2)$ 6. $E(-6, 5)$, $F(3, -3)$

For Exercises 7–9, use the graphic at the right. Note that the years are on the horizontal axis and the second homeownership is on the vertical axis.

7. Find the slope of the line representing the change from 1990 to 2000.

8. Find the slope of the line representing the change from 2000 to 2010.

9. Which part of the graph shows a greater rate of change? Explain.

USA TODAY Snapshots®

No place like home — both of them

Thanks to the rise in affluent, childless households and the aging of baby boomers, second-home ownership is projected to almost double from 1990 to 2010. Second-home ownership by decade:

9.8

(millions)   6.4

5.5

1990   2000   2010

Source: Census Bureau for 1990-2000; Peter Francese for 2010 projection

By In-Sung Yoo and Suzy Parker, USA TODAY

Practice and Applications

Find the slope of the line that passes through each pair of points.

10. $A(0, 1)$, $B(2, 7)$ 11. $C(2, 5)$, $D(3, 1)$ 12. $E(1, 2)$, $F(4, 7)$

13. $G(-6, -1)$, $H(4, 1)$ 14. $J(-9, 3)$, $K(2, 1)$ 15. $M(-2, 3)$, $N(7, -4)$

16. $P(4, -4)$, $Q(8, -4)$ 17. $R(-1, 5)$, $S(-1, -2)$ 18. $T(3, -2)$, $U(3, 2)$

19. $V(-6, 5)$, $W(3, -3)$ 20. $X(21, 5)$, $Y(17, 0)$ 21. $Z(24, 12)$, $A(34, 2)$

HOMEWORK HELP

For Exercises	See Examples
10–21	1–4

Extra Practice
See pages 643, 658.

TRAVEL For Exercises 22 and 23, use the following information.
After 2 hours, Kendra has traveled 110 miles. After 3 hours, she has traveled 165 miles. After 5 hours, she has traveled 275 miles.

22. Graph the information with the hours on the horizontal axis and miles traveled on the vertical axis. Draw a line through the points.

23. What is the slope of the graph? What does it represent?

GEOMETRY For Exercises 24 and 25, use the following information to show that each quadrilateral graphed is a parallelogram.
Two lines that are parallel have the same slope.

24.

25.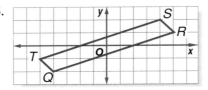

26. **CRITICAL THINKING** Without graphing, determine whether $A(5, 1)$, $B(1, 0)$, and $C(-3, -3)$ lie on the same line. Explain.

EXTENDING THE LESSON
For Exercises 27–29, use the graphs at the right.
The two lines in each graph are perpendicular.

Graph A 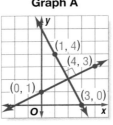   Graph B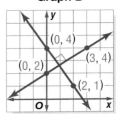

27. Find the slopes of the lines in graph A.

28. Find the slopes of the lines in graph B.

29. **Make a conjecture** about the slopes of perpendicular lines.

Standardized Test Practice and Mixed Review

FCAT Practice

30. **MULTIPLE CHOICE** Which graph has a slope of -2?

  A   B   C 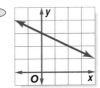 D

31. **SHORT RESPONSE** Draw a graph of a line with an undefined slope.

Graph each function. (Lesson 11-3)

32. $y = 5x$ 33. $y = x - 2$ 34. $y = 2x - 1$ 35. $y = 3x + 2$

36. **WEATHER** The function used to change a Celsius temperature (C) to a Fahrenheit temperature (F) is $F = \frac{9}{5}C + 32$. Change 25° Celsius to Fahrenheit. (Lesson 11-2)

GETTING READY FOR THE NEXT LESSON

PREREQUISITE SKILL Solve each equation. (Lesson 1-8)

37. $7 + a = 15$ 38. $23 = d + 44$ 39. $28 = n - 14$ 40. $t - 22 = -31$

Vocabulary and Concepts

1. **Explain** how to find the next three terms of the sequence 5, 8, 11, 14, 17, … . (Lesson 11-1)

2. **Explain** how to graph $y = -2x + 1$. (Lesson 11-3)

3. **Describe** the slopes of a horizontal line and a vertical line. (Lesson 11-4)

Skills and Applications

State whether each sequence is *arithmetic*, *geometric*, or *neither*. If it is arithmetic or geometric, state the common difference or common ratio. Then write the next three terms of the sequence. (Lesson 11-1)

4. 13, 17, 21, 25, 29, …   5. 64, −32, 16, −8, 4, …   6. 5, 6, 8, 11, 15, …

7. **PICNICS** Shelby is hosting a picnic. The cost to rent the shelter is $25 plus $2 per person. Write a function using two variables to represent the situation. Find the total cost if 150 people attend. (Lesson 11-2)

Graph each function. (Lesson 11-3)

8. $y = -2x$ 9. $y = x + 6$ 10. $y = 2x - 5$

Find the slope of the line that passes through each pair of points. (Lesson 11-4)

11. $A(2, 5), B(3, 1)$ 12. $C(-1, 2), D(-5, 2)$ 13. $E(5, 2), F(2, -3)$

Standardized Test Practice

FCAT Practice

A B C D

14. **MULTIPLE CHOICE** Which equation describes the function represented by the table? (Lesson 11-2)

x	f(x)
−2	−7
0	−3
2	1
4	5

   Ⓐ $f(x) = 2x - 3$

   Ⓑ $f(x) = x + 4$

   Ⓒ $f(x) = n - 3$

   Ⓓ $f(x) = 2x + 3$

15. **MULTIPLE CHOICE** Which graph has a negative slope? (Lesson 11-4)

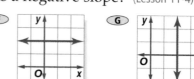

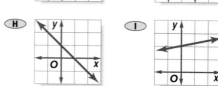

The Game Zone

A Place To Practice Your Math Skills

It's a Hit

● **GET READY!**

Players: two
Materials: large piece of paper, marker, grid paper

● **GET SET!**

• Use a marker to list the following functions on a piece of paper.

$y = -2x$	$y = x + 2$	$y = x - 2$
$y = -x + 2$	$y = 3 - x$	$y = 1 - 2x$
$y = 2x$	$y = -x - 1$	$y = x - 1$
$y = x + 1$	$y = 2x - 1$	$y = -x - 2$
$y = x - 3$	$y = 2x + 1$	$y = -x + 1$

• Each player makes two coordinate planes. Each plane should be on a 20-by-20 grid with the origin in the center.

● **GO!**

• Each player secretly picks one of the functions listed on the paper and graphs it on one of his or her coordinate planes.

• The first player names an ordered pair. The second player says *hit* if the ordered pair names a point on his or her line. If not, the player says *miss*.

• Then the second player names an ordered pair. It is either a *hit* or a *miss*. Players should use their second coordinate plane to keep track of their hits and misses. Players continue to take turns guessing.

• **Who Wins?** A player who correctly names the equation of the other player is the winner. However, if a player incorrectly names the equation, the other player is the winner.

Graphing Calculator Investigation

11-5a

A Preview of Lesson 11-5

Sunshine State Standards
MA.C.3.3.2-3, MA.D.1.3.1-1

What You'll Learn
Use a graphing calculator to graph families of lines.

Families of Linear Graphs

Families of graphs are graphs that are related in some manner. In this investigation, you will study families of linear graphs.

 ACTIVITY

Graph $y = -2x + 4$, $y = -2x + 1$, and $y = -2x - 3$.

STEP 1 **Clear any existing equations from the Y= list.**

Keystrokes: Y= CLEAR

STEP 2 **Enter each equation.**

Keystrokes: (–) 2 X,T,θ,*n* + 4 ENTER
(–) 2 X,T,θ,*n* + 1 ENTER
(–) 2 X,T,θ,*n* – 3 ENTER

STEP 3 **Graph the equations in the standard viewing window.**

Keystrokes: ZOOM 6

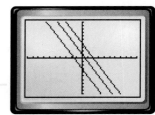

 EXERCISES

1. **Compare** the three equations.

2. **Describe** the graphs of the three equations.

3. **MAKE A CONJECTURE** Consider equations of the form $y = ax + b$, where the value of a is the same but the value of b varies. What do you think is true about the graphs of the equations?

4. **Use a graphing calculator** to graph $y = 2x + 3$, $y = -x + 3$, and $y = -3x + 3$.

5. **Compare** the three equations you graphed in Exercise 4.

6. **Describe** the graphs of the three equations you graphed in Exercise 4.

7. **MAKE A CONJECTURE** Consider equations of the form $y = ax + b$, where the value of a changes but the value of b remains the same. What do you think is true about the graphs of the equations?

8. **Write** equations of three lines whose graphs are a family of graphs. Describe the common characteristic of the graph.

 msmath3.net/other_calculator_keystrokes

Slope-Intercept Form

Sunshine State Standards
MA.B.1.3.2-2, MA.D.1.3.1-1, MA.D.1.3.2-4, MA.D.2.3.1-4, MA.D.2.3.1-5

What You'll Learn
Graph linear equations using the slope and y-intercept.

NEW Vocabulary
slope-intercept form

LINK To Reading
Everyday Meaning of Intercept: to intersect or cross

HANDS-ON Mini Lab

Materials
- grid paper

Work with a partner.

Graph each equation listed in the table at the right.

1. Use the graphs to find the slope and y-intercept of each line. Copy and complete the table.

Equation	Slope	y-intercept
$y = 3x + 2$		
$y = \frac{1}{4}x + (-1)$		
$y = -2x + 3$		

2. Compare each equation with the value of its slope. What do you notice?

3. Compare each equation with its y-intercept. What do you notice?

All of the equations in the table above are written in the form $y = mx + b$. This is called the **slope-intercept form**. When an equation is written in this form, m is the slope, and b is the y-intercept.

$$y = mx + b$$

slope ⤴   ⤴ y-intercept

EXAMPLES Find Slopes and y-intercepts of Graphs

State the slope and the y-intercept of the graph of each equation.

1 $y = \frac{2}{3}x - 4$

$y = \frac{2}{3}x + (-4)$ Write the equation in the form $y = mx + b$.

$y = mx + \quad b \qquad m = \frac{2}{3}, b = -4$

The slope of the graph is $\frac{2}{3}$, and the y-intercept is -4.

2 $x + y = 6$

$x + y = 6$ Write the original equation.

$\underline{-x \qquad\qquad -x}$ Subtract x from each side.

$y = 6 - x$ Simplify.

$y = -1x + 6$ Write the equation in the form $y = mx + b$.
                  Recall that $-x$ means $-1x$.

$y = \quad mx + b \quad m = -1, b = 6$

The slope of the graph is -1, and the y-intercept is 6.

You can use the slope-intercept form of an equation to graph the equation.

EXAMPLE **Graph an Equation**

3 Graph $y = -\frac{3}{2}x - 1$ using the slope and y-intercept.

Step 1 Find the slope and y-intercept.

$$y = -\frac{3}{2}x - 1$$

slope $= -\frac{3}{2}$ y-intercept $= -1$

Step 2 Graph the y-intercept $(0, -1)$.

Step 3 Write the slope $-\frac{3}{2}$ as $\frac{-3}{2}$. Use it to locate a second point on the line.

$$m = \frac{-3}{2} \quad \begin{array}{l} \leftarrow \text{change in } y\text{: down 3 units} \\ \leftarrow \text{change in } x\text{: right 2 units} \end{array}$$

Step 4 Draw a line through the two points.

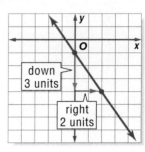

down 3 units

right 2 units

Your Turn Graph each equation using the slope and y-intercept.

a. $y = x + 3$ **b.** $y = \frac{1}{2}x - 1$ **c.** $y = -\frac{4}{3}x + 2$

EXAMPLES **Graph an Equation to Solve Problems**

ADVERTISING Student Council wants to buy posters advertising the school's carnival. The Design Shoppe charges \$15 to prepare the design and \$3 for each poster printed. The total cost y can be represented by the equation $y = 3x + 15$, where x represents the number of posters.

4 Graph the equation.

First find the slope and the y-intercept.

$$y = 3x + 15$$

slope y-intercept

Plot the point $(0, 15)$. Then locate another point up 3 and right 1. Draw the line.

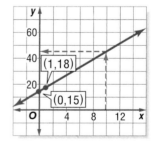

(1,18)

(0,15)

5 Use the graph to find the cost for 10 posters.

Locate 10 on the x-axis. Find the y-coordinate on the graph where the x-coordinate is 10. The total cost is \$45.

6 Describe what the slope and y-intercept represent.

The slope 3 represents the cost per poster, which is the rate of change. The y-intercept 15 is the one-time charge for preparing the design.

1. **Explain** how to graph a line with a slope of $-\frac{5}{4}$ and a y-intercept of -3.

2. **OPEN ENDED** Draw the graph of a line that has a y-intercept but no x-intercept. What is the slope of the line?

3. **Which One Doesn't Belong?** Identify the equation that has a graph with a different slope. Explain your reasoning.

$y = \frac{2}{3}x - 4$	$y = \frac{3}{2}x + 1$	$y = \frac{2}{3}x + 7$	$y = \frac{2}{3}x$

GUIDED PRACTICE

State the slope and the y-intercept for the graph of each equation.

4. $y = x + 2$
5. $y = -\frac{1}{6}x - \frac{1}{2}$
6. $2x + y = 3$

Graph each equation using the slope and the y-intercept.

7. $y = \frac{1}{3}x - 2$
8. $y = -\frac{5}{2}x + 1$
9. $y = -2x + 5$

MONEY MATTERS For Exercises 10–12, use the following information.
Lydia borrowed \$90 from her mother and plans to pay her mother \$10 per week. The equation for the amount of money y Lydia owes her mother is $y = 90 - 10x$, where x is the number of weeks after the loan.

10. Graph the equation.

11. What does the slope of the graph represent?

12. What does the x-intercept of the graph represent?

Practice and Applications

State the slope and the y-intercept for the graph of each equation.

13. $y = 3x + 4$
14. $y = -5x + 2$
15. $y = \frac{1}{2}x - 6$
16. $y = -\frac{3}{7}x - \frac{1}{7}$
17. $y - 2x = 8$
18. $3x + y = -4$

19. Graph a line with a slope of $\frac{1}{2}$ and a y-intercept of -3.

20. Graph a line with a slope of $-\frac{2}{3}$ and a y-intercept of 0.

21. Write an equation in slope-intercept form of the line with a slope of -2 and a y-intercept of 6.

22. Write an equation in slope-intercept form of the line with slope of 4 and a y-intercept of -10.

Graph each equation using the slope and the y-intercept.

23. $y = \frac{1}{3}x - 5$
24. $y = -x + \frac{3}{2}$
25. $y = -\frac{4}{3}x + 1$
26. $y = \frac{3}{2}x - 4$
27. $y = -2x - 3.5$
28. $y = 3x + 1.5$
29. $y - 3x = 5$
30. $5x + y = -2$

HOMEWORK HELP

For Exercises	See Examples
13–18, 21–22	1, 2
19–20, 23–30	3
31–36	4–6

Extra Practice
See pages 644, 658.

GEOMETRY For Exercises 31–33, use the information at the right.

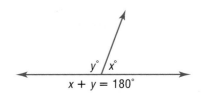

31. Write the equation in slope-intercept form.

32. Graph the equation.

33. Use the graph to find the value of y if $x = 70$.

SPACE TRAVEL For Exercises 34–36, use the following information.
From 4,074 meters above Earth, the space shuttle Orbiter glides to the runway. Let $y = 4{,}074 - 47x$ represent the altitude of the Orbiter after x seconds.

34. Graph the equation.

35. What does the slope of the graph represent?

36. What does the x-intercept of the graph represent?

37. **WRITE A PROBLEM** Write a real-life problem that involves a linear equation in slope-intercept form. Graph the equation. Explain the meaning of the slope and y-intercept.

38. Is it *sometimes*, *always*, or *never* possible to draw more than one line given a slope and a y-intercept? Explain.

39. **CRITICAL THINKING** Suppose the graph of a line is vertical. What is the slope and y-intercept of the line?

Standardized Test Practice and Mixed Review

FCAT Practice

40. **MULTIPLE CHOICE** What is the equation of the graph at the right?

   (A) $y = \frac{1}{2}x - 3$ (B) $y = -\frac{1}{2}x - 3$

   (C) $y = 2x - 3$ (D) $y = -2x - 3$

41. **MULTIPLE CHOICE** A taxi fare y can be determined by the equation $y = 3x + 5$, where x is the number of miles traveled. What does the slope of the graph of this equation represent?

   (F) the distance traveled   (G) the cost per mile

   (H) the initial fare   (I) none of the above

Find the slope of the line that passes through each pair of points. (Lesson 11-4)

42. $M(4, 3), N(-2, 1)$ 43. $S(-5, 4), T(-7, 1)$ 44. $X(-9, 5), Y(-2, 5)$

45. **MEASUREMENT** The function $y = 0.39x$ approximates the number of inches y in x centimeters. Make a function table. Then graph the function. (Lesson 11-3)

GETTING READY FOR THE NEXT LESSON

PREREQUISITE SKILL Graph each point on the same coordinate plane. (Page 614)

46. $A(5, 2)$ 47. $B(1.5, 2.5)$ 48. $C(2.3, 1.8)$ 49. $D(7.5, 3.2)$

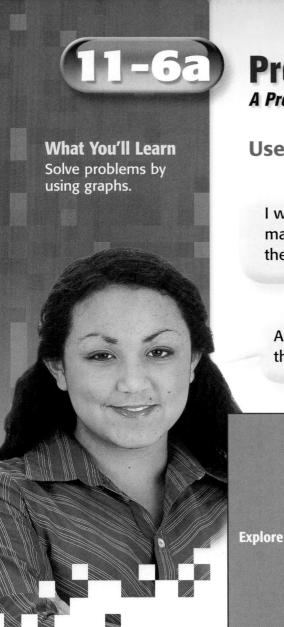

What You'll Learn
Solve problems by using graphs.

Use a Graph

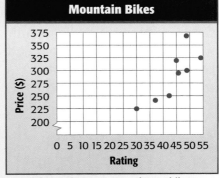

I want to buy a mountain bike. I made a graph with the ratings and the prices of 8 different bikes.

Are the highest rated bikes the most expensive bikes?

Explore	We have a graph. We want to know whether the highest rated bikes are the most expensive.

Mountain Bikes

Price ($): 375, 350, 325, 300, 275, 250, 225, 200

Rating: 0 5 10 15 20 25 30 35 40 45 50 55

Higher ratings represent better bikes.

Plan	Let's study the graph.
Solve	The graph shows that the highest rated bike is *not* the most expensive bike. Also the prices of the two bikes with the second highest rating vary considerably.
Examine	Look at the graph. The dot farthest to the right is *not* the highest on the graph.

Analyze the Strategy

1. **Explain** why the bike represented by (48, 300) might be the best bike to buy.

2. **Find** a graph in a newspaper, magazine, or the Internet. Write a sentence explaining the information contained in the graph.

Solve. Use a graph.

3. **TECHNOLOGY** Teenagers were asked which they spent more time using their computer, their video game system, or both equally. The graph shows the results of the survey. How many teenagers were surveyed?

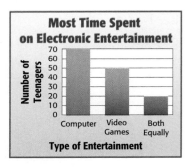

4. **BIRDS** A zoologist studied extinction times in years of birds on an island. Make a graph of the data. Does the bird with the greatest average number of nests have the greatest extinction time?

Bird	Average Number of Nests	Extinction Time (yr)
Cuckoo	1.4	2.5
Magpie	4.5	10.0
Swallow	3.8	2.6
Robin	3.3	4.0
Stonechat	3.6	2.4
Blackbird	4.7	3.3
Tree-Sparrow	2.2	1.9

Solve. Use any strategy.

5. **MULTI STEP** Canton's big brother has a full scholarship for tuition, books, and room and board for four years of college. The total scholarship is $87,500. Room and board cost $9,500 per year. His books cost about $750 per year. What is the cost of his yearly tuition?

EDUCATION For Exercises 6 and 7, use the table below.

Students per Computer in U.S. Public Schools			
Year	Students	Year	Students
1991	20	1996	10
1992	18	1997	7.8
1993	16	1998	6.1
1994	14	1999	5.7
1995	10.5	2000	5.4

Source: National Center for Education Statistics

6. Make a graph of the data.

7. Describe how the number of students per computer changed from 1991 to 2000.

8. **MONEY MATTERS** Francisco spent twice as much on athletic shoes as he did on a new pair of jeans. The total bill came to $120. What was the cost of his new jeans?

9. **STANDARDIZED TEST PRACTICE** FCAT Practice

The blue line shows the cost of producing T-shirts. The green line shows the amount of money received from the sales of the T-shirts. How many shirts must be sold to make a profit?

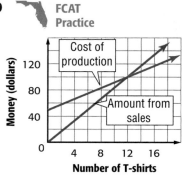

Ⓐ less than 12 T-shirts

Ⓑ exactly 12 T-shirts

Ⓒ more than 12 T-shirts

Ⓓ cannot be determined from the graph

You will use graphs in the next lesson.

Scatter Plots

Sunshine State Standards
MA.E.1.3.1-1, MA.E.1.3.1-2

What You'll Learn

Construct and interpret scatter plots.

NEW Vocabulary

scatter plot
best-fit line

HANDS-ON Mini Lab

Materials
• tape measure
• grid paper

Work with a partner.

Measure your partner's height in inches. Then ask your partner to stand with his or her arms extended parallel to the floor. Measure the distance from the end of the longest finger on one hand to the longest finger on the other hand. Write these measures as the ordered pair (height, arm span) on the chalkboard.

1. Graph each of the ordered pairs listed on the chalkboard.

2. Examine the graph. Do you think there is a relationship between height and arm span? Explain.

The graph you made in the Mini Lab is called a scatter plot. A **scatter plot** is a graph that shows the relationship between two sets of data. In this type of graph, two sets of data are graphed as ordered pairs on a coordinate plane. Scatter plots often show a pattern, trend, or relationship between the variables.

Concept Summary   **Types of Relationships**

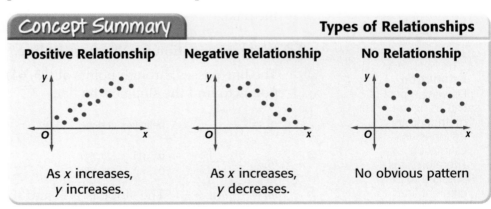

Positive Relationship	Negative Relationship	No Relationship
As *x* increases, *y* increases.	As *x* increases, *y* decreases.	No obvious pattern

EXAMPLE   **Identify a Relationship**

1. Determine whether a scatter plot of the data for the hours traveled in a car and the distance traveled might show a *positive*, *negative*, or *no* relationship.

As the number of hours you travel increases, the distance traveled increases. Therefore, the scatter plot shows a positive relationship.

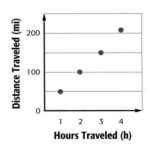

EXAMPLE Identify a Relationship

2 Determine whether a scatter plot of the data for the month of birth and birth weight show a *positive*, *negative*, or *no* relationship.

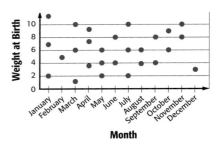

Birth weight does not depend on the month of birth. Therefore, the scatter plot shows no relationship.

If a scatter plot shows a positive relationship or a negative relationship, a best-fit line can be drawn to represent the data. A **best-fit line** is a line that is very close to most of the data points.

EXAMPLES Draw a Best-Fit Line

LAKES The water temperatures at various depths in a lake are given.

Water Depth (ft)	0	10	20	25	30	35	40	50
Temperature (°F)	74	72	71	64	61	58	53	53

3 Make a scatter plot using the data. Then draw a line that seems to best represent the data.

Graph each of the data points. Draw a line that best fits the data.

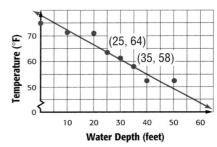

4 Write an equation for this best-fit line.

The line passes through points at (25, 64) and (35, 58). Use these points to find the slope of the line.

$m = \dfrac{y_2 - y_1}{x_2 - x_1}$ Definition of slope

$m = \dfrac{58 - 64}{35 - 25}$ $(x_1, y_1) = (25, 64)$, $(x_2, y_2) = (35, 58)$

$m = \dfrac{-6}{10}$ or $-\dfrac{3}{5}$ The slope is $-\dfrac{3}{5}$, and the y-intercept is 79.

Use the slope and y-intercept to write the equation.

$y = \quad mx \ + \ b$ Slope-intercept form

$y = -\dfrac{3}{5}x + 79$ The equation for the best-fit line is $y = -\dfrac{3}{5}x + 79$.

5 Use the equation to predict the temperature at a depth of 55 feet.

$y = -\dfrac{3}{5}x + 79$ Equation for the best-fit line

$y = -\dfrac{3}{5}(55) + 79$ or 46   The temperature will be about 46°F.

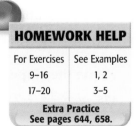

Writing
Math
Exercise 1

1. **Describe** how you can use a scatter plot to display two sets of related data.

2. **OPEN ENDED** Give an example of data that would show a negative relationship on a scatter plot.

3. **NUMBER SENSE** Suppose a scatter plot shows that as the values of *x* decrease, the values of *y* decrease. Does the scatter plot show a *positive*, *negative*, or *no* relationship?

GUIDED PRACTICE

Determine whether a scatter plot of the data for the following might show a *positive*, *negative*, or *no* relationship.

4. hours worked and earnings   5. miles per gallon and weight of car

EDUCATION For Exercises 6–8, use the following table.

Enrollment in U.S. Public and Private Schools (millions)					
Year	Students	Year	Students	Year	Students
1900	15.5	1940	25.4	1980	41.7
1910	17.8	1950	25.1	1990	40.5
1920	21.6	1960	35.2	2000	46.9
1930	25.7	1970	45.6		

6. Draw a scatter plot of the data and draw a best-fit line.

7. Does the scatter plot show a *positive*, *negative*, or *no* relationship?

8. Use your graph to estimate the enrollment in public and private schools in 2010.

Data Update What is the current number of students in school? Visit msmath3.net/data_update to learn more.

Practice and Applications

Determine whether a scatter plot of the data for the following might show a *positive*, *negative*, or *no* relationship.

9. length of a side of a square and perimeter of the square

10. day of the week and amount of rain

11. grade in school and number of pets

12. length of time for a shower and amount of water used

13. outside temperature and amount of heating bill

14. age and expected number of years a person has yet to live

15. playing time and points scored in a basketball game

16. pages in a book and copies sold

HOMEWORK HELP

For Exercises	See Examples
9–16	1, 2
17–20	3–5

Extra Practice
See pages 644, 658.

FOOD For Exercises 17–20, use the table.

17. Draw a scatter plot of the data. Then draw a best-fit line.

18. Does the scatter plot show a *positive*, *negative*, or *no* relationship?

19. Write an equation for the best-fit line.

20. Use your equation to estimate the number of fat grams in a muffin with 350 Calories.

21. **RESEARCH** Use the Internet or other resource to find the number of goals and assists for the players on one of the National Hockey teams for the past season. Make a scatter plot of the data.

Nutritional Information of Commercial Muffins		
Muffin (brand)	Fat (grams)	Calories
A	2	250
B	3	300
C	4	260
D	9	220
E	14	410
F	15	390
G	10	300
H	18	430
I	23	480
J	20	490

22. **CRITICAL THINKING** A scatter plot of skateboard sales and swimsuit sales for each month of the year shows a positive relationship.

   a. Why might this be true?

   b. Does this mean that one factor caused the other? Explain.

Standardized Test Practice and Mixed Review
FCAT Practice

23. **MULTIPLE CHOICE** Find the situation that matches the scatter plot at the right.

   Ⓐ adult height and year of birth

   Ⓑ number of trees in an orchard and the number of apples produced

   Ⓒ number of words written and length of pencil

   Ⓓ length of campfire and amount of firewood remaining

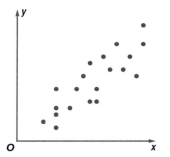

24. **MULTIPLE CHOICE** What type of graph is most appropriate for displaying the change in house prices over several years?

   Ⓕ scatter plot   Ⓖ circle graph

   Ⓗ bar graph   Ⓘ line plot

State the slope and the *y*-intercept for the graph of each equation. (Lesson 11-5)

25. $y = \frac{4}{5}x + 7$ 26. $y = -\frac{1}{6}x - 4$ 27. $4x + y = 2$

28. **GEOMETRY** The vertices of a triangle are located at $(-2, 1)$, $(1, 7)$, and $(5, 1)$. Find the slope of each side of the triangle. (Lesson 11-4)

GETTING READY FOR THE NEXT LESSON

PREREQUISITE SKILL Graph each equation. (Lessons 11-3 and 11-5)

29. $y = -x - 5$ 30. $y = \frac{1}{2}x + 3$ 31. $y = -\frac{1}{3}x + 1$ 32. $y = 5x - 3$

Graphing Calculator Investigation

11-6b

A Follow-Up of Lesson 11-6

Sunshine State Standards
MA.E.1.3.1-2

Scatter Plots

You can use a TI-83 Plus graphing calculator to create scatter plots.

The following table gives the results of a survey listing the number of vehicles owned by a family and the average monthly gasoline cost in dollars. Make a scatter plot of the data.

Number of Vehicles	1	3	5	2	5	1	2
Monthly Gasoline Cost ($)	19	59	90	55	115	35	58

Number of Vehicles	2	1	3	4	3	3	2
Monthly Gasoline Cost ($)	80	62	77	90	80	112	63

STEP 1 **Clear the existing data.**
Keystrokes: [STAT] [ENTER] [▲] [CLEAR] [ENTER]

STEP 2 **Enter the data.**
Input each number of vehicles in L1 and press [ENTER].
Then enter the monthly gasoline cost in L2.

STEP 3 **Turn on the statistical plot.**
Select the scatter plot, L1 as the Xlist, and L2 as the Ylist.
Keystrokes: [2nd] [STAT PLOT]
[ENTER] [ENTER] [▼] [ENTER] [▼]
[2nd] [L1] [ENTER] [2nd] [L2]
[ENTER]

STEP 4 **Graph the data.**
Keystrokes: [ZOOM] 9

Use the [TRACE] feature and the left and right arrow keys to move from one point to another.

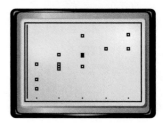

EXERCISES

1. **Describe** the relationship of the data.

2. **RESEARCH** Find some data to use in a scatter plot. Enter the data in a graphing calculator. Determine whether the data has a *positive*, *negative*, or *no* relationship.

Graphing Systems of Equations

Sunshine State Standards
MA.C.3.3.2-3, MA.D.1.3.1-1, MA.D.2.3.1-5

What You'll Learn
Solve systems of linear equations by graphing.

⇒NEW Vocabulary
system of equations
substitution

⊙REVIEW Vocabulary
solution: a value for the variable that makes an equation true **(Lesson 1-8)**

WHEN am I ever going to use this?

TRAVEL A storm is approaching a cruise ship. If x represents the number of hours, then $y = 6x$ represents the position of the storm, and $y = 5x + 2$ represents the position of the ship.

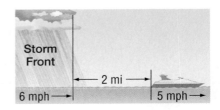

1. Graph both of the equations on a coordinate plane.

2. What are the coordinates of the point where the two lines intersect? What does this point represent?

The equations $y = 6x$ and $y = 5x + 2$ form a system of equations. A set of two or more equations is called a **system of equations**.

When you find an ordered pair that is a solution of all of the equations in a system, you have solved the system. The ordered pair for the point where the graphs of the equations intersect is the solution.

EXAMPLE One Solution

1 Solve the system $y = x + 3$ and $y = 2x + 5$ by graphing.

The graphs of the equations appear to intersect at $(-2, 1)$. Check this estimate.

Check $\quad y = x + 3 \qquad\qquad y = 2x + 5$

$\qquad\qquad 1 \stackrel{?}{=} -2 + 3 \qquad\quad 1 \stackrel{?}{=} 2(-2) + 5$

$\qquad\qquad 1 = 1 ✓ \qquad\qquad\quad 1 = 1 ✓$

The solution of the system is $(-2, 1)$.

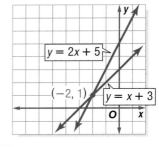

EXAMPLE Infinitely Many Solutions

2 Solve the system $y = x - 3$ and $y - x = -3$ by graphing.

Write $y - x = -3$ in slope-intercept form.

$\qquad y - x = -3 \qquad$ Write the equation.

$y - x + x = -3 + x \qquad$ Add x to each side.

$\qquad\qquad y = x - 3 \qquad$ Both equations are the same.

The solution of the system is all the coordinates of points on the graph of $y = x - 3$.

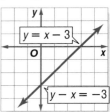

EXAMPLE No Solution

3 **MONEY MATTERS** The Buy Online Company charges $1 per pound plus $2 for shipping and handling. The Best Catalogue Company charges $1 per pound plus $3 for shipping and handling. For what weight will the shipping and handling for the two companies be the same?

Let x equal the weight in pounds of the item or items ordered.

Let y equal the total cost of shipping and handling.

Write an equation to represent each company's charge for shipping and handling.

Buy Online Company: $y = 1x + 2$ or $y = x + 2$

Best Catalogue Company: $y = 1x + 3$ or $y = x + 3$

Graph the system of equations.

$y = x + 2$

$y = x + 3$

The graphs appear to be parallel lines. Since there is no coordinate pair that is a solution of both equations, there is no solution of this system of equations.

For any weight, the Buy Online Company will charge less than the Best Catalogue Company.

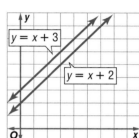

A more accurate way to solve a system of equations than by graphing is by using a method called **substitution**.

EXAMPLE Solve by Substitution

4 Solve the system $y = 2x - 3$ and $y = -1$ by substitution.

Since y must have the same value in both equations, you can replace y with -1 in the first equation.

$y = 2x - 3$	Write the first equation.
$-1 = 2x - 3$	Replace y with -1.
$-1 + 3 = 2x - 3 + 3$	Add 3 to each side.
$2 = 2x$	Simplify.
$\dfrac{2}{2} = \dfrac{2x}{2}$	Divide each side by 2.
$1 = x$	Simplify.

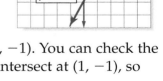

The solution of this system of equations is $(1, -1)$. You can check the solution by graphing. The graphs appear to intersect at $(1, -1)$, so the solution is correct.

Your Turn Solve each system of equations by substitution.

a. $y = x - 4$
    $y = 7$

b. $y = 3x + 4$
    $x = 2$

c. $y = -2x - 7$
    $y = 1$

1. **Explain** what is meant by a system of equations and describe its solution.

2. **OPEN ENDED** Draw a graph of a system of equations that has $(-2, 3)$ as its solution.

3. **NUMBER SENSE** Describe the solution of the system $y = \frac{1}{2}x + 1$ and $y = \frac{1}{2}x - 3$ without graphing. Explain.

GUIDED PRACTICE

Solve each system of equations by graphing.

4. $y = 2x + 1$
   $y = -x + 7$

5. $y = -2x + 4$
   $y = 2x$

6. $y = 3x + 1$
   $y = 3x - 1$

Solve each system of equations by substitution.

7. $y = 3x - 4$
   $y = 8$

8. $y = -2x + 1$
   $x = -3$

9. $y = 0.5x - 4$
   $y = 1$

JOBS For Exercises 10–12, use the information at the right about the summer jobs of Neka and Savannah.

	Weekly Salary	Starting Bonus
Neka	$300	$200
Savannah	$350	$100

10. Write an equation for Neka's total income y after x weeks.

11. Write an equation for Savannah's total income y after x weeks.

12. When will Neka and Savannah have earned the same total amount? What will that amount be?

Practice and Applications

Solve each system of equations by graphing.

HOMEWORK HELP

For Exercises	See Examples
13–20	1–3
21–28	4

Extra Practice
See pages 644, 658.

13. $y = x - 4$
    $y = -2x + 2$

14. $y = 2x - 3$
    $y = -x - 6$

15. $y = 3x - 2$
    $y = -\frac{1}{2}x + 5$

16. $y = \frac{1}{3}x + 1$
    $y = -2x + 8$

17. $x + y = -3$
    $x + y = 4$

18. $y - x = 0$
    $2x + y = 3$

19. Graph the system $y = x + 8$ and $y = -2x - 1$. Find the solution.

20. Graph the system $y = -2x - 6$ and $y = -2x - 3$. Find the solution.

Solve each system of equations by substitution.

21. $y = 3x + 4$
    $y = -5$

22. $y = -2x + 4$
    $y = -6$

23. $y = -3x - 1$
    $x = -4$

24. $y = 4x - 5$
    $x = 2$

25. $y = -2x + 9$
    $y = x$

26. $y = 5x - 8$
    $y = x$

27. Solve the system $y = -3x + 5$ and $y = 2$ by substitution.

28. Solve the system $y = 2x - 1$ and $y = 5$ by substitution.

CLUBS For Exercises 29 and 30, use the following information.
The Science Club wants to order T-shirts for their members. The Shirt Shack will make the shirts for a $30 set-up fee and then $12 per shirt. T-World will make the same shirts for $70 set-up fee and then $8 per shirt.

29. For how many T-shirts will the cost be the same? What will be the cost?

30. If the club wants to order 30 T-shirts, which store should they choose?

HOT-AIR BALLOONS For Exercises 31–34, use the information at the right about two ascending hot-air balloons.

Balloon	Distance from Ground (meters)	Rate of Ascension (meters per minute)
A	60	15
B	40	20

31. Write an equation that describes the distance from the ground y of balloon A after x minutes.

32. Write an equation that describes the distance from the ground y of balloon B after x minutes.

33. When will the balloons be at the same distance from the ground?

34. What is the distance of the balloons from the ground at that time?

35. **CRITICAL THINKING** One equation in a system of equations is $y = 2x + 1$.
   a. Write a second equation so that the system has (1, 3) as its only solution.
   b. Write an equation so that the system has no solutions.
   c. Write an equation so the system has many solutions.

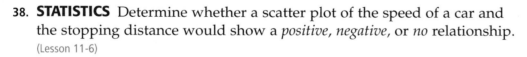

Standardized Test Practice and Mixed Review

FCAT Practice

36. **MULTIPLE CHOICE** Which ordered pair represents the intersection of lines ℓ and k?

   Ⓐ $(-3, 2)$ Ⓑ $(-2, 3)$ Ⓒ $(3, -2)$ Ⓓ $(2, -3)$

37. **GRID IN** The equation $c = 900 + 5t$ represents the cost c in cents that a long-distance telephone company charges for t minutes. Find the value of t if $c = 1,200$.

38. **STATISTICS** Determine whether a scatter plot of the speed of a car and the stopping distance would show a *positive*, *negative*, or *no* relationship. (Lesson 11-6)

Graph each equation using the slope and the y-intercept. (Lesson 11-5)

39. $y = -\frac{2}{3}x + 2$ 40. $y = \frac{2}{5}x - 1$ 41. $y = 4x - 3$

42. **MULTI STEP** Write an equation to represent *three times a number minus five is 16*. Then solve the equation. (Lesson 10-3)

GETTING READY FOR THE NEXT LESSON

PREREQUISITE SKILL Graph each inequality on a number line. (Lesson 10-5)

43. $x > -3$ 44. $x \leq 5$ 45. $x < 0$ 46. $x \geq -1$

Graphing Linear Inequalities

Sunshine State Standards
MA.D.1.3.2-2

What You'll Learn
Graph linear inequalities.

NEW Vocabulary

boundary
half plane

REVIEW Vocabulary

inequality: a mathematical sentence that contains $>$, $<$, $\neq$, $\leq$, or $\geq$ (Lesson 10-5)

WHEN am I ever going to use this?

MONEY MATTERS At a sidewalk sale, one table has a variety of CDs for $2 each, and another table has a variety of books for $1 each. Sabrina wants to buy some CDs and books.

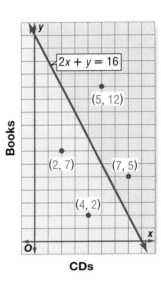

1. Use the graph at the right to list three different combinations of CDs and books that Sabrina can purchase for $16.

2. Suppose Sabrina wants to spend less than $16. Substitute (2, 7), (4, 2), (5, 12), and (7, 5) in $2x + y < 16$. Which values make the inequality true?

3. Which region do you think represents $2x + y < 16$?

4. Suppose Sabrina can spend more than $16. Substitute (2, 7), (4, 2), (5, 12), and (7, 5) in $2x + y > 16$. Which values make the inequality true?

5. Which region do you think represents $2x + y > 16$?

To graph an inequality such as $y < x + 1$, first graph the related equation $y = x + 1$. This is the **boundary**.

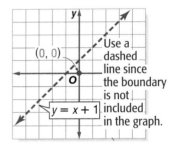

- If the inequality contains the symbol $\leq$ or $\geq$, a solid line is used to indicate that the boundary is included in the graph.

- If the inequality contains the symbol $<$ or $>$, a dashed line is used to indicate that the boundary is not included in the graph.

Next, test any point above or below the line to determine which region is the solution of $y < x + 1$. For example, it is easy to test (0, 0).

$y < x + 1$ Write the inequality.

$0 \overset{?}{<} 0 + 1$ Replace x with 0 and y with 0.

$0 < 1$ ✔   Simplify.

Since $0 < 1$ is true, (0, 0) is a solution of $y < x + 1$. Shade the region that contains this solution. This region is called a **half plane**. All points in this region are solutions of the inequality.

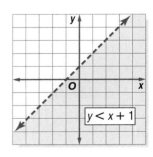

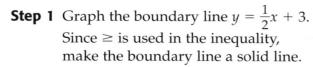 **EXAMPLE** Graph an Inequality

1 Graph $y \geq \frac{1}{2}x + 3$.

Step 1 Graph the boundary line $y = \frac{1}{2}x + 3$. Since $\geq$ is used in the inequality, make the boundary line a solid line.

Step 2 Test a point not on the boundary line, such as $(0, 0)$.

$y \geq \frac{1}{2}x + 3$ Write the inequality.

$0 \stackrel{?}{\geq} \frac{1}{2}(0) + 3$ Replace x with 0 and y with 0.

$0 \ngeq 3$ Simplify.

Step 3 Since $(0, 0)$ is not a solution of $y \geq \frac{1}{2}x + 3$, shade the region that does not contain $(0, 0)$.

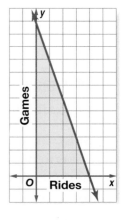

 **STUDY TIP**

Check You may want to check the graph in Example 1 by choosing a point in the shaded region. Do the coordinates of that point make the inequality a true statement?

The solution of an inequality includes negative numbers as well as fractions. However, in real-life situations, sometimes negative numbers and fractions have no meaning.

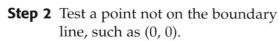 **EXAMPLE** Graph an Inequality to Solve a Problem

2 **FAIRS** At the local fair, rides cost $3 and games cost $1. Gloria has $12 to spend on the rides and games. How can she spend her money?

Let x represent the number of rides and y represent the number of games. Write an inequality.

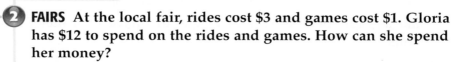

Words	Cost of rides plus cost of games is no more than 12.
Inequality	$3x$ + $1y$ $\leq$ 12

The related equation is $3x + y = 12$.

$3x + y = 12$ Write the equation.

$3x + y - 3x = 12 - 3x$ Subtract $3x$ from each side.

$y = -3x + 12$ Write in slope-intercept form.

Graph $y = -3x + 12$. Test $(0, 0)$ in the original inequality.

$3x + 1y \leq 12$ Write the inequality.

$3(0) + 1(0) \stackrel{?}{\leq} 12$ Replace x with 0 and y with 0.

$0 \leq 12$ ✔   Simplify.

Since Gloria cannot ride or play a negative number of times or a fractional number of times, the answer is any pair of integers represented in the shaded region. For example, she could ride 3 rides and play 2 games.

REAL-LIFE MATH

FAIRS Each year, there are more than 3,200 fairs held in the United States and Canada.

Source: *World Book*

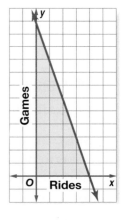

1. **OPEN ENDED** Write an inequality that has a graph with a dashed line as its boundary. Graph the inequality.

2. **FIND THE ERROR** Nathan and Micheal are graphing $y \geq \frac{2}{3}x - 2$. Who is correct? Explain.

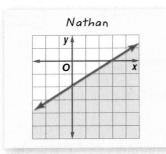

Nathan

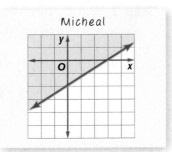

Micheal

GUIDED PRACTICE

Graph each inequality.

3. $y > 2x - 1$ 4. $y \leq \frac{4}{3}x - 2$ 5. $y \geq \frac{1}{2}x$

GEOMETRY For Exercises 6 and 7, use the following information.
A formula for the perimeter of an isosceles triangle where x is the length of the legs and y is the length of the base is $P = 2x + y$.

6. Make a graph for all isosceles triangles that have a perimeter greater than 8 units.

7. Give the lengths of the legs and the base of three isosceles triangles with perimeters greater than 8 units.

Practice and Applications

Graph each inequality.

8. $y > x - 4$ 9. $y \geq x + 5$ 10. $y \leq -3x + 3$

11. $y < -x - 2$ 12. $y > \frac{5}{2}x + 1$ 13. $y < \frac{3}{4}x - 1$

14. $y \leq -\frac{3}{2}x + 2$ 15. $y \geq -\frac{2}{5}x - 3$ 16. $y < 5x + 3$

17. $y \leq 4x - 1$ 18. $3x + y \geq 1$ 19. $y - 2x > 6$

20. Graph the inequality *the sum of two numbers is less than 6.*

21. Graph the inequality *the sum of two numbers is greater than 4.*

SCHOOL For Exercises 22 and 23, use the following information.
Alberto must finish his math and social studies homework during the next 60 minutes.

22. Make a graph showing all the amounts of time Alberto can spend on each subject.

23. Give three possible ways Alberto can spend his time on math and social studies.

HOMEWORK HELP

For Exercises	See Examples
8–21	1
22–25	2

Extra Practice
See pages 645, 658.

TRAVEL For Exercises 24 and 25, use the information below.

In the monetary system of the African country of Mauritania, five khoums equals one ouguiya. Heather is visiting Mauritania and wants to take at least an amount equal to 30 ouguiyas to the market. The inequality $\frac{1}{5}x + y \geq 30$, where x is the number of khoums and y is the number of ouguiyas, represents the situation.

24. Make a graph showing all the combinations of khoums and ouguiyas Heather can take to the market.

25. Give three possible ways Heather can take an appropriate amount to the market.

26. **CRITICAL THINKING** Graph the intersection of $y \leq -x - 3$ and $y > x + 2$.

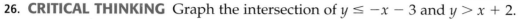
27. **MULTIPLE CHOICE** Which is the graph of $2x + y > 5$?

A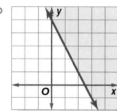

B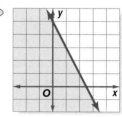

C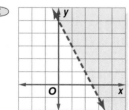

D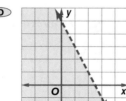

28. **MULTIPLE CHOICE** Which ordered pair is *not* a solution of $y + 3 \leq 2x$?

    F $(1, 2)$ **G** $(2, 1)$ **H** $(3, -2)$ **I** $(2, -3)$

Solve each system of equations by graphing. (Lesson 11-7)

29. $y = 2x - 5$
$y = -x + 1$

30. $y = x + 2$
$y = -x$

31. $y = -2x + 4$
$y = -2x - 2$

32. $y = -x + 4$
$y = 2x - 2$

33. **STATISTICS** Determine whether a scatter plot of the amount of studying and test scores would show a *positive*, *negative*, or *no* relationship. (Lesson 11-6)

Find the area of each figure. (Lesson 7-1)

34.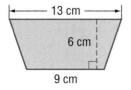

35.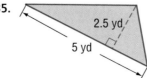

36.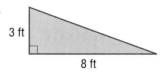

37. **MULTI STEP** The original price of a jacket is $58. Find the price of the jacket if it is marked down 25%. (Lesson 5-7)

Vocabulary and Concept Check

arithmetic sequence (p. 512)	function table (p. 518)	slope formula (p. 526)
best-fit line (p. 540)	geometric sequence (p. 513)	slope-intercept form (p. 533)
boundary (p. 548)	half plane (p. 548)	substitution (p. 545)
common difference (p. 512)	independent variable (p. 518)	system of equations (p. 544)
common ratio (p. 513)	linear function (p. 523)	term (p. 512)
dependent variable (p. 518)	range (p. 518)	x-intercept (p. 523)
domain (p. 518)	scatter plot (p. 539)	y-intercept (p. 523)
function (p. 517)	sequence (p. 512)	

Choose the correct term or number to complete each sentence.

1. The (domain, range) is the set of input values of a function.
2. The range is the set of (input, output) values of a function.
3. A (sequence, term) is an ordered list of numbers.
4. A geometric sequence has a (common difference, common ratio).
5. A(n) (arithmetic sequence, geometric sequence) has a common difference.
6. The (x-intercept, y-intercept) has the coordinates $(0, b)$.
7. The (half-plane, boundary) is the graph of the equation related to an inequality.
8. The slope formula is $m = \left(\dfrac{y_2 - y_1}{x_2 - x_1} , \dfrac{x_2 - x_1}{y_2 - y_1} \right)$.

Lesson-by-Lesson Exercises and Examples

11-1 Sequences (pp. 512–515)

State whether each sequence is *arithmetic*, *geometric*, or *neither*. If it is arithmetic or geometric, state the common difference or common ratio. Write the next three terms of the sequence.

9. 64, 32, 16, 8, 4, …   10. −7, −4, −1, 2, 5, …
11. 1, 2, 6, 24, 120, …   12. 1, −1, 1, −1, 1, …

13. **SAVINGS** Loretta has $5 in her piggy bank. Each week, she adds $1.50. If she does not take any money out of the piggy bank, how much will she have after 6 weeks?

Example 1 State whether the sequence is *arithmetic*, *geometric*, or *neither*. If it is arithmetic or geometric, state the common difference or common ratio. Write the next three terms of the sequence.

$$1, -2, 4, -8, 16, \ldots$$

1,   −2,   4,   −8,   16, …
× (−2) × (−2) × (−2) × (−2)

The terms have a common ratio of −2, so the sequence is geometric. The next three terms are 16(−2) or −32, −32(−2) or 64, and 64(−2) or −128.

11-2 Functions (pp. 517–520)

Find each function value.

14. $f(3)$ if $f(x) = 3x + 1$

15. $f(9)$ if $f(x) = 1 - 3x$

16. $f(0)$ if $f(x) = 2x + 6$

17. $f(-11)$ if $f(x) = -2x$

18. $f(-2)$ if $f(x) = x - 1$

19. $f(2)$ if $f(x) = \frac{1}{2}x - 4$

Example 2 Complete the function table for $f(x) = 2x - 1$.

x	$2x - 1$	$f(x)$
-2	$2(-2) - 1$	-5
0	$2(0) - 1$	-1
1	$2(1) - 1$	1
5	$2(5) - 1$	9

11-3 Graphing Linear Functions (pp. 522–525)

Graph each function.

20. $y = -2x + 1$

21. $y = x - 4$

22. $y = -3x$

23. $y = \frac{1}{2}x - 2$

24. **GEOMETRY** The function $y = 4x$ represents the perimeter y of a square with side x units long. Graph $y = 4x$.

Example 3 Graph $y = 3 - x$.

x	$3 - x$	y	(x, y)
-1	$3 - (-1)$	4	$(-1, 4)$
0	$3 - 0$	3	$(0, 3)$
2	$3 - 2$	1	$(2, 1)$
3	$3 - 3$	0	$(3, 0)$

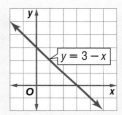

11-4 The Slope Formula (pp. 526–529)

Find the slope of each line that passes through each pair of points.

25. $A(-2, 3)$, $B(-1, 5)$

26. $E(-3, 2)$, $F(-3, 5)$

27. $G(6, 2)$, $H(1, 5)$

28. $K(2, 1)$, $L(-3, 1)$

Example 4 Find the slope of the line that passes through $A(-3, 2)$ and $B(5, -1)$.

$m = \dfrac{y_2 - y_1}{x_2 - x_1}$ Definition of slope

$m = \dfrac{-1 - 2}{5 - (-3)}$ or $-\dfrac{3}{8}$ $(x_1, y_1) = (-3, 2)$, $(x_2, y_2) = (5, -1)$

11-5 Slope-Intercept Form (pp. 533–536)

State the slope and y-intercept for the graph of each equation.

29. $y = 2x + 5$

30. $y = \frac{1}{2}x - 7$

31. $y = \frac{1}{5}x + 6$

32. $y = -3x - 2$

33. $y = -\frac{3}{4}x + 7$

34. $y = 3x + 7$

Example 5 State the slope and y-intercept of the graph of $y = -\frac{1}{2}x + 3$.

$y = -\frac{1}{2}x + 3$ Write the equation.

$y = \quad mx + b$

The slope of the graph is $-\frac{1}{2}$, and the y-intercept is 3.

11-6 Scatter Plots (pp. 539–542)

Determine whether a scatter plot of the data for the following might show a *positive*, *negative*, or *no* relationship.

35. number of people in the household and the cost of groceries

36. day of the week and temperature

37. child's age and grade level in school

38. temperature outside and amount of clothing

Example 6
Determine whether the graph at the right shows a *positive*, *negative*, or *no* relationship.

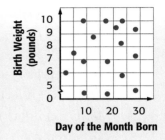

Since there is no obvious pattern, there is no relationship.

11-7 Graphing Systems of Equations (pp. 544–547)

Solve each system of equations by graphing.

39. $y = 2x$
$y = x + 1$

40. $y = 3x - 1$
$y = x - 3$

41. $y = 2x - 2$
$y = x + 2$

42. $y = x - 4$
$y = -x + 2$

Solve each system of equations by substitution.

43. $y = -2x + 7$
$x = 3$

44. $y = 3x - 5$
$x = 4$

Example 7 Solve the system of equations $y = 2x$ and $y = x - 1$ by graphing.

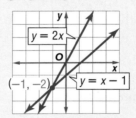

The graphs of the equation appear to intersect at $(-1, -2)$.

Check this estimate.

11-8 Graphing Linear Inequalities (pp. 548–551)

Graph each inequality.

45. $y > x$

46. $y < -2x$

47. $y \geq 2x + 3$

48. $y \leq -x + 5$

49. $y > 3x + 5$

50. $y \leq 2x + 1$

51. FESTIVALS At the Spring Festival, games cost $2 and rides cost $3. Nate wants to spend no more than $20 at the festival. Give three possible ways Nate can spend his money.

Example 8 Graph $y < 4x - 2$.

Graph the boundary line $y = 4x - 2$. Since the $<$ symbol is used in the inequality, make the boundary line a dashed line.

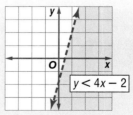

Test a point not on the boundary line such as $(0, 0)$. Since $(0, 0)$ is not a solution of $y < 4x - 2$, shade the region that does not contain $(0, 0)$.

Vocabulary and Concepts

1. **Describe** how you can tell that there is no solution when you graph a system of equations.

2. **Describe** two different ways to graph $y = 2x + 5$.

Skills and Applications

State whether each sequence is *arithmetic*, *geometric*, or *neither*. If it is *arithmetic* or *geometric*, state the common difference or common ratio. Then write the next three terms of the sequence.

3. $-10, -6, -2, 2, 6, \ldots$ 4. $\frac{1}{5}, 1, 5, 25, 125, \ldots$ 5. $61, 50, 40, 31, 23, \ldots$

Find each function value.

6. $f(-2)$ if $f(x) = \frac{x}{2} + 5$ 7. $f(3)$ if $f(x) = -2x + 6$

Graph each function or inequality.

8. $y = \frac{1}{3}x - 1$ 9. $y = -4x + 1$ 10. $y < -2x + 3$ 11. $y \geq x - 5$

Find the slope of the line that passes through each pair of points.

12. $A(-2, 5), B(-2, 1)$ 13. $C(0, 3), D(-5, 2)$ 14. $E(2, -1), F(5, -3)$

CHILD CARE For Exercises 15–17, use the following information.
The cost for a child to attend a certain day care center is \$35 a day plus a registration fee of \$50. The cost c for d days of child care is $c = 35d + 50$.

15. Graph the equation.   16. What does the y-intercept represent?

17. What does the slope of the graph represent?

18. Solve the system $y = x + 1$ and $y = 2x - 2$ by graphing.

19. **TRAVEL** Would a scatter plot of data describing the gallons of gas used and the miles driven show a *positive*, *negative*, or *no* relationship?

Standardized Test Practice

20. **MULTIPLE CHOICE** Which is the graph of $y = -3x$?

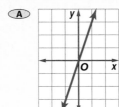 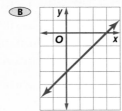 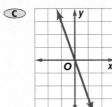 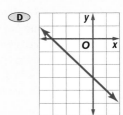

🐊 FCAT Practice

PART 1 Multiple Choice

Record your answers on the answer sheet provided by your teacher or on a sheet of paper.

1. If the following ordered pairs are plotted on a graph, which graph passes through all five points? (Prerequisite Skill, p. 614)

$(-2, 4), (-1, 1), (0, 0), (1, 1), (2, 4)$

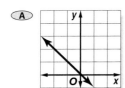
Ⓐ

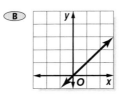
Ⓑ

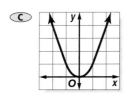
Ⓒ

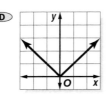
Ⓓ

2. Which could be the value of x if $0.6 < x < 68\%$? (Lesson 5-2)

Ⓕ $\dfrac{3}{5}$ Ⓖ $\dfrac{2}{3}$

Ⓗ $\dfrac{5}{7}$ Ⓘ $\dfrac{6}{8}$

3. Which of the following is *not* a quadrilateral? (Lesson 6-4)

Ⓐ

Ⓑ

Ⓒ

Ⓓ

TEST-TAKING TIP

Question 3 Read each question carefully so that you do not miss key words such as *not* or *except*. Then read every answer choice carefully. If allowed to write in the test booklet, cross off each answer choice that you know is not the answer, so you will not consider it again.

4. What is the value of $f(x)$ when $x = 5$? (Lesson 11-1)

x	$f(x)$
0	3
1	4
2	5
5	?

Ⓕ 7   Ⓖ 8

Ⓗ 9   Ⓘ 10

5. Which of the following ordered pairs is a solution of $y = \dfrac{1}{2}x - 4$? (Lesson 11-2)

Ⓐ $(-2, -3)$ Ⓑ $(4, 2)$

Ⓒ $(6, 1)$ Ⓓ $(8, 0)$

6. The graph shows the distance Kimberly has traveled. What does the slope of the graph represent? (Lesson 11-5)

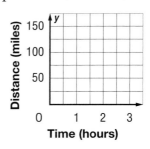

Ⓕ the distance Kimberly traveled

Ⓖ how long Kimberly has traveled

Ⓗ Kimberly's average speed

Ⓘ the time Kimberly will arrive

7. Which display would be most appropriate for the data in a table that shows the relationship between height and weight of 20 students? (Lesson 11-6)

Ⓐ scatter plot   Ⓑ line plot

Ⓒ bar graph   Ⓓ circle graph

PART 2 Short Response/Grid In

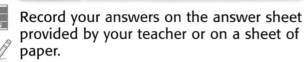

THINK
SOLVE
EXPLAIN

Record your answers on the answer sheet provided by your teacher or on a sheet of paper.

8. On a number line, how many units apart are −6 and 7? (Lesson 1-3)

9. Find the volume of the pyramid. (Lesson 7-6)

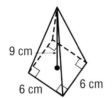

9 cm

6 cm   6 cm

10. Mr. Thomas is drawing names out of a hat in order to create class debating teams. There are 13 girls and 12 boys in the class, and the first name picked is a girl's name. What is the probability that a boy's name will be drawn next? (Lesson 8-5)

11. Molly received grades of 79, 92, 68, 90, 72, and 92 on her history tests. What measure of central tendency would give her the highest grade for the term? (Lesson 9-4)

12. A pair of designer jeans costs $98, which is $35 more than 3 times the cost of a discount store brand. The equation to find the cost of the discount store brand d is $3d + 35 = 98$. What is the price of the discount store brand of jeans? (Lesson 10-3)

13. Copy and complete the function table for $f(x) = 0.4x + 2$. (Lesson 11-2)

x	y
−5	
0	
5	
10	

14. Graph $y = \frac{1}{2}x - 2$. (Lesson 11-3)

15. Describe the relationship shown in the graph. (Lesson 11-6)

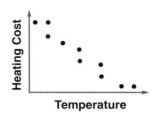

16. What ordered pair is a solution of both $y = 3x - 5$ and $y = x - 7$? (Lesson 11-7)

PART 3 Extended Response

Record your answers on a sheet of paper. Show your work.

THINK
SOLVE
EXPLAIN

17. Study the data below. (Lesson 11-6)

Date	Number of Customers	Ice Cream Scoops Sold
June 1	75	100
June 2	125	230
June 3	350	460
June 4	275	370
June 5	175	300
June 6	225	345
June 7	210	325

a. What type of display would be most appropriate for this data?

b. Graph the data.

c. Describe the relationship of the data.

18. In Major League Soccer, a team gets 3 points for a win and 1 point for a tie. (Lesson 11-8)

a. Write an inequality for the number of ways a team can earn more than 10 points.

b. Graph the inequality.

c. Compare the values graphed and those that actually satisfy the situation.

Algebra: Nonlinear Functions and Polynomials

"What does racing have to do with math?"

As a race car increases its speed, or accelerates, the distance it travels each second also increases. The relationship between these two quantities, however, is not linear. The distance d that a race car travels in time t, given the rate of acceleration a, can be described by the quadratic function $d = \frac{1}{2}at^2$.

You will solve a problem about racing in Lesson 12-2.

GETTING STARTED

Take this quiz to see whether you are ready to begin Chapter 12. Refer to the lesson number in parentheses if you need more review.

▶ Vocabulary Review

Choose the correct term to complete each sentence.

1. A function in which the graph of the solutions forms a line is called a (straight, **linear**) function. (Lesson 11-3)

2. The (**base**, exponent) is the number in a power that is multiplied. (Lesson 2-8)

▶ Prerequisite Skills

Identify the like terms in each expression. (Lesson 10-1)

3. $3x + 5 - x$ 4. $2 - 4n + 1 + 6n$

Rewrite each expression using parentheses so that the like terms are grouped together. (Lessons 1-2 and 10-1)

5. $(a + 2b) + (2a - 5b)$

6. $(8w + 7x) + (3w + 9x)$

Rewrite each expression as an addition expression by using the additive inverse.
(Lessons 1-4 and 1-5)

7. $3 - 5y$ 8. $2m - 7n$

Write each expression using exponents.
(Lesson 2-8)

9. $6 \cdot 6 \cdot 6 \cdot 6$ 10. $3 \cdot 7 \cdot 7 \cdot 3 \cdot 7$

Use the Distributive Property to rewrite each expression. (Lesson 10-1)

11. $9(d + 2)$ 12. $4(n - 1)$

13. $-2(a + 3)$ 14. $-8(f - 3)$

 **Nonlinear Functions**
Make this Foldable to help you organize your notes. Begin with 7 sheets of $8\frac{1}{2}" \times 11"$ paper.

STEP 1 Fold and Cut
Fold a sheet of paper in half lengthwise. Cut a 1" tab along the left edge through one thickness.

STEP 2 Glue and Label
Glue the 1" tab down. Write the title of the lesson on the front tab.

Linear & Nonlinear Functions

STEP 3 Repeat and Staple
Repeat Steps 1–2 for the remaining sheets of paper. Staple together to form a booklet.

Linear & Nonlinear Functions

Reading and Writing As you read and study the chapter, write notes, define terms, record concepts, and write examples under each tab.

12-1 Linear and Nonlinear Functions

Sunshine State Standards
MA.D.1.3.1-1, MA.D.1.3.1-2, MA.D.1.3.1-3

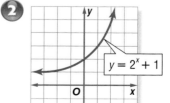

What You'll Learn
Determine whether a function is linear or nonlinear.

→NEW Vocabulary

nonlinear function

⟳REVIEW Vocabulary

function: a relationship where one quantity depends upon another (Lesson 11-2)

WHEN am I ever going to use this?

ROCKETRY The tables show the flight data for a model rocket launch. The first table gives the rocket's height at each second of its ascent, or upward flight. The second table gives its height as it descends back to Earth using a parachute.

Ascent		Descent	
Time (s)	Height (m)	Time (s)	Height (m)
0	0	7	140
1	38	8	130
2	74	9	120
3	106	10	110
4	128	11	100
5	138	12	90
6	142	13	80

1. During its ascent, did the rocket travel the same distance each second? Explain.

2. During its descent, did the rocket travel the same distance each second? Explain.

3. Graph the data whose ordered pairs are (time, height) for the rocket's ascent and descent on separate axes. Connect the points with a straight line or smooth curve. Then compare the two graphs.

In Lesson 11-3, you learned that linear functions have graphs that are straight lines. These graphs represent constant rates of change. **Nonlinear functions** do not have constant rates of change. Therefore, their graphs are not straight lines.

EXAMPLES Identify Functions Using Graphs

Determine whether each graph represents a *linear* or *nonlinear* function. Explain.

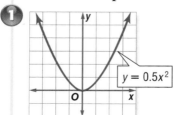

1 $y = 0.5x^2$

2 $y = 2^x + 1$

The graph is a curve, not a straight line. So it represents a nonlinear function.

This graph is also a curve. So it represents a nonlinear function.

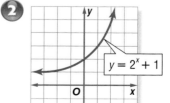

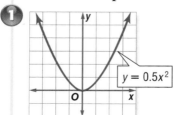

Since the equation for a linear function can be written in the form $y = mx + b$, where m represents the constant rate of change, you can determine whether a function is linear by examining its equation.

 Identify Functions Using Equations

Determine whether each equation represents a *linear* or *nonlinear* function. Explain.

3 $y = x + 4$

Since the equation can be written as $y = 1x + 4$, this function is linear.

4 $y = \dfrac{6}{x}$

Since x is in the denominator, the equation cannot be written in the form $y = mx + b$. So this function is nonlinear.

A nonlinear function does not increase or decrease at the same rate. You can use a table to determine if the rate of change is constant.

 Identify Functions Using Tables

Determine whether each table represents a *linear* or *nonlinear* function. Explain.

5

x	y
2	50
4	35
6	20
8	5

$+2$ / -15 (each step)

As x increases by 2, y decreases by 15 each time. The rate of change is constant, so this function is linear.

6

x	y
1	1
4	16
7	49
10	100

$+3$ / $+15$, $+33$, $+51$

As x increases by 3, y increases by a greater amount each time. The rate of change is not constant, so this function is nonlinear.

7 **BASKETBALL** Use the table to determine whether the number of teams is a linear function of the number of rounds of play.

Examine the differences between the number of teams for each round.

$4 - 2 = 2$ $8 - 4 = 4$ $16 - 8 = 8$ $32 - 16 = 16$

While there is a pattern in the differences, they are not the same. Therefore, this function is nonlinear.

Round(s) of play	Teams
1	2
2	4
3	8
4	16
5	32

REAL-LIFE MATH

BASKETBALL The NCAA women's basketball tournament begins with 64 teams and consists of 6 rounds of play.

Your Turn Determine whether each equation or table represents a *linear* or *nonlinear* function. Explain.

a. $y = 2x^3 + 1$ **b.** $y = 3x$

c.

x	0	5	10	15
y	20	16	12	8

Skill and Concept Check

1. **OPEN ENDED** Give an example of a nonlinear function using a table of values.

2. **Which One Doesn't Belong?** Identify the function that is not linear. Explain your reasoning.

$y = 2x$ $y = x^2$ $y - 2 = x$ $x - y = 2$

GUIDED PRACTICE

Determine whether each graph, equation, or table represents a *linear* or *nonlinear* function. Explain.

3.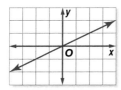

4.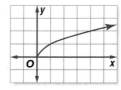

5. $y = \dfrac{x}{3}$

6. $y - x = 1$

7.
x	3	6	9	12
y	12	10	8	6

8.
x	1	2	3	4
y	1	4	9	16

Practice and Applications

Determine whether each graph, equation, or table represents a *linear* or *nonlinear* function. Explain.

HOMEWORK HELP

For Exercises	See Examples
9–14	1, 2
15–22	3, 4
23–28	5, 6
29–31	7

Extra Practice
See pages 645, 659.

9.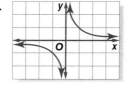

10.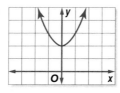

11.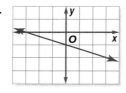

12.

13.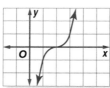

14.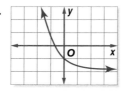

15. $xy = -9$

16. $y = 0.6x$

17. $y = x^3 - 1$

18. $y = 4x^2 + 9$

19. $y = 2^x$

20. $y = \dfrac{4}{x}$

21. $y = 7$

22. $y = \dfrac{3x}{2}$

23.
x	1	2	3	4
y	0	2	6	12

24.
x	−1	0	1	2
y	−4	1	6	11

25.
x	2	5	8	11
y	21	19	17	15

26.
x	−4	0	4	8
y	2	1	−1	−4

27.
x	4	6	8	10
y	4	13.5	32	62.5

28.
x	0.5	1	1.5	2
y	15	8	1	−6

29. FOOD The graphic shows the increase in garlic consumption from 1970 to 2000. Would you describe the growth as linear or nonlinear? Explain.

 Data Update Is the growth in the consumption of your favorite food linear or nonlinear? Visit msmath3.net/data_update to learn more.

GEOMETRY **For Exercises 30 and 31, use the following information.**
Recall that the circumference of a circle is equal to pi times its diameter and that the area of a circle is equal to pi times the square of its radius.

30. Is the circumference of a circle a linear or nonlinear function of its diameter? Explain.

31. Is the area of a circle a linear or nonlinear function of its radius? Explain.

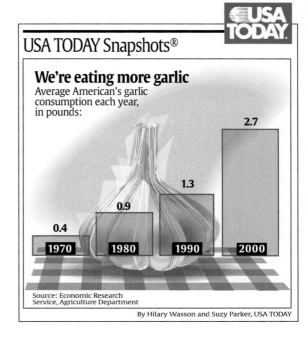

USA TODAY Snapshots®

We're eating more garlic
Average American's garlic consumption each year, in pounds:

2.7 — 2000
1.3 — 1990
0.9 — 1980
0.4 — 1970

Source: Economic Research Service, Agriculture Department

By Hilary Wasson and Suzy Parker, USA TODAY

32. CRITICAL THINKING *True* or *False*? All graphs of straight lines are linear functions. Explain your reasoning or provide a counterexample.

Standardized Test Practice and Mixed Review

FCAT Practice

33. MULTIPLE CHOICE Which equation represents a nonlinear function?

 Ⓐ $y = 3x + 1$ Ⓑ $y = \dfrac{x}{3}$ Ⓒ $2xy = 10$ Ⓓ $y = 3(x - 5)$

34. SHORT RESPONSE Water is poured at a constant rate into the vase at the right. Draw a graph of the water level as a function of time. Is the water level a linear or nonlinear function of time? Explain your reasoning.

COPYING **For Exercises 35 and 36, use the following information.**
Black-and-white copies at Copy Express cost $0.12 each, and color copies cost $1.00 each. Suppose you want to spend no more than $10 on copies of your club's flyers. (Lesson 11-8)

35. Write an inequality to represent this situation.

36. Graph the inequality and use the graph to determine three possible combinations of copies you could make.

Solve each system of equations by substitution. (Lesson 11-7)

37. $y = 2x + 1$
 $y = 3$

38. $y = -4x - 3$
 $y = 1$

39. $y = -5x + 8$
 $y = -2$

40. $y = 0.5x - 6$
 $y = -4$

GETTING READY FOR THE NEXT LESSON

PREREQUISITE SKILL Graph each function. (Lesson 11-3)

41. $y = 2x$ **42.** $y = x + 3$ **43.** $y = 3x - 2$ **44.** $y = \dfrac{1}{3}x + 1$

Graphing Calculator Investigation

A Preview of Lesson 12-2

What You'll Learn
Use a graphing calculator to graph families of quadratic functions.

Families of Quadratic Functions

In Lesson 11-5a, you discovered that families of linear functions share the same slope or y-intercept. Families of nonlinear functions also share a common characteristic. You can use a TI-83 Plus graphing calculator to investigate families of quadratic functions.

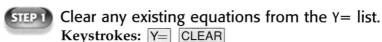

ACTIVITY

Graph $y = x^2$, $y = x^2 + 5$, and $y = x^2 - 3$ on the same screen.

STEP 1 Clear any existing equations from the Y= list.
Keystrokes: Y= CLEAR

STEP 2 Enter each equation.
Keystrokes: X,T,θ,n x^2 ENTER
X,T,θ,n x^2 + 5 ENTER
X,T,θ,n x^2 − 3 ENTER

STEP 3 Graph the equations in the standard viewing window.
Keystrokes: ZOOM 6

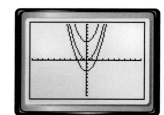

EXERCISES

1. **Compare and contrast** the three equations you graphed.

2. **Describe** how the graphs of the three equations are related.

3. **MAKE A CONJECTURE** How does changing the value of c in the equation $y = x^2 + c$ affect the graph?

4. **Use a graphing calculator** to graph $y = 0.5x^2$, $y = x^2$, and $y = 2x^2$.

5. **Compare and contrast** the three equations you graphed in Exercise 4.

6. **Describe** how the graphs of the three equations are related.

7. **MAKE A CONJECTURE** How does changing the value of a in the equation $y = ax^2$ affect the graph?

8. **Write** a family of three quadratic functions. Describe the common characteristic of their graphs.

 msmath3.net/other_calculator_keystrokes

12-2 Graphing Quadratic Functions

Sunshine State Standards
MA.D.1.3.2-3

What You'll Learn
Graph quadratic functions.

NEW Vocabulary
quadratic function

HANDS-ON Mini Lab

Materials
- graph paper

Work with a partner.

You know that the area A of a square is equal to the length of a side s squared, $A = s^2$. What happens to the area of a square as its side length is increased?

STEP 1 Copy and complete the table.

STEP 2 Graph the ordered pairs from the table. Connect them with a smooth curve.

s	s^2	(s, A)
0	0	(0, 0)
1	1	(1, 1)
2		
3		
4		
5		
6		

1. Is the relationship between the side length and the area of a square linear or nonlinear? Explain.

2. Describe the shape of the graph.

A **quadratic function** is a function in which the greatest power of the variable is 2.

EXAMPLES Graph Quadratic Functions: $y = ax^2$

1 Graph $y = x^2$.

To graph a linear function, make a table of values, plot the ordered pairs, and connect the points with a smooth curve.

x	x^2	y	(x, y)
−2	$(−2)^2 = 4$	4	(−2, 4)
−1	$(−1)^2 = 1$	1	(−1, 1)
0	$(0)^2 = 0$	0	(0, 0)
1	$(1)^2 = 1$	1	(1, 1)
2	$(2)^2 = 4$	4	(2, 4)

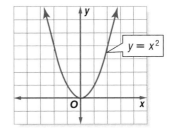

$y = x^2$

2 Graph $y = −2x^2$.

x	$−2x^2$	y	(x, y)
−2	$−2(−2)^2 = −8$	−8	(−2, −8)
−1	$−2(−1)^2 = −2$	−2	(−1, −2)
0	$−2(0)^2 = 0$	0	(0, 0)
1	$−2(1)^2 = −2$	−2	(1, −2)
2	$−2(2)^2 = −8$	−8	(2, −8)

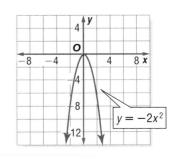

$y = −2x^2$

EXAMPLES Graph Quadratic Functions: $y = ax^2 + c$

3 Graph $y = x^2 + 2$.

x	$x^2 + 2$	y	(x, y)
-2	$(-2)^2 + 2 = 6$	6	$(-2, 6)$
-1	$(-1)^2 + 2 = 3$	3	$(-1, 3)$
0	$(0)^2 + 2 = 2$	2	$(0, 2)$
1	$(1)^2 + 2 = 3$	3	$(1, 3)$
2	$(2)^2 + 2 = 6$	6	$(2, 6)$

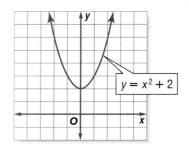

$y = x^2 + 2$

4 Graph $y = -x^2 + 4$.

x	$-x^2 + 4$	y	(x, y)
-2	$-(-2)^2 + 4 = 0$	0	$(-2, 0)$
-1	$-(-1)^2 + 4 = 3$	3	$(-1, 3)$
0	$-(0)^2 + 4 = 4$	4	$(0, 4)$
1	$-(1)^2 + 4 = 3$	3	$(1, 3)$
2	$-(2)^2 + 4 = 0$	0	$(2, 0)$

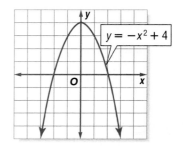

$y = -x^2 + 4$

Your Turn Graph each function.

a. $y = x^2 + 1$ **b.** $y = -2x^2 - 1$ **c.** $y = -x^2$

Many real-life situations can be described using quadratic functions.

EXAMPLE Graph a Function to Solve a Problem

5 **MONUMENTS** The function $h = 0.66d^2$ represents the distance d in miles you can see from a height of h feet. Graph this function. Then use your graph and the information at the left to estimate how far you could see from the top of the Eiffel Tower.

The equation $h = 0.66d^2$ is quadratic, since the variable d has an exponent of 2. Distance cannot be negative, so use only positive values of h.

d	$h = 0.66d^2$	(d, h)
0	$0.66(0)^2 = 0$	$(0, 0)$
5	$0.66(5)^2 = 16.5$	$(5, 16.5)$
10	$0.66(10)^2 = 66$	$(10, 66)$
15	$0.66(15)^2 = 148.5$	$(15, 148.5)$
20	$0.66(20)^2 = 264$	$(20, 264)$
25	$0.66(25)^2 = 412.5$	$(25, 412.5)$
30	$0.66(30)^2 = 594$	$(30, 594)$
35	$0.66(35)^2 = 808.5$	$(35, 808.5)$
40	$0.66(40)^2 = 1,056$	$(40, 1,056)$

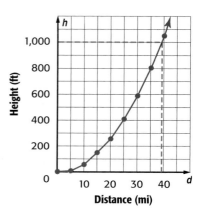

At a height of 986 feet, you could see approximately 39 miles.

1. **Explain** how to determine whether a function is quadratic.

2. **OPEN ENDED** Write a quadratic function of the form $y = ax^2 + c$ and explain how to graph it.

3. **Which One Doesn't Belong?** Identify the function whose graph does not have the same characteristic as the other three. Explain your reasoning.

| $y = 2x^2 + 1$ | $y = -5x^2$ | $y = 7x - 3$ | $y = 4x^2 - 2$ |

GUIDED PRACTICE

Graph each function.

4. $y = 3x^2$ 5. $y = -5x^2$ 6. $y = 0.5x^2$

7. $y = x^2 - 2$ 8. $y = -x^2 + 1$ 9. $y = -2x^2 + 2$

Practice and Applications

Graph each function.

10. $y = 4x^2$ 11. $y = -3x^2$ 12. $y = -1.5x^2$

13. $y = 3.5x^2$ 14. $y = x^2 + 6$ 15. $y = x^2 - 4$

16. $y = 2x^2 - 1$ 17. $y = 2x^2 + 3$ 18. $y = -x^2 + 2$

19. $y = -x^2 - 5$ 20. $y = -4x^2 - 1$ 21. $y = -3x^2 + 2$

22. Graph the function $y = 0.5x^2 + 1$.

23. Graph the function $y = \frac{1}{3}x^2 - 2$.

HOMEWORK HELP

For Exercises	See Examples
10–13	1, 2
14–23	3, 4
24–29	5

Extra Practice
See pages 645, 659.

RACING For Exercises 24–26, use the following information.

The function $d = \frac{1}{2}at^2$ represents the distance d that a race car will travel over an amount of time t given the rate of acceleration a. Suppose a car is accelerating at a rate of 5 feet per second every second.

24. Graph $d = \frac{1}{2}(5t^2)$.

25. Find the distance traveled after 10 seconds.

26. About how long would it take the car to travel 125 feet?

WATERFALL For Exercises 27–29, use the following information.
The quadratic equation $d = -16t^2 + h$ models the distance d in feet a falling object is from the ground or other surface t seconds after it is dropped from a beginning height of h feet. Suppose a drop of water descends from the 182-foot tall American Falls in New York, toward the river below.

27. Graph $d = -16t^2 + 182$.

28. How high is a drop of water after 2 seconds?

29. After about how many seconds will the drop of water reach the river below?

GEOMETRY For Exercises 30 and 31, write a function for each of the following. Then graph the function in the first quadrant.

30. the volume V of a cube as a function of the edge length a

31. the volume V of a rectangular prism as a function of a fixed height of 5 and a square base of varying length s

CRITICAL THINKING The graphs of quadratic functions may have exactly one highest point, called a *maximum*, or exactly one lowest point, called a *minimum*.

Graph each quadratic equation. Determine whether each graph has a maximum or a minimum. If so, give the coordinates of each point.

32. $y = 2x^2 + 1$ **33.** $y = -x^2 + 5$ **34.** $y = x^2 - 3$

EXTENDING THE LESSON Another type of nonlinear function is graphed at the right. A *cubic function*, such as $y = x^3$, is a function in which the greatest power is 3.

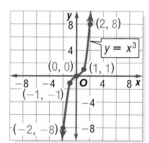

Graph each function. (*Hint*: You may need to let x represent decimal values.)

35. $y = 2x^3$ **36.** $y = x^3 + 1$ **37.** $y = 2x^3 + 2$

38. Graph the equations $y = x^2$ and $y = x^3$ on the same coordinate plane. Describe their similarities and differences.

Standardized Test Practice and Mixed Review

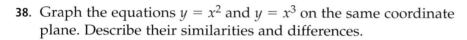

FCAT Practice

39. MULTIPLE CHOICE Which equation represents the graph at the right?

   A $y = 2x^2 - 2$ **B** $y = -0.5x^2 - 2$

   C $y = -x^2 + 2$ **D** $y = x^2 - 2$

40. MULTIPLE CHOICE Which equation represents a quadratic function?

   F $y = 2x$ **G** $y = \dfrac{2}{x}$

   H $y = x + 2$ **I** $y = -x^2 + 8$

Determine whether each equation represents a *linear* or *nonlinear* function. (Lesson 12-1)

41. $y = x - 5$ **42.** $y = 3x^3 + 2$ **43.** $x + y = -6$ **44.** $y = -2x^2$

Graph each inequality. (Lesson 11-8)

45. $y < 2x$ **46.** $y \geq x + 1$ **47.** $y > -x - 3$ **48.** $y \leq -3x + 4$

GETTING READY FOR THE NEXT LESSON

PREREQUISITE SKILL Identify the like terms in each expression. (Lesson 10-1)

49. $4a + 1 - 2a$ **50.** $2x + 3x + 5 - 1$ **51.** $-1 - 2d + 3 + d$ **52.** $x + 2 - 7x + 8$

12-3a HANDS-ON LAB
A Preview of Lesson 12-3

What You'll Learn
Model expressions using algebra tiles.

Materials
• algebra tiles

Modeling Expressions with Algebra Tiles

In a set of algebra tiles, the integer 1 is represented by a tile that is 1 unit by 1 unit. Notice that the area of this tile is 1 square unit. The opposite of 1, -1, is represented by a red tile with the same shape and size.

The variable x is represented by a tile that is 1 unit by x units. Notice that the area of this tile is x square units. The opposite of x, $-x$, is represented by a red tile with the same shape and size.

Similarly, the expression x^2 is represented by a tile that is x units by x units. A red tile with the same shape and size is used to represent $-x^2$.

You can use these tiles to model expressions like $2x^2 + 5x - 6$.

ACTIVITY

Use algebra tiles to model $2x^2 + 5x - 6$.

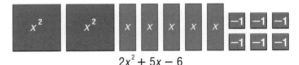

$2x^2 + 5x - 6$

Your Turn Use algebra tiles to model each expression.

a. $4x^2$
b. $-3x^2$
c. $3x^2 - 4x$
d. $-x^2 + 2x$
e. $x^2 - x + 1$
f. $-2x^2 + x - 5$
g. $2x^2 - 3x + 2$
h. $-5x^2 + 3x + 8$

Writing Math

1. **Name** the expression modeled below.

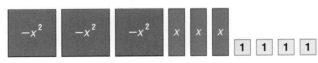

2. **MAKE A CONJECTURE** What might a model of the expression x^3 look like?

Simplifying Polynomials

Sunshine State Standards
MA.D.2.3.1-7, MA.D.2.3.2-1

What You'll Learn

Simplify polynomials.

➢NEW Vocabulary

monomial
polynomial

↻REVIEW Vocabulary

like terms: terms that contain the same variable (Lesson 10-1)

simplest form: an algebraic expression that has no like terms or parentheses (Lesson 10-1)

WHEN **am I ever going to use this?**

MONEY Suppose you need money to buy a drink and a snack. The table shows the number and type of coins you find in your backpack and in your pocket.

Coin Type	Number in Backpack	Number in Pocket
Quarter	3	0
Dime	5	2
Nickel	2	3
Penny	4	0

1. Let q, d, n, and p represent the value of a quarter, a dime, a nickel, and a penny, respectively. Write an expression for the total amount of money in your backpack.

2. Write an expression for the total amount of money in your pocket.

3. Write an expression for the total amount of money in all.

In Lesson 10-1, you learned that like terms, such as $-5d$ and $2d$, can be combined using the Distributive Property.

$$-5d + 2d = (-5 + 2)d \quad \text{Distributive Property}$$
$$= -3d \quad \text{Simplify.}$$

The terms of an expression are also called monomials. A **monomial** is a number, a variable, or a product of numbers and/or variables. An algebraic expression that is the sum or difference of one or more monomials is called a **polynomial**.

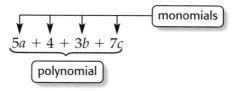

$$\underbrace{5a + 4 + 3b + 7c}_{\text{polynomial}} \quad \text{monomials}$$

You have already learned how to simplify polynomials like $3x + 4 + 2x - 8$ by combining like terms. You can use the same process to simplify polynomials containing more than one variable.

EXAMPLE **Simplify a Polynomial**

1 **Simplify $-5d + 2n + 4d - 3n$.**

The like terms in this expression are $-5d$ and $4d$, and $2n$ and $-3n$.

$-5d + 2n + 4d - 3n$	Write the polynomial.
$= -5d + 2n + 4d + (-3n)$	Definition of subtraction
$= (-5d + 4d) + [2n + (-3n)]$	Group like terms.
$= -1d + (-1n)$ or $-d - n$	Simplify by combining like terms.

The expression $2x^2$ is another example of a monomial, since it is the product of 2, x, and x. You can simplify expressions like $2x^2 + 4 - x^2$ using algebra tiles.

EXAMPLE Simplify Polynomials

2 **Simplify $2x^2 + 4 - x^2$.**

Use the definition of subtraction to write this polynomial as $2x^2 + 4 + (-x^2)$.

STUDY TIP

Look Back To review **zero pair**s, see Lesson 1-4.

Method 1 Use models.

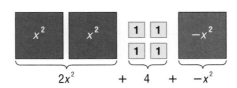

$$2x^2 \quad + \quad 4 \quad + \quad -x^2$$

Group tiles with the same shape and remove zero pairs.

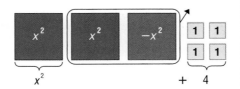

$$x^2 \qquad\qquad + \quad 4$$

Thus, $2x^2 + 4 - x^2 = x^2 + 4$.

Method 2 Use symbols.

Write the polynomial. Then group and add like terms.

$2x^2 + 4 + (-x^2)$
$= [2x^2 + (-x^2)] + 4$
$= [2x^2 + (-1x^2)] + 4$
$= 1x^2 + 4$
$= x^2 + 4$

From these examples, you can see that like terms must have the same variable and the same power. Thus, $2x^2$ and $3x^2$ are like terms, while $4x^2$ and $5x$ are not.

EXAMPLE Simplify Polynomials

3 **Simplify $x^2 - 1 - x + 3 + 2x$.**

$x^2 - 1 - x + 3 + 2x$ is equal to $x^2 + (-1) + (-1x) + 3 + 2x$.

Method 1 Use models.

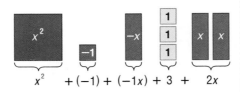

$$x^2 \quad +(-1)+ (-1x) + 3 \quad + \quad 2x$$

Group tiles with the same shape and remove zero pairs.

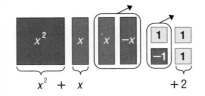

$$x^2 \quad + \quad x \qquad\qquad +2$$

Method 2 Use symbols.

Write the polynomial. Then group and add like terms.

$x^2 + (-1) + (-1x) + 3 + 2x$
$= x^2 + (-1x + 2x) + (-1 + 3)$
$= x^2 + 1x + 2$
$= x^2 + x + 2$

STUDY TIP

Standard Form When simplifying polynomials, it is customary to write the result in *standard form*; that is, with the powers of the variable decreasing from left to right.
$-5x^2 + 3x + 2$, *not* $3x - 5x^2 + 2$

Thus, $x^2 - 1 - x + 3 + 2x = x^2 + x + 2$.

1. **OPEN ENDED** Write a polynomial with four terms that simplifies to $5a - 9b$.

2. **Explain** why $6x$ and $3x^2$ are not like terms.

3. **Which One Doesn't Belong?** Identify the expression that is not a like term. Explain your reasoning.

$$2y^2 \qquad -x^2 \qquad 5x^2 \qquad -4x^2$$

GUIDED PRACTICE

Simplify each polynomial. If the polynomial cannot be simplified, write *simplest form*.

4. $4c + 5d + 6d + c$
5. $-7x - 8 - 2y$
6. $9g - 9h + 3g + 1$

7. $4x + x^2 + 2x$
8. $-x^2 + 5 + 3x - 1 + x^2$

9. $-5w^2 + 3w^2 - 8w$
10. $-9m^2 + 4m - m + 2$
11. $6g^2 + 5 - 7g + 3 - 8g^2$

Practice and Applications

Simplify each polynomial. If the polynomial cannot be simplified, write *simplest form*.

12. $6a + 8b - 7a + b$
13. $5x + 7y + 8 - z$
14. $-n + 4p + 5 - 6n$

15. $3f - 2g - 9g + 5f$
16. $-8c - d + 4c + 2$
17. $2j + 7 + k - 9$

18. $2x^2 - 3x + x$
19. $-3x^2 + 2x + x^2$

20. $-2x^2 + 4 - x - 3 + 2x$
21. $x^2 + x + 2 - x^2 - 3x - 5$

22. $m^2 + m - 3$
23. $a + 5a^2 - 7a$
24. $4 - 3x^2 + 6x + x^2$

25. $2w^2 - 6w - w + 1$
26. $3k^2 + 4 - 8k + k - 2$
27. $y^2 + 8y + 1 + 7y^2 - 4$

28. $a^2 + 3a^2 - 4a + a - 7 - 1$
29. $-z^2 + z^2 - 5z + 9z - 2 + 13$

30. $b^2 + 6b - 9 + b^2 - b + 3$
31. $r^2 - 3r + 8 + 2r^2 - 4r + 4$

32. $11 - 4n^3 - 8 + n^3 - 4n^3$
33. $-5t^3 - 8t^2 + 4t - 6 + 7t^3 + 3t$

34. $1.4x^2 - 3.8x + 1.2x^2 + 4.5x$
35. $\frac{3}{4}y^2 - 5y - \frac{1}{4}y + 5y$

HOMEWORK HELP

For Exercises	See Examples
12–17, 36	1
18–35, 37	2, 3

Extra Practice
See pages 646, 659.

36. COOKIES The table shows the number of boxes of each type of cookie Orlando and Emma bought from Science Club members. If m represents the cost of mint cookies, p the cost of peanut butter cookies, and c the cost of chocolate chip cookies per box, write an expression in simplest form for the total amount spent by Orlando and Emma on cookies.

Name	Mint	Peanut Butter	Chocolate Chip
Orlando	2 boxes	1 box	0 boxes
Emma	0 boxes	2 boxes	3 boxes

37. SAVINGS Shanté receives $50 each birthday from her aunt. Her parents put this money in a savings account with an interest rate of r. The table gives the account balance after each birthday. Write the balance of Shanté's account after her third birthday in simplest form.

Birthday	Balance ($)
1	50
2	$(50r + 50) + 50$
3	$(50r^2 + 100r + 50) + (50r + 50) + 50$

38. CRITICAL THINKING Determine whether $2x^2 + 3x = 5x^2$ is *sometimes*, *always*, or *never* true for all x. Explain your reasoning.

Standardized Test Practice and Mixed Review

FCAT Practice

39. MULTIPLE CHOICE Simplify $x^2 - 4x - 5x + 3 - 2x^2 + 9$.

   Ⓐ $3x^2 - 9x + 12$ Ⓑ $2x^2$

   Ⓒ $-2x^2 - 9x + 12$ Ⓓ $-x^2 - 9x + 12$

40. MULTIPLE CHOICE Write the perimeter of the figure in simplest form.

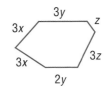

   Ⓕ $14xyz$ Ⓖ $15xyz$

   Ⓗ $6x + 5y + 3z$ Ⓘ $6x + 5y + 4z$

Graph each function. (Lesson 12-2)

41. $y = 5x^2$ **42.** $y = x^2 + 5$ **43.** $y = x^2 - 4$ **44.** $y = -x^2 - 3$

45. BIOLOGY The table shows how long it took for the first 400 bacteria cells to grow in a petri dish. Is the growth of the bacteria a linear function of time? Explain. (Lesson 12-1)

Time (min)	46	53	57	60
Number of Cells	100	200	300	400

GETTING READY FOR THE NEXT LESSON

PREREQUISITE SKILL Rewrite each expression using parentheses so that the like terms are grouped together. (Lessons 1-2 and 10-1)

46. $(a + 2) + (3a + 4)$ **47.** $(2n + 5) + (5n + 1)$

48. $(c + d) + (7c - 2d)$ **49.** $(x^2 + 4x) + (6x^2 - 8x)$

12-4 Adding Polynomials

What You'll Learn
Add polynomials.

HANDS-ON Mini Lab

Materials
- algebra tiles

Work with a partner.

Consider the polynomials $3x^2 - 2x + 1$ and $-x^2 + 3x - 4$ modeled below.

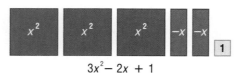

$3x^2 - 2x + 1$

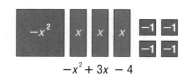

$-x^2 + 3x - 4$

Follow these steps to add the polynomials.

STEP 1 Combine the tiles that have the same shape.

STEP 2 Remove any zero pairs.

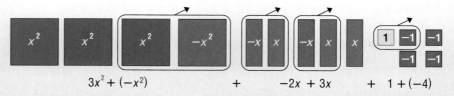

$3x^2 + (-x^2)$ $+$ $-2x + 3x$ $+$ $1 + (-4)$

1. Write the polynomial for the tiles that remain.
2. Use algebra tiles to find $(x^2 + x - 2) + (6x^2 - 5x - 1)$.

You can add polynomials horizontally or vertically by combining like terms.

EXAMPLES Add Polynomials

1 Find $(4x + 1) + (2x + 3)$.

Method 1 Add vertically.

$$\begin{array}{r} 4x + 1 \\ (+)\ 2x + 3 \quad \text{Align like terms.} \\ \hline 6x + 4 \quad \text{Add.} \end{array}$$

The sum is $6x + 4$.

Method 2 Add horizontally.

$(4x + 1) + (2x + 3)$ **Associative and**
$= (4x + 2x) + (1 + 3)$ **Commutative**
$= 6x + 4$ **Properties**

2 Find $(3x^2 + 5x - 9) + (x^2 + x + 6)$.

Method 1 Add vertically.

$$\begin{array}{r} 3x^2 + 5x - 9 \\ (+)\ x^2 + x + 6 \\ \hline 4x^2 + 6x - 3 \end{array}$$

Method 2 Add horizontally.

$(3x^2 + 5x - 9) + (x^2 + x + 6)$
$= (3x^2 + x^2) + (5x + x) + (-9 + 6)$
$= 4x^2 + 6x - 3$

The sum is $4x^2 + 6x - 3$.

574 Chapter 12 Algebra: Nonlinear Functions and Polynomials

Adding Vertically
When adding vertically, be sure to correctly identify the terms of each polynomial. For example, the last term of the polynomial $2x^2 - 3$ is -3, not 3.

EXAMPLES Add Polynomials

③ Find $(7y^2 + 2y) + (-5y + 8)$.

$(7y^2 + 2y) + (-5y + 8) = 7y^2 + (2y - 5y) + 8$ Group like terms.

$\qquad\qquad\qquad\qquad = 7y^2 - 3y + 8$ Simplify.

The sum is $7y^2 - 3y + 8$.

④ Find $(6x^2 - x + 5) + (2x^2 - 3)$.

$$
\begin{array}{r}
6x^2 - x + 5 \\
(+)\ 2x^2\quad\ - 3 \\
\hline
8x^2 - x + 2
\end{array}
$$

> Leave a space because there is no other term like $-x$.

The sum is $8x^2 - x + 2$.

Your Turn Add.

a. $(4x + 3) + (x - 1)$ b. $(10a^2 + 5a + 7) + (a^2 - 3)$

Polynomials are often used to represent measures of geometric figures.

EXAMPLE Use Polynomials to Solve a Problem

⑤ MULTIPLE-CHOICE TEST ITEM Find the measure of $\angle B$ in the figure at the right.

Ⓐ 41°   Ⓑ 63°   Ⓒ 76°   Ⓓ 166°

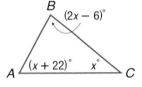

Read the Test Item The figure is a triangle. The sum of the measures of the angles of a triangle equals 180°. The measure of each angle is determined by the value of x.

Solve the Test Item

Write an equation to find the value of x.

$$\underbrace{(2x - 6) + (x + 22) + x}_{\text{measures of the angles}} \quad \underset{\text{equals}}{=} \quad \underset{180}{180}$$

The sum of the measures of the angles

$(2x - 6) + (x + 22) + x = 180$ Write the equation.

$(2x + x + x) + (-6 + 22) = 180$ Group like terms.

$\qquad\qquad\qquad\quad 4x + 16 = 180$ Simplify.

$\qquad\qquad\qquad\quad\ \underline{- 16 = - 16}$ Subtract 16 from each side.

$\qquad\qquad\qquad\quad 4x\qquad = 164$ Simplify.

$\qquad\qquad\qquad\quad \dfrac{4x}{4} = \dfrac{164}{4}$ Divide each side by 4.

$\qquad\qquad\qquad\quad\quad x = 41$ Simplify.

Find the measure of angle B.

$m\angle B = 2x - 6$ Write the expression for the measure of angle B.

$\quad\ = 2(41) - 6$ Replace x with 41.

$\quad\ = 82 - 6$ or 76   Simplify.

The measure of $\angle B$ is 76°. The answer is C.

Test-Taking Tip
The Princeton Review

Many standardized tests provide a list of common geometry facts and formulas. Be sure to find this list before the test begins so you can refer to it easily.

1. **OPEN ENDED** Write two polynomials whose sum is $4x - 5y$.

2. **FIND THE ERROR** Benito and Cleavon are adding $5a^2 - 7a$ and $3a^2 + 2$. Who is correct? Explain.

> Benito
> $5a^2 - 7a$
> $(+) \ 3a^2 + 2$
> ——————
> $8a^2 - 5a$

> Cleavon
> $5a^2 - 7a$
> $(+) \ 3a^2 \qquad + 2$
> ——————
> $8a^2 - 7a + 2$

GUIDED PRACTICE

Add.

3. $\quad h + 3$
   $(+) \ 2h + 1$

4. $\quad 2b^2 + 6b + 9$
   $(+) \ b^2 + 2b - 7$

5. $\quad 4t^2 + t + 1$
   $(+) \ 3t^2 \qquad - 5$

6. $(6g^2 - g + 3) + (-2g^2 - 3g + 1)$

7. $(8f^2 - 3f) + (5f - 9)$

GEOMETRY For Exercises 8–10, use the rectangle at the right.

8. Write an expression in simplest form for the perimeter of the rectangle.

9. Find the value of x if the perimeter of the rectangle is 48.

10. Find the measure of the length and the width of the rectangle.

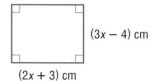

$(3x - 4)$ cm

$(2x + 3)$ cm

Practice and Applications

Add.

11. $\quad 5y + 6$
    $(+) \ 2y + 4$

12. $\quad 5p^2 + 3$
    $(+) \ 8p^2 + 1$

13. $\quad s^2 - s - 4$
    $(+) \ 4s^2 + 2s - 5$

HOMEWORK HELP	
For Exercises	See Examples
11–30	1–4
39–40	5

Extra Practice
See pages 646, 659.

14. $\quad k^2 + 6k - 2$
    $(+) \ 7k^2 - 3k - 1$

15. $\quad 4m^2 + m - 5$
    $(+) \ 3m^2 \qquad + 9$

16. $\quad 8x^2 - 6x - 7$
    $(+) \ -4x^2 - 6x$

17. $(2c - 4) + (3c + 3)$

18. $(9z + 6) + (-5z - 6)$

19. $(7j^2 + j + 1) + (j^2 - 5j - 2)$

20. $(4q^2 - 2q - 1) + (q^2 + 5q + 1)$

21. $(5d^2 - 6) + (3d^2 + 5)$

22. $(-9w^2 + 4) + (4w^2 - 9)$

23. $(4n^2 + 8) + (2n^2 - 5n + 1)$

24. $(-6r - 2) + (r^2 + 9r + 4)$

25. $(5v^2 - v + 1) + (v^2 + v + 1)$

26. $(6x^2 - 5x - 4) + (-x^2 - 8x - 9)$

27. $(5m^2 - 2) + (4m + 6)$

28. $(-3g - 10) + (6g^2 + 7g)$

29. $(-2b^2 - 3b - 7) + (5b + 2)$

30. $(-3a^2 - 2a - 9) + (-3a^2 - 5a - 3)$

Add. Then evaluate each sum if $x = 6$, $y = 3$, and $z = -5$.

31. $(6x + 2y) + (-4x - y)$

32. $(-3y + 5z) + (10y - 2z)$

33. $(-3x + 4z) + (5y - 2z)$

34. $(4x - 6y - 13z) + (-3x - 4y + 11z)$

WORK For Exercises 35–38, use the following information.

Wei-Ling works at a grocery store a few hours after school on weekdays and baby-sits on weekends. She makes the same hourly wage for both jobs. During one week, Wei-Ling worked 18 hours at the grocery store, and $9 was deducted for taxes. She worked 7 hours baby-sitting, and no taxes were deducted. Let x represent her hourly pay.

35. Write a polynomial expression to represent Wei-Ling's grocery store pay.

36. Write a polynomial expression to represent Wei-Ling's pay for baby-sitting.

37. Write a polynomial expression to represent Wei-Ling's total weekly pay.

38. Suppose Wei-Ling makes $5.50 an hour at both jobs. How much was her weekly pay after taxes?

GEOMETRY For Exercises 39 and 40, use the figure at the right.

39. Find the sum of the measures of the angles.

40. **MULTI STEP** The sum of the measures of the angles in any quadrilateral is 360°. Find the measure of each angle.

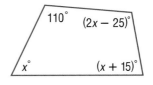

41. **CRITICAL THINKING** If $(3a - 5b) + (2a + 3b) = 5a - 2b$, then what is $(5a - 2b) - (3a - 5b)$? Explain.

Standardized Test Practice and Mixed Review

FCAT Practice

42. **MULTIPLE CHOICE** What is the sum of $14n + m$ and $n - 9m$?

   Ⓐ $13n - 8m$ Ⓑ $14n - 9m$ Ⓒ $15n - 10m$ Ⓓ $15n - 8m$

43. **SHORT RESPONSE** Find the measure of each angle in the figure at the right.

Simplify each polynomial. If the polynomial cannot be simplified, write *simplest form*. (Lesson 12-3)

44. $3t + 2s + s + 8t$

45. $7v - 10w + 2$

46. $6f + 9e - 2e + 16$

47. $4q^2 - q - 7 + 6q + 2$

48. **SKYDIVING** The distance d a skydiver falls in t seconds is given by the function $d = 16t^2$. Graph this function and estimate how far a skydiver will fall in 5.5 seconds. (Lesson 12-2)

Find the total amount in each account to the nearest cent. (Lesson 5-9)

49. $250 at 4% for 2.5 years

50. $760 at 5% for 10 months

51. $375 at 9.4% for 14 years

52. $1,200 at 2.2% for $3\frac{1}{3}$ years

GETTING READY FOR THE NEXT LESSON

PREREQUISITE SKILL Rewrite each expression as an addition expression by using the additive inverse. (Lessons 1-4 and 1-5)

53. $6 - 7$ **54.** $a^2 - 8$ **55.** $4x - 5y$ **56.** $(c + d) - 3c$

Vocabulary and Concepts

1. **Describe** the difference between the graphs of linear functions and nonlinear functions. (Lesson 12-1)

2. **OPEN ENDED** Write two polynomials whose sum is $5x - 3y$. (Lesson 12-4)

Skills and Applications

Determine whether each equation or table represents a *linear* or *nonlinear* function. Explain. (Lesson 12-1)

3. $3y = x$

4. $y = 5x^3 + 2$

5.

x	1	3	5	7
y	−5	−6	−7	−8

6.

x	−1	0	1	2
y	1	0	1	4

Graph each function. (Lesson 12-2)

7. $y = 2x^2$

8. $y = -x^2 + 3$

9. $y = 4x^2 - 1$

Simplify each polynomial. If the polynomial cannot be simplified, write *simplest form.* (Lesson 12-3)

10. $3x + 2 - 5x + 1$

11. $6a^2 + 5x - 2a$

12. $y^2 + 3y + 1 + 5y - 2y^2$

13. $3x^2 - 6x + 5x + 8$

Add. (Lesson 12-4)

14. $(3a + 6) + (2a - 5)$

15. $(-3x - 2) + (-2x + 5)$

16. $(3q^2 - 5) + (2q^2 - q)$

17. $(a^2 - 2a + 3) + (3a^2 - 5a + 6)$

18. **AMUSEMENT PARK RIDES** Your height h above the ground t seconds after being released at the top of a free-fall ride is given by the function $h = -16t^2 + 200$. Graph this function. After about how many seconds will the ride be 60 feet above the ground? (Lesson 12-2)

Standardized Test Practice

FCAT Practice

19. **MULTIPLE CHOICE** Which expression is *not* a monomial? (Lesson 12-3)

   Ⓐ -6

   Ⓑ x^3

   Ⓒ $4a$

   Ⓓ $\dfrac{4}{n}$

20. **SHORT RESPONSE** Find the measure of each angle in the figure below. (Lesson 12-4)

$(4x - 2)°$

$(3x + 1)°$

The Game Zone

A Place To Practice Your Math Skills

Math Skill
Adding Polynomials

Polynomial Challenge

GET READY!

Players: four
Materials: stopwatch, algebra tiles, scissors, 10 index cards

GET SET!

- Cut each index card in half to make 20 playing cards.

- Each player should write a polynomial with four or five terms that can be modeled with algebra tiles on each of five cards.

- At least two of the polynomials should contain one or more positive or negative x^2-terms.

$$2x - 4 + x^2 - 3$$

$$-5x^2 + 3x - 7 + 2x - 1$$

GO!

- Mix the cards and place the stack facedown on the table.

- The first player turns over the top two cards and lays them side-by-side on the table.

- The player then has one minute to model the two polynomials using algebra tiles and find the sum.

- If the player is correct, he or she scores one point. Those cards are then placed in a discard pile and it becomes the next player's turn.

- **Who Wins?** The first player to score 5 points wins.

12-5 Subtracting Polynomials

⟳ REVIEW Vocabulary

additive inverse: a number and its opposite (Lesson 1-4)

⟨H⟩ HANDS-ON Mini Lab

Materials
• algebra tiles

Work with a partner.

You can use algebra tiles to find $(x + 4) - (-2x + 3)$.

STEP 1 Model the polynomial $x + 4$.

STEP 2 To subtract $-2x + 3$, you need to remove 2 negative x-tiles and 3 1-tiles.

STEP 3 Since there are no negative x-tiles to remove, add 2 zero pairs of x-tiles. Then remove 2 negative x-tiles and 3 1-tiles.

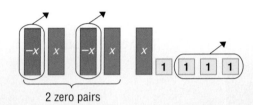

2 zero pairs

1. From the tiles that remain, determine the value of $(x + 4) - (-2x + 3)$.

2. Use algebra tiles to find $(2x^2 + 3x + 5) - (x^2 - x + 2)$.

As with adding polynomials, to subtract two polynomials, you subtract the like terms.

EXAMPLES Subtract Polynomials

Subtract.

① $(7a + 5) - (3a + 4)$

$$
\begin{array}{r}
7a + 5 \\
(-)\ 3a + 4 \\
\hline
4a + 1
\end{array}
$$

Align like terms.

Subtract.

The difference is $4a + 1$.

② $(5x^2 + 3x + 4) - (3x^2 - 2)$

$$
\begin{array}{r}
5x^2 + 3x + 4 \\
(-)\ 3x^2 \qquad - 2 \\
\hline
2x^2 + 3x + 6
\end{array}
$$

Align like terms.

Subtract.

The difference is $2x^2 + 3x + 6$.

Recall that you can subtract a number by adding its *additive inverse*. You can also subtract a polynomial by adding its additive inverse. To find the additive inverse of a polynomial, find the opposite of each term.

Polynomial	Terms	Opposites	Additive Inverse
$x + 5$	$x, 5$	$-x, -5$	$-x - 5$
$-x^2 - 4x + 2$	$-x^2, -4x, 2$	$x^2, 4x, -2$	$x^2 + 4x - 2$

EXAMPLES **Subtract Using the Additive Inverse**

3 **Find $(4x + 9) - (7x - 2)$.**

The additive inverse of $7x - 2$ is $-7x + 2$.

$(4x + 9) - (7x - 2)$

$= (4x + 9) + (-7x + 2)$ To subtract $(7x - 2)$, add $(-7x + 2)$.

$= (4x - 7x) + (9 + 2)$ Group like terms.

$= -3x + 11$ Simplify by combining like terms.

The difference is $-3x + 11$.

4 **Find $(6y^2 - 5) - (-3y + 4)$.**

The additive inverse of $-3y + 4$ is $3y - 4$.

$$\begin{array}{r} 6y^2 \quad\;\; - 5 \\ (-)\quad -3y + 4 \end{array} \quad\longrightarrow\quad \begin{array}{r} 6y^2 \quad\;\; - 5 \\ (+)\quad\;\; 3y - 4 \\ \hline 6y^2 + 3y - 9 \end{array}$$

The difference is $6y^2 + 3y - 9$.

Your Turn **Subtract.**

a. $(5p + 3) - (12p - 8)$ **b.** $(x^2 - 6x + 4) - (2x^2 - 7x - 1)$

EXAMPLE **Use Polynomials to Solve a Problem**

5 **CARS** Car A travels a distance of $4t^2 + 60t$ feet t seconds after the start of a soapbox derby. Car B travels $5t^2 + 55t$ feet. How far apart are the two cars 8 seconds after the start of the race?

Write an expression for the difference of the distances traveled by each car.

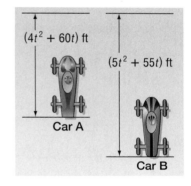

$(4t^2 + 60t)$ ft

$(5t^2 + 55t)$ ft

Car A

Car B

Words	car B's distance minus car A's distance
Variables	t = the time in seconds
Expression	$(5t^2 + 55t) - (4t^2 + 60t)$

$$\begin{array}{r} 5t^2 + 55t \\ (-)\;(4t^2 + 60t) \end{array} \quad\longrightarrow\quad \begin{array}{r} 5t^2 + 55t \\ (+)\;(-4t^2 - 60t) \\ \hline t^2 - \;\; 5t \end{array}$$

Now evaluate this expression for a time of 8 seconds.

$t^2 - 5t = (8)^2 - 5(8)$ Replace t with 8.

$= 64 - 40$ or 24   Simplify.

After 8 seconds, the cars are 24 feet apart.

1. **NUMBER SENSE** Write the opposite of each term in $4x^2 - 8x + 9$. Then write the additive inverse of this polynomial.

2. **OPEN ENDED** Write two polynomials whose difference is $3x - 8$.

3. **FIND THE ERROR** Karen and Yoshi are finding $(3a^2 - 3a + 5) - (2a^2 + a - 1)$. Who is correct? Explain.

Karen
$(3a^2 - 3a + 5) - (2a^2 + a - 1)$
$= (3a^2 - 3a + 5) + (-2a^2 - a + 1)$
$= a^2 - 4a + 6$

Yoshi
$(3a^2 - 3a + 5) - (2a^2 + a - 1)$
$= (3a^2 - 3a + 5) + (-2a^2 + a - 1)$
$= a^2 - 2a + 4$

GUIDED PRACTICE

Subtract.

4. $\quad 5z + 2$
$\underline{(-)\ 3z + 1}$

5. $\quad 7c^2 + c + 5$
$\underline{(-)\ 2c^2\quad\ \ + 4}$

6. $\quad 2m^2 + 6m + 8$
$\underline{(-)\ m^2 + 3m - 1}$

7. $(6p + 2) - (p - 1)$

8. $(x^2 - x - 4) - (-x + 1)$

9. $(5n^2 + n - 2) - (3n^2 + 2n - 1)$

10. $(r^2 + r - 1) - (2r^2 - r + 2)$

11. Find the difference of $-4a + 5$ and $a - 1$.

Practice and Applications

Subtract.

12. $\quad 3x + 6$
$\underline{(-)\ 2x + 5}$

13. $\quad 9w + 15$
$\underline{(-)\ 4w + 12}$

14. $\quad 8g^2 + 8g + 5$
$\underline{(-)\ 7g^2 + 5g + 1}$

15. $\quad 10b^2 - 4b + 9$
$\underline{(-)\ 5b^2 +\ b + 3}$

16. $\quad 4u^2 + 3u + 2$
$\underline{(-)\ 2u^2\qquad - 4}$

17. $\quad 7y^2 + y + 6$
$\underline{(-)\ 5y^2\qquad + 1}$

HOMEWORK HELP

For Exercises	See Examples
12–17	1, 2
18–25	3, 4
32–40	5

Extra Practice
See pages 646, 659.

18. $(10h + 4) - (2h - 3)$

19. $(6a + 6) - (-a + 8)$

20. $(4m^2 - 8) - (-3m + 2)$

21. $(5k^2 - 7) - (9k + 13)$

22. $(c^2 - 2c + 1) - (c^2 + c - 5)$

23. $(3r^2 + r - 1) - (r^2 - r + 3)$

24. Find the difference of $7x^2 + 12x - 9$ and $4x^2 - 3$.

25. What is $-z - 5$ subtracted from $16z - 7$?

Subtract. Then evaluate the difference if $x = -8$ and $y = 5$.

26. $(4x + 10) - (3x + 7)$

27. $(6y - 2) - (2y + 6)$

28. $(-3x - 8) - (y - 5)$

29. $(9x + 2y) - (8x - 4)$

30. $(x + 5y) - (-4x + 3y)$

31. $(-2x - y) - (-6x - 3y)$

32. **GEOMETRY** The measure of $\angle ABC$ is $(12x - 8)°$. Write an expression in simplest form for the measure of $\angle ABD$.

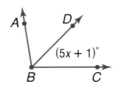

$(5x + 1)°$

FAST FOOD For Exercises 33–36, use the following information.
Khadijah ordered 3 burritos and 7 tacos from a fast-food drive through. When Khadijah looked at her receipt, she discovered that she had been charged for 5 burritos and 5 tacos.

Order # 368

5 burritos......$7.95
5 tacos.........$4.95

Total...........$12.90

33. If burritos cost b dollars and tacos cost t dollars, write an expression for the amount Khadijah was charged.

34. Write an expression for the cost of the food she ordered.

35. Write an expression for the amount Khadijah was overcharged.

36. If tacos cost $0.99 and burritos cost $1.59, how much was she overcharged?

FUND-RAISER For Exercises 37–40, use the following information.
Your club spends $200 on a pizza fund-raiser kit. Each pizza costs you $6.50 to make. You sell each pizza for $10.

37. Write a polynomial that models your total expenses for making x pizzas.

38. Write a polynomial that models your income from selling x pizzas.

39. Write a polynomial that models your profit from selling x pizzas.
(*Hint*: Profit = Income − Expenses)

40. How much profit will you make if you sell 150 pizzas?

41. **CRITICAL THINKING** Suppose A and B represent polynomials. If $A + B = 7x + 4$ and $A - B = 3x + 2$, find A and B.

Standardized Test Practice and Mixed Review

 FCAT Practice

42. **MULTIPLE CHOICE** Write the additive inverse of $n^2 - 2n + 3$.

   Ⓐ $n^2 + 2n - 3$ Ⓑ $-n^2 + 2n - 3$

   Ⓒ $-n^2 - 2n - 3$ Ⓓ $n^2 + 2n + 3$

43. **SHORT RESPONSE** The perimeter of the triangle is $16x - 7$ units. Write an expression for the missing length.

$7x - 2$ $4x + 3$

Add. (Lesson 12-4)

44. $(7b + 2) + (-5b + 3)$ 45. $(6v^2 - 4) + (v - 1)$ 46. $(t^2 - 8t) + (t^2 + 5)$

SCHOOL For Exercises 47 and 48, use the following information.
The drama club is selling flowers. The sales for the first two weeks are shown in the table. (Lesson 12-3)

Number of Flowers Sold		
Week	Carnations	Roses
1	54	38
2	65	42

47. The selling prices of a carnation and a rose are C and R respectively. Write a polynomial expression for the total sales.

48. If carnations cost $2 each and roses cost $5 each, what was the total amount of sales?

GETTING READY FOR THE NEXT LESSON

PREREQUISITE SKILL Write each expression using exponents. (Lesson 2-8)

49. $3 \cdot 3 \cdot 3 \cdot 3$ 50. $5 \cdot 4 \cdot 5 \cdot 5 \cdot 4$ 51. $7 \cdot (7 \cdot 7)$ 52. $(2 \cdot 2) \cdot (2 \cdot 2 \cdot 2)$

Multiplying and Dividing Monomials

Sunshine State Standards
MA.A.2.3.1-3

WHEN am I ever going to use this?

SCIENCE The pH of a solution describes its acidity. Neutral water has a pH of 7. Lemon juice has a pH of 2. Each one-unit decrease in the pH means that the solution is 10 times more acidic. So a pH of 8 is 10 times more acidic than a pH of 9.

pH	Times More Acidic Than a pH of 9	Written Using Powers
8	10	10^1
7	$10 \times 10 = 100$	$10^1 \times 10^1 = 10^2$
6	$10 \times 10 \times 10 = 1{,}000$	$10^1 \times 10^2 = 10^3$
5	$10 \times 10 \times 10 \times 10 = 10{,}000$	$10^1 \times 10^3 = 10^4$
4	$10 \times 10 \times 10 \times 10 \times 10 = 100{,}000$	$10^1 \times 10^4 = 10^5$

1. Examine the exponents of the factors and the exponents of the products in the last column. What do you observe?

Exponents are used to show repeated multiplication. You can use this fact to help find a rule for multiplying powers with the same base.

2 factors   4 factors

$$3^2 \cdot 3^4 = \underbrace{(3 \cdot 3)} \cdot \underbrace{(3 \cdot 3 \cdot 3 \cdot 3)} \text{ or } 3^6$$

6 factors

Notice the sum of the original exponents and the exponent in the final product. This relationship is stated in the following rule.

Key Concept   Product of Powers

Words   To multiply powers with the same base, add their exponents.

Symbols   Arithmetic   Algebra

$$2^4 \cdot 2^3 = 2^{4+3} \text{ or } 2^7 \qquad a^m \cdot a^n = a^{m+n}$$

EXAMPLE   Multiply Powers

1 Find $5^2 \cdot 5$. Express using exponents.

$5^2 \cdot 5 = 5^2 \cdot 5^1$ $5 = 5^1$ **Check** $5^2 \cdot 5 = (5 \cdot 5) \cdot 5$

$ = 5^{2+1}$ The common base is 5. $= 5 \cdot 5 \cdot 5$

$ = 5^3$ Add the exponents. $= 5^3$ ✔

What You'll Learn
Multiply and divide monomials.

MATH Symbols

$x^5 \leftarrow$ exponent
$\llcorner$ base

x^5 x to the fifth power

STUDY TIP

Common Error
When multiplying powers, do not multiply the bases.
$4^5 \cdot 4^2 = 4^7$, not 16^7.

EXAMPLE **Multiply Monomials**

2 Find $-3x^2(4x^5)$. Express using exponents.

$$-3x^2(4x^5) = (-3 \cdot 4)(x^2 \cdot x^5)$$ Commutative and Associative Properties
$$= (-12)(x^{2+5})$$ The common base is x.
$$= -12x^7$$ Add the exponents.

 Your Turn Multiply. Express using exponents.

a. $9^3 \cdot 9^2$ **b.** $y^4 \cdot y^9$ **c.** $-2m(-8m^5)$

There is also a rule for dividing powers that have the same base.

Key Concept   **Quotient of Powers**

Words   To divide powers with the same base, subtract their exponents.

Symbols   **Arithmetic**   **Algebra**

$$\frac{3^7}{3^3} = 3^{7-3} \text{ or } 3^4$$ $$\frac{a^m}{a^n} = a^{m-n}, \text{ where } a \neq 0$$

EXAMPLES **Divide Powers**

Divide. Express using exponents.

 $\dfrac{4^8}{4^2}$ **4** $\dfrac{n^9}{n^4}$

$\dfrac{4^8}{4^2} = 4^{8-2}$ The common base is 4. $\dfrac{n^9}{n^4} = n^{9-4}$ The common base is n.

$\phantom{\dfrac{4^8}{4^2}} = 4^6$ Simplify. $\phantom{\dfrac{n^9}{n^4}} = n^5$ Simplify.

 Your Turn Divide. Express using exponents.

d. $\dfrac{5^7}{5^4}$ **e.** $\dfrac{x^{10}}{x^3}$ **f.** $\dfrac{12w^5}{2w}$

EXAMPLE **Divide Powers to Solve a Problem**

5 **SOUND** The loudness of a conversation is 10^6 times as intense as the loudness of a pin dropping, while the loudness of a jet engine is 10^{12} times as intense. How many times more intense is the loudness of a jet engine than the loudness of a conversation?

To find how many times more intense, divide 10^{12} by 10^6.

$$\frac{10^{12}}{10^6} = 10^{12-6}$$ Quotient of Powers

$$= 10^6$$ Simplify.

The loudness of a jet engine is 10^6 or 1,000,000 times as intense as the loudness of a conversation.

Skill and Concept Check

1. **Determine** whether the following statement is *true* or *false*.
   If you change the order in which you multiply two monomials, the product will be different.
   Explain your reasoning or give a counterexample.

2. **OPEN ENDED** Write a multiplication expression whose product is 4^{15} and a division expression whose quotient is 4^{15}.

3. **NUMBER SENSE** Is $\frac{2^{100}}{2^{99}}$ greater than, less than, or equal to 2?

GUIDED PRACTICE

Multiply or divide. Express using exponents.

4. $4^5 \cdot 4^3$

5. $3^6 \cdot 3$

6. $n^2 \cdot n^9$

7. $-2a(3a^4)$

8. $\frac{7^6}{7}$

9. $\frac{9c^7}{3c^2}$

Practice and Applications

Multiply or divide. Express using exponents.

10. $6^8 \cdot 6^5$

11. $7^3 \cdot 7^3$

12. $2^9 \cdot 2$

13. $11 \cdot 11^4$

14. $n \cdot n^7$

15. $b^{13} \cdot b$

16. $2g \cdot 7g^6$

17. $(3x^8)(5x)$

18. $-4a^5(6a^5)$

19. $(8w^4)(-w^7)$

20. $(-p)(-9p^2)$

21. $-5y^3(-8y^6)$

22. $\frac{3^9}{3^2}$

23. $\frac{4^{10}}{4^5}$

24. $\frac{8^4}{8}$

25. $\frac{10^{12}}{10}$

26. $\frac{r^7}{r^2}$

27. $\frac{x^{14}}{x^8}$

28. $\frac{14n^6}{7n}$

29. $\frac{24k^3}{8k^2}$

30. $xy^2(x^3y)$

31. $4a^2b^3(7ab^2)$

32. $\frac{20a^5b}{4ab}$

33. $\frac{16x^3y^2}{2x^2y}$

HOMEWORK HELP

For Exercises	See Examples
10–13, 34–35	1
14–21	2
22–33, 36–37	3, 4
38–43	5

Extra Practice
See pages 647, 659.

34. the product of seven to the tenth power and seven cubed

35. the quotient of n to the sixth power and n squared

36. What is the product of 2^6, 2, and 2^3?

37. Find $x^3 \cdot x^9 \div x^5$.

EARTHQUAKES For Exercises 38 and 39, use the information in the table at the right and below.
For each increase on the Richter scale, an earthquake's vibrations, or *seismic waves*, are 10 times greater.

38. How many times greater are the seismic waves of an earthquake with a magnitude of 6 than an aftershock with a magnitude of 3?

39. How many times greater were the seismic waves of the 1906 San Francisco earthquake than the 1998 Adana, Turkey, earthquake?

Earthquake	Richter Scale Magnitude
San Francisco, 1906	8.3
Adana, Turkey, 1998	6.3

40. LIFE SCIENCE A cell culture contains 2^6 cells. By the end of the day, there are 2^{10} times as many cells in the culture. How many cells are there in the culture by the end of the day?

41. ASTRONOMY Venus is approximately 10^8 kilometers from the Sun. The gas giant Saturn is more than 10^9 kilometers from the Sun. About how many times farther away from the Sun is Saturn than Venus?

42. GEOMETRY Find the volume of the rectangular prism.

43. POPULATION The continent of North America contains approximately 10^7 square miles of land. If the population doubles, there will be about 10^9 people on the continent. At that point, how many people will be on each square mile of land?

44. CRITICAL THINKING What is half of 2^{30}? Write your answer using exponents.

CRITICAL THINKING Divide.

45. $\dfrac{a^8}{a^8}$ **46.** $\dfrac{n^2}{n^5}$ **47.** $\dfrac{6x^7y^4}{3x^3y^9}$ **48.** $\dfrac{y^6}{y^{-4}}$ **49.** $\dfrac{a^8b^{-1}c^{-5}}{a^3b^2c^{-3}}$

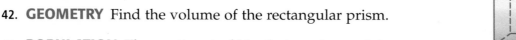

Standardized Test Practice and Mixed Review

FCAT Practice

50. MULTIPLE CHOICE Find the product of $-5x^2$ and $-6x^8$.

ⓐ $-11x^{10}$ ⓑ $-30x^{16}$ ⓒ $11x^{10}$ ⓓ $30x^{10}$

51. MULTIPLE CHOICE Find $\dfrac{(-2)^5}{(-2)^4}$.

ⓕ $(-2)^9$ ⓖ -2 ⓗ 1 ⓘ 2^9

Subtract. (Lesson 12-5)

52. $(3x + 8) - (5x + 1)$ **53.** $(5a - 2) - (3a - 4)$ **54.** $(6y^2 + 3y + 9) - (2y^2 + 8y + 1)$

SCHOOL For Exercises 55 and 56, use the following information and the table at the right.
Suppose your total number of grade points for the first semester was $2A + 2B + C$ and your total for the second semester was $A + 3B + D$.
(Lesson 12-4)

55. Add the polynomials to find your total grade points for the year.

56. Evaluate the sum by substituting the grade point value for each variable.

Grade	Grade Points
A	4
B	3
C	2
D	1
F	0

Find the mean, median, and mode of each set of data. Round to the nearest tenth if necessary. (Lesson 9-4)

57. 52, 57, 52, 33, 39, 43, 53   **58.** 19, 28, 25, 64, 64, 76, 18

GETTING READY FOR THE NEXT LESSON

PREREQUISITE SKILL Use the Distributive Property to write each expression as an equivalent algebraic expression. (Lesson 10-1)

59. $3(x + 4)$ **60.** $5(y - 2)$ **61.** $-2(n + 8)$ **62.** $-4(p - 6)$

What You'll Learn
Solve problems by making a model.

Make a Model

We need to arrange some of these square tables into a square that is open in the middle and has 10 tables on each side.

We have 35 tables. Do we have enough? Let's **make a model** using these tiles.

Explore	We want to know how many square tables it will take to make the outline of a 10-by-10 square.
Plan	Let's start by making a model of a 4-by-4 square and of a 5-by-5 square. Then, let's look for a pattern.
Solve	 4-by-4 square 2 groups of 4 and 2 groups of 2 5-by-5 square 2 groups of 5 and 2 groups of 3 For a 10-by-10 square we need $2 \times 10 + 2 \times 8$ or 36. We have 35 tables, so we need one more.
Examine	We get the same answer when we make 4 groups of tiles. Each group has 1 less tile than the length of the square. Since $4(10 - 1)$ is 36, our answer is reasonable.

4 groups of 3

Analyze the Strategy

1. **Explain** why building a model is an appropriate strategy for solving the problem.

2. **Draw a diagram** showing another way the students could have grouped the tiles to solve this problem. Use a 4 by 4 square.

3. **Write** a problem that can be solved by making a model. Describe the model. Then solve the problem.

Solve. Make a model.

4. **STICKERS** In how many different ways can three rectangular stickers be torn from a sheet of such stickers so that all three stickers are attached to each other? Draw each arrangement.

5. **GEOMETRY** A 10-inch by 12-inch piece of cardboard has a 2-inch square cut out of each corner. Then the sides are folded up and taped together to make an open box. Find the volume of the box.

Mixed Problem Solving

Solve. Use any strategy.

6. **PETS** Mrs. Harper owns both cats and canaries. Altogether her pets have thirty heads and eighty legs. How many cats does she have?

TOWER For Exercises 7 and 8, use the figure at the right.

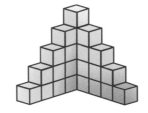

7. How many cubes would it take to build this tower?

8. How many cubes would it take to build a similar tower that is 12 cubes high?

9. **CARS** Yesterday you noted that the mileage on the family car read 60,094.8 miles. Today it reads 60,099.1 miles. Was the car driven about 4 or 40 miles?

10. **HOBBIES** Lorena says to Angela, "If you give me one of your baseball cards, I will have twice as many baseball cards as you have." Angela answers, "If you give me one of your cards, we will have the same numbers of cards." How many cards do each of the girls have?

11. **PARKING** Campus parking space numbers consist of three digits. They are typed on a slip of paper and given to students at orientation. Tara accidentally read her number upside-down. The number she read was 723 more than her actual parking space number. What is Tara's parking space number?

12. **SCIENCE** The light in the circuit will turn on if one or more switches are closed. How many combinations of open and closed switches will result in the light being on?

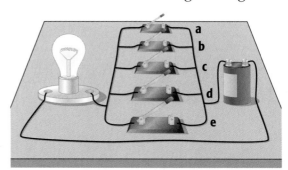

13. **CAMP** The camp counselor lists 21 chores on separate pieces of paper and places them in a basket. The counselor takes one piece of paper, and each camper takes one as the basket is passed around the circle. There is one piece of paper left when the basket returns to the counselor. How many people could be in the circle if the basket goes around the circle more than once?

14. **STANDARDIZED TEST PRACTICE** FCAT Practice

In how many different ways can five squares be arranged to form a single shape so that touching squares border on a full side? One arrangement is shown at the right.

Ⓐ 8

Ⓑ 12

Ⓒ 16

Ⓓ 20

You will use the make a model strategy in the next lesson.

12-7 Multiplying Monomials and Polynomials

HANDS-ON Mini Lab

Materials
• algebra tiles
• product mat

Algebra tiles can be used to form a rectangle whose length and width each represent a polynomial. The area of the rectangle is the product of the polynomials. Use algebra tiles to find $x(x + 3)$.

STEP 1 Use algebra tiles to mark off a rectangle with a width of x and a length of $x + 3$ on a product mat.

STEP 2 Using the marks as a guide, fill in the rectangles with algebra tiles.

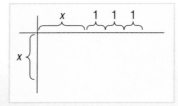 

1. What is $x(x + 3)$ in simplest form?

Use algebra tiles to find each product.

2. $x(x + 4)$ 3. $x(3x + 1)$ 4. $2x(x + 3)$

In Lesson 10-1, you learned how to rewrite an expression like $4(x + 3)$ using the Distributive Property. This is an example of multiplying a polynomial by a monomial.

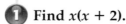

polynomial

monomial → $4(x + 3) = 4(x) + 4(3)$ Distributive Property

$= 4x + 12$ Simplify.

Often, the Distributive Property and the definition of exponents are needed to simplify the product of a monomial and a polynomial.

EXAMPLE Use the Distributive Property

1 Find $x(x + 2)$.

$x(x + 2) = x(x) + x(2)$ Distributive Property

$= x^2 + 2x$ $x \cdot x = x^2$

Use the Distributive Property

2 **Find $-5y(y + 8)$.**

$$-5y(y + 8) = -5y(y) + (-5y)(8) \quad \text{Distributive Property}$$
$$= -5y^2 + (-40y) \quad\quad -5 \cdot y \cdot y = -5y^2$$
$$= -5y^2 - 40y \quad\quad \text{Definition of subtraction}$$

Your Turn **Multiply.**

    a. $n(n - 9)$ **b.** $(10 + 2p)4p$ **c.** $-3x(6x - 4)$

Sometimes you may need to use the Product of Powers rule.

EXAMPLES **Use the Product of Powers Rule**

3 **Find $3n(n^2 - 7)$.**

$$3n(n^2 - 7) = 3n[n^2 + (-7)] \quad \text{Rewrite } n^2 - 7 \text{ as } n^2 + (-7).$$
$$= 3n(n^2) + 3n(-7) \quad \text{Distributive Property}$$
$$= 3n^3 + (-21n) \quad\quad 3n(n^2) = 3n^{1+2} \text{ or } 3n^3$$
$$= 3n^3 - 21n \quad\quad \text{Definition of subtraction}$$

4 **Find $2x(x^2 + 3x - 5)$.**

$$2x(x^2 + 3x - 5)$$
$$= 2x[x^2 + 3x + (-5)] \quad\quad \text{Rewrite } x^2 + 3x - 5 \text{ as } x^2 + 3x + (-5).$$
$$= 2x(x^2) + 2x(3x) + 2x(-5) \quad \text{Distributive Property}$$
$$= 2x^3 + 6x^2 + (-10x) \quad\quad \text{Simplify.}$$
$$= 2x^3 + 6x^2 - 10x \quad\quad \text{Definition of subtraction}$$

Your Turn **Multiply.**

    d. $5y(4y^2 - 2y)$ **e.** $a(a^2 - 4a + 6)$ **f.** $-4p(2p^2 - p + 3)$

Skill and Concept Check

Writing Math
Exercise 2

1. **OPEN ENDED** Write a polynomial with three terms and a monomial that contains a variable with a power of 1. Then find their product.

2. **FIND THE ERROR** Christopher and Stephanie are finding the product of $3x$ and $2x^2 - 3x + 8$. Who is correct? Explain.

     Christopher
     $3x(2x^2 - 3x + 8)$
     $= 6x^2 - 9x + 24$

     Stephanie
     $3x(2x^2 - 3x + 8)$
     $= 6x^3 - 9x^2 + 24x$

GUIDED PRACTICE

Multiply.

3. $m(m + 5)$ **4.** $(2w - 1)(3w)$ **5.** $-4x(x + 1)$

6. $k(k^2 - 7)$ **7.** $g(2g^2 - 5g + 9)$ **8.** $3z(4z^2 - 6z - 10)$

HOMEWORK HELP

For Exercises	See Examples
9–18, 25–26	1, 2
19–24	3, 4

Extra Practice
See pages 647, 659.

Multiply.

9. $r(r + 9)$

10. $t(t - 4)$

11. $(3b - 2)(3b)$

12. $(5x + 1)(2x)$

13. $-6d(d + 5)$

14. $-a(7a - 8)$

15. $6h(4 + 3h)$

16. $8w(1 - 7w)$

17. $11e(2e - 7)$

18. $10a(5a + 5)$

19. $4y(y^2 - 9)$

20. $-6g(2g^2 + 1)$

21. $t(t^2 + 5t + 9)$

22. $-n(3n^2 - 4n + 13)$

23. $-2r(4r^2 - r - 8)$

24. $11c(6c^2 - 8c + 1)$

25. **GARDENING** A square garden plot measures x feet on each side. Suppose you double the length of the plot and increase the width by 4 feet. Write two expressions for the area of the new plot.

26. **GEOMETRY** Write an expression in simplest form for the area of the figure.

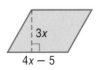

27. **CRITICAL THINKING** Draw a model showing how algebra tiles can be used to find the following product of two *binomials*, or polynomials with two terms: $(x + 2)(x + 3)$.

Standardized Test Practice and Mixed Review

FCAT
Practice

28. **MULTIPLE CHOICE** What is the product of $4x^2$ and $x^2 + 2x - 3$?

   Ⓐ $4x^2 + 8x - 12$

   Ⓑ $4x^4 + 8x^2 - 12x$

   Ⓒ $4x^4 + 8x^3 - 12x^2$

   Ⓓ $5x^2 + 6x + 1$

29. **SHORT RESPONSE** The length of a rectangle is twice its width. If the width is x, write an equation for the area A of the rectangle. Then graph the area as a function of x.

Multiply or divide. Express using exponents. (Lesson 12-6)

30. $5^2 \cdot 5$

31. $\dfrac{11^8}{11^5}$

32. $3x^3 \cdot 9x^3$

33. $\dfrac{21a^5}{3a^4}$

34. **BUSINESS** Allison's income from selling x beaded bracelets is $6.50x$. Her expenses are $4x + 35$. Write an expression for her profit. (Lesson 12-5)

WebQuest Interdisciplinary Project

Getting Down to Business

It's time to complete your project. Use the information and data you have gathered about the cost of materials and the feedback from your peers to prepare a video or brochure. Be sure to include a scatter plot with your project.

msmath3.net/webquest

Vocabulary and Concept Check

| monomial (p. 570) | nonlinear function (p. 560) | polynomial (p. 570) | quadratic function (p. 565) |

State whether each sentence is *true* or *false*. If *false*, replace the underlined word or number to make a true sentence.

1. The expression $x^2 - 3x$ is an example of a <u>monomial</u>.
2. A <u>nonlinear</u> function has a constant rate of change.
3. To <u>multiply</u> two polynomials, you combine like terms.
4. A quadratic function is a <u>nonlinear</u> function.
5. To divide powers with the same base, <u>subtract</u> the exponents.

Lesson-by-Lesson Exercises and Examples

12-1 Linear and Nonlinear Functions (pp. 560–563)

Determine whether each equation or table represents a *linear* or *nonlinear* function. Explain.

6. $y - 4x = 1$ 7. $y = x^2 + 3$

8.
x	2	3	4	5
y	1	3	7	12

Example 1 Determine whether the table represents a *linear* or *nonlinear* function.

x	y
−2	−3
−1	−1
0	1
1	3

As x increases by 1, y increases by 2. The rate of change is constant, so this function is linear.

12-2 Graphing Quadratic Functions (pp. 565–568)

Graph each function.

9. $y = -4x^2$ 10. $y = x^2 + 4$

11. **SCIENCE** A ball is dropped from the top a 36-foot tall building. The quadratic equation $d = -16t^2 + 36$ models the distance d in feet the ball is from the ground at time t. Graph the function. Then use your graph to find how long it takes for the ball to reach the ground.

Example 2 Graph $y = -x^2 - 1$.

Make a table of values. Then plot and connect the ordered pairs with a smooth curve.

x	$y = -x^2 - 1$	(x, y)
−2	$-(-2)^2 - 1$	(−2, −5)
−1	$-(-1)^2 - 1$	(−1, −2)
0	$-(0)^2 - 1$	(0, −1)
1	$-(1)^2 - 1$	(1, −2)
2	$-(2)^2 - 1$	(2, −5)

$y = -x^2 - 1$

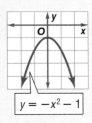

12-3 **Simplifying Polynomials** (pp. 570–573)

Simplify each polynomial. If the polynomial cannot be simplified, write *simplest form*.

12. $3a - b - 7a + 2 + 4b$

13. $8x - y + 1$

14. $3n^2 + 7n - 4n^2 + n$

Example 3 Simplify $8a^2 - 5a + 6 - 9a^2 - 6$.

$8a^2 - 5a + 6 - 9a^2 - 6$

$= 8a^2 + (-5a) + 6 + (-9a^2) + (-6)$

$= [8a^2 + (-9a^2)] + (-5a) + [6 + (-6)]$

$= -1a^2 + (-5a) + 0$ or $-a^2 - 5a$

12-4 **Adding Polynomials** (pp. 574–577)

Add.

15. $(3a^2 + 6a) + (2a^2 - 5a)$

16. $(b^2 - 2b + 4) + (2b^2 + b - 8)$

17. $(10m^2 + 5m - 9) + (-2m + 3)$

Example 4 Find $(3x^2 - 2) + (2x^2 + 5)$.

$(3x^2 - 2) + (2x^2 + 5)$

$= (3x^2 + 2x^2) + (-2 + 5)$ Group like terms.

$= 5x^2 + 3$ Simplify.

12-5 **Subtracting Polynomials** (pp. 580–583)

Subtract.

18. $(7g + 2) - (5g + 1)$

19. $(3c - 7) - (-3c + 4)$

20. $(7p^2 + 2p - 5) - (4p^2 + 6p - 2)$

21. $(6k^2 - 3) - (k^2 - 5k - 2)$

Example 5 Find $(5x - 1) - (6x + 4)$.

To subtract $6x + 4$, add $-6x - 4$.

$(5x - 1) - (6x + 4)$

$= (5x - 1) + (-6x - 4)$

$= [5x + (-6x)] + [-1 + (-4)]$

$= -1x + (-5)$

$= -x - 5$

12-6 **Multiplying and Dividing Monomials** (pp. 584–587)

Multiply or divide. Express using exponents.

22. $4 \cdot 4^5$

23. $-9y^2(-4y^9)$

24. $\dfrac{n^5}{n}$

25. $\dfrac{21c^{11}}{-7c^8}$

Example 6

Find $3a^3 \cdot 4a^7$.

$3a^2 \cdot 4a^7 = (3 \cdot 4)a^{3+7}$

$= 12a^{10}$

Example 7

Find $\dfrac{6^8}{6^3}$.

$\dfrac{6^8}{6^3} = 6^{8-3}$

$= 6^5$

12-7 **Multiplying Monomials and Polynomials** (pp. 590–592)

Multiply.

26. $a(a - 7)$

27. $(3y + 4)(3y)$

28. $-4n(n - 2)$

29. $p(p^2 - 6)$

30. $x(2x^2 + x - 5)$

31. $-2k(5k^2 - 3k + 8)$

Example 8 Find $-2x(5x + 3)$.

$-2x(5x + 3) = -2x(5x) + (-2x)(3)$

$= -10x^2 + (-6x)$

$= -10x^2 - 6x$

Vocabulary and Concepts

1. **OPEN ENDED** Write two polynomials whose difference is $-4n + 5$.

2. **State** whether the Quotient of Powers rule applies to $\frac{6^5}{3^2}$. Explain.

3. **Describe** the function $y = 3x^2$ using two different terms.

Skills and Applications

Determine whether each graph, equation, or table represents a *linear* or *nonlinear* function. Explain.

4.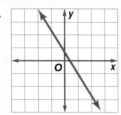

5.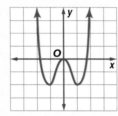

6. $2x = y$

7. $y = \frac{x}{7} + 3$

8.

x	-3	-1	1	3
y	2	10	18	26

9. Graph the function $y = -2x^2 + 3$.

Simplify each polynomial. If the polynomial cannot be simplified, write *simplest form*.

10. $-6x + 4y - 8 + y - 1$

11. $2a^2 + 4a + 3a^2 + 5a$

12. $10p + 7p^2 + 1$

Add or subtract.

13. $(4c^2 + 2) + (-4c^2 + 1)$

14. $(-x^2 + 2x - 5) + (4x^2 - 6x)$

15. $(9z^2 - 3z) - (5z^2 + 8z)$

16. $(5n^2 - 4n + 1) - (4n - 5)$

17. **GEOMETRY** Write an expression for the measure of $\angle JKM$. Then find the value of x.

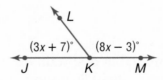

Multiply or divide. Express using exponents.

18. $15^3 \cdot 15^5$

19. $-5m^6(-9m^8)$

20. $\frac{3^{15}}{3^7}$

21. $\frac{-40w^8}{8w}$

Multiply.

22. $8n(n + 3)$

23. $g^3(6g - 5)$

24. $-4x(3x^2 - 6x + 8)$

FCAT Practice

Standardized Test Practice

Ⓐ Ⓑ Ⓒ Ⓓ

25. **MULTIPLE CHOICE** Find the value of ● in the equation $\frac{x^9}{x^●} = x^3$.

Ⓐ 3   Ⓑ 6   Ⓒ 12   Ⓓ 27

PART 1 Multiple Choice

Record your answers on the answer sheet provided by your teacher or on a sheet of paper.

1. Nolan wants to make a larger sail for his model boat. How long will the base of the new sail be in inches? (Lesson 4-5)

    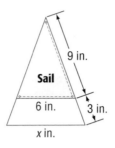

    A 2   B 8

    C 27   D 36

2. Which statement is false? (Lesson 5-5)

    F 42% of 60 is greater than 24.

    G 31% of 90 is greater than 30.

    H 79% of 250 is less than 200.

    I 3% of 80 is less than 3.

3. A group of dancers form a circle for a routine they are performing. The radius of their circle is 8 yards. If they increase the area of their circle by 4 times, what will be the radius, in yards, of the new circle? (Lesson 7-2)

    A 2   B 12   C 16   D 32

4. Fourteen dogs are enrolled in dog-training class. All the dogs weigh about 50 pounds, except for Spot, who weighs 35 pounds. How does Spot's weight affect the mean and median weights of the entire class? (Lesson 9-4)

    F Spot's weight affects the mean more.

    G Spot's weight affects the median more.

    H Spot's weight has an equal affect on the mean and median.

    I Spot's weight has no affect on the mean or median.

5. On the basis of the graph below, what relationship exists between the number of DVDs a person owns and the number of books they own? (Lesson 11-6)

    A As the number of DVDs owned increases, the number of books owned increases.

    B As the number of DVDs owned increases, the number of books owned decreases.

    C There is no relationship between the number of DVDs owned and the number of books owned.

    D For every DVD owned, there are 3 books owned.

6. Which is the graph of a nonlinear function? (Lesson 12-1)

    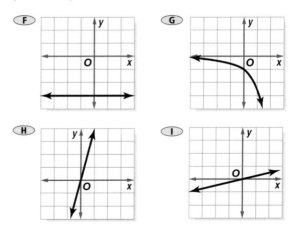

7. If $a = b^4$, which expression is equal to b^8? (Lesson 12-6)

    A a^2 B a^4 C a^{12} D a^{32}

FCAT Practice

PART 2 Short Response/Grid In

THINK
SOLVE
EXPLAIN

Record your answers on the answer sheet provided by your teacher or on a sheet of paper.

8. A square tile measures 9 inches by 9 inches. What is the least number of tiles needed to cover a rectangular floor measuring 21 feet by 27 feet? (Lesson 1-1)

9. There are 240 students in attendance at a student government conference held in Atlanta. Half of these students are from Georgia. Of the remaining students, $\frac{1}{5}$ are from Alabama, and $\frac{1}{4}$ are from Florida. All others are from Tennessee. How many students are from Tennessee? (Lesson 2-3)

10. The block shown below weighs 54 grams. What would be the weight of a block of the same material that measures 6 inches by 6 inches by 6 inches? (Lesson 7-5)

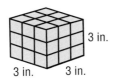

3 in.
3 in.   3 in.

11. Brad's Internet password is a permutation of his initials, B, W, and D, and the numbers 5, 8, and 2. How many different passwords does he have to choose from if no letter or number is used more than once? (Lesson 8-3)

12. What are the coordinates of the point where the graph of $9 + y = 4x$ intercepts the y-axis? (Lesson 11-5)

TEST-TAKING TIP

Question 12 Before graphing an equation, determine whether it is necessary to do so in order to answer the question. In Question 12, you can write $9 + y = 4x$ in slope-intercept form and identify its y-intercept to determine where its graph will cross the y-axis.

13. Copy and complete the table below so that it represents a linear function. (Lesson 12-1)

x	7	5			
y	−4	1			

14. Write an expression in simplest form for the perimeter of the figure at the right. (Lesson 12-3)

15. Find the measure of the value of x in the figure below. (Lesson 12-4)

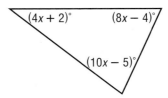

FCAT Practice

PART 3 Extended Response

Record your answers on a sheet of paper. Show your work.

THINK
SOLVE
EXPLAIN

You have 40 feet of fencing to make a rectangular kennel for your dog. You will use your house as one side. (Lessons 12-2 and 12-7)

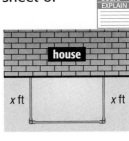

16. Write an algebraic expression for the kennel's length.

17. Write an algebraic expression in simplest form for the area of the kennel.

18. Write the area A of the kennel as a function of its width x.

19. Make a table of values and graph the function you wrote in Exercise 18.

20. Use your graph to determine the width that produces a kennel with the greatest area.

STUDENT HANDBOOK

598

HOW TO...

USE THE STUDENT HANDBOOK

A Student Handbook is the additional skill and reference material found at the end of books. The Student Handbook can help answer these questions.

What If I Forget What I Learned Last Year?

Use the **Prerequisite Skills** section to refresh your memory about things you have learned in other math classes. Here's a list of the topics.

- Estimation Strategies
- Displaying Data on Graphs
- Converting Measurements within the Customary System
- Converting Measurements within the Metric System
- Divisibility Patterns
- Prime Factorization
- Greatest Common Factor
- Simplifying Fractions
- Least Common Multiple
- Perimeter and Area of Rectangles
- Plotting Points on a Coordinate Plane
- Measuring and Drawing Angles

What If I Need More Practice?

You, or your teacher, may decide that working through some additional problems would be helpful. The **Extra Practice** section provides these problems for each lesson.

What If I Have Trouble with Word Problems?

The **Mixed Problem Solving** pages provide additional word problems that use the skills in each chapter.

What If I Forget a Vocabulary Word?

The **English-Spanish Glossary** provides a list of important, or difficult, words used throughout the textbook. It provides a definition in English and Spanish as well as the page number(s) where the word can be found.

What If I Need to Check a Homework Answer?

The answers to the odd-numbered problems are included in **Selected Answers**. Check your answers to make sure you understand how to solve all of the assigned problems.

What If I Need to Find Something Quickly?

The **Index** alphabetically lists the subjects covered throughout the entire textbook and the pages on which each subject can be found.

What If I Forget a Formula?

Inside the back cover of your math book is a list of **Formulas and Symbols** that are used in the book.

Need to Cover Your Book?
Inside the back cover are directions for a Foldable that you can use to cover your math book quickly and easily!

Prerequisite Skills

Estimation Strategies

Sometimes you do not need to know the exact answer to a problem, or you may want to check the reasonableness of an answer. In those instances, you can use **estimation**. There are several different methods of estimation. A common method is to use **rounding**.

EXAMPLES Estimate by Rounding

Estimate by rounding.

1 189.2×315.6

Round each number to the nearest hundred. Then multiply.

$$
\begin{array}{rcr}
189.2 & \to & 200 \\
\times\ 315.6 & \to & \times\ 300 \\
\hline
& & 60{,}000
\end{array}
$$

The product is about 60,000.

2 $453\frac{1}{5} + 68\frac{2}{3}$

Round each number to the nearest ten. Then add.

$$
\begin{array}{rcr}
453\frac{1}{5} & \to & 450 \\
+\ 68\frac{2}{3} & \to & +\ 70 \\
\hline
& & 520
\end{array}
$$

The sum is about 520.

You can use clustering to estimate sums. **Clustering** works best with numbers that all round to approximately the same number.

EXAMPLES Estimate by Clustering

Estimate by clustering.

3 $13\frac{1}{4} + 16\frac{2}{5} + 14\frac{5}{6} + 15\frac{3}{8}$

All of the numbers are close to 15. There are four numbers.

The sum is about 4×15 or 60.

4 $99.6 + 97.83 + 102.18 + 100.101 + 99.90$

All of the numbers are close to 100. There are five numbers.

The sum is about 5×100 or 500.

Compatible numbers are numbers that are easy to compute with mentally.

EXAMPLES Estimate by Using Compatible Numbers

Estimate by using compatible numbers.

5 $76.36 \div 24.73$

76.36 is close to 75, and 24.73 is close to 25.

$$
24.73\overline{)76.36} \quad \to \quad 25\overline{)75}^{\,3}
$$

The quotient is about 3.

6 $7\frac{3}{8} + 12 + 20\frac{2}{3}$

The fractions $\frac{3}{8}$ and $\frac{2}{3}$ are close to $\frac{1}{2}$.

$$
7\frac{1}{2} + 12 + 20\frac{1}{2} = (7 + 12 + 20) + \left(\frac{1}{2} + \frac{1}{2}\right)
$$

$$
= 39 + 1 \text{ or } 40
$$

The sum is about 40.

A strategy that works well for some addition and subtraction problems is **front-end estimation**. This strategy involves adding or subtracting the left-most column of digits. Then, add or subtract the next column of digits. Annex zeros for the remaining digits.

EXAMPLES Use Front-End Estimation

Use front-end estimation to find an estimate.

7 5,283 + 3,634

$$
\begin{array}{r} 5{,}283 \\ +\,3{,}634 \\ \hline 8 \end{array}
\quad\rightarrow\quad
\begin{array}{r} 5{,}283 \\ +\,3{,}634 \\ \hline 8{,}800 \end{array}
$$

The sum is about 8,800.

8 118.1 − 57.5

$$
\begin{array}{r} 118.1 \\ -\,57.5 \\ \hline 6 \end{array}
\quad\rightarrow\quad
\begin{array}{r} 118.1 \\ -\,57.5 \\ \hline 61.0 \end{array}
$$

The difference is about 61.

Exercises

Estimate by rounding.

1. $42\frac{1}{3} \times 59\frac{3}{4}$

2. $78.26 + 90.1 + 18.5$

3. $425\frac{2}{3} \times 10\frac{1}{8}$

4. 51.68×72.31

5. $18\frac{3}{4} + 32\frac{2}{5} + 53$

6. 96.88×31.98

Estimate by clustering.

7. $19.9 + 17.63 + 21.45 + 20.17 + 18.75$

8. $353\frac{1}{3} + 349\frac{2}{5} + 347\frac{3}{4} + 351$

9. $74\frac{1}{2} + 72 + 77\frac{2}{3} + 76\frac{3}{5}$

10. $3.12 + 2.75 + 2.89 + 3.25 + 2.9 + 3.05$

Estimate by using compatible numbers.

11. $105\frac{2}{3} \div 26\frac{1}{2}$

12. $69.3 \div 34.5$

13. $85\frac{2}{5} + 14\frac{1}{3}$

14. $2\frac{3}{5} + 7\frac{1}{2} + 15$

15. $85.1 \div 22.3$

16. $12.4 + 19 + 35.6$

Estimate by using front-end estimation.

17. $109.67 - 25.88$

18. $4{,}456 + 8{,}703$

19. $625.28 - 400.35$

20. $34\frac{5}{8} + 56\frac{1}{3} + 62\frac{3}{7}$

21. $99\frac{3}{8} - 15\frac{1}{2}$

22. $628 + 547 + 432$

Use any method to estimate.

23. $752.6 - 50.1$

24. 69.5×32

25. $88\frac{2}{3} + 2\frac{1}{3}$

26. $99.6 \div 18.25$

27. 700.45×2.1

28. $1{,}065.6 - 200.8$

29. $390 \div 52$

30. 9.5×2.3

31. $77\frac{3}{8} - 55\frac{2}{3}$

32. $1{,}208.85 \div 399.1$

33. $80\frac{1}{5} \div 9\frac{1}{3}$

34. $1{,}715.3 - 1{,}399.9$

35. **MONEY MATTERS** At an arts and crafts festival, Lena selected items priced at $5.98, $7.25, $3.25, $8.75, $9.85, $2.50, and $7.25. She has $50 in cash. How could she use estimation to see if she can use cash or if she needs to write a check?

Displaying Data in Graphs

Statistics involves collecting, analyzing, and presenting information, called **data**. Graphs display data to help readers make sense of the information.

- **Bar graphs** are used to compare the frequency of data. The bar graph below compares the average number of vacation days given by countries to their workers.

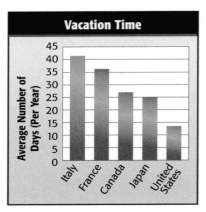

Source: *The World Almanac*

- **Double bar graphs** compare two sets of data. The double bar graph below shows the percent of men and women 65 and older who held jobs in various years.

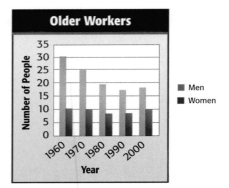

Source: *The World Almanac*

- **Line graphs** usually show how values change over time. The line graph below shows the number of people per square mile in the U.S. from 1800 through 2000.

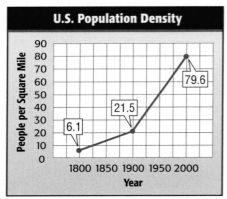

Source: *The World Almanac*

- **Double line graphs**, like double bar graphs, show two sets of data. The double line graph below compares the amount of money spent by both domestic and foreign U.S. travelers.

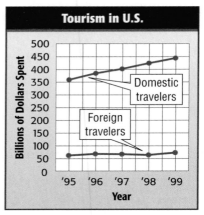

Source: *The World Almanac*

- **Stem-and-leaf plots** are a system used to condense a set of data where the greatest place value of the data is used for the **stems** and the next greatest place value forms the **leaves**. Each data value can be seen in this type of graph.

The stem-and-leaf plot below contains this list of mathematics test scores:

95 76 64 88 93 68 99 96 74 75 92 80 76 85 91 70 62 81

The least number has 6 in the tens place.

The greatest number has 9 in the tens place.

The stems are 6, 7, 8, and 9.

The leaves are ordered from least to greatest.

Stem	Leaf
6	2 4 8
7	0 4 5 6 6
8	0 1 5 8
9	1 2 3 5 6 9

6 | 2 = 62

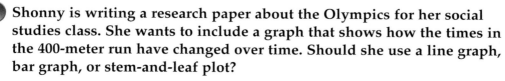

EXAMPLE   Choose a Display

1 Shonny is writing a research paper about the Olympics for her social studies class. She wants to include a graph that shows how the times in the 400-meter run have changed over time. Should she use a line graph, bar graph, or stem-and-leaf plot?

Since the data would show how the times have changed over a period of time, she should choose a line graph.

Exercises

Determine whether a bar graph, double bar graph, line graph, double line graph, or stem-and-leaf plot is the best way to display each of the following sets of data. Explain your reasoning.

1. how the income of households has changed from 1950 through 2000

2. the income of an average household in six different countries

3. the prices for a loaf of bread in twenty different supermarkets

4. the number of boys and the number of girls participating in six different school sports

Refer to the bar graph, double bar graph, line graph, double line graph, and stem-and-leaf plot on page 602.

5. Write several sentences to describe the data shown in the graph titled "Vacation Time." Include a comparison of the days worked for Canada and the U.S.

6. Write several sentences to describe the data shown in the graph titled "Older Workers." What other type or types of graphs could you use to display this data? Explain your reasoning.

7. Write several sentences to describe the data shown in the graph titled "Tourism in U.S." What other type or types of graphs could you use to display this data? Explain your reasoning.

8. Write several sentences to describe the data shown in the graph titled "U.S. Population Density." What other type or types of graphs could you use to display this data? Explain your reasoning.

9. Write several sentences to describe the data shown in the stem-and-leaf plot of mathematics test scores. What is an advantage of showing the scores in this type of graph?

For Exercises 10–14, use the stem-and-leaf plot at the right that shows the number of stories in the tallest buildings in Dallas, Texas.

10. How many buildings does the stem-and-leaf plot represent?

11. How many stories are there for the shortest building in the stem-and-leaf plot? the tallest building?

12. What is the median number of stories for these buildings?

13. What is the mean number of stories for these buildings?

14. Explain how the stem-and-leaf plot is useful in displaying the data.

Stem	Leaf
2	7 9 9
3	0 1 1 1 3 3 4 4 6 6 7
4	0 2 2 5 9
5	0 0 0 0 2 5 6 8
6	0
7	2 *2	7 means 27*

Converting Measurements within the Customary System

The units of length in the customary system are inch, foot, yard, and mile. The table shows the relationships among these units.

Customary Units of Length

1 mile (mi)	= 5,280 feet
1 foot (ft)	= 12 inches (in.)
1 yard (yd)	= 3 feet

- To convert from larger units to smaller units, multiply.
- To convert from smaller units to larger units, divide.

Larger Units	→	Smaller Units

7 ft = 7 × 12 = 84 in.

4 mi = 4 × 5,280 = 21,120 ft

There will be a greater number of smaller units than larger units.

Smaller Units	→	Larger Units

108 in. = 108 ÷ 12 = 9 ft

15 ft = 15 ÷ 3 = 5 yd

There will be fewer larger units than smaller units.

EXAMPLES Convert Customary Units of Length

Complete each sentence.

1 8 yd = __?__ ft

8 yd = (8 × 3) ft

= 24 ft

2 144 in. = __?__ ft

144 in. = (144 ÷ 12) ft

= 12 ft

3 7.5 mi = __?__ ft

7.5 mi = (7.5 × 5,280) ft

= 39,600 ft

The units of weight in the customary system are ounce, pound, and ton. The table at the right shows the relationships among these units. As with units of length, to convert from larger units to smaller units, multiply. To convert from smaller units to larger units, divide.

Customary Units of Weight

1 pound (lb)	= 16 ounces (oz)
1 ton (T)	= 2,000 pounds

EXAMPLES Convert Customary Units of Weight

Complete each sentence.

4 12,400 lb = __?__ T

12,400 lb = 12,400 ÷ 2,000 or 6.2 T

5 92 oz = __?__ lb

92 oz = 92 ÷ 16 or 5.75 lb

Capacity is the amount of liquid or dry substance a container can hold. Customary units of capacity are fluid ounces, cup, pint, quart, and gallon. The relationships among these units are shown in the table.

Customary Units of Capacity

1 cup (c)	= 8 fluid ounces (fl oz)
1 pint (pt)	= 2 cups
1 quart (qt)	= 2 pints
1 gallon (gal)	= 4 quarts

EXAMPLES Convert Customary Units of Capacity

Complete each sentence.

6 64 fl oz = __?__ c

64 fl oz = 64 ÷ 8 or 8 c

7 4.4 gal = __?__ qt

4.4 gal = 4.4 × 4 or 17.6 qt

EXAMPLE Convert Customary Units Using Two Steps

8 **12 pt = __?__ gal**

12 pt = (12 ÷ 2) qt   First, change pints to quarts. | 6 qt = (6 ÷ 4) gal   Then, change quarts to gallons.

    = 6 qt | = 1.5 gal

So, 12 pints = 1.5 gallons.

Units of time can also be converted. The table shows the relationships between these units

Units of Time	
60 seconds (s) = 1 minute (min)	7 days = 1 week
60 minutes = 1 hour (h)	52 weeks = 1 year
24 hours = 1 day	365 days = 1 year

EXAMPLES Convert Units of Time

Complete each sentence.

9 **84 h = __?__ days**
84 h = 84 ÷ 24 or 3.5 days

10 **5 weeks = __?__ days**
5 weeks = 5 × 7 or 35 days

EXAMPLE Adding Mixed Measures

11 **Find the sum of 4 feet 7 inches and 5 feet 10 inches. Simplify.**

   4 ft   7 in.   Line up like units and add.
+ 5 ft 10 in.
  9 ft 17 in. = 9 ft + (12 in. + 5 in.)   Separate 17 in. into 12 in. and 5 in.
           = 10 ft + 5 in.   Replace 12 in. with 1 ft and add like units.

Exercises

Complete each sentence.

1. 2 mi = __?__ ft
2. 48 oz = __?__ lb
3. 120 min = __?__ h
4. 8.5 T = __?__ lb
5. 5 days = __?__ h
6. 63,360 ft = __?__ mi
7. 150 ft = __?__ yd
8. 5 gal = __?__ qt
9. 128 fl oz = __?__ c
10. 20 weeks = __?__ days
11. 24 c = __?__ gal
12. 190,080 in. = __?__ mi
13. 5 T = __?__ oz
14. 36 h = __?__ days
15. 12 oz = __?__ lb
16. 10 pt = __?__ gal
17. 1 mi = __?__ yd
18. 12 gal = __?__ c
19. 14,080 yd = __?__ mi
20. 49 days = __?__ weeks
21. 1 day = __?__ s

Find each sum.

22.   15 ft 2 in.
    + 32 ft 7 in.

23.   5 gal 1 qt
    + 10 gal 2 qt

24.   12 h 15 min
    + 27 h 55 min

25.   45 lb 14 oz
    + 62 lb 12 oz

26.   4 yd 2 ft
    + 16 yd 1 ft

27.   12 days 7 h
    + 44 days 20 h

Converting Measurements within the Metric System

All units of length in the metric system are defined in terms of the meter (m). The diagram below shows the relationships between some common metric units.

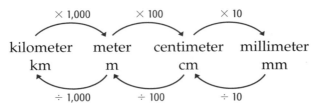

Comparing Metric and Customary Units of Length

1 mm ≈ 0.04 inch (height of a comma)
1 cm ≈ 0.4 inch (half the width of a penny)
1 m ≈ 1.1 yards (width of a doorway)
1 km ≈ 0.6 mile (length of a city block)

- To convert from larger units to smaller units, multiply.
- To convert from smaller units to larger units, divide.

There will be a greater number of smaller units than larger units.

Converting From Larger Units to Smaller Units	Converting From Smaller Units to Larger Units
1 km = 1 × 1,000 = 1,000 m	1 mm = 1 ÷ 10 = 0.1 cm
1 m = 1 × 100 = 1,000 cm	1 cm = 1 ÷ 100 = 0.01 m
1 cm = 1 × 10 = 1,000 mm	1 m = 1 ÷ 1,000 = 0.001 km

There will be fewer larger units than smaller units.

EXAMPLES · Convert Metric Units of Length

Complete each sentence.

1 7 km = __?__ m

7 km = (7 × 1,000) m
    = 7,000 m

2 123 cm = __?__ m

123 cm = (123 ÷ 100) m
    = 1.23 m

3 38.9 cm = __?__ mm

38.9 cm = (38.9 × 10) mm
    = 389 mm

The basic unit of capacity in the metric system is the liter (L). A liter and milliliter (mL) are related in a manner similar to meter and millimeter.

× 1,000

1 L = 1,000 mL

÷ 1,000

Comparing Metric and Customary Units of Capacity

1 mL ≈ 0.03 ounce (drop of water)
1 L ≈ 1 quart (bottle of ketchup)

EXAMPLES · Convert Metric Units of Capacity

Complete each sentence.

4 14.5 L = __?__ mL

14.5 L = 14.5 × 1,000 or 14,500 mL

5 750 mL = __?__ L

750 mL = 750 ÷ 1,000 or 0.75 L

The *mass* of an object is the amount of matter that it contains. The basic unit of mass in the metric system is the kilogram (kg). Kilogram, gram (g), and milligram (mg) are related in a manner similar to kilometer, meter, and millimeter.

1 kg = 1,000 g   1 g = 1,000 mg

Comparing Metric and Customary Units of Mass

1 g ≈ 0.04 ounce (one raisin)
1 kg ≈ 2.2 pounds (six medium apples)

EXAMPLES   Convert Metric Units of Mass

Complete each sentence.

6 53 kg = _?_ g
53 kg = 53 × 1,000 or 53,000 g

7 4,500 g = _?_ kg
4,500 g = 4,500 ÷ 1,000 or 4.5 kg

Sometimes you need to perform more than one conversion to get the desired unit.

EXAMPLES   Convert Metric Units Using Two Steps

Complete each sentence.

8 35,000 cm = _?_ km
35,000 cm = 35,000 ÷ 100 m
　　　　　 = 350 m
350 m = 350 ÷ 1,000 km
　　　 = 0.35 km
So, 35,000 cm = 0.35 km.

9 4.5 kg = _?_ mg
4.5 kg = 4.5 × 1,000 g
　　　 = 4,500 g
4,500 g = 4,500 × 1,000 mg
　　　　 = 4,500,000 mg
So, 4.5 kg = 4,500,000 mg.

Exercises

State which metric unit you would probably use to measure each item.

1. mass of an elephant
2. amount of juice in a pitcher
3. length of a room
4. distance across a state
5. mass of a small stone
6. length of a paper clip
7. height of a large tree
8. amount of water in a medicine dropper
9. width of a sheet of paper
10. diameter of the head of a pin
11. mass of a truck
12. cruising altitude of a passenger jet

Complete each sentence.

13. 45 mm = _?_ cm
14. 2,500 g = _?_ kg
15. 5,000 m = _?_ km
16. 7 L = _?_ mL
17. 8,000 mg = _?_ g
18. 10 km = _?_ m
19. 25 kg = _?_ g
20. 450 cm = _?_ mm
21. 6.4 m = _?_ cm
22. 8.25 kg = _?_ g
23. 655 mL = _?_ L
24. 982 cm = _?_ m
25. 79 m = _?_ km
26. 4,000 mm = _?_ m
27. 60,000 mg = _?_ kg
28. 82,500 cm = _?_ km
29. 5 km = _?_ cm
30. 12 kg = _?_ mg
31. 8 L = _?_ mL
32. 72.6 cm = _?_ mm
33. 0.45 L = _?_ mL
34. 0.625 km = _?_ m
35. 425,000 mg = _?_ kg
36. 1 km = _?_ mm

37. **RACES** Priscilla is running a five-kilometer race. How many meters long is the race?

38. **MEDICINE** A large container of medicine contains 0.5 liter of the drug. How many 25-milliliter doses of the drug are in this container?

Divisibility Patterns

If a number is a factor of a given number, you can also say the given number is **divisible** by the factor. For example, 144 is divisible by 9 since $144 \div 9 = 16$, a whole number. A number n is a factor of a number m if m is divisible by n.

A number is divisible by:

- 2 if the ones digit is divisible by 2.
- 3 if the sum of the digits is divisible by 3.
- 4 if the number formed by the last two digits is divisible by 4.
- 5 if the ones digit is 0 or 5.
- 6 if the number is divisible by both 2 and 3.
- 8 if the number formed by the last three digits is divisible by 8.
- 9 if the sum of the digits is divisible by 9.
- 10 if the ones digit is 0.

EXAMPLE   Use Divisibility Rules

1 **Determine whether 2,418 is divisible by 2, 3, 4, 5, 6, 8, 9, or 10.**

2: Yes; the ones digit, 8, is divisible by 2.

3: Yes; the sum of the digits, $2 + 4 + 1 + 8 = 15$, is divisible by 3.

4: No; the number formed by the last two digits, 18, is not divisible by 4.

5: No; the ones digit is not 0 or 5.

6: Yes; the number is divisible by 2 and 3.

8: No; 418 is not divisible by 8.

9: No; the sum of the digits, 15, is not divisible by 9.

10: No; the ones digit is not 0.

So, 2,418 is divisible by 2, 3, and 6, but not by 4, 5, 8, 9, or 10.

Exercises

Determine whether each number is divisible by 2, 3, 4, 5, 6, 8, 9, or 10.

1. 48
2. 153
3. 2,470
4. 56
5. 165
6. 323
7. 918
8. 1,700
9. 2,865
10. 12,357
11. 16,084
12. 50,070
13. 199
14. 999
15. 808,080
16. 117

17. Is 3 a factor of 777?
18. Is 5 a factor of 232?
19. Is 6 a factor of 198?
20. Is 795 divisible by 10?
21. Is 989 divisible by 9?
22. Is 2,348 divisible by 4?

23. The number $87a,46b$ is divisible by 6. What are possible values of a and b?

24. **FLAGS** Each star in the U.S. flag represents a state. If another state joins the Union, could the stars be arranged in a rectangular array? Explain.

Prerequisite Skills

Prime Factorization

When a whole number greater than 1 has *exactly* two factors, 1 and itself, it is called a **prime number**. When a whole number greater than 1 has more than two factors, it is called a **composite number**. The numbers 0 and 1 are *neither* prime *nor* composite. Notice that 0 has an endless number of factors and 1 has only one factor, itself.

EXAMPLES Identify Numbers as Prime or Composite

Determine whether each number is *prime*, *composite*, or *neither*.

1 33

The numbers 1, 3, and 11 divide into 33 evenly. So, 33 is composite.

2 59

The only numbers that divide evenly into 59 are 1 and 59. So, 59 is prime.

When a number is expressed as a product of factors that are all prime, the expression is called the **prime factorization** of the number. A **factor tree** is useful in finding the prime factorization of a number.

EXAMPLE Write Prime Factorization

3 Use a factor tree to write the prime factorization of 60.

You can begin a factor tree for 60 in several ways.

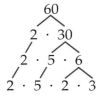 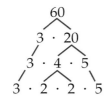 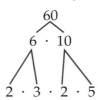

Notice that the bottom row of "branches" in every factor tree is the same except for the order in which the factors are written. So, $60 = 2 \cdot 2 \cdot 3 \cdot 5$ or $2^2 \cdot 3 \cdot 5$.

Every number has a unique set of prime factors. This property of numbers is called the **Fundamental Theorem of Arithmetic**.

Exercises

Determine whether each number is *prime*, *composite*, or *neither*.

1. 45	**2.** 23	**3.** 1	**4.** 13
5. 27	**6.** 96	**7.** 37	**8.** 0
9. 177	**10.** 233	**11.** 507	**12.** 511

Write the prime factorization of each number.

13. 20	**14.** 49	**15.** 225	**16.** 32
17. 25	**18.** 36	**19.** 51	**20.** 75
21. 80	**22.** 117	**23.** 72	**24.** 4,900

Greatest Common Factor

The greatest of the factors common to two or more numbers is called the **greatest common factor (GCF)** of the numbers. One way to find the GCF is to list the factors of the numbers.

EXAMPLE **Find the GCF**

1 Find the greatest common factor of 36 and 60.

Method 1 List the factors.

factors of 36: 1, 2, 3, 4, 6, 9, ⑫ 18, 36

factors of 60: 1, 2, 3, 4, 5, 6, 10, ⑫ 15, 20, 30, 60

Common factors of 36 and 60: 1, 2, 3, 4, 6, 12

The greatest common factor of 36 and 60 is 12.

Method 2 Use prime factorization.

$36 = \boxed{2} \cdot \boxed{2} \cdot \boxed{3} \cdot 3$

$60 = \boxed{2} \cdot \boxed{2} \cdot \boxed{3} \cdot 5$

Common prime factors of 36 and 60: 2, 2, 3

The GCF is $2 \cdot 2 \cdot 3$ or 12.

EXAMPLE **Find the GCF**

2 Find the greatest common factor of 54, 81, and 90.

Use a factor tree to find the prime factorization of each number.

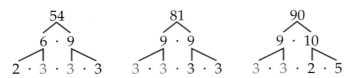

The common prime factors of 54, 81, and 90 are 3 and 3.
The GCF of 54, 81, and 90 is $3 \cdot 3$ or 9.

Exercises

Find the GCF of each set of numbers.

1. 45, 20
2. 27, 54
3. 24, 48
4. 63, 84
5. 40, 60
6. 32, 48
7. 30, 42
8. 54, 72
9. 36, 144
10. 3, 51
11. 24, 36, 42
12. 35, 49, 84

13. **DESIGN** Suppose you are tiling a tabletop with 6-inch square tiles. How many of these squares will be needed to cover a 30-inch by 24-inch table?

14. **SHELVING** Emil is cutting a 72-inch-long board and a 54-inch-long board to make shelves. He wants the shelves to be the same length while not wasting any wood. What is the longest possible length of the shelves?

Two or more numbers are relatively prime if their greatest common factor is 1. Determine whether each set of numbers is relatively prime.

15. 9, 19
16. 7, 21
17. 3, 51
18. 4, 28, 31

Simplifying Fractions

Fractions, mixed numbers, decimals, and integers are examples of **rational numbers**. When a rational number is represented as a fraction, it is often expressed in **simplest form**. A fraction is in simplest form when the GCF of the numerator and denominator is 1.

EXAMPLE Simplify Fractions

 Write $\frac{30}{45}$ in simplest form.

Method 1 Divide by the GCF.

$30 = \boxed{2} \cdot \boxed{3} \cdot 5$ Factor the numerator.

$45 = \boxed{3} \cdot \boxed{3} \cdot 5$ Factor the denominator.

The GCF of 30 and 45 is $3 \cdot 5$ or 15.

$\frac{30}{45} = \frac{30 \div 15}{45 \div 15}$ Divide numerator and denominator by the GCF, 15.

$= \frac{2}{3}$

Method 2 Use prime factorization.

$\frac{30}{45} = \frac{2 \cdot 3 \cdot 5}{3 \cdot 3 \cdot 5}$ Write the prime factorization of the numerator and denominator.

$= \frac{2 \cdot \cancel{3} \cdot \cancel{5}}{\cancel{3} \cdot 3 \cdot \cancel{5}}$ Divide the numerator and denominator by the GCF, $3 \cdot 5$.

$= \frac{2}{3}$ Simplify.

Exercises

Write each fraction in simplest form. If the fraction is already in simplest form, write _simplest form_.

1. $\frac{8}{72}$
2. $\frac{27}{45}$
3. $-\frac{60}{75}$
4. $\frac{36}{54}$
5. $-\frac{3}{9}$

6. $\frac{15}{25}$
7. $\frac{36}{81}$
8. $-\frac{18}{54}$
9. $\frac{14}{66}$
10. $\frac{24}{54}$

11. $-\frac{15}{24}$
12. $\frac{48}{72}$
13. $\frac{24}{120}$
14. $-\frac{66}{88}$
15. $\frac{72}{98}$

16. $-\frac{45}{100}$
17. $\frac{7}{91}$
18. $\frac{15}{100}$
19. $\frac{15}{60}$
20. $\frac{17}{51}$

21. $-\frac{6}{9}$
22. $\frac{16}{40}$
23. $-\frac{6}{16}$
24. $\frac{64}{68}$
25. $\frac{30}{80}$

26. $-\frac{2}{15}$
27. $\frac{90}{120}$
28. $\frac{75}{89}$
29. $-\frac{16}{96}$
30. $\frac{133}{140}$

31. $\frac{99}{300}$
32. $-\frac{50}{1,000}$
33. $-\frac{90}{6,000}$
34. $-\frac{150}{400}$
35. $\frac{10}{10,000}$

36. Both the numerator and the denominator of a fraction are even. Can you tell whether the fraction is in simplest form? Explain.

37. **WEATHER** The rainiest place on Earth is Waialeale, Hawaii. Of 365 days per year, the average number of rainy days is 335. Write a fraction in simplest form to represent these rainy days as a part of a year.

38. **OLYMPICS** In the 2000 Olympics, Brooke Bennett of the U.S. swam the 800-meter freestyle event in about 8 minutes. Express 8 minutes in terms of hours using a fraction in simplest form.

Least Common Multiple

A **multiple** of a number is the product of that number and any whole number.

> **EXAMPLE** List Multiples
>
> **1** **List the first six multiples of 15.**
>
> $0 \cdot 15 = 0, 1 \cdot 15 = 15, 2 \cdot 15 = 30, 3 \cdot 15 = 45, 4 \cdot 15 = 60, 5 \cdot 15 = 75$
>
> The first six multiples of 15 are 0, 15, 30, 45, 60, 75.

The least of the nonzero common multiples of two or more numbers is called the **least common multiple (LCM)** of the numbers. To find the LCM of two or more numbers, you can list the multiples of each number until a common multiple is found, or you can use prime factorization.

> **EXAMPLE** Find the LCM
>
> **2** **Find the LCM of 12 and 18.**
>
> **Method 1** List the multiples.
>
> multiples of 12: 0, 12, 24, 36, 48, …
> multiples of 18: 0, 18, 36, 72, 90, …
>
> The LCM of 12 and 18 is 36. Remember that the LCM is a *nonzero* number.
>
> **Method 2** Use prime factorization.
>
> $12 = 2 \cdot 2 \cdot 3$
> $18 = 2 \cdot 3 \cdot 3$ Write the prime factorization of each number.
>
> $2 \cdot 2 \cdot 3 \cdot 3$ Multiply the factors, using the common factors only once.
>
> The LCM is $2 \cdot 2 \cdot 3 \cdot 3$ or 36.

Exercises

List the first six multiples of each number.

1. 7	**2.** 11	**3.** 4	**4.** 5	**5.** 14
6. 25	**7.** 150	**8.** 2	**9.** 3	**10.** 6

Find the least common multiple (LCM) of each set of numbers.

11. 8, 20	**12.** 15, 18	**13.** 12, 16	**14.** 7, 12
15. 20, 50	**16.** 16, 24	**17.** 2, 7, 8	**18.** 2, 3, 5
19. 4, 8, 12	**20.** 7, 21, 5	**21.** 8, 28, 30	**22.** 10, 12, 14
23. 35, 25, 49	**24.** 24, 12, 6	**25.** 68, 170, 4	**26.** 45, 10, 6
27. 10, 100, 1,000	**28.** 100, 200, 300	**29.** 2, 3, 5, 7	**30.** 2, 15, 25, 36

31. CIVICS In the United States, a president is elected every four years. Members of the House of Representatives are elected every two years. Senators are elected every six years. If a voter had the opportunity to vote for a president, a representative, and a senator in 1996, what will be the next year the voter has a chance to make a choice for a president, a representative, and the same Senate seat?

Perimeter and Area of Rectangles

The distance around a geometric figure is called its **perimeter**. The perimeter P of a rectangle is twice the sum of the length ℓ and width w, or $P = 2\ell + 2w$. The measure of the surface enclosed by a figure is its **area**. The area A of a rectangle is the product of the length ℓ and width w, or $A = \ell w$.

EXAMPLES Find the Perimeter and Area of a Rectangle

1 Find the perimeter of the rectangle.

$P = 2\ell + 2w$ Write the formula.

$P = 2(27) + 2(12)$ Replace ℓ with 27 and w with 12.

$P = 54 + 24$ Multiply.

$P = 78$ Add.

The perimeter is 78 feet.

27 ft

12 ft

2 Find the area of the rectangle.

$A = \ell w$ Write the formula.

$A = 27 \cdot 12$ Replace ℓ with 27 and w with 12.

$A = 324$ Multiply.

The area is 324 square feet.

A square is a rectangle whose sides are all the same length. The perimeter P of a square is four times the side length s, or $P = 4s$. Its area A is the square of the side length, or $A = s^2$.

EXAMPLE Find the Perimeter and Area of a Square

3 Find the perimeter and area of a square with side length 5 yards.

$P = 4s$ Write the formula. $A = s^2$ Write the formula.

$P = 4(5)$ or 20   Replace s with 5. $A = 5^2$ or 25   Replace s with 5.

The perimeter is 20 yards.   The area is 25 square yards.

Exercises

Find the perimeter and area of each figure.

1. 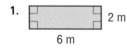 2 m   6 m

2. 5 yd   8 yd

3. 5.5 in.   6.5 in.

4. 7.5 cm   7.5 cm

5. rectangle: 3 mm by 5 mm

6. rectangle: 144 mi by 25 mi

7. square: side length, 75 ft

8. square: side length, 0.75 yd

9. rectangle: 4.3 cm by 2.7 cm

10. square: side length of 625 m

11. square: side length of 87 km

12. rectangle: 875.5 mm by 245.3 mm

Plotting Points on a Coordinate Plane

An ordered pair of numbers is used to locate any point on a coordinate plane. The first number is called the *x*-coordinate. The second number is called the *y*-coordinate.

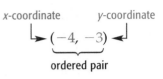

EXAMPLES Identify Ordered Pairs

1 **Write the ordered pair that names point A.**

Step 1 Start at the origin.

Step 2 Move left on the *x*-axis to find the *x*-coordinate of point *A*, which is -1.

Step 3 Move up along the *y*-axis to find the *y*-coordinate which is 4.

The ordered pair for point *A* is $(-1, 4)$.

2 **Write the ordered pair that names point B.**

The *x*-coordinate of *B* is 2. Since the point lies on the *x*-axis, its *y*-coordinate is 0.

The ordered pair for point *B* is $(2, 0)$.

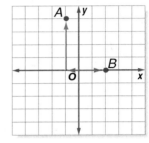

EXAMPLE Graph an Ordered Pair

3 **Graph and label the point $C(3, -2)$ on a coordinate plane.**

Step 1 Start at the origin.

Step 2 Since the *x*-coordinate is 3, move 3 units right.

Step 3 Since the *y*-coordinate is -2, move down 2 units. Draw and label a dot.

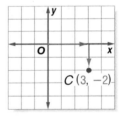

Exercises

Name the ordered pair for the coordinates of each point on the coordinate plane.

1. *Z*	2. *X*	3. *W*
4. *Y*	5. *T*	6. *V*
7. *U*	8. *S*	9. *Q*
10. *R*	11. *P*	12. *M*

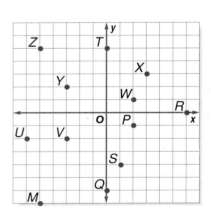

Graph each point on the same coordinate plane.

13. $A(4, 7)$	14. $C(1, 0)$	15. $B(0, 7)$
16. $E(-1, -2)$	17. $D(-4, -7)$	18. $F(-10, 3)$
19. $G(9, 9)$	20. $J(7, -8)$	21. $K(-6, 0)$
22. $H(0, -3)$	23. $I(4, 0)$	24. $M(2, 7)$
25. $N(8, -1)$	26. $L(-1, -1)$	27. $P(3, 3)$

Measuring and Drawing Angles

Two rays that have a common endpoint form an **angle** . The common endpoint is called the **vertex** , and the two rays that make up the angle are called the **sides** of the angle.

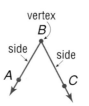

A circle can be divided into 360 equal sections. Each section is one **degree** . You can use a **protractor** to measure an angle in degrees and draw an angle with a given degree measure.

EXAMPLE   Measure an Angle

1 Use a protractor to measure ∠FGH.

Step 1 Place the center point of the protractor's base on vertex G. Align the straight side with side $\overrightarrow{GH}$ so that the marker for 0° is on one of the rays.

Step 2 Use the scale that begins with 0° at $\overrightarrow{GH}$. Read where the other side of the angle, $\overrightarrow{GF}$, crosses this scale.

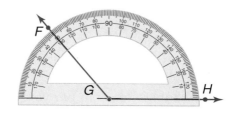

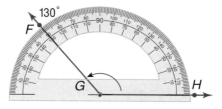

The measure of angle FGH is 130°. Using symbols, $m\angle FGH = 130°$.

EXAMPLE   Draw an Angle

2 Draw ∠X having a measure of 75°.

Step 1 Draw a ray. Label the endpoint X.

Step 2 Place the center point of the protractor's base on point X. Align the mark labeled 0 with the ray.

Step 3 Use the scale that begins with 0. Locate the mark labeled 75. Then draw the other side of the angle.

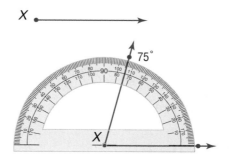

Exercises

Use a protractor to find the measure of each angle.

1. ∠XZY   2. ∠SZT   3. ∠SZY
4. ∠UZX   5. ∠TZW   6. ∠UZV

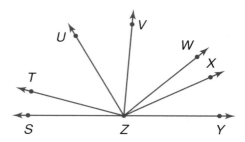

Use a protractor to draw an angle having each measurement.

9. 40°   10. 70°   11. 65°
12. 110°   13. 85°   14. 90°
15. 155°   16. 140°   17. 117°

Extra Practice

Use the four-step plan to solve each problem.

1. Joseph is planting bushes around the perimeter of his lawn. If the bushes must be planted 4 feet apart and Joseph's lawn is 64 feet wide and 124 feet long, how many bushes will Joseph need to purchase?

2. Find the next three numbers in the pattern 1, 3, 7, 15, 31, ...

3. At the bookstore, pencils cost $0.15 each and erasers cost $0.25 each. What combination of pencils and erasers can be purchased for a total of $0.65?

4. Cheap Wheels Car Rental rents cars for $50 per day plus $0.15 per mile. How much will it cost to rent a car for 2 days and to drive 200 miles?

5. Josie wants to fence in her yard. She needs to fence three sides and the house will supply the fourth side. Two of the sides have a length of 25 feet and the third side has a length of 35 feet. If the fencing costs $10 per foot, how much will it cost Josie to fence in her yard?

Evaluate each expression.

1. $15 - 5 + 9 - 2$
2. $(5^2 + 2) \div 3$
3. $12 + 20 \div 4 - 5$
4. $6 \times 3 \div 9 - 1$
5. $(4^2 + 2^2) \times 5$
6. $24 \div 8 - 2$
7. $\dfrac{25}{(3^2 - 4)}$
8. $(15 - 7) \cdot 6 + 2$
9. $3[15 - (2 + 7) \div 3]$

Evaluate each expression if $a = 3$, $b = 6$, and $c = 5$.

10. $2a + bc$
11. ba^2
12. $\dfrac{bc}{a}$
13. $3a + c - 2b$
14. $(2c + b) \cdot a$
15. $\dfrac{2(ac)^2}{b}$
16. abc
17. $(3b + a)c$

Name the property shown by each statement.

18. $2(a + b) = 2a + 2b$
19. $3 \times 5 = 5 \times 3$
20. $(2 + 6) + 5 = 2 + (6 + 5)$
21. $3(4 + 1) = (4 + 1)3$
22. $(7 \times 5)2 = 7(5 \times 2)$
23. $8(2x + 1) = 8(2x) + 8(1)$
24. $5(x + 2) = (x + 2)5$
25. $(3x + 2) + 0 = 3x + 2$
26. $5 \cdot 1 = 5$

Replace each ● with $>$, $<$, or $=$ to make a true sentence.

1. $-3 ● 0$
2. $-1 ● -2$
3. $-5 ● -4$
4. $6 ● -7$
5. $8 ● 10$
6. $-6 ● 6$
7. $-11 ● -20$
8. $-8 ● 2$
9. $-13 ● -12$
10. $|-2| ● |5|$
11. $|13| ● |-19|$
12. $|-6| ● |2|$
13. $|14| ● |-14|$
14. $|0| ● |-4|$
15. $|23| ● |-20|$
16. $|-12| ● -12$

Evaluate each expression.

17. $|-1|$
18. $|-92|$
19. $|3|$
20. $|160 + 32|$
21. $|80 + 100|$
22. $|0|$
23. $|7 - 3|$
24. $|3| + |-7|$
25. $|-161|$
26. $|150|$
27. $|102| - |-2|$
28. $|-116|$

Lesson 1-4

(Pages 23–27)

Add.

1. $-7 + (-7)$
2. $-36 + 40$
3. $18 + (-32)$
4. $47 + 12$
5. $-69 + (-32)$
6. $-120 + (-2)$
7. $-56 + (-4)$
8. $14 + 16$
9. $-18 + 11$
10. $-42 + 29$
11. $-13 + (-11)$
12. $95 + (-5)$
13. $-120 + 2$
14. $25 + (-25)$
15. $-4 + 8$
16. $-9 + (-6)$
17. $42 + (-18)$
18. $-33 + (-12)$
19. $7 + (-13) + 6 + (-7)$
20. $-6 + 12 + (-20)$
21. $4 + 9 + (-14)$
22. $-20 + 0 + (-9) + 25$
23. $5 + 9 + 3 + (-17)$
24. $-36 + 40 + (-10)$
25. $(-2) + 2 + (-2) + 2$
26. $6 + (-4) + 9 + (-2)$
27. $9 + (-7) + 2$
28. $100 + (-75) + (-20)$
29. $-12 + 24 + (-12) + 2$
30. $9 + (-18) + 6 + (-3)$

Lesson 1-5

(Pages 28–31)

Subtract.

1. $3 - 7$
2. $-5 - 4$
3. $-6 - 2$
4. $12 - 9$
5. $0 - (-14)$
6. $58 - (-10)$
7. $-41 - 15$
8. $-81 - 21$
9. $26 - (-14)$
10. $6 - (-4)$
11. $63 - 78$
12. $-5 - (-9)$
13. $72 - (-19)$
14. $-51 - 47$
15. $-99 - 1$
16. $8 - 13$
17. $-2 - 23$
18. $-20 - 0$
19. $55 - 33$
20. $84 - (-61)$
21. $-4 - (-4)$
22. $-2 - (-3)$
23. $65 - (-2)$
24. $0 - (-3)$
25. $0 - 5$
26. $-2 - 6$
27. $-4 - 7$
28. $-3 - (-3)$
29. $15 - 6$
30. $5 - 8$

Lesson 1-6

(Pages 34–38)

Multiply.

1. $5(-2)$
2. $-11(-5)$
3. $-5(-5)$
4. $-12(6)$
5. $2(-2)$
6. $-3(2)(-4)$
7. $(-4)(-4)$
8. $4(21)$
9. $-50(0)$
10. $3(-13)$
11. $2(2)$
12. $-2(-2)$
13. $5(-12)$
14. $2(2)(-2)$
15. $6(-4)$

Divide.

16. $4 \div (-2)$
17. $16 \div (-8)$
18. $-14 \div (-2)$
19. $-18 \div 3$
20. $-25 \div 5$
21. $-56 \div (-8)$
22. $81 \div 9$
23. $-55 \div 11$
24. $-42 \div (-7)$
25. $18 \div (-3)$
26. $0 \div (-1)$
27. $-32 \div 8$
28. $81 \div (-9)$
29. $18 \div (-2)$
30. $-21 \div 3$

Lesson 1-7
(Pages 39–42)

Write each verbal phrase as an algebraic expression.

1. 12 more than a number
2. 3 less than a number
3. a number divided by 4
4. a number increased by 7
5. a number decreased by 12
6. 8 times a number
7. 28 multiplied by m
8. 15 divided by a number
9. 54 divided by n
10. 18 increased by y
11. q decreased by 20
12. n times 41
13. 5 less than a number
14. the product of a number and 15

Write each verbal sentence as an algebraic equation.

15. 6 less than the product of q and 4 is 18.
16. Twice x is 20.
17. A number increased by 6 is 8.
18. The quotient of a number and 7 is 8.
19. The difference between a number and 12 is 37.
20. The product of a number and 7 is 42.

Lesson 1-8
(Pages 45–49)

Solve each equation. Check your solution.

1. $g - 3 = 10$
2. $b + 7 = 12$
3. $a + 3 = 15$
4. $r - 3 = 4$
5. $t + 3 = 21$
6. $s + 10 = 23$
7. $9 + n = 13$
8. $13 + v = 31$
9. $-4 + b = 12$
10. $z - 10 = -8$
11. $-7 = x + 12$
12. $7 + g = 91$
13. $63 + f = 71$
14. $a + 6 = -9$
15. $c - 18 = 13$
16. $23 = n - 5$
17. $j - 3 = 7$
18. $18 = p + 3$
19. $12 + p = 16$
20. $25 = y - 50$
21. $x + 2 = 4$
22. $r - (-8) = 14$
23. $m + (-2) = 6$
24. $5 + q = 12$
25. $t + 12 = 6$
26. $8 + p = 0$
27. $12 - x = 8$
28. $14 + t = 10$
29. $x + 5 = 7$
30. $2 = 3 + x$

Lesson 1-9
(Pages 50–53)

Solve each equation. Check your solution.

1. $4x = 36$
2. $39 = 3y$
3. $4z = 16$
4. $\frac{t}{5} = 6$
5. $100 = 20b$
6. $8 = \frac{w}{8}$
7. $10a = 40$
8. $\frac{s}{9} = 8$
9. $420 = 5s$
10. $8k = 72$
11. $2m = 18$
12. $\frac{m}{8} = 5$
13. $\frac{r}{7} = -8$
14. $\frac{w}{7} = 8$
15. $18q = 36$
16. $9w = 54$
17. $4 = p \div 4$
18. $14 = 2p$
19. $12 = 3t$
20. $\frac{m}{4} = 12$
21. $6h = 12$
22. $-2a = -8$
23. $0 = 6r$
24. $\frac{y}{12} = -6$
25. $3m = -15$
26. $\frac{c}{-4} = 10$
27. $-6f = -36$
28. $81 = -9w$
29. $6r = 42$
30. $\frac{x}{-2} = -15$

Lesson 2-1

(Pages 62–66)

Write each fraction or mixed number as a decimal.

1. $\frac{2}{5}$

2. $2\frac{3}{11}$

3. $-\frac{3}{4}$

4. $\frac{5}{7}$

5. $\frac{3}{4}$

6. $-\frac{2}{3}$

7. $\frac{7}{11}$

8. $\frac{1}{2}$

9. $\frac{5}{6}$

10. $1\frac{3}{5}$

11. $-2\frac{1}{4}$

12. $\frac{8}{9}$

Write each decimal as a fraction or mixed number in simplest form.

13. 0.5

14. $0.\overline{8}$

15. 0.32

16. -0.75

17. $2.\overline{2}$

18. $0.\overline{38}$

19. -0.486

20. 20.08

21. -9.36

22. $10.1\overline{8}$

23. 1.24

24. $-5.\overline{7}$

Lesson 2-2

(Pages 67–70)

Replace each ● with $<$, $>$, or $=$ to make a true sentence.

1. $-5.6 ● 4.2$

2. $4.256 ● 4.25$

3. $0.233 ● 0.\overline{23}$

4. $\frac{5}{7} ● \frac{2}{5}$

5. $\frac{6}{7} ● \frac{7}{9}$

6. $\frac{2}{3} ● \frac{2}{5}$

7. $\frac{3}{8} ● 0.375$

8. $-\frac{1}{2} ● 0.5$

9. $12.56 ● 12\frac{3}{8}$

10. $-0.25 ● -0.26$

11. $1.31 ● 1.\overline{31}$

12. $\frac{3}{5} ● \frac{2}{3}$

Order each set of rational numbers from least to greatest.

13. $0.24, 0.2, 0.245, 2.24, 0.25$

14. $0.\overline{3}, 0.3, 0.3\overline{4}, 0.\overline{34}, 0.33$

15. $\frac{2}{5}, \frac{2}{3}, \frac{2}{7}, \frac{2}{9}, \frac{2}{1}$

16. $\frac{1}{2}, \frac{5}{7}, \frac{2}{9}, \frac{8}{9}, \frac{6}{6}$

17. $0.25, 0.2, 0.02, 0.251, \frac{253}{1,000}$

18. $\frac{3}{10}, \frac{3}{2}, \frac{3}{5}, \frac{3}{1}, \frac{3}{8}, \frac{3}{7}, \frac{3}{4}$

19. $\frac{3}{5}, \frac{2}{3}, 0.61, 0.65, \frac{33}{50}$

20. $-\frac{3}{5}, -\frac{2}{3}, -\frac{1}{2}, -\frac{3}{4}, -\frac{5}{6}$

21. $\frac{4}{9}, 0.4, 0.44, \frac{3}{5}$

22. $7.5, 7\frac{2}{3}, 6\frac{5}{6}, 6.8$

23. $-\frac{2}{3}, \frac{1}{3}, 0.1, \frac{5}{6}$

24. $-0.5, 0.5, 0, 0.35, -0.51$

Lesson 2-3

(Pages 71–75)

Multiply. Write in simplest form.

1. $\frac{2}{11} \cdot \frac{3}{4}$

2. $4\left(-\frac{7}{8}\right)$

3. $-\frac{4}{7} \cdot \frac{3}{5}$

4. $\frac{6}{7}\left(-\frac{7}{12}\right)$

5. $\frac{7}{8} \cdot \frac{1}{3}$

6. $\frac{3}{4} \cdot \frac{4}{5}$

7. $-1\frac{1}{2} \cdot \frac{2}{3}$

8. $\frac{5}{6} \cdot \frac{6}{7}$

9. $8\left(-2\frac{1}{4}\right)$

10. $-3\frac{3}{4} \cdot \frac{8}{9}$

11. $\frac{10}{21}\left(-\frac{7}{8}\right)$

12. $-1\frac{4}{5}\left(-\frac{5}{6}\right)$

13. $5\frac{1}{4} \cdot 6\frac{2}{3}$

14. $-8\frac{3}{4} \cdot 4\frac{2}{5}$

15. $6 \cdot 8\frac{2}{3}$

16. $\left(\frac{3}{5}\right)\left(\frac{3}{5}\right)$

17. $-4\frac{1}{5}\left(-3\frac{1}{3}\right)$

18. $-8\left(\frac{3}{4}\right)$

19. $3\frac{2}{3}\left(-3\frac{1}{2}\right)$

20. $\left(-\frac{2}{5}\right)\left(-\frac{2}{5}\right)$

21. $4\frac{1}{2}\left(-1\frac{1}{3}\right)$

22. $-5\left(-3\frac{1}{5}\right)$

23. $4\frac{1}{3} \cdot 1\frac{1}{2}$

24. $-5\left(3\frac{1}{3}\right)$

25. $\frac{4}{5} \cdot \frac{7}{6}$

26. $-\frac{3}{8} \cdot \frac{7}{9}$

27. $-1\frac{3}{4}\left(-2\frac{1}{7}\right)$

28. $-8\left(5\frac{1}{4}\right)$

Lesson 2-4

(Pages 76–80)

Name the multiplicative inverse of each number.

1. 3

2. -5

3. $\frac{2}{3}$

4. $2\frac{1}{8}$

5. $\frac{1}{15}$

6. -8

7. $1\frac{1}{3}$

8. $-\frac{4}{5}$

Divide. Write in simplest form.

9. $\frac{2}{3} \div \frac{3}{4}$

10. $-\frac{4}{9} \div \frac{5}{6}$

11. $\frac{7}{12} \div \frac{3}{8}$

12. $\frac{5}{18} \div \frac{2}{9}$

13. $\frac{1}{3} \div 4$

14. $5\frac{1}{4} \div \left(-2\frac{1}{2}\right)$

15. $-6 \div \left(-\frac{4}{7}\right)$

16. $-6\frac{3}{8} \div \frac{1}{4}$

17. $\frac{6}{7} \div \frac{3}{5}$

18. $3\frac{1}{3} \div (-4)$

19. $2\frac{5}{12} \div 7\frac{1}{3}$

20. $\frac{5}{6} \div 1\frac{1}{9}$

21. $\frac{3}{8} \div (-6)$

22. $\frac{5}{8} \div \frac{1}{6}$

23. $4\frac{1}{4} \div 6\frac{3}{4}$

24. $4\frac{1}{6} \div 3\frac{1}{8}$

25. $8 \div \left(-1\frac{4}{5}\right)$

26. $-5 \div \frac{2}{7}$

27. $\frac{3}{5} \div \frac{6}{7}$

28. $4\frac{8}{9} \div \left(-2\frac{2}{3}\right)$

29. $8\frac{1}{6} \div 3$

30. $-\frac{3}{4} \div 9$

31. $1\frac{11}{14} \div 2\frac{1}{2}$

32. $-2\frac{1}{4} \div \frac{4}{5}$

Lesson 2-5

(Pages 82–85)

Add or subtract. Write in simplest form.

1. $\frac{17}{21} + \left(-\frac{13}{21}\right)$

2. $\frac{5}{11} + \frac{6}{11}$

3. $-\frac{8}{13} + \left(-\frac{11}{13}\right)$

4. $-\frac{7}{12} + \frac{5}{12}$

5. $\frac{13}{28} - \frac{9}{28}$

6. $-1\frac{2}{9} - \frac{7}{9}$

7. $\frac{15}{16} + \frac{13}{16}$

8. $2\frac{1}{3} - \frac{2}{3}$

9. $-\frac{4}{35} - \left(-\frac{17}{35}\right)$

10. $\frac{3}{8} + \left(-\frac{5}{8}\right)$

11. $\frac{8}{15} - \frac{2}{15}$

12. $-2\frac{4}{7} - \frac{3}{7}$

13. $-\frac{29}{9} - \left(-\frac{26}{9}\right)$

14. $2\frac{3}{5} + 7\frac{3}{5}$

15. $\frac{5}{18} - \frac{13}{18}$

16. $-2\frac{2}{7} + \left(-1\frac{6}{7}\right)$

17. $-\frac{3}{10} + \frac{7}{10}$

18. $\frac{4}{11} + \frac{9}{11}$

19. $\frac{1}{8} + 1\frac{7}{8}$

20. $\frac{5}{6} - \frac{7}{6}$

21. $5 - 3\frac{5}{7}$

22. $-3 - 4\frac{5}{8}$

23. $5\frac{3}{7} + 2\frac{6}{7}$

24. $-9\frac{3}{4} - \left(-2\frac{3}{4}\right)$

25. $4\frac{5}{9} - 1\frac{2}{9}$

26. $2\frac{5}{12} - 8\frac{7}{12}$

27. $-5\frac{1}{4} + 1\frac{3}{4}$

28. $6\frac{1}{5} - 2\frac{4}{5}$

Lesson 2-6

(Pages 88–91)

Add or subtract. Write in simplest form.

1. $\frac{7}{12} + \frac{7}{24}$

2. $-\frac{3}{4} + \frac{7}{8}$

3. $\frac{2}{5} + \left(-\frac{2}{7}\right)$

4. $-\frac{3}{5} - \left(-\frac{5}{6}\right)$

5. $\frac{5}{24} - \frac{3}{8}$

6. $-\frac{7}{12} - \frac{3}{4}$

7. $-\frac{3}{8} + \left(-\frac{4}{5}\right)$

8. $\frac{2}{15} + \left(-\frac{3}{10}\right)$

9. $-\frac{2}{9} - \left(-\frac{2}{3}\right)$

10. $-\frac{7}{15} - \frac{5}{12}$

11. $\frac{3}{8} + \frac{7}{12}$

12. $-2\frac{1}{4} + \left(-1\frac{1}{3}\right)$

13. $3\frac{2}{5} - 3\frac{1}{4}$

14. $\frac{3}{4} + \left(-\frac{4}{15}\right)$

15. $-1\frac{2}{3} + 4\frac{3}{4}$

16. $-\frac{1}{8} - 2\frac{1}{2}$

17. $3\frac{2}{5} - 1\frac{1}{3}$

18. $5\frac{1}{3} + \left(-8\frac{3}{7}\right)$

19. $\frac{3}{5} - \frac{2}{3}$

20. $1\frac{1}{3} - 2\frac{5}{6}$

21. $2\frac{1}{2} - \frac{3}{4}$

22. $5\frac{2}{3} + 3\frac{5}{6}$

23. $-5\frac{1}{7} - 3\frac{1}{5}$

24. $-8\frac{1}{2} + 1\frac{2}{3}$

25. $5\frac{3}{4} - 8\frac{1}{3}$

26. $1\frac{3}{4} + 3\frac{5}{8}$

27. $4\frac{1}{2} - \frac{5}{7}$

28. $3\frac{2}{3} + 9\frac{3}{4}$

Solve each equation. Check your solution.

1. $434 = -31y$
2. $6x = -4.2$
3. $\frac{3}{4}a = -12$
4. $-10 = \frac{b}{-7}$
5. $7.2 = \frac{3}{4}c$
6. $r + 0.4 = 1.4$
7. $-2.4n = 7.2$
8. $7 = \frac{1}{2}d$
9. $n - 0.64 = -5.44$
10. $\frac{t}{3} = 2$
11. $\frac{3}{8} = \frac{1}{2}x$
12. $\frac{1}{2}h = -14$
13. $k - 1.18 = 1.58$
14. $4\frac{1}{2}s = -30$
15. $\frac{2}{3}f = \frac{8}{15}$
16. $\frac{2}{3}m = 22$
17. $\frac{2}{3}g = 4\frac{5}{6}$
18. $7 = \frac{1}{3}v$
19. $\frac{g}{1.2} = -6$
20. $z - 4\frac{5}{8} = 15\frac{3}{8}$
21. $-12 = \frac{1}{5}j$
22. $a + 3.2 = 6.5$
23. $q - \frac{1}{5} = \frac{2}{3}$
24. $3.5z = 7$
25. $2.5x = \frac{1}{2}$
26. $c - 3 = 5\frac{2}{3}$
27. $\frac{2}{3}x = \frac{5}{6}$

Write each expression using exponents.

1. $4 \cdot 4 \cdot 4 \cdot 4$
2. $3 \cdot 3$
3. $7 \cdot 7 \cdot 7 \cdot 7 \cdot 7 \cdot 7$
4. $4 \cdot 4 \cdot 4 \cdot 4 \cdot 4 \cdot 5 \cdot 5 \cdot 5 \cdot 5 \cdot 5 \cdot 5 \cdot 5 \cdot 5$
5. $x \cdot y \cdot y \cdot y \cdot y \cdot x \cdot x \cdot x \cdot y \cdot x$
6. $b \cdot b \cdot b \cdot b \cdot c \cdot c \cdot c \cdot c \cdot c \cdot c$
7. $3 \cdot 2 \cdot 5 \cdot 5 \cdot 5 \cdot 2 \cdot 2 \cdot 2 \cdot 3 \cdot 5$
8. $a \cdot a \cdot a \cdot b \cdot b \cdot b \cdot a \cdot a \cdot a \cdot b$
9. $6 \cdot 6 \cdot 6 \cdot 6 \cdot 6 \cdot 6 \cdot 6 \cdot 6$
10. $x \cdot x \cdot x \cdot x \cdot x \cdot x \cdot x \cdot x \cdot x \cdot x$
11. $a \cdot b \cdot b \cdot b \cdot b \cdot b \cdot b \cdot b$

Evaluate each expression.

12. 4^3
13. 6^2
14. 2^6
15. $5^2 \cdot 6^2$
16. $3 \cdot 2^4$
17. $10^4 \cdot 3^2$
18. $5^3 \cdot 1^9$
19. $2^2 \cdot 2^4$
20. $2 \cdot 3^2 \cdot 4^2$
21. 7^3
22. $2^2 \cdot 5^2$
23. $3^5 \cdot 4^2$
24. $7^2 \cdot 3^4$
25. 3^{-3}
26. 2^{-4}
27. 5^{-2}

Write each number in standard form.

1. 4.5×10^3
2. 2×10^4
3. 1.725896×10^6
4. 9.61×10^2
5. 1×10^7
6. 8.256×10^8
7. 5.26×10^4
8. 3.25×10^2
9. 6.79×10^5
10. 3.1×10^{-4}
11. 2.51×10^{-2}
12. 6×10^{-1}
13. 2.15×10^{-3}
14. 3.14×10^{-6}
15. 1×10^{-2}

Write each number in scientific notation.

16. 720
17. 7,560
18. 892
19. 1,400
20. 91,256
21. 51,000
22. 0.012
23. 0.0002
24. 0.054
25. 0.231
26. 0.0000056
27. 0.000123

Lesson 3-1

(Pages 116–119)

Find each square root.

1. $\sqrt{9}$

2. $\sqrt{81}$

3. $-\sqrt{625}$

4. $\sqrt{36}$

5. $-\sqrt{169}$

6. $\sqrt{144}$

7. $\sqrt{961}$

8. $\sqrt{324}$

9. $-\sqrt{225}$

10. $-\sqrt{4}$

11. $\sqrt{529}$

12. $-\sqrt{484}$

13. $\sqrt{196}$

14. $\sqrt{729}$

15. $\sqrt{289}$

16. $-\sqrt{16}$

17. $\sqrt{1,024}$

18. $\sqrt{0.16}$

19. $\sqrt{0.04}$

20. $\sqrt{2.25}$

21. $\sqrt{0.01}$

22. $-\sqrt{0.09}$

23. $\sqrt{0.49}$

24. $\sqrt{1.69}$

25. $\sqrt{0.36}$

26. $\sqrt{\dfrac{289}{10,000}}$

27. $\sqrt{\dfrac{169}{121}}$

28. $-\sqrt{\dfrac{4}{9}}$

29. $-\sqrt{\dfrac{81}{64}}$

30. $\sqrt{\dfrac{25}{81}}$

Lesson 3-2

(Pages 120–122)

Estimate to the nearest whole number.

1. $\sqrt{229}$

2. $\sqrt{63}$

3. $\sqrt{290}$

4. $\sqrt{27}$

5. $\sqrt{333}$

6. $\sqrt{23}$

7. $\sqrt{96}$

8. $\sqrt{19}$

9. $\sqrt{200}$

10. $\sqrt{76}$

11. $\sqrt{17}$

12. $\sqrt{34}$

13. $\sqrt{137}$

14. $\sqrt{540}$

15. $\sqrt{165}$

16. $\sqrt{326}$

17. $\sqrt{52}$

18. $\sqrt{37}$

19. $\sqrt{79}$

20. $\sqrt{89}$

21. $\sqrt{71}$

22. $\sqrt{117}$

23. $\sqrt{410}$

24. $\sqrt{47}$

25. $\sqrt{1.30}$

26. $\sqrt{8.4}$

27. $\sqrt{18.35}$

28. $\sqrt{25.70}$

29. $\sqrt{1.41}$

30. $\sqrt{15.3}$

Lesson 3-3

(Pages 125–129)

Name all sets of numbers to which each real number belongs.

1. 6.5

2. $\sqrt{25}$

3. $\sqrt{3}$

4. -7.2

5. $-0.\overline{61}$

6. $\dfrac{1}{2}$

7. $\dfrac{16}{4}$

8. -102.1

9. $\sqrt{29}$

Estimate each square root. Then graph the square root on a number line.

10. $-\sqrt{12}$

11. $\sqrt{23}$

12. $\sqrt{2}$

13. $\sqrt{10}$

14. $-\sqrt{30}$

15. $\sqrt{5}$

16. $\sqrt{21}$

17. $-\sqrt{202}$

18. $-\sqrt{10}$

Replace each ● with <, >, or = to make a true sentence.

19. $\sqrt{7}$ ● 2.8

20. $2\dfrac{1}{3}$ ● $2.\overline{3}$

21. $\sqrt{121}$ ● 11

22. 5.6 ● $\sqrt{30}$

23. 9.45 ● $9.\overline{4}$

24. $\sqrt{5}$ ● 2.23

25. $\sqrt{6.25}$ ● $2\dfrac{1}{2}$

26. $5\dfrac{1}{3}$ ● $\sqrt{30}$

27. $4\dfrac{2}{3}$ ● $\sqrt{22}$

Lesson 3-4

(Pages 132–136)

Write an equation you could use to find the length of the missing side of each right triangle. Then find the missing length. Round to the nearest tenth if necessary.

1.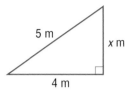

2.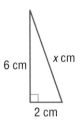

3.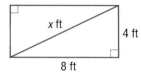

4. a, 6 cm; b, 5 cm

5. a, 12 ft; b, 12 ft

6. a, 8 in.; b, 6 in.

7. a, 20 m; c, 25 m

8. a, 9 mm; c, 14 mm

9. b, 15 m; c, 20 m

Determine whether each triangle with sides of given lengths is a right triangle.

10. 15 m, 8 m, 17 m

11. 7 yd, 5 yd, 9 yd

12. 5 in., 12 in., 13 in.

13. 9 in., 12 in., 16 in.

14. 10 ft, 24 ft, 26 ft

15. 2 ft, 2 ft, 3 ft

Lesson 3-5

(Pages 137–140)

Write an equation that can be used to answer each question. Then solve. Round to the nearest tenth if necessary.

1. How far apart are the boats?

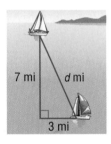

2. How high does the ladder reach?

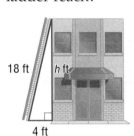

3. How long is each rafter?

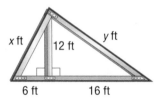

Lesson 3-6

(Pages 142–145)

Find the distance between each pair of points whose coordinates are given. Round to the nearest tenth if necessary.

1.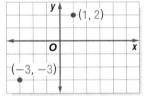

2.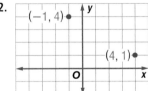

3.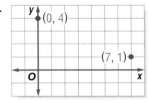

Graph each pair of ordered pairs. Then find the distance between the points. Round to the nearest tenth.

4. $(-4, 2), (4, 17)$

5. $(5, -1), (11, 7)$

6. $(-3, 5), (2, 7)$

7. $(7, -9), (4, 3)$

8. $(5, 4), (-3, 8)$

9. $(-8, -4), (-3, 8)$

10. $(2, 7), (10, -4)$

11. $(9, -2), (3, 6)$

12. $(2, 3), (-1, 6)$

13. $(-5, 1), (2, -3)$

14. $(0, 1), (5, 2)$

15. $(-1, 2), (-2, 3)$

Lesson 4-1

(Pages 156–159)

Express each ratio in simplest form.

1. 27 to 9
2. 4 inches per foot
3. 16 out of 48
4. 10:50
5. 40 minutes per hour
6. 35 to 15
7. 16 wins to 16 losses
8. 7 out of 13
9. 5 out of 50
10. 3 out of 5
11. 20 minutes per hour
12. 6 inches per foot

Express each rate as a unit rate.

13. 6 pounds gained in 12 weeks
14. $800 for 40 tickets
15. $6.50 for 5 pounds
16. 6 inches of rain in 3 weeks
17. 20 preschoolers to 2 teachers
18. 10 inches of snow in 2 days
19. $500 for 50 tickets
20. $360 for 100 dinners

Lesson 4-2

(Pages 160–164)

For Exercises 1–3, use the following information.

Time	1:00	2:00	2:30	3:00	3:15
Temperature	88°F	89°F	80°F	76°F	76°F

1. Find the rate of change between 2:00 and 2:30.
2. Find the rate of change between 1:00 and 3:00.
3. Find the rate of change between 3:00 and 3:15. Explain the meaning of this rate of change.

For Exercises 4–7, use the following information.

Time	6:00	6:30	6:45	7:00	7:10	7:30	8:00	8:15	8:30
Number of Tickets Sold	2	32	77	137	139	140	142	142	142

4. Find the rate of change between 6:45 and 7:00.
5. Was the rate of change between 8:00 and 8:15 positive, negative, or zero?
6. Find the rate of change between 6:00 and 8:30.
7. During which time period was the greatest rate of change?

Lesson 4-3

(Pages 166–169)

Find the slope of each line.

1.
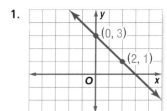

2.
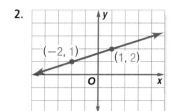

3.
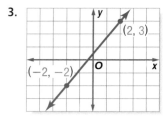

The points given in each table lie on a line. Find the slope of the line.

4.
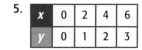

x	0	1	2	3
y	1	0	−1	−2

5.

x	0	2	4	6
y	0	1	2	3

6.
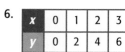

x	0	1	2	3
y	0	2	4	6

Lesson 4-4

(Pages 170–173)

Determine whether each pair of ratios forms a proportion.

1. $\dfrac{3}{5}, \dfrac{5}{10}$ 2. $\dfrac{8}{4}, \dfrac{6}{3}$ 3. $\dfrac{10}{15}, \dfrac{5}{3}$ 4. $\dfrac{2}{8}, \dfrac{1}{4}$

5. $\dfrac{6}{18}, \dfrac{3}{9}$ 6. $\dfrac{14}{21}, \dfrac{12}{18}$ 7. $\dfrac{4}{20}, \dfrac{5}{25}$ 8. $\dfrac{9}{27}, \dfrac{1}{3}$

9. $\dfrac{2}{3}, \dfrac{3}{2}$ 10. $\dfrac{2}{3}, \dfrac{14}{21}$ 11. $\dfrac{1}{2}, \dfrac{5}{10}$ 12. $\dfrac{3}{5}, \dfrac{15}{9}$

Solve each proportion.

13. $\dfrac{2}{3} = \dfrac{a}{12}$ 14. $\dfrac{7}{8} = \dfrac{c}{16}$ 15. $\dfrac{3}{7} = \dfrac{21}{d}$ 16. $\dfrac{2}{5} = \dfrac{18}{x}$

17. $\dfrac{3}{5} = \dfrac{n}{21}$ 18. $\dfrac{5}{12} = \dfrac{b}{5}$ 19. $\dfrac{4}{36} = \dfrac{2}{y}$ 20. $\dfrac{16}{8} = \dfrac{y}{12}$

21. $\dfrac{x}{3} = \dfrac{14}{21}$ 22. $\dfrac{2}{x} = \dfrac{8}{24}$ 23. $\dfrac{y}{15} = \dfrac{8}{60}$ 24. $\dfrac{1}{5} = \dfrac{a}{3}$

25. $\dfrac{27}{8} = \dfrac{z}{4}$ 26. $\dfrac{0.3}{0.2} = \dfrac{1.5}{c}$ 27. $\dfrac{3}{10} = \dfrac{z}{36}$ 28. $\dfrac{2}{3} = \dfrac{t}{4}$

Lesson 4-5

(Pages 178–182)

Determine whether each pair of polygons is similar. Explain your reasoning.

1.

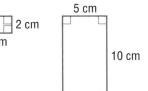

2.

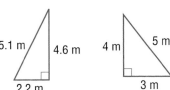

Each pair of polygons is similar. Write a proportion to find each missing measure. Then solve.

3.

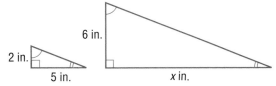

4.

Lesson 4-6

(Pages 184–187)

Solve.

1. The distance between two cities on a map is 3.2 centimeters. If the scale on the map is 1 centimeter = 50 kilometers, find the actual distance between the two cities.

2. A scale model of the Empire State Building is 10 inches tall. If the Empire State Building is 1,250 feet tall, find the scale of this model.

3. On a scale drawing of a house, the dimensions of the living room are 4 inches by 3 inches. If the scale of the drawing is 1 inch = 6 feet, find the actual dimensions of the living room.

4. Columbus, Ohio, is approximately 70 miles from Dayton, Ohio. If a scale on an Ohio map is 1 inch = 11 miles, about how far apart are the cities on the map?

Lesson 4-7

(Pages 188–191)

Write a proportion. Then determine the missing measure.

1. A road sign casts a shadow 14 meters long, while a tree nearby casts a shadow 27.8 meters long. If the road sign is 3.5 meters high, how tall is the tree?

2. Use the map to find the distance across Catfish Lake. Assume the triangles are similar.

3. A 7-foot tall flag stick on a golf course casts a shadow 21 feet long. A golfer standing nearby casts a shadow 16.5 feet long. How tall is the golfer?

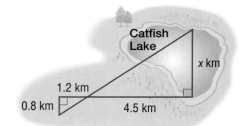

4. A building casts a shadow that is 150 feet. A tree casts a shadow that is 25 feet. If the tree is 150 feet tall, how tall is the building?

5. A tower casts a shadow that is 120 feet. A pole casts a shadow that is 5 feet. If the tower is 2,400 feet tall, how tall is the pole?

Lesson 4-8

(Pages 194–197)

Find the coordinates of the vertices of triangle $A'B'C'$ after triangle ABC is dilated using the given scale factor. Then graph triangle ABC and its dilation.

1. $A(-1, 0)$, $B(2, 1)$, $C(2, -1)$; scale factor 2

2. $A(4, 6)$, $B(0, -2)$, $C(6, 2)$; scale factor $\frac{1}{2}$

3. $A(1 -1)$, $B(1, 2)$, $C(-1, 1)$; scale factor 3

4. $A(2, 0)$, $B(0, -4)$, $C(-2, 4)$; scale factor $\frac{3}{2}$

In each figure, the green figure is a dilation of the blue figure. Find the scale factor of each dilation, and classify each dilation as an *enlargement* or as a *reduction*.

5.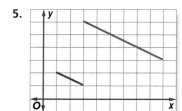

6.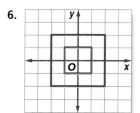

7.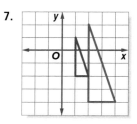

Lesson 5-1

(Pages 206–209)

Write each ratio or fraction as a percent.

1. 3 out of 5
2. $\frac{1}{4}$
3. $\frac{7}{10}$
4. 39:100

5. 11 out of 25
6. 72.5:100
7. 3 out of 4
8. $\frac{1}{2}$

9. $\frac{7}{20}$
10. 93:100
11. 2 out of 8
12. $\frac{9}{20}$

Write each percent as a fraction in simplest form.

13. 30%
14. 4%
15. 20%
16. 85%

17. 3%
18. 80%
19. 17%
20. 55%

21. 82%
22. 48%
23. 32%
24. 51%

Lesson 5-2

Write each percent as a decimal.

1. 2%
2. 25%
3. 29%
4. 6.2%
5. 16.8%
6. 14%
7. 23.7%
8. 42%

Write each decimal as a percent.

9. 0.35
10. 14.23
11. 0.9
12. 0.13
13. 6.21
14. 0.08
15. 0.036
16. 2.34

Write each fraction as a percent.

17. $\frac{2}{5}$
18. $\frac{49}{50}$
19. $\frac{21}{50}$
20. $\frac{1}{3}$
21. $\frac{81}{100}$
22. $\frac{2}{25}$
23. $\frac{11}{20}$
24. $\frac{9}{75}$
25. $\frac{33}{40}$
26. $\frac{1}{50}$
27. $\frac{41}{50}$
28. $\frac{39}{100}$

Lesson 5-3

(Pages 216–219)

Write a percent proportion to solve each problem. Then solve. Round answers to the nearest tenth if necessary.

1. 39 is 5% of what number?
2. What is 19% of 200?
3. 6 is what percent of 30?
4. 24 is what percent of 72?
5. 9 is $33\frac{1}{3}$% of what number?
6. Find 55% of 134.
7. 8 is what percent of 32?
8. What is 35% of 215?
9. 62 is 50% of what number?
10. 93 is what percent of 186?
11. 90 is 36% of what number?
12. 15 is 60% of what number?
13. What is 15% of 60?
14. 15 is 20% of what number?
15. 66 is 75% of what number?
16. 31 is what percent of 155?
17. 22 is 25% of what number?
18. What is 65% of 150?
19. 6 is 75% of what number?
20. 27 is what percent of 100?

Lesson 5-4

(Pages 220–223)

Compute mentally.

1. 10% of 206
2. 1% of 19.3
3. 20% of 15
4. 87.5% of 80
5. 50% of 46
6. 12.5% of 56
7. $33\frac{1}{3}$% of 93
8. 90% of 2,000
9. 30% of 70
10. 40% of 95
11. $66\frac{2}{3}$% of 48
12. 80% of 25
13. 25% of 400
14. 75% of 72
15. 37.5% of 96
16. 40% of 35
17. 60% of 85
18. 62.5% of 160
19. 90% of 205
20. 1% of 2,364
21. 20% of 85
22. 75% of 12
23. 12.5% of 800
24. 30% of 90
25. 1% of 70
26. 40% of 45
27. 62.5% of 88

Lesson 5-5

(Pages 228–231)

Estimate.

1. 33% of 12
2. 24% of 84
3. 39% of 50
4. 19% of 135
5. 21% of 50
6. 49% of 121
7. 72% of 101
8. 99% of 255
9. 25% of 41

Estimate each percent.

10. 11 out of 99
11. 28 out of 89
12. 9 out of 20
13. 25 out of 270
14. 5 out of 49
15. 7 out of 57
16. 2 out of 21
17. 12 out of 61
18. 7 out of 15

Estimate the percent of the area shaded.

19.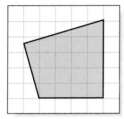
20.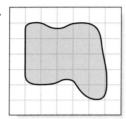
21.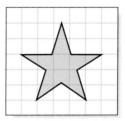

Lesson 5-6

(Pages 232–235)

Solve each equation using the percent equation.

1. Find 5% of 73.
2. What is 15% of 15?
3. Find 80% of 12.
4. What is 7.3% of 500?
5. Find 21% of 720.
6. What is 12% of 62.5?
7. Find 0.3% of 155.
8. What is 75% of 450?
9. Find 7.2% of 10.
10. What is 10.1% of 60?
11. Find 23% of 47.
12. What is 89% of 654?
13. 20 is what percent of 64?
14. Sixty-nine is what percent of 200?
15. Seventy is what percent of 150?
16. 26 is 30% of what number?
17. 7 is 14% of what number?
18. 35.5 is what percent of 150?
19. 17 is what percent of 25?
20. 152 is 2% of what number?

Lesson 5-7

(Pages 236–240)

Find each percent of change. Round to the nearest tenth if necessary. State whether the percent of change is an *increase* or a *decrease*.

1. original: 35
   new: 29
2. original: 550
   new: 425
3. original: 72
   new: 88
4. original: 25
   new: 35
5. original: 28
   new: 19
6. original: 46
   new: 55

Find the selling price for each item given the cost to the store and markup.

7. golf clubs: $250, 30% markup
8. compact disc: $17, 15% markup
9. shoes: $57, 45% markup
10. book: $26, 20% markup

Find the sale price of each item to the nearest cent.

11. piano: $4,220, 35% off
12. scissors: $14, 10% off
13. book: $29, 40% off
14. sweater: $38, 25% off

Lesson 5-8

(Pages 241–244)

Find the simple interest to the nearest cent.

1. $500 at 7% for 2 years
2. $2,500 at 6.5% for 36 months
3. $8,000 at 6% for 1 year
4. $1,890 at 9% for 42 months
5. $760 at 4.5% for $2\frac{1}{2}$ years
6. $12,340 at 5% for 6 months

Find the total amount in each account to the nearest cent.

7. $300 at 10% for 3 years
8. $3,200 at 8% for 6 months
9. $20,000 at 14% for 20 years
10. $4,000 at 12.5% for 4 years
11. $450 at 11% for 5 years
12. $17,000 at 15% for $9\frac{1}{2}$ years

Lesson 6-1

(Pages 256–260)

Find the value of x in each figure.

1.

2.
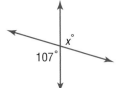

3.
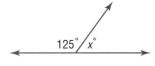

4.

5.

6.

For Exercises 7–10, use the figure at the right.

7. Find $m\angle 6$, if $m\angle 3 = 42°$.
8. Find $m\angle 4$, if $m\angle 3 = 71°$.
9. Find $m\angle 1$, if $m\angle 5 = 128°$.
10. Find $m\angle 7$, if $m\angle 2 = 83°$.

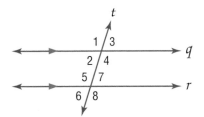

Lesson 6-2

(Pages 262–265)

Find the value of x in each triangle.

1.

2.
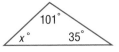

3.

Classify each triangle by its angles and by its sides.

4.
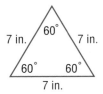

5.
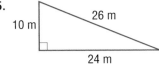

6.
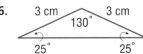

Lesson 6-3

Lesson 6-3 (Pages 267–270)

Find each missing length. Round to the nearest tenth if necessary.

1.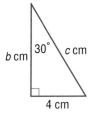

2.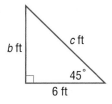

3.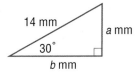

4.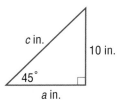

5.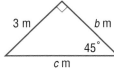

6.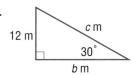

Lesson 6-4 (Pages 272–275)

Find the value of *x* in each quadrilateral.

1.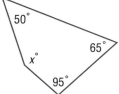

2.

3.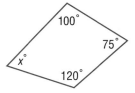

Classify each quadrilateral with the name that best describes it.

4.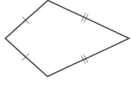

5.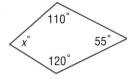

6.

Lesson 6-5 (Pages 279–282)

Determine whether the polygons are congruent. If so, name the corresponding parts and write a congruence statement.

1.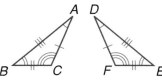

2.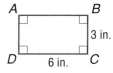

3.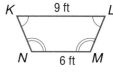

In the figure, quadrilateral *ABCD* is congruent to quadrilateral *EFGH*. Find each measure.

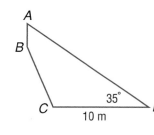

4. $m\angle A$

5. BC

6. GH

7. $m\angle H$

630 Extra Practice

Lesson 6-6

(Pages 286–289)

Complete parts a and b for each figure.

a. **Determine whether the figure has line symmetry. If it does, trace the figure and draw all lines of symmetry. If not write** *none.*

b. **Determine whether the figure has rotational symmetry. write** *yes* **or** *no.* **If** *yes,* **name the angle(s) of rotation.**

1.

2.

3.

4.

5.

6.

Lesson 6-7

(Pages 290–294)

Graph the figure with the given vertices. Then graph the image of the figure after a reflection over the given axis, and write the coordinates of its vertices.

1. triangle *CAT* with vertices $C(2, 3)$, $A(8, 2)$, and $T(4, -3)$; *x*-axis

2. trapezoid *TRAP* with vertices $T(-2, 5)$, $R(1, 5)$, $A(4, 2)$, and $P(-5, 2)$; *y*-axis

Name the line of reflection for each pair of figures.

3.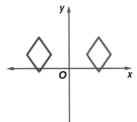

4.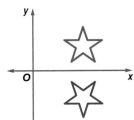

5.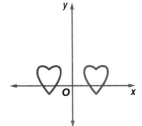

Lesson 6-8

(pages 296–299)

Graph the figure with the given vertices. Then graph the image of the figure after the indicated translation, and write the coordinates of its vertices.

1. rectangle *PQRS* with vertices $P(-7, 6)$, $Q(-5, 6)$, $R(-5, 2)$, and $S(-7, 2)$ translated 9 right and 1 unit down

2. pentagon *DGLMR* with vertices $D(1, 3)$, $G(2, 4)$, $L(4, 4)$, $M(5, 3)$ and $R(3, 1)$ translated 5 units left and 7 units down

3. triangle *TRI* with vertices $T(2, 1)$, $R(0, 3)$, and $I(-1, 1)$ translated 2 units left and 3 units down

4. quadrilateral *QUAD* with vertices $Q(3, 2)$, $U(3, 0)$, $A(6, 0)$ and $D(6, 2)$, translated 3 units left and 1 unit down

Lesson 6-9

(Pages 300–303)

Graph the figure with the given vertices. Then graph the image of the figure after the indicated rotation about the origin, and write the coordinates of its vertices.

1. triangle ABC with vertices $A(-2, -1)$, $B(0, 1)$, and $C(1, -1)$; 90° counterclockwise

2. rectangle $WXYZ$ with vertices $W(1, 1)$, $X(1, 3)$, $Y(6, 3)$, and $Z(6, 1)$; 180°

3. quadrilateral $QRST$ with vertices $Q(-2, 1)$, $R(3, 1)$, $S(3, 3)$, and $T(-2, 3)$; 90° counterclockwise

4. triangle PQR with vertices $P(1, 1)$, $Q(3, 1)$, and $R(1, 4)$; 90° counterclockwise

5. rectangle $ABCD$ with vertices $A(-1, 1)$, $B(-1, 3)$, $C(-4, 3)$, and $D(-4, 1)$; 180°

6. parallelogram $GRAM$ with vertices $G(1, -2)$, $R(2, -4)$, $A(2, -3)$, and $M(1, -1)$; 90° counterclockwise

7. triangle DEF with vertices $D(0, 2)$, $E(-3, 3)$, and $F(-3, 1)$; 180°

Lesson 7-1

(Pages 314–318)

Find the area of each figure.

1.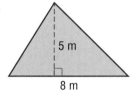
5 m
8 m

2.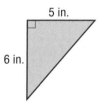
5 in.
6 in.

3.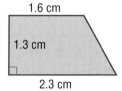
1.6 cm
1.3 cm
2.3 cm

4. triangle: base, $2\frac{1}{2}$ in.; height, 7 in.

5. triangle: base, 12 cm; height, 3.2 cm

6. trapezoid: bases, 5 ft and 7 ft; height, 11 ft

7. trapezoid: bases, $4\frac{1}{4}$ yd and $3\frac{1}{2}$ yd; height, 5 yd

8. parallelogram: base, 15 cm; height, 3 cm

9. parallelogram: base, 11.2 in.; height, 5 in.

10. triangle: base, 7 yd; height, 9 yd

11. trapezoid: bases, 9 cm and 10 cm; height, 5 cm

Lesson 7-2

(Pages 319–323)

Find the circumference and area of each circle. Round to the nearest tenth.

1.
20 mm

2.
3.5 m

3.
6 yd

4.
4 in.

5.
16 ft

6.
2.4 cm

7.
56 mm

8.
35 in.

9.
22.4 m

Find the area of each figure. Round to the nearest tenth of necessary.

1.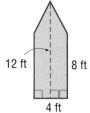

12 ft   8 ft

4 ft

2.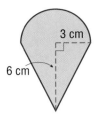

3 cm

6 cm

3.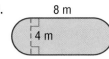

8 m

4 m

4.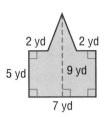

2 yd   2 yd

5 yd   9 yd

7 yd

5.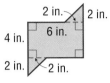

2 in.   2 in.

6 in.

4 in.

2 in.   2 in.

6.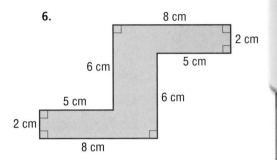

8 cm

2 cm

6 cm   5 cm

5 cm   6 cm

2 cm

8 cm

Identify each solid. Name the number and shapes of the faces. Then name the number of edges and vertices.

1.

2.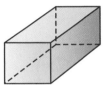

3.

4.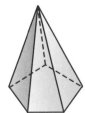

5.

6.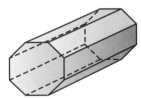

Find the volume of each solid. Round to the nearest tenth if necessary.

1.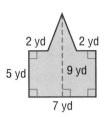

3 m

3 m

3 m

2.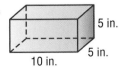

5 in.

10 in.   5 in.

3.

6 yd

11 yd

4.

26 cm

8 cm

5.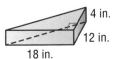

4 in.

12 in.

18 in.

6.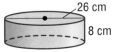

7 ft

30 ft

Lesson 7-6

Find the volume of each solid. Round to the nearest tenth if necessary.

1.
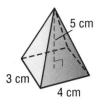
5 cm
3 cm
4 cm

2.

60 in.
60 in.
60 in.

3.

12 yd
7 yd

4.
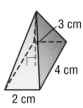
3 cm
4 cm
2 cm

5.
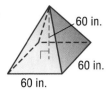
15 ft
11 ft

6.

8 ft
5 ft
8 ft

Lesson 7-7

(Pages 347–350)

Find the surface area of each solid. Round to the nearest tenth if necessary.

1.

2 ft
2 ft
2 ft

2.
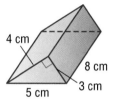
3 ft
4 ft
6 ft

3.
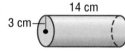
4 cm
8 cm
5 cm
3 cm

4.

8 in.
6 in.

5.
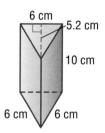
6 cm
5.2 cm
10 cm
6 cm   6 cm

6.
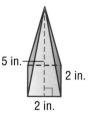
14 cm
3 cm

Lesson 7-8

(Pages 352–355)

Find the surface area of each solid. Round to the nearest tenth if necessary.

1.

12 ft
4 ft

2.
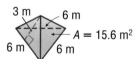
3 m   6 m
$A = 15.6$ m²
6 m   6 m

3.
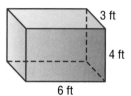
5 in.
2 in.
2 in.

4.

6.5 cm
3 cm

5.
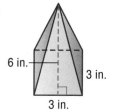
6 in.
3 in.
3 in.

6.

10 ft
12 ft

634 Extra Practice

Lesson 7-9

(Pages 358–362)

Determine the number of significant digits in each measure.

1. 18 min
2. 7.5 lb
3. 92.46 m
4. 7 ft
5. 0.067 kg
6. 61.7 cm
7. 8 mm
8. 6.02 cm

Find each sum or difference using the correct precision.

9. 9 L + 5.7 L
10. 15.27 in. − 3.16 in.
11. 3.67 ft + 2.1 ft
12. 5.612 m − 3.1 m
13. 7.1 mi − 5.421 mi
14. 0.81 kg + 5.1 kg

Find each product or quotient using the correct number of significant digits.

15. 3.257 ft ÷ 0.52 ft
16. 3.25 in · 0.2 in
17. 5.7 mm · 3 mm
18. 7.1 cm ÷ 2.1 cm
19. 18 kg ÷ 3.5 kg
20. 3.7 m · 20 m
21. 1.44 cm · 2.2 cm
22. 500 mL ÷ 3.5 mL
23. 100 mm · 73.2 mm

Lesson 8-1

(Pages 374–377)

A date is chosen at random from the calendar below. Find the probability of choosing each date. Write each probability as a fraction, a decimal, and a percent.

1. The date is the thirteenth.
2. The date is Friday.
3. It is after the twenty-fifth.
4. It is before the seventh.
5. It is an odd-numbered date.
6. The date is divisible by 3.
7. The date is Wednesday.
8. It is after the seventeenth.

			November			
S	**M**	**T**	**W**	**T**	**F**	**S**
		1	2	3	4	5
6	7	8	9	10	11	12
13	14	15	16	17	18	19
20	21	22	23	24	25	26
27	28	29	30			

Lesson 8-2

(Pages 380–383)

Draw a tree diagram to determine the number of outcomes.

1. A car comes in white, black, or red with standard or automatic transmission and with a 4-cylinder or 6-cylinder engine.

2. A customer can buy roses or carnations in red, yellow, pink, or white.

3. A bed comes in queen or king size with a firm or super firm mattress.

4. A pizza can be ordered with a regular or deep dish crust and with a choice of one topping, two toppings, or three toppings.

Use the Counting Principle to determine the number of outcomes.

5. A woman's shoe comes in red, white, blue, or black with a choice of high, medium, or low heels.

6. Sandwiches can be made with either ham or bologna, American or Swiss cheese, on wheat, rye, or white bread.

7. Sugar cookies, chocolate chip, or oatmeal raisin cookies can be ordered either with or without icing.

8. Susan can choose for her outfit a black or tan skirt, a white, pink, or cream shirt, black or tan shoes, and a red or black jacket.

Lesson 8-3

(Pages 384–387)

Find each value.

1. 8!
2. 10!
3. 0!
4. 7!
5. 6!
6. 5!
7. 2!
8. 11!
9. 9!
10. 4!
11. $P(5, 4)$
12. $P(3, 3)$
13. $P(12, 5)$
14. $P(8, 6)$
15. $P(10, 2)$
16. $P(6, 4)$

17. How many different ways can a family of four be seated in a row?

18. In how many different ways can you arrange the letters in the word *orange* if you take the letters five at a time?

19. How many ways can you arrange five different colored marbles in a row if the blue one is always in the center?

20. In how many different ways can Kevin listen to each of his ten CDs once?

Lesson 8-4

(Pages 388–391)

Find each value.

1. $C(8, 4)$
2. $C(30, 8)$
3. $C(10, 9)$
4. $C(7, 3)$
5. $C(12, 5)$
6. $C(17, 16)$
7. $C(24, 17)$
8. $C(9, 7)$

9. How many ways can you choose five compact discs from a collection of 17?

10. How many combinations of three flavors of ice cream can you choose from 25 different flavors of ice cream?

11. How many ways can you choose three books out of a selection of ten books?

Determine whether each statement is a *permutation* or a *combination*.

12. choosing a committee of 3 from a class.

13. placing 6 different math books in a line

Lesson 8-5

(Pages 396–399)

Two socks are drawn from a drawer which contains one red sock, three blue socks, two black socks, and two green socks. Once a sock is selected, it is not replaced. Find each probability.

1. P(a black sock and then a green sock)
2. P(a red sock and then a green sock)
3. P(two blue socks)
4. P(two green socks)
5. P(two black socks)
6. P(a black sock and then a red sock)
7. P(a red sock and then a blue sock)
8. P(a blue sock and then a black sock)

There are three quarters, five dimes, and twelve pennies in a bag. Once a coin is drawn from the bag it is not replaced. If two coins are drawn at random, find each probability.

9. P(a quarter and then a penny)
10. P(a nickel and then a dime)
11. P(a dime and then a penny)
12. P(two dimes)
13. P(two quarters)
14. P(two pennies in a row)
15. P(a quarter and then a dime)
16. P(a penny and then a quarter)

Lesson 8-6

(Pages 400–403)

FOOD For Exercises 1–6, use the survey results at the right.

Favorite Pizza Topping	
Topping	**Number**
pepperoni	45
sausage	25
green pepper	15
mushrooms	5
other	10

1. What is the probability that a person's favorite pizza topping is pepperoni?

2. Out of 280 people, how many would you expect to have pepperoni as their favorite pizza topping?

3. What is the probability that a person's favorite pizza topping is green pepper?

4. Out of 280 people, how many would you expect to have green pepper as their favorite pizza topping?

5. What is the probability that a person's favorite pizza topping is pepperoni or sausage?

6. Out of 280 people, how many would you expect to have pepperoni or sausage as their favorite pizza topping?

Lesson 8-7

(Pages 406–409)

Describe each sample.

1. To predict who will be the next mayor, a radio station asks their listeners to call one of two numbers to indicate their preferences.

2. To award prizes at a hockey game, four seat numbers are picked from a barrel containing individual papers representing each seat number.

3. To evaluate the quality of the televisions coming off the assembly line, the manufacturer takes one every half hour and tests it.

4. To determine what movies people prefer, people leaving a movie theater showing an action film are asked to give their preference.

5. To form a committee to discuss how the cafeteria can be improved, one student is picked at random from each second period class.

Lesson 9-1

(Pages 420–424)

ARCHITECTURE For Exercises 1–8, use the histogram at the right.

1. How large is each interval?

2. How many buildings are represented in the histogram?

3. Which interval represents the most number of buildings?

4. Which interval represents the least number of buildings?

5. How many buildings are taller than 70 feet?

6. How many buildings are less than 51 feet tall?

7. What is the height of the tallest building?

8. How does the number of buildings between 61 and 80 feet tall compare to the number of buildings between 31 and 50 feet tall?

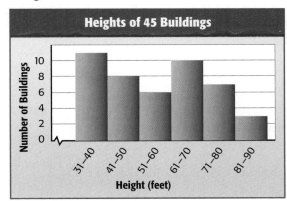

Lesson 9-2

(Pages 426–429)

Make a circle graph for each set of data.

1.

Sporting Goods Sales	
shoes	44%
apparel	30%
equipment	26%

2.

Energy Use in Home	
heating/cooling	51%
appliances	28%
lights	21%

3.

Household Income	
primary job	82%
secondary job	9%
investments	5%
other	4%

4.

Students in North High School	
freshmen	30%
sophomores	28%
juniors	24%
seniors	18%

5.

Number of Siblings	
0	25%
1	45%
2	20%
3	5%
4	2%
5+	3%

6.

Household Expenses	
food	45%
housing	30%
utilities	15%
other	10%

Lesson 9-3

(Pages 430–433)

Choose the most appropriate type of display for each situation.

1. number of televisions in homes compared to the total number of homes in the survey

2. the amount of sales by different sales people compared to the total sales

3. ages by intervals of amusement park attendees in marketing information for the park

4. average proficiency test score for five consecutive years

5. numbers of Americans who own motorcycles, boats, and recreational vehicles

6. percent of people who own a certain type of car compared to all car owners

7. a child's age and his or her height

8. amount of fat grams in intervals in various sandwiches

9. the number of students who have read each of three popular books

10. number of people filing tax returns electronically over the past ten years

Lesson 9-4

(Pages 435–438)

Find the mean, median, and mode for each set of data. Round to the nearest tenth if necessary.

1. 2, 7, 9, 12, 5, 14, 4, 8, 3, 10

2. 58, 52, 49, 60, 61, 56, 50, 61

3. 122, 134, 129, 140, 125, 134, 137

4. 36, 41, 43, 45, 48, 52, 54, 56, 56, 57, 60, 64, 65

5. 3, 9, 14, 3, 0, 2, 6, 11

6. 6, 3, 1, 8, 7, 2

7. 11, 15, 21, 11, 6, 10, 11

8. 21, 20, 19, 20, 18, 21, 23, 25

9. 1, 3, 2, 1, 1, 2, 2, 2, 3

10. 23, 35, 42, 26, 27, 29, 31, 29, 27

11. 32.1, 33.5, 31.5, 37.8

12. 25.5, 26.7, 20.9, 23.4, 26.8, 24.0, 25.7

13. 98.6, 97.9, 98.1, 100.1, 100.2

14. 10.1, 12.3, 11.4, 15.6, 7.3, 10.1

Lesson 9-5

(Pages 442–445)

Find the range, median, upper and lower quartiles, interquartile range, and any outliers for each set of data.

1. 15, 12, 21, 18, 25, 11, 17, 19, 20
2. 2, 24, 6, 13, 8, 6, 11, 4
3. 189, 149, 155, 290, 141, 152
4. 451, 501, 388, 428, 510, 480, 390
5. 22, 18, 9, 26, 14, 15, 6, 19, 28
6. 245, 218, 251, 255, 248, 241, 250
7. 46, 45, 50, 40, 49, 42, 64
8. 128, 148, 130, 142, 164, 120, 152, 202
9. 2, 3, 2, 6, 4, 14, 13, 2, 6, 3
10. 88, 84, 92, 93, 90, 96, 87, 97
11. 2, 3, 5, 4, 3, 3, 2, 5, 6
12. 6, 7, 9, 10, 11, 11, 13, 14, 12, 11, 12
13. 117, 118, 120, 109, 117, 117, 100
14. 12, 14, 17, 19, 13, 16, 17
15. 378, 480, 370, 236, 361, 394, 345, 328, 388, 339
16. 80, 91, 82, 83, 77, 79, 78, 75, 75, 88, 84, 82, 61, 93, 88, 85, 84, 89, 62, 79
17. 195, 121, 135, 123, 138, 150, 122, 138, 149, 124, 149, 151, 152

Lesson 9-6

(Pages 446–449)

Draw a box-and-whisker plot for each set of data.

1. 2, 3, 5, 4, 3, 3, 2, 5, 6
2. 6, 7, 9, 10, 11, 11, 13, 14, 12, 11, 12
3. 15, 12, 21, 18, 25, 11, 17, 19, 20
4. 2, 24, 6, 13, 8, 6, 11, 4
5. 22, 18, 9, 26, 14, 15, 6, 19, 28
6. 46, 45, 50, 40, 49, 42, 64
7. 2, 3, 2, 6, 4, 14, 13, 2, 6, 3
8. 88, 84, 92, 93, 90, 96, 87, 97
9. 80, 91, 82, 83, 77, 79, 78, 75, 75, 88, 84, 82, 61, 93, 88, 85, 84, 89, 62, 79
10. 195, 121, 135, 123, 138, 150, 122, 138, 149, 124, 149, 151, 152

ZOOS For Exercises 11 and 12, use the following box-and-whisker plot.

Area (acres) of Major Zoos in the United States

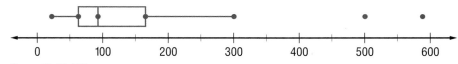

Source: *The World Almanac*

11. How many outliers are in the data?

12. Describe the distribution of the data. What can you say about the areas of the major zoos in the United States?

Lesson 9-7

(Pages 450–453)

FITNESS For Exercises 1 and 2, use the graphs at the right.

1. Do both graphs contain the same information? Explain.

2. Which graph would you use to indicate that many more eighth graders finished the obstacle course than sixth or seventh graders? Explain.

Graph A

Students Completing Obstacle Course

Number of Students: 130, 120, 110, 100, 90
6th grade, 7th grade, 8th grade

Graph B

Students Completing Obstacle Course

Number of Students: 120, 100, 80, 60, 40, 20, 0
6th grade, 7th grade, 8th grade

Lesson 9-8

State the dimension of each matrix. Then identify the position of the circled element.

1. $\begin{bmatrix} 2 \\ ③ \\ 1 \end{bmatrix}$

2. $\begin{bmatrix} 2 & ⑥ \\ -1 & 0 \end{bmatrix}$

3. $[3 \quad 4 \quad ⊖2 \quad 1]$

4. $\begin{bmatrix} 6 & 3 & 2 & 1 \\ 0 & ① & -3 & 5 \end{bmatrix}$

5. $\begin{bmatrix} ⑥ & 4 & 1 \\ 9 & 0 & 1 \\ 2 & -8 & 11 \end{bmatrix}$

6. $\begin{bmatrix} 10 & 9 \\ 2 & 8 \\ 15 & ⑬ \end{bmatrix}$

Add or subtract. If there is no sum or difference, write *impossible*.

7. $\begin{bmatrix} 2 & 0 \\ -6 & 4 \end{bmatrix} + \begin{bmatrix} 8 & 6 \\ 1 & -3 \end{bmatrix}$

8. $\begin{bmatrix} 8 & 5 & 2 \\ -1 & 3 & 11 \end{bmatrix} + \begin{bmatrix} -4 & 12 \\ 1 & 8 \\ 6 & -3 \end{bmatrix}$

9. $\begin{bmatrix} 1 & 2 \\ 6 & -3 \\ -5 & 7 \end{bmatrix} - \begin{bmatrix} -2 & 7 \\ 5 & 4 \\ -3 & 9 \end{bmatrix}$

10. $\begin{bmatrix} 0 & 8 & 6 \\ 7 & -2 & 4 \\ 0 & 0 & 9 \end{bmatrix} - \begin{bmatrix} -6 & 0 & -1 \\ 2 & 4 & 0 \\ -1 & 5 & 8 \end{bmatrix}$

11. $\begin{bmatrix} 13 & 2 \\ 7 & -1 \end{bmatrix} + \begin{bmatrix} 6 & -5 & 0 \\ 3 & 1 & -2 \\ 2 & 7 & 8 \end{bmatrix}$

12. $\begin{bmatrix} 13 & 5 & 7 \\ -6 & 12 & -1 \end{bmatrix} - \begin{bmatrix} 8 & -6 & 0 \\ -2 & 15 & 2 \end{bmatrix}$

Lesson 10-1

(Pages 469–473)

Use the Distributive Property to rewrite each expression.

1. $2(x + 3)$
2. $3(a + 7)$
3. $3(g - 6)$
4. $-2(a + 3)$
5. $-1(x - 6)$
6. $4(a - 5)$
7. $-6(x - 1)$
8. $3(2x + 5)$
9. $2(3x - 1)$
10. $-1(2x - 1)$
11. $5(-3x + 2)$
12. $-7(2x - 2)$

Simplify each expression.

13. $3x + 2x$
14. $6x - 3x$
15. $2a - 5a$
16. $5x - 6x$
17. $8a - 3a$
18. $a - 4a$
19. $3a + 2a - 6$
20. $6x + 2x - 3$
21. $5a - 3 + 2a$
22. $3x + 7 - 5x$
23. $x - 3 + 5x$
24. $6x - 3x - 2$
25. $a - 2a + 5$
26. $6x - 2 + 7x$
27. $5a - 7a + 2$
28. $4a + 2 - 7a - 5$
29. $3a - 2 + 5a - 7$
30. $5x - 3x + 2 - 5$

Lesson 10-2

(Pages 474–477)

Solve each equation. Check your solution.

1. $2x + 4 = 14$
2. $5p - 10 = 0$
3. $5 + 6a = 41$
4. $\frac{x}{3} - 7 = 2$
5. $18 = 6q - 24$
6. $18 = 4m - 6$
7. $3r - 3 = 9$
8. $2x + 3 = 5$
9. $0 = 4x - 28$
10. $3x - 1 = 5$
11. $3z + 5 = 14$
12. $3x - 15 = 12$
13. $9a - 8 = 73$
14. $2x - 3 = 7$
15. $3t + 6 = 9$
16. $2y + 10 = 22$
17. $15 = 2y - 5$
18. $3c - 4 = 2$
19. $6 + 2p = 16$
20. $8 = 2 + 3x$
21. $4b + 24 = 24$
22. $2x + 3x - 6 = 19$
23. $-2x - 6 = 14$
24. $3x - 9 = -18$
25. $2a - 3a + 1 = 15$
26. $5x - 3x + 6 = -10$
27. $3a - 5a + a = 11$
28. $5a - 3a - 5 + 1 = -10$
29. $3 = 7a - 6a + 2$
30. $3y + 5y - 1 = 15$

Extra Practice

Lesson 10-3

(Pages 478–481)

Translate each sentence into an equation. Then find each number.

1. The sum of a number and 7 is 11.

2. Seven more than the quotient of a number and -2 is 6.

3. The sum of a number and 6 is 21.

4. The difference of a number and 2 is 4.

5. Twice a number plus 5 is -3.

6. The product of a number and 3 is 18.

7. The product of a number and 4 plus 2 is 14.

8. Eight less than the quotient of a number and 3 is 5.

9. The difference of twice a number and 3 is 11.

10. The sum of 3 times a number and 7 is 25.

Lesson 10-4

(Pages 484–487)

Solve each equation. Check your solution.

1. $6x + 10 = 1x$
2. $2a - 5 = -3a$
3. $7a - 5 = 2a$
4. $3a + 7 = 10$
5. $8x + 3 = 2x$
6. $5x - 3 = -18$
7. $3a - 1 = 2a$
8. $7a - 2 = 12$
9. $3x + 6 = x$
10. $2x + 7 = 11 - 2x$
11. $8x + 10 = 3x$
12. $7a + 4 = 3a$
13. $7x + 8 = 11x$
14. $21x + 11 = 10x$
15. $5x + 5 = 14 + 2x$
16. $7b - 4 = 2b + 16$
17. $2y - 3 = 5 - 2y$
18. $3m = 2m + 7$
19. $9t + 1 = 4t - 9$
20. $-2a + 3 = a - 12$
21. $3x = 9x - 12$
22. $2c + 3 = 3c - 4$
23. $s - 3 = 5 - s$
24. $3w - 5 = 5w - 7$
25. $4x - 7 = 11 + x$
26. $5x + 2 = 10 + x$
27. $3x + 2 = 2x + 5$
28. $8a + 7 = 7a + 8$
29. $3a - 11 = 4a - 12$
30. $2a - 5 = 8a - 11$

Lesson 10-5

(Pages 492–495)

Write an inequality for each sentence.

1. A number is less than 10.
2. A number is greater than or equal -7.
3. A number is less than -2.
4. A number is more than 5.
5. A number is less than or equal to 11.
6. a number is no more than 8.

Graph each inequality on a number line.

7. $x > 5$
8. $y > 0$
9. $z < -2$
10. $a \geq 6$
11. $b \leq 2$
12. $x \geq 1$
13. $a \leq 3$
14. $b \geq 1$
15. $x < -2$
16. $n \geq -3$
17. $t > -1$
18. $y \leq -5$

Lesson 10-6

(Pages 496–499)

Solve each inequality. Check your solution.

1. $y + 3 > 7$
2. $c - 9 < 5$
3. $x + 4 \geq 9$
4. $y - 3 < 15$
5. $t - 13 \geq 5$
6. $x + 3 < 10$
7. $y - 6 \geq 2$
8. $x - 3 \geq -6$
9. $a + 3 \leq 5$
10. $c - 2 \leq 11$
11. $a + 15 \geq 6$
12. $y + 3 \geq 18$
13. $y + 16 \geq -22$
14. $x - 3 \geq 17$
15. $y - 6 > -17$
16. $y - 11 < 7$
17. $a + 5 \geq 21$
18. $c + 3 > -16$
19. $x - 12 \geq 12$
20. $x + 5 \geq 5$
21. $y - 6 > 31$
22. $a - 6 > 17$
23. $y + 7 > 3$
24. $a + 13 \geq -16$
25. $y - 6 > 5$
26. $y + 6 < -5$
27. $x - 17 \geq 34$
28. $y + 1 \leq 16$
29. $a - 14 \geq 16$
30. $x + 14 \leq 20$

Lesson 10-7

(Pages 500–504)

Solve each inequality and check your solution. Then graph the solution on a number line.

1. $5p \geq 25$
2. $4x < 12$
3. $15 \leq 3m$
4. $\frac{d}{3} > 15$
5. $8 < \frac{r}{7}$
6. $9g < 27$
7. $4p \geq 24$
8. $5p > 25$
9. $-4 > \frac{-k}{3}$
10. $\frac{-z}{5} > 2$
11. $-3x \leq 9$
12. $-5x > -35$
13. $\frac{a}{-6} < 1$
14. $\frac{x}{-5} \leq -2$
15. $-2x < 16$
16. $3p \geq 12$
17. $\frac{x}{-2} \leq -2$
18. $\frac{y}{6} \leq -5$
19. $5p \geq 100$
20. $-4x \leq 64$
21. $8x \geq 56$
22. $-2t < 14$
23. $18 > 3x$
24. $5x > 10$
25. $\frac{a}{3} > 1$
26. $14 \leq 2x$
27. $\frac{x}{2} > 0$
28. $2y \geq 22$
29. $35 \leq 5d$
30. $3x \geq 9$

Lesson 11-1

(Pages 512–515)

State whether each sequence is *arithmetic*, *geometric*, or *neither*. If it is arithmetic or geometric, state the common difference or common ratio. Write the next three terms of each sequence.

1. 1, 5, 9, 13, ...
2. 2, 6, 18, 54, ...
3. 1, 4, 9, 16, 25, ...
4. 729, 243, 81, ...
5. 2, −3, −8, −13, ...
6. 5, −5, 5, −5, ...
7. 810, −270, 90, −30, ...
8. 11, 14, 17, 20, 23, ...
9. 33, 27, 21, ...
10. 21, 15, 9, 3, ...
11. $\frac{1}{8}, -\frac{1}{4}, \frac{1}{2}, -1, \ldots$
12. $\frac{1}{81}, \frac{1}{27}, \frac{1}{9}, \frac{1}{3}, \ldots$
13. $\frac{3}{4}, 1\frac{1}{2}, 3, \ldots$
14. 2, 5, 9, 14, ...
15. $-1\frac{1}{4}, -1\frac{3}{4}, -2\frac{1}{4}, -2\frac{3}{4}, \ldots$
16. 9.9, 13.7, 17.5, ...
17. $\frac{1}{2}, 1\frac{1}{2}, 2\frac{1}{2}, 3\frac{1}{2}, \ldots$
18. 2, 12, 32, 62, ...
19. 3, −6, 12, −24, ...
20. 5, 7, 9, 11, 13, ...
21. −0.06, 2.24, 4.54, ...
22. 7, 14, 28, ...
23. −5.4, −1.4, 2.6, ...
24. −96, 48, −24, 12, ...
25. 4, 12, 36, ...
26. 20, 19, 18, 17, ...
27. 768, 192, 48, ...

Find each function value.

1. $f\left(\frac{1}{2}\right)$ if $f(x) = 2x - 6$

2. $f(-4)$ if $f(x) = -\frac{1}{2}x + 4$

3. $f(1)$ if $f(x) = -5x + 1$

4. $f(6)$ if $f(x) = \frac{2}{3}x - 5$

5. $f(0)$ if $f(x) = 1.6x + 4$

6. $f(2)$ if $f(x) = 2x - 8$

7. $f(-1)$ if $f(x) = -3x + 5$

8. $f\left(\frac{1}{2}\right)$ if $f(x) = 2x - 1$

9. $f(6)$ if $f(x) = \frac{2}{3}x + 4$

Copy and complete each function table.

10. $f(x) = -4x$

x	−4x	f(x)
−2		
−1		
0		
1		
2		

11. $f(x) = x + 6$

x	x + 6	f(x)
−6		
−4		
−2		
0		
2		

12. $f(x) = 3x + 2$

x	3x + 2	f(x)
−3		
−2		
−1		
0		
1		

Copy and complete the table. Then graph the function.

1. $y = 6x + 2$

x	6x + 2	y	(x, y)
−2			
−1			
0			
1			

2. $y = -2x + 3$

x	−2x + 3	y	(x, y)
−2			
−1			
0			
2			

Graph each function.

3. $y = -5x$

4. $y = 10x - 2$

5. $y = -2.5x - 1.5$

6. $y = 7x + 3$

7. $y = \frac{x}{4} - 8$

8. $y = 3x + 1$

9. $y = 25 - 2x$

10. $y = \frac{x}{6}$

11. $y = -2x + 11$

12. $y = 7x - 3$

13. $y = \frac{x}{2} + 5$

14. $y = 4 - 6x$

15. $y = -3.5x - 1$

16. $y = 4x + 10$

17. $y = 8x$

18. $y = \frac{x}{3} + 2$

Find the slope of the line that passes through each pair of points.

1. $A(2, 3), B(1, 5)$

2. $C(-6, 1), D(2, 1)$

3. $E(3, 0), F(5, 0)$

4. $G(-1, -3), H(-2, -5)$

5. $I(6, 7), J(11, 1)$

6. $K(5, 3), L(5, -2)$

7. $M(10, 2), N(-3, 5)$

8. $O(6, 2), P(1, 7)$

9. $Q(5, 8), R(-3, -2)$

10. $S(-1, 7), T(3, 8)$

11. $U(4, -1), V(-5, -2)$

12. $W(3, -2), X(7, -1)$

13. $Y(0, 5), Z(2, 1)$

14. $A(6, 5), B(-3, -5)$

15. $C(2, 1), D(7, -1)$

16. $E(-5, 2), F(0, 2)$

17. $G(-3, 5), H(-2, 5)$

18. $I(2, 0), J(3, 5)$

19. $K(11, 1), L(21, 3)$

20. $M(6, 5), N(-1, 7)$

21. $O(-2, 3), P(2, -1)$

22. $Q(5, 0), R(1, 1)$

23. $S(0, 0), T(3, -4)$

24. $U(5, 3), V(5, -2)$

Lesson 11-5

(Pages 533–536)

State the slope and *y*-intercept for the graph of each line.

1. $y = 3x - 5$

2. $y = 2x - 6$

3. $y = -6x + \frac{1}{2}$

4. $y = -7x + \frac{5}{2}$

5. $y = \frac{1}{2}x + 7$

6. $y = \frac{3}{4}x + 8$

7. $y = -\frac{2}{3}x - \frac{1}{3}$

8. $y = -\frac{1}{8}x - \frac{3}{8}$

9. $y = \frac{2}{3}x + 5$

10. $y = -\frac{2}{7}x - 1$

11. $3x + y = 6$

12. $y - 4x = 7$

Graph each equation using the slope and *y*-intercept.

13. $y = -2x + 5$

14. $y = -3x + 1$

15. $y = -x + 1$

16. $y = -x + 3$

17. $y = x - 3$

18. $y = x - 5$

19. $y = 3x - 6$

20. $y = \frac{5}{2}x - 1$

21. $y = \frac{1}{2}x + 3$

22. $y = -2x - 2$

23. $y - 4x = -1$

24. $2x + y = 3$

Lesson 11-6

(Pages 539–542)

Determine whether a scatter plot of the data for the following might show a *positive*, *negative*, or *no* relationship.

1. height and hair color

2. hours spent studying and test scores

3. income and month of birth

4. price of oranges and number available

5. size of roof and number of shingles

6. number of clouds and number of stars seen

7. child's age and height

8. age and eye color

9. number of hours worked and earnings

10. temperature outside and heating bill

11. length of foot and shoe size

12. number of candies eaten and number left in a bowl

13.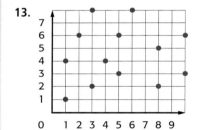

14.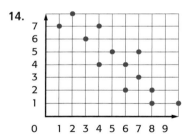

15.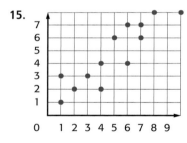

Lesson 11-7

(Pages 544–547)

Solve each system of equations by graphing.

1. $y = x - 1$
   $y = -x + 11$

2. $y = -x$
   $y = 2x$

3. $y = -x + 3$
   $y = x + 3$

4. $y = x - 3$
   $y = 2x$

5. $y = -x + 6$
   $y = x + 2$

6. $y = -x + 2$
   $y = x - 4$

7. $y = -3x + 6$
   $y = x - 2$

8. $y = 3x - 4$
   $y = -3x - 4$

9. $y = 2x + 1$
   $y = 3x$

10. $y = -x + 4$
    $y = x - 10$

11. $y = -x + 6$
    $y = 2x$

12. $y = x - 4$
    $y = -2x + 5$

Solve each system of equations by substitution.

13. $y = 2x - 5$
    $x = 5$

14. $y = -3x + 5$
    $x = -2$

15. $y = 5$
    $y = 2x + 5$

16. $y = -4$
    $y = 2x - 6$

Lesson 11-8

(Pages 548–551)

Graph each inequality.

1. $y < 3x$
2. $y < 2x$
3. $y > -2x$
4. $y < -5x$

5. $y \geq 3x + 1$
6. $y \leq 4x + 2$
7. $y \leq -2x$
8. $y \geq -x$

9. $y < x$
10. $y > -x$
11. $y < x + 1$
12. $y \geq -x - 1$

13. $y \leq \frac{1}{2}x + 2$
14. $y \geq \frac{2}{3}x + 5$
15. $y \leq \frac{1}{5}x - 2$
16. $y > \frac{3}{5}x$

17. $y > \frac{7}{8}x$
18. $y \leq -\frac{2}{5}x + 1$
19. $y \geq \frac{1}{4}x$
20. $y < \frac{1}{3}x + 4$

21. $y \geq 2x + 1$
22. $y < x - 3$
23. $y \leq x + 4$
24. $y > -x - 2$

25. $y \geq \frac{1}{2}x - 3$
26. $y < -\frac{1}{2}x + 3$
27. $y > -\frac{1}{3}x - 2$
28. $y \geq \frac{2}{3}x + 1$

Lesson 12-1

(Pages 560–563)

Determine whether each graph, equation, or table represents a *linear* or *nonlinear* function. Explain.

1.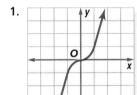
2.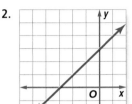
3.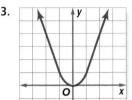

4. $y = 3x$
5. $y = \frac{2}{3}x$
6. $y = x^2 + 5$

7. $y = 4^x$
8. $y = -\frac{3}{x}$
9. $xy = -3$

10. $y = x^3 + 2$
11. $y = 2$
12. $y = 3x + 5$

13.

x	−1	0	1	2
y	2	0	2	8

14.

x	−1	0	1	2
y	−1	0	1	8

15.

x	−1	0	1	2
y	−3	0	3	6

16.

x	−1	0	1	2
y	−5	0	5	10

17.

x	−1	0	1	2
y	−1	0	−1	−4

18.

x	−1	0	1	2
y	1	0	1	16

Lesson 12-2

(Pages 565–568)

Graph each function.

1. $y = x^2 - 1$
2. $y = 1.5x^2 + 3$
3. $y = x^2 - x$
4. $y = 2x^2$

5. $y = x^2 + 3$
6. $y = -3x^2 + 4$
7. $y = -x^2 + 7$
8. $y = 3x^2$

9. $y = 3x^2 + 9x$
10. $y = -x^2$
11. $y = \frac{1}{2}x^2 + 1$
12. $y = 5x^2 - 4$

13. $y = -x^2 + 3x$
14. $y = 2.5x^2$
15. $y = -2x^2$
16. $y = 8x^2 + 3$

17. $y = -x^2 + \frac{1}{2}x$
18. $y = -4x^2 + 4$
19. $y = 4x^2 + 3$
20. $y = -4x^2 + 1$

21. $y = 2x^2 + 1$
22. $y = x^2 - 4x$
23. $y = 3x^2 + 5$
24. $y = 0.5x^2$

25. $y = 2x^2 - 5x$
26. $y = \frac{3}{2}x^2 - 2$
27. $y = 6x^2 + 2$
28. $y = 5x^2 + 6x$

Lesson 12-3

(Pages 570–573)

Simplify each polynomial. If the polynomial cannot be simplified, write
simplest form.

1. $2x - x^2 + x - 1$

2. $2x^2 + 1 + x^2$

3. $-2y + 3 + x^2 + 5y$ **4.** $m + m^2 + n + 3m^2$ **5.** $a^2 + b^2 + 3 + 2b^2$

6. $1 + a + b + 6$ **7.** $x + x^2 + 5x - 3x^2$ **8.** $-2y + 3 + y - 2$

9. $6x + 3y - 2x$ **10.** $5a + 2b - 7a - 3b$ **11.** $-2x + 5 - 3x + 7$

12. $5y - 2z - 6z - 1$ **13.** $x^2 - 3x + 2x^2 - 5x$ **14.** $y^2 + 2y + 1 - 3y^2 - 5y$

15. $4t + 3s - 2t + 7s$ **16.** $4d + 5 - 7d - 8$ **17.** $x^2 - 3x + 4x^2 - 4x + 7$

Lesson 12-4

(Pages 574–577)

Add.

1. $\begin{array}{r} 2x^2 - 5x + 7 \\ + x^2 - x + 11 \\ \hline \end{array}$ **2.** $\begin{array}{r} 2m^2 + m + 1 \\ + (-m^2) + 2m + 3 \\ \hline \end{array}$ **3.** $\begin{array}{r} 2a - b + 6c \\ + 3a - 7b + 2c \\ \hline \end{array}$

4. $\begin{array}{r} 5a + 3a^2 - 2 \\ + 2a + 8a^2 + 4 \\ \hline \end{array}$ **5.** $\begin{array}{r} 3c + b + a \\ + (-c) + b - a \\ \hline \end{array}$ **6.** $\begin{array}{r} -z^2 + x^2 + 2y^2 \\ + 3z^2 + x^2 + y^2 \\ \hline \end{array}$

7. $(5x + 6y) + (2x + 8y)$ **8.** $(4a + 6b) + (2a + 3b)$

9. $(7r + 11m) + (4m + 2r)$ **10.** $(-z + z^2) + (-2z + z^2)$

11. $(3x - 7y) + (3y + 4x + 1)$ **12.** $(5m + 3n - 3) + (8m + 6)$

13. $(a + a^2) + (3a - 2a^2)$ **14.** $(3s - 5t) + (8t + 2s)$

15. $(3x - 2y) + (5x + 3y)$ **16.** $(2a - 5b) + (-3a - 6b)$

17. $(x + 3) + (x^2 - 3x + 4)$ **18.** $(x^2 + 2x - 3) + (3x^2 - 5x - 6)$

Lesson 12-5

(Pages 580–583)

Subtract.

1. $\begin{array}{r} 5a - 6m \\ (-) 2a + 5m \\ \hline \end{array}$ **2.** $\begin{array}{r} 2a - 7 \\ (-) 8a - 11 \\ \hline \end{array}$ **3.** $\begin{array}{r} 9r^2 + r + 3 \\ (-) 11r^2 - r + 12 \\ \hline \end{array}$

4. $(9x + 3y) - (9y + x)$ **5.** $(3x^2 + 2x - 1) - (2x + 2)$

6. $(a^2 + 6a + 3) - (5a^2 + 5)$ **7.** $(5a + 2) - (3a^2 + a + 8)$

8. $(3x^2 - 7x) - (8x - 6)$ **9.** $(3m + 3n) - (m + 2n)$

10. $(3m - 2) - (2m + 1)$ **11.** $(x^2 - 2) - (x + 3)$

12. $(5x^2 - 4) - (3x^2 + 8x + 4)$ **13.** $(7z^2 + 1) - (3z^2 + 2z - 6)$

14. $(2x - 5) - (3x - 6)$ **15.** $(5a - 1) - (8a - 3)$

16. $(3y^2 - 5) - (5y + 6)$ **17.** $(2x^2 - 6) - (7x - 3)$

18. $(x^2 + 5x - 6) - (2x^2 - 3x + 5)$ **19.** $(a^2 - 3a - 5) - (5a^2 - 6a + 2)$

Lesson 12-6

(Pages 584–587)

Multiply or divide. Express using exponents.

1. $2^3 \cdot 2^4$

2. $5^6 \cdot 5$

3. $t^2 \cdot t^2$

4. $y^5 \cdot y^3$

5. $(-3x^3)(-2x^2)$

6. $b^{12} \cdot b$

7. $3^5 \cdot 3^8$

8. $(-2y^3)(5y^7)$

9. $(6a^5)(-3a^6)$

10. $(-x)(-6x^3)$

11. $(3x^2)(2x^5)$

12. $(-6y^2)(-2y^5)$

13. $(-3a)(-2a^6)$

14. $8a^9(5a^5)$

15. $(6x^2)(2x^{11})$

16. $\dfrac{x^{11}}{x^2}$

17. $\dfrac{a^6}{a^3}$

18. $\dfrac{b^9}{b^3}$

19. $\dfrac{7^9}{7^6}$

20. $\dfrac{2^5}{2^2}$

21. $\dfrac{11^{10}}{11}$

22. $\dfrac{16x^3}{4x^2}$

23. $\dfrac{25y^5}{5y^2}$

24. $\dfrac{-48y^3}{-8y}$

25. $\dfrac{12y^2}{3y^2}$

26. $\dfrac{39x^7y^5}{3x^3y}$

27. $\dfrac{21a^7b^2}{7ab^2}$

28. $\dfrac{22a^5b^3}{2a^2b}$

29. $\dfrac{15x^2y}{3xy}$

30. $\dfrac{20a^3b^2}{2a^2b}$

Lesson 12-7

(Pages 590–592)

Multiply.

1. $a(a + 2)$

2. $x(2x - 3)$

3. $t(3t + 1)$

4. $a(a + 4)$

5. $m(m - 7)$

6. $z^2(z + 3)$

7. $6x(x + 10)$

8. $3y(5 + y)$

9. $2d(d^3 + 1)$

10. $m^3(m^2 - 2)$

11. $p^5(3p - 1)$

12. $b(9 + 4b)$

13. $4t^3(t + 3)$

14. $2r^4(5r + 9)$

15. $3n^2(6 - 7n^3)$

16. $3x(x^2 + 2)$

17. $-2x(x^3 + 2x + 5)$

18. $-3x(-2x - 6)$

19. $-2(x - 5)$

20. $2x^2(3x + 5)$

21. $3a^2(2a - 6)$

22. $5a(7 - 3a)$

23. $-2a^2(8 - 5a^2)$

24. $3x^2(1 + x + 5x^2)$

25. $y(2y^2 + 3y - 1)$

26. $-a(a^2 - 3a + 5)$

27. $2x^2(2x^2 + 2x + 2)$

28. $n(n^2 - n + 3)$

29. $-3t^2(2t^2 - 2t)$

30. $-5p^2(4p + 10)$

Mixed Problem Solving

Chapter 1 Algebra: Integers

(pages 4–59)

1. **PATTERNS** Draw the next two figures in the pattern below. (Lesson 1-1)

TEMPERATURE For Exercises 2 and 3, use the following information.

The formula $F = \frac{9}{5}C + 32$ is used to convert degrees Celsius to degrees Fahrenheit. (Lesson 1-2)

2. Find the degrees Fahrenheit if it is 30°C outside.

3. A local newscaster announces that today is his birthday. Rather than disclose his true age on air, he states that his age in Celsius is 10. How old is he?

4. **SPORTS** In football, a penalty results in a loss of yards. Write an integer to describe a loss of 10 yards. (Lesson 1-3)

BILLS For Exercises 5 and 6, use the table below. (Lesson 1-4)

Description	Amount ($)
Beginning Balance	435
Gas Company	−75
Electric Company	−75
Phone Company	−100
Deposit	75
Rent	−200

5. How much is in the account?

6. Kirsten owes the cable company $65. Does she have enough to pay this bill?

7. **WEATHER** For the month of August, the highest temperature was 98°F. The lowest temperature was 54°F. What was the range of temperatures for the month? (Lesson 1-5)

8. **WEATHER** During a thunderstorm, the temperature dropped by 5 degrees per half-hour. What was the temperature change after 3 hours? (Lesson 1-6)

9. **AGE** Julia is 6 years older than Elias. Define a variable and write an expression for Julia's age. (Lesson 1-7)

HISTORY For Exercises 10 and 11, use the following information.
To be President of the United States, a person must be at least 35 years old. (Lesson 1-7)

10. If y is the year a person was born, write an expression for the earliest year that he or she could be president.

11. If a person became President this year, write an equation to find the latest year he or she could have been born.

12. **BANKING** After you withdraw $75 from your checking account, the balance is $205. Write and solve a subtraction equation to find your balance before the withdrawal. (Lesson 1-8)

13. **HISTORY** Dario gained 25 pounds during his junior year. By the end of his junior year, he weighed 160 pounds. Write and solve an addition equation to find out how much he weighed at the beginning of his junior year. (Lesson 1-8)

14. **MONEY** Janelle baby-sits and charges $5 per hour. Write and solve a multiplication equation to find how many hours she needs to baby-sit in order to make $55. (Lesson 1-9)

15. **PHYSICAL SCIENCE** Work is done when a force acts on an object and the object moves. The amount of work, measured in foot-pounds, is equal to the amount of force applied, measured in pounds, times the distance, in feet, the object moved. Write and solve a multiplication equation that could be used to find how far you have to lift a 45-pound object to produce 180 foot-pounds of work. (Lesson 1-9)

1. **HEALTH** A newborn baby weighs $6\frac{3}{4}$ pounds. Write this weight as a decimal. (Lesson 2-1)

MEASUREMENT For Exercises 2 and 3, use the figure below. (Lesson 2-1)

2. Write the length of the pencil as a fraction.
3. Write the length of the pencil as a decimal.

4. **SEWING** Which is the smallest seam: $\frac{1}{4}$ inch, $\frac{1}{2}$ inch, or $\frac{1}{8}$ inch? (Lesson 2-2)

Find the area of each rectangle. (Lesson 2-3)

5.
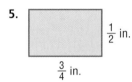
$\frac{1}{2}$ in.
$\frac{3}{4}$ in.

6.
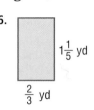
$1\frac{1}{5}$ yd
$\frac{2}{3}$ yd

7. **COOKING** Giovanni is increasing his double chocolate chip cookie recipe to $1\frac{1}{2}$ batches. If the original recipe calls for $3\frac{1}{2}$ cups of flour, how much flour does he need for $1\frac{1}{2}$ batches? (Lesson 2-3)

8. **MEDICINE** A baby gets 1 dropper of medicine for each $2\frac{1}{4}$ pounds of body weight. If a baby weighs $11\frac{1}{4}$ pounds, how many droppers of medicine should she get? (Lesson 2-4)

9. **LIBRARIES** Lucas is storing a set of art books on a shelf that has $11\frac{1}{4}$ inches of shelf space. If each book is $\frac{3}{4}$ inch wide, how many books can be stored on the shelf? (Lesson 2-4)

10. **HEIGHT** Molly is $64\frac{1}{4}$ inches tall. Minya is $62\frac{3}{4}$ inches tall. How much taller is Molly than Minya? (Lesson 2-5)

GEOMETRY Find the perimeter of each figure. (Lesson 2-5)

11.

$\frac{1}{2}$ ft
$1\frac{1}{2}$ ft

12.
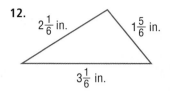
$2\frac{1}{6}$ in.
$1\frac{5}{6}$ in.
$3\frac{1}{6}$ in.

13. **ELECTIONS** In the student council elections, Janie won $\frac{1}{5}$ of the votes, and Jamal won $\frac{2}{3}$ of the votes. What fraction of the votes did the only other candidate receive? (Lesson 2-6)

14. **CONSTRUCTION** Three pieces of wood are $4\frac{3}{4}$, $5\frac{1}{8}$, and $7\frac{3}{16}$ inches long. If these pieces of wood are laid end to end, what is their total length? (Lesson 2-6)

FINANCES For Exercises 15 and 16, use the following information.
Jenna makes \$3.25 per hour delivering newspapers. (Lesson 2-7)

15. Write a multiplication equation you can use to determine how many hours she must work to earn \$35.75.

16. How many hours does Jenna need to work to earn \$35.75?

17. **BIOLOGY** If one cell splits in two every $\frac{1}{2}$ hour, how many cells will there be after $4\frac{1}{2}$ hours? (Lesson 2-8)

18. **EARTH SCIENCE** There are approximately 10^{21} kilograms of water on Earth. Write the number of kilograms of water on Earth. (Lesson 2-9)

19. **HAIR** There are an estimated 100,000 hairs on a person's head. Write this number in scientific notation. (Lesson 2-9)

20. **LIFE SCIENCE** A petri dish contains 2.53×10^{11} bacteria. Write the number of bacteria in standard form. (Lesson 2-9)

Mixed Problem Solving

1. **GARDENING** A square garden has an area of 576 square feet. What is the length of each side of the garden? (Lesson 3-1)

GEOMETRY The formula for the perimeter of a square is $P = 4s$, where s is the length of the side. Find the perimeter of each square. (Lesson 3-1)

2.
Area = 16 square meters

3.
Area = 144 square inches

SCIENCE The formula $t = \frac{\sqrt{h}}{4}$ represents the time t in seconds that it takes an object to fall from a height of h feet. (Lesson 3-2)

4. If a ball is dropped from a height of 100 feet, estimate how long it will take to reach the ground.

5. If a ball is dropped from a height of 500 feet, estimate how long it will take to reach the ground.

6. **WAVES** The speed s in knots of a wave can be estimated using the formula $s = 1.34\sqrt{\ell}$, where ℓ is the length of the wave in feet. Find the estimated speed of a wave of length 5 feet. (Lesson 3-3)

7. **GEOMETRY** To approximate the radius of a circle, you can use the formula $r = \sqrt{\frac{A}{3.14}}$, where A is the area of the circle. To the nearest tenth, find the radius of a circle that has an area of 60 square feet. (Lesson 3-3)

8. **GEOGRAPHY** In Ohio, a triangle is formed by the cities Cleveland, Columbus, and Toledo. From the distances given below, is this triangle a right triangle? Explain your reasoning. (Lesson 3-4)

9. **INTERIOR DESIGN** A room is 20 feet by 15 feet. Find the length of the diagonal of the room. (Lesson 3-4)

10. **TRAVEL** Plane A travels north 500 miles. Plane B leaves from the same location at the same time and travels east 250 miles. How far apart are the two planes? (Lesson 3-5)

11. **TELEVISION** A television screen has a diagonal measurement of 32 inches and a width of 15 inches. How long is the television? (Lesson 3-5)

12. **KITES** A kite string is 25 yards long. The horizontal distance between the kite and the person flying it is 12 yards. How high is the kite? (Lesson 3-5)

13. **REPAIRS** Shane is painting his house. He has a ladder that is 10 feet long. He places the base of the ladder 6 feet from the house. How far from the ground will the top of the ladder reach? (Lesson 3-5)

14. **ARCHEOLOGY** A dig uncovers an urn at $(1, 1)$ and a bracelet at $(5, 3)$. How far apart were the two items if one unit on the grid equals 1 mile? (Lesson 3-6)

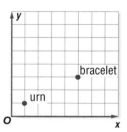

15. **TRAVEL** A unit on the grid below is 0.25 mile. Find the distance from point A to point B. (Lesson 3-6)

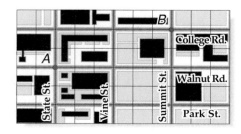

1. **SHOPPING** You can buy 3 tapes at The Music Shoppe for $12.99, or you can buy 5 of the same tapes for $19.99 at Quality Sounds. Which is the better buy? Explain your reasoning. (Lesson 4-1)

2. **TRAVEL** On a trip, you drive 1,565 miles on 100 gallons of gas. Find your car's gas mileage. (Lesson 4-1)

3. **WEATHER** The temperature is 88°F at 2 P.M. and 72°F at 3:30 P.M. What was the rate of change in temperature between these two time periods? (Lesson 4-2)

WEIGHTS **The table below gives the age and weight of a young child.** (Lesson 4-2)

Age (yr)	3	3.5	4	5	7
Weight (lb)	31	34	38	40	45

4. Between which two ages did the child's weight increase the most? Explain.

5. Between which two ages did the child's weight increase the least? Explain.

6. **LOANS** Find the slope of the line below and interpret its meaning as a rate of change. (Lesson 4-3)

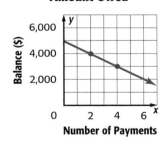

Amount Owed

7. **ELECTIONS** About $\frac{2}{3}$ of the eighth grade class voted for Dominic to be Student Council president. If there are 350 students in the eighth grade class, how many voted for Dominic? (Lesson 4-4)

8. **HEALTH** About $\frac{3}{5}$ of the babies born at Memorial Hospital are boys. If there are 250 babies born during the month of September, about how many are boys? (Lesson 4-4)

9. **PHOTOGRAPHY** Eva wants to enlarge the picture below and frame it. The scale factor from the original picture to the enlarged picture is to be 2:5. Find the dimensions of the enlarged picture. (Lesson 4-5)

6 in.

4 in.

10. **ARCHITECTURE** The Eiffel Tower is 986 feet tall. If a scale model is 6 inches tall, what is the scale of the model? (Lesson 4-6)

11. **CARS** A model is being built of a car. The car is 12 feet long and 9 feet wide. If the length of the model is 4 inches, how wide should the model be? (Lesson 4-6)

12. **FLAGPOLE** A 10-foot tall flagpole casts a 4-foot shadow. At the same time, a nearby tree casts a 25-foot shadow. Draw a diagram of this situation. Then write and solve a proportion to find the height of the tree. (Lesson 4-7)

13. **SURVEYING** Write and solve a proportion to find the distance across the river shown in the diagram below. (Lesson 4-7)

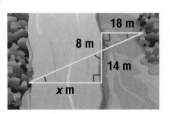

18 m

8 m

14 m

x m

MURAL **For Exercises 14 and 15, use the following information.**
A design 10 inches long and 7 inches wide is to be enlarged to appear as a wall mural that is 36 inches long. (Lesson 4-8)

14. What is the scale factor for this enlargement?

15. How wide will the mural be?

1. **SCHOOL** Two out of five children entering kindergarten can read. Write this ratio as a percent. (Lesson 5-1)

2. **ELECTIONS** About 25% of the school voted for yellow and red to be the school colors. Write this percent as a fraction. (Lesson 5-1)

3. **FOOD** About $\frac{17}{25}$ of Americans eat fast food at least two times a week. Write $\frac{17}{25}$ as a percent. (Lesson 5-2)

4. **GEOMETRY** What percent of the area of the rectangle is shaded? (Lesson 5-2)

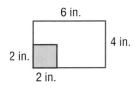

5. **EXAMS** Lexie answered 75% of the questions correctly on an exam. If she answered 30 questions correctly, how many questions were on the exam? (Lesson 5-3)

CANDY For Exercises 6–8, use the table below listing the number of each color of chocolate candies in a jar. (Lesson 5-3)

Color	Number
yellow	4
brown	12
red	2
green	5
orange	1
blue	1

6. What percent of the candies are brown?

7. What percent of the candies are green?

8. What percent of the candies are blue?

9. **RETAIL** Mr. Lewis receives a 10% commission on items he sells. What is his commission on a $35 purse? (Lesson 5-4)

10. **FARMING** A farmer receives 25% of the cost of a bag of flour. Determine the amount of money a farmer receives from a bag of flour that sells for $1.60. (Lesson 5-4)

MOVIES For Exercises 11 and 12, use the following information.
The results of a survey asking children ages 3 to 6 if they liked a recent animated movie are depicted below. (Lesson 5-4)

11. If 120 children were surveyed, how many said they liked the movie?

12. How many said they did not like the movie?

13. **LIFE SCIENCE** The table below lists the elements found in the human body. If Jacinta weighs 120 pounds, estimate how many pounds of each element are in her body. (Lesson 5-5)

Element	Percent of Body
Oxygen	63
Carbon	19
Hydrogen	9
Nitrogen	5
Calcium	1.5
Phosphorus and Sulfur	1.2

Source: *The New York Public Library Science Desk Reference*

14. **RETAIL** A pair of shoes costs $50. If a 5.75% sales tax is added, what is the total cost of the shoes? (Lesson 5-6)

15. **WEATHER** The average wind speed on Mount Washington is 35.3 miles per hour. The highest wind speed ever recorded there is 231 miles per hour. Find the percent of change from the average wind speed to the highest wind speed recorded. (Lesson 5-7)

16. **MONEY** Suppose $500 is deposited into an account with a simple interest rate of 5.5%. Find the total in the account after 3 years. (Lesson 5-8)

FURNITURE **For Exercises 1–3, use the following information.**
A single piece of wood is used for both the backrest of a chair and its rear legs. The inside angle that the wood makes with the floor is 100°, and the seat is parallel to the floor. (Lesson 6-1)

1. Find the values of *x* and *y*.

2. Classify the angle measuring *x*°.

3. Classify the angle pair measuring 100° and *y*° using all names that apply.

4. **URBAN PLANNING** Ambulances cannot safely make turns of less than 70°. The proposed site of a hospital's emergency entrance is to be at the northeast corner of Bidwell and Elmwood. Should this site be approved? Explain your reasoning. (Lesson 6-1)

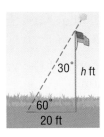

5. **CONSTRUCTION** A 12-foot ladder leans against a house. The base of the ladder rests on level ground and is 4 feet from the house. The top of the ladder reaches 11.3 feet up the side of the house. Classify the triangle formed by the house, the ground, and the ladder by its angles and by its sides. (Lesson 6-2)

6. **MEASUREMENT** At the same time that the sun's rays make a 60° angle with the ground, the shadow cast by a flagpole is 20 feet. To the nearest foot, find the height of the flagpole. (Lesson 6-3)

7. **CARPENTRY** Suppose you are constructing a doghouse with a triangular roof. Into what shape will you need to cut the boards labeled *A* and *B* in the diagram below? (Lesson 6-4)

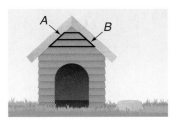

8. **GARDENING** Two triangular gardens have congruent shapes. If 36 bricks are needed to border the first garden how many bricks are needed to border the second garden? Explain your reasoning. (Lesson 6-5)

QUILT PATTERNS **For Exercises 9 and 10, use the diagrams below.** (Lesson 6-6)

a.   b.

9. Determine whether each pattern has line symmetry. If it does, trace the pattern and draw all lines of symmetry. If not, write *none*.

10. Which pattern has rotational symmetry? Name its angles of rotation.

ART **For Exercises 11–13, copy and complete the design shown at the right so that each finished four-paneled piece of art fits the given description.**

Upper Right Corner

11. The finished art has only a vertical line of symmetry. (Lesson 6-7)

12. The finished art shows translations of the first design to each of the other 3 panels. (Lesson 6-8)

13. The finished art has rotational symmetry. Its angles of rotation are 90°, 180°, and 270° about the bottom left corner. (Lesson 6-9)

1. **FLOORING** How much will it cost to tile the floor shown if the tile costs $2.55 per square foot? (Lesson 7-1)

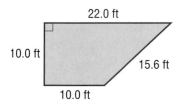

2. **FOOD** An apple pie has a diameter of 8 inches. If 1 slice is $\frac{1}{6}$ of the pie, what is the area of each slice? (Lesson 7-2)

3. **MONEY** The diameter of a dime is about 17.9 millimeters. If the dime is rolled on its edge, how far will it roll after one complete rotation? (Lesson 7-2)

4. **FURNITURE** The top of a desk is shown below. How much workspace does the desktop provide? (Lesson 7-3)

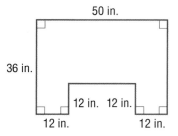

5. **STORAGE** Denise has a hatbox in the shape of a hexagonal prism. How many faces, edges, and vertices are on the hatbox? (Lesson 7-4)

ANT FARM For Exercises 6 and 7, use the following information.
A 3-foot by 2-foot by 1.5-foot terrarium is to be filled with dirt for an ant farm. (Lesson 7-5)

6. How much dirt will the terrarium hold?

7. If each bag from the store holds 3 cubic feet of dirt, how many bags will be needed to fill the terrarium?

8. **BATTERY** A size D battery is cylinder shaped, with a diameter of 33.3 millimeters and a height of 61.1 millimeters. Find the battery's volume in cubic centimeters. (*Hint*: 1 cm^3 = 1,000 mm^3) (Lesson 7-5)

9. **HATS** A clown wants to fill his party hat with confetti. Use the drawing below to determine how much confetti his hat will hold. (Lesson 7-6)

10. **PRESENTS** Viviana wants to wrap a gift in a box that is 5 inches by 3 inches by 3 inches. How much wrapping paper will she need? Assume that the paper does not overlap. (Lesson 7-7)

11. **PAINTING** A front of a government building has four columns that are each 15 feet tall and 6 feet in diameter. If the columns are to be painted, find the total surface area to be painted. (*Hint*: The tops and bottoms of the columns will not be painted.) (Lesson 7-7)

12. **ICE CREAM** Mr. Snow wants to wrap his ice cream cones in paper. If the radius of the base of the cone is 1.5 inches and the slant height is 5 inches, how much paper will he need to cover one cone? (Lesson 7-8)

13. **FAMOUS BUILDINGS** The front of the Rock and Roll Hall of Fame in Cleveland, Ohio, is a square pyramid made out of glass. The pyramid has a slant height of 120 feet and a base length of 230 feet. Find the lateral area of the pyramid. (Lesson 7-8)

SURVEYING For Exercises 14 and 15, use the following information.
A surveyor records that Mrs. Smith's yard is 62.5 feet by 30 feet. (Lesson 7-9)

14. Find the perimeter of the yard using the correct precision.

15. What is the area of the yard? Round to the correct number of significant digits.

1. **GAMES** To start the game of backgammon, each player rolls a number cube. The player with the greater number starts the game. Ebony rolls a 2. What is the probability that Cristina will roll a number greater than Ebony? (Lesson 8-1)

2. **WEATHER** The news reports that there is a 55% chance of snow on Monday. What is the probability that there will be *no* snow on Monday? (Lesson 8-1)

3. **MONEY** A dime, a penny, a nickel, and a quarter are tossed. How many different outcomes are there? (Lesson 8-2)

4. **PHONE NUMBERS** How many seven-digit phone numbers can be made using the numbers 0 through 9 if the first number cannot be 0? (Lesson 8-2)

5. **MUSIC** A disc jockey has 12 songs he plans to play in the next hour. How many ways can he pick the next 3 songs? (Lesson 8-3)

6. **CONSTRUCTION** A contractor can build 11 different model homes. She only has 4 lots. How many ways can she put a different house on each lot? (Lesson 8-3)

7. **GAMES** In the game Tic Tac Toe, players place an X or an O in any of the nine locations that are empty. How many different ways can the first 3 moves of the game occur if X goes first? (Lesson 8-3)

8. **CAPS** Austin wants to take 2 of his 5 baseball caps on his trip. How many different combinations of baseball caps can he take? (Lesson 8-4)

9. **MEDICINE** There are 8 standard classifications of blood types. An examination for prospective technicians requires them to correctly identify 3 different samples of blood. How many groups of samples can be set up for the examination? (Lesson 8-4)

ELECTRONICS For Exercises 10 and 11, use the following information.
The table below shows the percent of students at Midpark Middle School who have various electronic devices in their bedrooms. (Lesson 8-5)

Electronic Device	Percent
TV	60%
DVD Player	15%
Computer	20%
Game Station	75%

10. What is the probability that a student has both a TV and a computer?

11. What is the probability that a student has a TV, a DVD player, and a computer?

12. **CARDS** Two cards are drawn from a deck of 20 cards numbered 1 to 20. Once a card is selected, it is not returned. Find the probability of drawing two odd cards. (Lesson 8-5)

TELEVISION For Exercises 13 and 14, use the table below. (Lesson 8-6)

Television Show	Number Who Selected as Favorite Show
Show A	35
Show B	25
Show C	20
Show D	10
Show E	10

13. What is the probability a person's favorite prime-time TV show is Show A?

14. Out of 320 people, how many would you expect to say that Show A is their favorite prime-time TV show?

CONCERTS For Exercises 15 and 16, use the following information.
As they leave a concert, 50 people are surveyed at random. Six people say they would buy a concert T-shirt. (Lesson 8-7)

15. What percent say they would buy a T-shirt?

16. If 6,330 people attend the next concert, how many would you expect to buy T-shirts?

Mixed Problem Solving

ADVERTISING For Exercises 1 and 2, use the histogram below. (Lesson 9-1)

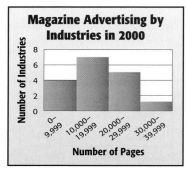

Magazine Advertising by Industries in 2000

Source: Publisher Information Bureau, Inc.

1. How many industries used 20,000 pages or more of magazine advertising?

2. How many industries used less than 30,000 pages of magazine advertising?

3. **AIR** Use the circle graph below to describe the makeup of the air we breathe. (Lesson 9-2)

The Air We Breathe

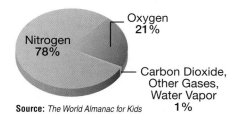

Source: *The World Almanac for Kids*

For Exercises 4 and 5, choose an appropriate type of display for each situation. Then make a display. (Lesson 9-3)

4. **MUSIC** A survey asked teens what they liked most about a song. 59% said the sound, and 41% said the lyrics.

5. **TAXES**

Tax Returns Filed Electronically			
Year	1990	1991	1992	1993
Percent	3.7%	6.6%	9.6%	11.0%
Year	1994	1995	1996	1997
Percent	12.2%	10.5%	12.6%	15.8%
Year	1998	1999	2000	2001
Percent	19.9%	23.3%	27.6%	30.7%

Source: Internal Revenue Service

6. **ANIMALS** What is the mean, median, and mode of the incubation periods of all the birds shown in the table below? (Lesson 9-4)

Bird	Incubation Period (days)
Australian King Parrot	20
Eclectus Parrot	26
Princess Parrot	21
Red Tailed Cockatoo	30
Red-Winged Parrot	21
Regent Parrot	21
Sulphur Crested Cockatoo	30
White Tailed Cockatoo	29
Yellow Tailed Cockatoo	29

Source: www.birds2grow.com

POPULATION For Exercises 7–9, use the following information.

The populations of the smallest countries in 2000 were 860, 10,838, 11,845, 18,766, 26,937, 31,693, and 32,204. (Lessons 9-5 and 9-6)

7. Find the range and median of the data.

8. Find the upper quartile, lower quartile, and interquartile range of the data.

9. Make a box-and-whisker plot for the data.

ENTERTAINMENT For Exercises 10 and 11, use the following information.

The average wait times at each of the major attractions at a theme park are 20, 25, 30, 45, 45, 45, 50, and 50 minutes. (Lesson 9-7)

10. Which measure of central tendancy would the theme park use to encourage people to attend the park? Explain.

11. Which measure of central tendancy would be more representative of the data?

12. **AUTO RACING** Make a matrix for the following information. (Lesson 9-8)

Daytona 500 Lap and Mileage Leaders			
Drivers	**Appearances**	**Laps**	**Miles**
R. Petty	32	4,860	12,150.0
D. Marcis	32	4,859	12,147.5
D. Waltrip	28	4,726	11,815.0

Source: *USA TODAY*

Mixed Problem Solving

1. **SCHOOL SUPPLIES** You buy two gel pens for x dollars each, a spiral-bound notebook for $1.50, and a large eraser for $1. Write an expression in simplest form for the total amount of money you spent on school supplies. (Lesson 10-1)

2. **ENTERTAINMENT** You buy x CDs for $15.99 each, a tape for $9.99, and a video for $20.99. Write an expression in simplest form for the total amount of money you spent. (Lesson 10-1)

3. **ZOO** Four adults took a trip to the zoo. If they spent $37 for admission and $3 for parking, solve the equation $4a + 3 = 37$ to find the cost of admission per person. (Lesson 10-2)

4. **POOLS** There were 820 gallons of water in a 1,600-gallon pool. Water is being pumped into the pool at a rate of 300 gallons per hour. Solve the equation $300t + 820 = 1,600$ to find how many hours it will take to fill the pool. (Lesson 10-2)

5. **FOOTBALL** In football, a touchdown and extra point is worth 7 points, and a field goal is worth 3 points. The winning team scored 27 points. The score consisted of two field goals, and the rest were touchdowns with extra points. Write and solve an equation to determine how many touchdowns the winning team scored. (Lesson 10-3)

6. **DIVING** In diving competitions where there are three judges, the sum of the judges' scores is multiplied by the dive's degree of difficulty. A diver's final score is the sum of all the scores for each dive. The degree of difficulty for Angel's final dive is 2.0. Her current score is 358.5, and the current leader's final score is 405.5. Write and solve an equation to determine what the sum of the judge's scores for Angel's last dive must be in order for her to tie the current leader for first place. (Lesson 10-3)

7. **GEOMETRY** Write an equation to find the value of x so that each pair of polygons has the same perimeter. Then solve. (Lesson 10-4)

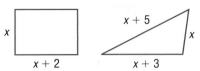

8. **MUSIC** One music club charges $35 a month plus $5 per CD. Another club charges $7 a month plus $9 per CD. Write and solve an equation to find the number of CD purchases that results in the same monthly cost. (Lesson 10-4)

For Exercises 9 and 10, write an inequality for each sentence. (Lesson 10-5)

9. **AMUSEMENT PARKS** Your height must be over 48 inches tall to ride the roller coaster.

10. **SHOPPING** You can spend no more than $500 on your vacation.

11. **GEOMETRY** The base of the rectangle shown is less than its height. Write and solve an inequality to find the possible positive values of x. (Lesson 10-6)

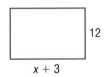

12. **STORMS** A hurricane has winds in excess of 75 miles per hour. A tropical storm currently has wind speeds of 68 miles per hour. Write and solve an inequality to find how much the wind speed must increase to be classified as a hurricane force. (Lesson 10-6)

13. **SWIMMING** A swimming pool charges $4 per adult per visit. They also offer a yearly pass for $112. Write and solve an inequality to find how many times a person must go to the pool so that the yearly pass is less expensive than paying per visit. (Lesson 10-7)

EARNINGS For Exercises 1–3, use the following information.
Annie earns $6.50 per hour at her job as a veterinarian's assistant. (Lesson 11-1)

1. Make a list of the total amount of money earned for 1, 2, 3, 4, and 5 hours.

2. State whether the sequence is *arithmetic*, *geometric*, or *neither*.

3. How much money would she earn for working 7 hours?

4. **SPORTS** Tyree's adjusted bowling score can be found using the function $f(x) = x + 30$. In the function, x is his actual score, and 30 is his handicap. Make a function table to show Tyree's adjusted scores if he bowled 153, 144, 161, 163, and 166 in his first five games of the season. (Lesson 11-2)

GEOMETRY For Exercises 5–7, use the following information.
A regular pentagon is a polygon with five sides of equal length. (Lesson 11-3)

5. Write a function for the perimeter of a regular pentagon.

6. Graph the function.

7. Determine the perimeter of a regular pentagon with sides 3 units long.

WATER FLOW For Exercises 8–10, use the following information.
An empty Olympic-sized swimming pool is being filled with water. The table below shows the amount of water in the pool after the indicated amount of time. (Lesson 11-4)

Time (h)	Volume (m³)
2	144
3	216
5	360

8. Graph the information with the hours on the horizontal axis and cubic meters of water on the vertical axis. Draw a line through the points.

9. What is the slope of the graph?

10. What does the slope represent?

For Exercises 11 and 12, use the following information.
Chen is saving for an $850 computer. He plans to save $50 each month. The equation $f(x) = 850 - 50x$ represents the amount Chen still needs to save. (Lesson 11-5)

11. Graph the equation.

12. What does the slope of the graph represent?

LIFE EXPECTANCY For Exercises 13 and 14, use the following table. (Lesson 11-6)

Year Born	Life Expectancy
1900	47.3
1910	50.0
1920	54.1
1930	59.7
1940	62.9
1950	68.2
1960	69.7
1970	70.8
1980	73.7
1990	75.4
2000	77.1

Source: U.S. Census Bureau

13. Draw the scatter plot for the data.

14. Does the scatter plot show a *positive*, *negative*, or *no* relationship?

RENTALS For Exercises 15 and 16, use the following information.
Company A charges $25 plus $0.10 per mile to rent a car. Company B charges $15 plus $0.20 per mile. (Lesson 11-7)

15. Write equations for the cost of renting a car from Company A and from Company B.

16. When will the costs be the same?

PACKAGING For Exercises 17 and 18, use the following information.
The weight limit on a certain package is 80 pounds. Two items are to go in this package. (Lesson 11-8)

17. Graph all of the possible combinations of weights for the two items.

18. Give three possible weight combinations.

1. **GEOMETRY** Recall that the volume V of a sphere is equal to four-thirds pi times the cube of its radius. Is the volume of a sphere a *linear* or *nonlinear* function of its radius? Explain. (Lesson 12-1)

2. **PRODUCTION** The table lists the cost of producing a specific number of items at the ABC Production Company. Does this table represent a *linear* or *nonlinear* function? Explain. (Lesson 12-1)

Number of Items	Cost ($)
2	2,507
4	2,514
6	2,521
8	2,528

SCIENCE For Exercises 3–5, use the following information.
A ball is dropped from a 200-foot cliff. The quadratic equation $h = -16t^2 + 200$ models the height of the object t seconds after it is dropped. (Lesson 12-2)

3. Graph the function.

4. How high is the ball after 2 seconds?

5. After about how many seconds will the ball reach the ground?

6. **LANDSCAPING** Write a polynomial that represents the perimeter of the garden below in feet. (Lesson 12-3)

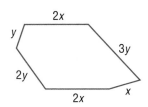

7. **GEOMETRY** Find the measure of each angle in the figure below. (Lesson 12-4)

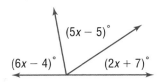

8. **CARPENTRY** To build the top cupboard below, $8x + 6$ square feet of wood is required. The bottom cupboard will require $2x^2 + 12x$ square feet of wood. Find the total square feet of wood required for both cupboards, assuming that there is no waste. (Lesson 12-4)

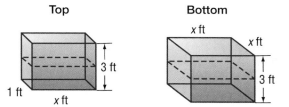

Top   Bottom

MONEY MATTERS For Exercises 9–11, use the following information.
Alan borrowed $200 each year for college expenses. The amount he owes the bank at the beginning of his second and third years is $(400 + 200r)$ and $(600 + 600r + 200r^2)$ respectively, where r is the interest rate. (Lesson 12-5)

9. Find how much his debt increased between his second and third years.

10. Evaluate the increase for $r = 6\%$.

11. Evaluate the increase for $r = 8\%$.

12. **GEOMETRY** Find the volume of a box that is x inches by $3x$ inches by $5x$ inches. (Lesson 12-6)

13. **LIFE SCIENCE** The number of cells in a petri dish starts at 2^5. By the end of the day, there are 2^{12} cells in the dish. About how many times more cells are in the dish at the end of the day than at the beginning? (Lesson 12-6)

HOME IMPROVEMENT For Exercises 14 and 15, use the following information.
A patio's length is to be 6 feet longer than its width. (Lesson 12-7)

14. Write a simplified expression for the area of the patio.

15. Evaluate the expression you wrote in Exercise 14 to find the area of the patio if its width is 24 feet.

Glossary/Glosario

Cómo usar el glosario en español:
1. Busca el término en inglés que desees encontrar.
2. El término en español, junto con la definición, se encuentran en la columna de la derecha.

English

Español

abscissa (p. 142) The first number of an ordered pair. The *x*-coordinate.

abscisa El primer número de un par ordenado. La coordenada *x*.

absolute value (p. 19) The distance a number is from zero on the number line.

valor absoluto Número de unidades en la recta numérica que un número dista de cero.

acute angle (p. 256) An angle with a measure greater than 0° and less than 90°.

ángulo agudo Ángulo que mide más de 0° y menos de 90°.

acute triangle (p. 263) A triangle having three acute angles.

triángulo acutángulo Triángulo que tiene tres ángulos agudos.

Addition Property of Equality (p. 46) If you add the same number to each side of an equation, the two sides remain equal.

propiedad de adición de la igualdad Si sumas el mismo número a ambos lados de una ecuación, los dos lados permanecen iguales.

additive inverse (p. 25) Two integers that are opposite of each other are called additive inverses. The sum of any number and its additive inverse is zero.

inverso aditivo Dos enteros que son opuestos mutuos reciben el nombre de inversos aditivos. La suma de cualquier número y su inverso aditivo es cero.

adjacent angles (p. 256) Angles that have the same vertex, share a common side, and do not overlap. In the figure, ∠1 and ∠2 are adjacent angles.

ángulos adyacentes Ángulos que comparten el mismo vértice y un común lado, pero no se sobreponen. En la figura, ∠1 y ∠2 son adyacentes.

algebraic expression (p. 11) A combination of variables, numbers, and at least one operation.

expresión algebraica Una combinación de variables, números y por lo menos una operación.

alternate exterior angles (p. 258) In the figure, transversal *t* intersects lines ℓ and *m*. ∠1 and ∠7, ∠2 and ∠8 are alternate exterior angles. If lines ℓ and *m* are parallel, then these angles are congruent.

ángulos alternos externos En la figura, la transversal *t* interseca las rectas ℓ y *m*. ∠1 y ∠7, ∠2 y ∠8 son ángulos alternos externos. Si las rectas ℓ y *m* son paralelas, entonces estos ángulos son congruentes.

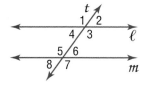

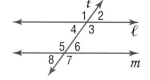

alternate interior angles (p. 258) In the figure at the bottom of page 660, transversal *t* intersects lines ℓ and *m*. $\angle 3$ and $\angle 5$, $\angle 4$ and $\angle 6$ are alternate interior angles. If lines ℓ and *m* are parallel, then these angles are congruent.

altitude (p. 314) A line segment perpendicular to the base of a figure with endpoints on the base and the vertex opposite the base.

angle of rotation (p. 287) The degree measure of the angle through which a figure is rotated.

arithmetic sequence (p. 512) A sequence in which the difference between any two consecutive terms is the same.

ángulos alternos internos En la figura al pie de la página 660, la transversal *t* interseca las rectas ℓ y *m*. $\angle 3$ y $\angle 5$, $\angle 4$ y $\angle 6$ son ángulos alternos internos. Si las rectas ℓ y *m* son paralelas, entonces estos ángulos son congruentes.

altura Segmento de recta perpendicular a la base de una figura y con extremos en la base y el vértice opuesto a la base.

ángulo de rotación La medida en grados del ángulo a través del cual se rota una figura.

sucesión aritmética Sucesión en la cual la diferencia entre dos términos consecutivos es constante.

B

bar notation (p. 63) In repeating decimals, the line or bar placed over the digits that repeat. Another way to write 2.6363636… is $2.\overline{63}$.

base (p. 98) In a power, the number used as a factor. In 10^3, the base is 10. That is, $10^3 = 10 \times 10 \times 10$.

base (p. 216) In a percent proportion, the number to which the percentage is compared.

base (p. 314) The base of a parallelogram or a triangle is any side of the figure. The bases of a trapezoid are the parallel sides.

base (p. 331) The bases of a prism are the two parallel congruent faces.

best-fit line (p. 540) A line that is very close to most of the data points in a scatter plot.

biased sample (p. 407) A sample drawn in such a way that one or more parts of the population are favored over others.

boundary (p. 548) The line which separates the solutions from the points that are not solutions in the graph of a linear inequality.

box-and-whisker plot (p. 446) A diagram that summarizes data using the median, the upper and lower quartiles, and the extreme values. A box is drawn around the quartile values and whiskers extend from each quartile to the extreme data points.

notación de barras En decimales periódicos, la línea o barra que se coloca sobre los dígitos que se repiten. Otra forma de escribir 2.6363636… es $2.\overline{63}$.

base Número que se usa como factor en un potencia. En 10^3, la base es 10. Es decir, $10^3 = 10 \times 10 \times 10$.

base En una proporción porcentual, el número con que se compara el porcentaje.

base La base de un paralelogramo o de un triángulo es cualquier lado de la figura. Las bases de un trapecio son sus lados paralelos.

base Las bases de un prisma son las dos caras congruentes paralelas.

recta de óptimo ajuste Recta que mejor aproxima a los puntos de los datos de una gráfica de dispersión.

muestra sesgada Muestra en que se favorece una o más partes de una población.

frontera La recta que separa las soluciones de los puntos que no son soluciones en la gráfica de una desigualdad lineal.

diagrama de caja y patillas Diagrama que resume información usando la mediana, los cuartiles superior e inferior y los valores extremos. Se dibuja una caja alrededor de los cuartiles y se trazan patillas que los unan a los valores extremos respectivos.

C

center (p. 319) The given point from which all points on a circle are the same distance.

centro Un punto dado del cual equidistan todos los puntos de un círculo.

center of rotation (p. 300) The fixed point a rotation of a figure turns or spins around.

circle (p. 319) The set of all points in a plane that are the same distance from a given point called the center.

circle graph (p. 426) A type of statistical graph used to compare parts of a whole. The entire circle represents the whole.

circumference (p. 319) The distance around a circle.

coefficient (p. 470) The numerical factor of a term that contains a variable.

column (p. 454) In a matrix, numbers stacked on top of each other in a vertical arrangement form a column.

combination (p. 388) An arrangement or listing in which order is not important.

common difference (p. 512) The difference between any two consecutive terms in an arithmetic sequence.

common ratio (p. 513) The quotient between any two consecutive terms in a geometric sequence.

compatible numbers (p. 228) Two numbers that are easy to add, subtract, multiply, or divide mentally.

complementary angles (p. 256) Two angles are complementary if the sum of their measures is 90°.

complementary events (p. 375) The events of one outcome happening and that outcome not happening are complementary events. The sum of the probabilities of complementary events is 1.

complex figure (p. 326) A figure that is made up of two or more shapes.

complex solid (p. 337) An object made up of more than one type of solid.

compound event (p. 396) An event which consists of two or more simple events.

cone (p. 343) A three-dimensional figure with one circular base. A curved surface connects the base and the vertex.

centro de rotación El punto fijo alrededor del cual se lleva a cabo la rotación de un figura.

círculo Conjunto de todos los puntos en un plano que equidistan de un punto dado llamado centro.

gráfica circular Tipo de gráfica estadística que se usa para comparar las partes de un todo. El círculo completo representa el todo.

circunferencia La distancia alrededor de un círculo.

coeficiente Factor numérico de un término que contiene una variable.

columna En una matriz, los números colocados uno encima de otro en un arreglo vertical forman una columna.

combinación Arreglo o lista de objetos en que el orden no es importante.

diferencia común La diferencia entre cualquier par de términos consecutivos en una sucesión aritmética.

razón común El cociente entre cualquier par de términos consecutivos en una sucesión geométrica.

números compatibles Dos números que son fáciles de sumar, restar, multiplicar o dividir mentalmente.

ángulos complementarios Dos ángulos son complementarios si la suma de sus medidas es 90°.

eventos complementarios Se dice de los eventos de un resultado que ocurren y el resultado que no ocurre. La suma de las probabilidades de eventos complementarios es 1.

figura compleja Figura compuesta de dos o más formas.

sólido complejo Cuerpo compuesto de más de un tipo de sólido.

evento compuesto Evento que consta de dos o más eventos simples.

cono Figura tridimensional con una base circular. Una superficie curva conecta la base con el vértice.

congruent (p. 179) Have the same measure.

congruent polygons (p. 279) Polygons that have the same size and shape.

conjecture (p. 7) An educated guess.

constant (p. 470) A term without a variable.

convenience sample (p. 407) A sample which includes members of the population that are easily accessed.

converse (p. 134) The converse of the Pythagorean Theorem can be used to test whether a triangle is a right triangle. If the sides of the triangle have lengths a, b, and c, such that $c^2 = a^2 + b^2$, then the triangle is a right triangle.

coordinate (p. 18) A number associated with a point on the number line.

coordinate plane (p. 142) A plane in which a horizontal number line and a vertical number line intersect at their zero points.

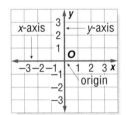

corresponding angles (p. 258) Angles that have the same position on two different parallel lines cut by a transversal. $\angle 1$ and $\angle 5$, $\angle 2$ and $\angle 6$, $\angle 3$ and $\angle 7$, $\angle 4$ and $\angle 8$ are corresponding angles.

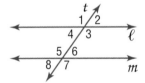

corresponding parts (p. 178) Parts of congruent or similar figures that match.

counterexample (p. 13) A statement or example that shows a conjecture is false.

cross products (p. 170) The products of the terms on the diagonals when two ratios are compared. If the cross products are equal, then the ratios form a proportion. In the proportion $\frac{2}{3} = \frac{8}{12}$, the cross products are 2×12 and 3×8.

cylinder (p. 336) A solid whose bases are congruent, parallel circles, connected with a curved side.

congruente Que tienen la misma medida.

polígonos congruentes Polígonos que tienen la misma medida y la misma forma.

conjetura Suposición informada.

constante Término sin variables.

muestra de conveniencia Muestra que incluye miembros de una población fácilmente accesibles.

recíproco El recíproco del Teorema de Pitágoras puede usarse para averiguar si un triángulo es un triángulo rectángulo. Si las longitudes de los lados de un triángulo son a, b, y c, tales que $c^2 = a^2 + b^2$, entonces el triángulo es un triángulo rectángulo.

coordenada Número asociado con un punto en la recta numérica.

plano de coordenadas Plano en que una recta numérica horizontal y una recta numérica vertical se intersecan en sus puntos cero.

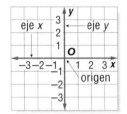

ángulos correspondientes Ángulos que ocupan la misma posición en dos rectas paralelas distintas cortadas por una transversal. $\angle 1$ y $\angle 5$, $\angle 2$ y $\angle 6$, $\angle 3$ y $\angle 7$, $\angle 4$ y $\angle 8$ son ángulos correspondientes.

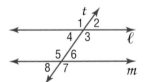

partes correspondientes Partes de figuras congruentes o semejantes que coinciden.

contraejemplo Ejemplo o enunciado que demuestra que una conjetura es falsa.

productos cruzados Productos que resultan de la comparación de los términos de las diagonales de dos razones. Si los productos son iguales, las razones forman una proporción. En la proporción $\frac{2}{3} = \frac{8}{12}$, los productos cruzados son 2×12 y 3×8.

cilindro Sólido cuyas bases son círculos congruentes y paralelos, conectados por un lado curvo.

defining a variable (p. 39) Choosing a variable and a quantity for the variable to represent in an expression or equation.

dependent events (p. 397) Two or more events in which the outcome of one event does affect the outcome of the other event or events.

dependent variable (p. 518) The variable for the output of a function.

diameter (p. 319) The distance across a circle through its center.

dilation (p. 194) A transformation that results from the reduction or enlargement of an image.

dimensional analysis (p. 73) The process of including units of measurement when you compute.

dimensions (p. 454) A description of a matrix by the number of rows and columns it has. The number of rows is always stated first. For example, a matrix with 3 rows and 5 columns has dimensions 3 by 5.

discount (p. 238) The amount by which a regular price is reduced.

Division Property of Equality (p. 50) If you divide each side of an equation by the same nonzero number, the two sides remain equal.

domain (p. 518) The set of input values in a function.

definir una variable El elegir una variable y una cantidad que esté representada por la variable en una expresión o en una ecuación.

eventos dependientes Dos o más eventos en que el resultado de uno de ellos afecta el resultado de los otros eventos.

variable dependiente La variable para el valor de salida de una función.

diámetro La distancia a través de un círculo pasando por el centro.

dilatación Transformación que resulta de la reducción o ampliación de una imagen.

análisis dimensional Proceso que incorpora las unidades de medida al hacer cálculos.

dimensiones Descripción de una matriz según el número de filas y columnas que tiene. El número de filas siempre se escribe primero. Por ejemplo, las dimensiones de una matriz con 3 filas y 5 columnas es 3 por 5.

descuento La cantidad de reducción del precio normal.

propiedad de división de la igualdad Si cada lado de una ecuación se divide entre el mismo número no nulo, los dos lados permanecen iguales.

dominio Conjunto de valores de entrada de una función.

edge (p. 331) The intersection of faces of a three-dimensional figure.

element (p. 454) Each number in a matrix is called an element.

equation (p. 13) A mathematical sentence that contains the equal sign (=).

equilateral triangle (p. 263) A triangle that has three congruent sides.

equivalent expressions (p. 469) Expressions which have the same value regardless of the value(s) of the variable(s).

arista La intersección de las caras de una figura tridimensional.

elemento Cada número en una matriz se llama un elemento.

ecuación Un enunciado matemático que contiene el signo de igualdad (=).

triángulo equilátero Triángulo con tres lados congruentes.

expresiones equivalentes Expresiones que poseen el mismo valor, sin importar los valores de la(s) variable(s).

evaluate (p. 11) To find the value of an expression by replacing the variables with numerals.

experimental probability (p. 400) An estimated probability based on the relative frequency of positive outcomes occurring during an experiment.

exponent (p. 98) In a power, the number of times the base is used as a factor. In 10^3, the exponent is 3.

evaluar Calcular el valor de una expresión sustituyendo las variables por números.

probabilidad experimental Probabilidad de un evento que se estima basándose en la frecuencia relativa de los resultados favorables al evento en cuestión, que ocurren durante un experimento.

exponente En una potencia, el número de veces que la base se usa como factor. En 10^3, el exponente es 3.

F

face (p. 331) Any surface that forms a side or a base of a prism.

factorial (p. 385) The expression $n!$ is the product of all counting numbers beginning with n and counting backward to 1.

function (p. 517) A relation in which each element of the input is paired with exactly one element of the output according to a specified rule.

function table (p. 518) A table organizing the input, rule, and output of a function.

Fundamental Counting Principle (p. 381) Uses multiplication of the number of ways each event in an experiment can occur to find the number of possible outcomes in a sample space.

cara Cualquier superficie que forma un lado o una base de un prisma.

factorial La expresión $n!$ es el producto de todos los números naturales, comenzando con n y contando al revés hasta 1.

función Relación en que cada elemento de entrada se relaciona con un único elemento de salida, según una regla específica.

tabla de funciones Tabla que organiza las entradas, la regla y las salidas de una función.

principio fundamental de contar Método que usa la multiplicación del número de maneras en que cada evento puede ocurrir en un experimento, para calcular el número de resultados posibles en un espacio muestral.

G

geometric sequence (p. 513) A sequence in which the quotient between any two consecutive terms is the same.

sucesión geométrica Sucesión en la cual el cociente entre cualquier par de términos consecutivos es la misma.

H

half plane (p. 548) The region which contains the solutions in the graph of a linear inequality.

histogram (p. 420) A special kind of bar graph that displays the frequency of data that has been organized into equal intervals. The interval covers all possible values of data, therefore there are no spaces between the bars of the graph.

hypotenuse (p. 132) The side opposite the right angle in a right triangle.

semiplano Región que contiene las soluciones en la gráfica de una desigualdad lineal.

histograma Tipo especial de gráfica de barras que exhibe la frecuencia de los datos que han sido organizados en intervalos iguales. El intervalo cubre todos los valores posibles de datos, sin dejar espacios entre las barras de la gráfica.

hipotenusa El lado opuesto al ángulo recto de un triángulo rectángulo.

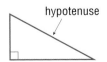

hypotenuse

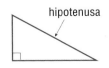

hipotenusa

independent events (p. 396) Two or more events in which the outcome of one event does not affect the outcome of the other event(s).

independent variable (p. 518) The variable for the input of a function.

indirect measurement (p. 188) A technique using proportions to find a measurement.

inequality (p. 18) A mathematical sentence that contains $<$, $>$, $\neq$, $\leq$, or $\geq$.

integers (p. 17) The set of whole numbers and their opposites.
..., -3, -2, -1, 0, 1, 2, 3, ...

interest (p. 241) The amount of money paid or earned for the use of money.

interquartile range (p. 442) The range of the middle half of the data. The difference between the upper quartile and the lower quartile.

inverse operations (p. 46) Pairs of operations that undo each other. Addition and subtraction are inverse operations. Multiplication and division are inverse operations.

irrational number (p. 125) A number that cannot be expressed as $\frac{a}{b}$, where a and b are integers and $b \neq 0$.

isosceles triangle (p. 263) A triangle that has at least two congruent sides.

eventos independientes Dos o más eventos en los cuales el resultado de uno de ellos no afecta el resultado de los otros eventos.

variable independiente La variable correspondiente al valor de entrada de un función.

medición indirecta Técnica que usa proporciones para calcular una medida.

desigualdad Enunciado matemático que contiene $<$, $>$, $\neq$, $\leq$ o $\geq$.

enteros El conjunto de los números enteros y sus opuestos.
..., -3, -2, -1, 0, 1, 2, 3, ...

interés Cantidad que se cobra o se paga por el uso del dinero.

amplitud intercuartílica El rango de la mitad central de un conjunto de datos. La diferencia entre el cuartil superior y el cuartil inferior.

operaciones inversas Pares de operaciones que se anulan mutuamente. La adición y la sustracción son operaciones inversas. La multiplicación y la división son operaciones inversas.

números irracionales Un número que no se puede expresar como $\frac{a}{b}$, donde a y b son enteros y $b \neq 0$.

triángulo isósceles Triángulo que tiene por lo menos dos lados congruentes.

lateral area (p. 352) The sum of the areas of the lateral faces of a pyramid.

lateral face (p. 352) A triangular side of a pyramid.

legs (p. 132) The two sides of a right triangle that form the right angle.

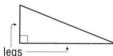

like fractions (p. 82) Fractions that have the same denominator.

like terms (p. 470) Terms that contain the same variable(s).

linear function (p. 523) A function in which the graph of the solutions forms a line.

line of reflection (p. 290) The line a figure is flipped over in a reflection.

área lateral La suma de las áreas de las caras laterales de una pirámide.

cara lateral Un lado triangular de una pirámide.

catetos Los dos lados de un triángulo rectángulo que forman el ángulo recto.

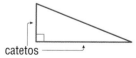

fracciones semejantes Fracciones que tienen el mismo denominador.

términos semejantes Términos que contienen la(s) misma(s) variable(s).

función lineal Función en la cual la gráfica de las soluciones forma un recta.

línea de reflexión Línea a través de la cual se le da vuelta a una figura en una reflexión.

Glossary/Glosario

line of symmetry (p. 286) A line that divides a figure into two halves that are reflections of each other.

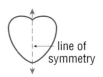
line of symmetry

line symmetry (p. 286) Figures that match exactly when folded in half have line symmetry.

lower quartile (p. 442) The median of the lower half of a set of data. Represented by LQ.

eje de simetría Recta que divide una figura en dos mitades que son reflexiones una de la otra.

eje de simetría

simetría lineal Exhiben simetría lineal las figuras que coinciden exactamente al doblarse una sobre otra.

cuartil inferior La mediana de la mitad inferior de un conjunto de datos, la cual se denota por CI.

markup (p. 238) The amount the price of an item is increased above the price the store paid for the item.

matrix (p. 454) A rectangular arrangement of numerical data in rows and columns.

mean (p. 435) The sum of the numbers in a set of data divided by the number of pieces of data in the data set.

measures of central tendency (p. 435) Numbers or pieces of data that can represent the whole set of data.

measures of variation (p. 442) Numbers used to describe the distribution or spread of a set of data.

median (p. 435) The middle number in a set of data when the data are arranged in numerical order. If the data has an even number, the median is the mean of the two middle numbers.

mode (p. 435) The number(s) or item(s) that appear most often in a set of data.

monomial (p. 570) A number, a variable, or a product of a number and one or more variables.

Multiplication Property of Equality (p. 51) If you multiply each side of an equation by the same number, the two sides remain equal.

multiplicative inverse (p. 76) A number times its multiplicative inverse is equal to 1. The multiplicative inverse of $\frac{2}{3}$ is $\frac{3}{2}$.

margen de utilidad Cantidad de aumento en el precio de un artículo por encima del precio que paga la tienda por dicho artículo.

matriz Arreglo rectangular de datos numéricos en filas y columnas.

media La suma de los números de un conjunto de datos dividida entre el número total de datos.

medidas de tendencia central Números o fragmentos de datos que pueden representar el conjunto total de datos.

medidas de variación Números que se usan para describir la distribución o separación de un conjunto de datos.

mediana El número central de un conjunto de datos, una vez que los datos han sido ordenados numéricamente. Si hay un número par de datos, la mediana es el promedio de los dos datos centrales.

moda El número(s) o artículo(s) que aparece con más frecuencia en un conjunto de datos.

monomio Un número, una variable o el producto de un número por una o más variables.

propiedad de multiplicación de la igualdad Si cada lado de una ecuación se multiplica por el mismo número, los lados permanecen iguales.

inverso multiplicativo Un número multiplicado por su inverso multiplicativo es igual a 1. El inverso multiplicativo de $\frac{2}{3}$ es $\frac{3}{2}$.

negative number (p. 17) A number that is less than zero.

nonlinear function (p. 560) A function that does not have a constant rate of change. The graph of a nonlinear function is not a straight line.

numerical expression (p. 11) A mathematical expression that has a combination of numbers and at least one operation. 4 + 2 is a numerical expression.

número negativo Número menor que cero.

función no lineal Función que no tiene una tasa constante de cambio. La gráfica de una función no lineal no es una recta.

expresión numérica Expresión matemática que tiene una combinación de números y por lo menos una operación. 4 + 2 es una expresión numérica.

obtuse angle (p. 256) An angle that measures greater than 90° but less than 180°.

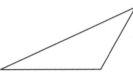

obtuse triangle (p. 263) A triangle having one obtuse angle.

open sentence (p. 13) An equation that contains a variable.

opposites (p. 25) Two numbers with the same absolute value but different signs. The sum of opposites is zero.

order of operations (p. 11) The rules to follow when more than one operation is used in an expression.
1. Do all operations within grouping symbols first; start with the innermost grouping symbols.
2. Evaluate all powers before other operations.
3. Multiply and divide in order from left to right.
4. Add and subtract in order from left to right.

ordered pair (p. 142) A pair of numbers used to locate a point in the coordinate plane. The ordered pair is written in this form: (x-coordinate, y-coordinate).

ordinate (p. 142) The second number of an ordered pair. The y-coordinate.

origin (p. 142) The point of intersection of the x-axis and y-axis in a coordinate plane.

outcome (p. 374) One possible result of a probability event. For example, 4 is an outcome when a number cube is rolled.

outlier (p. 443) Data that is more than 1.5 times the interquartile range from the upper or lower quartiles.

ángulo obtuso Ángulo que mide más de 90° pero menos de 180°.

triángulo obtuso Triángulo que tiene un ángulo obtuso.

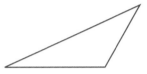

enunciado abierto Ecuación que contiene una variable.

opuestos Dos números con el mismo valor absoluto, pero distintos signos. La suma de opuestos es cero.

orden de las operaciones Reglas a seguir cuando se usa más de una operación en una expresión.
1. Primero ejecuta todas las operaciones dentro de los símbolos de agrupamiento.
2. Evalúa todas las potencias antes que las otras operaciones.
3. Multiplica y divide en orden de izquierda a derecha.
4. Suma y resta en orden de izquierda a derecha.

par ordenado Par de números que se utiliza para ubicar un punto en un plano de coordenadas. Se escribe de la siguiente forma: (coordenada x, coordenada y).

ordenada El segundo número de un par ordenado. La coordenada y.

origen Punto en que el eje x y el eje y se intersecan en un plano de coordenadas.

resultado Uno de los resultados posibles de un evento probabilístico. Por ejemplo, 4 es un resultado posible cuando se lanza un dado.

valor atípico Dato o datos que dista(n) de los cuartiles respectivos más de 1.5 veces la amplitud intercuartílica.

parallel lines (p. 257) Lines in the same plane that never intersect or cross. The symbol ∥ means parallel.

rectas paralelas Rectas que yacen en un mismo plano y que no se intersecan. El símbolo ∥ significa paralela a.

parallelogram (p. 273) A quadrilateral with both pairs of opposite sides parallel and congruent.

part (p. 216) The number that is being compared to the whole quantity in a percent proportion.

percent (p. 206) A ratio that compares a number to 100.

percent equation (p. 232) An equivalent form of the percent proportion in which the percent is written as a decimal. Part = Percent · Base

percent of change (p. 236) A ratio that compares the change in a quantity to the original amount.

percent of decrease (p. 237) The percent of change when the new amount is less than the original.

percent of increase (p. 237) The percent of change when the new amount is greater than the original.

percent proportion (p. 216) Compares part of a quantity to the whole quantity using a percent. $\frac{part}{base} = \frac{percent}{100}$

perfect square (p. 116) A rational number whose square root is a whole number. 25 is a perfect square because its square root is 5.

permutation (p. 384) An arrangement or listing in which order is important.

perpendicular lines (p. 257) Two lines that intersect to form right angles.

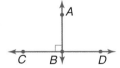

pi (p. 319) The ratio of the circumference of a circle to its diameter. The Greek letter π represents this number. The value of pi is always 3.1415926… .

plane (p. 331) A two-dimensional flat surface that extends in all directions.

polygon (p. 178) A simple closed figure in a plane formed by three or more line segments.

polyhedron (p. 331) A solid with flat surfaces that are polygons.

polynomial (p. 570) The sum or difference of one or more monomials.

population (p. 406) The entire group of items or individuals from which the samples under consideration are taken.

paralelogramo Cuadrilátero con ambos pares de lados opuestos paralelos y congruentes.

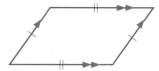

parte El número que se compara con la cantidad total en una proporción porcentual.

por ciento Razón que compara un número con 100.

ecuación porcentual Forma equivalente de proporción porcentual en la cual el por ciento se escribe como un decimal. Parte = Por ciento · Base

porcentaje de cambio Razón que compara el cambio en una cantidad, con la cantidad original.

porcentaje de disminución El porcentaje de cambio cuando la nueva cantidad es menos que la cantidad original.

porcentaje de aumento El porcentaje de cambio cuando aumenta la nueva cantidad es mayor que la cantidad original.

proporción porcentual Compara parte de una cantidad con la cantidad total mediante un por ciento. $\frac{parte}{base} = \frac{por\ ciento}{100}$.

cuadrados perfectos Número racional cuya raíz cuadrada es un número entero. 25 es un cuadrado perfecto porque su raíz cuadrada es 5.

permutación Arreglo o lista donde el orden es importante.

rectas perpendiculares Dos rectas que se intersecan formando ángulos rectos.

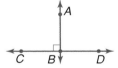

pi Razón de la circunferencia de un círculo al diámetro del mismo. La letra griega π representa este número. El valor de pi es siempre 3.1415926… .

plano Superficie plana bidimensional que se extiende en todas direcciones.

polígono Figura simple y cerrada en el plano formada por tres o más segmentos de recta.

poliedro Sólido cuyas superficies planas son polígonos.

polinomio La suma o la diferencia de uno o más monomios.

población El grupo total de individuos o de artículos del cual se toman las muestras bajo estudio.

powers (p. 12) Numbers written using exponents. Powers represent repeated multiplication. The power 7^3 is read seven to the third power, or seven cubed.

precision (p. 358) The precision of a measurement depends on the unit of measure. The smaller the unit, the more precise the measurement is.

principal (p. 241) The amount of money invested or borrowed.

principal square root (p. 117) A positive square root.

prism (p. 331) A polyhedron with two parallel, congruent faces called bases.

probability (p. 374) The chance that some event will happen. It is the ratio of the number of ways a certain event can occur to the number of possible outcomes.

property (p. 13) An open sentence that is true for any numbers.

proportion (p. 170) A statement of equality of two ratios.

pyramid (p. 331) A polyhedron with one base that is a polygon and faces that are triangles.

Pythagorean Theorem (p. 132) In a right triangle, the square of the length of the hypotenuse c is equal to the sum of the squares of the lengths of the legs a and b. $c^2 = a^2 + b^2$

Pythagorean triple (p. 138) A set of three integers that satisfy the Pythagorean Theorem.

potencias Números que se expresan usando exponentes. Las potencias representan multiplicación repetida. La potencia 7^3 se lee siete a la tercera potencia, o siete al cubo.

precisión El grado de exactitud de una medida, lo cual depende de la unidad de medida. Entre más pequeña es una unidad, más precisa es la medida.

capital Cantidad de dinero que se invierte o que se toma prestada.

raíz cuadrada principal Una raíz cuadrada positiva.

prisma Poliedro con dos caras congruentes y paralelas llamadas bases.

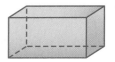

probabilidad La posibilidad de que suceda un evento. Es la razón del número de maneras en que puede ocurrir un evento al número total de resultados posibles.

propiedad Enunciado abierto que se cumple para cualquier número.

proporción Un enunciado que establece la igualdad de dos razones.

pirámide Poliedro cuya base tiene forma de polígono y caras en forma de triángulos.

Teorema de Pitágoras En un triángulo rectángulo, el cuadrado de la longitud de la hipotenusa es igual a la suma de los cuadrados de las longitudes de los catetos. $c^2 = a^2 + b^2$

triplete pitagórico Conjunto de tres enteros que satisfacen el Teorema de Pitágoras

quadrants (p. 142) The four regions into which the two perpendicular number lines of the coordinate plane separate the plane.

quadratic function (p. 565) A function in which the greatest power of the variable is 2.

quadrilateral (p. 272) A polygon that has four sides and four angles.

quartiles (p. 442) Values that divide a set of data into four equal parts.

cuadrantes Las cuatro regiones en que las dos rectas numéricas perpendiculares dividen el plano de coordenadas.

función cuadrática Función en la cual la potencia mayor de la variable es 2.

cuadrilátero Un polígono con cuatro lados y cuatro ángulos.

cuartiles Valores que dividen un conjunto de datos en cuatro partes iguales.

radical sign (p. 116) The symbol used to indicate a nonnegative square root. √

radius (p. 319) The distance from the center of a circle to any point on the circle.

random (p. 374) Outcomes occur at random if each outcome is equally likely to occur.

range (p. 442) The difference between the greatest number and the least number in a set of data.

range (p. 518) The set of output values in a function.

rate (p. 157) A ratio of two measurements having different units.

rate of change (p. 160) A rate that describes how one quantity changes in relation to another.

ratio (p. 156) A comparison of two numbers by division. The ratio of 2 to 3 can be stated as 2 out of 3, 2 to 3, 2:3, or $\frac{2}{3}$.

rational number (p. 62) Numbers of the form $\frac{a}{b}$, where a and b are integers and $b \neq 0$.

real numbers (p. 125) The set of rational numbers together with the set of irrational numbers.

reciprocals (p. 76) The multiplicative inverse of a number. The product of reciprocals is 1.

rectangle (p. 273) A parallelogram with four right angles.

reflection (p. 290) A type of transformation in which a mirror image is produced by flipping a figure over a line.

repeating decimal (p. 63) A decimal whose digits repeat in groups of one or more. Examples are 0.181818… and 0.8333… .

rhombus (p. 273) A parallelogram with four congruent sides.

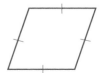

signo radical Símbolo que se usa para indicar una raíz cuadrada no negativa. √

radio Distancia desde el centro de un círculo hasta cualquier punto del mismo.

aleatorio Un resultado ocurre al azar si la posibilidad de ocurrir de cada resultado es equiprobable.

rango La diferencia entre el número mayor y el número menor en un conjunto de datos.

rango El conjunto de valores de salida en una función.

tasa Razón que compara dos cantidades que tienen distintas unidades de medida.

tasa de cambio Tasa que describe cómo cambia una cantidad con respecto a otras.

razón Comparación de dos números mediante división. La razón de 2 a 3 puede escribirse como 2 de cada 3, 2 a 3, 2:3, o $\frac{2}{3}$.

número racional Números de la forma $\frac{a}{b}$, donde a y b son enteros y $b \neq 0$.

número real El conjunto de números racionales junto con el conjunto de números irracionales.

recíproco El inverso multiplicativo de un número. El producto de recíprocos es 1.

rectángulo Un paralelogramo que tiene cuatro ángulos rectos.

reflexión Tipo de transformación en que se produce una imagen especular al darle vuelta de campana a una figura por encima de una línea.

decimal periódico Decimal cuyos dígitos se repiten en grupos de uno o más. Por ejemplo: 0.181818… y 0.8333… .

rombo Paralelogramo que tiene cuatro lados congruentes.

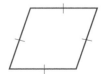

right angle (p. 256) An angle that measures 90°.

ángulo recto Ángulo que mide 90°.

right triangle (p. 132) A triangle having one right angle.

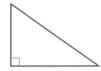

triángulo rectángulo Triángulo que tiene un ángulo recto.

rise (p. 166) The vertical change between any two points on a line.

rotation (p. 300) A transformation involving the turning or spinning of a figure around a fixed point.

rotational symmetry (p. 287) A figure has rotational symmetry if it can be turned less than 360° about its center and still look like the original.

row (p. 454) In a matrix, the numbers side by side horizontally form a row.

run (p. 166) The horizontal change between any two points on a line.

elevación El cambio vertical entre cualquier par de puntos en una recta.

rotación Transformación que involucra girar una figura en torno a un punto central fijo.

simetría rotacional Una figura posee simetría rotacional si se puede girar menos de 360° en torno a su centro sin que esto cambie su apariencia con respecto a la figura original.

fila En una matriz, los números que están horizontalmente uno al lado del otro.

carrera El cambio horizontal entre cualquier par de puntos en una recta.

sample (p. 406) A randomly selected group chosen for the purpose of collecting data.

sample space (p. 374) The set of all possible outcomes of a probability experiment.

scale (p. 184) The ratio of a given length on a drawing or model to its corresponding length in reality.

scale drawing (p. 184) A drawing that is similar, but either larger or smaller than the actual object.

scale factor (p. 179) The ratio of the lengths of two corresponding sides of two similar polygons.

scale model (p. 184) A replica of an original object that is too large or too small to be built at actual size.

scalene triangle (p. 263) A triangle with no congruent sides.

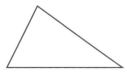

scatter plot (p. 539) A graph that shows the general relationship between two sets of data.

muestra Grupo escogido al azar o aleatoriamente que se usa con el propósito de recoger datos.

espacio muestral Conjunto de todos los resultados posibles de un experimento probabilístico.

escala Razón de una longitud dada en un dibujo o modelo a su longitud real.

dibujo a escala Dibujo que es semejante, pero más grande o más pequeño que el objeto real.

factor de escala La razón de las longitudes de dos lados correspondientes de dos polígonos semejantes.

modelo a escala Una replica del objeto original, el cual es demasiado grande o demasiado pequeño como para construirlo de tamaño natural.

triángulo escaleno Triángulo que no tiene lados congruentes.

diagrama de dispersión Gráfica que muestra la relación general entre dos conjuntos de datos.

scientific notation (p. 104) A way of expressing numbers as the product of a number that is at least 1 but less than 10 and a power of 10. In scientific notation, 5,500 is 5.5×10^3.

selling price (p. 238) The amount the customer pays for an item.

sequence (p. 512) An ordered list of numbers, such as 0, 1, 2, 3, or 2, 4, 6, 8.

significant digits (p. 358) All of the digits of a measurement that are known to be accurate plus one estimated digit.

similar (p. 178) Polygons that have the same shape are called similar polygons.

simple event (p. 374) A specific outcome or type of outcome.

simple random sample (p. 406) A sample where each item or person in the population is as likely to be chosen as any other.

simplest form (p. 471) An algebraic expression which has no like terms and no parentheses.

simplifying the expression (p. 471) Using properties to combine like terms.

slant height (p. 352) The altitude or height of each lateral face of a pyramid.

slope (p. 166) The rate of change between any two points on a line. The ratio of vertical change to horizontal change.

slope formula (p. 526) The slope m of a line passing through two points is the ratio of the difference in the y-coordinates to the corresponding difference in the x-coordinates.
$$m = \frac{y_2 - y_1}{x_2 - x_1}$$

slope-intercept form (p. 533) An equation written in the form $y = mx + b$, where m is the slope and b is the y-intercept.

solid (p. 331) A three-dimensional figure formed by intersecting planes.

solution (p. 45) The value for the variable that makes an equation true. The solution for $10 + y = 25$ is 15.

solve (p. 45) Find the value of the variable that makes the equation true.

square (p. 273) A parallelogram with four congruent sides and four right angles.

notación científica Manera de expresar números como el producto de un número que es al menos igual a 1, pero menor que 10, por una potencia de 10. En notación científica, 5,500 es 5.5×10^3.

precio de venta Cantidad de dinero que paga un consumidor por un artículo.

sucesión Lista ordenada de números, tales como 0, 1, 2, 3 ó 2, 4, 6, 8.

dígitos significativos Todos los dígitos de una medición que se sabe son exactos, más un dígito aproximado.

semejante Los polígonos que tienen la misma forma se llaman polígonos semejantes.

evento simple Un resultado específico o un tipo de resultado.

muestra aleatoria simple Muestra de una población que tiene la misma probabilidad de escogerse que cualquier otra.

forma reducida Expresión algebraica que carece de términos semejantes y de paréntesis.

simplificar una expresión El uso de propiedades para combinar términos semejantes.

altura oblicua La longitud de la altura de cada cara lateral de una pirámide.

pendiente Razón de cambio entre cualquier par de puntos en una recta. La razón del cambio vertical al cambio horizontal.

fórmula de la pendiente La pendiente m de una recta que pasa por dos puntos es la razón de la diferencia en la coordenada y a la diferencia correspondiente en la coordenada x.
$$m = \frac{y_2 - y_1}{x_2 - x_1}$$

forma pendiente intersección Ecuación de la forma $y = mx + b$, donde m es la pendiente y b es la intersección y.

sólido Figura tridimensional formada por planos que se intersecan.

solución El valor de la variable de una ecuación que hace que se cumpla la ecuación. La solución de $10 + y = 25$ es 15.

resolver Proceso de encontrar la variable que satisface una ecuación.

cuadrado Paralelogramo con cuatro lados congruentes y cuatro ángulos rectos.

square root (p. 116) One of the two equal factors of a number. If $a^2 = b$, then a is the square root of b. A square root of 144 is 12 since $12^2 = 144$.

straight angle (p. 256) An angle that measures 180°.

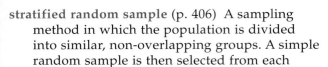

stratified random sample (p. 406) A sampling method in which the population is divided into similar, non-overlapping groups. A simple random sample is then selected from each group.

substitution (p. 545) A method used for solving a system of equations that replaces one variable in one equation with an expression derived from the other equation.

Subtraction Property of Equality (p. 45) If you subtract the same number from each side of an equation, the two sides remain equal.

supplementary angles (p. 256) Two angles are supplementary if the sum of their measures is 180°.

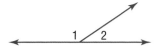

surface area (p. 347) The sum of the areas of all the faces of a three-dimensional figure.

systematic random sample (p. 406) A sampling method in which the items or people are selected according to a specific time or item interval.

system of equations (p. 544) A set of two or more equations considered together.

raíz cuadrada Uno de dos factores iguales de un número. Si $a^2 = b$, la a es la raíz cuadrada de b. Una raíz cuadrada de 144 es 12 porque $12^2 = 144$.

ángulo llano Ángulo que mide 180°.

muestra aleatoria estratificada Método de muestreo en que la población se divide en grupos semejantes que no se sobreponen. Luego se selecciona una muestra aleatoria simple de cada grupo.

sustitución Método que se usa para resolver un sistema de ecuaciones en que se reemplaza una variable en una ecuación con una expresión derivada de la otra ecuación.

propiedad de sustracción de la igualdad Si sustraes el mismo número de ambos lados de una ecuación, los dos lados permanecen iguales.

ángulos suplementarios Dos ángulos son suplementarios si la suma de sus medidas es 180°.

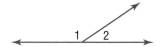

área de superficie La suma de las áreas de todas las caras de una figura tridimensional.

muestra aleatoria sistemática Muestra en que los elementos de la muestra se escogen según un intervalo de tiempo o elemento específico.

sistema de ecuaciones Conjunto de dos o más ecuaciones consideradas juntas.

T

term (p. 470) A number, a variable, or a product of numbers and variables.

term (p. 512) A number in a sequence.

terminating decimal (p. 63) A decimal whose digits end. Every terminating decimal can be written as a fraction with a denominator of 10, 100, 1,000, and so on.

theoretical probability (p. 400) Probability based on known characteristics or facts.

transformation (p. 290) A mapping of a geometric figure.

translation (p. 296) A transformation in which a figure is slid horizontally, vertically, or both.

término Un número, una variable o un producto de números y variables.

término Un número en una sucesión.

decimal terminal Decimal cuyos dígitos terminan. Todo decimal terminal puede escribirse como una fracción con un denominador 10, 100, 1,000, etc.

probabilidad teórica Probabilidad que se basa en características o hechos conocidos.

transformación Movimiento de una figura geométrica.

traslación Transformación en que una figura se desliza horizontal o verticalmente o de ambas maneras.

transversal (p. 258) A line that intersects two or more other lines to form eight angles.

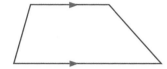

transversal Recta que interseca dos o más rectas formando ocho ángulos.

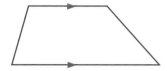

trapezoid (p. 273) A quadrilatFeral with exactly one pair of parallel opposite sides.

trapecio Cuadrilátero con un único par de lados opuestos paralelos.

tree diagram (p. 380) A diagram used to show the total number of possible outcomes in a probability experiment.

diagrama de árbol Diagrama que se usa para mostrar el número total de resultados posibles en experimento probabilístico.

triangle (p. 262) A figure formed by three line segments that intersect only at their endpoints.

triángulo Figura formada por tres segmentos de recta que sólo se intersecan en sus extremos.

two-step equation (p. 474) An equation that contains two operations.

ecuación de dos pasos Ecuación que contiene dos operaciones.

U

unbiased sample (p. 406) A sample that is selected so that it is representative of the entire population.

muestra no sesgada Muestra que se selecciona de modo que sea representativa de la población entera.

unit rate (p. 157) A rate with a denominator of 1.

razón unitaria Una tasa con un denominador de 1.

unlike fractions (p. 88) Fractions whose denominators are different.

fracciones con distinto denominador Fracciones cuyos denominadores son diferentes.

upper quartile (p. 442) The median of the upper half of a set of data. Represented by UQ.

cuartil superior La mediana de la mitad superior de un conjunto de números, denotada por CS.

V

variable (p. 11) A symbol, usually a letter, used to represent a number in mathematical expressions or sentences.

variable Un símbolo, por lo general, una letra, que se usa para representar números en expresiones o enunciados matemáticos.

vertex (p. 331) The vertex of a prism is the point where three or more planes intersect.

vértice El vértice de un prisma es el punto en que se intersecan dos o más planos del prisma.

vertical angles (p. 256) Opposite angles formed by the intersection of two lines. Vertical angles are congruent. In the figure, the vertical angles are ∠1 and ∠3, and ∠2 and ∠4.

ángulos opuestos por el vértice Ángulos congruentes que se forman de la intersección de dos rectas. En la figura, los ángulos opuestos por el vértice son ∠1 y ∠3, y ∠2 y ∠4.

volume (p. 335) The number of cubic units needed to fill the space occupied by a solid.

volumen El número de unidades cúbicas que se requieren para llenar el espacio que ocupa un sólido.

voluntary response sample (p. 407) A sample which involves only those who want to participate in the sampling.

muestra de respuesta voluntaria Muestra que involucra sólo aquellos que quieren participar en el muestreo.

x-**axis** (p. 142) The horizontal number line that helps to form the coordinate plane.

x-**coordinate** (p. 142) The first number of an ordered pair.

x-**intercept** (p. 523) The value of *x* where the graph crosses the *x*-axis.

eje *x* La recta numérica horizontal que ayuda a formar el plano de coordenadas.

coordenada *x* El primer número de un par ordenado.

intersección *x* El valor de *x* donde la gráfica cruza el eje *x*.

y-**axis** (p. 142) The vertical number line that helps to form the coordinate plane.

y-**coordinate** (p. 142) The second number of an ordered pair.

y-**intercept** (p. 523) The value of *y* where the graph crosses the *y*-axis.

eje *y* La recta numérica vertical que ayuda a formar el plano de coordenadas.

coordenada *y* El segundo número de un par ordenado.

intersección *y* El valor de *y* donde la gráfica cruza el eje *y*.

Glossary/Glosario

Selected Answers

Chapter 1 Algebra: Integers

Page 5 Chapter 1 Getting Started
1. multiply 3. 77 5. 79.5 7. 152 9. 2.6 11. 30 13. 72
15. 1,220 17. 32 19. 0.4 21. < 23. >

Pages 9–10 Lesson 1-1
1. Explore—Identify what information is given and what you need to find. Plan—Estimate the answer and then select a strategy for solving. Solve—Carry out the plan and solve. Examine—Compare the answer to the estimate and determine if it is reasonable. If not, make a new plan.
3. The numbers increase by 1, 2, 3, and so on; 25. 5. 6 lb
7. almost 3 jars/s

9.   11. 20 13. 529,920,000 acres 15. No; $8 + $12 + $12 > $30 17. $6 per pair 19. C 21. 60 23. 14.4

Pages 14–15 Lesson 1-2
1. The everyday meaning of variable is something that is likely to change or vary, and the mathematical meaning of a variable is a placeholder for a value that can change or vary. 3. Sample answer: $4 \cdot 5 = 5 \cdot 4$ 5. 12 7. 32 9. 29
11. Commutative ($\times$) 13. false; $6 + 0 \not> 6$ 15. 22 17. 20
19. 17 21. 12 23. 20 25. 2 27. 57 29. 11 31. 11 33. 144
35. 6 37. 5 39. 35 41. about 6,031 43. Distributive
45. Commutative (+) 47. Identity (+) 49. 6(4 + 3) 51. true
53. false; $(24 \div 4) \div 2 \neq 24 \div (4 \div 2)$ 55. Sample answer: The equals sign was first introduced by Robert Recorde in 1557. 57. C 59. $4.38 61. = 63. >

Pages 20–21 Lesson 1-3
1. Sample answer: $4 > -5$; $-5 < 4$ 3. +10

5. [number line from −4 to 6] 7. > 9. 5 11. 28
   −4−3−2−1 0 1 2 3 4 5 6 13. 4 15. −2

17. +4 19. +60 21. [number line from −6 to 4]
   −6 −4 −2 0 2 4

23. [number line from −10 to 0] 25. < 27. <
   −10−9−8−7−6−5−4−3−2−1 0 29. > 31. =

33. {−37, −23, −12, 0, 45, 55} 35. {−17, −11, −5, −2, 6}
37. 14 39. 25 41. 15 43. 3 45. −3 47. 14 49. −5, −2, −2, −2, −1, −1, 0, +1, +1, +2 51. helium 53. 2 55. 20 57. 11
59. Never; the absolute value of a positive number is always positive. 61. Sometimes; $5 > -4$ and $\left|5\right| > \left|-4\right|$, but $-4 > -5$ and $\left|-4\right| \not> \left|-5\right|$. 63. H 65. 26
67. about 2 h 69. 43 71. 65

Pages 26–27 Lesson 1-4
1. To add numbers with different signs, subtract the absolute values of the numbers. Then use the sign of the number with the greater absolute value. 3. 45 and 54; All of the other numbers are additive inverses. 5. 4 7. −13
9. −20 11. −26 13. 22 15. −15 17. 15 19. 41 21. 24
23. −30 25. −32 27. −11 29. 8 + (−5); 3 31. −1 + 7; 6
33. 7 35. 2 37. −2 39. 3 41. −6 43. Rock: 25, Rap/Hip Hop: 13, Pop: 11, Country: 11 45. Sometimes; If x and y have different signs, then $\left|x + y\right| \neq \left|x\right| + \left|y\right|$. If x and y have the same sign, then $\left|x + y\right| = \left|x\right| + \left|y\right|$.

47. F 49. > 51. > 53. Sample answer: about 500 million.
55. 17 57. 4

Pages 30–31 Lesson 1-5
1. Sample answer: $7 - (-3)$; $7 + 3$ 3. −5 5. −19 7. 4 9. −7
11. 2 13. −7 15. −12 17. −16 19. −20 21. 14 23. 16
25. −11 27. 7 29. −2 31. 589 m 33. Lake Michigan is 37 m deeper than Lake Ontario. 35. 5 37. −23 39. 6 41. −9
43. 10°F 45. Sample answer: You have $26 in your checking account. Find the balance in your account after you write a check for $30. 47. false; $3 - 2 \neq 2 - 3$ 49. I 51. 17
53. Identity (+) 55. 135 57. 540

Pages 37–38 Lesson 1-6
1a. positive 1b. negative 1c. negative 3a. Positive; the product of two negative numbers is always a positive number. 3b. Negative; the product of three negative numbers is always a negative number. 3c. Positive; the product of four negative numbers is always a positive number. 3d. Negative; the product of five negative numbers is always a negative number. 5. 14 7. 140 9. −4
11. −3 13. −4 15. −56 17. 24 19. 60 21. −72 23. 64
25. 84 27. −125 29. −240 31. −32 33. −35°C 35. −10
37. −20 39. 5 41. 17 43. −9 45. 7 47. 3 h 49. −21 51. 1
53. −6 55. −25 57. −4 59. −25 61. If the number of negative factors is odd, then the product is negative. If the number of negative factors is even, then the product is positive. 63. true 65. D 67. −6 69. −23 71. −11
73. Sample answer: in all 75. Sample answer: of

Pages 41–42 Lesson 1-7
1. Sample answer: the sum of a number and 4; 4 more than a number. 3. $18 + t$ 5. $\frac{n}{9}$ 7. $p \div 5 = 3$ 9. $n - 6$
11. $n + (-9)$ 13. $\frac{n}{-3}$ 15. $n - 20$ 17. $t + 4$ 19. k = the year Kentucky became a state; $k + 4$ 21. g = gallons of gasoline; $\frac{260}{g}$ 23. $150n$ 25. $n - 8 = 15$ 27. $-14 = 2n$
29. $10 - a = 3.50$ 31. Sample answer: If the year is 2004, then the equation is $y + 25 = 2004$. 33. B 35. −90 37. −2
39. −$1,800 41. −9 43. −27

Pages 47–49 Lesson 1-8
1. If you replace x with 4 in the equation $x + 3 = 7$ you get a true statement, $4 + 3 = 7$. 3. $m - 6 = 4$; this equation can be solved using the Addition Property of Equality, and the others can be solved using the Subtraction Property of Equality. 5. −5 7. 9 9. 3 11. $n - 20 = -14$; 6 13. −2
15. −15 17. 17 19. 18 21. 16 23. −10 25. −59 27. 76
29. −8 31. 8 33. $n - 8 = -14$; −6 35. $n + 30 = 9$; −21
37. $b + 50 = 124$; $74 39. $s - 6 = -5$; 1 or 1 over par
41. $p = 16.8 - 0.6$; 16.2 43. $x + 0.40 = 5.15$; $4.75 per hour
45. Sample answer: While playing a computer trivia game, you answer a question correctly and your score is increased by 60 points. If your score is now 20 points, what was your original score? Answer: −40 47. B 49. $7m$ 51. −940 tigers per year 53. −36 55. −60

Pages 52–53 Lesson 1-9
1. Division Property of Equality; the inverse operation of multiplication is division. 3. Sample answer: it is greater than 300. 5. −2 7. −5 9. −36 11. −56 13. $\frac{n}{4} = -16$; −64

15. 8 **17.** −2 **19.** 7 **21.** −15 **23.** 70 **25.** −99 **27.** −72 **29.** 60 **31.** 6 **33.** −3n = 39; −13 **35.** $\frac{n}{7} = -14$; −98 **37.** 12f = 288; 24 **39.** 5,280m = 26,400; 5 **41.** 500d = 3,000; 6 days **43.** about 114 h **45.** $\frac{d}{r}$ **47.** C **49.** x + 10 = 4

Pages 54–56 Chapter 1 Study Guide and Review
1. e **3.** h **5.** a **7.** d **9.** c **11.** 21.4 g **13.** 8 **15.** 1 **17.** 15 **19.** −80 **21.** > **23.** 5 **25.** −33 **27.** −34 **29.** 5 **31.** 3 **33.** −4 **35.** 100 **37.** −60 **39.** 34 **41.** −12 **43.** n + 7 **45.** 4x = 48 **47.** 13 **49.** −22 **51.** 39 **53.** −5 **55.** −13 **57.** −72

Chapter 2 Algebra: Rational Numbers

Page 61 Chapter 2 Getting Started
1. opposites, additive inverses **3.** −9 **5.** 3 **7.** −14 **9.** −28 **11.** −84 **13.** 43 **15.** −12 **17.** 24 **19.** 30 **21.** 48

Pages 65–66 Lesson 2-1
1. Sample answer: $0.\overline{12}$; Since $0.\overline{12} = \frac{4}{33}$, it is a rational number. **3.** $\frac{1}{2}$; It cannot be written as a repeating decimal. **5.** 4.375 **7.** $7.\overline{15}$ **9.** $-1\frac{11}{20}$ **11.** $2\frac{1}{9}$ **13.** 0.875 in. **15.** 0.2 **17.** −0.22 **19.** 5.3125 **21.** $-0.\overline{2}$ **23.** $-0.\overline{54}$ **25.** $7.\overline{24}$ **27.** $0.0\overline{4}$ **29.** $\frac{1}{2}$ **31.** $-\frac{7}{20}$ **33.** $7\frac{8}{25}$ **35.** $-\frac{4}{9}$ **37.** $2\frac{7}{9}$ **39.** $-3\frac{8}{11}$ **41.** $\frac{19}{50}$; $\frac{191,919}{500,000}$ **43.** $\frac{11}{100}$ oz **45.** $1\frac{5}{8}$ lb **47a.** $\frac{1}{2} = 0.5$, $\frac{1}{4} = 0.25$, $\frac{1}{5} = 0.2$, $\frac{1}{8} = 0.125$ **47b.** $\frac{1}{3} = 0.\overline{3}$, $\frac{1}{6} = 0.1\overline{6}$, $\frac{1}{7} = 0.\overline{142857}$, $\frac{1}{9} = 0.\overline{1}$ **49.** G **51.** −22 **53.** −4 **55.** 15 **57.** 24

Pages 69–70 Lesson 2-2
1. Since 0.28 = 0.28000000 and $0.\overline{28} = 0.28282828...$, 0.28 is less than $0.\overline{28}$. **3.** Greatest to least; since the numerators are the same, the values of the fractions decrease as the denominators increase. **5.** < **7.** < **9.** $-0.68, -\frac{2}{3}, 0.7, \frac{3}{4}$ **11.** $\frac{5}{32}, \frac{1}{4}, \frac{3}{8}, \frac{7}{16}, \frac{9}{16}$ **13.** < **15.** > **17.** > **19.** < **21.** < **23.** = **25.** $6\frac{3}{4}, 6.8, 7\frac{1}{5}, 7.6$ **27.** $-\frac{3}{5}, -0.5, 0.45, \frac{4}{7}$ **29.** −2.95, $-2\frac{13}{14}, -2.9, -2\frac{9}{11}$ **31.** $\frac{2}{5}$ **33.** $\frac{1}{125}$ s **35.** Florida State University **37.** D **39.** 0.875 in. **41.** −15 **43.** −27 **45.** −96 **47.** 115

Pages 74–75 Lesson 2-3
1. Since $\frac{1}{2} \cdot 1 = \frac{1}{2}$ and $\frac{7}{8}$ is less than 1, $\frac{1}{2} \cdot \frac{7}{8}$ is less than $\frac{1}{2}$. **3.** Enrique; to multiply mixed numbers, you must first rename them as fractions. **5.** $-\frac{1}{18}$ **7.** $\frac{16}{25}$ **9.** $\frac{2}{5}$ **11.** $\frac{1}{21}$ **13.** $-\frac{3}{5}$ **15.** $1\frac{1}{2}$ **17.** $3\frac{1}{2}$ **19.** $-14\frac{1}{6}$ **21.** $\frac{4}{9}$ **23.** $\frac{3}{20}$ **25.** $-\frac{2}{9}$ **27.** $\frac{4}{27}$ **29.** $2\frac{1}{16}$ in. **31.** $\frac{6}{7}$ **33.** 3.78 **35.** C **37.** < **39.** < **41.** $\frac{11}{20}$ **43.** 27 **45.** 15

Pages 79–80 Lesson 2-4
1. If the product of the two numbers is 1, the two numbers are multiplicative inverses. **3.** Sample answer: $\frac{2}{3} \div \frac{5}{6}$ **5.** $\frac{7}{5}$ **7.** $-\frac{4}{11}$ **9.** $-1\frac{1}{4}$ **11.** 7 times longer **13.** $-\frac{8}{5}$ **15.** $\frac{1}{18}$ **17.** $\frac{15}{7}$ **19.** $\frac{8}{33}$ **21.** $\frac{9}{16}$ **23.** $-\frac{4}{5}$ **25.** $\frac{3}{8}$ **27.** $3\frac{4}{7}$ **29.** $-\frac{3}{14}$ **31.** $-3\frac{3}{16}$

33. $\frac{8}{9}$ **35.** $\frac{9}{10}$ **37.** $\frac{1}{2}$ **39.** $1\frac{4}{5}$ **41a.** $\frac{43}{76}$ **41b.** $\frac{53}{72}$ **43.** −a **45.** $\frac{1}{3}$ **47.** $2\frac{1}{6}$ **49.** 4x − 8,000,000 **51.** 8 **53.** −18

Pages 84–85 Lesson 2-5
1.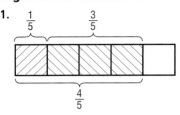
3. Allison; to add like fractions, add the numerators and write the sum over the denominator. **5.** $-\frac{1}{2}$ **7.** $-\frac{1}{2}$ **9.** $\frac{6}{7}$ **11.** $\frac{6}{7}$ **13.** $\frac{1}{6}$ **15.** $-1\frac{3}{4}$ **17.** $-\frac{1}{2}$ **19.** $-1\frac{2}{5}$ **21.** $11\frac{1}{4}$ **23.** $5\frac{1}{5}$ **25.** $-5\frac{2}{3}$ **27.** $1\frac{3}{5}$ **29.** $-4\frac{3}{8}$ **31.** $7\frac{2}{3}$ **33.** Since $\frac{1}{2} \cdot \frac{3}{4} + \frac{1}{2} \cdot \frac{1}{4} = \frac{1}{2}\left(\frac{3}{4} + \frac{1}{4}\right)$ or $\frac{1}{2}(1)$, the answer is $\frac{1}{2}$. **35.** $1\frac{1}{2}$ in. **37.** Since $\frac{2}{3} + \frac{1}{3} = 1$, $\frac{2}{5} + \frac{3}{5} = 1$, and $\frac{1}{6} + \frac{5}{6} = 1$, add 3 to the sum of the whole numbers; 15. **39.** F **41.** $\frac{5}{16}$ **43.** $\frac{3}{4}$ **45.** 42 **47.** 36

Pages 90–91 Lesson 2-6
1. Rename the fractions so they have a common denominator. **3.** Greater than; since both fractions are greater than $\frac{1}{2}$, the sum will be greater than $\frac{1}{2} + \frac{1}{2}$ or 1. **5.** $\frac{1}{8}$ **7.** $-1\frac{7}{30}$ **9.** $-\frac{13}{15}$ **11.** $1\frac{5}{24}$ **13.** $\frac{7}{12}$ **15.** $-\frac{11}{21}$ **17.** $14\frac{13}{14}$ **19.** $-10\frac{3}{8}$ **21.** $4\frac{2}{3}$ **23.** $-5\frac{3}{10}$ **25.** $-3\frac{23}{24}$ **27.** $\frac{421}{1,496}$ **29.** $15\frac{1}{4}$ **31.** $-1\frac{1}{8}$ **33.** $12\frac{1}{8}$ **35.** about $\frac{25}{168}$ **37.** $\frac{1}{3}$ **39.** $1\frac{7}{8}$ min **41a.** $\frac{3}{4} \cdot \frac{2}{3} = \frac{1}{2}$ **41b.** $\frac{2}{3} + \frac{3}{4} = 1\frac{5}{12}$ **41c.** $\frac{2}{3} - \frac{3}{4} = -\frac{1}{12}$ **41d.** $\frac{2}{3} \div \frac{3}{4} = \frac{8}{9}$

43. $5\frac{1}{6} + 4\frac{2}{9}$
$$= (5 + 4) + \left(\frac{1}{6} + \frac{2}{9}\right)$$
$$= (5 + 4) + \left(\frac{3}{18} + \frac{4}{18}\right)$$
$$= 9 + \frac{7}{18}$$
$$= 9\frac{7}{18}$$

45. $-\frac{11}{15}$ **47.** −4 **49.** −15 **51.** −80

Pages 94–95 Lesson 2-7
1. Sample answer: $x + \frac{3}{4} = 1$ **3.** −4.37 **5.** −54 **7.** $\frac{16}{45}$ **9.** 11.9r = 59.5 **11.** 0.84 **13.** $\frac{1}{6}$ **15.** −28 **17.** 6 **19.** $-1\frac{1}{9}$ **21.** 0.85 **23.** −14.4 **25.** $2\frac{3}{10}$ **27.** $-\frac{1}{4}$ **29.** $-1\frac{3}{5}$ **31.** $5.60 **33.** 4.5 million visitors **35.** A **37.** $\frac{13}{42}$ **39.** $-12\frac{3}{10}$ **41.** $106\frac{3}{16}$ in. **43.** 17 + p **45.** 32 **47.** 125

Pages 100–101 Lesson 2-8
1. Sample answer: 3^{-2}; $3^{-2} = \frac{1}{3^2}$ or $\frac{1}{9}$ **3.** 3^6 **5.** $s^3 \cdot r^5$ **7.** 288 **9.** $\frac{1}{216}$ **11.** 300 stars **13.** 30,000 stars **15.** 8^3 **17.** p^6 **19.** $3^2 \cdot 4^3$ **21.** $4^3 \cdot 7^5$ **23.** $a^4 b^4$ **25.** $7^4 \cdot 15^2$ **27.** 8 **29.** 243 **31.** 225 **33.** 4,000 **35.** $\frac{1}{625}$ **37.** $\frac{8}{49}$ **39.** 224 **41.** 64 bacteria **43.** 8 in³; 216 in³ **45a.** 10^5 **45b.** $5 \cdot 10^7$ **45c.** $3 \cdot 10^9$ **45d.** $6 \cdot 10^4$ **47.** $(2 \cdot 2 \cdot 2) \cdot (6 \cdot 6)$; 288 **49.** $\frac{11}{18}$ **51.** $2\frac{5}{18}$ **53.** x + 12 **55.** 320

1. Sometimes; if the decimal is greater than or equal to 1 and less than 10, the value is in scientific notation. If the fraction is less than 1 or greater than or equal to 10, the value is not in scientific notation. 3. 1.2×10^6; 1.2×10^5 is only 120,000, but 1.2×10^6 is just over one million. 5. 993,100 7. 0.000602 9. 8.785×10^9 11. 5.24×10^{-1} 13. 208 15. 71,130,000 17. 0.0078 19. 0.000873 21. 1,046,000 23. 0.000006299 25. 14,000 lb 27. 6.3×10^5 29. 6.7×10^3 31. 5.23×10^7 33. 3.7×10^{-2} 35. 7.07×10^{-6} 37. 1×10^{-24} s 39. 1×10^{100} 41. Wrigley Field, Network Associates Coliseum, H.H.H. Metrodome, The Ballpark in Arlington, Yankee Stadium 43. $\dfrac{(9 \times 10^4)(1.6 \times 10^{-3})}{(2 \times 10^5)(3 \times 10^4)(1.2 \times 10^{-4})} = 2 \times 10^{-4}$ 45. Huron 47. $-\dfrac{5}{6}$ 49. 3.12

1. exponent 3. Like fractions 5. base 7. rational number 9. $1.\overline{3}$ 11. 5.26 13. -2.3 15. $\dfrac{3}{10}$ 17. $-2\dfrac{3}{4}$ 19. $4\dfrac{1}{3}$ 21. $<$ 23. $=$ 25. $-\dfrac{3}{4}, -\dfrac{1}{2}, 0, 0.75$ 27. $\dfrac{4}{9}$ 29. $\dfrac{5}{11}$ 31. $2\dfrac{1}{3}$ 33. $2\dfrac{1}{5}$ 35. 1 37. $-\dfrac{3}{4}$ 39. $-\dfrac{1}{15}$ 41. $-11\dfrac{1}{6}$ 43. $11\dfrac{11}{20}$ 45. 3.2 47. $1\dfrac{1}{6}$ 49. 40 51. $2^2 \cdot 5^3$ 53. $4^2 \cdot 9^2$ 55. 432 57. 128 59. 67,100 61. 0.015 63. 3.51×10^{-4} 65. 7.41×10^6

Chapter 3   Algebra: Real Numbers and the Pythagorean Theorem

1. true

3–6.
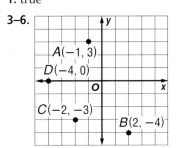

7. 20 9. 164 11. 32 13. 62 15. $\dfrac{2}{3}$ 17. $\dfrac{3}{8}$ 19. 36, 49

1. $\sqrt{16} = 4$, which is what golfers yell to warn other players that the ball is coming. 3. Sample answer: $x^2 = 100$ 5. 5 7. $-\dfrac{4}{9}$ 9. 6 or -6 11. 30 or -30 13. 16 or -16 15. 9 17. -6 19. -12 21. 18 23. $-\dfrac{3}{7}$ 25. 1.2 27. -20 29. 10 or -10 31. 12 or -12 33. 50 or -50 35. 31 or -31 37. $\dfrac{3}{8}$ or $-\dfrac{3}{8}$ 39. 1.1 or -1.1 41. 2.01 or -2.01 43. 44 in. 45. 24 m 47a. 36 47b. 81 47c. 21 47d. x 49. C 51. 25,000,000,000,000 mi 53. $2^4 \cdot 3^2$ 55. $s^4 \cdot t^3$ 57. 49, 64 59. 25, 36

1.

3. Julia; $7^2 = 49$ or about 50, but $25^2 = 625$. 5. 5 7. 12 9. 9 or -9 11. 4 13. 5 15. 10 17. 7 19. 5 21. 6 23. 13 25. 12 27. 30 29. 28 31. 10 or -10 33. 5, $\sqrt{38}, 7, \sqrt{91}$ 35. about 3.5 s 37. B 39. 90 or -90 41. $\dfrac{3}{20}$ 43. $\dfrac{1}{3}$

1. Sample answer: $\sqrt{4}$ 3. $\sqrt{25}$; since $\sqrt{25} = 5$, it is not an irrational number like the others. 5. integer, rational 7. rational 9. -4.2

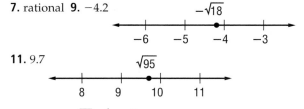

11. 9.7

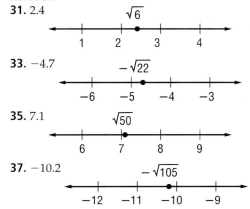

13. $=$ 15. $\sqrt{30}, 5\dfrac{1}{2}, 5.\overline{5}, 5.56$ 17. whole, integer, rational 19. integer, rational 21. rational 23. irrational 25. rational 27. rational 29. Always; an integer can always be written as the integer over 1, so an integer is always a rational number.
31. 2.4

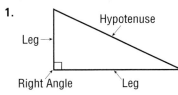

33. -4.7

35. 7.1

37. -10.2

39. $<$ 41. $>$ 43. $=$ 45. $\sqrt{3}, 2.\overline{2}, \sqrt{5}, 2.25$ 47. $\sqrt{17}, 4.01, -4.1, -4.\overline{1}$ 49. about 53.3 mph 51. always 53. 76 ft 55. 5 or -5 57. 0.8 or -0.8 59. 34 61. 202

1.

3. Morgan; the hypotenuse (8) and a leg (5) are given. The correct equation to solve the problem is $8^2 = a^2 + 5^2$.

5. $12^2 = a^2 + 8^2$; 8.9 yd 7. $6^2 = 5^2 + b^2$; 3.3 ft 9. $10^2 = a^2 + 4^2$; 9.2 yd 11. yes 13. $c^2 = 5^2 + 12^2$; 13 in. 15. $18^2 = 8^2 + b^2$; 16.1 m 17. $x^2 = 14^2 + 6^2$; 15.2 in. 19. $c^2 = 48^2 + 55^2$; 73 yd 21. $c^2 = 23^2 + 18^2$; 29.2 in. 23. $12.3^2 = a^2 + 5.1^2$; 11.2 m 25. about 11.5 ft 27. no 29. no 31. no 35. at knots 3 and 7 from unstaked end

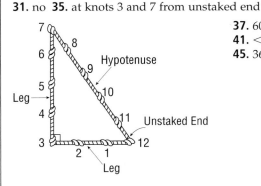

37. 60 in. 39. $>$ 41. $<$ 43. 33 45. 36

1. The Pythagorean Theorem relates the lengths of the three sides of a right triangle. If you know the lengths of two

Selected Answers

sides of a right triangle, you can substitute the values into the Pythagorean Theorem and solve for the missing length. **3.** 5-7-9; $9^2 \neq 5^2 + 7^2$ **5.** $d^2 = 7^2 + 10^2$; 12.2 mi **7.** about 5.7 in. **9.** $d^2 = 60^2 + 150^2$; 161.6 yd **11.** $24^2 = 18^2 + \ell^2$; 15.9 mi **13.** $20^2 = 19.5^2 + h^2$; 4.4 m **15.** about 28.5 in. **17.** about 2.6 cm **19.** about 15.3 cm **21.** about 0.5 ft **23.** I **25.** $6.\overline{6}$, 6.7, $\sqrt{45}$, 6.75 **27.** 27 **29.** 1,600,000

31.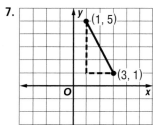

33.

Pages 144–145   Lesson 3-6

1. Pythagorean Theorem **3.** Sample answer: (1, 2) and (4, 6) **5.** 5.7 units

7.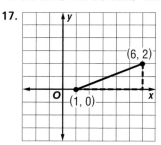

4.5 units

9.

8.6 units

11. 5.4 units **13.** 6.3 units **15.** 5.8 units

17.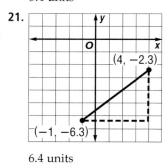

5.4 units

19.

7.6 units

21.

6.4 units

23. about 842.7 mi **25.** For horizontal lines, the x-coordinate is half the sum of the x-coordinates of the endpoints and the y-coordinate is the y-coordinate of the endpoints. For vertical lines, the x-coordinate is the x-coordinate of the endpoints and the y-coordinate is half the sum of the y-coordinates of the endpoints. **27.** $c^2 = 4^2 + 2^2$ **29.** 23.4 cm

Pages 146–148   Chapter 3   Study Guide and Review

1. false; cannot **3.** true **5.** false; vertical axis **7.** true **9.** 9 **11.** 8 **13.** $-\frac{2}{3}$ **15.** 17 rows of 17 trees in each row **17.** 6

19. 10 **21.** 3 **23.** 4 **25.** irrational **27.** rational **29.** irrational **31.** $c^2 = 24^2 + 18^2$; 30 in. **33.** $c^2 = 8^2 + 5^2$; 9.4 ft **35.** $6^2 = 5^2 + b^2$; 3.3 in. **37.** $25^2 = 20^2 + h^2$; 15 ft **39.** $\ell^2 = 8^2 + 5^2$; 9.4 ft **41.** about 13.9 m

43.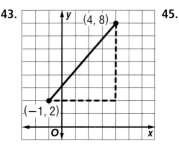

7.8 units

45.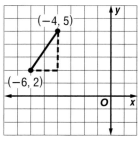

3.6 units

47.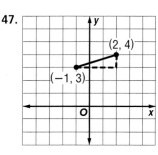

3.2 units

Chapter 4   Proportions, Algebra, and Geometry

Page 155   Chapter 4   Getting Started

1. variable **3.** ordered pair **5.** $\frac{11}{13}$ **7.** $\frac{7}{13}$ **9.** $\frac{3}{4}$ **11.** 6 **13.** -7 **15.** 7 **17.** 1 **19.** 14 **21.** 1.75 **23.** 10.5

Pages 158–159   Lesson 4-1

1. Sample answer: $\frac{16 \text{ yellow}}{10 \text{ red}}$; $\frac{8}{5}$; for every 8 yellow marbles, 5 are red. **3.** 1:15 **5.** 12 to 1 **7.** \$12.50/day **9.** Ben's Mart; the cost at Ben's Mart is about 23.8¢ per apple, while at SaveMost it is about 24.8¢. **11.** 7:8 **13.** $\frac{7}{25}$ **15.** 2 to 9 **17.** 17:2 **19.** \$5.65/h **21.** 1,225 tickets/theater **23.** about 1.8 lb/wk **25.** 6 cans for \$1; 6 for \$1 costs about 16.7¢/can and 10 for \$1.95 cost about 19.5¢/can. **27.** 2 liters for \$1.39; 2 liters for \$1.39 costs about 2.1¢/oz and 12 12-ounce cans for \$3.49 costs about 2.4¢/oz. **29.** 384 **31.** $\frac{1,662,269}{9}$ **33.** about \$471,000/in^2 **35.** 18 **37.** Darnell

39.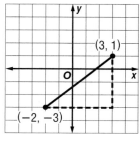

8.1 units

41.

6.4 units

43. 6 **45.** -3

Pages 163–164   Lesson 4-2

1. cost of postage for a 1-oz letter over a period of 4 months in which the cost did not change **3.** 6°/h; about −6.3°/h

5.

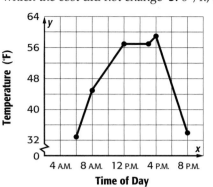

7. 4.1 flyers/min
9. between 1:25 and 1:30
11. 100 eagle pairs/yr
13. between 1984 and 1994

15. Between 2000 and 2002; reading the graph from left to right, the segment connecting 2000 and 2002 is steeper than the segment connecting 1980 and 1990. **17.** $22 billion/yr

19. 18 **21.** B **23.** 3 to 8 **25.** $\frac{3}{2}$ **27.** $\frac{1}{3}$

Pages 168–169   Lesson 4-3

1.

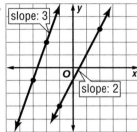

the line with slope 3
3. $\frac{4}{3}$

5. 2;

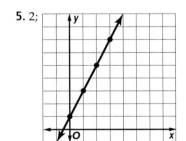

7. $-\frac{3}{4}$ **9.** 0 **11.** $\frac{3}{2}$

13. $\frac{2}{3}$;

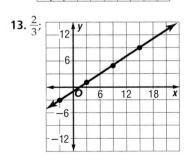

15. 10; $10 increase in cost for each pizza delivered **17.** $\frac{5}{11}$; pressure increases 5 lbs/in² for every 11 feet increase in depth **19.** Pedro; he is saving $12.50 per week, while Jenna is only saving $6 per week.

21. 2/3 **23.** 1.8 in./min **25.** 12 **27.** 7.5

Pages 172–173   Lesson 4-4

1. Sample answer: $\frac{3}{10}, \frac{6}{20}, \frac{9}{30}, \frac{24}{80}$ **3.** yes **5.** yes **7.** 16.4
9. $\frac{18}{15} = \frac{b}{60}$; 72 **11.** no **13.** yes **15.** yes **17.** yes **19.** 4
21. 120 **23.** 7.2 **25.** 1.68 **27.** 1.4 **29.** 22 **31.** $\frac{6}{1} = \frac{96}{p}$; 16

33. $\frac{3}{2,000} = \frac{x}{3,500}$; 5.25 **35.** 24 **37.** about 8.5 in.
39. $\frac{12}{x} = \frac{1}{2.54}$; 30.48 **41.** $\frac{2}{x} = \frac{3.78}{1}$; 0.53 **43.** Proportional; each statement can be written as a ratio equivalent to 3:2.

45. D **47.** $\frac{8}{3}$;

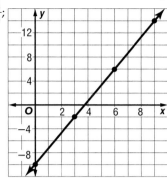

49. $\overline{AB}, \overline{BC}, \overline{CA}$ **51.** $\overline{LM}, \overline{MN}, \overline{NP}, \overline{PL}$

Pages 181–182   Lesson 4-5

1. If 2 polygons have corresponding congruent angles and have corresponding sides that are in proportion, then the polygons are similar. **3.** No; the corresponding angles are congruent, but $\frac{5}{3} \neq \frac{13}{5}$. **5.** $\frac{x}{6} = \frac{2}{1}$; 12 **7.** Yes; the corresponding angles are congruent and $\frac{3}{5} = \frac{3}{5}$. **9.** No; the corresponding angles are congruent, but $\frac{5}{4} \neq \frac{8}{6}$.
11. $\frac{x}{4.8} = \frac{5}{4}$; 6 **13.** $\frac{26}{x} = \frac{1.6}{1}$; 16.25 **15.** 46.67 mm **17a.** $\frac{1}{4}$
17b. $\frac{1}{9}$ **17c.** $\frac{1}{16}$ **17d.** $\frac{1}{25}$; The ratio of the area is the square of the scale factor. **19.** Always; all corresponding angles between squares are congruent since all four angles in a square are right angles. In addition, all sides in a square are congruent. Therefore, all four ratios of corresponding sides are equal. **21.** $3\frac{3}{4}$ in.

23.

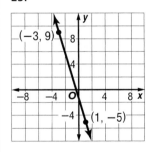

$-\frac{7}{2}$

25.

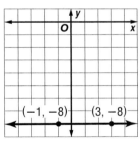

0

27. $\frac{4}{5} = \frac{5}{x}$; 6.25

Pages 186–187   Lesson 4-6

1. Sample answer: 1 in. = 10 ft; 1:120 **3.** 720 mi
5. 1 in. = 60 ft **7.** 12 ft **9.** 8.4 ft **11.** $13\frac{1}{2}$ ft **13.** $\frac{1}{72}$
15. 1 cm = 0.0015 mm **17.** Sample answer: 1 cm = 1.25 m; 11.25 m **19.** A tennis ball; if the diameter of the model is d, then $\frac{d}{11,000} = \frac{3}{4,000}$, so $d = 8.25$. **21.** The model built on the 1:75 scale since $\frac{1}{75} > \frac{1}{100}$. **23.** D **25.** Yes; the corresponding angles are congruent, and $\frac{3}{4.8} = \frac{2}{3.2} = \frac{2}{3.2} = \frac{1.5}{2.4}$.
27. 12.5 **29.** $\angle A \cong \angle D, \angle B \cong \angle E, \angle ACB \cong \angle DCE$

Pages 189–191   Lesson 4-7

1.
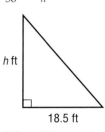
$$\frac{14}{5} = \frac{h}{6}$$

3. $\frac{0.6}{56} = \frac{1.5}{h}$; 140 m

5.
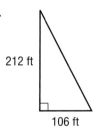
$$\frac{h}{18.5} = \frac{10}{7}$$; about 26 ft

7. $\frac{248}{h} = \frac{186}{9}$; 12 ft   **9.** $\frac{135}{252} = \frac{204}{d}$; 380.8 ft

11.

$$\frac{3}{106} = \frac{h}{212}$$; 6 ft

13.
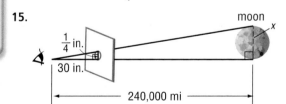
$$\frac{3}{x} = \frac{5}{62.5}$$; 37.5 ft

15.
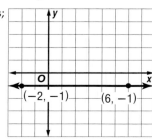

$$\frac{\frac{1}{4}}{x} = \frac{30}{240,000}$$; 2,000 mi

17. $\frac{h}{ED} = \frac{BC}{DC}$ **19.** B   **21.** 10 mi   **23.** 32.5 mi   **25.** $\frac{4.5}{m} = \frac{3}{8}$; 12

27. $-4\frac{1}{10}$ **29.** $-1\frac{1}{3}$ **31.** 4.3×10^6 **33.** 2.0×10^{-7}

35. 8 units;
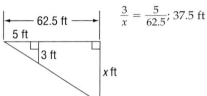

37. 13 units;
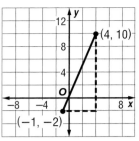

Pages 196–197   Lesson 4-8

1. Sample answer:
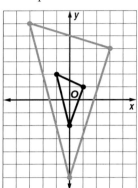

3. $A'(-1, 3)$, $B'\left(-\frac{1}{2}, -1\right)$, $C'\left(2, \frac{3}{2}\right)$;
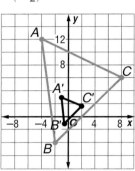

5. $\frac{3}{2}$; enlargement   **7.** $H'(-6, 3)$, $J'\left(4\frac{1}{2}, 3\right)$, $K'\left(4\frac{1}{2}, -3\right)$, $L'(-6, -3)$;
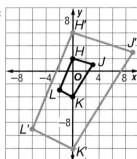

9. $H'(0, 6)$, $J'(9, 3)$, $K'(0, -12)$, $L'(-6, -9)$;
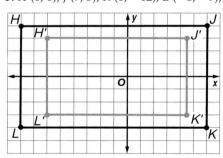

11. $\frac{3}{5}$; reduction   **13.** $\frac{4}{3}$; enlargement

15. Sample answer:
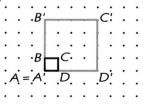

682   Selected Answers

17. 3; enlargement **19.** $\frac{1}{2}$; reduction **21.** 2.5

23.

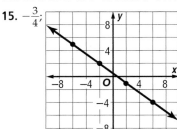

vanishing point

25. The figure is enlarged and rotated 180°. **27.** G
29. 5 in. = 7 ft

Pages 198–200   Chapter 4   Study Guide and Review

1. f **3.** a **5.** d **7.** b **9.** 1 for 8 **11.** 1 out of 12 **13.** $\frac{5}{2}$

15. $-\frac{3}{4}$;

17. 3.5 **19.** 2.85
21. $\frac{13}{x} = \frac{5}{1}$; 2.6
23. $13\frac{1}{3}$ in. **25.** 22.5 mi
27. 1 in. = 12 in. or 1 in. = 1 ft
29. $\frac{5.5}{x} = \frac{2.25}{6}$; $14\frac{2}{3}$ ft
31. $\frac{2}{5}$; reduction

Chapter 5   Percent

Page 205   Chapter 5   Getting Started
1. equation **3.** proportion **5.** 322 **7.** 32 **9.** 0.875 **11.** 0.375
13. 0.25 **15.** 450 **17.** 0.35 **19.** 0.5 **21.** 31.5 **23.** 15.6

Pages 208–209   Lesson 5-1
1. 30%; $\frac{3}{10}$ **3.** $\frac{20}{100}$; All the other numbers equal 40%.
5. 237% **7.** 45% **9.** $\frac{1}{2}$ **11.** $\frac{1}{250}$ **13.** 23% **15.** 0.3% **17.** 60%
19. 32% **21.** 34% **23.** 195% **25.** 12% **27.** $\frac{29}{100}$ **29.** $\frac{2}{5}$ **31.** $\frac{9}{20}$
33. $\frac{16}{25}$ **35.** $1\frac{1}{4}$ **37.** $\frac{1}{500}$ **39.** $\frac{1}{25}$ **41.** $\frac{9}{20}$; $\frac{3}{20}$; $\frac{1}{10}$; $\frac{2}{25}$; $\frac{7}{100}$
43. 37% **45.** 25% **47.** Since $\frac{43}{50} = \frac{86}{100}$, a student would receive a 86% on the test if he or she answered 43 out of the 50 questions correctly. **49.** H **51.** 4; enlargement **53.** 0.6
55. 0.625

Pages 212–214   Lesson 5-2
1. $\frac{13}{25}$; 52%; 0.52 **3.** Aislyn; 0.7 is 7 tenths, not 7 hundredths.
5. 0.16 **7.** 0.003 **9.** 123% **11.** 72.5% **13.** 87.5% **15.** $83.\overline{3}$%
17. 0.9 **19.** 0.15 **21.** 1.72 **23.** 0.275 **25.** 0.07 **27.** 0.082
29. 0.55 **31.** 54% **33.** 37.5% **35.** 0.7% **37.** 40% **39.** 275%
41. 21% **43.** 85% **45.** 2.5% **47.** 160% **49.** 0.5% **51.** $44.\overline{4}$%
53. 32% **55.** $83.\overline{3}$% **57.** > **59.** < **61.** = **63.** 31.25%
65. 2%, $\frac{3}{20}$, 0.2, $\frac{1}{4}$ **67.** more **69.** $\frac{7}{25}$ **71.** 160% **73.** H **75.** 0.6%
77. 66% **79.** {−12, −5, −1, 5, 13} **81.** {−65, −61, −58, 57, 64}
83. 1.6 **85.** 2.4

Pages 218–219   Lesson 5-3
1. Percent means per 100 and p is the number out of 100.
3. Roberto; the base b is unknown. **5.** $\frac{a}{90} = \frac{60}{100}$; 54
7. $\frac{7}{49} = \frac{p}{100}$; 14.3% **9.** $\frac{125}{b} = \frac{30}{100}$; 416.7 **11.** $\frac{16}{64} = \frac{p}{100}$; 25%

13. $\frac{a}{200} = \frac{35}{100}$; 70 **15.** $\frac{95}{b} = \frac{95}{100}$; 100 **17.** $\frac{120}{360} = \frac{p}{100}$; 33.3%
19. $\frac{a}{350} = \frac{17.2}{100}$; 60.2 **21.** $\frac{225}{b} = \frac{95}{100}$; 236.8 **23.** $\frac{a}{42} = \frac{5.8}{100}$; 2.4
25. $\frac{12}{27} = \frac{p}{100}$; 44.4% **27.** $\frac{57}{b} = \frac{13.5}{100}$; 422.2 **29.** 12.5%
31. 1,840,000 households **33.** 760,000 households
35. 680,000 households **39.** Hurt; 7 out of 13 is about 53.8% which is less than 56%. **41.** 75% **43.** 12% **45.** 173.5%
47. 211 **49.** 72.2

Pages 222–223   Lesson 5-4
1. Since 75% equals $\frac{3}{4}$, find $\frac{3}{4}$ of 40. $\frac{1}{4}$ of 40 is 10. So $\frac{3}{4}$ of 40 is $3 \cdot 10$ or 30. **3.** Candace; 10% of 95 = $0.1 \cdot 95$ or 9.5. **5.** 20
7. 0.52 **9.** 126 **11.** 11 **13.** 8 **15.** 14 **17.** 80 **19.** 5.7 **21.** 0.283
23. 3.9 **25.** 120 **27.** 7.2 **29.** $422 **31.** < **33.** = **35.** 5
37. 450 Calories **39.** 700 women **41.** Sample answer:
$a = 300$, $b = 100$; Since 10% is $\frac{1}{3}$ of 30%, a must equal $3b$.
43. B **45.** about 22.9% **47.** 87.5% **49.** $22.\overline{2}$% **51.** Sample answer: $\frac{2}{3}$ of 90 or 60 **53.** Sample answer: $\frac{2}{7}$ of 70 or 20

Pages 230–231   Lesson 5-5
1. 26% is about 25% or $\frac{1}{4}$. $98.98 is about $100. $\frac{1}{4}$ of 100 is 25. So, 26% of $98.98 is about $25. **3.** 51% of 120; 24% of 240 is less than $\frac{1}{4}$ of 240 or 60. 51% of 120 is greater than $\frac{1}{2}$ of 120 or 60. **5–39. Sample answers given. 5.** $\frac{2}{3}$ of 21 or 14
7. $\frac{6}{35} \approx \frac{7}{35}$ or 20% **9.** $\frac{17.5}{23} \approx \frac{18}{24}$ or 75% **11.** $\frac{3}{10}$ of 50 or 15
13. $\frac{1}{5}$ of 75 or 15 **15.** $\frac{1}{5}$ of 70 or 14 **17.** $\frac{1}{2}$ of 160 or 80
19. $\frac{2}{3}$ of 9 or 6 **21.** $1\frac{1}{4}$ of 40 or 50 **23.** $\frac{1}{3}$ of 120 or 40
25. $\frac{7}{29} \approx \frac{7}{28}$ or 25% **27.** $\frac{4}{21} \approx \frac{4}{20}$ or 20% **29.** $\frac{8}{13} \approx \frac{8}{12}$ or
$66\frac{2}{3}$% **31.** $\frac{150,078}{299,065} \approx \frac{150,000}{300,000}$ or 50% **33.** $\frac{6}{25} \approx \frac{5}{25}$ or 20%
35. $\frac{13}{36} \approx \frac{12}{36}$ or $33\frac{1}{3}$% **37.** $\frac{8,008,278}{18,976,457} \approx \frac{8,000,000}{20,000,000}$ or 40%
39. $\frac{2,896,016}{12,419,293} \approx \frac{3,000,000}{12,000,000}$ or 25% **41.** always
43. sometimes **45.** G **47.** $\frac{7}{70} = \frac{p}{100}$; 10% **49.** $\frac{42}{b} = \frac{35}{100}$; 120
51. 0.4 **53.** 0.1

Pages 234–235   Lesson 5-6
1. $32 = n(40)$ **3.** $80 = n(60)$; The solution of $80 = n(60)$ is $1\frac{1}{3}$, while the solution of each of the others is $\frac{3}{4}$. **5.** 4% **7.** 0.25%
9. 36 **11.** 30% **13.** 200 **15.** 0.12% **17.** 201.6 **19.** 500
21. $2,222\frac{2}{9}$ **23.** $133\frac{1}{3}$% **25.** 11.5 **27.** 1,131 attempts
29. New York **31.** B **33.** Sample answer: $\frac{19}{30} \approx \frac{20}{30}$ or $66\frac{2}{3}$%
35. 60 **37.** 90 **39.** 87 **41.** 229

Pages 239–240   Lesson 5-7
1. Find the amount of change. **3.** Sample answer: original: 10, new: 30; The percent of change is 200%.
5. 20%; decrease **7.** 23.1%; increase **9.** $116.00 **11.** $29.96
13. 20%; decrease **15.** 25%; decrease **17.** 53.1%; increase
19. $144.00 **21.** $17.76 **23.** 50% **25.** $339.15 **27.** $439.08
29. 30% **31.** 308 hours or 12 days and 20 hours **33.** Sample answer: There were 25 students in the math class. Two more students enrolled in the class. What is the percent of change? Answer: 8%

35. Sample answer:
$27.9 - 22.4 = 5.5$ Find the difference.
$\frac{a}{b} = \frac{p}{100} \rightarrow \frac{5.5}{22.4} = \frac{p}{100}$ Replace a with 5.5 and b with 22.4.
$5.5 \cdot 100 = 22.4 \cdot p$ Find the cross products.
$550 = 22.4p$ Multiply.
$\frac{550}{22.4} = \frac{22.4p}{22.4}$ Divide each side by 22.4.
$24.6 \approx p$ Simplify.
The percent of change is about 24.6%.

37. about $0.54 **39.** Sample answer: $\frac{1}{4}$ of 84 or 21
41. Sample answer: $\frac{1}{3}$ of 96 or 32 **43.** 3 **45.** 0.04

Pages 243–244   Lesson 5-8
1. In the formula $I = prt$, I represents the interest, p represents the principal, r represents the simple interest rate written as a decimal, and t represents the time in years. **3.** Yes; Yoshiko will earn half of the interest which is 3.5% or 0.035. **5.** $18.40 **7.** $421.38 **9.** $48.75 **11.** $90.70 **13.** $187.50 **15.** $112.50 **17.** $2,621.25 **19.** $636.09 **21.** $14,925.00 **23.** $1,016.75 **25.** $72,500 **27.** 14 yr **29.** $540 = 750 \cdot r \cdot 6$; 12% **31.** 25%

Pages 246–248   Chapter 5   Study Guide and Review
1. percent **3.** percent proportion **5.** markup **7.** principal **9.** 4 **11.** 80% **13.** 16.5% **15.** 20% **17.** $1\frac{1}{5}$ **19.** 0.043 **21.** 0.13 **23.** 1.47 **25.** 65.5% **27.** 70% **29.** 1.5% **31.** 87.5% **33.** 96% **35.** $\frac{15}{b} = \frac{30}{100}$; 50 **37.** $\frac{75}{250} = \frac{p}{100}$; 30% **39.** 90 **41.** 16 **43.** 2.43 **45.** Sample answer: $\frac{1}{8}$ of 80 or 10 **47.** Sample answer: $\frac{2}{5}$ of 40 or 16 **49.** Sample answer: $\frac{33}{98} \approx \frac{33}{99}$ or $33\frac{1}{3}$% **51.** 4,620 **53.** 12.3 **55.** 50%; increase **57.** 20%; decrease **59.** 48% **61.** $17.00 **63.** $68.25

Chapter 6   Geometry

Page 255   Chapter 6   Getting Started
1. false; $a^2 + b^2 = c^2$ **3.** 86 **5.** 98 **7.** 11.4 ft **9.** 9.5 yd **11.** No; the angles do not have the same measure.

Pages 259–260   Lesson 6-1
1. Sample answer:
3. straight **5.** adjacent **7.** 153 **9.** 43° **11.** 126° **13.** acute **15.** vertical **17.** adjacent, complementary
19. 140 **21.** 36 **23.** 45 **25.** 73 **27.** 20 **29.** 70° **31.** 111° **33.** 63° **35.** 59° **37.** 82°
39. They are supplementary.
Sample answer: In the diagram, and $\angle 1$ and $\angle 2$ are supplementary. Since $\angle 1$ and $\angle 3$ are alternate interior angles, $\angle 1 \cong \angle 3$. Therefore, replacing $\angle 1$ with $\angle 3$, $\angle 3$ and $\angle 2$ are supplementary.
41. A **43.** 35%; increase **45.** 95%; decrease **47.** 81 **49.** 45

Pages 264–265   Lesson 6-2
1. Sample answer: a baseball pennant **3.** 74 **5.** 61 **7.** acute scalene **9.** obtuse isosceles **11.** 40 **13.** 134 **15.** 27 **17.** acute equilateral **19.** right scalene **21.** right isosceles **23.** obtuse isosceles

25.    **27.**

29. not possible **31.** sometimes **33.** The sum of the measures of the angles of a triangle is 180°. If two of the angles of a triangle were greater than or equal to 90°, then the sum of these angles would already be greater than or equal to 180°. **35.** B **37.** 95° **39.** 85° **41.** 9.4 ft **43.** 11.5 in.

Pages 269–270   Lesson 6-3
1. The length of the hypotenuse is twice the length of the leg opposite the 30° angle. **3.** $a = 10$ in., $b \approx 17.3$ in. **5.** $b = 9$ m, $c \approx 12.7$ m **7.** $a = 11$ in., $b \approx 19.1$ in. **9.** $c = 50$ ft, $b \approx 43.3$ ft **11.** $b = 18$ yd, $c \approx 25.5$ yd **13.** 11.6 cm **15.** 7.5 ft, about 10.6 ft **17.** about 3.5 in. **19.** Sample answer: A flowerbed is in the shape of a 45°-45° right triangle. The length of one leg is 6 feet. What is the length of the other two sides of the triangle? 6 ft and about 8.5 ft **21.** B **23.** acute isosceles **25.** acute **27.** alt. exterior **29.** $-\frac{3}{10}$ **31.** -6 **33.** 80

Pages 274–275   Lesson 6-4
1. A square is a parallelogram with four congruent sides. **3.** Trapezoid; the others are all examples of parallelograms. **5.** 30 **7.** square **9.** trapezoid **11.** 95 **13.** 65 **15.** 142 **17.** trapezoid **19.** parallelogram **21.** square **23.** trapezoid **25.** 120 **27.** trapezoid **29.** rhombus, square **31.** true

33. False;

35. C **37.** 13.9 ft **39.** acute, scalene **41.** obtuse, scalene **43.** Yes, the angles have the same measure.

Pages 281–282   Lesson 6-5
1.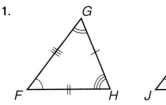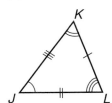

3. yes; $\angle A \cong \angle G$, $\angle C \cong \angle H$, $\angle E \cong \angle F$, $\overline{AC} \cong \overline{GH}$, $\overline{CE} \cong \overline{HF}$, $\overline{AE} \cong \overline{GF}$; $\triangle ACE \cong \triangle GHF$ **5.** 73° **7.** 7 yd **9.** yes; $\angle H \cong \angle P$, $\angle K \cong \angle Q$, $\angle J \cong \angle M$, $\overline{HK} \cong \overline{PQ}$, $\overline{KJ} \cong \overline{QM}$, $\overline{HJ} \cong \overline{PM}$; $\triangle HJK \cong \triangle PMQ$ **11.** no **13.** yes; $\angle A \cong \angle E$, $\angle B \cong \angle D$, $\angle C \cong \angle F$, $\overline{AB} \cong \overline{ED}$, $\overline{BC} \cong \overline{DF}$, $\overline{AC} \cong \overline{EF}$; $\triangle ABC \cong \triangle EDF$ **15.** 6 m **17.** 45° **19.** 90° **21.** 11 in. **23.** 2.5 m **25.** a and d **27.** trapezoid **29.** quadrilateral **31.** A

Pages 288–289   Lesson 6-6
1.

3a.

5a.

3b. no

5b. no

7a.

7b. yes; 180°

9a.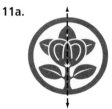

9b. yes; 72°, 144°, 216°, 288°

11a.

11b. no

13a. none **13b.** yes; 120°, 240° **15.** Isosceles and equilateral triangles; equilateral triangles

17a.

17b.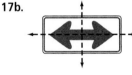

17c. none **17d.** none
19. Sample answers:

line symmetry

line symmetry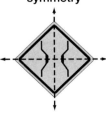

line and rotational symmetry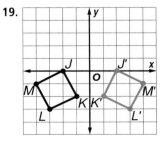

21. true **23.** B **25.** The 4 triangles that form the large triangle in the center appear to be congruent. Three of these smaller triangles are divided into 3 smaller triangles. These smaller triangles appear to be congruent.

27.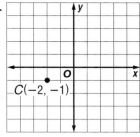

29.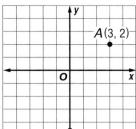

Pages 292–294   Lesson 6-7

1. Sample answer: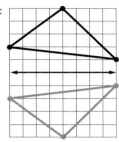

3. The third transformation; all the transformations are reflections of the original figure about the given line. The image of the tip of the dog's tail is not directly across from the tip of the original dog's tail.
5. $Q'(-3, -3)$, $R'(2, -4)$, $S'(3, -2)$, $T'(-2, -1)$;

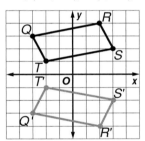 **7.**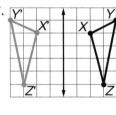

9. 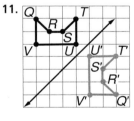 **11.**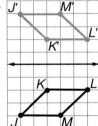

13. No; the image of the tip of the balloon's tail is not directly across from the tip of the original balloon's tail.
15. Yes; each point on the image of the cup is directly across from each corresponding point on the original cup.

17.

19.

$A'(-1, 1)$, $B'(-2, 4)$, $C'(-4, 1)$

$J'(2, 0)$, $K'(1, -2)$, $L'(3, -3)$, $M'(4, -1)$

21. x-axis **23.** y-axis

25. yes;

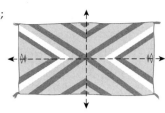

27. H, I, M, O, T, U, V, W, X, Y **29.** x-axis; The x-coordinates are the same, but the y-coordinates are opposites.
31. The two pieces are reflections of each other and they are congruent. **33.** yes; 180° **35.** no **37.** 20 **39.** −2 **41.** 0

Pages 298–299   Lesson 6-8
1. The fourth transformation; the other transformations are translations, but in the fourth, the figure is turned, so it is not a translation.

3.

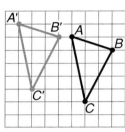

5.

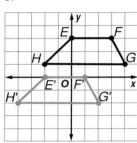

E'(−2, 0), F'(1, 0),
G'(2, −2), H'(−4, −2)

7.

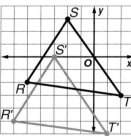

9.

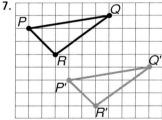

R'(−6, −5), S'(−3, 0),
T'(1, −6)

11.

A'(3, 5), B'(1, 2),
C'(3, 0), D'(5, 3)

13.

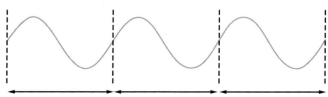

15. S'(−14, 2), T'(0, 9) **17.** (−4, 2) **19.** yes **21.** yes; 180°
23. no

Pages 302–303   Lesson 6-9
1. Sample answer: fan blade, Ferris wheel, car tire

3.

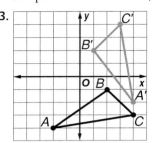

A'(4, −2), B'(1, 2), C'(3, 4)

5.

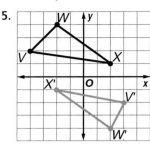

V'(4, −2), W'(2, −4),
X'(−2, −1)

7.

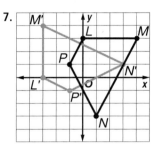

L'(−3, 0), M'(−3, 4), N'(3, 1), P'(−1, −1)

9. Yes; the figure in green is a rotation of the figure in blue 180° about the origin.
11. No; the figure in green is a reflection of the figure in blue over the y-axis.

13.

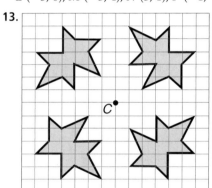

15.

17. rotation **19.** reflection **21.** The 4 hearts at the bottom of the tie are translations of the first heart at the top of the tie.

23. 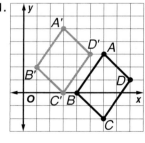 **25.** acute equilateral; yes

Pages 306–308   Chapter 6   Study Guide and Review
1. false; obtuse **3.** false; perpendicular **5.** true **7.** true
9. 137 **11.** 135° **13.** 23 **15.** a = 2 cm, c ≈ 3.5 cm
17. b = 10 ft, c ≈ 14.1 ft **19.** 102° **21.** 124° **23.** 11 cm

25.  **27.** none

29.

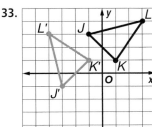

$Q'(2, -5)$, $R'(4, -5)$,
$S'(3, -1)$, $T'(1, -1)$

31.

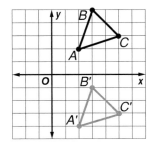

$A'(2, -4)$, $B'(3, -1)$,
$C'(5, -3)$

$J'(-3, -1)$, $K'(-1, 1)$,
$L'(-4, 3)$

33.

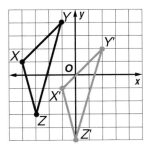

Chapter 7 Geometry: Measuring Area and Volume

Page 313 Chapter 7 Getting Started
1. trapezoid **3.** 32 **5.** 34.0 **7.** 1.8 **9.** 47.1 **11.** 153.9
13. quadrilateral **15.** pentagon

Pages 317–318 Lesson 7-1
1. They are the same; $A = bh$. **3.** Malik; The area of a trapezoid is half the product of the height and the sum of the bases. **5.** 180 m² **7.** 25.8 km² **9.** 14.04 m²
11. 25 ft² **13.** 28 in² **15.** 32.4 cm² **17.** 14.4 cm² **19.** 7 cm
21. 120,000 km² **23.** 112,500 km² **25.** Tennessee: 109,158 km², Arkansas: 137,741 km², Virginia: 109,391 km², North Dakota 183,123 km² **27.** The area is doubled. **29.** A
31.

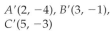

$X'(-1, -1)$, $Y'(2, 2)$,
$Z'(0, -5)$

33.

$X'(4, -1)$, $Y'(1, -4)$,
$Z'(3, 3)$

35. 58.4 **37.** 176.7

Pages 322–323 Lesson 7-2

1. Sample answer:

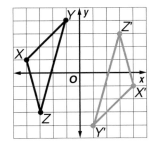

3. 75.4 yd; 452.4 yd² **5.** 66.0 ft; 346.4 ft² **7.** 16.7 mi; 22.1 mi²
9. 62.8 in.; 314.2 in² **11.** 119.4 mi; 1,134.1 mi² **13.** 60.9 m;

295.6 m² **15.** 22.0 cm; 38.5 cm² **17.** 32.6 ft; 84.5 ft² **19.** about 7,854 in. or 654.5 feet **21.** about 70,686 yd² **23.** 254.5 in²; 153.9 in²; 78.5 in² **25.** 13.3 ft **27.** 88.0 cm² **29.** 18.2 m²
31. Sample answer:

33. C **35.** 17.4 cm²

37. $W'(3, 1)$, $X'(1, 3)$, $Y'(2, 4)$;

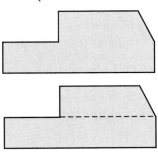

39. 25.1 **41.** 300.15

Pages 328–329 Lesson 7-3
1. Sample answer:

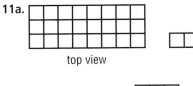

3. 68 cm² **5.** 216 in²
7. 240 yd² **9.** 87.5 m²
11. 121.2 cm² **13.** 103.8 m²
15. about 480.5 units²
17. 3 cans; The area to be painted is 857.5 ft², so 857.5 ÷ 350 or about 2.5 cans of paint are needed. Since you cannot buy half a can, you must buy 3 cans.

19. No, the area of the field is about 60,638.3 ft², so it will take 60,638.3 ÷ 1,750 or about 35 minutes to mow the field. The grounds crew only has 30 minutes. **21.** C **23.** 95.8 m
25. 220 ft² **27.** quadrilateral **29.** hexagon

Pages 333–334 Lesson 7-4
1. a: vertex; b: face; c: base; d: edge **3.** rectangular prism; 6 faces, all rectangles; 12 edges; 8 vertices **5.** rectangular pyramid; 5 faces, 1 rectangle and 4 triangles; 8 edges; 5 vertices **7.** triangular pyramid; 4 faces, all triangles; 6 edges; 4 vertices **9.** triangular prism; 5 faces, 2 triangles and 3 rectangles; 9 edges; 6 vertices

11a.

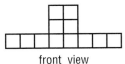

top view   front view

side view

11b. $1\frac{1}{2}$ ft **11c.** $1\frac{1}{2}$ ft² **13.** Sometimes; three planes can intersect in a line or not intersect at all if two or more are parallel. **15.** Sometimes; a rectangular prism has 5 vertices, but a triangular prism has 4. **17a.** 2 square pyramids

top view   side view

19. $E = 2n$
21. $\overline{EF}$ and $\overline{AD}$
23. top: rectangular pyramid; bottom: rectangular prism
25. 146.3 ft^2
27. 161.1 in^2
29. 15 in^2
31. 27.5 cm^2

Pages 337–339 Lesson 7-5

1. The area of the base B of a rectangular prism equals the length ℓ times the width w. Replacing B with ℓw in the formula $V = Bh$, gives another formula for the volume of a rectangular prism, $V = (\ell w)h$ or $V = \ell wh$. **5.** 539 m^3
7. 420 ft^3 **9.** 216 mm^3 **11.** 768 m^3 **13.** 55.4 m^3 **15.** 297.5 ft^3
17. 576 mm^2 **19.** 14,790 cm^3 **21.** 891.3 yd^3 **23.** 6 in.
25. 330 ft^3 **27.** 1,728 **29.** 1,000,000 **31.** 40 ft **33.** volume doubles **35.** volume is multiplied by 8 **37.** B **39.** 126 ft^2
41. $\frac{3}{2,500}$ **43.** $1\frac{7}{20}$ **45.** 20 **47.** 48

Pages 344–345 Lesson 7-6

1. Doubling its radius; doubling the radius means the volume of the cone is multiplied by 4, while doubling the height multiplies the volume of the cone by 2. **3.** 183.3 m^3
5. 14 ft^3 **7.** 1,731.8 mm^3 **9.** 43.3 in^3 **11.** 261.3 m^3
13. 230.9 in^3 **15.** 175 cm^3 **17.** 654.5 ft^3 **19.** 13 m^3
21. Sample answer: 250 cm^3 **23.** Sample answer: A paper cup is shaped like a cone. If the cup is 6 cm wide and 10 cm tall, find the volume of water the cup will hold; 94.2 cm^3 **25.** 113.1 in^3 **27.** 523.6 m^3 **29.** It multiplies it by 8; replacing r with $2r$ in the formula for the volume of a sphere gives $\frac{4}{3}\pi(2r)^3$ or $8 \cdot \frac{4}{3}\pi r^3$. **31.** 45 cm^3 **33.** trapezoidal prism; 6 faces, 2 trapezoids, 4 rectangles; 12 edges; 8 vertices **35.** 17.3 ft **37.** 23.9 cm

Pages 349–351 Lesson 7-7

1. False; a rectangular prism 2 ft long, 4 ft wide, and 6 feet high has the same volume as a prism 2 ft long, 2 ft wide, and 12 ft high, 48 ft^3. The surface area of the first prism is 88 ft^2, but the surface area of the second prism is 104 ft^2.
3. Sample answer: 2 ft $\times$ 2 ft $\times$ 11 ft **5.** 216 in^2 **7.** 467.3 cm^2
9. 168.7 cm^2 **11.** 360 ft^2 **13.** 1,154.5 yd^2 **15.** 864 m^2
17. 725.7 in^2 **19.** 805 ft^2 **21.** 574.7 in^2 **23.** Double the radius; consider the expression for the surface area of a cylinder, $2\pi r^2 + 2\pi rh$. If you double the height, you will double the second addend. If you double the radius, you will quadruple the first addend and double the second addend. **25.** 12
27. 1 **29.** rectangle **31.** triangle **33.** D **35.** 392 m^3 **37.** No; the volume of the refrigerator is 12,852 in^3 or about 7.4 ft^3, which is less than 8 ft^3. **39.** 115 **41.** 35

Pages 354–355 Lesson 7-8

1. The slant height is the height of each lateral face of the pyramid and the height is the perpendicular distance from the vertex to the base of the pyramid. **3.** 64 ft^2 **5.** 115.0 cm^2
7. 47.3 ft^2 **9.** 140.4 mm^2 **11.** 659.6 yd^2 **13.** 149.5 cm^2 **15.** 5; The surface area of the roof is 502.7 ft^2. $502.7 \div 120 \approx 4.2$. Since you cannot buy a fraction of a roll, 5 rolls of roofing material are needed. **17.** 254.5 in^2 **19.** $\sqrt{18}$ in. **21.** 113.1 m^2
23. 804.2 ft^2 **25.** B **27.** 1,278.6 cm^2 **29.** 62.6 **31.** 25.7

Pages 360–362 Lesson 7-9

1. 74.8 oz; 2 lbs is measured to the nearest lb, 74 ounces is measured to the nearest ounce, and 74.8 ounces is measured

to the nearest 0.1 ounce, therefore 74.8 oz is most precise.
3. 375.0; all of the other numbers have 3 significant digits, while 375.0 has 4 significant digits. **5.** 2°F **7.** 2 **9.** 5
11. 1.5 m **13.** 1.4 **15.** $\frac{1}{8}$ in. **17.** $\frac{1}{4}$ pound **19.** 3 **21.** 2 **23.** 3
25. 2 **27.** 9.39 L **29.** 190 m **31.** 5.2 s **33.** 13 ft^2 **35.** 80 **37.** 40
39. 10,600 m^2 **41.** 3 **43.** 150 cm^2 **45.** D **47.** 74.6 cm^2
49. −3.31 **51.** $\frac{2}{3}$

Pages 363–366 Chapter 7 Study Guide and Review

1. b **3.** d **5.** i **7.** c **9.** g **11.** $140\frac{1}{4}$ in^2 **13.** 106.4 m^2
15. 18.8 cm; 28.3 cm^2 **17.** 16.3 m; 21.2 m^2 **19.** 57.5 mm^2
21. 200.5 in^2 **23.** hexagonal pyramid; 7 faces, 1 hexagon and 6 triangles; 12 edges; 7 vertices **25.** 660 yd^3 **27.** 163.3 ft^3
29. 445.3 yd^3 **31.** 612 m^2 **33.** 95 ft^2 **35.** 520.7 cm^2
37. about 175.9 ft^2 **39.** 2 **41.** 3 **43.** 45.3 lb **45.** 16.7

Chapter 8   Probability

Page 373 Chapter 8 Getting Started

1. proportion **3.** $\frac{2}{3}$ **5.** $\frac{7}{33}$ **7.** 7,920 **9.** 3,024 **11.** 35 **13.** 70
15. $\frac{1}{2}$ **17.** $\frac{7}{18}$ **19.** 22.4

Pages 376–377 Lesson 8-1

1.

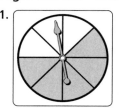

3. Masao; a 2 is only 1 out of 6 possibilities when rolling a number cube. The probability is $\frac{1}{6}$. **5.** $\frac{1}{2}$; 0.5; 50% **7.** $\frac{7}{8}$; 0.875; 87.5% **9.** $\frac{3}{4}$; 0.75; 75% **11.** $\frac{6}{25}$; 0.24; 24% **13.** $\frac{14}{25}$; 0.56; 56%

15. 1; 1; 100% **17.** $\frac{1}{4}$ **19.** No; P(greater than 3) $= \frac{1}{2}$ and P(less than 3) $= \frac{1}{3}$, but $\frac{1}{2} + \frac{1}{3} \neq 1$. **21.** 60% **25.** 0.225
27. 17 red crayons **29.** 5:1 **31.** C **33.** The measurement is to the nearest centimeter. There is 1 significant digit. The greatest possible error is 0.5 cm, and the relative error is $\frac{0.5}{8}$ or about 0.063. **35.** The measurement is to the nearest 0.01 m. There are 3 significant digits. The greatest possible error is 0.005 m, and the relative error is $\frac{0.005}{4.83}$ or about 0.0010. **37.** about 267 in^2 **39.** 200 **41.** 112

Pages 382–383 Lesson 8-2

1. With a tree diagram, you can see all the different outcomes. However, with the Fundamental Counting Principle, you only know how many outcomes there are.
3. 4 more outfits **5.** $\frac{1}{9}$ **7.** 24 pizzas

9.

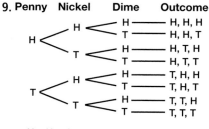

H = Heads
T = Tails

8 outcomes

11.

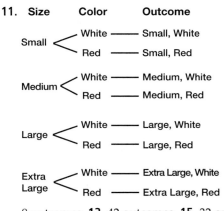

Size	Color	Outcome
Small	White	Small, White
	Red	Small, Red
Medium	White	Medium, White
	Red	Medium, Red
Large	White	Large, White
	Red	Large, Red
Extra Large	White	Extra Large, White
	Red	Extra Large, Red

8 outcomes **13.** 42 outcomes **15.** 32 outcomes

17.

First Spinner	Second Spinner	Outcome
Green	Red	Green, Red
	Blue	Green, Blue
	White	Green, White
Blue	Red	Blue, Red
	Blue	Blue, Blue
	White	Blue, White
Yellow	Red	Yellow, Red
	Blue	Yellow, Blue
	White	Yellow, White
Red	Red	Red, Red
	Blue	Red, Blue
	White	Red, White

12 outcomes

19. $\frac{1}{2}$ **21.** $\frac{1}{100,000}$ **23.** 45,697,600 plates **25.** D **27.** $\frac{2}{11}$
29. 1 **31.** 3 significant digits **33.** 5,040 **35.** 1,680

Pages 386–387 Lesson 8-3
1. $9! = 9 \cdot 8 \cdot 7 \cdot 6 \cdot 5 \cdot 4 \cdot 3 \cdot 2 \cdot 1$ and $P(9, 5) =$
$9 \cdot 8 \cdot 7 \cdot 6 \cdot 5$ **3.** Bailey; $P(7, 3)$ means to start with 7 and
use 3 factors. **5.** 840 **7.** 40,320 **9.** 720 ways **11.** 120 **13.** 120
15. 240,240 **17.** 303,600 **19.** 2 **21.** 39,916,800 **23.** 24 ways
25. 3,024 passwords **27.** $\frac{1}{5}$ **29.** 120 ways **31.** 15,600 ways
33. 3,628,800; $10! = 10 \cdot 9!$ **35.** B **37.** 44 different ways
39. $\frac{1}{2}$ **41.** $\frac{2}{3}$ **43.** 20 **45.** 190

Pages 390–391 Lesson 8-4
1. Sample answers: selecting a committee of 5 people;
selecting a president and vice president of a club **3.** 15
5. 7 **7.** permutation **9.** 210 squads **11.** 36 **13.** 9 **15.** 126
17. 3,060 **19.** combination **21.** permutation **23.** permutation
25. 220 pizzas **27.** 6,840 ways **29.** 2,598,960 hands
31. 259,459,200 ways **33.** Sometimes; they are equal if $y = 1$.
35. 4,249,575 committees **37.** 32,760 **39.** 5,040 **41.** $\frac{3}{10}$ **43.** $\frac{1}{8}$

Pages 398–399 Lesson 8-5
1. Both independent events and dependent events are
compound events. Independent events do not affect each
other. Dependent events affect each other. **3.** Evita;
spinning the spinner twice represents two independent
events. The probability of getting an odd number is $\frac{3}{5}$
each time. **5.** $\frac{1}{4}$ **7.** $\frac{1}{18}$ **9.** $\frac{1}{30}$ **11.** $\frac{1}{10}$ **13.** $\frac{2}{15}$ **15.** $\frac{1}{8}$ **17.** $\frac{1}{19}$

19. $\frac{3}{95}$ **21.** $\frac{33}{95}$ **23.** 68.9% **25.** $\frac{4}{13}$ **27.** 4 **29.** 2/17 **31.** 21
33. 84 **35.** $\frac{13}{30}$ **37.** $\frac{7}{10}$

Pages 402–403 Lesson 8-6
1. Each experimental probability will be different.
Theoretical probability tells you approximately what
should happen. **3.** $\frac{13}{25}$ **5.** $\frac{1}{2}$ **7.** The theoretical probability
is about the same as the experimental probability.
9. about 70 cars **11.** about 67 errors **13.** $\frac{4}{15}$ **15.** $\frac{4}{5}$
17. about 80 teens **21.** about 200 times **23.** $\frac{3}{2,500}$
25. The experimental probability is $\frac{15}{75}$ or $\frac{1}{5}$. The theoretical
probability is $\frac{1}{4}$. The experimental probability is less than
the theoretical probability. **27.** $\frac{5}{44}$ **29.** 31.5 **31.** 16.2

Pages 408–409 Lesson 8-7
1. Taking a survey is one way to determine experimental
probability. **3.** This is a biased sample, since people in other
states would spend much more than those in Arizona. The
sample is a convenience sample since all the people are
from the same state. **5.** 48% **7.** This is an unbiased,
systematic random sample. **9.** This is a biased sample,
since only voluntary responses are used. **11.** This is an
unbiased, simple random sample. **13.** Sample answer: Get
a list of all the students in the school and contact every 20th
student on the list. **15.** about 240 containers **19.** No; the
survey should be representative of the whole school.
21. Sample answer: If the questions are not asked in a
neutral manner, the people may not give their true opinion.
For example, the question "You really don't like Brand X,
do you?" might not get the same answer as the question
"Do you prefer Brand X or Brand Y?" Also, the question
"Why would anyone like rock music?" might not get the
same answer as the question "What do you think about
rock music?" **23.** I **25.** $\frac{1}{20}$

Pages 410–412 Chapter 8 Study Guide and Review
1. sample space **3.** multiplying **5.** compound event
7. Theoretical probability **9.** $\frac{6}{25}$; 0.24; 24% **11.** $\frac{18}{25}$; 0.72;
72% **13.** $\frac{18}{25}$; 0.72; 72% **15.** $\frac{1}{6}$ **17.** $\frac{1}{8}$ **19.** $\frac{3}{8}$ **21.** 6 **23.** 60
25. 720 **27.** 120 numbers **29.** 4 **31.** 126 **33.** 21 **35.** $\frac{1}{12}$ **37.** $\frac{1}{6}$
39. $\frac{1}{6}$ **41.** $\frac{25}{102}$ **43.** $\frac{4}{15}$ **45.** $\frac{23}{30}$ **47.** This is a biased sample,
since only people leaving a concert are surveyed. This is a
convenience sample. **49.** about 120 people

Chapter 9 Statistics and Matrices

Page 417 Chapter 9 Getting Started
1. false; biased **3.**

5.

7. −12 **9.** 4 **11.** −12 **13.** 0.23, 0.32, 2.03
15. 0.01, 0.10, 1.01, 1.10 **17.** 187.2 **19.** 50.4

Pages 422–424 Lesson 9-1
1. Sample answer: 2, 3, 5, 6, 7, 8, 9, 10, 11, 12, 13, 15, 16, 17

3. Record High Temperatures for Each State

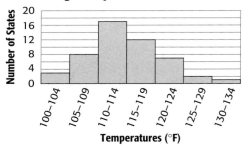

5. New Broadway Productions for Each Year from 1960-2001

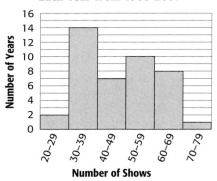

7. Sample answer:

Calories of Various Types of Frozen Bars

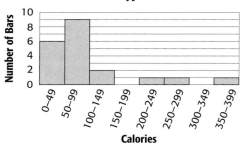

9. 200–399 **11.** 7 states **13.** 37 states **15.** 8 courts
17. 16 courts **19.** Vermont **21.** 6 counties **25.** G **27.** $\frac{3}{4}$
29. $\frac{11}{20}$ **31.** $\frac{33}{50}$ **33.** 5 **35.** 12.6 **37.** 190.8

Pages 428–429   Lesson 9-2

1. Both graphs show how the ages of the signers of the Declaration of Independence were distributed. The bar graph shows how many were in each age interval. The circle graph shows what percent of the signers were in each age interval.
3. Sample answer:

My Day

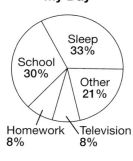

5.   Hawaiian Counties

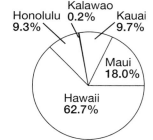

7. Flowers and Plants Purchased for Mother's Day

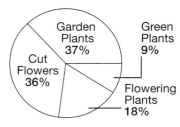

9. U.S. Population by Age

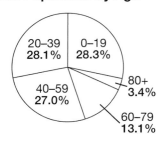

11. Half of the homes are heated with piped gas. About a third of the homes are heated with electricity. The rest of the homes are heated with fuel oil, bottled gas, wood, or something else. **13.** C

15. Sample answer:

Calories of Single Serving, Frozen Pizzas

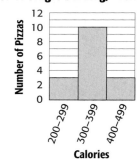

17.

Pages 432–433   Lesson 9-3

1. Both bar graphs and histograms use bars to show how many things are in each category. A histogram shows the frequency of data that has been organized into equal intervals. There is no space between the bars in a histogram.
3. circle graph

5. Sample answer: histogram   **7.** table, bar graph, or pictograph **9.** histogram
11. line graph

Grams of Carbohydrates in a Serving of Various Vegetables

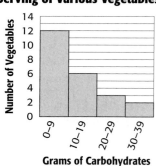

13. Sample answer: line graph;

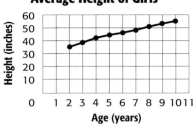

Average Height of Girls

17. Sample answer:

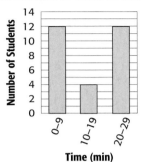

Time Needed to Walk to School

19. 20.5 **21.** 6.5

Pages 437–438   Lesson 9-4
1. No; the mode must always be a member of the set of data, but the mean and median may or may *not* be a member of the set of data. **3.** Erica; you must first order the numbers from least to greatest. **5.** 9; 9; no mode **7.** The median; the mean is affected by the extreme value of 74, and the mode is the least number in the set of data. **9.** 12.8; 14; no mode **11.** 34; 34; 34 **13.** 1.5; 1.6; no mode **15.** 0.5; 0.6; 0.6 **17.** Median; the mode is the least number in the set of data and the mean is affected by the very large number 53. **19.** Sample answer: 1, 1, 1, 1, 14, 15, 18 **21.** G
23. histogram **25.** 2.89, 2.9, 3.1, 3.2, 3.25 **27.** 15.01, 15.1, 16.79, 16.8, 17.4

Pages 444–445   Lesson 9-5
1. Sample answer: {1, 50, 50, 60, 60, 70, 70, 80} **3.** 9; 59; 62, 58; 4; no outliers **5.** 41.1 million **7.** 33.0 million, 9.1 million **9.** no outliers **11.** 38; 52; 57, 48; 9; 22 **13.** 6.3; 16.6; 18.7, 14.55; 4.15; no outliers **15.** 0.7; 0.55; 0.65, 0.25; 0.4; no outliers **17.** 38; 46.5, 30.5 **19.** Philadelphia **21.** 54; 70; 39
23. The interquartile range for San Francisco is only 10°F, while the interquartile range for Philadelphia is 31°F.
25a. Sample answer: {1, 1, 2, 2, 2, 5, 9, 9, 9, 10, 10} and {1, 4, 4, 4, 4, 5, 5, 5, 9, 10, 10} **25b.** Sample answer: {1, 2, 5, 7, 9, 10, 12, 14, 15, 17, 22} and {0, 2, 5, 7, 9, 10, 12, 14, 15, 17, 27} **27.** 1.35 **29.** 7; 7; 3

31.

33.

Pages 448–449   Lesson 9-6
1. The box represents the spread of the middle half of the data. **3.** Joseph; 64 is an outlier.

5.

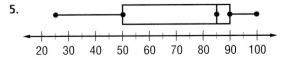

7. 260 **9.**

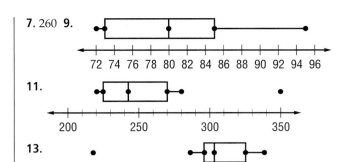

11.

13.

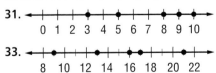

15. domestic **17.** 75% **19.** Sample answer: {20, 20, 20, 30, 35, 40, 50, 60, 70, 70, 70} **21.** 40 **23.** 49; 81; 88, 74; 14; 50
25. Since this is a systematic random sample of the entire population concerned with the park, it is an unbiased sample.

Pages 451–453   Lesson 9-7
1. The scale may have a break or may have different sized intervals. The graph may show a larger area than the actual increase. **3.** Graph B; since there is a break in the vertical scale, the number of medals for Norway appears to be much greater than the number of medals for the Soviet Union. **5.** Graph B; the area of the house indicates a much greater median income for the male householder than a female householder. **7.** The advertising is not false. In the last survey, 3 out of 4 people liked Tasty Treats better than Groovy Goodies. However, it is misleading, because combining all the survey results shows only half of the people liked Tasty Treats better than Groovy Goodies.
9. Mean; it is greater than the median and they will want to appear to pay more money. **11.** Mean; it is greater than the median and they will want to appear to pay more money.

13a. **Number of Admissions to Movie Theaters** **13b.** **Number of Admissions to Movie Theaters**

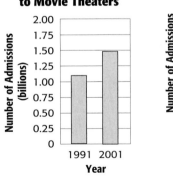 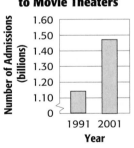

15. G **17.**

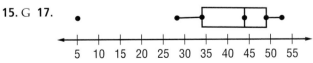

19. 7.02×10^4 **21.** 4.56×10^{-4} **23.** 3 **25.** -2

Pages 456–457   Lesson 9-8
1. A 3-by-2 matrix has 3 rows and 2 columns, and a 2-by-3 matrix has 2 rows and 3 columns. **3.** 3 by 1; third row, first column **5.** 2 by 5; second row, fourth column

7. $\begin{bmatrix} -2 & 13 & -8 \\ 11 & 3 & -18 \\ -5 & -8 & 8 \end{bmatrix}$ **9.** 1 by 3; first row, third column

11. 3 by 3; third row, first column **13.** 3 by 4; third row, second column

15. $\begin{bmatrix} 2 & 1 & 2 \\ 3 & -5 & -1 \\ 1 & -8 & 8 \end{bmatrix}$ **17.** impossible **19.** $\begin{bmatrix} 12 & 8 & 12 \\ 21 & 17 & 11 \\ 16 & 16 & 15 \end{bmatrix}$

21. C **23.**

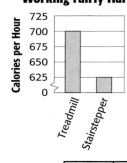

Calories Burned Working Fairly Hard

25.

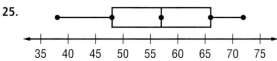

Pages 458–460 Chapter 9 Study Guide and Review
1. true **3.** true **5.** false; median **7.** false; dimensions
9. 10 students
11.

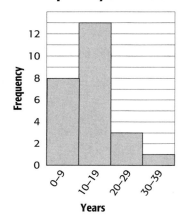

Life Expectancy of Animals

13. circle graph **15.** 16.4, 15, 15 **17.** 7.7, 8, 8
19. 11; 3; 5, 2; 3; 12 **21.** 8; 6.5; 8.5, 4.5; 4; no outliers

23.

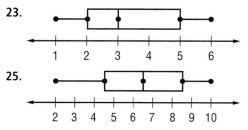

25.

27. median **29.** median **31.** $\begin{bmatrix} 9 & 7 \\ -2 & 2 \end{bmatrix}$ **33.** impossible

Chapter 10 Algebra: More Equations and Inequalities

Page 467 Chapter 10 Getting Started
1. algebraic **3.** true **5.** false **7.** $10 + x = -8$ **9.** $2x - 4 = 26$
11. -17 **13.** 19 **15.** 14 **17.** -6 **19.** 7 **21.** -84

Pages 472–473 Lesson 10-1
1. terms that contain the same variable or are constants
3. $5(x - 3)$; $5(x - 3)$ is equivalent to $5x - 15$, while the other three expressions are equivalent to $5x - 3$.
5. $-3a - 27$ **7.** terms: $8a$, 4, $-6a$; like terms: $8a$ and $-6a$; coefficients: 8, -6; constant: 4 **9.** terms: $5n$, $-n$, 3, $-2n$; like terms: $5n$, $-n$, and $-2n$; coefficients: 5, -1, -2; constant: 3
11. $9n$ **13.** $11c$ **15.** $5x - 3$ **17.** $7m + 42$ **19.** $-7n - 14$
21. $-8c + 64$ **23.** $-4x + 24$ **25.** $4x - 4y$ **27.** $-12x - 20$
29. $12(x - 7)$; $12x - 84$ **31.** $9(x - 3)$; $9x - 27$ **33.** terms: 7, $-5x$, 1; like terms: 7, 1; coefficients: -5; constants: 7, 1
35. terms: n, $4n$, $-7n$, -1; like terms: n, $4n$, $-7n$; coefficients: 1, 4, -7; constant: -1 **37.** terms: 9, $-z$, 3, $-2z$; like terms: 9 and 3, $-z$ and $-2z$; coefficients: -1, -2; constants: 9, 3 **39.** $6n$ **41.** $-3k$ **43.** $14x + 4$ **45.** $6 - 3c$ **47.** 7
49. -8 **51.** $5\frac{1}{2}y + 100$ **53.** $2t + 24$ **55.** $12x + 6$ **57.** 1 by 2; first row, first column **59.** 3 by 1; third row, first column
61. Graph B **63.** -6 **65.** -8

Pages 476–477 Lesson 10-2
1. You identify the order in which operations would be performed on the variable, then you undo each operation using its inverse operation in reverse order. **3.** Tomás; Alexis did not undo the operations in reverse order. **5.** 1
7. 28 **9.** 8 **11.** -3 **13.** 6 **15.** 3 **17.** -8 **19.** 4 **21.** 27 **23.** 8
25. \$6 **27.** -3 **29.** 5 **31.** -9 **33.** -6 **35.** 4 **37.** 4 **39.** -1
41. -1 **43.** 7 **45.** Sample answer: $4x + 3 = -17$
47. $13 + 3x = 25$; 4 **49.** $-3x - 15$ **51.** $-8p + 56$
53. $4n + 5 = 17$

Pages 480–481 Lesson 10-3
1. multiplication by 2 **3.** $3n + 1 = 7$; 2 **5.** $\frac{n}{5} - 10 = 3$; 65
7. $5n - 4 = 11$; 3 **9.** $4n + 8 = -12$; -5 **11.** $\frac{n}{3} + 9 = 14$; 15
13. $3n - 10 = 17$; 9 **15.** $4x + 25 = 75$; \$12.50 each
17. $0.07m + 3.95 = 12.63$; 124 min **19.** $4.45s + 105.34 = 216.59$; 25 **21.** $n + 2n + (2n + 5) = 200$; \$37, \$74, \$89 **23.** 46
25. 9 **27.** -2 **29.** 6 **31.** 2 **33.** 5 **35.** 8 **37.** $3 + 5y$

Pages 486–487 Lesson 10-4
1. Addition Property of Equality **3.** -3 **5.** -4 **7.** 5
9. $n = $ number; $3n - 18 = 2n$; 18 **11.** 8 **13.** -9 **15.** 10 **17.** 1
19. 5 **21.** 3.6 **23.** -0.25 **25.** -1.8 **27.** 8 **29.** $n = $ number; $4n + 2 = n - 7$; -3 **31.** $60x = 8x + 26$; 0.5
33. $14 + 0.8x = x$; \$70 **35.** $5 + 0.10(10x) + 8x = 10x$; 5 mugs
37. C **39.** $4n + 8 = 60$; 13 **41.** -3 **43.** 18 **45.** false **47.** true

Pages 494–495 Lesson 10-5
1. Sample answer: $n \geq 9$; you will earn at least \$9. **3.** $a < 6$
5. false **7.** true

9.
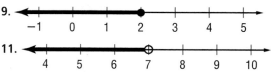

11.

13. $s > 100$ **15.** $\ell \geq 4$ **17.** $c \leq 25$ **19.** true **21.** true **23.** false

25.
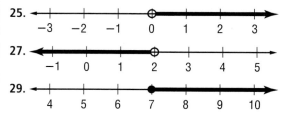

27.

29.

31.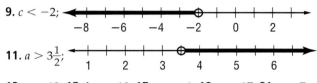

33.

35. $8n \geq 24$ **37.** $n - 4 < 12$ **39.** $t \leq 900$ **41.** $n < -2$
43. Sample answer: These symbols were introduced by the editor of Thomas Harriot's work *Artis Analyticae Praxis ad Aequationes Algebraicas Resolvendas.* **45.** D **47.** 4 **49.** −5
51. $-3 + 6h = 21$; 4 h **53.** −11 **55.** −9

Pages 498–499   Lesson 10-6
1. The same quantity can be subtracted from each side of the equation or inequality without changing the truth of the statement. **3.** $b > 4$ **5.** $g \geq -13$ **7.** $k \geq 7$

9. $c < -2$;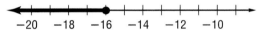

11. $a > 3\frac{1}{2}$;

13. $x \leq 13$ **15.** $k > -10$ **17.** $c < -1$ **19.** $g \geq 17$ **21.** $s \leq 7$
23. $w \geq -2$ **25.** $q \leq -1.3$ **27.** $p > -1.2$ **29.** $f < 3\frac{3}{4}$
31. $n - 11 < 8$; $n < 19$ **33.** $n + 17 \leq 6$; $n \leq -11$

35. $n > 4$;

37. $x \leq -16$;

39. $g < -7$;

41. $h \geq 2$;

43. $b \leq 7.75$;

45. $w < 4\frac{2}{3}$;

47. $99.2 + t > 101$; $t > 1.8$; more than 1.8°F **49.** $15 > x - 3$; $x < 18$; x is less than 18 cm **51.** Never; subtracting x from each side gives $0 > 1$, which is never true. **53.** I **55.** false
57. true **59.** 133° **61.** 9 **63.** −100

Pages 503–504   Lesson 10-7
1. Sample answer: $\frac{x}{-6} > 3$

3. $x \leq -9$;

5. $p \leq 8$;

7. $g > 14$;

9. $a < -8$ **11.** $m \geq 27$

13. $n \leq 5$;

15. $g < -4$;

17. $y < -11$;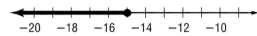

19. $r < -3$;

21. $c \leq 1$;

23. $a \leq -15$;

25. $n < -98$;

27. $t \geq 10$;

29. $k < 20$;

31. $5.25c \geq 42$; $c \geq 8$; at least 8 h **33.** $4,000,000 > 8d$; $d < 500,000$; less than 500,000 mi **35.** $k \leq -1$ **37.** $n < -9$
39. $c > 28$ **41.** $x \leq -3$ **43.** $\frac{n}{-5} \leq 7$; $n \geq -35$
45. $-2n > -18$; $n < 9$ **47.** D **49.** $y < 2$ **51.** $j \geq -4$
53. $b > 100,000$

Pages 505–506   Chapter 10   Study Guide and Review
1. d **3.** a **5.** f **7.** $4a + 12$ **9.** $-7n + 35$ **11.** $7p$ **13.** 6 **15.** −7
17. 35 **19.** $2n + 6 = -4$; −5 **21.** $\frac{x}{8} - 2 = 5$; 56 **23.** −7 **25.** 3

27. $g \geq 92$;

29. $y \leq -2$ **31.** $d > 14$ **33.** $c \leq -2$ **35.** $m < -3$

Chapter 11   Algebra: Linear Functions

Page 511   Chapter 11   Getting Started
1. false; vertical **3–6.**
7. 18 **9.** 20
11. 12 **13.** 5
15. 4 **17.** −5
19. 3

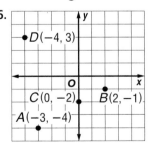

Pages 514–515   Lesson 11-1
1. When the quotient between any two consecutive terms is the same, the sequence is geometric. **3.** 5, 10, 15, 20, 25, …; It is an arithmetic sequence and the others are geometric sequences. **5.** neither; −14, −16, −17 **7.** 1, 2, 3, 4, 5
9. 6 cans **11.** geometric; 10; 100,000, 1,000,000, 10,000,000
13. arithmetic; −3; 73, 70, 67 **15.** neither; 26, 37, 50
17. geometric; −3; −1,215, 3,645, −10,935 **19.** arithmetic; $2\frac{1}{2}$; $16\frac{1}{2}$, 19, $21\frac{1}{2}$ **21.** geometric; $-\frac{1}{4}$; $-\frac{1}{64}$, $\frac{1}{256}$, $-\frac{1}{1,024}$
23. arithmetic; $-\frac{1}{3}$; $2\frac{5}{6}$, $2\frac{1}{2}$, $2\frac{1}{6}$ **25.** 100, −600, 3,600, −21,600 **27.** 4, 8, 12, 16, …; arithmetic **29.** arithmetic

31. Yes; the common ratio is 1. **33.** 0, 5, 10, 15, 20, 25, …
35. I **37.** $b > 17$ **39.** $-7 \le t$ **41.** 18 **43.** 26

Pages 519–520   Lesson 11-2
1. domain; range **3.** Tomi; the input is the value of x, not $f(x)$. **5.** -7

7.

x	$5x + 1$	$f(x)$
-2	$5(-2) + 1$	-9
0	$5(0) + 1$	1
1	$5(1) + 1$	6
3	$5(3) + 1$	16

9. 35 **11.** 11 **13.** 19 **15.** 2

17.

x	$6x - 4$	$f(x)$
-5	$6(-5) - 4$	-34
-1	$6(-1) - 4$	-10
2	$6(2) - 4$	8
7	$6(7) - 4$	38

19.

x	$7 + 3x$	y
-3	$7 + 3(-3)$	-2
-2	$7 + 3(-2)$	1
1	$7 + 3(1)$	10
6	$7 + 3(6)$	25

21. $P = 4s$ **23.** $c = 45.00 + 3.50p$ **25.** Sample answer: Mr. Jones is traveling on the interstate at an average speed of 55 miles per hour. Write a function to determine the distance he travels in h hours. How far will he travel in 3 hours? $d = 55h$; 165 mi **27.** A **29.** geometric; -2; -96, 192, -384 **31.** neither; 17, 23, 30 **33.** 28

34–37.

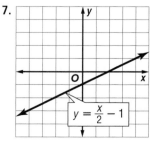

Pages 524–525   Lesson 11-3
1. Make ordered pairs using the x value and its corresponding y value. Then graph the ordered pairs on a coordinate plane. Draw a line that the points suggest.
3. $(0, 3)$; 3 does not equal $2(0) - 3$ or -3.

5.

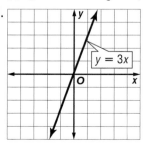

7.

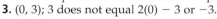

9. 100 gal

11. $y = 2x$

x	$2x$	y	(x, y)
-2	$2(-2)$	-4	$(-2, -4)$
0	$2(0)$	0	$(0, 0)$
1	$2(1)$	2	$(1, 2)$
2	$2(2)$	4	$(2, 4)$

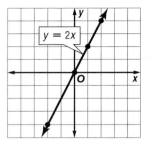

13.

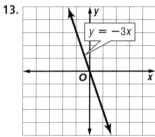

15.

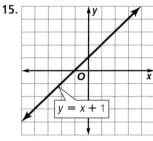

17.

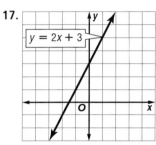

19.

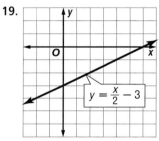

21.

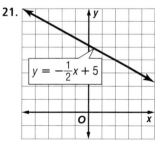

23.

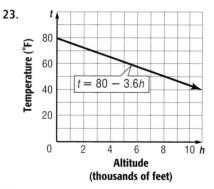

25a. Sample answer: $(-2, -4)$, $(0, -2)$, $(2, 0)$, $(4, 2)$; $y = x - 2$ **25b.** Sample answer: $(-1, 4)$, $(0, 3)$, $(1, 2)$, $(3, 0)$; $y = 3 - x$ **27.** A **29.** 39 **31.** -1 **33.** $\frac{1}{3}$ **35.** 4

Selected Answers

1. If $x_2 = x_1$, then $\frac{y_2 - y_1}{x_2 - x_1} = \frac{y_2 - y_1}{0}$ and division by 0 is undefined. Therefore the slope is undefined. **3.** Dylan; Martin did not use the x-coordinates in the same order as the y-coordinates. **5.** 0 **7.** $\frac{9}{100}$ **9.** The second half; it has a greater slope. **11.** -4 **13.** $\frac{1}{5}$ **15.** $-\frac{7}{9}$ **17.** undefined **19.** $-\frac{8}{9}$ **21.** -1 **23.** 55; her average speed **25.** Slope of $\overline{QR}$: $m = \frac{1 - (-2)}{5 - (-4)}$ or $\frac{1}{3}$; Slope of $\overline{RS}$: $m = \frac{2 - 1}{4 - 5}$ or -1; Slope of $\overline{ST}$: $m = \frac{-1 - 2}{-5 - 4}$ or $\frac{1}{3}$; Slope of $\overline{TQ}$: $m = \frac{-1 - (-2)}{-5 - (-4)}$ or -1; Therefore, $\overline{QR} \parallel \overline{ST}$ and $\overline{RS} \parallel \overline{TQ}$, and quadrilateral $QRST$ is a parallelogram. **27.** $-2, \frac{1}{2}$ **29.** The product of the slopes of two perpendicular lines is -1.

31. Sample answer:

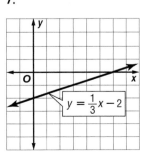

33.

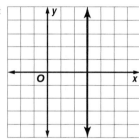

35.

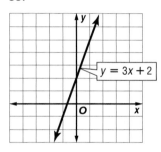

37. 8 **39.** 42

1. Locate the first point at $(0, -3)$. From this point, go down 5 and right 4 to locate the second point. Draw a line through the two points. **3.** $y = \frac{3}{2}x + 1$; The slope of this line is $\frac{3}{2}$, but the slope of each of the other lines is $\frac{2}{3}$.

5. $-\frac{1}{6}; -\frac{1}{2}$

7.

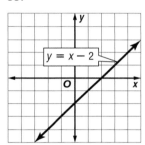

9.

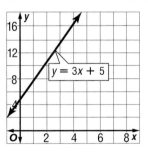

11. the amount paid each week **13.** 3; 4 **15.** $\frac{1}{2}$; -6
17. 2; 8

19.

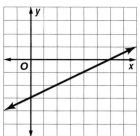

21. $y = -2x + 6$

23.

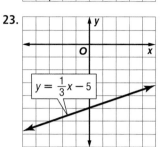

25.

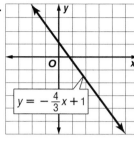

27.

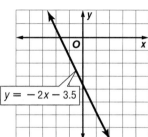

29.

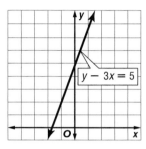

31. $y = -x + 180$ **33.** 110 **35.** the rate of descent
37. Sample answer: Jim is on a long hike. He has already gone 5 miles. He plans to hike 3 miles each hour. The distance y he has traveled in x hours can be determined by $y = 3x + 5$. Draw a graph of the function.

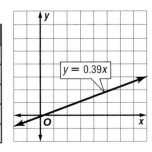

What does the slope represent? (the miles per hour) What does the y-intercept represent? (the distance already traveled) **39.** The slope is undefined. There is no y-intercept unless the graph of the line is the y-axis.
41. G **43.** $\frac{3}{2}$

45.

x	0.39x	y	(x, y)
0	0.39(0)	0	(0, 0)
1	0.39(1)	0.39	(1, 0.39)
2	0.39(2)	0.78	(2, 0.78)
3	0.39(3)	1.17	(3, 1.17)
4	0.39(4)	1.56	(4, 1.56)

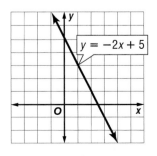

46–49.

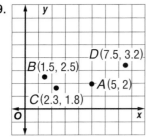

Pages 541–542 Lesson 11-6

1. Let one set of data be the x values and the other set of data be the y values. Pair the corresponding x and y values to form ordered pairs. Graph the ordered pairs to form a scatter plot. **3.** positive **5.** negative
7. positive **9.** positive **11.** no relationship
13. negative **15.** positive

17.

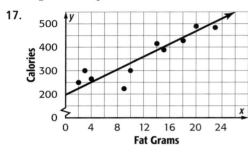

19. Sample answer: $y = 12x + 210$
23. B **25.** $\frac{4}{5}$; 7 **27.** -4; 2

29.

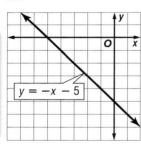

31.

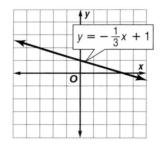

Pages 546–547 Lesson 11-7

1. A system of equations is a set of 2 or more equations. The solution of a system of equations is the solution or solutions that solve all the equations in the system.
3. The system has no solution. Since the graphs of the equations have the same slope and different y-intercepts, the graphs are parallel lines. Therefore, they do not intersect.

5.

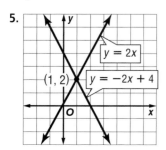

$(1, 2)$

7. $(4, 8)$ **9.** $(10, 1)$ **11.** $y = 350x + 100$

13.

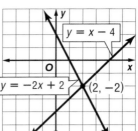

$(2, -2)$

15.

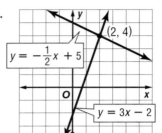

$(2, 4)$

17.

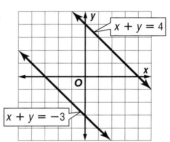

no solution

19.

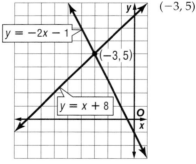

$(-3, 5)$

21. $(-3, -5)$ **23.** $(-4, 11)$ **25.** $(3, 3)$ **27.** $(1, 2)$ **29.** 10 shirts; $150 **31.** $y = 15x + 60$ **33.** 4 min **35a.** Sample answer: $y = -x + 4$ **35b.** Sample answer: $y = 2x - 1$ **35c.** Sample answer: $y - 2x = 1$ **37.** 60

39.

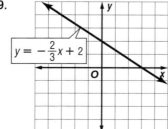

41.

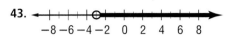

(graph) $y = 4x - 3$

43.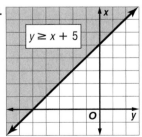
$-8 -6 -4 -2 \ 0 \ 2 \ 4 \ 6 \ 8$

45.
$-8 -6 -4 -2 \ 0 \ 2 \ 4 \ 6 \ 8$

Pages 550–551   Lesson 11-8

1. Sample answer: $y > x - 3$

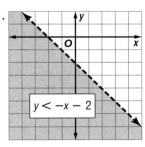
$y > x - 3$

3.

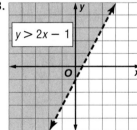
$y > 2x - 1$

5.

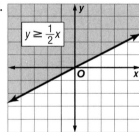
$y \geq \frac{1}{2}x$

7. Sample answers: legs: 5 units, base: 2 units; legs: 7 units, base: 5 units; legs: 9 units, base: 1 unit

9.

$y \geq x + 5$

11.

$y < -x - 2$

13.

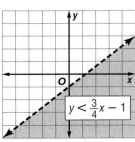
$y < \frac{3}{4}x - 1$

15.

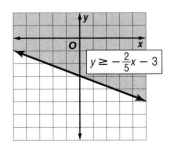
$y \geq -\frac{2}{5}x - 3$

17.

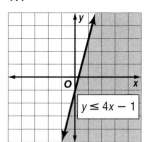
$y \leq 4x - 1$

19.

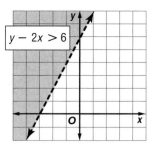
$y - 2x > 6$

21.

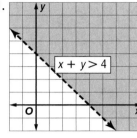
$x + y > 4$

23. Sample answers: math: 30 min, social studies: 20 min; math: 25 min, social studies: 25 min; math: 20 min, social studies: 35 min **25.** Sample answers: khoums: 5, ouguiya: 29; khoums: 25, ouguiya: 30; khoums: 10, ouguiya: 35 **27.** C

29. $(2, -1)$

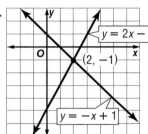
$y = 2x - 5$
$(2, -1)$
$y = -x + 1$

31. no solution

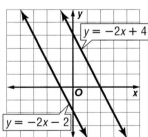
$y = -2x + 4$
$y = -2x - 2$

33. positive **35.** 6.25 yd^2 **37.** \$43.50

Pages 552–554   Chapter 11   Study Guide and Review
1. domain **3.** sequence **5.** arithmetic sequence
7. boundary **9.** geometric; $\frac{1}{2}$; 2, 1, $\frac{1}{2}$ **11.** neither; 720, 5,040, 40,320 **13.** \$14 **15.** -26 **17.** 22 **19.** -3

21.

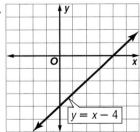
$y = x - 4$

23.

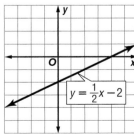
$y = \frac{1}{2}x - 2$

25. 2 **27.** $-\frac{3}{5}$ **29.** 2; 5 **31.** $\frac{1}{5}$; 6 **33.** $-\frac{3}{4}$; 7 **35.** positive
37. positive

39.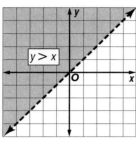

$y = x + 1$ (1, 2)
$y = 2x$

(1, 2)

41.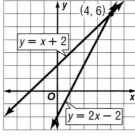

(4, 6)
$y = x + 2$
$y = 2x - 2$

(4, 6)

43. (3, 1)

45.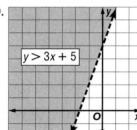

$y > x$

47.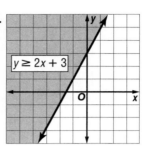

$y \geq 2x + 3$

49.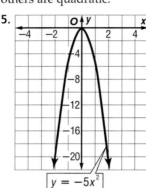

$y > 3x + 5$

51. Sample answers: 3 games, 3 rides; 5 games, 3 rides; 10 games, 0 rides

Chapter 12   Algebra: Nonlinear Functions and Polynomials

Page 559   Chapter 12   Getting Started
1. linear **3.** $3x$, $-x$ **5.** $(a + 2a) + (2b - 5b)$ **7.** $3 + (-5y)$
9. 6^4 **11.** $9d + 18$ **13.** $-2a - 6$

Pages 562–563   Lesson 12-1

1.

x	1	2	3	4
y	3	5	9	15

3. linear; graph is a straight line **5.** linear; can be written as $y = \frac{1}{3}x + 0$ **7.** linear; rate of change is constant, as x increases by 3, y decreases by 2 **9.** nonlinear; graph is two curves **11.** linear; graph is a straight line **13.** nonlinear; graph is a curve **15.** nonlinear; when solved for y, x appears in denominator so the equation cannot be written in the form $y = mx + b$ **17.** nonlinear; power of x is greater than 1 **19.** nonlinear; x is an exponent, so the equation cannot be written in the form $y = mx + b$ **21.** linear; can be written in the form $y = 0x + 7$ **23.** nonlinear; rate of change is not constant **25.** linear; rate of change is constant, as x increases by 3, y decreases by 2 **27.** nonlinear; rate of change is not constant **29.** Nonlinear; the points (year, pounds) would lie on a curved line, not on a straight line and the rate of change is not constant. **31.** Nonlinear; the power of r in the function $A = \pi r^2$ is greater than 1. **33.** C

35. $0.12x + y \leq 10$ **37.** (1, 3) **39.** (2, −2)

41.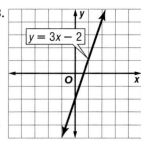

$y = 2x$

43.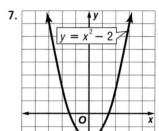

$y = 3x - 2$

Pages 567–568   Lesson 12-2
1. A function is quadratic if the greatest power of the variable is 2. **3.** $y = 7x - 3$, it is a linear function, while the others are quadratic.

5.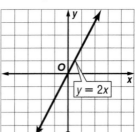

$y = -5x^2$

7.

$y = x^2 - 2$

9.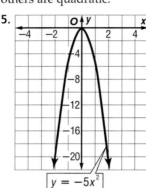

$y = -2x^2 + 2$

11.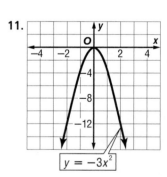

$y = -3x^2$

13.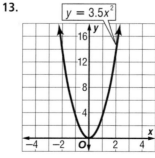

$y = 3.5x^2$

15.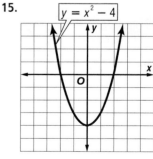

$y = x^2 - 4$

17.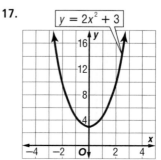

$y = 2x^2 + 3$

19.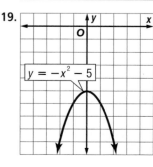

$y = -x^2 - 5$

21.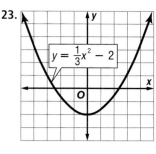
$$y = -3x^2 + 2$$

23.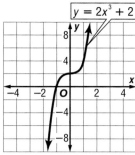
$$y = \frac{1}{3}x^2 - 2$$

25. 250 ft

27.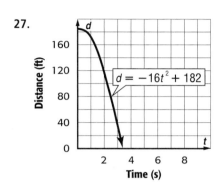
$$d = -16t^2 + 182$$

29. about 3.4 s

31. $V = 5s^2$;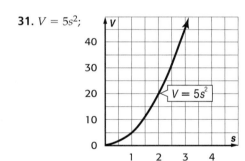
$$V = 5s^2$$

33. 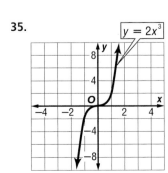 maximum; (0, 5)
$$y = -x^2 + 5$$

35.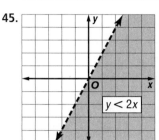
$$y = 2x^3$$

37.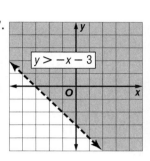
$$y = 2x^3 + 2$$

39. B **41.** linear
43. linear

45.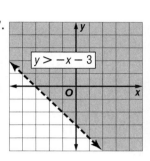
$$y < 2x$$

47.
$$y > -x - 3$$

49. $4a$ and $-2a$ **51.** -1 and 3; $-2d$ and d

Pages 572–573   Lesson 12-3
1. Sample answer: $6a - a + b - 10b$ **3.** $2y^2$; the others are like terms. **5.** simplest form **7.** $x^2 + 6x$ **9.** $-2w^2 - 8w$
11. $-2g^2 - 7g + 8$ **13.** simplest form **15.** $8f - 11g$
17. $2j + k - 2$ **19.** $-2x^2 + 2x$ **21.** $-2x - 3$ **23.** $5a^2 - 6a$
25. $2w^2 - 7w + 1$ **27.** $8y^2 + 8y - 3$ **29.** $4z + 11$
31. $3r^2 - 7r + 12$ **33.** $2t^3 - 8t^2 + 7t - 6$ **35.** $\frac{3}{4}y^2 - \frac{1}{4}y$
37. $50r^2 + 150r + 150$ **39.** D

41.
$$y = 5x^2$$

43.
$$y = x^2 - 4$$

45. No; the difference between the times varies, so the growth is not constant. **47.** $(2n + 5n) + (5 + 1)$
49. $(x^2 + 6x^2) + (4x - 8x)$

Pages 576–577   Lesson 12-4
1. Sample answer: $x - 3y, 3x - 2y$ **3.** $3h + 4$ **5.** $7t^2 + t - 4$
7. $8f^2 + 2f - 9$ **9.** 5 **11.** $7y + 10$ **13.** $5s^2 + s - 9$
15. $7m^2 + m + 4$ **17.** $5c - 1$ **19.** $8j^2 - 4j - 1$ **21.** $8d^2 - 1$
23. $6n^2 - 5n + 9$ **25.** $6v^2 + 2$ **27.** $5m^2 + 4m + 4$
29. $-2b^2 + 2b - 5$ **31.** $2x + y$; 15 **33.** $-3x + 5y + 2z$; -13
35. $18x - 9$ **37.** $25x - 9$ **39.** $4x + 100$ **41.** $2a + 3b$; the sum of two numbers minus one of its addends is equal to the other addend. **43.** 132°; 48° **45.** simplest form
47. $4q^2 + 5q - 5$ **49.** \$275 **51.** \$868.50 **53.** $6 + (-7)$
55. $4x + (-5y)$

Pages 582–583   Lesson 12-5
1. $-4x^2, 8x, -9$; $-4x^2 + 8x - 9$ **3.** Karen; Yoshi did not add the additive inverse of the entire second polynomial.
5. $5c^2 + c + 1$ **7.** $5p + 3$ **9.** $2n^2 - n - 1$ **11.** $-5a + 6$

13. $5w + 3$ **15.** $5b^2 - 5b + 6$ **17.** $2y^2 + y + 5$ **19.** $7a - 2$
21. $5k^2 - 9k - 20$ **23.** $2r^2 + 2r - 4$ **25.** $17z - 2$
27. $4y - 8$; 12 **29.** $x + 2y + 4$; 6 **31.** $4x + 2y$; -22
33. $5b + 5t$ **35.** $2b - 2t$ **37.** $6.5x + 200$ **39.** $3.5x - 200$
41. $A = 5x + 3$, $B = 2x + 1$ **43.** $5x - 8$ units
45. $6v^2 + v - 5$ **47.** $119C + 80R$ **49.** 3^4 **51.** 7^3

Pages 586–587   Lesson 12-6
1. false; $4x(5x^2) = 5x^2(4x) = 20x^3$ **3.** equal **5.** 3^7 **7.** $-6a^5$
9. $3c^5$ **11.** 7^6 **13.** 11^5 **15.** b^{14} **17.** $15x^9$ **19.** $-8w^{11}$ **21.** $40y^9$
23. 4^5 **25.** 10^{11} **27.** x^6 **29.** $3k$ **31.** $28a^3b^5$ **33.** $8xy$ **35.** n^4
37. x^7 **39.** 10^2 or 100 times greater **41.** about 10 times
43. 10^2 or 100 people/mi^2 **45.** 1 **47.** $2x^4y^{-5}$ or $\frac{2x^4}{y^5}$
49. $a^5b^{-3}c^{-2}$ or $\frac{a^5}{b^3c^2}$ **51.** G **53.** $2a + 2$ **55.** $3A + 5B + C + D$
57. 47; 52; 52 **59.** $3x + 12$ **61.** $-2n - 16$

Pages 591–592   Lesson 12-7
1. Sample answer: $5x^2 + 4x + 1$ and $2x$; $10x^3 + 8x^2 + 2x$
3. $m^2 + 5m$ **5.** $-4x^2 - 4x$ **7.** $2g^3 - 5g^2 + 9g$ **9.** $r^2 + 9r$
11. $9b^2 - 6b$ **13.** $-6d^2 - 30d$ **15.** $24h + 18h^2$ **17.** $22e^2 - 77e$
19. $4y^3 - 36y$ **21.** $t^3 + 5t^2 + 9t$ **23.** $-8r^3 + 2r^2 + 16r$
25. $2x(x + 4)$; $2x^2 + 8x$

27.

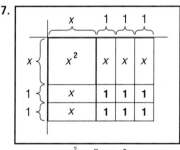

$x^2 + 5x + 6$

29. $A = 2x^2$;

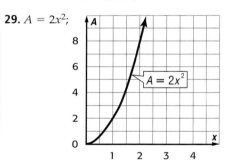

31. 11^3 **33.** $7a$

Pages 593–594   Chapter 12   Study Guide and Review
1. false; polynomial **3.** false; add **5.** true **7.** nonlinear; power of x is greater than 1

9.

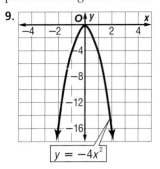

$y = -4x^2$

11.

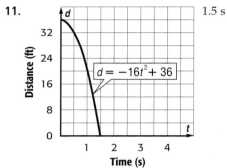

1.5 s

$d = -16t^2 + 36$

13. simplest form **15.** $5a^2 + a$ **17.** $10m^2 + 3m - 6$
19. $6c - 11$ **21.** $5k^2 + 5k - 1$ **23.** $36y^{11}$ **25.** $-3c^3$
27. $9y^2 + 12y$ **29.** $p^3 - 6p$ **31.** $-10k^3 + 6k^2 - 16k$

Photo Credits

Index

Index

Index

Index

x-intercept, 523
y-intercept, 523

Line symmetry, 286, 287

Logical reasoning, 276

Lower quartile (LQ), 442

Magnifications. *See* Dilations

Markup, 238

Mass, changing metric units, 607

Matrix (matrices), 454–457
  adding and subtracting, 455
  columns, 454
  dimensions, 454
  elements, 454
  rows, 454

M.C. Escher, 304

Mean, 36, 435

Measurement, 53. *See also*
  Customary System, Metric
  System
  angles, 256–258, 615
  applying the Pythagorean
    Theorem, 133, 134, 137, 138, 143
  area
    circles, 320, 321
    complex figures, 326–329
    parallelograms, 314
    rectangles, 613
    squares, 613
    trapezoids, 315
    triangles, 315
  circumference, 319, 320
  customary system conversions,
    53, 604–605
  effect of changing dimensions,
    73, 78, 178–180, 184, 185, 194,
    195, 338, 356
  greatest possible error, 362
  indirect, 188, 189
  perimeter, 178–180
    rectangles, 613
    squares, 613
  precision, 358–360
  significant digits, 358–360
  similar polygons, 178–180
  surface area, 347
    cones, 353
    cylinders, 348, 349
    pyramids, 352, 353
    rectangular prisms, 347, 348
  trigonometry, 192
  volume, 335
    complex solids, 337
    cones, 342, 343

cube, 101
cylinders, 336
prisms, 335, 336
pyramids, 342, 343
spheres, 345

Measures of central tendency,
  435–438
  mean, 435
  median, 435
  mode, 435
  summary, 436
  using appropriate measures, 436,
    451

Measures of variation, 442–445
  interquartile range, 442
  lower quartile (LQ), 442
  outliers, 443
  quartiles, 442
  range, 442
  upper quartile (UQ), 442

Median, 69, 435

Mental Math. *See also* Number
  Sense
  25, 63, 73, 78, 104, 127, 133, 160,
    188, 211, 220–221, 238, 375, 397,
    401, 407

Metric System, 606–607
  capacity units, 606
  choosing appropriate units, 607
  compared to Customary System,
    606
  conversions, 606–607
  length units, 606
  mass units, 606

Mid-Chapter Practice Test, 32, 86,
  130, 174, 224, 284, 340, 394, 440,
  490, 530, 578

Misleading statistics, 451–453

Mixed numbers, 62
  adding, 83, 89
  dividing, 78
  multiplying, 72
  subtracting, 83, 89
  written as decimals, 63

Mode, 435

Monomials, 570
  dividing, 585
  multiplying, 584, 585

Multiple, 612

Multiplication, 34
  Associative Property of, 13
  Commutative Property of, 13
  Identity Property of, 13
  integers, 34, 35
  Inverse Property of, 76
  monomials, 584, 585

phrases indicating, 39
polynomials and monomials, 590
powers, 584
with powers of 10, 104
rational numbers, 71–73
as repeated addition, 34
solving equations, 50
symbols, 12

**Multiplication Property of
  Equality,** 51

Multiplicative inverses, 76

Mutually exclusive events, 399

Negative exponents, 99

Negative numbers, 17

Negative square roots, 116, 117

Nets, 342, 346

Nonlinear functions. *See* Functions

Nonproportional relationship, 173

Number line, 17
  absolute value, 19
  comparing integers, 18
  comparing rational numbers, 68
  graphing inequalities, 548–549
  graphing integers, 18
  graphing irrational numbers, 120,
    141
  integers, 17
  ordering integers, 18
  real numbers, 126

Numbers
  compatible, 228
  composite, 609
  irrational, 125
  prime, 609
  real, 125
  scientific notation, 105
  standard form, 104

Number Sense, 9, 37, 52, 69, 74, 79,
  90, 100, 106, 121, 163, 172, 187,
  230, 243, 322, 325, 344, 349, 382,
  428, 480, 489, 494, 541, 546, 582,
  586

Numerical expressions, 11

Obtuse angle, 256

Obtuse triangle, 253

random, 374
sample space, 374
simulations, 404–405
theoretical, 400, 401
tree diagrams, 380

Problem solving, 6–8
four-step plan, 6

Problem-Solving Strategy
determine reasonable answers, 226
draw a diagram, 176
guess and check, 488
look for a pattern, 96
make a model, 588
make an organized list, 378
make a table, 418
solve a simpler problem, 324
use a graph, 537
use a Venn diagram, 123
use logical reasoning, 276
work backward, 43

Product, 34. *See also* Multiplication

Product of powers, 584

Projects. *See* WebQuest

Properties, 13
Addition Property of Equality, 46
Additive Inverse Property, 25
Associative Property of Addition, 13
Associative Property of Multiplication, 13
Closure Property, 38
Commutative Property of Addition, 13
Commutative Property of Multiplication, 13
Distributive Property, 13
Division Property of Equality, 50
of geometric figures, 262–263, 267–268, 273, 278
Identity Property of Addition, 13
Identity Property of Multiplication, 13
Inverse Property of Multiplication, 76
Multiplication Property of Equality, 51
of parallel lines, 257
of perpendicular lines, 529
Substitution Property of Equality, 13
Subtraction Property of Equality, 45

Proportional reasoning
circumference, 319, 320
distance on the coordinate plane, 142, 143
golden ratio, 121
golden rectangle, 121

indirect measurement, 188, 189
percent equation, 232, 233
percent proportion, 216, 217
proportions, 170, 171
rate of change, 160–162
scale drawings, 184, 185
scale factors, 179, 184, 195, 356
shadow reckoning, 188
similarity, 178–180
slope, 166, 167
theoretical probability, 400, 401
trigonometry, 192–193
unit rates, 157

Proportions, 170–173
cross products, 170
identifying, 171
property of, 170
solving, 170, 171

Pyramids, 331–334
lateral area, 352
lateral face, 352
slant height, 352
surface area, 352–355
vertex, 352
volume, 342–345

Pythagorean Theorem, 132–140
applying, 133, 134, 137, 138
converse, 134
distance, 143
identifying, 134
with special right triangles, 267, 268

Pythagorean triples, 138

Quadrants, 142

Quadratic functions. *See* Functions

Quadrilaterals, 272–275
classifying, 273
parallelogram, 273
rectangle, 273
rhombus, 273
square, 273
sum of angles, 272
trapezoid, 273

Quartiles, 442

Quotient, 35. *See also* Division

Quotient of powers, 585

Radical sign, 116

Radius, 319

Random, 374

Range, 442

Range (for a function), 518

Rate of change, 160–162. *See also* Rates
constant, 165
negative, 160, 161
slope, 166, 167
zero, 162

Rates, 156, 157
interest, 241
population density, 157
speed, 157
unit, 157

Rational numbers, 62, 125. *See also* Fractions; Percents
adding, 82, 89
comparing, 67, 68
dividing, 76–80
multiplying, 71–75
on a number line, 68
ordering, 68
solving equations, 92
subtracting, 83, 88
unit fractions, 66
writing as decimals, 63
writing as percents, 207, 211

Ratios, 156, 157, 206, 207
simplest form, 156
writing as percents, 206, 207

Reading, Link to, 62, 132, 194, 216, 266, 272, 290, 330, 352, 442, 469, 533

Reading Math
and so on, 513
angle measure, 257
interior and exterior angles, 258
isosceles trapezoid, 273
at least, 421
matrices, 455
naming triangles, 262
notation for combinations, 389
notation for permutations, 385
notation for segments, 280
notation for the image of a point, 290
parallel and perpendicular lines, 257
probability notation, 375
proportional, 171
ratios, 157
repeating decimals, 64
square roots, 117
subscripts, 161, 315
word problems, 8

Real-Life Careers
automotive engineer, 581